# THE CHEMICAL PROCESS INDUSTRIES INFRASTRUCTURE

# CHEMICAL INDUSTRIES

A Series of Reference Books and Textbooks

*Consulting Editor*

HEINZ HEINEMANN

### ADDITIONAL VOLUMES IN PREPARATION

*Transport Phenomena Fundamentals,* Joel Plawsky

# THE CHEMICAL PROCESS INDUSTRIES INFRASTRUCTURE

## FUNCTION AND ECONOMICS

**James R. Couper**
*University of Arkansas*
*Fayetteville, Arkansas*

**O. Thomas Beasley**
*CMAI, Inc.*
*Houston, Texas*

**W. Roy Penney**
*University of Arkansas*
*Fayetteville, Arkansas*

CRC Press
Taylor & Francis Group
Boca Raton  London  New York

CRC Press is an imprint of the
Taylor & Francis Group, an **informa** business

Portions of this book were adapted or reprinted from *The Structure of the Chemical Process Industries* by J. Wei, T. W. Fraser Russell, and M. W. Schwartzlander (McGraw-Hill, 1979).

First published 2001 by Marcel Dekker, Inc.

Published 2023 by CRC Press
Taylor & Francis Group
6000 Broken Sound Parkway NW, Suite 300
Boca Raton, FL 33487-2742

ISBN 13: 978-0-8247-0435-3 (hbk)

# Preface

We have developed this text based on the book *The Structure of the Chemical Process Industries*, by J. Wei, T. W. Fraser Russell, and M. W. Schwartzlander, (McGraw-Hill, 1979). Although the chapter titles are essentially the same, the material content has been updated, revised, or completely rewritten to reflect the global changes that have occurred in the chemical process industries since the late 1970s. Further, we have included an excerpt from the 1979 book because we endorse the philosophy in that edition. The story about the Caliph of Baghdad is particularly appropriate today:

> The Caliph of Baghdad in disguise saw three men working and asked them what they were doing. The first man said, "I work for the Caliph for a few dinars a day." The second man said, "I am an expert rock breaker." The third man said, "We are building a road from Baghdad to the seaport." The Caliph said to his Grand Vizier, "The first man is a clock watcher, who is only interested in his pay in dinars. The second man has pride in his skill and is the backbone and sinew of my realm. But the third man understands the goal of the project and may be able to contribute ideas on how to improve the road." The old wise Vizier replied, "The third man knows what we are doing but not why it should be done. An even wiser worker would have said that the broader mission is to improve transportation of goods and people, which improves the quality of life for all our subject people. Our wisest worker would even consider whether flying carpets are better than roads to fulfill the mission. How few are those who understand the needs of the people and how to serve them!"

If the Caliph were to return today, he might exclaim that the workers now have wondrous machines and skills but that the progeny of the first worker are more numerous than ever. Even those with pride in their skills often do not fully know the purpose and worth of their work; their lives may be without meaning outside of technical accomplishments, and they may be unaware of their broader contributions. This book is designed to broaden the minds of the Caliph's workers and to produce more knowledgeable and productive chemical professionals, at all levels of specialization.

With the restructuring of companies, chemical professionals are called on to be more knowledgeable about the threats and opportunities in the chemical process industries (CPI). These people educated in highly specialized fields of technology must now wear more hats and be able to interface with customers, marketing, purchasing, and manufacturing personnel, as well as upper-level administration. The chemicals professionals are becoming more generalists than specialists. They need to know more about global and domestic competition, marketing strategies, reduction of operating expenses, environmental regulations, safety requirements, and so on. Also, with the restructuring of companies, new challenges and opportunities are occurring in companies that are not normally considered a part of the CPI but that interface with these industries by providing materials, equipment, and services. A student embarking on a career as a chemical professional needs to have a broad knowledge of the structure and how the CPI functions. This text should help students understand how the CPI is affected by both domestic and global economic forces. The practicing chemical professional who is positioned to move into a management position, on the other hand, needs a broader concept of the pressures affecting management decisions in the CPI.

Chemical companies have realized that the production of major chemicals that were profitable from the 1950s to the 1980s may not even return the cost of capital to investors. About a decade and a half ago, Wei and Amundsen wrote a "frontiers" paper for the American Institute of Chemical Engineers predicting certain changes in the coming decades, and most of these have occurred. To perpetuate growth and improve the "bottom line," companies realigned their operations, sold off low-profitability ventures or shut them down, moved into more profitable specialty chemical production, and streamlined their operations with the latest process control and computer technology. In order to grow, many of the major companies moved into agroscience and medicinal and pharmaceutical areas. Still others produced special catalysts, electronic chemicals, and unique materials to ensure company growth as the industry prepared to enter the new millennium. There does not appear to be any strategic plan for growth except to subject all decisions to the test of maximizing company and stockholder value.

One should recognize that change is inevitable and in the CPI it is occurring at an ever-increasing pace. To stay somewhat current in the field, the chemical professional should read weekly publications such as: *Chemical and Engineering News, Chemical Week,* and the *Chemical Marketing Reporter,* as well as the *Wall Street Journal, Fortune,* and *Forbes* at a minimum. Also, we recommend that the chemical professional be cognizant of such new books on the CPI as *Chemicals and Long-Term Economic Growth* by Landau et al., to gain different perspective and insights into the industry.

In this volume, we have retained the format and chapter titles of the 1979 book by Wei et al. The material within the chapters has been updated or revised to conform to current practice. In Chapter 2, the data for the GNP and the GDP have been extended to 1995 with appropriate deflators. With respect to the AL-GOL-DELOS problem, Example 2.1, a number of intermediate equations were omitted in the development of the equations presented in the earlier book.

Chapter 3 was completely revised to reflect current practice with respect to accounting and financial reporting. The accounting example for Delchem Corporation was shortened, and financial reporting and financial analysis were augmented and updated. Financial ratios were presented and applied to Archem operations. Finally, a Z-score indicator that determines the potential for bankruptcy was introduced and also applied to Archem. Excerpts from Monsanto's 1994 annual report were included to illustrate a real-world annual report.

Chapter 4 (Input–Output Analysis) has been significantly revised to reflect the latest data from the 1987 transaction, direct-requirements, and total-requirements tables, Tables 4.4a, 4.4br, 4.10, and 4.15, which are found in Appendix C. The mathematical derivations and the supporting explanations have been modified to make for easier understanding.

As noted earlier, many companies have been acquired and merged with other firms or have ceased to exist. As a result, Chapters 5 and 7 have been updated and revised to reflect the names and identities of the surviving firms.

In Chapter 6, the product report on methanol has changed to reflect the drastic changes that have taken place in the market for this product. It has become a mature market, and the chemical production is on the decline. All the plots within this chapter have been updated. Other commercially significant products such as cumene-phenol and VCM have been included.

The structures of Chapters 8, 9, and 10 have been maintained and the content updated to the 1995–1998 time period. On the international scene, emerging nations in the Far East as well as others in Europe, Mideast, and Central and South America have become important in the global market, with even more significant potential in the future.

The appendices were expanded to include a List of Tables (Appendix D), a List of Figures (Appendix E), and a Quick Reference Guide for library materials (Appendix F).

The reader should be aware that some of the statistical data included in this revision come from government data sources that may not be updated yearly. Company financial data are based on the latest available annual report, which is 1998.

We express our sincere appreciation to Dean J. Wei and Professor T. W. F. Russell for their helpful suggestions and advice. Also, many professional colleagues have offered comments and provided information that added significantly to the text. Dr. Allen J. Lenz of the Chemical Manufacturers Association graciously provided financial and statistical data on the industry. We are particularly grateful to CMAI, Houston, Texas, for much of the marketing data on olefins. Dr. Robin Roberts, College of Business Administration, University of Central Florida, assisted by providing information and suggestions regarding accounting techniques. The University of Arkansas Mullins Library staff, especially Kinne Colpitts and Donald Batson of the reference department and Cathy Hooper of government publications, were most helpful in sorting through the maze of data available.

We also recognize the many helpful suggestions and contributions that the seniors and graduate students at the University of Arkansas have made. They have taken the associated course, solved the problems, gathered data, and assisted the instructor(s) in clarifying the text material.

We sincerely appreciate the efforts of B. J. Clark of Marcel Dekker, who helped expedite the publication of this book.

Lastly, the authors express their appreciation to our wives, Mary Couper, Betty Beasley, and Annette Penney, for their patience and for the many hours we were taken away from family activities during the preparation of this book.

The material in this text is suitable for a senior- or graduate-level course for chemical professionals, a continuing education course for these same professionals employed in the CPI, or as a text for self-study to broaden one's knowledge of the CPI.  .

*James R. Couper*

*O. Thomas Beasley*

*W. Roy Penney*

# Contents

*Appendices:*

# THE CHEMICAL
# PROCESS INDUSTRIES
# INFRASTRUCTURE

# 0

## Reader's Guide

### 0.1 ENRICHING AND UPDATING INFORMATION

The usefulness of this text will be greatly enhanced if you supplement it with information from current periodicals, basic references, statistical collections, abstracting services, indexes, and electronic data obtained from the World-Wide Web. This will allow you to update the data given in the text and—even more important—provide practice in finding the information chemical professionals need in day-to-day analysis and problem solving. The most important sources in this section are labeled with two asterisks and the next most important with one asterisk. If you are using the book for self-directed study, you should subscribe or have access to the doubled-asterisked periodicals and read the single-asterisked periodicals,[1] become acquainted with the contents of the references with asterisks, locate a library with national and international references, and obtain a copy of the *Statistical Abstract of the United States* (Sections 0.1.1 and 0.1.2).

### 0.1.1 Current Periodicals

*Chemical and Engineering News.*\*\*[2] This is the most important journal and is published weekly by the American Chemical Society. Anyone seriously interested in the chemical industry should subscribe to it.

---

[1]Subscription prices range from $20 to $120 per year.
[2]American Chemical Society, 1155 Sixteenth Street, NW, Washington, DC 20036.

Much of the statistical information in this text has been taken from the Facts and Figures issues of *Chemical and Engineering News* published in the months of June and July. Of equal importance are the news items, detailed studies, and articles that appear throughout the year (LCN TP1 .C35). **Note:** A Library of Congress number (LCN) is included in the paragraph about each periodical or book.

*Chemical Week.*\*[3] A subscription to this publication is very valuable. Like *Chemical and Engineering News,* it reports on business, economic, and technical aspects of the chemical industry globally. Frequently, longer studies on a particular product or the chemical industry in various countries are published. The *Annual Buyers' Guide* lists the names and addresses of more than 2600 companies and the manufacturers of 16,500 products they produce (LCN TP1 .C37).

*Chemical Engineering Progress.*[4] This monthly publication is particularly useful for information on new processes and technology, capital costs, and operating expenses of processes and new equipment of potential use in the CPI. It also provides issues of importance to chemical engineers (LCN TP1 .A6).

*CHEMTECH.*[5] This periodical published monthly by the American Chemical Society contains articles with new and refreshing ideas about technical as well as business aspects of the CPI (LCN TP1 .I612).

*Fortune.*[6] This twice-monthly publication deals with significant issues in business and economics. It publishes two very useful directories, *Fortune 500* and *Global 500,* which show the role of the chemical industry in the total business picture. Occasionally, there are articles on the CPI, which will enliven and expand the material in this text (LCN HF5001 .F7).

*Forbes.*[7] This monthly publication contains business articles written from the standpoint of investors. Frequent evaluations of the operations of various companies are an important feature of this publication. These critical evaluations provide more balanced information than the generally optimistic rhetoric in company annual reports. Occasionally, there are articles concerning the CPI giving insights not provided by other publications (LCN HF5001 .F6).

---

[3]Subscription address: Fulfillment Manager, P.O. Box 748, Mt. Morris, IL 61054-0748.

.. . . . .. . . . .. .. . . . .. . . .. . .. . . ... 10017.

NY 10020.

*Chemical Marketing Reporter.*\*[8] This weekly publication is in newspaper format. It is the best source of market prices of chemicals as well as news articles about the chemical industry (LCN HD9650.1 .C486).

*Chemical Engineering.*[9] This biweekly publication emphasizes technical over business aspects of the chemical industry. It is a good source of information about new processes and technology (LCN TN1 .M45).

*Hydrocarbon Processing.*[10] More technology- than business-oriented, this monthly publication places more emphasis on petroleum refining, petrochemical manufacturing, and the oil industry than *Chemical Engineering* does (LCN TP690 .A1 P4).

*Oil and Gas Journal.*[11] Written for the oil and gas industry, this biweekly journal contains useful statistics on oil and gas production and prices as well as articles and petroleum industry news (LCN TN860 .O4).

*Monthly Energy Review.*[12] Published by the Department of Energy (DOE/EIA-0035 ISSN 0095 7356), Washington D.C.

*National Petroleum Refiners Association publications.*[13] The NPRA publications of the U.S. International Trade Commission, at the Department of Commerce, in Washington, D.C., are the best source of monthly up-to-date production figures on many U.S. chemicals and petrochemicals. In some instances, prices are given.

*Economic Indicators.*[14] This monthly publication gives the most recent statistical data on total economic output, income, spending, employment, wages, business activity, prices, credit, and federal finance (LCN HC101.A186).

*Survey of Current Business.*[15] A monthly publication prepared by the Bureau of Economic Analysis of the Department of Commerce. Like *Economic Indicators,* it contains statistical macroeconomic data and articles of general business interest (LCN HC101 .A13).

*The Wall Street Journal, Business Week,* and *U.S. News & World Report.* The first publication is published Monday–Friday, providing daily business news. The other two are weekly business journals. All three often contain articles pertaining to the chemical industry.

---

[8]Schnell Publishing Company, Inc., 100 Church Street, New York, NY 10011.

[9]Subscription address: Fulfillment Manager, *Chemical Engineering,* P.O. Box 430, Hightstown, NJ 08520.

[10]Gulf Publishing Company, 3301 Allen Parkway, Houston, TX 77019-1805.

[11]Penwell Publishing Company, P.O. Box 1260, Tulsa, OK 74101.

[12]Superintendent of Documents, U.S. Government Printing Office, Washington, DC 20402.

[13]Superintendent of Documents, U.S. Government Printing Office, Washington, DC 20402.

[14]Superintendent of Documents, U.S. Government Printing Office, Washington, DC 20402.

[15]Superintendent of Documents, U.S. Government Printing Office, Washington, DC 20402.

*Chemistry and Industry.*[16] This biweekly publication reports on technical and business news internationally in the chemical business, with strong emphasis on Great Britain and Europe. There is some cross-over of news about the United States and Asia chemical business (LCN TP1 .S6332).

*Economic Report of the President.*[17] This annual report to Congress on the state of the national economy is written by the Council of Economic Advisers. It covers major economic issues and contains valuable statistical data (LCN HC 106.5 .A272).

*Harvard Business Review.*[18] A widely read and well-respected monthly journal of business management containing timely articles and excellent business case studies (LCN HF5001 .H3).

*Foreign Affairs.*[19] A quarterly journal of world events written by world leaders in government, education, foundations, and business (LCN D410 .F6).

*The Economist.*[20] A highly respected international business journal. It occasionally presents longer studies using statistics and analysis (LCN HG11 .E2).

*Far Eastern Economic Review.*[21] A weekly business magazine specializing in the Far East (LCN HC 411 .F18).

## 0.1.2  Basic References

There are a number of good introductory texts in basic economics. The following authors have been found to be particularly helpful:

Paul A. Samuelson, *Economics,* 16th ed. McGraw-Hill Book Company, New York, 1997.

James L. Riggs and Thomas M. West, *Engineering Economics,* 3rd ed. McGraw-Hill Book Company, New York, 1986.

Clarence B. Nickerson, *Accounting Handbook for Non-Accountants.* Van Nostrand Reinhold, New York, 1983.

Leopold A. Bernstein, *Financial Statement Analysis: Theory, Application & Interpretation,* 5th ed. Richard D. Irwin, Homewood, IL, 1992.

Erich A. Helfert, *Techniques of Financial Analysis,* 9th ed. Richard D. Irwin, Homewood, IL, 1996.

---

[16]Turpin Distribution, Blackhorse Road, Letchworth, Hertsniure SG6 1HN, England.

[17]Superintendent of Documents, U.S. Government Printing Office, Washington, DC 20402.

[18]Subscriber service: P.O. Box 52623, Boulder, CO 80322-2623.

[19]58 East 68th Street, New York, NY 10021.

[20]25 St. James Street, London SWIA 1HG, England.

[21]P.O. Box 160, Hong Kong.

The following out-of-print paperback books provide historical information and perspectives on the chemical industry up to 1974. They may be available in university libraries or in used-book stores:

Jules Backman, *The Economics of the Chemical Industry.* Manufacturing Chemists Association, Washington, DC, 1970. Valuable for the historical perspective and analysis of the chemical business; the statistics are out of date, however.

*Chemistry in the Economy.* American Chemical Society, Washington DC, 1974. Presents the social and economic impact of the chemical industry, accomplishments of chemistry, a good history of the development of important chemicals, and the expected future developments as of 1974.

The following paperback should be purchased each year:

*Statistical Abstract of the United States,*** Bureau of Census, Department of Commerce, published by the Government Printing Office, Washington, DC 20402. The price is about $50 (LCN HA202 GOVREF). The most valuable one-volume source of current statistics (population, labor force, income, and other types of information with which you should become familiar).

The following references should be available through a library:

George T. Austin, *Shreve's Chemical Process Industries,* 5th ed. McGraw-Hill Book Company, New York, 1984. (LCN TP145 .S5)

James A. Kent, *Riegel's Handbook of Industrial Chemistry,* 7th ed. Van Nostrand Reinhold, New York, 1992. (LCN TP145 .R54)

J. I. Kroschwitz, ed., *Kirk-Othmer Encyclopedia of Chemical Technology,* 4th ed. John Wiley & Sons, New York, 1996. (LCN TP9 .E685)

F. A. Lowenheim and M. K. Moran, *Faith, Keyes & Clark's Industrial Chemicals,* 4th ed. Wiley-Interscience, New York, 1975. A one-volume summary of process technology with a history of sales volume and price. This is of historical interest because the data are out of date. (LCN TP200 .F3)

J. J. McKetta, ed., *Encyclopedia of Chemical Processing Design,* Marcel Dekker, New York, 1998. (LCN TP 9 .E66)

*Moody's Industrial Manual.* Contains company financial data that are updated yearly. Moody's Investors Service, Inc., New York, NY. (LCN HG 4961 .M67)

*Standard & Poor's, Stock Reports.* Standard & Poor's Corporation, Division of McGraw-Hill Companies, New York. (LCN HG 4905 .S443)

*Standard & Poor's, Industry Surveys.* Standard and Poor's Corporation, Division of McGraw-Hill Companies, New York. It includes industry sector survey and some financial data, updated several times a year. (LCN HC 106.6 .S74)

*The Value Line Investment Survey.* Value Line, Inc., New York. This survey is a weekly report of stocks. (LCN HG4501 .V26).

**Note:** The *Chemical Economics Handbook* published by SRI International, Menlo Park, CA, is an extremely important reference presenting capacity, production, and sales data for chemicals, updated yearly. It is available only to subscribers, not to the general public.

*Chemical Market Associates, Inc. (CMAI),* Houston, TX. Provides market intelligence with such information as worldwide capacities, production prices, and process economics of various petrochemicals in either single- or multiclient studies. Multiclient studies are available to subscribers.

### 0.1.3  Statistical Collection

*U.S. Industry & Trade Outlook '98.* Government Printing Office, Washington, DC 20402. The current edition of an annual publication in both a narrative and statistical format. (LCN HC 106.5 .A17)

*Census of Manufacturers.* Subscription Services, Economics Statistics Administration, Government Printing Office, Washington, DC 20402. Contains detailed financial data on manufacturing establishments, specific products, and specific industries organized by SIC sector; includes industry reports on the CPI and related industries (published yearly). (LCN HD 9724 .U52)

*Predicasts,* Predicasts, Inc., Cleveland, OH. A quarterly abstract dealing with market data for manufacturing industries arranged by SIC code. (LCN HC 101 .P7)

*Synthetic Organic Chemicals,* International Trade Commission, Government Printing Office, Washington, DC 20402. An annual update with monthly supplements providing statistics on the organic chemical industry.

### 0.1.4  Electronic Sources

Much valuable information may be obtained by the use of the electronic information sources. For example, company annual reports, as well as 10K and 10Q reports containing a wealth of business and financial information of companies, may be retrieved on the Internet from www.investquest.com. As an alternative, the search engine AltaVista (www.altavista.digital.com) can be used to obtain similar information or to augment not only financial but also product information. Company information may be obtained by accessing a company's website (e.g., www.dow.com). A listing of selected Internet addresses may be found in *Chemical Week* and other literature sources. It is recommended that searches for financial or product information begin with electronic sources.

### 0.1.5 Abstracting Services and Indexes

Standard sources like *Chemical Abstracts, Applied Science and Technology Index, Technical Abstracts,* and the *Engineering Index* will frequently be needed, and the following will be needed and useful:

> *Chemical Market Abstracts,* Predicasts, Inc., Cleveland, OH. A monthly publication arranged by SIC code.
>
> *F and S Index of Corporations and Industries,* Predicasts, Inc., Cleveland, OH. Annual, quarterly, and weekly editions covering companies, products and other data. (LCN HG 4961 .F8)
>
> Many libraries also have available the Dialog, BRS, Orbit, and STN International Retrieval Services, which are online computer information retrieval services. All the Predicast data, the *Applied Science and Technology Index,* CA Search, Chemical Industry Notes, CSCORP, and Kirk-Othmer are searchable through these services.

Readers not enrolled in a formal course should read the next section, particularly the discussion on course objectives, skills to be developed, and problem assignments to get the maximum benefit from self-study.

Appendix F offers a quick reference table to assist the reader in locating desired information in Standard & Poor's, Moody's, and other sources.

## 0.2 DEVELOPING AN EFFECTIVE COURSE

### 0.2.1 Instructor's Guide

This text provides the necessary fundamentals and required statistical information for developing a course. In this section we discuss other materials essential for making the course effective.

The preparation and planning of this course differs from those activities as performed in most engineering or science courses. Much of the material discussed needs to be current (no one today knows which problems will be the crucial issues tomorrow). This text contains the relatively timeless fundamental economic and accounting principles, examples of economic analysis, and a detailed discussion of the structure of the CPI in the United States. Experience has shown that one can develop a lively, interesting, up-to-date, and effective course by expanding upon and adding to the textual material in a number of ways.

**0.2.1.1 Extensive Use of Current Periodicals.** These are listed in Section 0.1.1. We recommend a personal subscription to at least *Chemical and Engineering News* and one other periodical on the list such as *Chemical Week* or *CHEMTECH*. Both short news items and longer feature articles can be used to expand upon material in the text to stimulate class discussion and to generate effective homework assignments. For instance:

1. *Chemical Week* reports on timely topics like mergers and acquisitions, custom chemical manufacturing, and fine chemicals, as well as the chemical industry in various regions and of the world like Southeast Asia, Brazil, and China. These topics can provoke intense classroom discussions especially in connection with Chapters 6, 8, 9, and 10.
2. *Chemical and Engineering News* has a regular feature on key chemicals in which updated production figures, appear, as well as price ranges in the United States, Europe, and the Asia/Pacific area, market demand, and technology updates. This information is very useful for expanding upon and updating material in Chapter 5 and 6.
3. *CHEMTECH* occasionally has provocative articles on the management of research and development, technology, intellectual property, and other subjects that make excellent classroom discussion topics.
4. *Fortune* frequently publishes special articles on energy and oil crises, as well as corporate topics like mergers and acquisitions and company alliances. Further, on occasion, there are articles on segments of industry or on companies in the pharmaceutical, plastics, petrochemical, or petroleum industries.
5. *Forbes* will publish feature articles on companies and their business areas.

In almost every class, items from at least one of the periodicals should be discussed. Most universities have subscriptions to many of the periodicals mentioned heretofore. The instructor or students will clip or copy articles of interest from those journals for immediate discussion. As an alternate, some of the articles are filed until a particular topic is covered.

**0.2.1.2 Use of Case Studies.** Case studies are a particularly effective way of meeting the course objectives. Case studies also serve as examples, good and bad, of how a problem can be handled. Critical analysis of case studies helps develop the student's ability to detect flawed reasoning, weaknesses in logic, and inappropriate interpretation of facts and events. There are three good sources of case studies, as explained in Sections 0.2.1.2.1–0.2.1.2.3.

*0.2.1.2.1 Published Case Studies.* The Harvard Graduate School of Business publishes case studies on a variety of topics. These case studies are well prepared and add significantly to the course. There is no way a priori of determining which case studies are appropriate for a given course. The instructor should obtain from Harvard a list of case studies available and order those that are appropriate well in advance of the course, perhaps the semester or quarter prior to the course offering, to allow time to review the studies and select the ones to be used. In the past, the following case studies have been successful:

1. *Industrial Chemicals, Inc.* This Harvard Business School case examines research and development in a company by studying the personalities of key people involved, how they interacted with each other, and their career progression. This case study is well done and relatively timeless.

2. *Mobil Chemical Company.* This Harvard case study deals with the launching of a new product, marketing problems, and the company position in the manufacture of plastic bottles, as well as key management decisions that had to be made. Although the case study is dated, it serves as a basis for identifying the topic areas that needed to be addressed by the company. There is information in the literature regarding the outcome of this proposed venture. Much can be gained from this case study.

3. *Reichold Chemicals, Inc.* This University of Alabama case study deals with wastewater treatment problems at the company's Tuscaloosa plant. In addition to the emphasis on the methods for treating wastewater, economic, social, and community issues are raised. Although dated, this study provides a basis for considering alternate methods for the manufacture of phenol, the associated pollution problems, and community concerns.

*0.2.1.2.2 Case Studies Prepared by Industrial Concerns.* There is no central index of such case studies, many of which contain proprietary information and are for internal company use, hence are not made public. However, material presented by industrial guest speakers for seminars and from joint industry–university teaching ventures may make effective case study material.

*0.2.1.2.3 Case Studies Prepared by the Instructor.* Many topics come to the fore as one teaches a course and, although the work involved is not trivial, an instructor can prepare short case studies with the help of term papers and class assignments. If a research effort accompanies the classwork, one can prepare case studies of sufficiently high quality to meet thesis requirements of the master's degree.

0.2.1.3 Use of Speakers from Institutions Outside the University. This essential and rewarding part of the course allows students to hear and question people who are actually involved with the issues dealt with in the text, the supplementary sources, or the case studies. About 10–15% of the time should be devoted to outside speakers. Industrial concerns and government agencies are most cooperative, especially if you talk to people at the highest level. The speaker and the topic must be selected to fit into the course structure. Request background material from the speaker in the form of handouts or published articles, and make sure the students have read the material prior to the presentation. Time should be provided for questions and answers. Also informal meetings between the speaker and the students in the class is a forum for further discussion.

**0.2.1.4 Assignment as a Student Specialist.** After several years of experimentation, we have found that the following procedures greatly enliven class discussions, allow the more reticent student to participate more easily, and provide an effective means of motivating the students to become familiar with the basic references and to give practice in researching the economic literature of the chemical industry.

*0.2.1.4.1 Product Specialist Assignment.* Detailed instructions are given in Section 0.2.2. The plan is to have each student be the course specialist on at least one chemical product and to prepare a comprehensive term paper. The products must be carefully selected by the instructor, who should have a plan for class discussion utilizing the detailed information collected by the student.

Typical product assignments are found in Table 0.1. Normally the products discussed in Chapter 6 are not used as product assignments. The 50 largest-volume chemicals from the *Chemical and Engineering News* Facts and Figures issue are prime candidates for product specialist assignments. High market value chemicals that are not necessarily in the largest-volume list also make good product assignments. Any chemical products that are particularly timely (e.g., chiral drugs, bioherbicides) make interesting product assignments. The instructor should always ask student specialists to comment when their products appear in the news.

*0.2.1.4.2 Company Specialist Assignment.* Detailed instructions are found in Section 0.2.2. This assignment requires each student to become a class expert on a CPI company and to prepare a comprehensive term paper. The student is expected to provide both statistical and qualitative information on the company. The instructor must select the company assignments carefully, and it is essential to make use of the student's knowledge in both a planned and spontaneous fash-

**TABLE 0.1**  Typical Product Assignments

| | |
|---|---|
| Oxygen | Acetic acid |
| Methyl *tert*-butyl ether | Carbon black |
| Terephthalic acid | Color toners |
| Bisphenol A | Chlorine |
| Toluene diisocyanate | ABS |
| Adipic acid | Ethylene dichloride |
| Ethylene glycol | Maleic anhydride |
| Nylon 6 | Sulfuric acid |

TABLE 0.2   Typical Company Assignments

| | |
|---|---|
| Albemarle Corporation | Witco |
| Ciba-Geigy | Praxair |
| Lonza, Inc. | General Foods |
| Church & Dwight | FMC Corporation |
| Occidental Chemical Corporation | Drew Chemical |
| Millenium | Rohm & Haas |
| PPG | Biogen |
| Great Lakes Chemical | Hercules, Inc. |
| Alcoa | Stepan Company |

ion. Any of the firms listed in Appendix B make suitable assignments. The firms selected need not be large—perhaps specialty and fine chemical manufacturers. A typical list of assignments is found in Table 0.2.

A list of both product and company specialists should be given to all students in the class and to invited speakers, who should be encouraged to call upon the student experts.

*0.2.1.4.3 Special-Situation Specialist.* This assignment may be made instead of a product or company assignment to one or more students, or it may be an additional term paper for some or all students. It can be a very interesting and satisfying assignment if the term papers are properly researched and clearly written. They can be the basis for a case study or may lead to a graduate-level research program.

If there is time, it is worthwhile to have students present the reports, generally allotting one or two class periods to this activity. A speaker's performance may be evaluated on the following points:

> *Oral presentation:* quality of the presentation (clear, understandable, and well organized); competence in the preparation and use of training aids (slides, graphs, blackboard, or computer projection); ability to answer questions (does speaker understand the question?); unwillingness to admit inability to answer (trying to bluff).
> *Content:* comprehensiveness (covers all the necessary points); critical analysis (active independent criticism rather than just reporting the facts); use of concepts of finance and accounting principles developed in the course; originality and scope (instructor assesses overall impact of the talk and the general impression made).

**0.2.1.5   Development and Preparation of Supplemental Materials to Augment the Text.**   Each student should be given a package of material at the start of the course that contains as a minimum the following:

1. Facts and Figures issue of *Chemical and Engineering News.* As mentioned earlier in this chapter, much of the statistical material in this text came from tables presented in this periodic feature of the publication.
2. One company annual and 10K report (see Chapter 3). These are essential for review and discussion in Chapters 3, 5, and 7. Enough copies can be obtained for each member of the class by writing to the controller or treasurer of a company. This information may also be available on the Internet at www.investquest.com or at the individual company's website.
3. Case studies. Plan on two or three.
4. Reprints from current periodicals. These must be decided well before the class begins and ordered from the specific publications.
5. *Statistical Abstract of the United States.* As noted in Section 0.1.2, this is the standard summary of statistics on the political, economic, and social organizations of the United States. It needs to be available to update sections of the book. One or two copies of this expensive publication should be available in a "reserve" area of the departmental library for easy access.

Of course, there are many other sources of materials such as company house publications, government reports, and foundation and industry association reports, as well as those from other agencies like banks.

The data in the text and the "distribution package" above can be overwhelming if the class is not given chance to become familiar with the materials during or before the corresponding lecture. In Chapters 5–7, where by necessity there are a large number of tables, we have included Questions for Discussion designed to make students more aware of the significance of the data by requiring them to reorder the material, to compute ratios, to compare one table with another, and to interpret the tabular data. By using the questions in this special section as homework or as a short in-class assignment, the instructor can easily develop significant discussions. For example, the class can be asked to rank the companies in Table 5.9 by return on investment and by profit margin. This simple exercise can easily generate 15 minutes of lively questions and discussion, which can be related to the material already covered in Chapter 3 and to various company reports. Asking the class to find the companies in Table 5.10 with significant chemical sales can also generate good class discussion.

0.2.1.6 Planning. It cannot be emphasized strongly enough that to have a successful, interesting course requires considerable advance planning by the instructor. This course is not like the usual undergraduate lecture course because of the

---

[20]Reprints may be obtained from Special Issue Sales, American Chemical Society, 1155 Sixteenth St., N.W., Washington, D.C. 20036.

need to obtain additional information and outside speakers. This course cannot be "winged," an approach that would be disastrous and not beneficial to the student. To prepare for this course we try to follow a schedule like that of Table 0.3. In the three-credit course (42–45 hours), all 10 chapters can be adequately covered with time for guest speakers, case studies, and discussion of current problems of interest to the CPI. A typical flexible in-class schedule is presented in Table 0.4.

## 0.2.2 Student's Guide

This course should be studied differently from most engineering or science courses. The basic structure is more empirical and inductive, the kinds of information and the skills you need to develop are different and more diverse,

**TABLE 0.3** Planning Schedule

| Months before course begins | Phases to be completed | Comment |
|---|---|---|
| 12–10 | Collect articles which will expand upon and complement text; decide on reprints students should have. | Begin to rough out in-class course schedule. |
| 6–4 | Order text. | Not knowing class registration can be troublesome; class enrollment should have an upper limit to ensure adequate class discussions and interaction. |
| 4–2 | Order reprints, Facts and Figures issue of *Chemical and Engineering News*, case studies, company reports, etc. | |
| 2 | Decide upon companies and products to assign to student specialists; make a list in order of importance. | In-class schedule should now be fairly well decided. |
| 1½ | Invite outside speakers. | The last bit of information needed to firm up the class schedule. |
| 1 | Prepare package of supplemental material. | |

TABLE 0.4

| Chap. | Hours |
|---|---|
| 1 | 1–2 |
| 2 | 5–7 |
| 3 | 1–3 |
| 4 | 2–4 |
| 5 | 1–2 |
| 6 | 1–2 |
| 7 | 1–2 |
| 8 | 4–6 |
| 9 | 2–4 |
| 10 | 2–4 |
| Guest speakers | 3–5 |
| Case studies | 3–5 |

and there is seldom one correct answer to any of the problems posed. There must be active, informed, and well-thought-out discussion if you are to understand the economics of the complex CPI and the role chemical professionals play and must learn to play in the years to come. Tomorrow's engineers and chemists must understand more of the overall picture. This *does not mean* that they should discard their technical and scientific skills. Without them they become one of those drones of society who can only talk about problems. Chemical professionals must strengthen their technical expertise by obtaining additional economic and sociopolitical skills. We need chemical professionals who can *solve* real problems in which the technical component may not be the critical factor.

To develop the required skills you need to:

1. Read carefully the text, assigned case studies, reprints, journals, and references.
2. Participate in class discussion both by listening and by talking.
3. Do problems developed to demonstrate comprehension of concepts and important facts.
4. Do problems by applying economic and accounting tools to their analysis.
5. Do problems that give practice in library research for economic data.
6. Prepare two term papers that will develop your skills in data searching, logical analysis, and construction of convincing economic arguments.
7. Take examinations that will test comprehension and analytical skill.

Read Section 0.2.1 to understand the planning and scheduling for the course. Section 0.1 must constantly be consulted for appropriate sources. Your contribution is needed in and out of class in at least four ways.

**0.2.2.1  As a Product Specialist.**   Much can be learned about the chemical industry by studying the products it makes and the companies making up the industry. The task of the product specialist is to become a class expert on at least one important product in the chemical industry. Your instructor will assign or let you select the product(s) based upon the overall plan of the course. You will be given approximately 2–3 weeks to prepare a term paper on the product. You will be graded by your ability to respond to class questions and discussion regarding your product(s) as well as the quality of the written report.

To become the kind of expert required, it is necessary but not sufficient to consider:

1.  Total annual sales for the last decade
2.  Selling price structure over the last decade
3.  Principal uses and percentages of product consumed by each use
4.  Companies involved and as much information as possible on capacity and plant sites
5.  Process flow diagram and technical details of major and alternative manufacturing methods
6.  Some historical perspective
7.  Environmental and safety problems

The term paper should have at least three major sections:

1.  *Marketing.* The following questions need to be answered. Who are the customers? What do they use the product for? What other products compete with it? Who are the competitors? What is the sales history? How much has been sold each year for the past 10–20 years? What is the price history? How is the product classified (Chapter 5)? Are there any new uses or new markets to consider? What external threats exist? Can they be avoided?

2.  *Manufacturing.* This is the technical section of the report. It must be quantitative and up to date. Answers are needed to the following questions. What process or processes are used? (A simplified process flow diagram with some basic mass and energy balances is necessary at the very least.) How old is the process? Have there been any recent innovations? What are the major capital costs? What are the major operating expenses? Is the raw material supply secure? Are alternate raw materials available? Can productivity be improved (by, e.g., a new process, more control, or more research and development)? Which companies manufacture the product? What are approximate plant capacities?

3.  *Environmental impact.* Is a nuisance produced by the manufacturing process? Is there a problem in transportation? Do manufacture and use of the

product cause enviromnental or health problems? Do the benefits outweigh any harm? Can the bad effects be controlled? What is the cost of such control? Does the product present a potential customer liability problem?

### 0.2.2.2 As a Company Specialist.

You will also be required to become the class expert on one important company in the chemical industry. The instructor will assign or permit you to select the company based on the course plan, but it is your responsibility to gather the kinds of information that will encourage active and informed class discussion. As a minimum, you must consider the following for your company term paper:

1.  The company's financial structure (total sales, earnings per share, net income, stock price, assets, debts, etc.)
2.  The products made and their importance to the company business
3.  The company's planning for the future: expenditures for research and development, capital investment, innovations, and plans for dealing with a changing world

An annual report provides the minimum information needed. Financial data are also available from *Moody's, Valueline,* or *Standard & Poor's.* Current periodicals often provide interesting insights into company operations (particularly *Forbes*). The organization's website will provide an excellent source of company information.

A company term paper should consider the following questions:

1.  What business is it in? What products does it make and what customers does it serve? Will customers prosper in the future? Is there a mix of products? How vulnerable is the company to changes in market demand?

2.  What is the image of the company? Is it an aggressive, growth-oriented company making unusual chemicals for a high price, or a mature company making commodity chemicals in bulk for a low markup? Is innovation of new products, new markets, or new processes an important part of the company?

3.  What is the past record? Look at its past 10 years' record of sales, earnings per share, and stock price. What patterns are present: rising or falling? What are the causes behind these changes, and how does this company compare with other chemical companies manufacturing similar products, with industrial companies in general, with the GDP? Apply financial analysis.

4.  What are external threats and opportunities? What forces are gathering that may cause grief (a new process by a competitor, declining market for the product, regulation of pollution, price increase, or unavailability of raw materials) or prosperity?

Use numerical data to quantify your discussions as much as possible. Additional sources of information are brokerage houses and financial analysts, *The*

*Wall Street Journal, Moody's Handbook,* and company annual reports. Be critical and skeptical of your sources. Company reports tend to stress only the positive, anticompany critics of course stress the negative, and investor services such as *Moody's* and *Forbes* tend to be neutral. You need to consider all these evaluations as a broad base for your evaluations.

**0.2.2.3  As a Specialist on a Specific Situation or Problem.**  The instructor may assign, or you may decide on, an investigation of some special situation that affects the CPI. The possibilities are infinite, and much can be gained by both the student and the class if this type of term paper is well done.

Special-situation term papers must provide background information with as much quantitative data as possible. Be sure that both sides of any controversy are well presented. Principal arguments and logic should be analyzed. International data can be obtained in United Nations statistical tables, from major international or domestic banks operating in that nation, and often from the commerce attaché or commercial officer at the consulate.

Topics that have been successfully prepared include the following:

1. Environmental impact of synthetic versus natural fibers
2. Chemical plant buildup in the Far East
3. The chemical industry in China or South American countries
4. Recycling waste (e.g., plastic materials)
5. Experience curve for by-product chemicals
6. Pricing policy in the CPI

**0.2.2.4  As a Person Informed by Reading the Current Periodicals.** Sources of information on the chemical industry are listed in Section 0.1. You should read at least two of the current periodicals on a regular basis and raise issues in class which you feel should be discussed.

Most of your work requires well-organized library research in sources outside the usual technical area. You will greatly broaden your professional skills by doing these assignments effectively and by reading chemical industry literature.

Obtain your information from as many sources as possible to bring out different facets and interpretations. Every writer has prejudices and blind spots, and and a single individual cannot be depended upon to be comprehensive. The whole picture can emerge only after you have looked at the world through many pairs of eyes.

## 0.3  ADDITIONAL REFERENCES

The following books are recommended for reading because they give a sweeping view of the changing scene in the chemical industry:

J. Wei, *Frontiers in Chemical Engineering: Research Needs and Opportunities*. National Academy Press, Washington, DC, 1988.

Louis Hegedus, ed., *Chemical Technologies: The Role of Chemistry and Chemical Engineering*, National Academy Press, Washington, DC, 1992.

A. Arora, R. Landau, and N. Rosenberg, eds. *Chemicals and Long-Term Economic Growth: Insight from the Chemical Industry*. John Wiley & Sons, New York, 1998.

Robert Stobaugh, *Innovation and Competition: The Global Management of Petrochemical Products*. Harvard Business School Press, Cambridge, MA, 1988.

Paul Krugman, *The Age of Diminished Expectations: U.S. Economic Policy in the 1990s*. MIT Press, Cambridge, MA, 1990.

# 1

---

# Introduction

## 1.1 GENERAL REMARKS

The chemical professionals, chemical engineers, and chemists, devote most of their formal educational efforts to the study of technology and science. From courses in chemistry, physics, thermodynamics, kinetics, transport phenomena, unit operations, and design, the engineer or chemist learns about the laws of nature and how to describe physical phenomena in useful ways. To function in a truly useful manner, however, today's chemical professional must understand much more than science and technology. The dynamic professional must understand the complex economic and sociopolitical factors which affect the application of technical and scientific expertise.

Figure 1.1 graphically depicts the interactions between the decisive factors that affect the launching and success of a venture. We can use the diagram to illustrate the interactions in a typical problem facing our society today: the need to use alternate sources of energy and become less dependent on the crude oil sources outside the United States. Suppose that as a society we wish to manufacture a significant amount of a synthetic fuel from coal. Solution of the major technical and scientific problems is a necessary first step before any economic issues can be considered. What is the basic chemistry? What are the kinetic, thermodynamic, and transport factors affecting equipment design? What are the engineering process design problems? What materials of construction are needed? What processes have already been tested and demonstrated?

As a number of technical alternatives are developed, economic issues must be considered. At today's prices and availability of supplies, which process is more efficient in resource utilization? Are raw material prices and availability

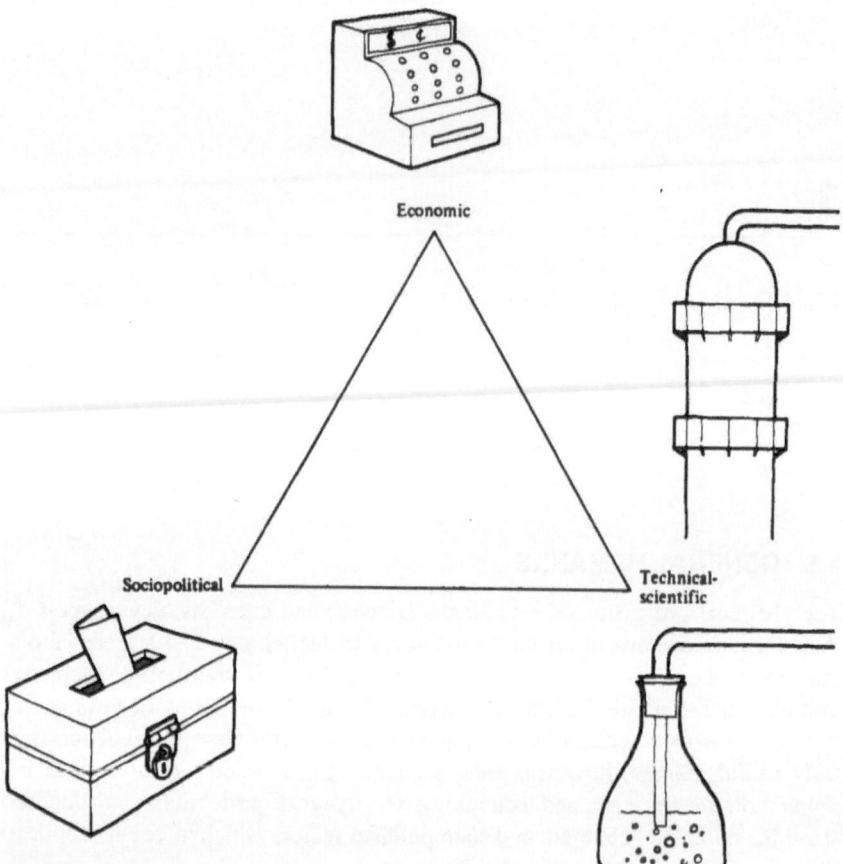

FIGURE 1.1   Decisive factors.

likely to be stable or to change dramatically? What size plant should be built? How many plants should be built?

Solutions that are technically and economically feasible must be acceptable to society. The preference of the people, expressed through the political process, is shaped by their hopes and fears as well as by their choices between economic efficiency and their quality of life. What environmental impact is acceptable to the public? What are the waste streams, and are there recoverable materials in them? What level of independence from imports strikes the people as the best balance between national security and consumer price increases? The solution path for solving a technical problem is illustrated in Figure 1.2.

Today, most of the technical problems in coal conversion are solvable, and

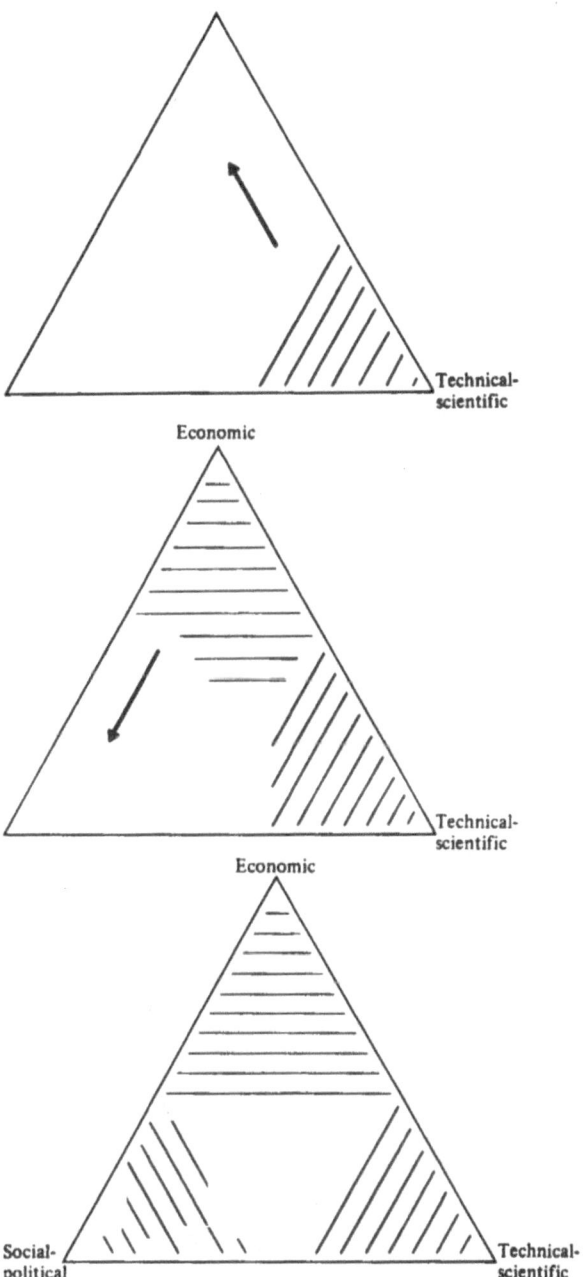

FIGURE 1.2  Problem solution path using decisive factors diagram.

the operating expenses and capital costs of coal processing plants are reasonably well known. The problems that must now be solved are political and social at both the national and international levels. How will the price of a barrel of imported crude oil vary over the next decade? Can a safe supply of imported crude oil be assumed? Should and will the United States government subsidize alternate energy processes? Should such ventures be partially supported by government and partially by a consortium of industry? Do we need a floor price on crude oil, and if so, what should it be? What political moves need to be made to achieve this?

We can also use the decisive factors diagram (Figure 1.3) to illustrate the structure of most educational programs and see how well prepared the chemical professional is to solve the questions posed in the preceding paragraph. Figure 1.3 shows examples of the types of course available in a typical university program in chemical engineering or chemistry. In addition to the technical and scientific courses, engineering students have a number of design-oriented courses, which help them see the economic constraints affecting the application of technology. Process design courses usually show how to determine which combination of technologies should be used to perform a chemical transformation in the most economical manner.

As a student, the future chemical professional takes courses in basic economics and social science as well as perhaps business courses that develop skills and give insight into economic and sociopolitical factors. To be used effectively, this nontechnical knowledge should be related to the technical and scientific knowledge while the student is in school.

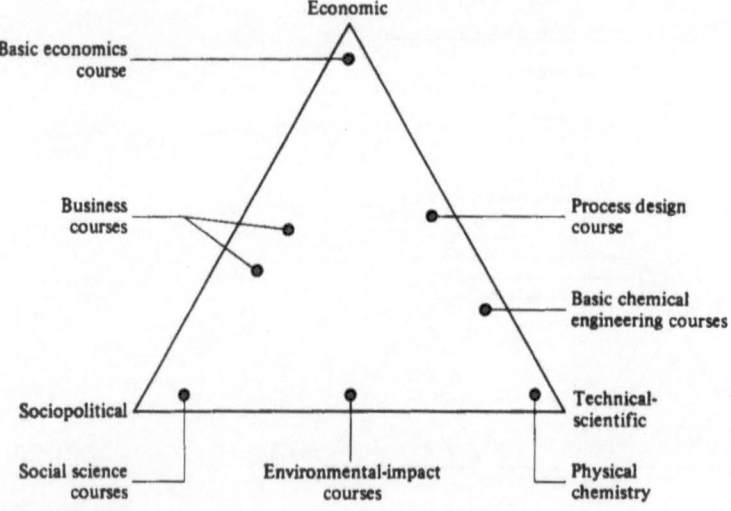

FIGURE 1.3   Decisive factors and university courses.

This text is written to help bridge the gap between the technical and nontechnical education. The text material emphasizes those aspects along the technical–economic side of the triangular diagram, with frequent departures into the sociopolitical aspects where appropriate. Case studies and outside collateral readings may be used to illustrate interactions between all three decisive factors. For appropriate case studies, one should consider those published by the Harvard Business School.

All chemical professionals should know how these decisive factors affect the ventures in which they are involved. Recently graduated chemical professionals may initially receive assignments that are mostly technical; however, they will quickly be introduced to problems in which economics play decisive roles. Within the first 5–10 years, the chemical professional will move into positions wherein economic factors play a dominant role.

After, say, 10–20 years, some professionals will move to upper-level managerial positions in which the main requirements will necessitate the handling of both economic and sociopolitical issues. A typical career can be sketched on the decisive factors triangle as shown in Figure 1.4. A chemical professional must be prepared to make the necessary transition in roles.

This text and the course based on it are designed to achieve the following goals:

1. To expand the mental horizon of the chemical profession beyond science and engineering and to show the economic purposes of the chemical process industries (CPI) and how the CPI benefits society.

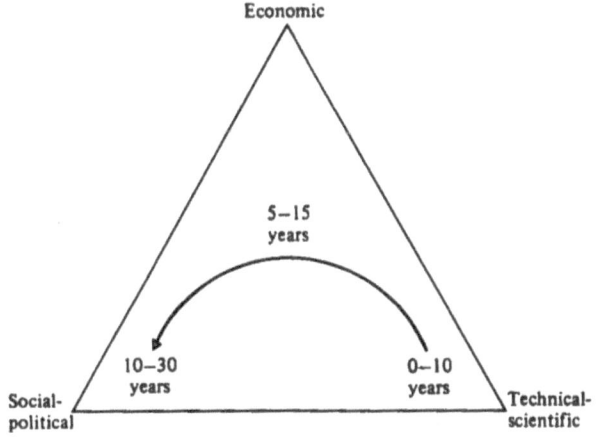

FIGURE 1.4   A typical career path.

2.  To help chemical professionals understand how their work relates to the goals of their company and of society.
3.  To develop in the chemical professional an appreciation for the potential impact of new developments in technology, marketing, finance, politics, or international affairs as threats and opportunities.
4.  To teach the chemical professional how to influence an organization to move in new directions by making fact-filled, comprehensive, and convincing economic studies.

To meet these objectives, certain *concepts and tools* should be mastered:

1.  Basic economic, financial, and accounting principles are covered in Chapters 2–4.
2.  External and internal factors that affect the CPI and the companies within them are identified and discussed in Chapters 4, 8, and 9.
3.  Future factors that are threats and opportunities and will affect innovation and growth are discussed in Chapter 10.

There are certain *facts* you must be familiar with if the objectives are to be attained:

1.  The major products and selected major CPI technologies are discussed in Chapters 1, 5, and 6.
2.  Major companies of the CPI are identified in Chapter 5 and some are discussed in more detail in Chapter 7.
3.  The major markets and suppliers of the CPI are discussed in Chapters 4 and 8.
4.  Employment in the CPI is dealt with in Chapters 1 and 8.
5.  International manufacturing and trade are covered in Chapter 9.
6.  The relation between the CPI and government is covered in Chapters 2, 4, and 10.

The effective chemical professional needs to master certain *skills:*

1.  The ability to read, analyze, and comprehend economic, financial, and accounting statements and studies to understand their significance, and to use them as a rational basis for decisions
2.  The ability to detect substantive weaknesses in a study (e.g., important facts and factors omitted, flaws in logic, inappropriate interpretations of facts and events)
3.  The ability to locate and select economic data, to supply appropriate logic and analysis, and to make convincing economic studies
4.  The ability to perceive threats and opportunities, to devise imaginative and effective measures for thwarting the one and taking advan-

tage of the other, and to persuade key people and organizations to effect needed changes

If you are studying this book without being enrolled in a course, you must follow the current literature outlined in Section 0.1 and do many of the problems designed to help develop the skills listed above. The chemical professional also should develop certain *habits* to keep in touch with important issues, events, and ideas being developed in the profession, the nation, and the world. The list that follows is illustrative.

1. Regular reading of newspapers and journals that carry national and international news as well as developments in politics, economics, and social events
2. Regular reading of professional journals that carry news on the economics of the industry and on the development of new processes and products
3. Periodic attendance at professional meetings, lectures, and symposia on broad issues that affect the profession
4. Periodic reading of fiction and nonfiction to gain an insight into changing values of society
5. Meeting people outside the profession in social and service capacities as well as paying particular attention to their concerns and their priorities
6. Periodic attendance in continuing education courses to enhance knowledge and capabilities in technical, business, management, and economic areas

## 1.2 THE CHEMICAL PROCESSING INDUSTRIES

### 1.2.1 Definition

The chemical processing industries produce many different and unrelated products and serve a myriad of different markets. Some authors define the CPI to include all industries in which a chemical reaction takes place. Using such an approach, the CPI would also include wineries, paper mills, steel mills, tanneries, petroleum refineries, sugar refineries, and petrochemical plants. *Chemical Engineering* takes this broad approach (Table 1.1).

An attempt to cover in detail a group of such diverse activities as those found in Table 1.1 would require many volumes. In this text we shall concentrate our efforts on Standard Industrial Classification (SIC) group 28 (chemicals and allied products) and group 29 (petroleum refining and related industries) (1).

The U.S. Bureau of the Budget publishes the SIC code and updates it about every 5 years. It defines SIC 28 as follows:

**TABLE 1.1**   The CPI as Defined by *Chemical Engineering*

| Industry segment | SIC code |
| --- | --- |
| 1. Industrial inorganic chemicals | **281** |
| Alkalies and chlorine | 2812 |
| Industrial gases | 2813 |
| Cyclic intermediates and crudes | 2865 |
| Inorganic pigments | 2816 |
| Industrial organic chemicals not otherwise classified | 2869 |
| Industrial inorganic chemicals not elsewhere classified | 2819 |
| Synthetic rubber (vulcanizable elastomers) | 2822 |
| Gum and wood chemicals | 2861 |
| Chemicals and chemical preparations not elsewhere classified | 2899 |
| 2. Drugs | **283** |
| Medicinal chemicals and botanical products | 2833 |
| Pharmaceutical preparations | 2834 |
| In vitro and in vivo diagnostic substances | 2836 |
| 3. Fats and oils | **207** |
| Cottonseed oil mills | 2074 |
| Soybean oil mills | 2075 |
| Vegetable oil mills except cottonseed and soybean mills | 2076 |
| Animal and marine fats and oils including grease and tallow | 2077 |
| Shortening, table oils, margarine, and other fats and oils not otherwise classified | 2079 |
| 4. Agricultural chemicals | **287** |
| Nitrogenous fertilizers | 2873 |
| Phosphatic fertilizers | 2874 |
| Fertilizers, mixing only | 2875 |
| Pesticides and agricultural chemicals not otherwise classified | 2879 |
| 5. Foods and beverages | **202** |
| Condensed and evaporated milk | 2023 |
| Wet corn milling | 2046 |
| Cane sugar, except refining only | 2061 |
| Cane sugar refining | 2062 |
| Beet sugar | 2063 |
| Malt liquors | 2082 |
| Malt | 2083 |
| Wines, brandy, and brandy spirits | 2084 |
| Distilled, rectified, and blended liquors | 2085 |
| Bottled and canned soft drinks and carbonated water | 2086 |
| Flavoring extracts and syrups not elsewhere classified | 2087 |
| Roasted coffee | 2095 |
| Food preparations not elsewhere classified | 2099 |
| 6. Leather tanning and finishing | **3111** |

TABLE 1.1   Continued

| Industry segment | SIC code |
|---|---|
| 7. Lime and cement | |
| Cement, hydraulic | **3241** |
| Lime | 3274 |
| 8. Metallurgical and metal products | |
| Electromagnetic products | 3313 |
| Primary smelting and refining of copper | 3331 |
| Primary smelting and refining of lead | 3339 |
| Primary smelting and refining of zinc | 3338 |
| Primary production of aluminum | 3334 |
| Primary smelting and refining of nonferrous metals not elsewhere classified | 3339 |
| Secondary smelting and refining and alloying of nonferrous metals and alloys | 3341 |
| Enameled iron and metal sanitary wear | 3431 |
| Electroplating, plating, polishing, anodizing, and coloring | 3471 |
| Coating, engraving, and allied services not elsewhere classified | 3479 |
| 9. Miscellaneous chemical products | **289** |
| Adhesives and sealants | 2891 |
| Explosives | 2892 |
| Printing ink | 2893 |
| Carbon blacks | 2895 |
| Chemicals and chemical preparations not elsewhere classified | 2899 |
| 10. Paints, varnishes, lacquers, enamels, and allied products | **2851** |
| 11. Petroleum, refining, and allied products | **29** |
| Petroleum refining | 2911 |
| Asphalt paving mixtures and blocks | 2951 |
| Asphalt felts and coatings | 2952 |
| Lubricating oils and greases | 2992 |
| Products of petroleum and coal, not elsewhere classified | 2999 |
| Coke and by-products including coke ovens part of | 3312 |
| 12. Plastic materials and synthetic resins, synthetic rubbers, cellulosic and other man-made fibers, except glass | **282** |
| Plastic materials, synthetic resins, and nonvulcanizable elastomers | 2821 |
| Synthetic rubber (vulcanizable elastomers) | 2822 |
| Cellulosic man-made fibers | 2823 |
| Man-made organic fibers except cellulosic | 2824 |
| 13. Rubber and miscellaneous plastic products | **30** |
| Tires and inner tubes | 3011 |
| Rubber and plastic footwear | 3021 |
| Rubber and plastic hose and belting | 3052 |
| Gaskets, packing, and sealing devices | 3053 |

**TABLE 1.1** Continued

| Industry segment | SIC code |
|---|---|
| Molded, extruded, and lathe-cut mechanical rubber products | 3061 |
| Fabricated rubber products not elsewhere classified | 3069 |
| Unsupported plastic film and sheet | 3081 |
| Unsupported plastic profile shapes | 3082 |
| Laminated plastic plate, sheet, and profile shapes | 3083 |
| Plastic pipe | 3084 |
| Plastic bottles | 3085 |
| Plastic foam products | 3086 |
| Custom compounding of purchased plastic resins | 3087 |
| Plastic plumbing fixtures | 3088 |
| Plastic products not elsewhere classified | 3089 |
| 14. Soap, detergents, and cleaning preparations, perfumes, cosmetics, and other toilet preparations | **284** |
| Soaps and other detergents except specialty cleaners | 2841 |
| Specialty cleaning, polishing, and sanitary preparations | 2842 |
| Surface active agents, finishing agents, sulfonated oils and assistants | 2843 |
| Perfumes, cosmetics, and other toilet preparations | 2844 |
| 15. Stone, clay, glass, and ceramics | **32** |
| Flat glass | 3211 |
| Glass containers | 3221 |
| Pressed and blown glass and glassware not elsewhere classified | 3229 |
| Brick and structural clay tile | 3251 |
| Ceramic wall and floor tile | 3253 |
| Clay refractories | 3255 |
| Structural clay products not elsewhere classified | 3259 |
| Pottery and related products | 3261–3264, 3269 |
| Gypsum products | 3275 |
| Abrasive products | 3291 |
| Asbestos products | 3292 |
| Minerals and earth, ground or otherwise classified | 3295 |
| Mineral wool | 3296 |
| Nonclay refractories | 3297 |
| Nonmetallic mineral products | 3299 |
| 16. Wood, pulp, paper, and board | **24, 26** |
| Wood preserving | 2491 |
| Pulp mills | 2611 |
| Paper mills | 2621 |
| Paperboard mills | 2631 |
| Paper coating and glazing | 2672 |
| Building-paper and building-board mills | 2621 |
| 17. Other chemically processed products | |
| Part of broad-woven fabric mills including dyeing and finishing | 2231 |

TABLE 1.1 Continued

| Industry segment | SIC code |
| --- | --- |
| Finishers of broad-woven fabrics of cotton | 2261 |
| Finishers of broad-woven fabrics of man-made fiber and silk | 2262 |
| Finishers of textiles not elsewhere classified | 2269 |
| Artificial leather, oilcloth, and other impregnated and coated fabrics except rubberized | 2295 |
| Glue and gelatin | 2891 |
| Storage batteries | 3691 |
| Lead pencils, crayons, and artists' materials | 3952 |
| Carbon paper and inked ribbons | 3955 |
| Linoleum, asphalted-felt-base and other hard-surface floor coverings not elsewhere classified | 3996 |
| Candles and matches (manufacturing industries not elsewhere classified) | 3999 |

Source: Chemical Engineering and Ref. 1.

This major group includes establishments producing basic chemicals, and establishments manufacturing products by predominantly chemical processes. Establishments classified in this major group manufacture three general classes of products: 1) basic chemicals such as acids, alkalies, salts and organic chemicals; 2) chemical products to be used in further manufacture such as raw materials to produce synthetic fibers, plastics, materials, dye colors and pigments; 3) finished chemical products to be used for ultimate consumption such as drugs, cosmetics and soaps; or to be used as materials or supplied in other industries such as paints, fertilizers and explosives.

SIC group 28 is broken into eight three-digit groups of chemical and allied products

| SIC no. | Subgroups |
| --- | --- |
| 281 | Industrial inorganic chemicals |
| 282 | Plastic materials and synthetics (plastics materials and resins, synthetic rubber, cellulosic man-made fibers, organic fibers, noncellulosic) |
| 283 | Drugs |
| 284 | Soaps, cleaners, and toilet goods (soap and other detergents, polishes and sanitation goods, surface active agents, and toilet preparations) |
| 285 | Paints and allied products |
| 286 | Industrial organic chemicals |
| 287 | Agricultural chemicals |
| 289 | Miscellaneous chemical products |

The major group of SIC 29 is defined by the U.S. Bureau of the Budget as including "establishments primarily engaged in petroleum refining, manufacturing paving and roofing materials and compounding lubricating oils from purchased materials."

SIC group 29 is broken down into three groups of petroleum refining and related industries:

| SIC no. | Subgroups |
|---------|-----------|
| 291 | Petroleum refining |
| 295 | Paving and roofing materials |
| 299 | Miscellaneous products of petroleum and coal |

Each of these three-digit subgroups is further subdivided into four-digit subgroups such as 2891, adhesives and gelatins; 2892, explosives; 2893, printing ink; and 2895 carbon black. Likewise, the four-digit subgroups are broken into five-digit subgroups, and so on. The complete SIC breakdown as used by the Census of Manufactures (1) may be found in Appendix A (Table A.1). To obtain data for any $n$-digit SIC group, the statistics for the $(n + 1)$-digit groups are included in that particular $n$-digit group are summed. For instance, the volume of sales for SIC 29 is the sum of the sales in SIC 291, 295, and 299.

The economic activities of many major corporations encompass operations in more than one SIC group. Exxon is engaged in oil and gas production (SIC 13), oil and gas transportation (SIC 40–46), oil refining (SIC 29), oil marketing (SIC 55), and petrochemicals (SIC 28). Many major companies that are not usually considered to be chemical companies have significant chemical production; examples include General Electric, Food Machinery Corporation, Borden, and USX (formerly U.S. Steel).

In addition to the SIC, where each plant or establishment receives an individual code, the federal government uses the Enterprise Standard Industrial Classification (ESIC) (2), where each company receives a single code according to the company's dominant activity. In the ESIC system, the real estate transactions of Shell Oil are all classed as ESIC 29 and the chemical operations of firms like General Electric, Borden, and Food Machinery Corporation are not part of ESIC 28. This system is useful in compiling and analyzing financial and related data (e.g., income, expenses, and profits that may be available only on a company basis). The existence of two systems requires care and discrimination in analyzing economic data.

The federal government also uses a third system of classification for input–output analysis (see Chapter 4) (3). For future reference, input–output classification numbers, titles, and relations to the SIC code numbers are shown in Table 1.2 for chemicals and selected chemicals products as well as several other industries. These tables are useful in business planning for 1-year, 5-year, and ·

TABLE 1.2   Partial Listing of Input–Output Classifications, 1987

| Input–output industry number and title | Related 1987 SIC codes |
|---|---|
| 24 Paper and allied products, except containers | |
| 24 0100 Pulp mills | 261 |
| 24.0400 Envelopes | 2677 |
| 24.0500 Sanitary paper products | 2676 |
| 24.0701 Paper coating and gluing | 2671–2 |
| 24.0702 Bags, except textile | 2673–4 |
| 24.0703 Die-cut paper and paperboard and cardboard | 2675 |
| 24.0705 Stationery, tablets, and related products | 2678 |
| 24.0706 Converted paper products, not elsewhere covered | 2679 |
| 24.0800 Paper and paperboard mills | 262–3 |
| 27A Industrial and other chemicals | |
| 27.0100 Industrial inorganic and organic chemicals | 281 |
| 27.0401 Gum and wood chemicals | 2861 |
| 27.0402 Adhesives and sealants | 2891 |
| 27.0403 Explosives | 2892 |
| 27.0404 Printing ink | 2893 |
| 27.0405 Carbon black | 2895 |
| 27.0406 Chemicals and chemical preparations, not elsewhere covered | 2899 |
| 27B Industrial and other chemicals | |
| 27.0201 Nitrogenous and phosphatic fertilizers | 2873–4 |
| 27.0202 Fertilizers, mixing only | 2875 |
| 27.0300 Pesticides and agricultural chemicals, not elsewhere covered | 2879 |
| 28 Plastics and synthetic materials | |
| 28.0100 Plastics materials and resins | 2821 |
| 28.0200 Synthetic rubber | 2822 |
| 28.0300 Cellulosic man-made fibers | 2823 |
| 28.0400 Man-made organic fibers, except cellulosic | 2824 |
| 29A Drugs | |
| 29.0100 Drugs | 283 |
| 29B Cleaning and toilet preparations | |
| 29.0201 Soap and other detergents | 2841 |
| 29.0202 Polishes and sanitation goods | 2842 |
| 29.0203 Surface active agents | 2843 |
| 29.0300 Toilet preparations | 2844 |
| 30 Paints and allied products | |
| 30.0000 Paint and allied products | 285 |
| 31 Petroleum refining and refined products | |
| 31.0101 Petroleum refining | 291 |
| 31.0102 Lubricating oils and greases | 2992 |
| 31.0103 Products of petroleum and coal, not elsewhere covered | 2999 |

**TABLE 1.2**  Continued

| Input–output industry number and title | Related 1987 SIC codes |
|---|---|
| 31.0200 Asphalt paving mixtures and blocks | 2951 |
| 31.0300 Asphalt felts and coatings | 2952 |
| 32 Rubber and miscellaneous plastics products | |
| 32.0100 Tires and inner tubes | 301 |
| 32.0200 Rubber and plastics footwear | 302 |
| 32.0300 Fabricated rubber products, not elsewhere covered | 306 |
| 32.0400 Miscellaneous plastics products, not elsewhere covered | 308 |
| 32.0500 Rubber and plastics hose and belting | 3052 |
| 32.0600 Gaskets, packing and sealing devices | 3053 |
| 35 Glass and glass products | |
| 35.0100 Glass and glass products, except containers | 3229, 3231 |
| 35.0200 Glass containers | 3221 |
| 36 Stone and clay products | |
| 36.0100 Cement, hydralic | 324 |
| 36.0200 Brick and structural clay tile | 3251 |
| 36.0300 Ceramic wall and floor tile | 3253 |
| 36.0400 Clay refractories | 3255 |
| 36.0500 Structural clay products, not elsewhere covered | 3259 |
| 36.0600 Vitreous china plumbing fixtures | 3261 |
| 36.0701 Vitreous china table and kitchenware | 3262 |
| 36.0702 Fine earthenware table and kitchenware | 3263 |
| 36.0800 Porcelain electrical supplies | 3264 |
| 36.0900 Pottery products, not elsewhere covered | 3269 |
| 36.1000 Concrete block and brick | 3271 |
| 36.1100 Concrete products, except block and brick | 3272 |
| 36.1200 Ready-mixed concrete | 3273 |
| 36.1300 Lime | 3274 |
| 36.1400 Gypsum products | 3275 |
| 36.1500 Cut stone and stone products | 3288 |
| 36.1600 Abrasive products | 3291 |
| 36.1700 Asbestos products | 3292 |
| 36.1900 Minerals, ground or treated | 3295 |
| 36.2000 Mineral wool | 3296 |
| 36.2100 Nonclay refractories | 3297 |
| 36.2200 Nonmetallic mineral products, not elsewhere covered | 3299 |
| 37 Primary iron and steel manufacturing | |
| 37.0101 Blast furnace and steel mills | 3312 |
| 37.0102 Electrometallurgical products, except steel | 3313 |
| 37.0103 Steel wiredrawing and steel nails and spikes | 3315 |
| 37.0104 Cold-rolled steel sheet, strip, and bars | 3316 |
| 37.0105 Steel pipe and tubes | 3317 |

TABLE 1.2    Continued

| Input–output industry number and title | Related 1987 SIC codes |
|---|---|
| 37.0200 Iron and steel foundries | 332 |
| 37.0300 Iron and steel forgings | 3462 |
| 37.0401 Metal heat treating | 3398 |
| 37.0402 Primary metal products, not elsewhere covered | 3399 |
| 38        Primary nonferrous metals manufacturing | |
| 38.0100 Primary smelting and refining of copper | 3331 |
| 38.0400 Primary aluminum | 3334, 2819 |
| 38.0501 Primary nonferrous metals, not elsewhere covered | 3339 |
| 38.0600 Secondary nonferrous metals | 334 |
| 38.0700 Rolling, drawing, and extruding of copper | 3351 |
| 38.0800 Aluminum rolling and drawing | 3363–5 |
| 38.0900 Nonferrous rolling and drawing, not elsewhere covered | 3356 |
| 38.1000 Nonferrous wiredrawing and insulating | 3357 |
| 38.1100 Aluminum castings | 3363, 3365 |
| 38.1200 Copper foundries | 3366 |
| 38.1300 Nonferrous castings, not elsewhere covered | 3364, 3369 |
| 38.1400 Nonferrous forgings | 3463 |

Source: Ref. 3, p. 94.

10-year scenarios. Of the various tabulations, the chemical professional will find the SIC table the most useful.

Trade statistics for the United States are compiled by the Department of Commerce, which uses the United Nations Standard International Trade Classification (SITC) (4). (Note: There have been changes in the code, with the result that data after 1994 are not comparable with those of previous years.) The following list of SITC classes gives the principal products in each category:

SITC 51, Organic Chemicals: hydrocarbons and their derivatives, alcohols, phenols, and other organic compounds

SITC 52, Inorganic Chemicals: chlorine and other inorganic elements, acids, metallic salts, and other inorganic compounds

SITC 53, Dyeing, Tanning, and Coloring: dyes, color lakes, tanning extracts, paints, pigments, varnishes, inks, and other related preparations

SITC 54, Medicinals and Pharmaceutical Products: vitamins, antibiotics, hormones, veterinary and other medicaments, and related preparations

SITC 55, Essential Oils and Perfume Materials: essential oils, perfume materials, perfumes, cosmetics, toilet preparations, polishes, soaps, and related cleansing preparations

*SITC 56, Fertilizers:* nitrogenous, phosphatic, and other chemical and mineral fertilizers

*SITC 57, Plastics in Primary Form:* polyethylene, PVC, polystyrene, and other plastic resins

*SITC 58, Plastics in Non-Primary Form:* tubes, pipes, hoses, plates, sheets, film, foil, strips, and other shapes of plastic

*SITC 59, Chemical Materials and Products not elsewhere classified:* pesticides, starches, glues, explosives, and pyrotechnic products, fuel additives, plasticizers, textile specialties, and other miscellaneous products and preparations

### 1.2.2 The Purpose of a Firm

Why is a CPI firm needed by society? Modern chemical technology requires a high concentration of technical and managerial experts, a complex body of technical expertise, and large amounts of capital to be invested in equipment and plants. At one point in time, assurance of the firm's continuity beyond the mortality of the founders was felt to be essential; however, that concept has changed with the mergers and acquisitions, not only nationally but internationally, which have been prevalent since the 1980s. A chemical company can be a closely held private firm, such as Huntsman Corporation or Koch Industries, a publicly owned firm like Monsanto, or a government corporation like the Tennessee Valley Authority, as long as it is organized to provide economic goods and services through chemistry.

One of the purposes of a private firm is to make a profit, whereas a government corporation should at the very least minimize losses, which are burdens to taxpayers. Some economists suggest that a firm's only purpose is to maximize profit, but this is a simplistic view of a complex issue although profit is a measure of a company's performance. Other measures of performance are equally important. Still other economists suggest that the purpose of a firm is to survive and operate as an independent entity. While a firm sometimes serves society in an altruistic manner as a "good" citizen by contributions to charitable and humanitarian causes (e.g., donations to public radio and television, support of the arts, museums hospitals, and educational institutions), its main purpose is to serve society by an exchange of goods and services with various groups of people and institutions. To appreciate contributions of a firm, it is useful to consider how well, or indeed whether, certain human needs could be satisfied if the chemical firm did not exist. CPI firms are important to five segments of the public.

*Employees.* A firm provides employment, paychecks, tools with which to work, an opportunity to learn skills, an opportunity to pursue career development, personal fulfillment possibilities, and opportunities to participate in the production of useful goods and services. If modern firms did not exist, large-scale chemical activities would not be possible. Unemployment and unfulfilled careers, especially for the chemical professional, would result.

*Stockholders and lenders.* The public firm receives savings from individuals as well as pension and insurance funds to invest. In return, the stockholder receives dividends or interest and stock appreciation. The government firm can return a surplus to the government treasury or perhaps constitute a burden to taxpayers. If the firm did not exist, this capital would be invested elsewhere, perhaps in home mortgages, municipal bonds, commercial paper, or government securities.

*Customers.* The firm delivers chemical goods and services needed for the customers to achieve their goals. If the firm did not exist, goods and services essential for the customer's requirements would not be produced. A lack of fertilizers and pesticides would mean a reduction in food production and a threat to many lives. So, too, automobiles cannot be produced without metal or plastics, tires, batteries, and paint.

*Suppliers.* The firm pays its suppliers for delivered equipment, raw materials, supplies, and services. If the firm did not exist, suppliers would suffer a business decline and would have to find alternate buyers. Presumably no one but CPI firms buys such industrial goods as fluidized catalysts, reactors, and distillation columns.

*Government.* The firm pays corporate income and excise taxes to the federal, state, and local governments, increase the general economic activity and prosperity in the nation, and cooperates with the nation in extending power and influence abroad. If the chemical firm did not exist, lower payrolls would mean lower personal income tax revenues, and lower exports would mean less ability to acquire resources from abroad. Lower government revenues would mean a decrease in the ability to carry out national programs—loans and aid to foreign nations, and domestic security and entitlement programs.

A firm in the CPI is organized to conduct large-scale economic activities through chemistry. Unlike such permanent activities as the post office and state universities, a chemical firm is allowed to continue in existence only if it performs well enough to satisfy all five segments of its public. An inefficient firm may have a workforce with low job qualifications, poor dedication, and low morale, perhaps as a result of poor training or education, as well as inadequate plants and facilities. The firm may produce high-cost, low-quality goods by antiquated processes, or equipment or goods that are not in demand. It may make unwise decisions by investing in ill-advised business ventures and by having too little cash on hand to transact business, or too much. The company may give the customer inferior goods or unreliable services and may fail to deliver on time. The firm may be unwise in its choice of suppliers or may alienate suppliers by unethical practices. It may offend the government and the public by illegal practices such as attempted price fixing, collusion, or bribery, or by damaging the environment, producing an unsafe product, and/or compromising the health of its employees and the public. When a firm loses the confidence of the public, all five segments can deal it punishing blows. Qualified key personnel may leave the company for better opportunities, and capable replacements may prove very difficult to recruit. Investors may sell their stocks and bonds and demand settlement on their bonds and loans. The company may not be able to raise necessary capital from banks or investment bankers. Customers may seek alternative sources of chemicals. Suppliers may demand cash payment on delivery of goods and services instead of the usual time payment. The government may intervene through regulations, fines, and injunctions, ultimately forcing the company into bankruptcy, with the result that the company must break up or cease to exist.

## 1.3   THE ROLE OF THE CHEMICAL PROFESSIONAL IN THE CPI

The chemical process industries represent one of the most technically complex sectors of our society. To fulfill the purposes outlined in Section 1.2, both chemical engineers and chemists at all degree levels are needed. These chemical professionals work in both fundamental and applied research in process and product development, in process design and engineering, in manufacturing, in sales and in administration, frequently at the highest level. A distribution of chemical professionals according to job function is presented in Table 1.3.

There were approximately 260,000 chemical professionals in the United

**TABLE 1.3**   Distribution by Function (self-identified)

| Chemical engineers | Percent | Chemists | Percent |
|---|---|---|---|
| R & D | 15.8 | Basic research | 9.6 |
| Design | 2.2 | Applied research | 29.1 |
| Process | 20.3 | Research administration | 14.6 |
| Administration | 9.6 | General administration | 8.7 |
| Manufacturing | 10.1 | Education | 19.4 |
| Sales, marketing | 4.2 | Sales, marketing | 5.0 |
| Planning, economics | 1.5 | Manufacturing | 6.8 |
| Consulting | 4.8 | Consulting | 1.1 |
| Education | 4.6 | Other | 5.7 |
| Other | 16.9 | | |

*Sources:* Ref. 5, and Facts and Figures issue, *Chemical and Engineering News*, July 1996.

States in 1995, about 60,000 chemical engineers (3) and about 200,000 chemists. In comparison, there were about 700,000 scientists and 2 million engineers in the 80-million-person labor force in the United States in 1997 (5,6). These numbers are approximate because of the difficulties in defining "a chemist" or "a chemical engineer." Statistics are gathered by the Department of Labor and professional societies on the basis of self-identification. Since there are obvious difficulties with either of these approaches, care is needed in interpretation and use of these statistics. Other groups have gathered statistics based on initial college degrees awarded.

Almost without exception, chemical professionals start their careers at institutions of higher education. Table 1.4 gives the distribution of the chemical professionals by academic degree. There is a clear trend toward greater numbers completing Ph.D. degrees in both areas, although there was a slight reversal in the late 1970s. The percentage of chemical engineers seeking the master's degree increased at first in response to the expanding job market in 1970–1980 and in the 1990s because of the lack of jobs at the bachelor's level. Member surveys frequently show different distributions by degree level, but the discrepancy is probably due to the high percentage of bachelors in chemical engineering and chemistry who do not practice in their fields. They may go on to study law, business, or medicine. Those who obtain advanced degrees tend to remain in the profession and be active in it. For comparison, about 8% of all engineers hold the Ph.D. degree, whereas almost 30% of the scientists hold that degree (7). It is interesting to note in Table 1.4 that the percentage of chemical engineers and chemists at various degree levels has not changed significantly over the years.

Not all the 250,000 chemical professionals are employed in the CPI. A high percentage of chemists and chemical engineers are employed outside the

**TABLE 1.4**  Distribution of Chemical Professionals (%) by Academic Degree in Selected Years, 1975–1998

|                    | 1975 | 1996 | 1998 |
|--------------------|------|------|------|
| Chemical engineers |      |      |      |
| B.S.               | 56   | 49   | 51   |
| M.S.               | 26   | 21   | 21   |
| Ph.D.              | 18   | 20   | 18   |
| MBA                | NA   | 10   | 10   |
| Chemists           |      |      |      |
| B.S.               | 55   |      | 53   |
| M.S.               | 17   |      | 16   |
| Ph.D.              | 28   |      | 31   |

Source: Ref. 5.

traditional sectors of the CPI in such industries as mining, primary metals, and government regulatory agencies, as well as in medical and dental laboratories. Petroleum refining is the only industry sector employing more chemical engineers than chemists.

Not all chemical professionals working in industry are employed by large companies. A recent member survey by the American Institute of Chemical Engineers (AIChE) indicated that 64% worked for companies employing fewer than 5000 people (see Table 1.5).

The impact of chemical professionals in the CPI has been profound. Many have become top-level managers or executives in CPI companies, the high point being the 1970s and 1980s. Today this has changed, with many executives not possessing technical degrees, although the opportunity for advancement to the top level still exists. This trend is discussed more fully in Chapter 8.

After receiving a bachelor's degree, the chemical professional starts in a technical position as a trainee under the guidance of an experienced professional. A chemical professional's development needs to be a lifelong learning process leading to greater opportunities and challenges as well as avoiding technical stagnation and obsolescence. One may elect to increase his or her formal education by working toward an M.S. or Ph.D. degree in chemistry or chemical engineering. Chemical engineers should become registered, because such certification is a measure of professionalism. Some chemical professionals take courses leading to an MBA or law degree. Some feel that obtaining an MBA will enhance their advancement up the management ladder, but this is not necessarily true. Other high-level managers feel that an advanced technical degree in a technical field is a plus but that a continuing education program to increase one's knowledge not only in technical subjects but business and sociopolitical subjects may be beneficial to the professional's long-term goals.

A chemical professional who wishes to succeed as a technical manager needs to have not only technical skills but also administrative, conceptual, and

**TABLE 1.5**   Size of Companies Employing Chemical Engineers

| Size of company (employees) | AIChE members (%) |
| --- | --- |
| 1–99 | 11.1 |
| 100–199 | 6.4 |
| 200–499 | 10.8 |
| 500–999 | 10.1 |
| 1000–4999 | 24.8 |
| 5000 or more employees | 36.0 |

Source: Ref. 5.

interpersonal skills. Such a person needs to be aware that companies today also want a professional with a great deal of scientific–technical knowledge that is integrated with a knowledge of economics, management, and the sociopolitical arena, to be able to cope with the ever-changing and ever-growing competitive technological environment of industry not only nationally but globally. In most companies today, the technical-ladder and the managerial-ladder routes are blended. Project, business, or management teams are formed with people of different talents, backgrounds, and experience to solve company problems. With the downsizing of companies, the original dual-ladder model is less evident, being replaced by the team approach. For the dual-ladder approach, see Figure 1.5.

Some chemical professionals aspire to becoming technical managers. Since technology management is of profound importance for a corporation to grow and be competitive, it is necessary to identify those individuals who have the knowledge, the attitude, and the professional skills to be competent mangers. The technology manager must have a body of knowledge including business management. This body of knowledge may be obtained by formal classes, on-the-job training, and personal development both formally and informally. Attitude toward the job and the company is indeed important. For a person to become an effective manager, three types of professional skill are needed: technical, administrative, and interpersonal. As a manager's responsibility increases, technical skills become less important as interpersonal and administrative skills become dominant. Being a competent manager requires the desire and ability to work effectively with people. Further, a manager must have strong communicative skills, be able to motivate and influence people (6).

Woodruff (7) conducted a research project to determine "why technical employees fail as managers and how to prevent technical managers from failing." From the surveys, he found that technical people fail as managers for the following reasons:

1. Poor people skills (i.e., leadership, motivation, interpersonal)
2. Lack of training, development opportunities, and support
3. Failure to delegate effectively
4. Inability to make the transition from details and tasks to people
5. Poor communicative skills
6. Lack of management skills (i.e., planning and organizing)
7. Poor relationships with peers and others
8. Poor time management/lack of organization
9. Did not really want the job

The following were keys to preventing such failures from occurring:

1. Effective general supervisory and management training
2. Specialized supervisory management training

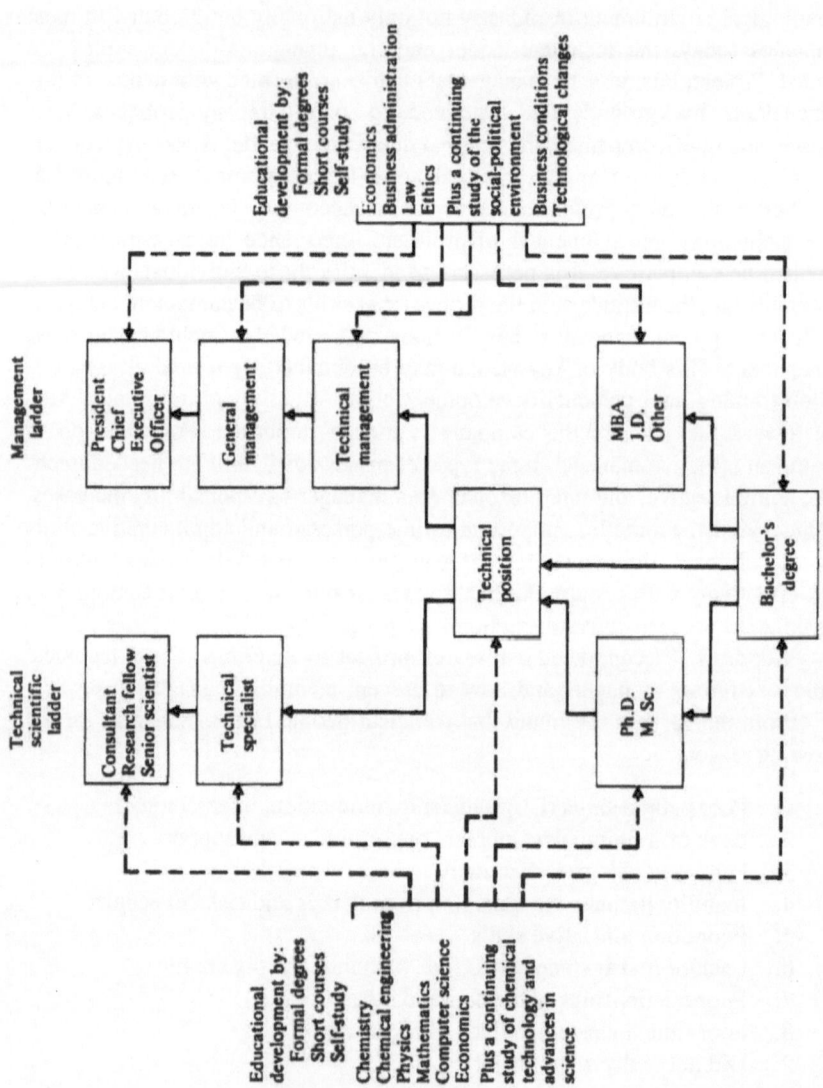

FIGURE 1.5  Developing a career: the dual-ladder approach.

3. Selection process that matches people with job
4. Provision of mentors/coaches
5. Explanation of roles/expectations and provision of feedback
6. Provision of training and development opportunities while future managers are still in technical positions
7. Internal support and resources
8. Allowing time for development
9. Time management/scheduling/planning

In summary, technical managers fail because they were not trained to manage and because interpersonal as well as social skills were not developed.

Table 1.6 lists some chemical professionals who have achieved top-level executive positions in North American corporations. About half have their basic education in chemistry or chemical engineering.

**TABLE 1.6** Chemical Professionals as Top Executives in North American Corporations, 1998[a]

| Company | Executive | Education |
|---|---|---|
| Monsanto Company | Robert B. Shapiro | Law |
| Dow Chemical | W. S. Stavropoulos | Pharmaceutical chemistry, medical chemistry |
| Du Pont | John A. Krol | Physical chemistry |
| Archer-Daniels-Midland | M. D. Andreas | Business |
| Union Carbide | William H. Joyce | Chemist, B.S. MBA, Ph.D. |
| Air Products | Howard A. Wagner | Mechanical engineering |
| Exxon | Lee R. Raymond | Chemical engineering |
| Exxon USA | Ansel Condray | Chemical engineering |
| Merck | R. V. Gilmartin | Electrical engineering, MBA |
| Amoco | H. L. Fuller | Chemical engineering |
| PPG Industries | R. W. LeBoeuf | Business |
| Eastman Chemical | E. W. Deavenport Jr. | Chemical engineering |
| Eli Lilly | R. L. Tobias | Marketing |
| Chevron | K. T. Derr | Mechanical engineering, MBA |
| Phillips Petroleum | W. W. Allen | Mechanical engineering |
| Alcoa | Paul H. O'Neill | Business, public administration |
| Procter & Gamble | John E. Pepper | Business |
| International Paper | John T. Dillon | Liberal arts, law |
| Goodyear | Samir G. Gibara | Business |

[a]Position held is either CEO or president.
*Sources: The Wall Street Journal, Fortune, Who's Who in America.*

TABLE 1.7 Compensation of the Top 10 CPI Executives in 1996

| Company | Name | Position | Salary | Bonus | Other | Totals Short-term | Totals Long-term | Totals Combined |
|---|---|---|---|---|---|---|---|---|
| Praxair | William Lichtenberger | Chairman and CEO | $668,300 | $1,200,000 | $93,332 | $1,961,632 | $13,635,024 | $15,596,656 |
| Allied Signal | L. A. Bossidy | Chairman and CEO | $2,000,000 | $2,800,000 | $6,576 | $4,806,576 | $8,033,914 | $12,840,490 |
| Praxair | Edgar G. Hotard | President | $415,000 | $700,000 | $15,555 | $1,130,555 | $7,603,483 | $8,734,038 |
| Lyondell Petrochemical | Robert G. Gower | Chairman and former CEO | $714,000 | $1,270,142 | $134,736 | $2,118,878 | $6,566,860 | $8,665,738 |
| Occidental Petroleum | Ray Irani | Chairman and CEO | $1,900,000 | $872,000 | $1,236,958 | $4,008,958 | $2,750,570 | $6,759,528 |
| Cabot | Samuel W. Bodman | Chairman | $675,000 | $650,000 | $0 | $1,325,000 | $4,556,086 | $5,881,086 |
| Du Pont (Conoco) | A. W. Dunham | Executive vice president | $600,000 | $1,020,000 | $0 | $1,620,000 | $4,164,167 | $5,784,167 |
| Praxair | John A. Clerico | Vice president and CFO | $356,300 | $450,000 | $40,000 | $846,300 | $4,772,190 | $5,618,490 |
| Hercules | T. L. Gossage | Chairman (retired) | $933,338 | $399,000 | $165,375 | $1,497,713 | $3,537,756 | $5,035,469 |
| Crompton & Knowles | V. A. Calarco | Chairman, president and CEO | $607,916 | $650,000 | $0 | $1,257,916 | $3,427,000 | $4,684,916 |

Source: Chemical Week, August 6, 1997, p 3, 21–29.

As one succeeds as a manager, he or she finds that more time must be dedicated to the job and the company. Less time is available for family and leisure pursuits. This is the downside to managerial achievements. On the upside, the manager is recognized as a leader by industry and the profession. Compensation at the upper administrative levels has escalated in recent years to be consistent with the responsibility. Table 1.7 lists the top 10 CPI executives according to the compensation received in 1998. The total compensation these administrators received is related to their respective companies' performance—a better performance means higher total compensation. In the news media, we often see that high achievers have resigned from the top-paying jobs to devote more time to family and other pursuits, having weighed the sacrifices just noted against the achievement, recognition, and compensation that accompany status as CEO, CFO, or other top manager.

Whether they are in the plant, laboratory, engineering department, or executive suite, chemical professionals in the CPI play an important role. Some of what they have accomplished and the challenges they must face in the future are described in this book.

## REFERENCES

1. *Standard Industrial Classification Manual.* Government Printing Office, Washington, DC, 1987.
2. *Enterprise Standard Industrial Classification Manual.* Government Printing Office, Washington, DC, 1974.
3. A complete list of all input–output classification groups and relationships to the SIC Code for the 1987 input–output tables is found in *Survey of Current Business,* April 1994, pp. 93–97.
4. United Nations Standard International Trade Classification, United Nations, New York, 1997.
5. *AIChE Salary Survey.* American Institute of Chemical Engineers, New York, 1998.
6. National Society of Professional Engineers, *Engineering News,* November 1997, p. 2.
7. M. K. Badawy, *Chemtech,* 27(1): 14–18, January 1997.
8. D. M. Woodruff, *Hydrocarbon Processing,* July 1995, pp. 109–113.

## PROBLEMS

1.1. Using Table 1.1 as a guide. prepare a list of 10 companies within a 200-mile radius of you that would be classified as part of the CPI. From this list, indicate which of these companies could be classified as SIC 28 or 29.

1.2. Use a basic reference (Moody or Standard & Poor, Section 1.1.2) to find the products made by each company. Find the SIC codes for each major product of the companies in your list regardless of whether the items are made in your area.

1.3. Your neighbor tells you that he objects to the presence of local chemical plants or refineries. Can you explain why the plant is needed and who needs it? What alternatives to the plant can you suggest? How would the local people be affected in terms of employment, safety, additional tax revenues, and so on? Discuss whether the cost to your local society are greater than the benefits.

1.4. Some critics say that the goods provided by the CPI are not needed and should not be produced. Examples of supposedly unnecessary products are synthetic fibers (vs. cotton or wool), plastic containers, disposable diapers, artificial or synthetic coloring and flavors in foods, food preservatives, fertilizers, and insecticides. Since these products are purchased, there must be some benefit to the consumer. Does the public have such poor judgment and taste that critics need to dictate what people should do? Select three CPI products and list their advantages and disadvantages to society as a whole.

1.5. Examine the job functions listed in Table 1.3. What types of skill and knowledge might be of particular value in administration or sales and marketing?

1.6. Job interviewers frequently ask the question, What would you like to be doing 5 years, 10 years, and 20 years from now? To prepare for such an interview, write down a potential career path that might suit your own goals, including what you might like to be doing in these time periods.

1.7. Investigate the backgrounds of two top executives of several of the largest CPI companies in the United States (Table 5.14). What were their college majors? What backgrounds might you expect for the top executives in General Electric, General Motors, Boeing, Chemical Bank, and Duke Power? Company annual reports, Moody's, or Standard & Poor's will yield the names of company officials. *Who's Who in America* and *American Men and Women of Science* are appropriate places to start investigating backgrounds.

1.8. Are more chemists than chemical engineers involved in research? What are the approximate percentages and numbers? Why are more chemists engaged in education than chemical engineers?

1.9. Is it worthwhile to obtain a master's degree in chemical engineering or chemistry? A Ph.D.? Prepare an economic analysis for each degree. Salary data are reported once a year in *Chemical and Engineering News* in the June, July, or August issue. What noneconomic factors affect a decision to pursue a M.S. or Ph.D?

# 2

## Basic Economics

Economics is concerned with the most efficient allocation of scarce resources to various uses and with the creation and distribution of wealth. The complete economic picture can best be understood by first discussing how the individual consumer or household chooses to allocate a limited budget and how these individual choices subsequently lead to demand for a product. Section 2.1 reviews of some important aspects of consumer economics.

When a product is in demand, numerous firms attempt to produce it at a profit. What processing unit should be built to produce the product and how the unit should be operated to make a profit are important aspects of a subject known as the *economics of the firm*. Section 2.2 discusses portions of this important subject relevant to the CPI.

The design of a chemical processing unit to meet a specific consumer demand and the comparison of alternative process costs (by return-on-investment and discounted-cash-flow analysis) are well covered in texts on process economics and are not treated in this book.

The financial well-being of the firm depends on the economic performance of the countries in which it operates. Section 2.3 discusses selected parts of national and international economics.

### 2.1 ECONOMICS OF THE CONSUMER

The individual consumer ultimately determines what is produced by the chemical industry (Figure 2.1). Some products made by the chemical industry, (e.g., antifreeze, plastic food wrapping, tires, drugs, cleaning agents) satisfy the *direct*

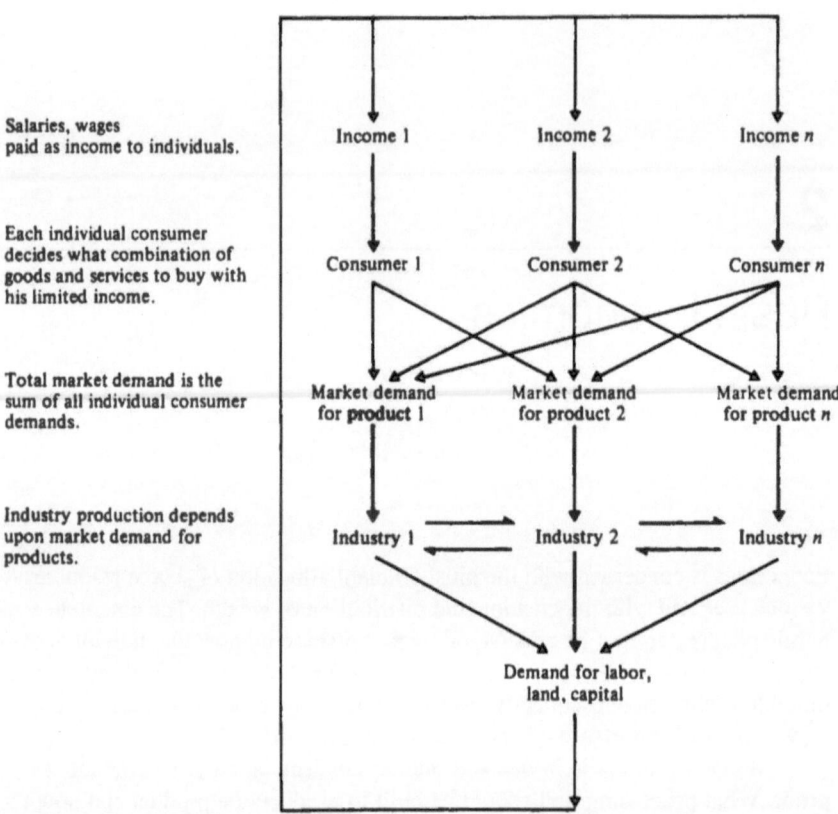

Salaries, wages
paid as income to individuals.

Each individual consumer
decides what combination of
goods and services to buy with
his limited income.

Total market demand is the
sum of all individual consumer
demands.

Industry production depends
upon market demand for
products.

Income 1          Income 2          Income $n$

Consumer 1        Consumer 2        Consumer $n$

Market demand     Market demand     Market demand
for product 1     for product 2     for product $n$

Industry 1        Industry 2        Industry $n$

Demand for labor,
land, capital

FIGURE 2.1   Derivation of demand and flow of dollars.

*demand* of consumers. Many other products of the chemical industry contribute indirectly to final products made for the consumer. These satisfy the *derived* demand of the consumer. Examples of chemical products that satisfy derived demand include (a) all chemical intermediates (e.g., monomers eventually used in polymer products), (b) fertilizers for growing food, (c) synthetic fibers for making textiles and carpets, and (d) pulping chemicals and dyes ultimately required for making books and newspapers; the list is almost endless. But whether the demand is direct or derived, it is always the consumers' needs that determine the well-being of the CPI. It is fitting that we begin with an analysis of consumer economics in our consideration of the economic structure of the CPI.

Economists maintain that a consumer derives a certain amount of *utility* from each final product bought. Utility is a relative measure of the physical and psychological satisfaction a person derives from a final product. Things urgently needed to maintain life (e.g., emergency health care) may provide the consumer with the most utility at some particular time. The consumer needs a mix of numerous products to sustain a satisfactory life, and doubling any one of them does not necessarily double satisfaction or utility. The functionality can only be expressed in a general way:

$$U = U(q_1, q_2, q_3, \ldots, q_n) \qquad (2.1)$$

where $U$ is total utility and $q_1, q_2, q_3, \ldots, q_n$ are the quantities of final product the consumer purchases. Rational consumers, aware of all options avail able to them, are assumed to purchase a combination of products in such a way as to maximize $U$. Of course, this maximization must be done within the constraints of a limited budget, stated as follows:

$$I = \Sigma p_i q_i \qquad (2.2)$$

where $I$ = consumer's income available for purchases
$p$ = price of product
$n$ = total number of products bought

It is extremely difficult to design an experiment for the individual consumer that would generate sufficient data to determine the functional relationship between utility $U$ and quantity of products $q_1, \ldots, q_n$. Nevertheless, the concept of utility is useful because it permits us to describe a reasonably complicated problem with the shorthand of mathematical symbols and significantly extends our ability to describe the economics of the consumer.

As Leavitt (1) explained in 1960, the firm supplying the consumer must focus on human needs or utility rather than the product lines that satisfy the need. If people need transportation, they can take electric street cars or they can buy bicycles. While human needs are relatively constant, the product lines required to satisfy those needs may change rapidly as substitutions are found.

### 2.1.1 Consumer Demand Curve

The amount of any one product $q_i$ a consumer will buy is a function of the person's income $I$ and the prices of all products $p_1, \ldots, p_n$

$$q_1 = D_1(p_1, \ldots, p_n, I) \qquad (2.3)$$

This functional relationship, referred to as the consumer's demand function, in theory can be derived from a utility–function–maximization analysis.

As an illustrative example, consider the simple case in which consumers can only buy two products, both of which are necessary to them, say food and housing. It may reasonably be assumed that utility will be zero if either of the products is not available. The utility function for this simple situation can be represented by $U = q_1 q_2$. The consumer's budget constraint is $I = p_1 q_1 + p_2 q_2$. Thus, it is necessary to maximize

$$U = q_1 q_2$$

subject to

$$I = p_1 q_1 + p_2 q_2$$

This can be done using the Lagrange multiplier procedure for determining the maximum value of a function with constraints. A function $V$ is created as follows:

$$V = U + \lambda (I - p_1 q_1 - p_2 q_2) = q_1 q_2 + \lambda(I - p_1 q_1 - p_2 q_2)$$

where $\lambda$ is an undetermined parameter. Taking the partial derivative of $V$ with respect to $q_1$, $q_2$, and $\lambda$ and equating each partial derivative to zero gives

$$\frac{\partial V}{\partial q_1} = q_2 - \lambda p_1 \quad \frac{\partial V}{\partial q_2} = q_1 - \lambda p_1 \quad \frac{\partial V}{\partial \lambda} = 1 - p_1 q_1 - p_2 q_2$$

Solving for $q_1$ and $q_2$ yields the functional form for Eq. (2.3) (the consumer's demand function):

$$q_1 = \frac{1}{2p_1} \quad q_2 = \frac{1}{2p_2}$$

Thus, with the function $U = q_1 q_2$ the consumer will maximize utility by spending half his or her income on each product. The quantity of each product bought is inversely proportional to its price and directly proportional to income. In this case, the demand for product 1 depends only on $p_1$.

For example, a family of consumers that includes several teenaged drivers will consider its utility for transportation to be equal to the number of vehicles times the miles the family can drive; that is, $q_1$ is miles, and $q_2$ is the number of vehicles the family owns. The cost of fuel, depreciation, insurance, and maintenance, $p_1$, is 30 cents per mile. The cost of a secondhand vehicle $p_{2a}$ is $4000 per unit, and that of a new car $p_{2b}$ is $8000 per unit. If the family has $16,000 to spend on transportation, it can maximize its utility by using the procedure described above:

$$q_1 = \frac{1}{2p_1} = \frac{16,000}{2 \times 0.30} = 26,667$$

$$q_2 = \frac{1}{2p_a} = \frac{16,000}{2 \times 4000} = 2$$

$$q_2 = \frac{1}{2p_{2b}} = \frac{16,000}{2 \times 8000} = 1$$

the utility $U$ is the product of $q_1$ and $q_2$:

$$U_{2a} = q_1 q_{2a} = 53,332 \quad \text{and} \quad U_{2b} = q_1 q_{2b} = 26,667$$

This family would decide to buy two secondhand cars and drive them a total of 26,667 miles.

If the consumer's income $I$ and all other prices $p_1, \ldots, p_n$ may be assumed to remain constant, the demand function of Eq. (2.3) simplifies to

$$q_1 = D_1(p_1) \tag{2.4}$$

At a given price, the sum of all the consumers' demands gives the market demand for the particular product at that price. If all the consumer demand functions were known, they could be added to give the *total market demand function.* The total market demand curve is a plot of price $p$ versus total amount of a given product required by all consumers $Q$. This function usually has a negative slope: the lower the price, the greater the quantity demanded.

**2.1.1.1 Total Market Demand.** When the market demand curve for a particular product is developed, it will take any of several downward-sloping forms (Figure 2.2). If the demand for a product is relatively independent of price, the demand is *inelastic;* if demand is highly dependent on price, the demand is *elastic.* This property is measured by the *price elasticity E,* which is simply the percentage change in demand for a product divided by the percentage change in the price causing the change in demand $Q$:

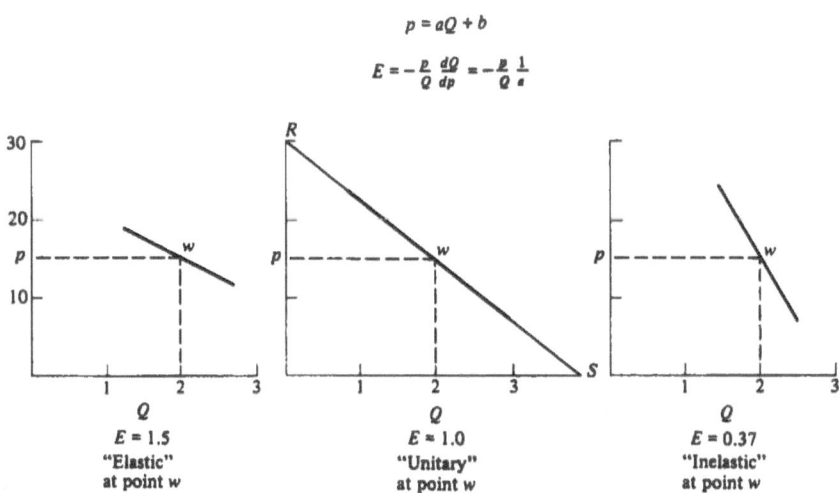

FIGURE 2.2 Price elasticity.

$$\text{Price elasticity} = -\frac{\dfrac{dQ}{Q}}{\dfrac{dp}{p}} = -\left(\frac{p}{Q}\right)\left(\frac{dQ}{dp}\right) \tag{2.5}$$

With this definition, $E$ is almost always a positive number.

Price elasticities $E$ were used to develop Figure 2.2. Assumed $E$ values were used to determine the slope of the curve for $p$ versus $Q$ at the point $Q = 2$ and $p = 15$ (i.e., point $w$ on the curves). The slope of $p$ versus $Q$ is $dp/dQ = -(1/E)(p/Q)$; for example, for $E = 1$, $dp/dQ = -(1/1)(15/2) = -7.5$. Thus, the equation of the demand function plotted in the middle figure (for $E = 1$) is $p = -7.5Q + 30$. Note that when a linear relationship is assumed between $p$ and $Q$, we write

$$p = aQ + b$$

and then

$$E = -\frac{(aQ+b)}{aQ} = -1 - \left(\frac{b}{a}\right)\left(\frac{1}{Q}\right)$$

For the unitary case ($E = 1.0$ at point $w$ on the middle figure), the elasticity is $\infty$ at point $R$ ($Q \Rightarrow 0$) and is 0 at point $S$ ($Q \Rightarrow +b/a$). In practice we find linearity between $p$ and $Q$ only over a limited range of $Q$. The only function with a constant elasticity is $Q = ap^{-n}$, where $E = n$.

One important determinant of price elasticity is the effect of availability of a reasonable substitute for the product. If consumers consider plastic milk cartons and paper milk cartons to be close substitutes for each other, with almost equal utility, the price elasticity for either product will be large ($E > 1$) when the price of the other product remains the same. Gasoline, on the other hand, has no readily available substitutes, and thus price changes have only a small effect on demand in the short run.

If the price of gasoline were to stay high for a long time, consumers could switch to smaller cars, develop a public transportation system, or learn to drive less. Short-run price elasticities are generally small because no alternative courses of action are readily available. Established equipment and entrenched production habits limit flexibility. However, long-run price elasticities may be much greater. Old equipment deteriorates, and habits change, allowing alternatives to be developed.

Total revenue to an industry is given by $TR = pQ$. The effect of a price change on revenue can be computed as follows:

$$\frac{d(TR)}{dp} = p\left(\frac{dQ}{dp}\right) + Q = Q\left(\frac{p}{Q}\right)\left(\frac{dQ}{dp}\right) + Q = (1 - E) \tag{2.6}$$

Thus, for an elastic product, $E > 1$, a price increase would lead to a loss of revenue. For an inelastic product, $E < 1$, a price increase would lead to an increase of revenue. If $E$ is 1 (unitary elasticity), changing price will have no effect on revenue.

Some products are *complementary* and are used jointly, so that a decrease in price for one product may increase the demand for another. For example, automobiles and antifreeze are *complementary* products. Other products are *substitutes;* that is, an increase in price for one product may increase the demand for the other. For example, nylon and polyester shirts are *substitute* products. The influence of one product's price $P_y$ on another product's demand $Q_x$ is measured by the *cross-elasticity:*

$$\text{Cross-elasticity CE} = \frac{\dfrac{dQx}{Qx}}{\dfrac{dp_y}{P_y}} = \left(\frac{P_y}{Q_x}\right)\left(\frac{dQ_x}{dp_x}\right) \tag{2.7a}$$

For *substitute* products, CE is positive; for *complementary* products, CE is negative.

The effect of fluctuations in personal income on the demand for a product is measured by *income elasticity:*

$$\text{Income elasticity, IE} = \frac{\dfrac{dQ_x}{Q_x}}{\dfrac{d_I}{I}} = \left(\frac{I}{Q_x}\right)\left(\frac{dQ_x}{dI}\right) \tag{2.7b}$$

Luxury items (e.g., travel abroad for pleasure) usually have high IE values, while common things (e.g., salt) have low IE values. Inferior items (e.g., cheap cuts of meat) may have negative IE values.

## 2.1.2 Determination of Total Market Demand

It is impractical to build up a *total market demand* curve for a product by estimating each individual consumer demand and then adding these demands. However, a qualitative understanding of utility and elasticities combined with statistical analysis of past market behavior, projected consumer incomes, and sales of major items does permit total demand for a particular product to be estimated. The ability to arrive at such an estimate is a crucial step in projecting the profitability of a business venture in the CPI. Two approaches may be used to derive a total market demand curve.

1. *Using past market data.* For many existing products, the historical price/quantity sold information is available and can be analyzed to derive total market demand versus price. For example, the amount of electricity sold per

household versus price collected from various parts of the country can be used to construct a crude market demand curve for electricity (Figure 2.3).

2. *Using projected market and sales data.* Considerable effort must be expended to make a sales forecast and to establish a selling price. There is much more art than science in this task.

Sales forecasting should begin with an estimation of consumer purchasing power as measured by gross national product (GNP) and disposable personal income. An end-use study, particularly if the product is a new one, must be made. For each end use, a value-in-use analysis must be carried out as an essential first step in estimating total sales and selling price. Value in use is the price the user must pay for the best alternative product plus any associated costs to derive the same performance or utility. For example, if the product is a new paint requiring only one coat for applications currently requiring two coats, the value in use for the new paint is equal to the cost of two coats of the old paint *plus* the cost of labor for the second coat, and so on. Knowing the value in use allows one to assume a selling price and to determine the market opportunity for which the product is both technically and economically advantageous. When the price of a product drops below the value in use of a potentially important customer, this product becomes the preferred one to buy. Lowering the price will broaden the market, and raising the price will price the product out of the reach of many customers. The demand curve ($p$ vs. $Q$) can be built up by estimating total possible

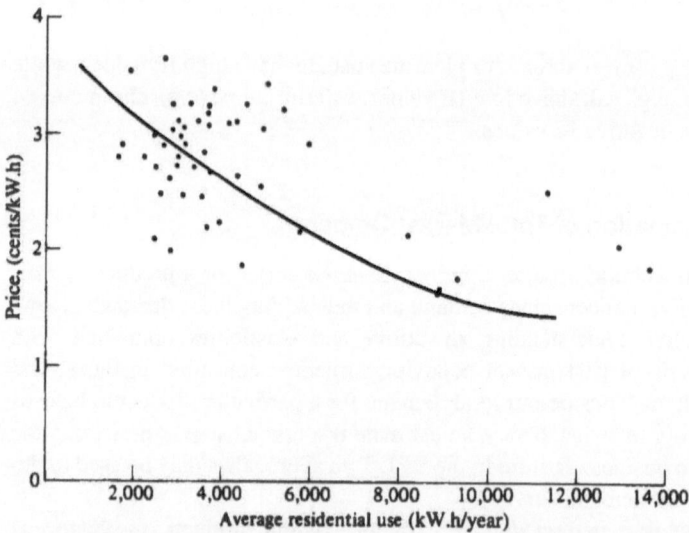

FIGURE 2.3 Demand curve for electric power. (From various Federal Power Commission reports in Ref. 2.)

sales $Q$ over a range of selling prices $p$. The fraction of the market that can be penetrated by any specific firm is affected by several nonprice factors:

1. Product information and awareness of the customer
2. Reluctance of satisfied present customers to change
3. Ability of the seller to deliver in a timely and regular manner
4. Ability of the seller to provide technical services and continued product improvement to the customer
5. Competition

An understanding of these factors is necessary to estimate $q$, the individual firm's share of the market.

For the most part, the CPI are not concerned with meeting final demand of consumer goods but with supplying other industries, which, in turn, furnish products to final demand. For the intermediate CPI products, or *industrial goods,* any demand is a *derived demand.* Consumer satisfaction enters the picture for such products several stages removed from the CPI. For example, consumers may demand garbage removal and choose garbage bags; the result is a derived demand for ethylene. Consumers may demand automobiles; the result is derived demand for steel, which, in turn, causes derived demand for oxygen to make the steel and hydrochloric acid to pickle the steel. If citizens demand certain government services; the result may be a derived demand for highways and bombers, leading in turn to a derived demand for asphalt and jet fuel. It is therefore necessary to estimate the final demand of consumers before the resulting derived demand for CPI products can be estimated.

## 2.2 ECONOMICS OF THE FIRM

Consumer demand for a product will generally be met by a business firm if it can make a reasonable profit by supplying the product without violating the law or arousing negative public opinion. As the product price increases, the firm will be willing to expand production by using higher cost methods (e.g., inefficient older equipment, inexperienced new labor, overtime premium for old labor) and by operating a process beyond the optimum capacity for best efficiency. The total market supply curve is the sum of individual supply curves of the various producers and generally increases with increasing price. These supply curves, or production functions, can be generated from engineering and technical data.

A supply curve for natural gas, for instance, is determined by analyzing each field to determine production costs. The cost is low for a large shallow field near a city. The cost is high for a small deep field in the Arctic. As the price increases, more fields of higher production costs are opened. A detailed example illustrating the development of a supply curve for a simple chemical processing unit is given later, in Example 2.2.

Supply curves can be classified with regard to elasticity. Figure 2.4 illustrates three situations of price elasticity of supply:

$$E_s = -\frac{\frac{p}{Q}}{\frac{dQ}{dp}} \tag{2.8}$$

where $Q$ is the total industry output. In a perfectly idealistic situation ($E_s \to 0$, the supply remains constant regardless of market price.

The elasticity of supply of helium, which is a by-product of natural gas processing, is inelastic. Regardless of the market price of helium, its production is limited by the production of the natural gas. Some petroleum products provide examples of a large elasticity of supply. A refinery is designed to take several different cuts form the barrel of crude oil. The processing equipment is versatile enough to vary the proportion of oil going to any particular cut. Thus, if the market price of gasoline increases in the summer because people drive more then,

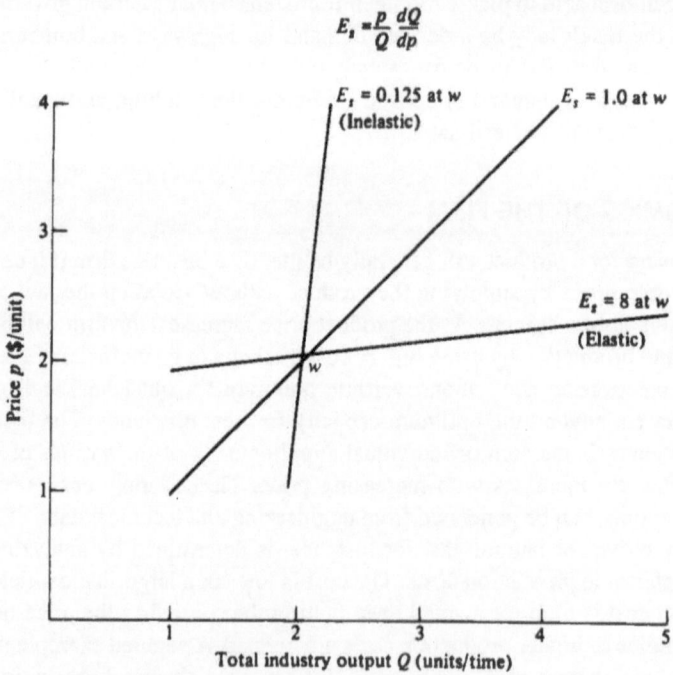

FIGURE 2.4  Supply elasticity.

refineries can quickly switch to producing more gasoline and less fuel oil. They reverse the operation in winter, when more home heating oil is needed.

Human need is infinite and can never be completely satisfied by the finite productive capacity of any economy. A free market is efficient in regulating supply and demand, so what is produced just meets what is demanded with "hard dollars." Since the market allocates a limited resource to those who can pay the price, some needs will not be satisfied. The free market differs from a centrally planned economy, where the government allocates a limited resource not according to individual purchasing power but according to overall political goals. Neither the free market nor the centrally planned economy can satisfy all needs.

A typical supply–demand situation for an industry in which there is perfect competition is qualitatively illustrated in Figure 2.5a. The supply and demand curves intersect at an equilibrium price, representing a stable situation in which supply equals demand. If supply temporarily exceeds the equilibrium value, the market price will have to be lowered to sell off any excess product (step 1 in Figure 2.5b). This lowered price in turn will cause production $Q$ to decrease in the next time period (step 2). A decrease in $Q$ below the equilibrium point will cause a shortage and induce the price to rise (step 3), which will cause total production to increase in the next period (step 4). But this increase will in turn cause a drop in demand. This postulated "cobweb" process will continue until the equilibrium is reestablished. The adjustment process requires the existence of:

1. Marginal producers who will either curtail production or go out of business when the price drops

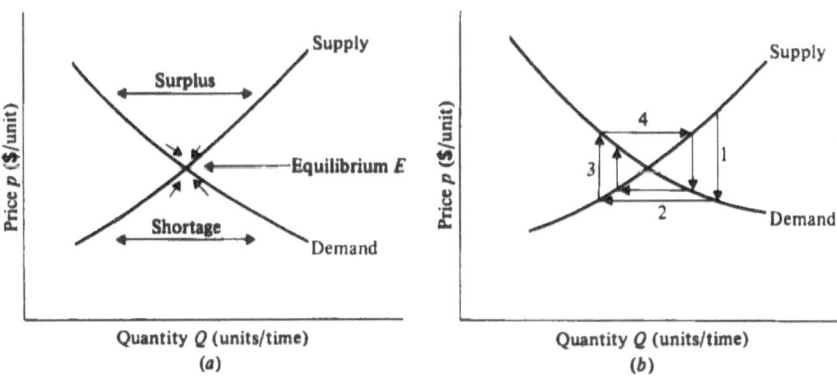

**FIGURE 2.5** (a) A supply–demand relationship. For the demand curve $Q$ is total demand. (b) Reestablishment of equilibrium for a small change in supply. For the supply curve $Q$ is total industry output.

2. Marginal consumers who will reduce consumption or do without entirely when the price rises

If the price is such that firms supplying the product make more than a "normal" profit, new firms will be induced to shift their investments to enter the market, causing the supply curve to shift from $S_1$ to $S_2$, as illustrated in Figure 2.6a. A new, lower price will then be established at the equilibrium point $E_2$. The lower price may cause some marginally profitable suppliers to stop making the product, whereupon the supply curve will shift back to the left. Under ideal conditions, the profits for every product would be the same "normal" profit, so that there is no incentive for any firm to enter or leave a market.

Shifts in the demand curve can also cause the equilibrium price to change. When a new use is found for the product, the demand curve may move from $D_1$ to $D_2$, as shown in Figure 2.6b. If the demand shift causes the equilibrium price to shift upward, new firms may be attracted into the industry or existing firms may increase production capability, causing a supply shift and readjustment of price downward. Alternatively, a decrease in demand (causing the equilibrium price to shift downward) may cause firms to withdraw from the industry and the price to readjust upward.

The time necessary to complete supply or demand shifts depends on the product and the industry. In a technologically simple industry, readjustments toward equilibrium may take place over a few months. In complex and capital-intensive industries like the CPI, in which the planning, design, and construction of a plant take several years, changes leading to equilibrium may likewise take several years. In the short run, both supply and demand tend to be inelastic, since

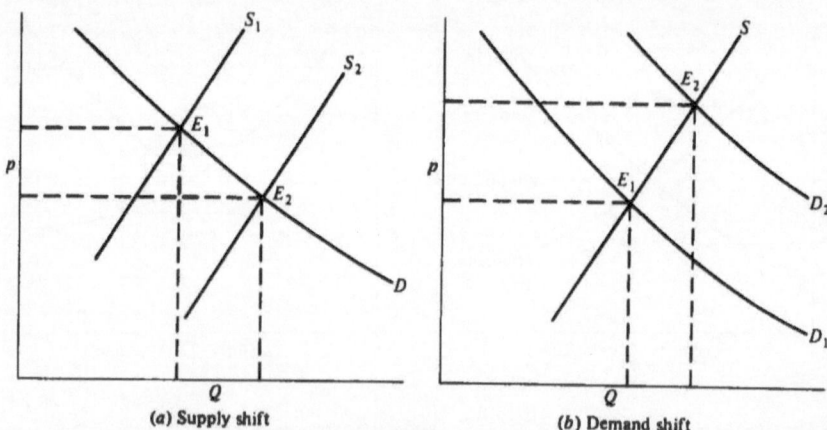

FIGURE 2.6   Qualitative illustration of shifts in supply and demand curves. For the demand curve $Q$ is total demand. For the supply curve $Q$ is total industry output.

it is not easy to find substitutes, shift suppliers, modify and build plants, and increase production rapidly. In the long run, all these adaptive movements can be accomplished, so that supply and demand both become more elastic. The free entrance of firms into the production of an item of extraordinary profit and the exit of firms from the production of an item of low profit tend to make the long-run profit of any industry stable. An especially high profit should be a transient reward for innovation (opening a new market, introducing a new product, developing a more efficient process) until other firms learn about the new development and move in to lower profits. A particularly low profit should also be a transient phenomenon. Marginal firms will have to phase out or close down operations, supply will dwindle, and price will rise, so that the remaining firms will eventually enjoy normal profits again.

This ideal balance is limited by the problem of *barriers to entry*. An entrepreneur who sees an extraordinary profit in producing an item may be unable to enter for several reasons:

1. Lack of a resource supply (e.g., crude oil for a refinery, natural gas for chemical feedstocks, a diamond mine for a jewelry operation).
2. Lack of technology (e.g., that needed for a complex chemical operation or for a nuclear power plant).
3. Lack of capital. In the CPI, enormous plants are often required. Units that produce a billion pounds per year, as for ethylene and methanol, are not uncommon. Plants of this size cannot be built unless a firm can raise substantial sums of money.
4. Exclusion by patent protection (e.g., as in many chemical processes or instant photography).
5. Exclusion by contract (e.g., a soft-drink bottling franchise).
6. Lack of a government franchise (e.g., that needed for a pipeline, telephone, or television broadcasting operation).

Many economists believe that the barriers to entry are bad because they restrict the free movement and most efficient use of resources. Some economists believe that some barriers to entry are good, since innovators need a few years of extraordinary profit, free from competition, to justify the risk and expenditure in sticking their necks out in a new area.

Similarly there is a *barrier to exit* for marginal and inefficient producers. Singly or in combination, ignorance and bad advice, stubbornness and pride, and political pressure to keep workers employed and regional economies from declining may keep a firm in production beyond the point at which a reasonable profit can be made.

The conceptual analysis of a firm is similar in a number of respects to the conceptual analysis of the consumer. But while a consumer attempts to maximize utility, it is usually the basic assumption that the firm attempts to maximize total profit, a quantity much easier to determine than utility. Some economists believe that owner-managers of small enterprises may try to maximize profit but hired managers of giant corporations are more interested in increasing their power and prerogatives by increasing the size of their respective organizations. Thus it may be more accurate to say that modern corporations try to maximize sales subject to adequate profit. For a firm to determine its profit for making any particular product it must know (a) the relation between production cost and quantity produced and (b) the relation between market price of product and quantity produced. The first item, the *production function*, can be estimated by engineering and cost analysis. The second item is determined by the market concentration power of the firm.

### 2.2.1   Determining Cost of Production

An analysis of the production of a firm must begin with a study of both capital and production costs. The term *cost is* often used in different senses by economists, engineers, and accountants. To an economist, "cost" means all the costs of producing a product that an engineer or accountant would consider *plus* a reasonable return for the capital needed to build (fixed capital) and operate (working capital plus expenses) the production facilities.

The total annual cost of producing a product is traditionally considered to have two components, fixed cost and variable cost. A list of such costs is provided by Peters and Timmerhaus (3). Fixed cost (FC) does not depend on the rate of production in the time period. The major components of any fixed cost are as follows:

1. Fixed charges on plants and property
   a. Depreciation
   b. Property taxes
   c. Insurance
   d. Rent paid to others
2. Overhead costs: medical, safety and protection, general plant overhead, payroll overhead, cafeteria, recreation, control laboratories, supervision
3. Administrative expenses: executive salaries, engineering salaries, research and development, general office expenses

Variable costs (VC) are those costs associated with producing and selling the product, which depend upon the rate of production. The major components of the VC are as follows:

1. Direct production costs
   a. Raw materials
   b. Operating labor
   c. Power and utilities

    d. Maintenance

    e. Operating supplies (i.e., catalysts, solvents, etc.)

    f. Laboratory charges

2. Distribution and marketing expenses: sales offices, salesperson wages and expenses, shipping, advertising, marketing, and technical sales

In the short run may costs are fixed. In the long run all costs are variable. Short-range fixed costs such as executive salaries, engineering salaries, research and development salaries, and expenses may be reduced in a prolonged business decline.

Variable costs usually increase sharply with production rate as plant design capacity is reached and exceeded. Chemical conversions and yields may sharply decrease, utility and labor costs may increase, and special operating measures may have to be undertaken

It is frequently useful to express the variable cost $VC$ as a function of the plant annual production rate $q$. A useful functional relationship often used is

$$VC(q) = aq + bq^n \qquad (2.9)$$

where the second term on the right-hand side represents variable costs that increase faster than they would if they were directly proportional to $q$.

The total cost of production $TC$ for producing a particular amount $q$ of a product is simply the sum of the fixed and variable costs at that production rate:

$$TC = FC + VC(q) \qquad (2.10)$$

The total cost of production is usually computed on an annual basis. A typical relationship between $TC$ and $q$ is shown in Figure 2.7; this curve was obtained by analyzing a chemical processing operation and is typical of the shape of curves for processing operations in the CPI. Both the average or unit cost $AC$ and the marginal cost $MC$ are shown in Figure 2.7. The average cost is defined as follows:

$$AC(q) = \frac{TC}{q} \qquad (2.11)$$

The marginal cost $MC$ is the incremental cost of producing one additional unit:

$$MC(q) = \frac{d(TC)}{dq} \qquad (2.12)$$

The essential elements of a production cost analysis can be illustrated with a simple example. In particular we will show how production costs are related to the chemistry, the reactor analysis, and the process design.

**Example 2.1** Suppose that laboratory bench-scale experiments have shown that it is feasible to manufacture a specialty chemical, Delos, by the simple isomerization of a readily available raw material, Algol. The chemistry is

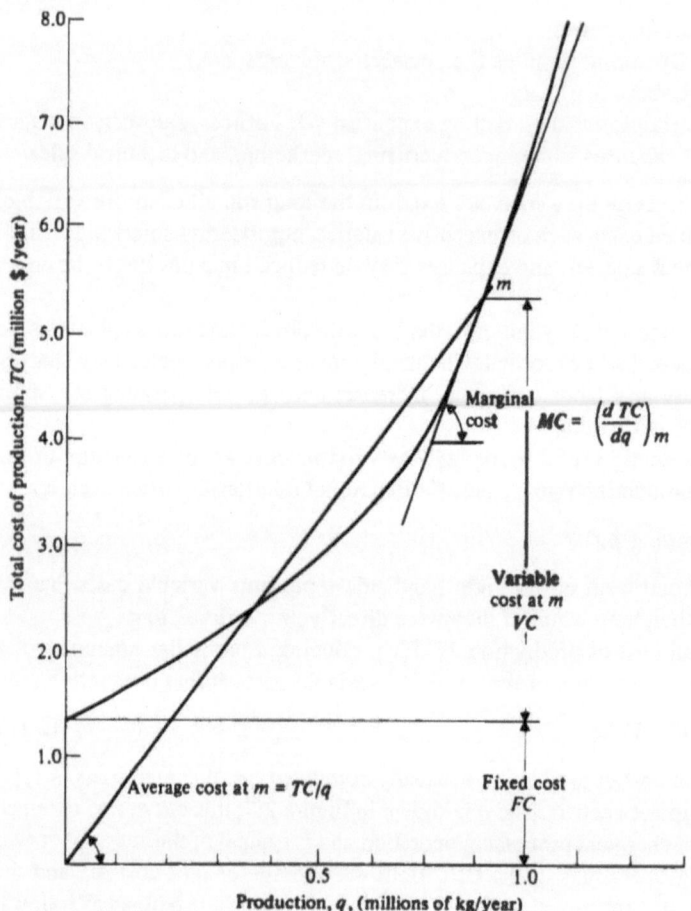

FIGURE 2.7  Total cost of production versus production.

### Algol ⇒ Delos

Enough laboratory experimentation and analysis have been carried out to determine that the rate at which Algol reacts to form Delos can be represented by the kinetic rate expression

$$r_A = kC_A$$

where $k$ is the reaction rate constant reciprocal minutes and $C_A$ is the concentration of Algol in moles per liter. At the processing temperature of interest (150°C) $k$ has a value of 0.005 min$^{-1}$.

The process to produce Delos is simple yet effective. The two key pieces of equipment are a well-stirred tank-type reactor and a separation column. A simplified flow diagram is shown in Figure 2.8.

The reactor performance is described by the following mass balance relationships:

Algol mass balance: $0 = WC_{AF} - WCA - kC_A L$ (2.13)

Delos mass balance: $0 = -WCD + kC_A L$ (2.14)

where $C_{AF}$ = concentration of Algol in reactor feed
$C_A$, $C_D$ = concentration of Algol and Delos in reactor effluent, respectively
$W$ = process flow rate (L/min)
$L$ = reactor volume (liters)

Equations (2.13) and (2.14) can be rearranged to yield various process operating parameters of interest:

1. Relationship between concentrations

$$C_{AF} = C_A + C_D$$ (2.15)

2. Concentration of Algol in the reactor effluent $C_A$ as a function of Delos produced $WC_D$ and reactor volume $L$.

$$C_A = \frac{WC_D}{kL}$$ (2.16)

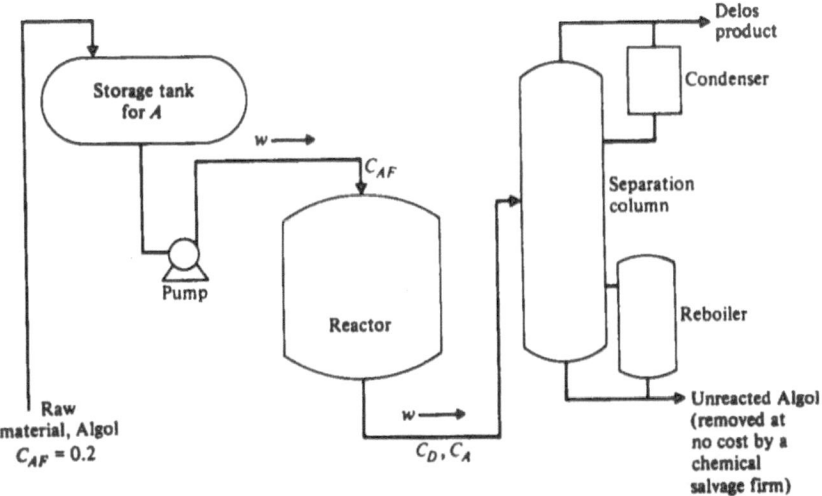

FIGURE 2.8 Simplified process flow diagram for manufacture of Delos.

$$C_A = \frac{C_{AF}}{1 + kL/W} \tag{2.17}$$

where $L/W$ is the reactor holding time.
3.  Process unit throughput $W$ in terms of the reactor volume $L$ and Delos produced $WC_D$:

$$W = \frac{WC_D}{\dfrac{C_{AF} - WC_D}{kL}} \tag{2.18}$$

The separation column design and operation are more complex and cannot be summarized as readily. The column height is essentially determined by the relative volatilities of Algol and Delos and any inert material present. The diameter of the column is set by total unit throughput $W$, as is the design of the reboiler and condenser. It is not necessary to consider in detail the column design to meet the purposes of this example, and we will assume that capital and operating costs for the column can be directly related to the reactor design. Product Delos is distilled overhead, and the unreacted Algol is produced as a bottom product and removed at no cost by a chemical salvage firm.

The total production cost can be computed with the following assumptions:

1.  The cost of the raw material, Algol, is $\$RC$/mol.
2.  Algol is available at a concentration $C_{AF}$ of 0.2 mol/L.
3.  Variable production costs such as utilities, power, operating supplies, and operating labor arc assumed proportional to unit throughput $W$ and equal to $\$PC$/L.
4.  Fixed costs, the most important of which are depreciation on the storage tank, pump, reactor, separation column, reboiler, and condenser, are assumed to be proportional to the reactor volume $L$ and equal to $\$45L$ per year.

Total production cost on an annual basis can be obtained by plugging the $FC$ value of $\$45$ into Eq. (2.10):

$$TC = 45L + VC(q)$$

The variable costs on an annual basis $VC$ are related to $q$, the total amount of Delos produced, by means of reactor design analysis. For the process being considered there are two components of the variable cost: raw material costs and variable production costs.

Delos has a molecular weight of 100. With a 350-day (504,000-min) operating year for the process, $q$ is related to the process parameters as follows:

$$q = WC_D \text{ mol/min} \times 504{,}000 \text{ min/year} \times 0.10 \text{ kg/mol} = 50{,}400 WC_D$$

Raw material cost can be expressed as a function of $q$ by using Eq. (2.18):

$$\text{Raw material cost} = (W)(C_{AF})(RC)(504,000)$$

$$= \frac{WC_D C_{AF}(RC)(504,000)}{\frac{C_{AF} - WC_D}{kL}}$$

$$= \frac{10(RC)q}{\left(\frac{1-q}{B}\right)}$$

where $B = 50,400 C_{AF} kL$.

For a specific reactor design, parameter $B$ is constant. Raw material costs increase faster than $q$ when $q/B > 0.1$. This is caused by increased feed rate $W$ and increased unconverted Algol in the effluent. The variable production cost can also be expressed in terms of the annual Delos production $q$:

$$\text{Variable operating cost} = \frac{PC(W)C_D 504,000)}{\frac{C_{AF} - WC_D}{kL}}$$

$$= \frac{10(PC)q}{C_{AF}\left(\frac{1-q}{B}\right)}$$

Total annual cost becomes

$$TC = 45L + \left(\frac{\frac{10q}{1-q}}{B}\right)\left(\frac{PC}{C_{AF} + RC}\right) \tag{2.19}$$

Figure 2.7 was prepared by using Eq. (2.19) for a reactor volume of $L = 30,000$ liters. For operating costs, $PC = 0.01$ and for raw material cost $RC = 0.15$. With these values for $L$, $PC$, and $RC$ substituted into Eq. (2.19), the following expression is obtained for $TC$ as a function of $B$ and $q$:

$$TC = \frac{45L + 2q}{\frac{1-q}{B}} \tag{2.20}$$

The cost curve in Figure 2.7 is specific for a particular plant at some particular time. The same processing unit can be made more efficient as operating experience is gained, leading to lower total costs at some later time. A firm may also build additional units, which will generally have a lower cost function owing to economies of scale and/or economies of experience. This phenomenon of decreasing cost due to increased experience can be presented in the form of a *learning curve* of unit cost versus time or as an *experience curve* of unit cost (corrected for inflation) versus a firm's accumulated production (defined as the sum of all production of the chemical up to a particular time). A log–log plot of deflated cost versus total accumulated production for a chemical often shows a linear decline in cost as accumulated production increases (Figure 2.9).

### 2.2.2  Optimizing a Firm's Profit

As an essential first step toward optimizing profit, a firm must decide how much product to sell and at what price. The income from sales or total revenue $TR$ of firm for selling a product is simply $pq$. The marginal revenue $MR$ is $d(TR)/dq$. The profit $\pi$ is the difference between the firm's total revenue and total cost:

$$\pi(q) = TR - TC \tag{2.21}$$

Assuming that the firm wishes to maximize profits, we have

$$\frac{d\pi}{dq} = \frac{\dfrac{d(TR)}{dq} - D(TC)}{dq} = MR - MC = 0 \tag{2.22}$$

Thus, to maximize profit, a firm will set production so that $MR = MC$. This means that to maximize profit, a firm will increase (or decrease) production unit the revenue from one additional unit sold equals the additional cost of making it. The value of $FC$ has no influence on the optimal production level once the plant has been designed and built. It is thus considered *sunk cost.*

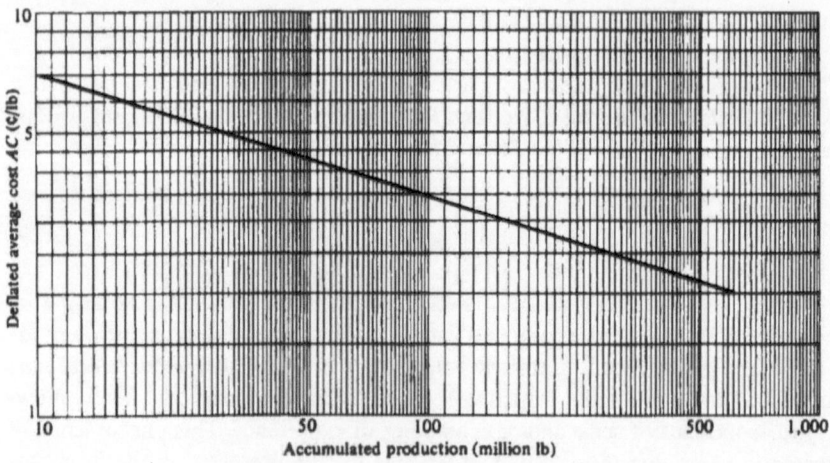

FIGURE 2.9  Typical experience curve.

The marginal cost $MC$ of production is the incremental cost of producing the product and is defined by Eq. (2.12) and shown in Figure 2.7:

$$MC = \frac{d(TC)}{dq}$$

The average or unit cost of production $AC$, as defined by Eq. (2.11), is plotted versus production $q$ in Figure 2.10 for a 30,000-liter reactor producing Delos ($C_{AF} = 0.2$, $PC = 0.01$, $RC = 0.15$).

To maximize profit for any type of operation, the expression for total revenue for the operation, $TR = pq$, is substituted into Eq. (2.22) and an expression relating price $p$, the operation's production $q$, and the marginal cost $MC$ is obtained:

$$\frac{d\pi}{dq} = \frac{d(pq)}{dq} - MC = p + q\left(\frac{dp}{dq}\right) - MC \tag{2.23}$$

To maximize profit with respect to production, $d\pi/dq$ must be zero:

$$p + q\left(\frac{dp}{dq}\right) - MC = 0$$

In using Eq. (2.23) one must take into consideration the relationship between $p$ and $q$. This depends on the type of competition in the industry. It is necessary to identify three extreme cases of competition:

1. *Perfect or atomistic competition.* In this situation the output of a product from each firm is a very small part of the total supply, and production decisions by any one firm have no influence on total supply and the product's market price ($dp/dq = 0$). The firm may decide to produce and to sell at any volume, but this choice has no influence on price.

2. *Monopoly.* In this situation one firm supplies all of a product ($q = Q =$ total industry production). Its decisions with regard to output have a decisive effect on price ($dp/dq = dp/dQ$, which is the slope of the total demand curve).

3. *Oligopoly.* In this intermediate situation a few firms supply most of a product, and a single firm's decision on production volume has some effect on price ($dp/dq$ not necessarily zero). The exact relationship between $p$ and $q$ *is* very difficult if not impossible to predict, since a price increase by one firm may be matched or ignored by the other firms.

We now consider these types of competition in turn, concluding with some remarks on capital allocation and design.

2.2.2.1 Atomistic, or Perfect, Competition.    If there is perfect competition, the amount produced by the atomistic firm will not affect $p$; $dp/dq = 0$. Then

$$\frac{d\pi}{dq} = p - MC = 0 \tag{2.24}$$

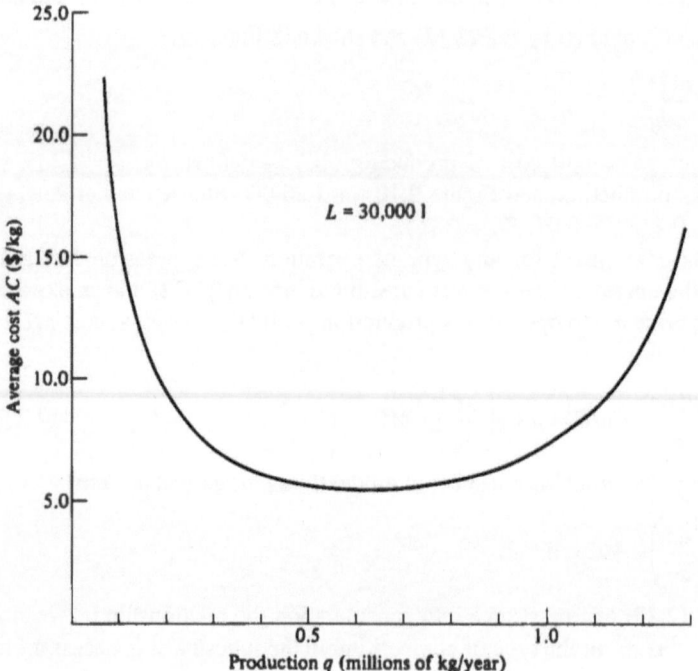

FIGURE 2.10   Average cost of production versus production.

To maximize profit, the firm will produce the quantity of product at which the marginal cost equals the externally determined price, $MC = p$. In cases of perfect competition, the extra revenue derived from an additional unit of production is simply the market price ($MC = MR = p$). This is not affected by the action of the individual atomistic firm.

As an example, assume the following relationships, where $p = \$1$ per unit of $q$ sold:

$$FC = \$60$$
$$VC = \$(0.25 \times 10^{-6}q^4)$$
$$TC = FC + VC = \$(60 + 0.25 \times 10^{-6}\ q^4)$$

To maximize profit

$$MC = d(TC)/dq = 10^{-6}q^3 = p$$

Therefore, the amount to produce is

$$q^3 = 10^6 p = 10^6 \times 1 \quad q = 100 \text{ units}$$

Figure 2.11 illustrates this example. When $q$ is below the break even point, profit is negative, when $q$ is at the optimum level; the profit $\pi = pq - TC$ is $15. A small firm in plastics processing may approach this ideal case.

### 2.2.2.2 Monopoly.

In a monopolistic situation one firm supplies all of a product: $q = Q$. The firm also sets the price $p$, and the total market demand function provides the $Q$ in response. Thus, $d\pi/dQ$ is not zero in Eq. (2.23). One needs the function relating $p$ and $q$ to determine the optimum $p$ for maximizing profits. To illustrate the operation of the monopoly firm, we assume the same total cost function as before but assume that for some reason all firms but one have dropped out of business. We also assume the following relationship for the demand curve:

$$p = 2 - 10^{-2}Q$$

Since the monopolistic firm is the sole producer ($q = Q$), Eq. (2.23) becomes

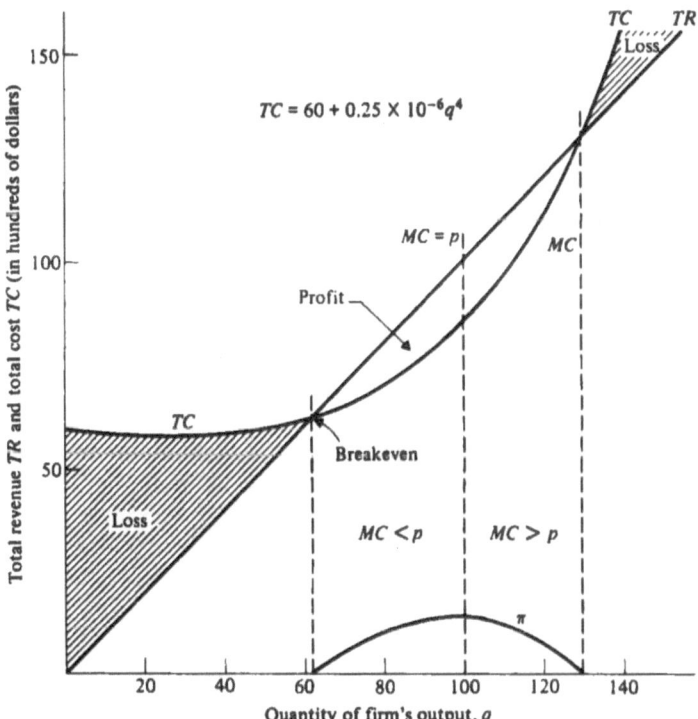

FIGURE 2.11  Determining production to maximize profit for the atomistic firm.

$$\frac{d\pi}{dq} = MR - MC = 0$$

Differentiating the demand curve relationship and substituting gives

$$MR = p + Q\left(\frac{dp}{dQ}\right) = p\left[1 + \left(\frac{Q}{p}\right)dp\right] = p\left(\frac{1-l}{E}\right)$$

For an elastic demand, the marginal revenue $MR$ is positive, and for an inelastic demand the $MR$ is negative. Finding the equation for marginal cost gives

$$MC = \frac{dC}{dQ} = \frac{d}{dQ}(60 + 0.25 \times 10^{-6}Q^4) = 10^{-6}Q^3$$

and setting $MC = MR$ leads to

$$10^{-6}Q_3 = 2 - 2 \times 10^{-2}Q$$

A solution to this equation yields $Q = 76.6$. At this production rate $p = \$1.24$ per unit and $\pi = \$25.6$. Figure 2.12 illustrates the example. The optimum in an unregulated monopoly yields a lower production level but a higher price and profit than under atomistic competition. In reality, government-sanctioned monopolies, such as an electric utility, have regulated prices. Such firms often try to maximize efficiencies and to increase market size but do not emphasize profit maximization.

**2.2.2.3  Oligopoly.**  Many CPI products are produced in an oligopolistic situation. Often, three or four firms produce over 50% of a chemical. The oligopoly has a set of characteristics that are not as easily defined as for perfect competition or monopoly.

Equation (2.23) can be used to compute the profit of any firm:

$$MR = p + q\left(\frac{dp}{dq}\right) = MC$$

While the atomistic firm has no influence on price and the monopolistic firm can set price unilaterally, the oligopolistic firm may announce a price increase, set a production figure, and then wait for the response of other firms. If the other firms do not follow with a similar price increase, the oligopolistic firm may find all its customers defecting and will have to rescind the price increase; if the price increase is adopted by the other firms, the market shares will remain the same in a shrunken market. If the oligopolistic firm starts a price war by decreasing the price in an effort to capture a larger market share, the other firms may or may not follow suit. The value of $dp/dq$ is determined by the business strategy of the other firms.

**2.2.2.4  Capital Allocation and Design.**  In the analysis leading to expressions for total production cost and strategies for maximizing the profit of a firm,

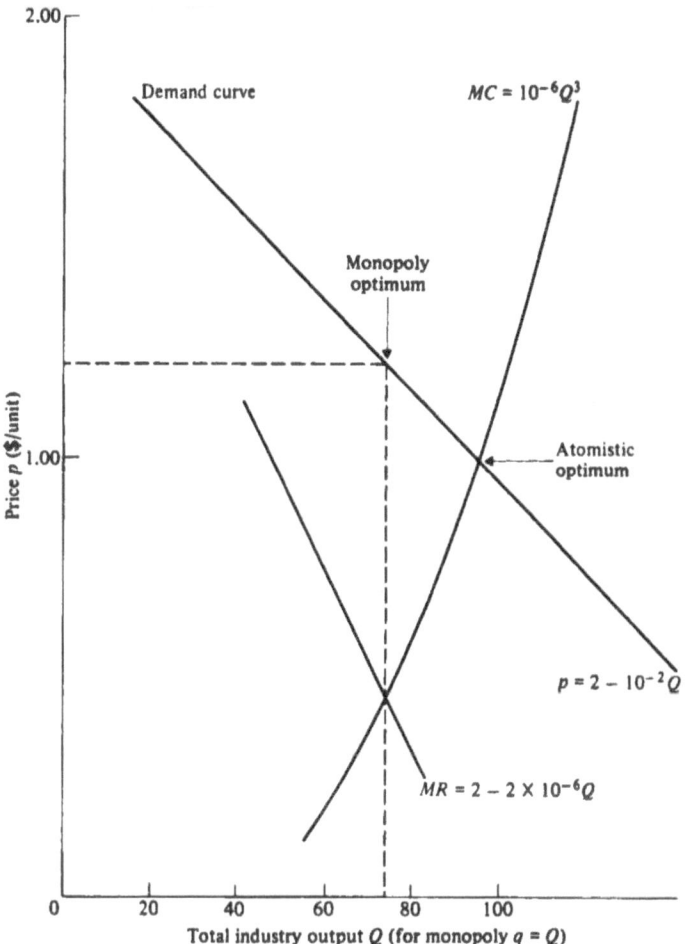

FIGURE 2.12 Determining production to maximize profit of a monopoly.

we have assumed that the operating plant exists and that firms are required to make decisions on production volume. The existing plant represents a sunk cost due to irreversible past decisions. Production level is governed only by the marginal cost. In the example of this section if the fixed cost $FC$ is $100 instead of $60, the optimum production remains at $q = 100$ units but the profit $\pi$ becomes -$45. Closing the plant down would be worse. The fixed cost $FC$ must be paid in any case, and the profit becomes -$100 (an even worse loss). Because of past errors, the best strategy now is to operate at $q = 100$ units and accept a loss of $45.

However, nothing would induce one to make a new plant investment of a similar nature again unless there were a sufficiently large positive profit. Projected return on investment (ROI) is one of the most important determinants of new plant investment. A firm must make sufficient *ROI* to justify plant expansion, and a firm with a risky investment would require even higher projected *ROI* to proceed with a new plant.

A firm has an annual capital investment budget, generated from internal cash flow of profit, depreciation, external borrowing, and new stock sales. A rational firm ranks all capital project proposals according to descending projected *ROI* and probable risk, and funds all projects with sufficient ROI subject to the budget constraint. An aggressive firm may have far more meritorious projects than funds available, so that many would stay on the planning board. A less aggressive firm with more funds than good projects may lend excess funds to other companies. In each case, a company tries to maximize its return on investment by putting its funds in investments with the highest return with safety.

### 2.2.3 Developing a Supply Curve

A supply curve for a particular product can be developed if the engineering and technical information is available for each firm's operation. This information can often be generated and a supply curve estimated for CPI products by a chemical professional.

The process for obtaining the supply curve is most conveniently demonstrated by expanding upon Example 2.1. The following steps are taken:

1.  From the process design an analytical or graphical relationship between total cost *TC* and production *q* for each operating unit must be developed.
2.  Using Eq. (2.23), one determines the production rate *q* at which each firm will operate at given price *p* for an expected range of prices:

$$MC = \frac{d(TC)}{dq} = p + q\left(\frac{dp}{dq}\right)$$

    Unless *dp/dq* is known, this part of the analysis may be carried out assuming *dp/dq* = 0 (the perfect competition situation).
3.  The supply curve, a plot of price *p* versus total product production *Q*, is made by summing all the individual firm productions *q* at one particular price.

**Example 2.2** If we consider that the Delos business consists of 10 firms each operating with a 30,000-liter reactor, 10 firms each operating with a 40,000-liter reactor, and 10 firms each operating with a 50,000-liter reactor, a supply curve can be generated using the simple reactor design and process economic parameters developed in Section 2.2.1. The total cost for any one firm is given according to Eq. (2.19), repeated here for convenience:

$$TC = 45L + \left( \frac{10q}{\frac{1-q}{B}} \right) \left( \frac{PC}{C_{AF} + RC} \right) \tag{2.19}$$

For the industry under consideration in our example, we can obtain a cost for each of the firm in the groups if we assume that $C_{AF}$ is 0.2 and $k$ is 0.005 for all firms.

These restrictions can easily be removed at the expense of some additional arithmetic, and the effect of different values for $C_{AF}$ and $k$ can be examined. It would also be easy to examine the effect of different raw material and operating costs for different firms, since they are reflected in the $PC$ and $RC$ terms in Eq. (2.19).

The simplified equation for $TC$ when $PC$ is 1 cent per liter and $RC$ is 15 cents per mole is

$$TC = \frac{45L - 2q}{\frac{q}{B}}$$

The total costs for the various groups are as follows:

Group I : $L_I = 30,000$

$$TC_I = 1.35 \times 10^6 + \frac{2q}{\frac{1-q}{B}} \qquad B = 1.51 \times 10^6$$

Group II : $L_{II} = 40,000$

$$TC_{II} = 1.81 \times 10^6 + \frac{2q}{\frac{1-q}{B}} \qquad B = 2.22 \times 10^6$$

Group III : $L_{III} = 50,000$

$$TC_{III} = 2.25 \times 10^6 + \frac{2q}{\frac{1-q}{B}} \qquad B = 2.52 \times 10^6$$

Figure 2.13 plots the total cost for a typical company in each group $TC$ versus the company's production $q$.

Any group's production can be obtained at any group total cost simply by multiplying the firm's $q$ or $TC$ by 10.

Equation (2.19), giving the total cost $TC$ in terms of reactor volume $L$ and

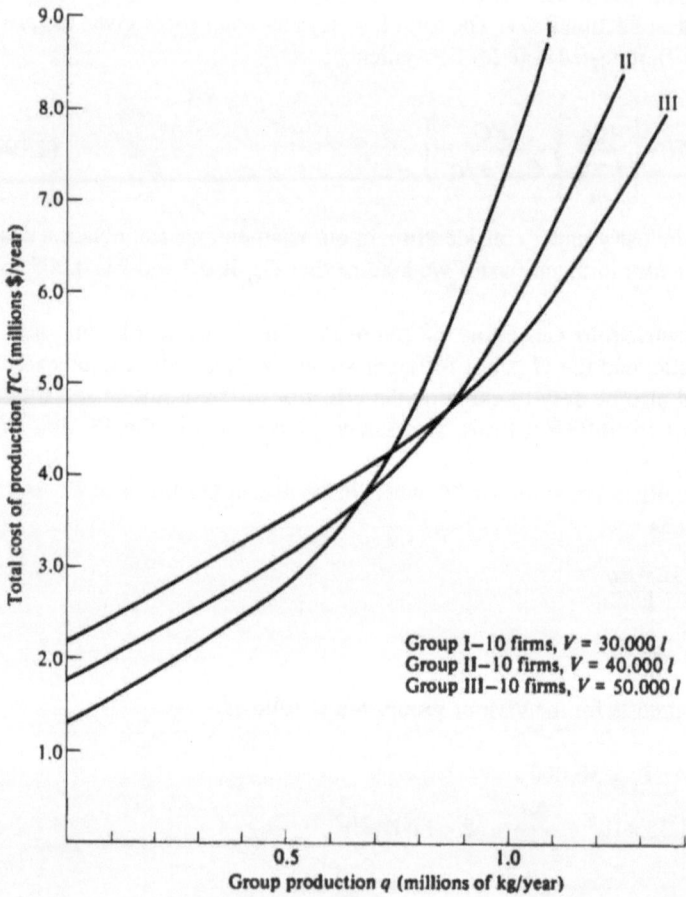

**FIGURE 2.13**  Total cost of production versus production for a typical firm producing Delos raw material.

production rate $q$, can be differentiated to obtain the marginal cost $MC$ (for $C_{AF} = 0.2$) as follows:

$$\frac{d(TC)}{dq} = \frac{d}{dq}\left[\frac{45L + 2q}{\dfrac{l-q}{B}}\right] = MC$$

$$(2.25)$$

$$MC = \frac{2}{\dfrac{l-q^2}{B}}$$

The marginal cost is related to the price $p$, as shown before [Eq. (2.23)]:

$$MC = \frac{p2}{l} = p$$
$$\frac{q}{B}$$

$$q = b\left[1 - \left(\frac{2}{p}\right)^{1/2}\right]$$ (2.26)

$$q = 50,400 C_{AF} kL\left[1 - \left(\frac{2}{p}\right)^{1/2}\right]$$

For CAF = 0.2 and $k = 0.005$, we have

$$q = 50.4L\left[1 - \left(\frac{2}{q}\right)^{1/2}\right]$$ (2.27)

Equation (2.27) gives the optimal production rate of a firm producing Delos for any price $p$. Note that when $p$ is 2 or less, there is no production.

The industry supply curve is generated by obtaining the functional relationship between total Delos production of the 30 firms $Q$ and $p$:

$$Q = \Sigma = q_i 50.5[10(30,000 + 40,000 + 50,000)]\left[1 - \left(\frac{2}{q}\right)^{1/2}\right]$$ (2.28)

$$= 60.48 \times 10^6\left[1 - \left(\frac{2}{q}\right)^{1/2}\right]$$

A plot of $p$ versus $Q$ is shown in Figure 2.14. This is the supply curve for Delos. Its intersection with a demand curve establishes the production rate for each firm manufacturing Delos. For example, if the price is $10/k, 33.4 million k of Delos will be produced. The production rate for each firm can be established using Eq. (2.27):

Group I:
$q = 0.836 \times 10^6$ kg/yr
Group II:
$q = 1.11 \times 10^6$ kg/yr
Group III:
$q = 1.39 \times 10^6$ kg/yr

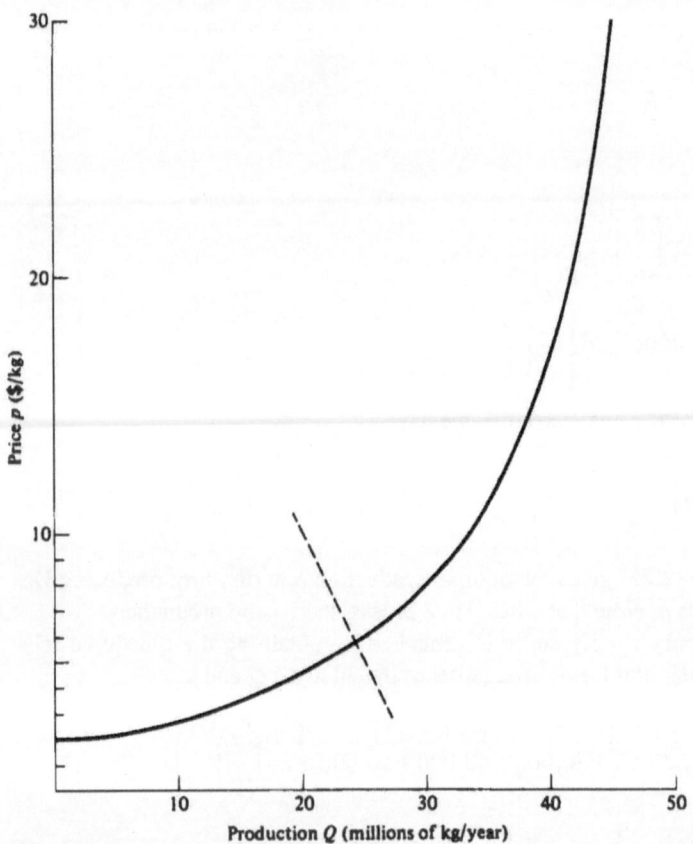

**FIGURE 2.14**  Supply curve for Delos.

Table 2.1 shows the profit for each firm in the various groups as a function of price for $PC = 1$ cent per liter and $RC = 15$ cents per mole:

$$\pi = pq - 45L - \frac{2q}{\frac{1-q}{B}} \tag{2.29}$$

At the price of about \$5.57, all firms will have a zero profit. With the simple cost functions used in this example, no one firm can make a profit while another loses money. In reality this may not always be the case, as can easily be shown if we assume some economy of scale for fixed cost:

$$FC = 45L^{0.8}$$

TABLE 2.1    Profit of Firms Manufacturing Delos [from Eqs. (2.27) and (2.29)]

| | Group I (L = 30,000) | | Group II (L = 40,000) | | Group III (L = 50,000) | |
|---|---|---|---|---|---|---|
| Price ($/kg) | Production q (10³ kg/yr) | Profit π (10³ $/yr) | Production q (10³ kg/yr) | Profit π (10³ $/yr) | Production q (10³ kg/yr) | Profit π (10³ $/yr) |
| 2 | 0 | −1350 | 0 | −1800 | 0 | −2250 |
| 4 | 442 | −831 | 591 | −1108 | 738 | −1358 |
| 5 | 556 | −328 | 741 | −438 | 926 | −548 |
| 6 | 639 | 270 | 852 | 360 | 1065 | 450 |
| 8 | 756 | 1674 | 1008 | 2232 | 1260 | 2790 |
| 10 | 836 | 3270 | 1114 | 4360 | 1393 | 5450 |

TABLE 2.2    Profit for FC = $45L^{0.8}$

| Group | Production q (kg/yr) | Profit π ($/yr) |
|---|---|---|
| I | 294,000 | −11,000,000 |
| II | 378,000 | −1,000,000 |
| III | 473,000 | 10,000,000 |

This expression for fixed costs assumes that there is an advantage to building a larger reactor. The profit equation now becomes

$$\pi = pq - 45L^{0.8} \tag{2.30}$$

At a price of $3.03/kg and with the economic constraints above, each firm in the three respective groups makes the profit shown in Table 2.2. The firms in group III with the larger reactors (50,000 liter) are able to make a small profit at this price, thanks to economy of scale, while firms in groups I and II lose money.

## 2.3  ECONOMICS OF A NATION

The well-being of the chemical industry depends on the economic performance and buying power of the nation in which it operates. To determine a nation's economic performance, the flow of goods, services, and money between consumers, government, business, and foreign interests must be known. In the United States, the Department of Commerce has the responsibility of maintaining the data, and the estimation of various important quantities such as GNP and national income is a difficult and nontrivial task. Some idea of the complexity of the national accounts as well as some of the key concepts used in the bookkeeping operation can be seen in Figure 2.15. The cash flows between blocks show where the money comes from and where it goes as it circulates through the U.S. and world economies. These are summed to give the measured quantity in the block. For example, personal income is the sum of compensation of employees, proprietors' income, interest, dividends, rental income of persons, transfer payments, and government and consumer interest payments. Personal income is distributed as personal savings, personal outlays, and personal tax and nontax (social security, etc.) payments.

A one-figure indication of the health of a nation's economy is provided by

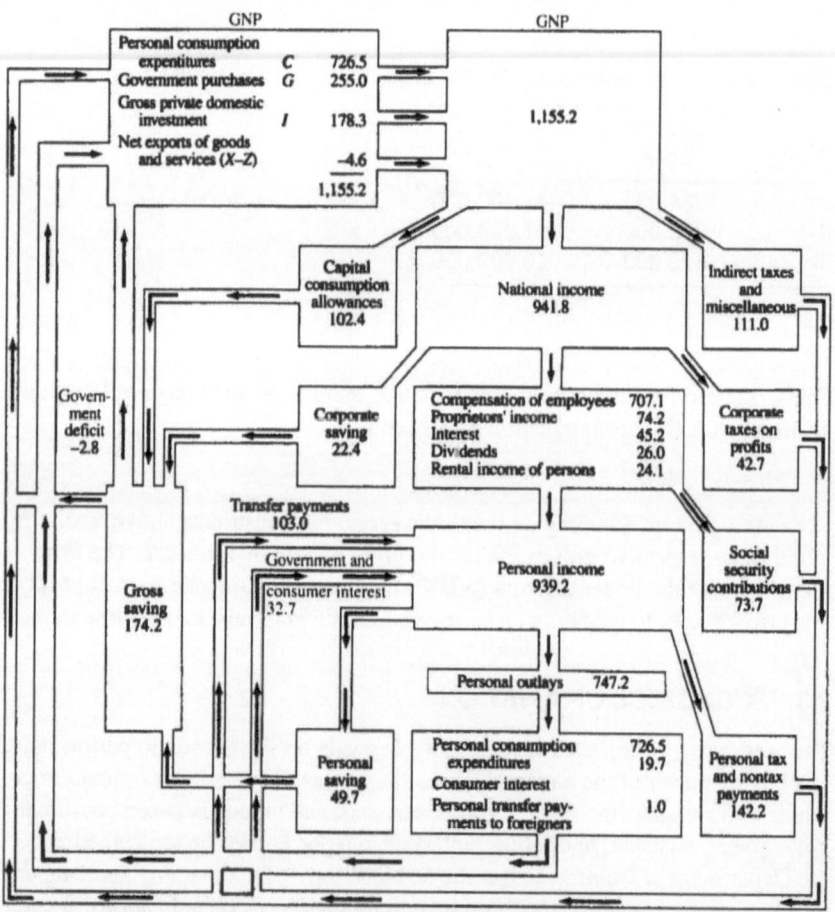

FIGURE 2.15   Flow of income and expenditures in the United States. (Original format of the figure by SRI, Menlo Park, CA, and 1994 data from Ref. 4.)

the Gross Domestic Product [GDP] (prior to 1997, the gross national product (GNP) was used as the indicator of national economic activity), the most common measure of economic activity, which has been used since. It is defined as the flow rate of final goods and services produced by a nation in any one year, or the rate at which money is spent in any one year. It is usually measured in terms of the market value of such production per year. The GNP can be computed as follows:

$$GDP = GNP + P$$
$$GDP = C + G + I + (X - Z) + P \tag{2.31}$$

where   $C$ = expenditure by consumers for final goods and services
   $G$ = market value of all government purchases
   $I$ = market value of investment goods produced by nation
   $X - Z$ = market value of exports minus imports from one nation to other countries
   $P$ = payments to rest of world

Consumer goods and services make up the largest part of GDP. These can be consumer durables (automobiles, major appliances, furniture, etc.), nondurables (food, clothing, fuel); and services (medical, insurance, garbage collection, etc.). The purchase of goods and services by government serves as a measure of the productive economic activity consumed in the public sector. Investments include capital equipment and building construction by businesses, new house construction, and all business inventory changes. The GNP and its individual components, as flow of goods and services for the United States, are presented for selected years from 1929 to 1994 in Table 2.3.

The National Income Division of the Department of Commerce computes the GDP (and GNP) regularly, and a monthly summary of the National Income Accounts is provided by *Economic Indicators*, a publication issued by the GPO. Data for computing GDP are provided by (a) periodic censuses, (b) payroll and employment statistics as developed from social security information, (c) tax information collected by the Internal Revenue Service, and (d) fiscal records of the federal government. More detail on sources of information and methods of estimation are given in Refs. 5 and 6.

The GNP values shown in Table 2.3 are expressed in dollars for the year calculated. When current dollars are used to measure economic activity, however, real growth is obscured by the effects of inflation, which decrease the real worth of the dollar. To correct the GDP for inflationary effects, price indexes using 1987 prices as the base year are computed by the Department of Commerce. The components of GNP are then converted into constant dollars by using the appropriate price index. In Table 2.4 the GNP has been corrected for the effects of inflation. The GNP deflator is an implicit price deflator derived from the rate of inflation.

**TABLE 2.3** GNP and Its Major Components (billions of dollars) for the United States in Selected Years, 1929–1994

|  |  | 1929 | 1930 | 1940 | 1950 | 1960 | 1970 | 1980 | 1990 | 1994 |
|---|---|---|---|---|---|---|---|---|---|---|
| GNP[a] |  | 103.4 | 90.7 | 100.0 | 286.2 | 506.0 | 982.4 | 2742.1 | 5567.8 | 6726.9 |
| By type of expenditure: |  |  |  |  |  |  |  |  |  |  |
| Personal consumption | C | 77.3 | 69.9 | 71.0 | 192.0 | 324.9 | 618.8 | 1748.1 | 3761.2 | 4628.4 |
| Government purchases | G | 8.8 | 9.5 | 14.2 | 38.5 | 100.3 | 218.9 | 507.1 | 1047.4 | 1175.3 |
| Investment goods | I | 16.2 | 10.2 | 13.1 | 53.8 | 75.4 | 140.8 | 467.6 | 808.9 | 1032.9 |
| Net exports | X − Z | 1.1 | 1.0 | 1.7 | 1.9 | 4.4 | 3.9 | −14.7 | −71.4 | −98.2 |
| Net payments row[b] |  | NA | NA | NA | NA | NA | NA | 34.1 | 21.7 | −11.5 |

[a] Some GNP figures are not exact totals of expenditures listed because of rounding.
[b] NA, not available
*Source:* Economic Report of the President, 1977, Table B-1; U.S. Bureau or the Census, *Statistical Abstracts of the United States, 1976,* 97th ed., Table 6–4; 1994, 114th ed., Tables 690.

In 1992 the Department of Commerce replaced the constant-dollar deflator with a 1987 constant-dollar deflator. The 1958, 1972, and 1987 deflators are reported in the Table 2.4 for comparison.

The GNP can also be estimated, by means of a procedure based on how the money is earned:

GNP = personal income + social security contributions + corporate saving + capital consumption allowances + corporate taxes on profits + indirect taxes and miscellaneous

Another measure of economic health is the national income, the amount of money paid to the owners of economic resources (land, labor, capital) in return for supplying the services of these resources. National income can be computed by subtracting indirect taxes and some minor charges and capital consumption charges (depreciation) from GNP (Figure 2.15). It can also be derived by adding all compensation of employees, proprietors' income, interest, dividends, rental income of persons, corporate taxes on profits, and corporate saving. The past record of GNP in the United States, as a flow of earnings, is shown in Table 2.5. By using this approach, one can estimate the contribution to the national income by individual industrial sectors. Table 2.6 shows the U.S. national income for 1950–1974 and the contribution to it by the chemical industry.

Individual households derive income from working or from selling (or lending) raw materials, land, and capital. The economic health of the household sector can be measured by personal income per person and by disposable income

**TABLE 2.4** U.S. GNP Corrected for Inflation in Selected Years, 1958–1987

| | GNP (constant dollars $\times 10^{12}$ | | | Deflators | | |
|---|---|---|---|---|---|---|
| | 1958 | 1972 | 1987 | 1958 = 100 | 1972 = 100 | 1987 = 100 |
| 1929 | 203.6 | 314.7 | 827.2 | 50.64 | 32.87 | 12.5 |
| 1933 | 141.5 | 222.1 | 587.4 | 39.29 | — | 9.5 |
| 1940 | 227.2 | 43.6 | 909.1 | 43.87 | 29.10 | 11.0 |
| 1950 | 355.3 | 533.5 | 1416.8 | 80.16 | 53.64 | 20.2 |
| 1958 | 447.3 | 679.5 | 1740.0 | 100.0 | 66.06 | 25.8 |
| 1960 | 487.7 | 736.8 | 1946.2 | 103.29 | 68.67 | 26.0 |
| 1970 | 722.5 | 1075.3 | 2740.9 | 135.24 | 91.36 | 35.2 |
| 1972 | 792.5 | 1171.1 | 3018.3 | 146.12 | 100.0 | 38.8 |
| 1980 | — | — | 3842.4 | — | — | 71.7 |
| 1987 | — | — | 4544.5 | — | — | 100.0 |
| 1990 | — | — | 4914.2 | — | — | 113.3 |
| 1994 | — | — | 5342.5 | — | — | 126.1 |

*Sources: Statistical Abstracts of the United States, 1994, 114th ed., Table 757; Economic Report of the President, February 1995, Table B-23.*

**TABLE 2.5**  National Income of the United States (billions of dollars) in Selected Years, 1929–1994

|                    | 1929  | 1940  | 1950  | 1960  | 1970  | 1980   | 1990   | 1994   |
|--------------------|-------|-------|-------|-------|-------|--------|--------|--------|
| National income    | 84.8  | 79.7  | 236.2 | 412.0 | 798.4 | 2198.2 | 4491.0 | 5458.4 |
| Employee           |       |       |       |       |       |        |        |        |
|   compensation     | 51.1  | 52.1  | 154.8 | 294.9 | 609.2 | 1644.4 | 3297.6 | 4004.6 |
| Proprietors'       |       |       |       |       |       |        |        |        |
|   income           | 14.9  | 12.9  | 38.4  | 47.0  | 65.1  | 171.8  | 363.3  | 473.7  |
| Rental income      | 4.9   | 2.7   | 7.1   | 13.8  | 18.6  | 13.2   | −14.2  | 27.7   |
| Corporate          |       |       |       |       |       |        |        |        |
|   profit[a]        | 10.5  | 9.8   | 37.6  | 48.9  | 66.4  | 177.7  | 380.6  | 542.7  |
| Net interest       | 4.7   | 3.3   | 2.3   | 9.8   | 37.5  | 191.2  | 463.7  | 409.7  |
| Capital            |       |       |       |       |       |        |        |        |
|   consumption      | 9.7   | 9.0   | 23.9  | 47.7  | 90.8  | 311.9  | 602.7  | 715.3  |
| Indirect tax       | 7.1   | 10.1  | 23.4  | 45.4  | 94.0  | 232.0  | 474.1  | 553.1  |
| GNP                | 103.4 | 100.0 | 286.2 | 506.0 | 982.4 | 2742.1 | 5567.8 | 6726.9 |

[a]Profits with inventory valuation adjustment and without capital consumption adjustment.
*Sources:* Economic Reports of the President, 1977, Tables B-17 and B-19; 1995, Table B-25.

**TABLE 2.6**  National Income for the United States in Selected Years, 1950–1993, and Contribution by Selected Industrial Sectors (billions of dollars)

|                                   | 1950  | 1960  | 1970  | 1980   | 1990   | 1993   |
|-----------------------------------|-------|-------|-------|--------|--------|--------|
| National income                   | 242.8 | 418.0 | 804.4 | 2198.2 | 4491.0 | 5131.4 |
| Contribution by:                  |       |       |       |        |        |        |
| Chemical and allied products      |       |       |       |        |        |        |
|   (SIC 28)                        | 4.9   | 9.1   | 16.0  | 40.0   | 84.2   | 96.4   |
| Petroleum and coal products       |       |       |       |        |        |        |
|   (SIC 29)                        | 3.4   | 4.4   | 6.6   | 19.5   | 32.4   | 38.6   |
| Rubber and miscellaneous          |       |       |       |        |        |        |
|   plastics products (SIC 30)      | 1.3   | 2.8   | 5.8   | 13.8   | 28.3   | 33.5   |
| Stone, glass, and clay (SIC 32)   | 2.8   | 4.6   | 6.9   | 14.6   | 20.2   | 21.0   |

*Sources: Statistical Abstracts of the United States,* 1976, 97th ed., Table 640; *Survey of Current Business, April 1995.*

per person (Figure 2.15). Personal income is simply the total current income before taxes of all persons (Figure 2.15). It equals national income *minus* social security tax, corporate taxes on profits, and corporate saving *plus* transfer payments made by the government (social security payments, welfare, etc.) and interest payments. Disposable income is what remains of personal income after taxes have been paid. These quantities are shown per person for the United States consumer, 1929–1994, both in current and constant dollars, in Table 2.7.

TABLE 2.7   Income Per Person in the United States in Selected Years
1929–1994

|      | Personal income per person | | Disposable income per person | |
|------|--------------------|------------------------|-------------------|------------------------|
|      | Current dollars | Constant 1987 dollars | Current dollars | Constant 1987 dollars |
| 1929 | 696   | 5569   | 675    | 5,400  |
| 1930 | 618   | 5,150  | 598    | 4,983  |
| 1940 | 589   | 5,354  | 569    | 5,173  |
| 1950 | 1,491 | 7,381  | 1,355  | 6,708  |
| 1960 | 2,212 | 8,508  | 1,934  | 7,438  |
| 1970 | 3911  | 11,111 | 3,348  | 9,511  |
| 1980 | 9,948 | 13,875 | 8,576  | 12,005 |
| 1990 | 18,666| 16,475 | 16,173 | 14,274 |
| 1994 | 21,846| 17,324 | 18,963 | 15,038 |

*Source: Survey of Current Business*, July 1976, 1991, 1994.

TABLE 2.8   Personal Consumption Expenditures Per Person in the United
States in Selected Years, 1929–1994

|      | Durable goods | | Nondurable goods | | Services | |
|------|-----------------|-----------------|-----------------|-----------------|-----------------|-----------------|
|      | Current dollars | 1987 dollars | Current dollars | 1987 dollars | Current dollars | 1987 dollars |
| 1929 | 76    | 387  | 309   | 1945 | 249    | 2,220 |
| 1933 | 28    | 192  | 177   | 1622 | 160    | 1,726 |
| 1940 | 59    | 360  | 280   | 2220 | 198    | 1,889 |
| 1950 | 203   | 625  | 648   | 2607 | 415    | 2,433 |
| 1960 | 239   | 637  | 836   | 2873 | 723    | 3,025 |
| 1970 | 418   | 901  | 1,292 | 3427 | 1,314  | 4,119 |
| 1980 | 963   | 1190 | 2992  | 3768 | 4,653  | 5,672 |
| 1990 | 1873  | 1773 | 4918  | 4244 | 8,257  | 7,077 |
| 1994 | 2,218 | 1,990| 5285  | 4215 | 10,139 | 7,455 |

*Sources: Survey of Current Business*, July 1991, 1994; Economic Report of the President,
1995, Table B-4.

The U.S. consumer's income has grown at a steady pace since the 1930's. Income per person in constant dollars has nearly tripled in 60 years.

   After providing for savings, consumers use disposable income to buy services and goods (collectively termed final products). Personal consumption expenditures per person for durable goods, nondurable goods, and services are
shown in Table 2.8.

**TABLE 2.9** Gross National Product, by Country: 1985–1991

| | Current Dollars | | | | Constant (1991) Dollars | | | | Per capita (dollars) | | |
|---|---|---|---|---|---|---|---|---|---|---|---|
| | 1985 | 1989 | 1990 | 1991 | 1985 | 1989 | 1990 | 1991 | 1985 | 1990 | 1991 |
| United States | 4,054 | 5,267 | 5,543 | 5,695 | 5,057 | 5,714 | 5,765 | 5,695 | 21,140 | 23,020 | 22,550 |
| Algeria | 34 | 39 | 40 | 41 | 42 | 42 | 41 | 41 | 1,890 | 1,620 | 1,564 |
| Argentina | 95 | 110 | 116 | 130 | 119 | 119 | 120 | 130 | 3,920 | 3,742 | 3,984 |
| Australia | 204 | 264 | 274 | 288 | 254 | 286 | 285 | 288 | 16,080 | 16,690 | 16,600 |
| Austria | 109 | 139 | 152 | 162 | 136 | 151 | 158 | 162 | 18,040 | 20,480 | 20,760 |
| Bangladesh | 14 | 19 | 21 | 23 | 18 | 21 | 22 | 23 | 177 | 192 | 195 |
| Belgium | 131 | 171 | 183 | 196 | 163 | 185 | 191 | 196 | 16,530 | 19,150 | 19,580 |
| Brazil | 287 | 384 | 384 | 404 | 358 | 416 | 399 | 404 | 2,596 | 2,618 | 2,601 |
| Bulgaria | 41 | 50 | 45 | 37 | 51 | 54 | 47 | 37 | 5,750 | 5,254 | 4,140 |
| Burma | 21 | 22 | 24 | 322 | 26 | 23 | 25 | 22 | 692 | 610 | 531 |
| Canada | 402 | 531 | 549 | 563 | 501 | 576 | 571 | 563 | 19,890 | 21,450 | 20,840 |
| Chile | 15 | 24 | 26 | 30 | 19 | 26 | 27 | 30 | 1,602 | 2,065 | 2,215 |
| China | | | | | | | | | | | |
| Mainland | 774 | 1,242 | 1,369 | 1,528 | 965 | 1,347 | 1,424 | 1,528 | 920 | 1,256 | 1,327 |
| Taiwan | 91 | 153 | 168 | 187 | 114 | 166 | 175 | 187 | 5,903 | 8,551 | 9,068 |
| Colombia | 25 | 34 | 37 | 40 | 31 | 37 | 39 | 40 | 1,038 | 1,173 | 1,187 |
| Cuba | 30 | 35 | 34 | 27 | 37 | 38 | 35 | 27 | 3,652 | 3,314 | 1,584 |
| Czechoslovakia | 98 | 123 | 124 | 109 | 122 | 134 | 129 | 109 | 7,842 | 8,248 | 6,914 |
| Denmark | 92 | 112 | 119 | 125 | 114 | 121 | 123 | 125 | 22,320 | 24,000 | 24,230 |
| Egypt | 20 | 27 | 28 | 30 | 25 | 29 | 29 | 30 | 533 | 544 | 543 |
| El Salvador | 4 | 5 | 5 | 6 | 5 | 5 | 6 | 6 | 1,064 | 1,060 | 1,075 |
| Ethiopia | 4 | 6 | 6 | 7 | 6 | 7 | 7 | 7 | 130 | 129 | 124 |
| Finland | 88 | 119 | 124 | 120 | 110 | 129 | 129 | 120 | 22,480 | 25,970 | 24,110 |

| | | | | | | | | | | | |
|---|---|---|---|---|---|---|---|---|---|---|---|
| France | 17 | 1,068 | 1,136 | 1,191 | 1,019 | 1,158 | 1,182 | 1,191 | 18,470 | 20,840 | 20,900 |
| Germany | 1,052 | 1,346 | 1,472 | 1,586 | 1,313 | 1,461 | 1,531 | 1,586 | 21,510 | 24,260 | 19,830 |
| Ghana | 4 | 5 | 6 | 6 | 5 | 6 | 6 | 6 | 366 | 394 | 401 |
| Greece | 50 | 63 | 66 | 69 | 63 | 68 | 68 | 69 | 6,317 | 6,813 | 6,878 |
| Hungary | 53 | 65 | 63 | 59 | 66 | 70 | 65 | 59 | 6,160 | 6,281 | 5,727 |
| India | 156 | 228 | 251 | 263 | 195 | 247 | 261 | 263 | 253 | 306 | 303 |
| Indonesia | 61 | 90 | 100 | 111 | 76 | 97 | 104 | 111 | 442 | 550 | 577 |
| Iran | 77 | 77 | 89 | 100 | 96 | 83 | 92 | 100 | 2,067 | 1,619 | 1,689 |
| Iraq | 33 | 34 | 25 | 13 | 41 | 37 | 26 | 13 | 2,617 | 1,400 | 705 |
| Israel | 37 | 50 | 55 | 62 | 46 | 54 | 57 | 62 | 11,680 | 13,290 | 13,590 |
| Italy | 777 | 1,014 | 1,078 | 1,134 | 970 | 1,100 | 1,121 | 1,134 | 16 | 19,440 | 19,630 |
| Japan | 2,071 | 2,841 | 3,117 | 3,386 | 2,584 | 3,082 | 3,242 | 3,386 | 21,400 | 26,240 | 27,300 |
| Kenya | 5 | 7 | 7 | 8 | 6 | 7 | 8 | 8 | 295 | 318 | 310 |
| Kuwait | 26 | 33 | 26 | 16 | 32 | 36 | 27 | 16 | 17,100 | 12,800 | 19,510 |
| Madagascar | 2 | 2 | 3 | 3 | 2 | 3 | 3 | 3 | 236 | 227 | 205 |
| Malaysia | 23 | 34 | 40 | 45 | 29 | 37 | 41 | 45 | 1,835 | 2,354 | 2,506 |
| Mexico | 195 | 232 | 254 | 276 | 243 | 251 | 264 | 276 | 3,079 | 2,988 | 3,051 |
| Morocco | 16 | 22 | 24 | 27 | 20 | 24 | 25 | 27 | 878 | 987 | 1,014 |
| Mozambique | 1 | 1 | 1 | 1 | 1 | 1 | 1 | 1 | 68 | 73 | 74 |
| Nepal | 2 | 2 | 3 | 3 | 2 | 3 | 3 | 3 | 123 | 137 | 141 |
| Netherlands | 199 | 252 | 273 | 290 | 248 | 274 | 284 | 290 | 17,140 | 19,010 | 19,310 |
| Nigeria | 19 | 24 | 26 | 30 | 23 | 26 | 27 | 30 | 227 | 230 | 242 |
| North Korea | 26 | 30 | 30 | 23 | 33 | 33 | 31 | 23 | 1,674 | 1,443 | 1,068 |
| Norway | 75 | 91 | 97 | 103 | 94 | 99 | 101 | 103 | 22,620 | 23,810 | 24,150 |
| Pakistan | 26 | 37 | 40 | 44 | 33 | 40 | 42 | 44 | 328 | 363 | 369 |
| Peru | 39 | 45 | 45 | 47 | 48 | 49 | 47 | 47 | 2,459 | 2,147 | 2,090 |
| Philippines | 28 | 40 | 44 | 46 | 35 | 44 | 46 | 46 | 615 | 709 | 694 |
| Poland | 144 | 175 | 163 | 160 | 180 | 190 | 170 | 160 | 4,841 | 4,458 | 4,185 |

**TABLE 2.9** Continued

| | Current Dollars | | | | Constant (1991) Dollars | | | | Per capita (dollars) | | |
|---|---|---|---|---|---|---|---|---|---|---|---|
| | 1985 | 1989 | 1990 | 1991 | 1985 | 1989 | 1989 | 1990 | 1985 | 1990 | 1991 |
| Romania | 39 | 113 | 105 | 94 | 124 | 123 | 109 | 94 | 5,455 | 4,705 | 4,064 |
| Saudi Arabia | 94 | 93 | 109 | 118 | 118 | 101 | 114 | 118 | 8,718 | 6,982 | 7,151 |
| South Africa | 76 | 96 | 100 | 105 | 94 | 104 | 104 | 105 | 2,726 | 2,629 | 2,753 |
| South Korea | 124 | 219 | 249 | 281 | 155 | 237 | 259 | 281 | 3,797 | 5,997 | 6,430 |
| Soviet Union | 2,118 | 2,645 | 2,660 | 2,531 | 2,642 | 2,870 | 2,767 | 2,531 | 9,475 | 9,509 | 8,639 |
| Spain | 328 | 454 | 491 | 522 | 409 | 493 | 510 | 522 | 10,630 | 13,100 | 13,370 |
| Sri Lanka | 6 | 7 | 8 | 9 | 7 | 8 | 9 | 9 | 455 | 502 | 518 |
| Sudan | 19 | 25 | 28 | 28 | 24 | 27 | 27 | 28 | 1,031 | 1,027 | 1,016 |
| Sweden | 170 | 216 | 226 | 230 | 213 | 234 | 235 | 230 | 25,430 | 27,490 | 28,900 |
| Switzerland | 172 | 220 | 233 | 242 | 214 | 239 | 242 | 242 | 32,740 | 35,920 | 35,590 |
| Syria | 17 | 19 | 23 | 25 | 21 | 20 | 23 | 25 | 1,957 | 1,842 | 1,909 |
| Tanzania | 2 | 2 | 2 | 3 | 2 | 2 | 2 | 3 | 98 | 95 | 96 |
| Thailand | 43 | 72 | 82 | 92 | 53 | 78 | 86 | 92 | 1,028 | 1,531 | 1,618 |
| Turkey | 62 | 87 | 1,030 | 105 | 77 | 93 | 104 | 105 | 1,518 | 1,819 | 1,790 |
| Uganda | 2 | 2 | 2 | 3 | 2 | 2 | 2 | 3 | 98 | 95 | 96 |
| United Kingdom | 711 | 943 | 985 | 1,002 | 887 | 1,023 | 1,025 | 1,002 | 15,660 | 17,850 | 17,400 |
| Venezuela | 33 | 39 | 46 | 53 | 41 | 43 | 48 | 53 | 2,378 | 2,413 | 2,609 |
| Yugoslavia | 91 | 107 | 105 | (NA) | 113 | 117 | 109 | (NA) | 4,892 | 4,545 | (NA) |
| Zaire | 7 | 9 | 9 | (NA) | 8 | 9 | 9 | (NA) | 268 | 244 | (NA) |

Source: Statistical Abstracts of the United States, 1994, 114th ed., p. 862.

To help you understand foreign economic activities and buying power, GNP values for other countries are shown in Table 2.9 for 1985–1991. Also shown is GNP per person. There is a tremendous range of GNP per capita, from $35,590 for Switzerland to $74 for Mozambique, or more than a 500-fold difference in material wealth and comfort. The wealthy nations tend to be the industrial nations in temperate climates, and the poor nations tend to be those relying on agricultural economies in tropical climates. Mineral resources are the basis of wealth for Kuwait, Qatar, and the United Arab Emirates. Among industrial nations, the per-capita wealth of the United States is surpassed by six countries listed in Table 2.9.

## REFERENCES

1.  Theodore Leavitt, "Marketing Myopia", *Harvard Business Review* July–August 1960.
2.  J. Wei, T.W.F. Russell, and M.W. Swartzlander, *The Structure of the Chemical Process Industries*, 1st ed. McGraw-Hill, New York, 1979.
3.  M.S. Peters and K.D. Timmerhaus, *Plant Design and Economics for Chemical Engineers*. McGraw-Hill, New York, 1991.
4.  *U.S. Department of Commerce, Bureau of Economic Analysis*, Survey of Current Business, January and May, 1991.
5.  S. Rosen, *National Income and Other Social Accounts*. Holt, New York, 1972.
6.  R. Eisner, *The Total Income System of Accounts*. University of Chicago Press, Chicago, 1989.

## NOTATIONS

| | |
|---|---|
| $a,b$ | parameters in variable cost [Eq. (2.9)] |
| $C$ | concentration of component (mol/L) |
| $CA$ | average cost, $TC/q$ |
| $CE$ | cross-elasticity $= (p_y/Q_x)(dQ_x/dp_x)$ [Eq. (2.7a)] |
| $E$ | price elasticity $= -(p/Q)(dQ/Dp)$ [Eq. (2.5)] |
| $E_s$ | supply elasticity $-(p/Q)(dQ/dp)$ [Eq. (2.8)] |
| $FC$ | fixed production cost |
| $I$ | consumers' income |
| $IE$ | income elasticity $= (I/Q_x)(dQ_x/dI)$ [Eq. (2.7b)] |
| $MR$ | marginal revenue $d(TR)/dq$ |
| $q$ | with number subscript, quantity of product desired by consumer; with letter subscript or with no subscript, quantity of product produced by a single firm |
| $Q$ | total demand for product or total amount supplied by firms engaged in manufacturing product (at equilibrium supply equals demand) |
| $RC$ | raw material cost |

| | |
|---|---|
| *TC* | total annual production cost |
| *TR* | total revenue for industry $pQ$, or firm $pq$ |
| *U* | consumers' utility |
| *VC* | variable production cost |
| *W* | volumetric process flow rate (L/min) |
| π | profit |

## PROBLEMS

2.1.   List the five most important products or services that influence your utility function. What qualities do these products possess that affect your utility function? How could these qualities be measured? Would it be satisfactory to use price to define your utility? Will the same products be important in determining your utility next year? In 5 years? By age 65?

2.2.   Propose a functional form for $U$ and two products different from that in Section 2.1.1. Justify the functional form with a paragraph of discussion.

2.3.   If $U = q_1 q_2$, find the consumer's demand functions if the usual budget constraint applies.

2.4.   Most customers of firms in the CPI are other CPI firms. For instance, a manufacturer of ethanolamines needs ethylene oxide and ammonia. It is a trivial matter to determine the required amounts of these raw materials if the total pounds of ethanolamines needed is known. Of course ethylene oxide is also used for a number of other products. What information must you obtain to construct a demand curve for ethylene oxide?

2.5.   For the following total market demand functions, develop relationships for the price elasticity $E$. How does $E$ depend on $p$? Plot $P$ versus $Q$ and $E$ versus $P$ for each case.

    a.   $P = aQ + b$

    b.   $P = aQ^b$

    c.   $P = ae^{-bQ}$

2.6.   Nuclear Power: Issues and Choices, a study prepared by the Ford Foundation and MITRE Corporation, a strategy study center, claims that the income elasticity of demand for energy was between 0.8 and 0.9 for the United States in 1975.

    a.   Use the data in Table 2.7 to predict the change in U.S. energy consumption for the time period 1970–1980. How would you expect the value for the income elasticity to change over the time period 1980–1995?

    b.   The same report predicted that price elasticity of demand for energy would shortly reach 0.35. If the United States wished to reduce energy consumption by 25%, how much would the price have to rise?

2.7.   The total annual cost $TC$ of Delos production depends on several factors, including the fixed cost $FC$ (Section 2.2.1). Show the effect of changes in fixed cost on the total annual cost by varying the fixed costs over a range of ⊕30% for a company with a 30,000-liter reactor. Assume that the price for Delos is $5.50/kg, the production cost $PC$ is 1 cent per liter, the raw material cost $RC$ is 15 cents per mole, and the functional form of $FC$ remains as a constant multiplied by $L$.

    a.  Suppose you, as a Delos producer, have reason to question your stability with respect to production costs. Show the dependence of the total annual Delos cost on production cost $PC$, if $PC$ varies up or down by as much as 30%.

    b.  Leading economic indicators have just revealed that a price increase for the raw material Algol is anticipated, but information about its magnitude is not available. As a concerned Delos competitor, prepare a plot to show the effect of a raw material price fluctuation of about ±30% on the total annual production cost.

2.8.   The following questions are based on The Last Word in *CHEMTECH* (August 1972).

    a.  What is the price elasticity of salt in this story?

    b.  Why does Sheik El Wadi receive more benefit than Abdullah in this new route innovation even when he has done nothing to deserve it?

    c.  It takes time and effort to teach villagers to use salt for food preservation. Does this effort benefit Abdullah exclusively, or does it benefit all salt sellers?

    d.  Would it be desirable for the sheik to grant Abdullah and progenies a perpetual monopoly for the new route as a reward?

2.9.   Old Smooky McCarthy is the last of the dipstick artists in wildcatting. or looking for oil in unpromising places. After a year with his faithful dipstick, he found the oil field he named Thornberry. All told, the cost of finding, drilling, and connecting was $5.2 million. Thornberry is a good well and flows at a steady rate of 200 bbl/day. In the last few years each barrel that flows contains more water and less oil. Old Smooky has to pay for hired help, utilities, upkeep of his pump, "and severance tax" to the county, all of which sums to $120 per day. It does not take a college education (Old Smooky got through third grade) to figure out that with oil selling at a steady $3 per barrel and the oil yield dropping steadily, the day will come when it pays to plug and abandon the well. But when? Professor *Sharppencil* was called in to consult. After a bit of digging into Smooky's records, which amount to a few scraps of paper and some scribbles on the back of an envelope, Professor *Sharppencil* did some thinking before reporting back.

*Sharppencil:* Despite the fact that you had 200 bbl/day all these years, only a declining fraction is oil. The fraction $f$ can be represented by $f = 0.800 - 0.60t$, where $t$ is the number of years since first production.

*Old Smooky:* Since I was knee high to a grasshopper I hated all them revenooers. I ain't gonna pay no dime to no Feds, so you can forget all about depletion allowance and income tax. Just tell me when to quit producing and start for the hills with my trusty dipstick again.

*Sharppencil:* According to sound financial policy, you should continue to produce till the day your marginal cost per barrel of oil equals your marginal revenue. On the other hand, a sound accounting practice is to depreciate your capital cost of $200,000 evenly over the life of the well; so if the well lasts 20 years, your depreciation cost would be $10,000 per year.

*Old Smooky:* Say, young fella. I heard a wild rumor a week ago that the price of oil may be going up, might even hit five. You reckon I could maybe squeeze a few more barrels?

*Sharppencil:* The answer to your question is in the affirmative.

*Old Smooky:* You sure talk smart!

*Sharppencil:* At $1000 per day consulting fee, I better talk smart.

    a.   What is the average cost of a barrel of oil produced over the life of the well?

    b.   Assuming straight-line depreciation of the capital cost, what is the annual profit in year 1?

    c.   What is the annual profit in the last year?

    d.   If the price of oil goes to $4 per barrel, what would be the new value of $t$ at which to plug and abandon the well?

    e.   At $4 per barrel, how much more oil would be produced?

    f.   What is the price elasticity of supply for this well?

    g.   The asset value at $t$ is $200,000. Forgetting the time value of money, if Old Smooky wants to sell Thornberry at $t = 0$, what would a reasonable customer who knows all the facts be willing to pay?

2.10.   Back in 1973, there were tales in Detroit of an aborted conspiracy to consolidate all U.S. automobile production in the Universal Motor Company. This budding monopoly was very concerned about the clamor, which started in the early 1970s, for putting antipollution devices on cars, costing as much as $500 per car. Mr. Bigelow, the president of UMC, wondered whether he could pass along this extra cost to the customers or whether UMC would have to take this extra cost out of profit. Thus, being just as shrewd a businessperson as Old Smooky McCarthy, he called in his favorite economics consultant, Professor Sharppencil. Recently released FBI tapes, available under the Freedom of Information Act, reveal the conversation between Mr. Bigelow and Professor Sharppencil.

*Bigelow:* In my job, there is a new headache every month. I can't wait till I reach retirement age next year. What do you think about splitting the $500 fairly with the consumers: we swallow $50 and increase the sticker price by $450. After all, we have to do our share for clean air too.

*Sharppencil:* Things are a lot more complicated than you think, Mr. Big!

*Bigelow:* I was afraid you'd say that.

*Sharppencil:* Now you are selling 10 million cars a year at $3000 a car. If you set the new car price at $3500, you won't sell 10 million. There'll be a lot of cheapskates who would say that with this price rise they'll stretch that old jalopy another year. Fortunately, I did a consumer study last year, and I can now say with confidence that the sales will drop to 8 million.

*Bigelow:* That's horrible! You're not suggesting that we leave the price at $3000 and take the $500 out of my hide!

*Sharppencil:* In that case, your profit would drop from $5 billion a year to zero. I figured all this out on a formula that even your vice president for manufacturing does not know about:

Total cost = 2.5 billion + $2000Q$ + $(0.25 \times 10^{-18})Q^4$ (before + $1,000)

New $TC$ = 2.5 billion + $2500Q$ + $(0.25 \times 10^{-18})Q^4$ (after + $1,000)

where $Q$ is the number of cars sold per year.

*Bigelow:* Don't keep me guessing. What should I do?

*Sharppencil:* I'll go home and do a few calculations and give you the pricing strategy to yield the maximum profit. You'd also like to know the new price, the forecast of volume, and your new profit picture. I'll throw in other goodies, such as the price elasticity of demand, the cost elasticity of production, the average cost formula, the marginal cost formula . . .

*Bigelow:* No formulas please, just the facts next week.

*Sharppencil* [to himself]: My students love to grind out formulas. I'll turn in the formulas to Bigelow anyway. He may not understand mathematics, but he is easily impressed.

Provide the analysis Sharppencil promised.

2.11.   The supply curve for Delos is defined by Eq. (2.28). If the demand curve can be defined as $p = 30 - 10^{-6}Q$, find the profit for each company in each of the groups.

2.12.   Suppose all the companies in group I producing Delos set up a cooperative research effort, which results in an increase in the reaction rate constant $k$ of 50%. Determine the new supply curve and the new profit for each company. Determine market share and compare both profit and market share with that when $k$ is the same for all firms (0.005 min$^{-1}$) How much should each firm in group I be willing to pay for the research effort?

2.13.   Suppose all the firms producing Delos can obtain their raw material at a concentration 50% greater than 0.2 mol/L and at the same price. Determine the new supply curve, the new share of market, and the new profit for each firm.

2.14.   If just one firm in group I can reduce its cost of raw material by 50%, how will its profit be affected?

2.15.   Using the data from Table 2.3, determine the percentage growth in personal consumption $C$ and government purchases $G$ for each of the years shown. What conclusions can you draw?

2.16.   Compare the disposable personal income in both current and 1958 dollars with total national income and prepare a graph showing disposable income as a percentage of total income versus year.

2.17.   As the United States grows more affluent, has the investment for the future become a larger or smaller percentage of the total GNP since 1929? Has the share of net income for employee compensation or corporate profit been growing?

2.18.   Table 2.9 shows that per capital GNP varies from $8,444 (Kuwait) to $71 (Rwanda) in a ratio of over 100. What can account for this huge disparity of productive capabilities? Is it land, resources, capital, technology, labor, or luck? Do you believe these differences should or will continue? How might they be narrowed?

2.19.   The definition of product in the legal sense may differ from the everyday sense. In 1956 the Supreme Court ruled that DuPont does not have a monopoly in cellophane, since the relevant product market under question is the "flexible packaging material market," in competition with many other materials. There is high enough cross-elasticity to prevent a monopoly. According to this definition, indicate which of the following chemical products have reasonable substitutes and could be included in larger market definitions: gasoline, sulfuric acid, titanium dioxide, aspirin, vitamin C. Discuss.

2.20.   A factory produces a product by a series of machine operations. The operators are paid at a piece rate of 20 cents per unit produced. The operation has an efficiency that depends on the production rate, creating rejects that have no commercial value. The relation between production and required feed rate is $f = q^{1.3}$, where $f$ and $q$ are in units per hour. The raw material cost is 20 cents per unit of feed, and the power cost is $I$ cents per unit. The plant operates 24 h/day, 7 days/week, and 52 weeks/yr. The first cost of plant and machines is $33,000. Annual fixed charges (interest, taxes, depreciations, etc.) are 32% of first cost. Heating and light is $500 per year, and general overhead is $3000 per year.

   a.   What is the fixed cost per $FC$ per year, variable cost $VC$ per year as a function of $q$, and total cost $TC$ per year as a function of $q$?

   b.   At a given price, ranging from zero to $1 per unit, what is the optimum production volume, the total cost $TC$ per year, the total revenue $TR$ per year, and the profit per year?

   c.   At what selling price would it pay to shut down the plant?

# 3

# Accounting and Financial Statements

Some basic knowledge of accounting is necessary before a chemical professional can develop and obtain the data needed to analyze a firm's operations, discover whether the firm is making a profit, and predict whether a company will continue to make a profit. It is also essential to know how a firm's operations are reported to determine its role in a particular industry or in the national economy. Financial reports of the individual firm are one of the most important sources of data for the national accounting summaries presented in Chapter 2 and for the input–output tables in Chapter 4.

There are differences of opinion concerning how much information about the bookkeeping process a chemical professional needs to know to understand accounting reports and financial statements. The average chemical professional interfaces with the accounting department in the budgeting and control function, operating performance of a department and, in some instances, with input and feedback during stages of plant design and construction. Accounting records also provide a file of prior information of actual projects that may be of value to the chemical professional in developing estimates.

The conventions governing accounting are fairly simple, but their detailed application is complex, requiring many years of study and experience. In this chapter, we acquaint the nonfinancial professional with the basic concepts of financial reporting by using simple examples and by analyzing a typical balance sheet and income statement from a standard company report.

Accounting systems have as input business transactions in their original form receipts and invoices. These business events are entered chronologically in a *journal* and then classified and posted in an appropriate account in a *ledger.* Pe-

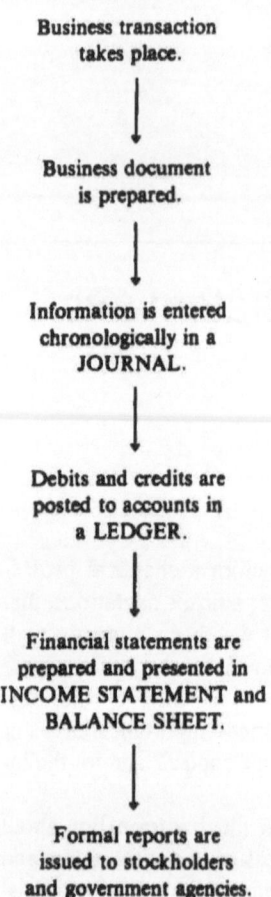

Business transaction
takes place.

Business document
is prepared.

Information is entered
chronologically in a
JOURNAL.

Debits and credits are
posted to accounts in
a LEDGER.

Financial statements are
prepared and presented in
INCOME STATEMENT and
BALANCE SHEET.

Formal reports are
issued to stockholders
and government agencies.

FIGURE 3.1  Flow of information through an accounting system.

riodically, perhaps, once a month but at least once a year, the accounts are closed
and a summary is issued as an *income statement* and a *balance sheet*. Data from
the financial statements are used by management, owners, creditors, and others
interested in the firm as well as local, state, and federal governments. A simple
informational flow diagram is found in Figure 3.1.

## 3.1  ACCOUNTING CONCEPTS AND CONVENTIONS

Various accounting texts deal with the fundamental concepts of accounting giving concise definitions well-suited to our purpose (1, 2).

Accounting methods in use today had their origin in the fourteenth century in what is now Italy (3). Fibonacci introduced the dual-aspect concept, and the Medicis made the system more efficient. The basis of these methods is called *double-entry bookkeeping,* which in the simplest terms is expressed as follows:

ASSETS = EQUITIES

*Assets* are economic resources that are owned by the firm and are expected to benefit future operations. Assets are items of value and may be tangible (equipment, buildings, furniture, etc.). Or, assets may be intangible (franchises, patents, trademarks, etc.).

*Equities* are claims against the firm. They may be divided into liabilities and owners' equity. Therefore, the equation above becomes:

ASSETS = LIABILITIES + OWNERS' EQUITY

*Liabilities* are outside claims against the assets of the firm. They are obligations to convey assets or to perform services, and they require settlement in the future. If the liabilities are deducted from the assets, the difference is the amount belonging to the firm's owners (e.g., stockholders) and is called *owners' equity.*

Any transaction that takes place causes changes in the accounting equation. An increase in assets must be accompanied by one of the following:

An increase in liabilities (e.g., money was borrowed to buy an item of equipment)

An increase in the stockholders' equity (e.g., they provided the money to purchase the equipment)

A decrease in assets (money was taken out of cash to buy the equipment: in this case, total assets do not change, but there is a change in the distribution of assets)

A change in one part of the equation due to an economic event must be accompanied by an equal change in another place—hence the term *double-entry bookkeeping.*

### 3.1.1 Debits and Credits

Economic events that change the accounting equation are recorded in books. The left side of the account book page has been arbitrarily designated the *debit* side and the right side the *credit* side. This is true regardless of the type of account.

### 3.1.2 Data Recording

Most of the accounting today is done by entering data into a computer and using software packages to record, accumulate, and classify data. The accounting equation, business transactions, debits and credits, accounts, journals, and ledgers are all part of the modern computer accounting system.

### 3.1.3 The General Journal

All transactions are recorded chronologically in a *general journal* of the form shown in Table 3.1 for Delchem Corporation in Example 3.1. In the first column is the date of the business transaction. An account title and a brief description of the transaction are found in the second column. The ledger page for each transaction is placed in the third column and serves as a cross-reference between the general journal and the various ledger accounts. The numbers in the column indicate the account to which the debit or credit has been transferred. The amount of each individual debit and credit entry is listed in the next two columns. Again, the left column contains the debits and the right column the credits.

### 3.1.4 The Ledger

Journal entries are transferred to a ledger. This process is called *posting*. Separate ledger accounts may be set up for each major type of transaction—asset account, liability account, revenue account, expense account, and so on. The number of ledger accounts depends on the information management needs to make decisions. Each debit entry to a ledger account is matched by a credit entry to another account. There is a one-to-one correspondence between journal entries and entries to the ledger, again reflecting the term *double-entry bookkeeping*. The ledger page (LP) is the cross-reference. Occasionally, the ledger must be enlarged to accommodate new types of account.

Periodically, perhaps on a monthly basis but certainly yearly, the ledger sheets are closed and balanced. The ledger sheets are used as intermediate documents between journal records and balance sheets, income statements, and retained earnings statements, as well as information for various government reports. For example, a consolidated income statement can be prepared from the ledger revenue and expense accounts. From the asset and liability accounts, a company's balance sheet is prepared. Table 3.2 is the ledger obtained from the general journal, Table 3.1.

### 3.2 JOURNAL AND LEDGER EXAMPLE

The basic concepts of accounting are illustrated by a simple example. This example demonstrates the flow of information as depicted in Figure 3.1 will be demonstrated.

**TABLE 3.1**   Delchem Corporation: Page 1 of General Journal

| Date | Account titles and explanation | LP[a] | Debit | Credit |
|---|---|---|---|---|
| 1/1 | Cash | 10 | 5000 | |
| | Stockholders' equity | 50 | | 5000 |
| | Capital invested by Armstrong | | | |
| 1/1 | Cash | 10 | 5000 | |
| | Stockholders' equity | 50 | | 5000 |
| | Capital invested by Bigelow | | | |
| 1/1 | Raw materials | 11 | 1000 | |
| | Stockholders' equity | 50 | | 1000 |
| | Raw materials from Custer | | | |
| 1/1 | Equipment | 12 | 3000 | |
| | Stockholders' equity | 50 | | 3000 |
| | Reactor and mixing vessels from Custer | | | |

[a]The ledger page (LP) column is used as a cross-reference between the general journal and the various ledger accounts. The number in the column indicates the account to which the debit or credit has been transferred.

**TABLE 3.2**   Delchem Corporation: Ledger

| Date | | JP[a] | Debit | Credit |
|---|---|---|---|---|
| | Asset accounts[b] | | | |
| 1/1 | Cash (10) | 1 | 5,000 | |
| | | | 5,000 | |
| | Ending balance | | 10,000 | |
| 1/1 | Raw materials account (11) | 1 | 1,000 | |
| | Ending balance | | 1,000 | |
| 1/1 | Equipment account (12) | 1 | 3,000 | |
| | Ending balance | | 3,000 | |
| | Liability accounts | | | |
| 1/1 | Stockholders' equity(50) | 1 | | 5,000 |
| | | 1 | | 5,000 |
| | | 1 | | 1,000 |
| | | 1 | | 3,000 |
| | Ending balance | | | 14,000 |

[a]Journal page.
[b]Ledger pages in parentheses.

On January 1, 20XX, Armstrong, Bigelow, and Custer agreed to start a business to manufacture of a specialty solvent, Cleansolv. The partners named the company Delchem, Inc., and their respective contributions to the company formation were as follows:

> Armstrong: $5000 cash
> Bigelow: $5000 cash
> Custer: basic process development information, a small reactor and mixing
>   vessels, and some raw materials

The three decide to distribute 1000 shares of stock as follows:

> Armstrong: 300 shares
> Bigelow: 300 shares
> Custer: 400 shares

All these initial transactions are recorded in the general ledger of the form shown in Table 3.1. Each journal transaction appears twice, once as a credit and once as a debit.

A ledger, Table, 3.2, was set up containing the necessary accounts to record the transactions of January 1, 20XX. Again, the number of accounts in the ledger depends on the information required by management to make decisions. Initially, Delchem, Inc. requires only asset and liability accounts. However, as the firm grows, more accounts will be established as necessary to record the business transactions.

Table 3.3 illustrates how information from the general journal and the ledger accounts is used to prepare a consolidated balance sheet. Manufacturing began during the month of January, so new asset and liability accounts were created to accommodate the new type of transaction. The general journal, Table 3.4, reflects the debits and credits with appropriate explanations and ledger page entries. For example, on January 26, there was a transfer of assets, namely, the transfer of $5000 in raw materials to $5000 of finished goods.

Temporary revenue and expense accounts are used to classify changes affecting stockholders' equity. Expense accounts covering legal expense (LP 40), depreciation expense (LP 41), and interest expense (LP 42) were created. A revenue account was not needed in January because there was no income. These ac-

**TABLE 3.3**  Delchem Corporation-Consolidated Balance Sheet as of January 1, 20XX

| Assets | | Liabilities and stockholders' equity | |
|---|---|---|---|
| Cash (10) | $10,000 | Current liabilities | $     0 |
| Inventory (raw materials) (11) | 1,000 | Stockholders' equity | |
| Plant and equipment (12) | 3,000 | (50) | $14,000 |
| | $14,000 | | $14,000 |

**TABLE 3.4**  Delchem Corporation: Page 2 of General Journal

| Date | Account titles and explanation | LP | Debit | Credit |
|------|-------------------------------|-----|-------|--------|
| 1/3 | Legal expense | 40 | 1,000 | |
| |   Cash | 10 | | 1,000 |
| | Paid lawyer to set up corporation | | | |
| 1/4 | Finished goods | 14 | 1,000 | |
| |   Accrued wages payable | 22 | | 1,000 |
| | Hired Davis as production labor and promised | | | |
| |   to pay him $1000 on 2/1[a] | | | |
| 1/4 | Prepaid expenses | 15 | 2,000 | |
| |   Cash | 10 | | 2,000 |
| | Cash downpayment on equipment to be | | | |
| |   delivered later | | | |
| 1/10 | Raw materials | 11 | 10,000 | |
| |   Cash | 10 | | 4,000 |
| |   Accounts payable | 24 | | 6,000 |
| | Purchased raw materials, paid $4000, balance | | | |
| |   of $6000 due in February | | | |
| 1/17 | Cash | 10 | 2,000 | |
| |   Bank loan | 21 | | 2,000 |
| | Obtained year loan from bank, interest at 12% | | | |
| |   per year | | | |
| 1/26 | Finished goods | 14 | 5,000 | |
| |   Raw materials | 11 | | 5,000 |
| | 5000 liters of Cleansolv manufactured | | | |
| |   using $5000 of raw materials[b] | | | |
| | **Adjusting entries** | | | |
| 1/31 | Depreciation expense: Equipment | 41 | 25 | |
| |   Equipment | 12 | | 25 |
| | $3000 \times 1/10 \times 1/12 = \$25$ per month | | | |
| 1/31 | Interest expense | 42 | 20 | |
| |   Interest payable | 23 | | 20 |
| | To record bank loan interest for January | | | |
| 1/31 | Income summary | 55 | 1,045 | |
| |   Legal expense | 40 | | 1,000 |
| |   Depreciation expense: equipment | 41 | | 25 |
| |   Interest expense | 42 | | 20 |
| | To close the expense and revenue accounts | | | |
| |   for the period | | | |
| 1/31 | Stockholders' equity | 50 | 1,045 | |
| |   Income summary | 55 | | 1,045 |
| | To close the income summary and transfer | | | |
| |   the gain (or loss) to the equity account | | | |

[a]Notice from 1/26 entry that we have now in the inventory 5000 liters of Cleansolv, incorporating $1000 of labor and $5,000 of raw materials.

[b]Given the costs of labor (see 1/4 entry), inventory value of this batch of Cleansolv is $6000/(5000 liters) or $1.20 per liter.

counts are used to prepare an income statement. The balance of revenue and expense accounts is reduced to zero through an *income summary* account at the end of the month. Table 3.4 is the general journal for the month of January, and Table 3.5 is the corresponding ledger. A consolidated income statement, Table 3.6, is developed from the income and expense accounts. Note that this statement reflects no income and that there was a loss of $1045 during that month. Table 3.7 is the consolidated balance sheet as of February 1, 20XX. If this balance sheet is compared with the January 1, 200 sheet, Table 3.3, it will be noted that the stockholders' equity decreased on the February statement by $1045, reflecting the operational deficit during January.

The same procedure is followed for each succeeding month, with each transaction being entered in the general journal then posted to the appropriate ledger account. At the end of February, an income statement and balance sheet may be prepared. In this manner, information for an annual report is assembled.

Today transactions are entered into a computer, ledger accounts are assigned, and the data manipulated electronically. Modern firms no longer keep manual ledgers.

Up to this point in this chapter, we have presented what is called "traditional" cost/managerial accounting. The traditional methods have helped finance departments monitor operations and value inventory, but some people feel that rather than providing accurate picture of a company's costs, this approach focused on direct costs and relied on arbitrary cost allocations such as labor-based overhead rates.

With the restructuring and downsizing of companies in the late 1980s, new management tools were introduced. With these new tools, new accounting concepts were developed (4). One of these new accounting systems is believed to provide useful information about direct and indirect costs of a production unit or a service, provide tracking of cost-contributing activities, as well as separating and identifying value-added activities from non-value-added ones that contribute to current costs. Although only about 10% of major corporations in the United States are using this system, proponents believe that it will allow managers to make better decisions about what products and services to offer and what the "real" costs are. This new approach may affect how accounting information is handled and perhaps alter company financial reporting. It will be interesting to see what happens in the next few years, that is, whether traditional accounting will withstand the test of time.

## 3.3 FINANCIAL REPORTS

A financial report, sometimes called an *annual report*, contains a large amount of information and is designed to tell the reader how well a company performed in the preceding year and how this performance compared against various standards.

TABLE 3.5 Delchem Corporation: Ledger

| Date | | JP | Debit | Credit |
|---|---|---|---|---|
| | Asset accounts | | | |
| | Cash (10) | | | |
| 1/1 | Starting balance | 1 | 10,000 | |
| 1/3 | | 2 | | 1,000 |
| 1/4 | | 2 | | 2,000 |
| 1/10 | | 2 | | 4,000 |
| 1/17 | | 2 | 2,000 | |
| | Ending balance | | 5,000 | |
| | Raw materials (11) | | | |
| 1/1 | Starting balance | 1 | 1,000 | |
| 1/10 | | 2 | 10,000 | |
| 1/26 | | 2 | | 5,000 |
| | Ending balance | | 6,000 | |
| | Equipment (12) | | | |
| 1/1 | Starting balance | 1 | 3,000 | |
| 1/31 | | 2 | | 25 |
| | Ending balance | | 2,975 | |
| | Finished goods (14) | | | |
| 1/1 | Starting balance | | 0 | |
| 1/4 | | 2 | 1,000 | |
| 1/26 | | 2 | 5,000 | |
| | Ending balance | | 6,000 | |
| | Prepaid expenses (15) | | | |
| 1/1 | Starting balance | | 0 | |
| 1/4 | | 2 | 2,000 | |
| | Ending balance | | 2,000 | |
| | Liability accounts | | | |
| | Bank loan (21) | | | |
| 1/1 | Starting balance | | | 0 |
| 1/17 | | 2 | | 2,000 |
| | Ending balance | | | 2,000 |
| | Accrued wages payable (22) | | | |
| 1/1 | Starting balance | | | 0 |
| 1/4 | | 2 | | 1,000 |
| | Ending balance | | | 1,000 |
| | Interest payable (23) | | | |
| 1/1 | Starting balance | | | 0 |
| 1/30 | | 2 | | 20 |
| | Ending balance | | | 20 |

TABLE 3.5   Continued

| Date | | Ref | | |
|---|---|---|---|---|
| | Accounts payable (24) | | | |
| 1/1 | Starting balance | | | 0 |
| 1/10 | | 2 | | 6,000 |
| | Ending balance | | | 6,000 |
| | Stockholders' equity (50) | | | |
| 1/1 | Starting balance | 1 | | 14,000 |
| 1/31 | | 2 | 1,045 | |
| | Ending balance | | | 12,955 |

| Expense accounts |
|---|

| Date | | Ref | | |
|---|---|---|---|---|
| | Legal expense (40) | | | |
| 1/1 | Starting balance | | 0 | |
| 1/3 | | 2 | 1,000 | |
| 1/31 | | 2 | | 1,000 |
| | Ending balance | | 0 | |
| | Depreciation expense: Equipment (41) | | | |
| 1/1 | Starting balance | | 0 | |
| 1/31 | | 2 | 25 | |
| 1/31 | | 2 | | 25 |
| | Ending balance | | 0 | |
| | Interest expense (42) | | | |
| 1/1 | Starting balance | | 0 | |
| 1/31 | | 2 | 20 | |
| 1/31 | | 2 | | 20 |
| | Ending balance | | 0 | |
| | Income summary (55) | | | |
| 1/1 | Starting balance | | | 0 |
| 1/31 | | 2 | 1,000 | |
| 1/31 | | 2 | 25 | |
| 1/31 | | 2 | 20 | |
| 1/31 | | 2 | | 1,045 |
| | Ending balance | | | 0 |

We have included parts of an annual report of a fictitious company, Archem, Inc., which we use to explain the terminology and construction of a balance sheet, an income statement, and a retained earnings statement.

The contents of a financial report may be classified into three distinct parts. The written part, mostly prose, is cast in simple language and is understandable. There may be some words or phrases new to the reader, but any financial jargon is explained in the text in this section, activities for all company divisions are

**TABLE 3.6**  Delchem Corporation:
Consolidated Income Statement, January
1–31, 20XX

| | |
|---|---|
| Revenue | $0 |
| Legal expenses | $1,000 |
| Depreciation expense: Equipment | 25 |
| Interest expense | 20 |
| Earnings (loss) | ($1,045) |

**TABLE 3.7**  Delchem Corporation: Consolidated Balance Sheet, February 1,
20XX

| Assets | | Liabilities and stockholders' equity | |
|---|---|---|---|
| Cash (10) | $ 5,000 | Accrued wages (22) | $ 1,000 |
| Prepaid expense (15) | 2,000 | Short-term borrowing: | |
| Inventory: | | Accounts payable (24) | 6,000 |
| Raw Materials (11) | 6,000 | Bank loan (21) | 2,000 |
| Finished Goods (14) | 6,000 | Interest payable (23) | 20 |
| Plant and | | Total liabilities | $ 9,020 |
| equipment (12) | 2,975 | Stockholders' equity | $12,955 |
| | | Total liabilities and | |
| Total assets | $21,975 | stockholders' equity | $21,975 |

presented—for example, any new product ventures undertaken as well as old
ones that were discontinued or sold to other companies. There will also be statements regarding how the company is meeting environmental, safety, health, and
product liability problems. The purpose of these statements is to demonstrate
that the company is a good corporate citizen. The second part of the report consists of the photographs that augment the prose and show what the equipment,
buildings, plants, and company personnel do for the company. The third part
contains the financial figures, which are usually the most difficult part of the report for the average reader to comprehend.

One of the most important parts of the third section is the footnotes that are
referenced and appended to the financial figures. When reading a financial report, one should always read the footnotes because they explain the sources from
which the numbers are derived. Although the style of a financial report has
changed through the years, the major three sections have remained intact.

A financial report contains two significant documents: the balance sheet and
the income statement. Two ancillary documents often included are the accumulated
retained earnings and the changes in working capital statements. In some annual

reports, the accumulated retained earnings are included in the statement of consolidated stockholders' equity. All four documents are discussed in the following sections. Tables 3.8 and 3.9 will help greatly to illustrate the text of Section 3.3.1.

### 3.3.1  The Balance Sheet

The balance sheet represents the financial status of the company on a particular date. The date frequently used is December 31 of any given year—as if the company's operation were "frozen" in time on that date (5).

The term "consolidated" balance sheet, introduced in Table 3.8, means that all the financial data for the parent company operations as well as financial data from all subsidiary companies, if there are any, are brought together in this document.

A balance sheet contains some real figures (e.g., cash and marketable securities), some estimated amounts or allowances (e.g., inventories and accounts receivable), as well as some fictitious figures (e.g., intangibles for which numbers are difficult to assess).

The balance sheet consists of two parts: the *assets,* which are what a company owns, and the *liabilities* and *stockholders' equity,* which comprise what a company owes. For both sides of the balance sheet to balance, the total assets must equal the total liabilities plus the stockholders' equity.

**3.3.1.1  Assets.**  The assets of a company are divided into three broad categories: *current assets, fixed assets, and intangibles.*

*3.3.1.1.1  Current Assets.*  The *current assets* are those that may be converted to cash within a year from the date of the balance sheet. The current assets include *cash* such as petty cash and money on deposit in a bank, while *marketable securities* are usually commercial paper and government bonds that can be readily converted to cash. *Accounts receivable* are goods sold to customers on a 30-, 60-, or 90-day basis for which full payment has not been received as of the date of the balance sheet. An allowance is made for uncollected bills because some customers are unable to pay. *Inventories* consist of raw materials on hand, goods in process, supplies, and finished goods ready for shipment to customers. Raw materials and supplies are carried at cost and goods in process at the raw material cost plus one-half the conversion cost; finished goods are valued at market price. Frequently, inventory costs are carried at slightly less than these figures to allow for deterioration, decline in prices, obsolescence, and so on.

*Prepaid expenses* include prepaid insurance premiums as well as leases for equipment, computers, and office machinery. These expenses are listed under current assets because although full benefit has not been received, the company has paid for the assets and expects to receive full benefit within the year.

**TABLE 3.8** Archem, Inc.: Consolidated Balance Sheet as of December 31[a]

| Assets | 19X6 | 19X5 |
|---|---|---|
| Current assets | | |
| Cash | $ 63,000 | $ 51,000 |
| Marketable securities | 41,000 | 39,000 |
| Accounts receivable[b] | 135,000 | 126,000 |
| Inventories | 149,000 | 153,000 |
| Prepaid expenses | 3,200 | 2,500 |
| Total current assets | $391,200 | $371,500 |
| Fixed assets | | |
| Land | 35,000 | 35,000 |
| Buildings | 101,000 | 97,500 |
| Machinery | 278,000 | 221,000 |
| Office equipment | 24,000 | 19,000 |
| Total fixed assets | $438,000 | $372,500 |
| Less accumulated depreciation | 128,000 | 102,000 |
| Net fixed assets | $310,000 | $270,500 |
| Intangibles | 4,500 | 4,500 |
| **Total assets** | $705,700 | $646,500 |

| Liabilities | 19X6 | 19X5 |
|---|---|---|
| Current liabilities | | |
| Accounts payable | $ 92,300 | $ 81,300 |
| Notes payable | 67,500 | 59,500 |
| Accrued expenses payable | 23,200 | 26,300 |
| Federal income taxes payable | 18,500 | 17,500 |
| Total current liabilities | $201,500 | $184,600 |
| Long-term liabilities | | |
| Debenture bonds, 10.3% due in 2015 | 110,000 | 110,000 |
| Debenture bonds, 11.5% due in 2007 | 125,000 | 125,000 |
| Deferred income taxes | 11,600 | 10,000 |
| **Total liabilities** | $448,100 | $429,600 |
| **Stockholders' equity** | | |
| Preferred stock, 5% cumulative | | |
| $5 par value—200,000 shares | $ 10,000 | $ 10,000 |
| Common stock, $1 par value | | |
| 19X5 28,000,000 shares | 32,000 | 28,000 |
| 19X6 32,000,000 shares | | |
| Capital surplus | 8,000 | 6,000 |
| Accumulated retained earnings | 207,600 | 172,900 |
| **Total stockholders' equity** | $257,600 | $216,900 |
| **Total liabilities and stockholders' equity** | $705,700 | $646,500 |

[a]All amounts in thousands of dollars.
[b]Includes an allowance for doubtful accounts.

*3.3.1.1.2   Total Current Assets.*   The sum of cash, marketable securities, inventories, accounts receivable, and prepaid expenses is called *total current assets.*

*3.3.1.1.3   Fixed Assets.*   A company's fixed assets include the land, buildings, manufacturing equipment, office equipment, automobiles, trucks, and so on that the company owns. These items are carried on the books at cost less the accumulated depreciation. Land value is entered at the same value year to year. The sum of these items is called *net fixed assets.*

Other assets include *intangibles.* They are assets that have substantial value to the company (patents, licenses, franchises, trademarks, goodwill, etc.). There is no consistent way to evaluate these assets, so the company often balances both sides of the balance sheet by making this value "close" the sheet. On occasion, other assets, such as investments in affiliates or deferred charges for which full benefit has not been received, may be included before the total assets are summed.

**3.3.1.2  Total Assets.**   The sum of current assets, fixed assets, deferred charges, and intangibles is called the *total assets.*

**3.3.1.3  Liabilities.**   The *liabilities* are what a company owes, divided into current and long-term liabilities.

*3.3.1.3.1   Current Liabilities.*   *Current liabilities* are debts that must be paid within a year from the date of the balance sheet. They are paid from the current assets. Current liabilities include accounts payable, notes payable, accrued expenses payable, and income taxes payable.

*Accounts payable* are such items as the invoices for raw materials and supplies that a company has purchased from suppliers for which payment is usually due in 30, 60, or 90 days.

*Notes payable* include the money owed to banks and other creditors. Promissory notes are in this category.

*Accrued expenses payable* are in addition to accounts payable. They may include such items as salaries, wages, interest on borrowed funds, insurance premiums, and pensions.

The liability known as *income taxes payable* is the debt due to various taxing authorities such as federal, state, and local governments. It is common practice to isolate this item from other expenses. These taxes are usually paid quarterly.

*3.3.1.3.2   Total Current Liabilities.*   The sum of the accounts payable, notes payable, accrued expenses payable, and income taxes payable is called *total current liabilities.*

*3.3.1.3.3   Long-Term Liabilities.*   *Long-term liabilities* are debts due more than one year from the date of the financial report.

*3.3.1.3.4 Bonds and Loans.* *First mortgage bonds* are issued at a stated interest rate due in a stated year. They are backed by the company's property. *Debenture bonds,* on the other hand, are backed by the general credit of the company rather than by company property. *Long-term loans* from insurance companies and investment houses are another form of long-term liability.

*3.3.1.3.5 Deferred Income Taxes.* *Deferred income taxes* are encouraged by the government as a tax incentive that will benefit the economy. An example of such an incentive is accelerated depreciation, which provides the rapid write-off in the early years of an investment. The net effect is to reduce what the company will pay in current taxes, but the full amount must be paid in the future. To smooth out wide fluctuations in a company's earnings, an entry is made for deferred taxes. This entry shows what the taxes would be without accelerated depreciation write-offs.

### 3.3.1.4 Total Liabilities.

The sum of current and long-term liabilities constitutes *total liabilities.*

### 3.3.1.5 Stockholders' Equity.

This is the total interest that the stockholders have in the business. The *stockholders' equity* is the net worth of the company, namely, the total assets minus the total liabilities. For convenience, stockholders' equity is divided into three categories: capital stock, capital surplus, and accumulated retained earnings.

*3.3.1.5.1 Capital Stock.* *Capital stock* is classified into two broad groups: preferred stock and common stock.

*Preferred stock* has a preference over other shares regarding dividends and/or the distribution of assets. Some preferred stock is called *cumulative,* which means that if in any given year the company does not pay dividends, the unpaid dividends accumulate, and when these obligations are paid, the holders receive of this stock dividends before the holders of common stock. Preferred stockholders do not normally have a voice in company affairs or voting rights unless the company fails to pay them dividends. The preferred stock is carried on the books at a stated par value.

On the other hand, there are no limitations on the dividends paid to the holders of *common stock.* If the company's earnings are high, dividends are paid, but if earnings are low, dividends may not be paid at all. Common stock is valued at a stated par value.

*3.3.1.5.2 Capital Surplus.* *Capital surplus* is the amount of money that the stockholders paid for the stock over and above the par value of the stock.

*3.3.1.5.3 Accumulated Retained Earnings.* This term is sometimes used synonymously with *earned surplus.* The *accumulated retained earnings* are calculated by subtracting the dividends paid to the stockholders from the net profit. If all the profits in one year are not distributed, they are retained by the company and added to the next year's earnings. They may be used, for example, for research and development activities, and/or for the purchase of capital equipment.

**3.3.1.6 Total Stockholders' Equity.** The *total stockholders' equity* is the sum of the preferred stock, common stock, capital surplus, and accumulated retained earnings.

**3.3.1.7 Total Liabilities and Stockholders' Equity.** The sum of the total liabilities and the stockholders' equity is what the company owes. For the balance sheet to "balance," it must equal the total assets.

### 3.3.2 The Income Statement

The income statement may also be known as the statement of profit and loss, the earnings statement, or the statement of operations. It displays the operating activities of a company for the year and may be an indication of company's future performance. A typical statement will show the figures for the current year as well as one or two previous year's activities. Frequently, an annual report will include a 5- or 10-year summary near the end of the report. The term "consolidated" will appear, indicating that all the financial activities of the company and its subsidiary operations are reported in a single statement (Table 3.9) (5).

**3.3.2.1 Net Sales.** The *net sales* is the amount of money received for the goods sold less the amount for returned goods and allowances for reduction in prices (e.g., allowing freight on goods shipped).

**3.3.2.2 Cost of Goods Sold and Operating Expenses.** This item includes all the expenses in converting raw materials into finished product, including depreciation, as well as sales, administration, research, and engineering expenses.

*3.3.2.2.1 Cost of Goods Sold.* The *cost of goods sold* represents the cash operating expenses for raw materials, labor, utilities, supplies, supervision, maintenance, waste disposal, plant indirect expanses, and so on.

*3.3.2.2.2 Depreciation, Amortization, and Depletion.* The federal government allows a company to charge off a portion of an asset due to wear and tear as well as obsolescence each year as an operating expense. This is a paper transaction and is not a cash item. Amortization is the decline in useful value of an intangible asset such as a patent. Depletion is the diminution of a natural resource, such as coal in a mine. All these paper allowances appear as one item in most income statements.

*3.3.2.2.3 Sales, Administration, Research and Engineering Expenses (SARE).* These are the expenses associated with the maintaining sales offices, paying the corporate officers and their staffs as well as research and engineering expenses not attributable to a specific project.

**TABLE 3.9**  Archem, Inc.: Consolidated Income Statement as of December 31[a]

|  | 19X6 | 19X5 |
|---|---|---|
| Net sales (revenue) | $932,000 | $850,000 |
| Cost of sales and operating expenses |  |  |
| Cost of goods sold | 692,000 | 610,000 |
| Depreciation and amortization | 40,000 | 36,000 |
| Selling, general and administrative expenses | 113,500 | 110,000 |
| Operating profit | $ 86,500 | $ 94,000 |
| Other income (expenses) |  |  |
| Dividends and interest income | 10,000 | 7,000 |
| Interest expense | (22,000) | (22,000) |
| Income before provision for income taxes | $ 74,500 | $ 79,000 |
| Provision for federal income taxes | 24,500 | 26,000 |
| **Net profit for year** | $ 50,000 | $ 53,000 |
| Accumulated retained earnings statement[a] |  |  |
| Balance as of January 1 | $172,900 | $141,850 |
| Net profit for year | 50,000 | 53,000 |
| Total for year | $222,900 | $194,850 |
| Less dividends paid on |  |  |
| Preferred stock | 700 | 700 |
| Common stock | 14,600 | 21,250 |
| **Balance December 31** | $207,600 | $172,900 |

[a]All amounts in thousands of dollars.

**3.3.2.3  Operating Profit (Operating Income).**   This entry is the difference between net sales and all operating expenses.

*3.3.2.3.1  Other Income.*   Other income may be derived from dividends or interest received by the company in other investments, income from patents and licenses, and additional sources.

*3.3.2.3.2  Income Before the Provision for Federal Income Taxes.*   When other income is subtracted from the operating profit, the result is the *income before the provision for federal income taxes.*

*3.3.2.3.3  Federal Income Taxes.*   Every company has a basic tax rate that it must pay. Because of tax incentives, tax credits, depreciation write-offs, capital gains, and so on, however, the actual taxes paid may be less than the basic rate.

**3.3.2.4 Net Profit for the Year After Income Taxes.** This entry is obtained by subtracting the provision for federal income taxes from the income before provision for federal income taxes.

### 3.3.3 Accumulated Retained Earnings

*Accumulated retained earnings* is an important part of the financial report because it shows how much money the company has retained for its growth and how much is paid out as dividends to stockholders. When the accumulated retained earnings increase, the company has more value (5).

To obtain the value of the retained earnings, the company starts at the beginning of a year with the previous year's balance. To that figure the *net profit for the year after taxes* is added. The dividends paid to the preferred and common stockholders are then subtracted. The result is the retained earnings at the end of the year.

### 3.3.4 Changes in Financial Position

If one looks at the balance sheet and the income statement, it is easy to see how much money passed through the company. How much profit was made? Did the working capital change and, if so, where did it go? (Working capital is the difference between the total current assets and the total current liabilities.) How were the funds provided from such sources as net profit, depreciation, and sale of common stock used? How did cash generated affect the company operations? By careful tracking through the changes in financial position, the reader can ascertain how the company managed its funds.

We have included excerpts of the 1994 Monsanto Company Annual Report as an example of a typical report from a chemical company. (See Tables 3.10–3.12.) It is useful to compare the entries in this report with the fictitious company, Archem, Inc. You will note that there are some differences in nomenclature and entries, but overall the income statement and balance sheets are similar in style.

### 3.3.5 Independent Accountants' Certification

When reading an annual report, one should look at the independent accountant's statement (see Table 3.13). The certificate will state that the auditing steps used in the verification of the account meet the accounting world's approved standards of practice, and that the financial statements contained in the annual report were prepared in conformance with generally accepted accounting practices (5).

These two statements assure the reader that the figures in the annual report

**TABLE 3.10** Monsanto Balance Sheet for 1994: Statement of Consolidated Financial Position[a–c]

| | As of Dec. 31, | |
| --- | --- | --- |
| Assets | 1994 | 1993 |
| Current assets | | |
| Cash and cash equivalents | $ 507 | $ 273 |
| Trade receivables, net of allowances of $57 in 1994 and $51 in 1993 | 1,530 | 1,445 |
| Miscellaneous receivables and prepaid expenses | 313 | 388 |
| Deferred income tax benefit | 321 | 342 |
| Inventories | 1,212 | 1,224 |
| Total current assets | 3,883 | 3,672 |
| Property, plant, and equipment | | |
| Land | 102 | 107 |
| Buildings | 1,268 | 1,237 |
| Machinery and equipment | 5,916 | 5,793 |
| Construction in progress | 269 | 245 |
| Total property, plant and equipment | 7,555 | 7,382 |
| Less accumulated depreciation | 4,738 | 4,580 |
| Net property, plant, and equipment | 2,817 | 2,802 |
| Investments in affiliates | 279 | 227 |
| Intangible assets, net of accumulated amortization of $522 in 1994 and $450 in 1993 | 1,134 | 1,189 |
| Other assets | 778 | 750 |
| Total assets | $8,891 | $8,640 |
| Liabilities and shareowners' equity | | |
| Current liabilities | | |
| Accounts payable | $ 629 | $ 538 |
| Wages and benefits | 343 | 299 |
| Income and other taxes | 150 | 140 |
| Restructuring reserves | 129 | 255 |
| Miscellaneous accruals | 872 | 840 |
| Short-term debt | 312 | 223 |
| Total current liabilities | 2,435 | 2,295 |
| Long-term debt | 1,405 | 1,502 |
| Deferred income taxes | 65 | 54 |
| Postretirement liabilities | 1,341 | 1,256 |
| Other liabilities | 697 | 678 |

**TABLE 3.10**  Continued

|                                                                                  | As of Dec. 31, | |
| Assets                                                                           | 1994    | 1993    |
| --- | --- | --- |
| Shareowners' equity                                                              |         |         |
| Common stock (authorized, 200,000,000 shares, par value $2)                      |         |         |
| Issued, 164,394,194 shares in 1994 and 1993                                      | 329     | 329     |
| Additional contributed capital                                                   | 849     | 826     |
| Treasury stock, at cost (52,859,031 shares in 1994 and 48,418,545 shares in 1993) | (2,744) | (2,348) |
| Reserve for employees' stock ownership plan (ESOP) debt retirement               | (199)   | (218)   |
| Unrealized investment holding gain                                               | 19      |         |
| Accumulated currency adjustment                                                  | 33      | (59)    |
| Reinvested earnings                                                              | 4,661   | 4,325   |
| Total shareowners' equity                                                        | 2,948   | 2,855   |
| Total liabilities and shareowners' equity                                        | $8,891  | $8,640  |

[a]Amounts in millions except per share.
[b]The original document specifies that the statement was to be read in conjunction with notes section of the annual report.
[c]Direct export sales from the United States to third-party customers outside the United States were $399 million in 1994.
*Source:* Ref. 9.

fairly represent the data they purport to describe. Occasionally, the words "subject to" or "except for" appear. This should alert the reader to dig further to find out what necessitated these qualifying statements. The answers are often found in the notes of the annual report.

3.3.6  Relationship Between the Balance Sheet and the Income Statement.  Someone inexperienced in reading a financial report may believe that there is no relationship between the balance sheet and the income statement. This is not the case, however, because information obtained from each is used to calculate the return on assets and the return on equity, which are of significance to financial people, as we shall see in Section 3.4, Financial Ratios. Figure 3.2, the operating profitability tree, contains the fixed and variable expenses as reported in internal reports such as the manufacturing cost sheet. This information is fed into the income statement as the cost of goods sold along with selling expenses, general overhead expenses, and depreciation, resulting in the operating expenses. The operating profit is calculated by subtracting the operating expenses from the sales. (The total assets are obtained from the balance sheet. Some of the data in the income statement flowsheet boxes may be found

TABLE 3.11    Monsanto Statement of Consolidated Income for 1994[a]

|  | 1994 | 1993 | 1992 |
|---|---|---|---|
| Net sales | $8,272 | $7,902 | $7,763 |
| Cost of goods sold | 4,774 | 4,564 | 4,710 |
| Gross profit | 3,498 | 3,338 | 3,053 |
| Marketing expenses | 1,191 | 1,199 | 1,115 |
| Administrative expenses | 589 | 548 | 487 |
| Technological expenses | 674 | 695 | 720 |
| Amortization of intangible assets | 81 | 81 | 237 |
| Restructuring expenses—net | 40 | 5 | 436 |
| Operating income | 923 | 810 | 58 |
| Interest expense | (131) | (129) | (169) |
| Interest income | 81 | 40 | 43 |
| Other income (expense)—net | 22 | 8 | (106) |
| Income (loss) from continuing operations before income taxes | 895 | 729 | (174) |
| Income taxes | 273 | 235 | (48) |
| Income (loss) from continuing operations | 622 | 494 | (126) |
| Discontinued operations |  |  |  |
| Income from fisher controls |  |  | 24 |
| Gain on sale of fisher controls |  |  | 554 |
| Income from discontinued operations |  |  | 578 |
| Income before accounting changes | 622 | 494 | 452 |
| Cumulative effect of accounting changes: |  |  |  |
| Postretirement benefits other than pensions |  |  | (658) |
| Income taxes |  |  | 118 |
| Net income (loss) | $ 622 | $ 494 | $ (88) |
| Earnings per share |  |  |  |
| Income (loss) from continuing operations | $ 5.32 | $ 4.10 | $(1.01) |
| Discontinued operations |  |  | 4.68 |
| Accounting changes |  |  | (4.38) |
| Net income (loss) | $ 5.32 | $ 4.10 | $(0.71) |
| Key financial statistics | 1994 | 1993 | 1992 |
| As a percent of net sales |  |  |  |
| Gross profit | 42% | 42% | 39% |
| Marketing, administrative, and technological expenses | 30 | 31 | 30 |
| Research and development expenses | 7 | 8 | 8 |
| Operating income | 11 | 10 | 1 |
| Income (loss) from continuing operations | 8 | 6 | (2) |
| Net income (loss) | 8 | 6 | (1) |
| Effective income tax rate | 31 | 32 | (28) |
| Return on shareowners' equity | 21.4 | 16.9 | (2.6) |

[a]See notes a and b of Table 3.10.
*Source:* Ref. 9.

**TABLE 3.12** Monsanto Statement of Consolidated Shareowners' Equity for 1994[a,b]

|  | 1994 | 1993 | 1992 |
|---|---|---|---|
| Common stock | | | |
| Balance, Jan. 1 and Dec. 31 | $ 329 | $ 329 | $ 329 |
| Additional contributed capital | | | |
| Balance, Jan. 1 | $ 826 | $ 820 | $ 726 |
| Employee stock plans and ESOP | 23 | 6 | 94 |
| Balance, Dec. 31 | $ 849 | $ 826 | $ 820 |
| Treasury stock | | | |
| Balance, Jan. 1 | $ (2,248) | $ (2,029) | $ (1,797) |
| Shares purchased (6,170,016; 5,795,600; and 6,732,300 shares in 1994, 1993, and 1992, respectively) | (478) | (380) | (417) |
| Shares issued under employee stock plans and ESOP (1,729,530; 1,306,882; and 4,269,180 shares in 1994, 1993, and 1992, respectively) | 82 | 61 | 185 |
| Balance, Dec. 31 | $(2,744) | $(2,348) | $(2,029) |
| Reserve for ESOP debt retirement | | | |
| Balance, Jan. 1 | $ (218) | $ (233) | $ (250) |
| Allocation of ESOP shares | 19 | 15 | 17 |
| Balance, Dec. 31 | $ (199) | $ (218) | $ (233) |
| Unrealized investment holding gain | | | |
| Balance, Jan. 1 | $ 15 | | |
| Net change in market value | 4 | | |
| Balance, Dec. 31 | $ 19 | | |
| Accumulated currency adjustment | | | |
| Balance, Jan. 1 | $ (59) | $ 15 | $ 187 |
| Translation adjustments | 92 | (74) | (172) |
| Balance, Dec. 31 | $ 33 | $ (59) | $ 15 |
| Reinvested earnings | | | |
| Balance, Jan. 1 | $ 4,325 | $ 4,103 | $ 4,459 |
| Net income (loss) | 622 | 494 | (88) |
| Dividends (net of ESOP tax benefits) | (286) | (272) | (268) |
| Balance, Dec. 31 | $ 4,661 | $ 4,325 | $ 4,103 |
| **Key Financial statistics** | **1994** | **1993** | **1992** |
| Stock price[c] | | | |
| High | $ 86 1/2 | $ 75 | 71 1/4 |
| Low | 66 1/2 | 48 7/8 | 49 3/4 |
| Year-end | 70 1/2 | 73 3/8 | 57 5/8 |
| Per share | | | |
| Dividends | 2.47 | 2.30 | 2.20 |
| Shareowners' equity | 26.43 | 24.62 | 24.95 |
| Average daily share trading volume (thousands of shares) | 376 | 335 | 392 |

[a]See notes a and b of Table 3.10.
[b]Monsanto maintained its current "A" debt rating in 1994.
[c]Based on daily reported high and low stock prices
*Source:* Ref. 9.

**TABLE 3.13** Independent Auditors' Statement from the 1994 Monsanto Annual Report

## Independent Auditors' Opinion

To the shareowners of Monsanto Company:

We have audited the accompanying statement of consolidated financial position of Monsanto Company and subsidiaries as of Dec. 31, 1994 and 1993, and the related statements of consolidated income, shareowners' equity and cash flow for each of the three years in the period ended Dec. 31, 1994. These financial statements are the responsibility of the company's management. Our responsibility is to express an opinion on these financial statements based on our audits.

We conducted our audits in accordance with generally accepted auditing standards. Those standards require that we plan and perform the audit to obtain reasonable assurance about whether the financial statements are free of material misstatement. An audit includes examining, on a test basis, evidence supporting the amounts and disclosures in the financial statements. An audit also includes assessing the accounting principles used and significant estimates made by management, as well as evaluating the overall financial statement presentation. We believe that our audits provide a reasonable basis for our opinion.

In our opinion, such consolidated financial statements present fairly in all material respects the financial position of Monsanto Company and subsidiaries as of Dec. 31, 1994 and 1993, and the results of their operations and their cash flows for each of the three years in the period ended Dec. 31, 1994, in conformity with generally accepted accounting principles.

As discussed in the Notes to Financial Statements, in 1992 Monsanto changed its methods of accounting for postretirement benefits other than pensions and for income taxes.

Deloitte & Touche LLP
St. Louis, Missouri

Feb. 24, 1995

*Source:* Ref. 9.

in the income statement, Table 3.9.) The return on assets is found by dividing the net income (after taxes) by the total assets.

For a broader perspective, Figure 3.3, the financial family tree, ties together the income statement and the balance sheet. The upper part of the figure includes information obtained from the income statement, while the lower part contains information from the balance sheet, namely, total assets and total liabilities. The difference between these two figures is the stockholders' equity from which the return on equity can be calculated. The net income found in the income statement divided by the stockholders' equity is the return on equity. Figures 3.2 and 3.3, therefore, depict how the income statement and the balance sheet are related.

### 3.3.7  The 10K Report

Every year, every public corporation in the United States is required to file a report, known as the 10K report with the Securities and Exchange Commission. When one requests a copy of a company's annual report, the 10K report frequently accompanies or is bound with it. The 10K report is a more thorough presentation and discussion of the details found in the annual report. Over the past decade, the format of the 10K report has been modified to eliminate redundancy and repetition of data included in the annual report.

The 10K report consists of four parts. In Part 1, the company presents the corporate structure, describing the nature of the business and providing details about acquisitions, mergers, raw material sources, environmental affairs, patents, and licenses, as well as research and development activities. Data on foreign, domestic, and discontinued operations are included. Properties, plants leases, and are also presented, including transportation and marketing information, if the company is engaged in petroleum operations. Any legal proceedings brought by states, the federal government, foreign governments, or individuals are also noted in Part 1. The subject of any litigation is stated in clear and concise terms. Also in this part of the report, matters submitted to security holders are discussed. The concluding section of Part 1 of the 10K report contains the name, position, and date of appointment of each corporate officer.

Part 2 consists of the financial position, capital expenditures, funds availability, and results of operations. Consolidated balance sheets, income statements, and selected quarterly data, as well as stock prices and dividends per share, may be found in Part 2. Any disagreements on accounting and financial disclosure are explained in narrative format.

Part 3 contains a listing of the nominees for election as directors as distributed in a proxy statement for the annual stockholders' meeting. Security ownership of voting securities and the principal holders are found in the proxy

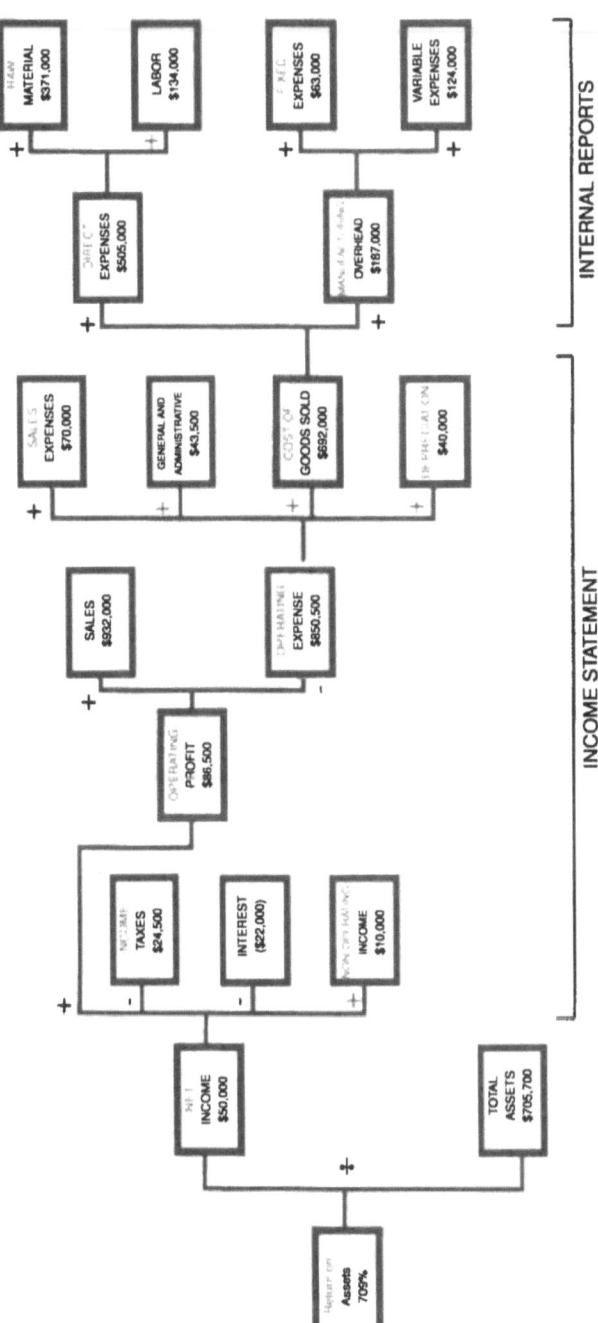

FIGURE 3.2 Operating profitability tree.

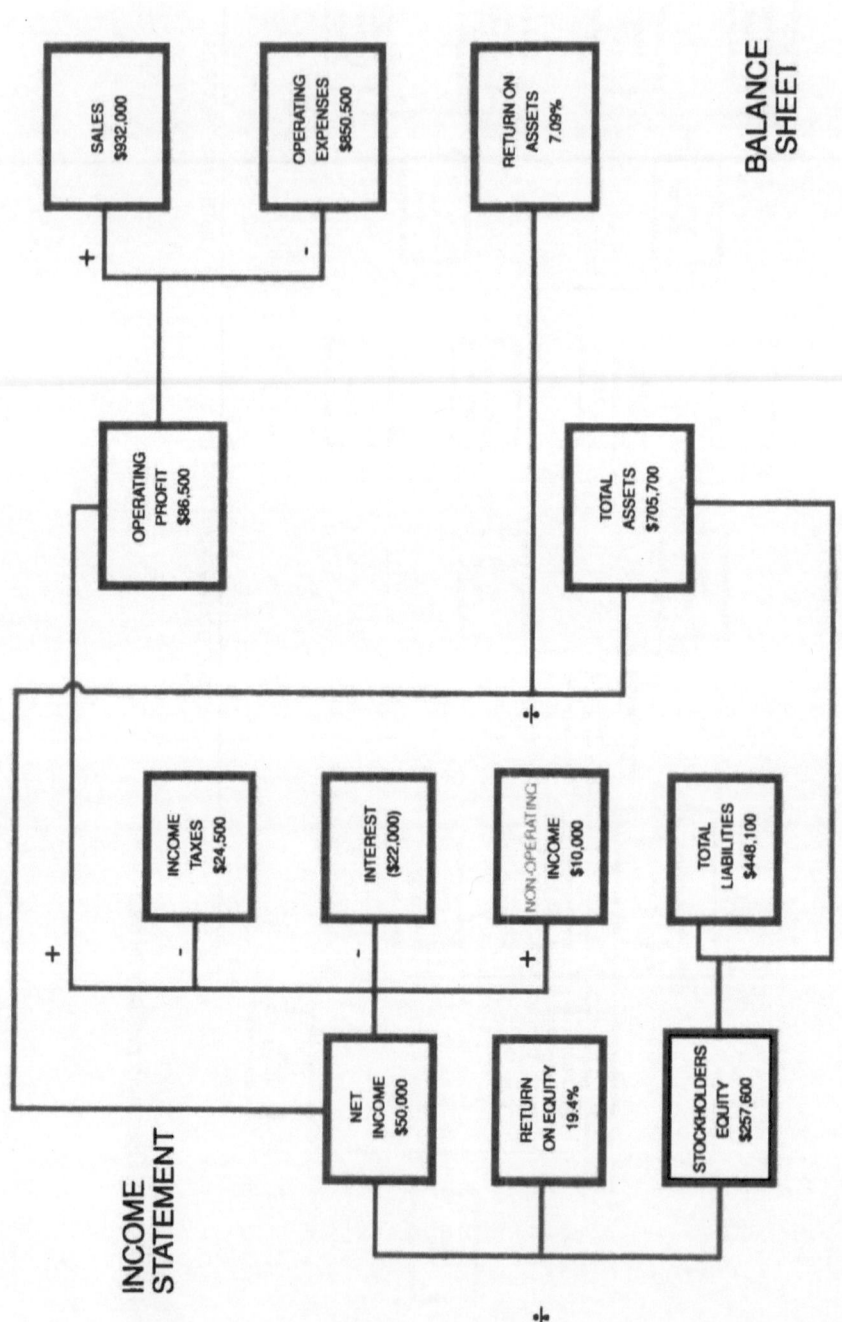

FIGURE 3.3   Financial family tree.

statement and only referenced in Part 3. Any related financial transaction not included in one or more of the foregoing sections is presented here.

Part 4 contains all financial statements and pertinent schedules referring to specific pages in the annual report. A statement concerning whether the registrant has filed any other reports is mentioned. A statement by the certified public accountants concerning their consent to include the annual report with the 10K report is made. An index to financial statements and financial schedules over the past three years is included as follows:

Properties, plants, and equipment
Accumulated depreciation, depletion, and amortization
Valuation of accounts and reserves
Short-term borrowing

Finally, a signature page for all corporate officers to sign is included.

## 3.4 FINANCIAL RATIOS

The sports and business worlds use statistics as a measure of performance. There are many financial ratios of interest to financial analysts. A summary of the definitions of financial ratios used in this text along with CPI average values is found in Table 3.14. Only the following common financial ratios are presented in this section:

*Liquidity ratios:* a measure of a company's ability to pay its short-term debts when due
*Leverage ratios:* a measure of a company's overall debt burden
*Activity ratios:* a measure of how effectively a company manages its assets
*Profitability ratios:* an indication of a firm's total operating performance, which is a combination of asset and income management

### 3.4.1 Liquidity Ratios

Two measures of a company's liquidity are the current ratio and the *cash (quick) ratio.*

The *current ratio* is defined as the current assets divided by the current liabilities. It is a measure of the company's overall ability to meet obligations from current assets. Today a "comfortable" level of between 1.5 and 2 is considered adequate. (*Note:* Numbers presented in this section are typical as of 1999 but will vary depending on the company's style of management.)

The *cash* or *quick ratio* expresses the ability of a company to cover from its assets an emergency. It is the cash plus marketable securities divided by the current liabilities. A typical figure is greater than 1.0.

TABLE 3.14   Summary of Selected Financial Ratios

| Ratio | Equation for calculation | Industry average |
|---|---|---|
| Liquidity | | |
| Current | current assets/current liabilities | 1.5–2.0 times |
| Cash or quick | current assets – inventory/current liabilities | 1.0–1.5 times |
| Leverage | | |
| Debt-to-total assets | total debt/total assets | 30–35% |
| Times interest earned | profit before taxes plus interest charges/interest charges | 7.0–8.0 times |
| Fixed-charge coverage | income available for meeting fixed charges/fixed charges | 5.5 times |
| Activity | | |
| Inventory turnover | sales or revenue/inventory | 7.0 times |
| Average collection period | receivables/sales per day | 45–60 days |
| Fixed assets turnover | sales/fixed assets | 2–3 times |
| Total assets turnover | sales/total assets | 1–2 times |
| Profitability | | |
| Gross profit margin | net sales – cost of goods sold/sales (revenue) | Varies |
| Net operating margin | net operating profit before taxes/sales (revenue) | Varies |
| Profit margin on sales | net profit after taxes/sales (revenue) | 5–8% |
| Return on net worth (return on equity) | net profit after taxes/net worth | 15% |
| Return on total assets | net profit after taxes/total assets | 10% |

### 3.4.2   Leverage Ratios

The leverage ratios summarize a company's overall debt burden.

The *debt-to-total assets ratio* is determined by dividing the total debt by the total assets expressed as a percentage. The all-industry CPI average is about 35%. Another measure is the *debt-to-equity ratio,* expressed as a percentage. Both these ratios are used to measure the amount of debt that a company maintains. The higher the ratio, the greater the financial risk. If an economic turndown occurred, it might be difficult for a company with high debt levels to meet creditors' demands.

The *times-interest-earned ratio* is a measure of the extent to which profits could decline before a company is unable to pay interest charges. This ratio is calculated by dividing the earnings before interest and taxes (EBIT) by the interest charges.

The *fixed-charge coverage* is obtained by dividing the income available for meeting fixed charges by the fixed charges. Many firms lease assets and incur long-term obligations under lease contracts. These lease contracts are part of the fixed costs and the cost of doing business. The numerator of this ratio is the operating profit before deducting the interest expense, lease costs, and income taxes divided by lease costs and interest expenses.

### 3.4.3 Activity Ratios

Activity ratios are measures of how effectively a company manages its assets. They are based on the assumption that there are proper relationships between a company's assets and the sales and income that the assets generate. Different methods are used to generate these ratios, and they depend on how a person wants to use the ratios. Most analysts compile average ratios from balance sheet data, which are end-of-year data. In this section and in the financial ratio example that follows, this method is used.

There are two *inventory/turnover ratios* in common use today. The *inventory/sales ratio* is found by dividing the sales by the inventory. Another method is to divide the cost of sales by the inventory. In either case, all-industry averages are between 7.0 and 9.0.

The *average collection period* measures the average number of days that customers' invoices remain unpaid. This figure is found by dividing the annual sales by 365 days to obtain the average daily sales and then dividing this figure into the accounts receivable balance. The average period for the CPI is about 45 days.

*Fixed-assets* and *total-assets turnover* indicate how well the fixed and total assets are being used. These figures are determined by dividing annual sales by the fixed assets or the total assets, respectively.

### 3.4.4 Profitability Ratios

Activity ratios are used to determine the company's management of assets, while profitability ratios help to evaluate its income management.

The *gross profit margin* is found by dividing the gross profits by the net sales, and the result is expressed as a percentage. This ratio is an indication of the effectiveness of a company's pricing, purchasing, and production policies.

The *net operating margin* is equal to earnings before interest and taxes (EBIT) divided by the net sales, again expressed as a percentage. This is a measure of a company's income performance before interest and taxes.

Another measure is the *profit margin on sales,* which is calculated by dividing the net profit after taxes by the net sales, expressed as a percentage. Industry averages vary, but 5% is a reasonable figure.

The *return on total assets ratio* is the net profit after taxes divided by the total assets expressed as a percentage. It reflects the overall return that a company has earned on its assets. The average for all industries is 10%, but figures vary widely.

The *return-on-equity ratio* is the net income after taxes and interest divided by stockholders' equity. It measures the return on the equity capital invested in the firm. Since one of management's objectives is to earn the highest return for the stockholders, this ratio is probably the best measure of management's performance.

In evaluating a company's performance, one should compare it with data from other companies in the same line of business. Thus a chemical company's performance should be compared with that of another chemical company, and one oil company should be compared with another oil company. Companies in the same line of business frequently use the same style of management.

### 3.4.5 Economic Value Added

In the mid-1990s, a new management concept, *economic value added* (EVA), was moved from a buzzword to an important financial tool. EVA, the after-tax net operating profit minus the cost of capital, measures whether a business earns more than the cost of capital. EVA is being used as a performance measure, as an analytical financial tool, and as a management discipline measure. Various economic sectors, manufacturing industries, health care companies, and the postal service are among the enterprises that employ EVA to help manage their operations. It is an indication of how efficient management is at turning investors' money, namely capital, into profits (6, 7).

In periods when the economy is strong and sales are growing, the bottom line still might not show good results. EVA analysis helps companies identify waste and inefficiency in daily operations and in the use of capital. It aids in identifying high inventories as well as the need to reduce the accounts receivable. In other words, EVA is a tool to improve overall efficiency. EVA analysis in many instances has demonstrated that debt capital is cheaper than equity capital. It also gives managers a clearer idea of whether they are increasing or decreasing shareholders' wealth. Stockholders, can benefit from EVA analysis if it results in higher dividends and permits stock share repurchases. Further, many companies are tying EVA to salaries and bonuses of upper-level company executives.

### 3.4.6 Other Financial Information

This section presents additional terms and information found frequently in financial reports.

*Net worth*, the difference between total assets and total liabilities, is the same as stockholders' equity.

*Working capital* is calculated by subtracting the total current liabilities from the total current assets.

Frequently, the *earnings per share* is of interest to owners of common stock and stockbrokers. This amount is found by dividing the net profit after taxes by the number of shares of common stock.

*After-tax cash flow* is defined as the net profit after taxes plus the depreciation. This item is an important to management because it is the amount of money available for operating the company. With the advent of accelerated depreciation, the cash flow in the early years of an asset tend to increase, when a plant is being brought on stream and markets are being developed.

*LIFO-FIFO* valuation of raw material and supplies inventory is an important consideration for management. *LIFO* means last in, first out, referring to the purchase of materials inventory. *FIFO* means first in, first out. The method selected is reflected not only in the operating expenses but also in the income statement and the balance sheet. At different times it may be more advantageous financially to use one instead of the other.

*Depreciation* methods in use today are either *straight-line* or *accelerated*. In the former method, the cost of an asset is divided by the asset's life. For example, if an equipment item cost $7000 and its life is 7 years, then $1000 per year for 7 years is charged as the depreciation charge on the operating expense sheet.

The accounting practice known as *accelerated* depreciation was introduced in the 1950s to stimulate capital spending. In the intervening years, there have been several revisions to the tax laws, with associated changes in the handling of depreciation. The net effect of accelerated depreciation was to provide a larger amount of depreciation in the early years of an investment, thereby increasing the cash flow, when a plant is coming on stream and the market for a product is being established.

As of 1999, the modified accelerated cost recovery system (MACRS) was in force. The depreciation rates applied to an asset are as follows for most CPI assets with a 7-year life:

| Year | Rate (%) |
|------|----------|
| 1 | 14.29 |
| 2 | 24.49 |
| 3 | 17.49 |
| 4 | 12.49 |
| 5 | 8.93 |
| 6 | 8.92 |
| 7 | 8.93 |
| 8 | 4.46 |

Note that the depreciation for a 7-year asset is spread over an 8-year period. It is assumed that full benefit will not be received during the first year of the life of an asset. So a half-year convention is adopted, and the remaining recovery is made in the eighth year.

### 3.4.7  Z Scores

Charles Kyd(8) developed an equation that measures how closely a firm's financial statements resemble those of companies that have gone bankrupt. The equation is as follows:

$$Z = 1.2 \text{ (working capital/total assets)} + 1.4 \text{ (retained earnings/total assets)}$$
$$+ 3.3 \text{ (EBIT/total assets)} + 0.6 \text{ (market value of equity/total liabilities)}$$
$$+ \text{ (sales/total assets)}$$

The five ratios displayed help one to evaluate how changes in the ratios produce changes in the value of Z. A Z score of less than 1.80 implies that the firm is economically bankrupt. A firm on a sound financial footing generally will have a score greater than 3.0. A score between these two values indicates some uncertainty about the company's prospects.

### 3.5  A FINANCIAL RATIO EXAMPLE

### 3.5.1  The Problem

To illustrate the use of the balance sheet and income statement in determining the financial soundness of a company, we present the example of Archem, Inc., a fictitious company supposedly formed over a period of about 15 years through acquisitions, mergers, and buyouts of several small companies. Archem manufactures primarily specialty and fine chemicals. We shall use their balance sheet and income statement for 19X6 as a typical set of statements, Tables 3.8 and 3.9. Determine the liquidity, leverage, activity, and profitability performance of Archem. Calculate the Z score for this company. You may assume that the stock is selling at $10 per share. What do you conclude about Archem's financial position?

### 3.5.2  The Solution

The financial ratios for Archem are found in Table 3.15. If these are compared with the CPI averages found in Table 3.14, one can draw the following conclusions:

> From the liquidity ratios, the company is able to pay its short-term debts when due.
> From a leverage standpoint, the "fixed-charge coverage" and "times interest earned" are a bit low. The debt-to-asset ratio is about average. The company needs to reduce fixed or interest charges or increase profit or income.
> The activity ratios are average for a company in the CPI, and the company manages its assets effectively.

**TABLE 3.15**  Financial Ratios for Archem, Inc.

Liquidity

Current ratio = $391,200/$201,500 = 1.94
Cash ratio = $391,200 − 149,000/$201,500 = 1.20

Leverage

Debt-to-assets = $448,100 − 201,500/$705,700 × 100 = 35%
Times interest earned = $74,500 − 22,000/$22,000 = 4.39
Fixed-charge coverage = $86,500/$22,000 = 3.93

Activity

Inventory turnover = $932,000/$149,000 = 6.25
Average collection period = $135,000/($932,000/365) = 52.8 days
Fixed-assets turnover = $932,000/$438,000 = 2.13
Total-assets turnover = $932,000/$705,700 = 1.32

Profitability

Gross profit margin = $932,000 − 692,000/$932,000 × 100 = 25.8%
Net operating margin = $74,500/$932,000 × 100 = 7.99%
Profit margin on sales = $50,000/$932,000 × 100 = 5.36%
Return on net worth (return on equity) = $50,000/$705,700 - 448,100 × 100 = 19.4%
Return on total assets = $50,000/$705,700 × 100 = 7.09%

$Z$ Score

$Z$ = 1.2 (working capital/total assets)
+ 1.4 (retained earnings/total assets)
+ 3.3 (EBIT/total assets)
+ 0.6 (market value of equity/total liabilities)
+ (sales/total assets)
$Z$ = 1.2 ($391,200 − $201,500)/$705,700
+ 1.4 ($268,400/$705,700) + 3.3 ($96,500/$705,700)
+ 0.6 ($320,000/$448,100) + $932,000/$705,700
$Z$ = 0.323 + 0.532 + 0.451 + 0.429 + 1.321 = 3.056

The profitability is about average, although the return on total assets is low. Overall, the operating performance appears to be satisfactory.

Overall, the company seems to be operated by management satisfactorily, despite some areas as cited above that bear watching.

The $Z$ score for Archem, Inc., obtained by substituting the appropriate information in Equation (3.3), is 3.06. See Table 3.15. The company is safe from bankruptcy at present but management must monitor the $Z$ score continually.

## 3.6  SUMMARY

This chapter introduced the basic elements of accounting, which readers are likely to encounter in industry. From accounting records and internal reports, the financial data ultimately become part of the balance sheet or the income statement. The various terms in the balance sheet and the income statement were defined, and an illustration of these documents for a typical company was presented. This information should allow readers to determine how well a company has done and is doing. All annual reports contain footnotes, which explain how changes in the financial information and pending litigation could affect the financial performance of the company. The 10K report was mentioned because it provides the source of the detailed backup financial information not in the annual report.

Financial ratios often used by analysts were introduced because they are indicators of how well the company uses its assets and how well management performs. An equation for determining the Z score was introduced. It measures how the position of a company with respect to the brink of bankruptcy.

The principles presented in this chapter will appear many times throughout the text to explain how financial data affect the marketing, manufacturing, engineering, research and development, and administrative functions of a company.

## REFERENCES

1.  Robert N. Anthony, *Management Accounting*. Richard D. Irwin, Homewood, IL, 1964.
2.  Clarence B. Nickerson, *Accounting Handbook for Non-Accountants*, 2nd ed. CBI Publishing Company, Boston, 1979.
3.  James, Burke. *The Day The Universe Changed*. Little, Brown & Company, Boston, 1985, p. 61.
4.  J. A. Ness and T. G. Cucuzza. *Harvard Business Review*, July–August 1995, pp. 130–138.
5.  Adapted from *How To Read an Annual Report*, 5th ed. Merrill Lynch, Pierce, Fenner & Smith, Inc., New York, 1984.
6.  *Fortune*, September 9, 1996, pp. 173–174.
7.  *Fortune*, November 10, 1997, pp. 265–276.
8.  Charles Kyd, "Calculating Financial Ratios," *Lotus*, January 1987, pp. 44–47.
9.  Monsanto Annual Report, St. Louis, MO, April 1994.

## PROBLEMS

3.1.   From the transactions in the general journal, prepare a ledger, a balance sheet, and an income statement for the month of February for Delchem Corporation. See (Table 3.16).

3.2.   From the following transactions for Delchem Corp. (see Table 3.17), con-

**TABLE 3.16** Page 3 of Delchem General Journal: For Problem 3.1

| Date | Account titles and explanation | LP | Debit | Credit |
|------|-------------------------------|-----|-------|--------|
| 2/1 | Accrued wages payable | 22 | 1,000 | |
| | Cash | 10 | | 1,000 |
| | Finished goods | 14 | 1,000 | |
| | Accrued wages payable | 22 | | 1,000 |
| | Back wages to Davis, owe for February | | | |
| 2/7 | Equipment | 12 | 5,000 | |
| | Prepaid expense | 15 | | 2,000 |
| | Cash | 10 | | 3,000 |
| | Equipment arrives, $3,000 balance paid in cash | | | |
| 2/10 | Accounts payable | 24 | 6,000 | |
| | Cash | 10 | | 6,000 |
| | Pay outstanding raw materials account upon delivery | | | |
| 2/20 | Finished goods | 14 | 4,000 | |
| | Raw materials | 11 | | 4,000 |
| | Make 5,000 liters of Cleansolv using $4,000 of raw materials[a] | | | |
| 2/25 | Cash | 10 | 2,000 | |
| | Bank loan | 21 | | 2,000 |
| | Take out another $2,000 loan at 12% interest; promise to pay back after 1 year | | | |
| 2/26 | Cost of goods sold | 44 | 1,000 | |
| | Finished goods | 14 | | 1,000 |
| | Cash | 10 | 3,000 | |
| | Revenue | 30 | | 3,000 |
| | Sell 1,000 liters of Cleansolv at $3 per liter[b] | | | |
| 2/28 | Salaries expense | 43 | 1,500 | |
| | Salaries payable | 26 | | 1,500 |
| | Owe Custer his salary | | | |
| | **Adjusting entries** | | | |
| 2/28 | Depreciation expense: equipment | 41 | 66.67 | |
| | Equipment | 12 | | 66.67 |
| | $8,000 × 1/12 = $66.67 per month | | | |
| 2/28 | Interest expense | 42 | 40.00 | |
| | Interest payable | 23 | | 40.00 |
| | To record bank loan interest for February | | | |
| 2/28 | Revenue | 30 | 3,000.00 | |
| | Income summary | 55 | | 3,000.00 |
| | To close the revenue account | | | |

**TABLE 3.16**  Continued

| 2/28 | Income summary | 55 | 2,606.67 | |
|------|----------------|----|----------|---|
| | Salaries expense | 43 | | 1,500.00 |
| | Depreciation expense: equipment | 41 | | 66.67 |
| | Interest expense | 42 | | 40.00 |
| | Cost of goods sold | 44 | | 1,000.00 |
| | To close the expense accounts | | | |
| 2/28 | Income summary | 55 | 393.33 | |
| | Stockholders' equity | 50 | | 393.33 |
| | To close out the income summary | | | |

ªThe book value of the second batch of finished goods is $5,000/(5,000 liters) or $1 per liter.
ᵇThis firm uses the LIFO method, where the goods sold are evaluated at the cost of the last batch manufactured, or $1 per liter. So the credit side of the "Finished goods" account shows a withdrawal of 1,000 liters at $1 per liter. The debit side of the "Cost of goods sold" account records the $1,000 cost of the 1,000 liters.

**TABLE 3.17**  March Delchem Data for Problem 3.2

| Date | Account titles, and explanation |
|------|--------------------------------|
| 3/1 | Pay $1,000 to Davis, give him a raise, owe him $1,500 for March |
| 3/10 | Pay $360 for advertisement in a chemical trade journal |
| 3/11 | Buy $5,500 of raw materials, pay $4,000 now, owe $1,500 due in April |
| 3/15 | Borrow $5,000 from bank at 12% interest, promise to pay back after 1 year |
| 3/15 | Sell 3,000 liters of Cleansolv at $3.50 per liter |

struct the journal, ledger, balance sheet, and income statement for March. Do not forget to include previous commitments (interest, depreciation, Custer's salary, etc.).

3.3.  In April, the price of Cleansolv drops to 50 cents per liter. Delchem decides not to sell any product but continues to produce Cleansolv at 1000 liters per month. To keep everything running, Custer makes a deal with a bank to borrow $5000 a month on a continuous basis. Would the income statement and the balance sheets look very bad? What kind of troubles would Delchem eventually encounter?

3.4. Old Smooky McCarthy was drilling wildcat holes in northern Michigan when he struck oil. The find is estimated at 2 million producible barrels of oil. The cost of drilling the well was $1.5 million; royalty to the owner is $2.50 per barrel plus a bonus of $5000 for signing the agreement last year. Oil currently sells for $18 per barrel. What should the asset value of this property be on the balance sheet?

3.5. Chuck Adamson built a new indigo plant at a cost of $2 million to capitalize on the worldwide craze for blue denim. He managed to convince the IRS that the craze would last only two years so that he should be able to depreciate the plant in 2 years by straight line, even though the plant will physically last 15 years. Let's assume the blue denim craze continued strongly into the third year. Will Adamson operate without a plant listed on the balance sheet? Will he have no capital depreciation expenses on his product cost? Can he undersell any would-be newcomers into the market?

3.6. Sharppencil & Associates, Inc., is a consulting firm operating from rented office space and furniture. It spent $200,000 in the past 2 years developing a set of computer programs for the control of batch-processing chemical plants. It hopes to sell them to the domestic as well as foreign CPI firms. How should the firm list this set of programs on the balance sheet? Suppose two buyers, each paying $500,000, could be found each year for the next 3 years. How should these transactions be listed in the books? Should the firm amortize the programs over a period of time? On the other hand, suppose no buyers can be located for 3 years. How would the programs be entered then?

3.7. Inflation has suddenly doubled the cost of all Cleansolv raw materials and tripled machinery costs. How is this likely to affect Delchem's evaluation of assets under raw materials, plants, and finished goods? The price of Cleansolv has also doubled. What will happen to the profit margin? Should the company use LIFO or FIFO accounting? Is it dangerous and unrealistic to keep depreciation charges at the original machine cost once replacement machinery cost has tripled?

3.8. From the 1994 annual report of Monsanto (Tables 3.10–3.12), answer as many of the following questions as possible:

    a.  What was the year's income?

    b.  What was the net income as a percentage of sales?

    c.  What was the income tax as a percentage of pretax income?

    d.  What percentage of the cost of plants and properties was shown as depreciation and obsolescence?

    e.  What percentage of sales was shown as depreciation and obsolescence?

    f.  What percentage of sales comprised selling, general, and administrative expenses?

    g.  What was the total revenue from net sales and other income?

    h.  What percentage of total revenue was spent for materials and services?

i.   What percentage of total revenue was spent for wages, salaries, and related expenses?

j.   What percentage of total revenue was paid out as dividends?

k.   How much was retained for use of the business, including depreciation "set-asides"?

l.   How much did it cost to construct (or acquire in other ways) the company's existing plants and properties?

m.   How much of the cost of plants and properties had been charged to date as an operating cost?

n.   What percentage of the cost of plants has been charged to cost?

o.   What percentage of total assets is in each of the following: cash, marketable securities, accounts receivable, and inventories?

p.   What were the total liabilities, excluding stockholders' equity?

q.   What was the stockholders' equity?

r.   How much cash and marketable securities were on hand at the beginning of the year?

s.   During the year, the cash and marketable securities increased and decreased. What was the increase due to the sum of net income and depreciation?

t.   What were the dividends as a percentage of net income plus depreciation?

u.   What were construction expenditures as a percentage of net income plus depreciation?

v.   How much cash and marketable securities were left at the end of the year?

3.9.   Using data from the 1994 annual report of Monsanto, calculate the CPI company's liquidity, leverage, activity, and profitability ratios. What was the Z score?

# 4

# Input–Output Analysis

The ultimate goal of all economic activity is to supply final products (including goods and services) for consumption. To produce a final product, input from several industries is required. The *net contribution to production* by a particular industry is termed *value added*. It can be measured as the difference between the industry's total sales and the cost of goods and services it purchases from other industries. The summation of all value-added contributions made by all industries as well as by individuals and the government equals the *gross national product* (GNP). The GNP is then the value measured in dollars of all such production for any particular year.

The *net contribution to ultimate consumption* by a particular industry is termed *final demand*. This is the difference between industry's total sales and its sales to other industries. This difference directly benefits ultimate consumption (i.e., consumption by individual households, by governments, by capital investors, and by exports). The summation of all final demands made by all industries as well as by individuals and the government also equals the GNP. See Table 4.1.

Although the GNP was a measure of output in the past, in the last two decades, the *gross domestic product* (GDP) has come to be viewed as a more realistic measure of the economic output of the United States or any nation. By definition, the GDP is the total output of goods and services produced by labor and property located in the United States or any nation and valued at market prices. For the economy as a whole, the total of all final uses of commodities equals the total value added by all industries, or GDP.

The role of the CPI is usually more one of supplying other industries with

**TABLE 4.1** Transaction Equations Written in Tabular Form

| Output from | Input to 1 | 2 | 3 | $n$ | Intermediate output $O$ | Final demand $F$ | Total output |
|---|---|---|---|---|---|---|---|
| 1 | $X_{11}$ | $X_{12}$ | $X_{13}$ | $X_{1n}$ | $O_1$ | $F_1$ | $X_1$ |
| 2 | $X_{21}$ | $X_{22}$ | $X_{23}$ | $X_{2n}$ | $O_2$ | $F_2$ | $X_2$ |
| 3 | $X_{31}$ | $X_{32}$ | $X_{33}$ | $X_{3n}$ | $O_3$ | $F_3$ | $X_3$ |
| $n$ | $X_{n1}$ | $X_{n2}$ | $X_{n3}$ | $X_{nn}$ | $O_n$ | $F_n$ | $X_n$ |
| Intermediate input | $I$ | $I_1$ | $I_2$ | $I_3$ | $I_n$ | | |
| Value added $V$ | $V_1$ | $V_2$ | $V_3$ | $V_n$ | ... | $W$ | |
| Total input | $X_1$ | $X_2$ | $X_3$ | $X_n$ | ... | | |

intermediate products than of supplying the ultimate consumer directly. Thus the GNP or GDP alone is not a significant indicator of the interindustry flow of CPI products. To account for interindustry interaction and to segregate the value-added contribution by industry, the powerful bookkeeping technique of *input–output analysis* has been developed.

Although the idea of input–output analysis dates back to eighteenth-century France, the modern version of such analysis was pioneered in the 1930s by Wassily Leontief at Harvard University. During World War II, the Bureau of Labor Statistics of the Department of Labor became interested in input–output analysis and applied it to the economy of the United States. Since then the Bureau of Economic Analysis of the Department of Commerce has been given the responsibility of updating the analysis for the U.S. economy. The transaction, direct-requirement, and total-requirement tables were developed for 1947, 1958, 1963, 1967, 1982, and 1987. The goal is to prepare benchmark tables every five years. Annual tables using the same procedures as the benchmark tables are prepared, but they are less comprehensive and a less reliable source of data. Computer diskettes of these tables may be obtained from Interindustry Economics Division, BE-51, Bureau of Economic Analysis, U.S. Department of Commerce, Washington, D.C. 20230.

Input–output analysis can be applied to a wide variety of problems, but its major contribution is that it reflects changes in final demand to be measured quantitatively for a great many industrial sectors. For example, the effect on CPI output of an increase in automobile sales can be determined by using input–output analysis. The techniques of input–output analysis can also be used to determine the effect on the CPI of increasing the amount of plastics in cars and decreasing the amount of steel. The effect of changes in the GNP on the CPI and on any other sector also can be estimated with input–output analysis. Also the economic impact of shortages (e.g., energy shortages and their implications) may be studied by input–output analysis. The basis of input–output analysis is a set of three tables, namely, the transactions table (now called the use table), the direct-requirements table, and the total-requirements table. Each of these tables is described in the following sections. Readers who want to understand this technique better should consult References 1–7.

## 4.1 THE USE (TRANSACTIONS) TABLE

The total *output* (sales) of any particular industry can be broken down into sales to other industries and sales to final demand:

$$X_k = (X_{k1} + X_{k2} + X_{k3} + \cdots + X_{kn}) + F_k \tag{4.1}$$

where $X_k$ = total dollar output of industry $k$
$X_k$ = sales of industry $k$ to industry $j$
$n$ = total number of industries
$F_k$ = sales of industry $k$ directly to final demand (personal consumption, government expenses, investment, and export)

For each industry of interest such an equation can be written as follows:

$$
\begin{aligned}
X_1 &= (X_{11} + X_{12} + X_{13} + \cdots + X_{1n}) + F_1 \\
X_2 &= (X_{21} + X_{22} + X_{23} + \cdots + X_{2n}) + F_2 \\
&\;\;\vdots \\
X_n &= (X_{n1} + X_{n2} + X_{n3} + \cdots + X_{nn}) + F_n
\end{aligned} \tag{4.2}
$$

The *output* $X_{kj}$ from industry $k$ to industry $j$ is also the *input* from $k$ to $j$ necessary for $j$ to make its products. Equation (4.2) can be presented in tabular form (Table 4.1), to illustrate once again the principle of double-entry bookkeeping (total output is equal to total input). Summing all industrial inputs $X_{ik}$ to any industry $k$ in a column yields the total cost of all intermediate inputs, $I_k$, required by industry $k$ to produce its total output, $X_k$.

The difference between the cost of all intermediate inputs to industry $k$ and its total sales measured by its total output $X_k$, is the *value added* to total gross production by industry $k$:

$$X_k = (X_{1k} + X_{2k} + X_{3k} + \cdots + X_{nk}) + V_k = I_k + V_k \tag{4.3}$$

Summing all industry outputs $X_{kj}$ from any industry $k$ in a row yields the cost of all intermediate outputs, $O_k$, sold to other industrial customers. The difference between the total sales $X_k$ and the intermediate output $O_k$ is the final demand of industry $k$:

$$X_k = (X_{k1} + X_{k2} + \cdots + X_{kn}) + F_k = O_k + F_k \tag{4.4}$$

Another item, $W$, of nonindustry contribution to the GNP represents transactions such as government payroll to cabinet members, government employees, and so on. The GNP is the sum

$$GNP = \sum_{1}^{n} V_j + W \tag{4.5}$$

or

$$GNP = \sum_{1}^{n} F_i + W \tag{4.6}$$

A simplified four-sector transactions table (Table 4.2) is used to illustrate these equations for value added: Eq. (4.5) or final demand, Eq. (4.6).

The sector designated as the consumer industry sells $60 billion worth of goods to itself, $7 billion to the basic industry sector, nothing to the energy sector, $19 billion to the service sector, and $180 billion to final demand. Its total output is $266 billion [Eq. (4.3) and (4.4)]. The consumer industry sector buys $60 billion from itself, $80 billion from the basic industry sector, $2 billion from the energy sector, and $26 billion from the service sector. Its net contribution to the production or its value added is $98 billion.

As would be expected, final demand is a high percentage of total output for the consumer industry sector, and value added is a high percentage of total output for the service sector. Basic industry sells most of its output to itself and also buys more from itself than from other sectors. This pattern is repeated for all sectors except the consumer industry sector, which buys more from basic industry than from itself: $W$ is $87 billion.

The Department of Commerce prefers to create *special industries* to represent the nonindustry sectors of individual households, government, investment, and export. For an economy with $n$ special industries, the use table is shown as Table 4.3 and the GNP equation is Eq. (4.7).

**TABLE 4.2**   Transactions Table for a Four-Sector Economy (billions of dollars)

| | Consumer industry | Basic industry | Energy industry | Service | Final demand | GNP | Total output |
|---|---|---|---|---|---|---|---|
| Consumer industry | $60 | $7 | $0 | $19 | $180 | | $266 |
| Basic industry | $80 | $102 | $1 | $23 | $39 | | $245 |
| Energy industry | $2 | $7 | $23 | $11 | $22 | | $65 |
| Service | $26 | $28 | $12 | $115 | $271 | | $452 |
| Value added | $98 | $101 | $29 | $285 | $87 | $600 | — |
| GNP | | | | | | | |
| Total ouput | $266 | $245 | $65 | $453 | $599 | | $1,028 |

TABLE 4.3 Transactions Table Including Special Industries

| Output from | Input to | | | | | | Intermediate output $O$ | Final demand $F$ | Total output |
|---|---|---|---|---|---|---|---|---|---|
| | 1 | 2 | 3 | n | n+1 | n + m | | | |
| 1 | $X_{11}$ | $X_{12}$ | $X_{13}$ | $X_{1n}$ | 0 | 0 | $O_1$ | $F_1$ | $X_1$ |
| 2 | $X_{21}$ | $X_{22}$ | $X_{23}$ | $X_{2n}$ | 0 | 0 | $O_2$ | $F_2$ | $X_2$ |
| 3 | $X_{31}$ | $X_{32}$ | $X_{33}$ | $X_{3n}$ | 0 | 0 | $O_3$ | $F_3$ | $X_3$ |
| n | $X_{n1}$ | $X_{n2}$ | $X_{n3}$ | $X_{nn}$ | 0 | 0 | $O_n$ | $F_n$ | $X_n$ |
| n + 1 | 0 | 0 | 0 | 0 | 0 | 0 | 0 | $F_{n+1}$ | $X_{n+1}$ |
| n + m | 0 | 0 | 0 | 0 | 0 | 0 | 0 | $F_{n+m}$ | $X_{n+m}$ |
| Intermediate input $I$ | $I_1$ | $I_2$ | $I_3$ | $I_n$ | 0 | 0 | | | |
| Value added $V$ | $V_1$ | $V_2$ | $V_3$ | $V_n$ | $V_{n+1}$ | $V_{n+m}$ | | GNP | |
| Total input | $X_1$ | $X_2$ | $X_3$ | $X_n$ | $X_{n+1}$ | $X_{n+m}$ | | | |

It is not a trivial matter to prepare the use (transactions) tables. Data and statistics from a great many diverse sources are needed, as well as guided estimation. The construction of the table is approached on an industry-to-industry basis. Output data from an industry form the basis for industry analysis. A complete technological analysis of the industry must be undertaken to estimate the inputs of raw materials, energy, and so on. Other inputs required are also estimated and, if possible, checked with limited output data from various sources. Marketing data and analyses are used in estimating outputs for an industry. Some valuable insights into the problems of constructing the use (transactions) tables are given in Ref. 1. Use tables for the U.S. economy have been developed by the federal government at irregular time periods since 1947. The general layout of the use table is found in Table 4.4. One should examine the contents of this table before attempting to use the more detailed table, Table 4.4a (Appendix C). The most recent use table based on 1987 data was published in 1994 (Table 4.4a).

The U.S. economy is divided into 85 industries. The input–output CPI sectors are related to the SIC code found in Table 1.24. Relationships of other input–output industries to the SIC code can be found in the *Survey of Current Business*. Another table, Table 4.4b (also found in Appendix C), is a summary of the input components, namely, value added and total intermediate inputs, to give the total industry output by industry number. In this table, the industries are shown in rows and total output, total intermediate inputs, and the components of value added are found in the columns (7). For example, the total output of industry 55 (electric lighting and wiring equipment) is $17.615 billion, of which $5.249 billion was employee compensation, $0.158 billion was intermediate business and nontax liability, $3.675 billion was other value added, and $8.532

TABLE 4.4  Use Table—Commodities Used by Industries and Final Uses, 1987

| | | INDUSTRIES | | | | | | | | | | FINAL USES (GDP) | | | | | | | TOTAL COMMODITY OUTPUT |
|---|---|---|---|---|---|---|---|---|---|---|---|---|---|---|---|---|---|---|---|
| | | Agricul- ture | Mining | Construc- tion | Manufac- turing | Transpor- tation | Trade | Finance | Services | Other* | Total inter- mediate use | Personal consumption expenditures | Gross private fixed investment | Changes in business inventories | Exports of goods and services | Imports of goods and services | Government purchases | GDP | |
| COMMODITIES | Agricultural products | | | | | | | | | | | | | | | | | | | |
| | Minerals | | | | | | | | | | | | | | | | | | | |
| | Construction | | | | | | | | | | | | | | | | | | | |
| | Manufactured products | | | | | | | | | | | | | | | | | | | |
| | Transportation | | | | | | | | | | | | | | | | | | | |
| | Trade | | | | | | | | | | | | | | | | | | | |
| | Finance | | | | | | | | | | | | | | | | | | | |
| | Services | | | | | | | | | | | | | | | | | | | |
| | Other* | | | | | | | | | | | | | | | | | | | |
| | Noncomparable imports | | | | | | | | | | | | | | | | | | | |
| | Total inter- mediate inputs | | | | | | | | | | | | | | | | | | | |
| VALUE ADDED | Compensation of employees | | | | | | | | | | | | | | | | | | | |
| | Indirect business tax and nontax liability | | | | | | | | | | | | | | | | | | | |
| | Other value added** | | | | | | | | | | | | | | | | | | | |
| | Total | | | | | | | | | | | | | | | | | | | |
| TOTAL INDUSTRY OUTPUT | | | | | | | | | | | | | | | | | | | | |

☐ TOTAL COMMODITY OUTPUT
▨ PRIMARY PRODUCT OF THE INDUSTRY
▨ TOTAL INDUSTRY OUTPUT

*The I-O accounts use two classification systems, one for industries and another for commodities, but both generally use the same I-O numbers and fees. "Other" includes government enterprises and I-O special industries; for more information, see "Appendix B,-Classification of the 1987 Benchmark Input–Output Accounts."
**For most industries, this item includes consumption of fixed capital, proprietors' income, corporate profits, and business transfer payments. For banking and for credit agencies other than banks, it also includes net interest. For owner-occupied dwellings and for real estate agents, managers, operators, and lessors, it also includes rental income. For the six industries covering the federal government and state and local government enterprises, it also includes current surplus less government subsidies. U.S. Department of Commerce, Bureau of Economic Analysis.

billion was total intermediate inputs. The column totals for industries in Table 4.4a equal the right-hand row totals in Table 4.4b. The column totals for industry 55 in Table 4.4a equals the row total for that industry in Table 4.4b, namely, $17.615 billion. Then, in this figure, the sum of the industry column yields the "total intermediate inputs." To this value, the components of "value added" (e.g., employee compensation, indirect business tax, and nontax liability) are summed to yield the total value added. If the "total intermediate inputs" and the "total value added" are summed, the result is "total industry output."

$$\text{GNP} = \sum_{1}^{n+m} V_j = \sum_{1}^{n+m} F_i \tag{4.7}$$

In Table 4.4a, the sum of industry rows yields the "total intermediate use." Final uses include the following (industry numbers given in brackets):

Personal consumption expenditures [91]
Gross private fixed investment [96]
Changes in business inventories [93]
Export of goods and services [94]
Import of goods and services [95]
Government purchases [96–99]

The sum of the final uses results in the *gross domestic product* (GDP). If the total intermediate use is summed with the GDP, the result is the *total commodity output*.

In Table 4.4a for a particular industry, each row shows the dollar value of goods and services (output) going to each of the 85 industries and to final demand. Each column shows the dollar value of an industry's consumption (input) of intermediate goods and services from every industry and its value added. The 85 industries can be grouped into 11 categories as shown in Table 4.5.

Special industries include government [82]; rest-of-the-world industry [83 not included in Table 4.4a]; household industry [84]. Industries 78 and 79 measure the income originating in federal, state, and local government excluding such government enterprises like TVA. Note that industry 82 has no output to other industries. Its output goes only to final demand and mainly reflects em-

TABLE 4.5   Input–Output Classifications

| Category | Industries included |
|---|---|
| Agriculture | 1–4 |
| Mining | 5–10 |
| Construction | 11, 12 |
| Manfacturing | 13–64 |
| Transportation, communication, electric, gas, and sanitary services | 65–68 |
| Trade | 69 |
| Finance, insurance, and real estate | 70, 71 |
| Services | 72–77 |
| Government enterprises | 78, 79, 82 |
| Household industries | 84 |
| Inventory valuation adjustment | 85 |

*Source:* Benchmark Input–Output Accounts of the United States, 1987 Bureau of Economic Analysis, U.S. Department of Commerce, published by U.S. Document Printing Office, November 1994.

ployee compensation. Rest-of-the-world industry [83] mostly reflects U.S. income originating in foreign nations. Industry 84 measures income originating in households, (e.g., employee compensation of domestic servants). Industry 85 also has no outputs to other industries.

In "Other Value Added" (Table 4.4b) is property-type income, which is a composite of several factors, including proprietor's income, rental income of persons, corporate profits, inventory valuation adjustment, net interest, business transfer payments, surplus of government enterprises less subsidies, and capital consumption allowances (depreciation, depletion, and amortization). As expected, property-type income is highest for sector 71B, real estate. The depreciation part of property-type income accounts for the high values of some of the CPI sectors (e.g., 27 and 29). The depletion part of property-type income accounts for the moderately high values of sector 8, crude petroleum and natural gas.

Because depreciation needs to be part of value added, the flow of capital expenditures in the use table requires special consideration. As shown earlier (see Section 2.1), gross private domestic investment is a component of the GNP. To make the summation of the final demand inputs equal to the GNP, it is necessary to consider output of capital goods from any industry $k$ as input to final demand, not as input to the industry to which the capital goods actually go.

The final demand in Table 4.4a is broken into several components, including gross private fixed capital formation to account for capital goods and output from each industry. As a consequence, interindustry flows of capital goods are not explicitly given by the transactions table. For example, if a firm in sector 27a (chemicals and selected chemical products) buys a compressor from sector 43 (engines and turbines), the amount paid appears as part of the final demand figure in row 43 of Table 4.4a. The depreciation on the compressor then becomes part of column 27A's value added under property-type income. In this capital goods transaction table, the compressor sale would be recorded as an output of sector 43 and an input to sector 27.

In addition to gross private fixed-capital formation, Table 4.4 presents several other components of final demand.

The 85 × 85 matrix in Table 4.4a contains a wealth of information useful to the chemical professional. Tables 4.6–4.8, which illustrate several important aspects of the sales patterns found in the United States, are typical examples. There is a wide variation among industries in the proportion of total output going to final demand. Some industries (e.g., construction [11 and 12], food and kindred products [14], apparel [18], furniture and fixtures [22 and 23]) sell most of their products directly to final demand. Other industries (e.g., nonmetallic minerals [9 and 10], lumber and wood products [20 and 21], metallic ores mining [5] metal containers [39]) sell practically all their output to intermediate consumers. The CPI and related industries fit mostly in the latter category (Table 4.6).

With notable exception of drugs, cleaning, and toilet preparations [29A

TABLE 4.6 Sales of CPI and Related Industries to Intermediate Consumers and to Final Demand, 1987

| Industry | Category | Sales to intermediate consumers (millions) | Sales to final demand (millions) | Total sales going to final demand (%)[a] |
|---|---|---|---|---|
| Crude petroleum and natural gas | 8 | $97,326 | −$29,318 | — |
| Nonmetallic minerals, mining | 9 + 10 | $11,974 | −$90 | — |
| Paper and allied products except containers | 24 | $69,529 | $11,432 | 14 |
| Industrial and other chemicals | 27A | $79,565 | $10,287 | 11 |
| Agricultural fertilizers and chemicals | 27B | $12,543 | $823 | 6 |
| Plastics and synthetic materials | 28 | $39,534 | $3,872 | 9 |
| Drugs | 29A | $11,676 | $24,186 | 67 |
| Cleaning and toilet preparations | 29B | $6,952 | $25,920 | 79 |
| Paints and allied products | 30 | $11,365 | $718 | 6 |
| Petroleum refining and related industries | 31 | $70,488 | $67,111 | 49 |
| Rubber and miscellaneous plastics products | 32 | $78,219 | $8,632 | 10 |
| Glass and glass products | 35 | $15,366 | $970 | 6 |
| Primary iron and steel manufacturing | 37 | $74,182 | −$7,981 | — |
| Primary nonferrous metal manufacturing | 38 | $57,761 | −$2,015 | — |

Source: Ref. 7.
[a]Undertermined percentages indicated by dashes (—).

and 29B], the connection between production and final demand of all industries in Table 4.6 can be traced only through intermediate sales to industries.

Table 4.7 illustrates this point for CPI in the industrial and chemicals sector [27A]. The first column of Table 4.7 shows the sales of nine industries buying the largest amounts (in dollar value) from industry [27A]. Also shown is the amount sold to final demand. Note that the largest industry buyer is industry [27A] itself. The next seven largest buyers, all in the CPI, are plastics and synthetic materials [28], agricultural fertilizers and chemicals [27B], cleaning and toilet preparations [29B], petroleum refining and related industries [31], paints and allied products [30], and paper and allied products except

TABLE 4.7  Derived Demand for Chemicals and Selected Chemical Products

| Seller | Category | Total sold (millions) | Major buyers [category] | Total bought from seller (millions) |
|---|---|---|---|---|
| Industrial and other chemicals | 27A | $89,852 | Industrial and other chemicals [27A] | $18,226 |
| | | | Agricultural fertilizers and chemicals [27B] | $1,371 |
| | | | Plastics and synthetic materials [28] | |
| | | | Other agricultural products [2] | $64 |
| | | | Drugs [29A] | $697 |
| | | | Cleaning and toilet preparations [29B] | $2,971 |
| | | | Petroleum refining and related industries [31] | $1,758 |
| | | | Paper and allied products except containers [24] | $3,444 |
| | | | Paints and allied products [30] | $2,334 |
| | | | Total final demand [27A] | $10,287 |
| Agricultural fertilizers and chemicals | 27B | | Industrial and other chemicals [27A] | $366 |
| | | | Agricultural fertilizers and chemicals [27B] | $2,533 |
| | | | Plastics and synthetic materials [28] | $95 |
| | | | Other agricultural products [2] | $4,607 |
| | | | Drugs [29A] | $54 |
| | | | Paper and allied products except containers [24] | $205 |
| | | | Paints and allied products [30] | $252 |
| | | | Total final demand [27A] | $823 |
| Plastics and synthetic materials | 28 | | Rubber and miscellaneous plastics products [32] | $15,955 |
| | | | Broad and narrow fabrics, yarn and thread mills [16] | $5,251 |
| | | | Miscellaneous textile goods and floor covering [17] | $3,825 |
| | | | Industrial and other chemicals [27A] | $757 |
| | | | Apparel [18] | $1,526 |
| | | | Paints and allied products [30] | $1,441 |
| | | | Plastics and synthetic materials [28] | $1,470 |
| | | | Total final demand [27A] | $3,872 |
| Other agricultural products | 2 | | Livestock and livestock products [1] | $23,778 |

TABLE 4.7   Continued

| Seller | Category | Total sold (millions) | Major buyers [category] | Total bought from seller (millions) |
|---|---|---|---|---|
| | | | Food and kindred products [14] | $22,262 |
| | | | Tobacco manufacture [15] | $1,707 |
| | | | Real estate and royalties [71B] | $19 |
| | | | Broad and narrow fabrics, yarn and thread mills [16] | $3,192 |
| | | | Other agricultural products [2] | $3,855 |
| | | | Agriculture, forestry, and fishery services [4] | $2,089 |
| | | | Total final demand [27A] | $23,152 |
| Drugs | 29A | | Drugs [29A] | $3,758 |
| | | | Cleaning and toilet preparations [29B] | $0 |
| | | | Health, educational, and social services, and membership organizations [77A + 77B] | $6,591 |
| | | | Industrial and other chemicals [27A] | $18,226 |
| | | | Agricultural fertilizers and chemicals [27B] | $1,371 |
| | | | Wholesale and retail trade [69A + 69B] | $65 |
| | | | Other business and pro-fessional services, except medical [73C] | $47 |
| | | | Total final demand [27A] | $24,186 |
| Cleaning and toilet preparations | 29B | | Health, educational, and social services, and membership organizations [77A + 77B] | $1,028 |
| | | | Drugs [29A] | $21 |
| | | | Cleaning and toilet preparations [29B] | $1,535 |
| | | | Hotels; personal repair services except auto [72A + 72B] | $1,140 |
| | | | Wholesale and retail trade [69A + 69B] | $65 |
| | | | Real estate and royalties [71B] | $36 |
| | | | Computer, data processing, legal, engineering, and other business and professional services [73A + 73B] | $764 |
| | | | Total final demand [27A] | $25,920 |
| Petroleum refining and related industries | 31 | | Petroleum refining and related industries [31] | $9,933 |

TABLE 4.7   Continued

| Seller | Category | Total sold (millions) | Major buyers [category] | Total bought from seller (millions) |
|---|---|---|---|---|
| | | | Industrial and other chemicals [27A] | $1,012 |
| | | | Agricultural fertilizers and chemicals [27B] | $69 |
| | | | New construction; maintenance, and repair construction [11 + 12] | $11,220 |
| | | | Wholesale and retail trade [69A + 69B] | $4,871 |
| | | | Other agricultural products [2] | $1,175 |
| | | | Real estate and royalties [71B] | $389 |
| | | | Total final demand [27A] | $67,111 |
| Paper and allied products except containers | 24 | | Newspapers, periodicals, printing, and publishing [26A + 26B] | $21,116 |
| | | | Paper and allied products except containers [24] | |
| | | | Paperboard containers and boxes [25] | $11,548 |
| | | | Wholesale and retail trade [69A + 69B] | $4,694 |
| | | | Food and kindred products [14] | $3,142 |
| | | | Computer, data processing, legal, engineering, and other business and professional services [73A + 73B] | $1,902 |
| | | | Plastics and synthetic materials [28] | $463 |
| | | | Total final demand [27A] | $8,631 |
| Paints and allied products | 30 | | New construction; maintenance and repair construction [11 + 12] | $4,688 |
| | | | Motor vehicles, trucks, bus bodies, and motor vehicles parts [59A + 59B] | $1,837 |
| | | | Heating, plumbing, and fabricated structural metal products [40] | $243 |
| | | | Metal containers [39] | $196 |
| | | | Industrial and other chemicals [27A] | $408 |
| | | | Furniture and fixtures [22 + 23] | $376 |
| | | | Total final demand [27A] | $719 |

Source: Ref.7.

TABLE 4.8  Diversity of Sales Distribution Patterns of the CPI and Related Industries, 1987

| Industry | Category | Number of industrial buyers[a] | Portion of total sales sold to the largest buyer (%) |
|---|---|---|---|
| Crude petroleum and natural gas | 8 | 10 | 78 |
| Chemical and fertilizer mineral mining | 10 | 27 | 40 |
| Paper and allied products except containers | 24 | 89 | 21 |
| Industrial and other chemicals | 27A | 90 | 23 |
| Plastics and synthetic materials | 28 | 45 | 40 |
| Drugs | 29A | 10 | 32 |
| Cleaning and toilet preparations | 29B | 53 | 22 |
| Paints and allied products | 30 | 58 | 41 |
| Petroleum refining and related industries | 31 | 89 | 16 |
| Rubber and miscellaneous plastic products | 32 | 90 | 10 |
| Glass and glass products | 35 | 72 | 26 |
| Primary iron and steel manufacturing | 37 | 75 | 14 |
| Primary nonferrous metal manufacturing | 38 | 53 | 30 |

[a]Buyers of $1 million or more annually
Source: Ref. 7.

containers [24]. The column headed "Major buyers", gives the industries buying the largest amounts from the industries listed in the "Seller" column. The demand for industry 27's products is thus derived through supplying other industries' needs.

There is great diversity in intermediate sales distribution patterns. This is especially true for the CPI and related industries (Table 4.8). For example, both industrial and other chemical products [27A] and petroleum refining and related industries [31] have sales over $1 million to over 75 industries. However, 20% of total sales of industry 27A goes to its largest buyer industry, while 8.2% of total sales of industry 31 goes to its largest buyer industry. Plastics and synthetic materials [28], as well as drugs, cleaning, and toilet preparations [29 and 29B] both sell to 42 industries. Industry 28, however, sells 37% of its total output to its largest buyer industry, while industry 29A sells 51% of its total output to its largest buyer.

## 4.2    THE DIRECT-REQUIREMENTS TABLE

The dollar flow matrix in Tables 4.2 and 4.4a can be normalized to yield a table of coefficients that represents the fractions of a dollar required by a sector to produce a dollar of output. This is done as follows: dividing any element Xkj in the use table by the output of a sector gives the dollar input from industry k required for $1 output from industry j (6). This is defined as the direct coefficient:

$$D_{kj} = \frac{X_{kj}}{X_j}$$                                          (4.8)

Equation (4.2) can be rewritten as follows:

$$X_1 = (D_{11}X_1 + D_{12}X_2 + D_{13}X_3 + \cdots + D_{1n}X_n) + F_1$$
$$X_2 = (D_{21}X_1 + D_{22}X_2 + D_{23}X_3 + \cdots + D_{2n}X_n) + F_2$$
$$\vdots$$                                                               (4.9)
$$X_n = (D_{n1}X_1 + D_{n2}X_2 + D_{n3}X_3 + \cdots + D_{nn}X_n) + F_n$$

In vector notation, Eqs. (4.9) are

$$X = DX + F$$                                                          (4.10)

where $X$ and $F$ are vectors and $D$ is a matrix.

In tabular form, the direct coefficients of an entire economy are referred to as direct-requirements tables. Table 4.9 is the direct-requirements table for the four-sector economy illustrated in Table 4.2. Each column shows the inputs to the industry named at the top of the column required from the industry in each row at the left for $1 of column industry output

| Industry j | |
|---|---|
| Industry k | Dollar input from k needed to produce $1 output from industry j |

Table 4.9 is obtained by using Eq. (4.8). For example, the direct requirement of the energy industry for service is 0.185, which is obtained by dividing $12 billion of service industry input by the total output of the energy sector, $65 billion. Table 4.9 shows that in addition to service costs of 18.5 cents, the energy sector needs 35.4 cents of input from itself, 44.6 cents of value added (mostly employee compensation), and 1.5 cents from basic industry to produce $1 of output.

Table 4.10, which may be found in Appendix C, is the direct-requirements table of the U.S. economy in 1987. Note that special industries 82–85 are not

**TABLE 4.9** Direct-Requirement Table (four-sector economy)[a]

|  | Consumer | Basic | Energy | Service | Final demand |
|---|---|---|---|---|---|
| Consumer | 0.226 | 0.029 | 0.000 | 0.042 | 0.300 |
| Basic | 0.301 | 0.416 | 0.015 | 0.051 | 0.065 |
| Energy | 0.008 | 0.029 | 0.354 | 0.024 | 0.037 |
| Service | 0.098 | 0.114 | 0.185 | 0.254 | 0.452 |
| Value added | 0.368 | 0.412 | 0.446 | 0.628 | 0.147 |
| Totals | 1.001 | 1.000 | 1.000 | 0.999 | |

[a]Some figures do not total 1.00 because of rounding.

listed. Except for the larger number of sectors, Table 4.10 is completely analogous to Table 4.9.

For example, to produce $1 output, the plastics and synthetic materials industry 28 requires input from 58 industries. It requires the most input from industrial and other chemicals (33.2 cents), paper and allied products (1.1 cents), and wholesale and retail trade (4.3 cents). Table 4.11 lists the five largest suppliers of sectors 8, 9 + 10, 24, 27A, and 28, as well as 29A and 29B, in terms of dollar of direct input per total dollar output.

The direct-requirements table permits the estimation of changes in output directly required of any industry $k$ due to changes in total output of industries supplied by industry $k$. Assuming that the direct coefficients remain constant, this is accomplished as follows:

$$\Delta X_k = \sum_{j=1}^{n} D_{kj} \Delta X_j \qquad (4.11)$$

Table 4.10 shows that motor vehicles sector [59A] requires $0.0047 of input from industrial and other chemicals sector [27A] to produce $1 of output

$$D_{27A, 59A} = 0.0047$$

If the final demand for sector [59A] increases by $10 million, we can estimate the change in direct requirements for sector 27A as follows:

$$\Delta X_{59} = \Delta F_{59} = \$10,000,000$$
$$\Delta X_{27} = 0.0047 \times \$10,000,000 = \$47,000$$

If other industries that are supplied by sector 27A increased their output, the total direct effect could be estimated by summing the individual amounts using Equation 4.11.

**TABLE 4.11**  Major Suppliers of the CPI and Related Industries

| Buyer [category] | Major suppliers [category] | Input per total dollar output |
|---|---|---|
| Crude petroleum and natural gas [8] | Owner-occupied dwellings, real estate and royalties [71A + 71B] | 0.17414 |
| | Noncomparable imports [80] | 0.00940 |
| | Maintenance and repair construction [12] | 0.02189 |
| | Crude petroleum and natural gas [8] | 0.03738 |
| | Computer, data processing, legal, engineering, accounting, other business services except medical, advertising (73A + 73B + 73C + 73D) | 0.02140 |
| Nonmetallic minerals, mining [9+10] | Noncomparable imports [80] [9 + 10] | 0.00001 |
| | Nonmetallic minerals, mining [9 + 10] | 0.03484 |
| | Electric services, gas production and distribution services, water and sewer services [68A + 68B + 68C] | 0.07745 |
| | Owner-occupied dwellings; real estate and royalties [71A + 71B] | 0.01185 |
| | Computer, data processing, legal, engineering, accounting, other business services except medical; advertising [73A + 73B + 73C + 73D] | 0.02637 |
| Paper and allied products, except containers [24] | Paper and allied products, except containers [24] | 0.16027 |
| | Noncomparable imports [80] | 0.00103 |
| | Lumber and wood products [20 + 21] | 0.05859 |
| | Industrial and other chemicals [27A] | 0.04201 |
| | Railroads, motor freight, water, air, pipelines, and other related services [65A + 65B + 65C + 65D + 65E] | 0.03608 |
| Industrial and other chemicals [27A] | Industrial and other chemicals [27A] | 0.21601 |
| | Petroleum refining and related industries [31] | 0.01199 |
| | Electric, gas, water, and sanitary services [68A + 68B + 68C] | 0.06586 |
| | Computers, data processing, legal, engineering, accounting, and related services advertising [73A + 73B + 73D + 73E] | 0.04819 |
| | Noncomparable imports [80] | 0.00753 |
| Plastics and synthetic materials [28] | Industrial and other chemicals [27A] | 0.33213 |
| | Wholesale and retail trade [69A + 69B] | 0.04310 |

TABLE 4.11 Continued

| Buyer [category] | Major suppliers [category] | Input per total dollar output |
|---|---|---|
| | Paper and allied products, except containers [24] | 0.01138 |
| | Computers, data processing, legal, engineering, accounting and related services, advertising [73A + 73B + 73D + 73E] | 0.04650 |
| | Plastics and synthetic materials [28] | 0.03615 |
| Drugs [29A] | Computers, data processing, legal, engineering, accounting and related services, advertising [73A + 73B + 73D + 73E] | 0.04650 |
| | Industrial and other chemicals [27A] | 0.01937 |
| | Drugs [29A] | 0.10434 |
| | Cleaning and toilet preparations [29B] | 0.00059 |
| | Rubber and miscellaneous plastic products [32] | 0.02081 |
| | Owner-occupied dwellings; real estate and royalties [71A + 71B] | 0.00585 |
| Cleaning and toilet preparations [29B] | Computers, data processing, legal, engineering, accounting and related services, advertising [73A + 73B + 73D + 73E] | 0.00666 |
| | Industrial and other chemicals [27A] | 0.08941 |
| | Cleaning and toilet preparations [29B] | 0.04618 |
| | Rubber and miscellaneous plastic products [32] | 0.05156 |
| | Owner-occupied dwellings; real estate and royalties [71A + 71B] | 0.00451 |
| Paint and allied products [30] | Industrial and other chemicals [27A] | 0.19337 |
| | Plastics and synthetic materials [28] | 0.11933 |
| | Metal containers [39] | 0.04422 |
| | Computers, data processing, legal, engineering, accounting and related services; advertising [73A + 73B + 73D + 73E] | 0.01427 |
| | Wholesale and retail trade [69A + 69B] | 0.02842 |
| Petroleum refining and related industries (31) | Crude petroleum and natural gas [8] | 0.55103 |
| | Petroleum refining and related industries [31] | 0.07204 |

**TABLE 4.11** Continued

| Buyer [category] | Major suppliers [category] | Input per total dollar output |
|---|---|---|
| | Railroads, motor freight, water, air, pipelines, and other related services (65A + 65B + 65C + [65D + 65E] | 0.05191 |
| | Noncomparable imports [80] | 0.00284 |
| | Computers, data processing, legal, engineering, accounting and related services, advertising [73A + 73B + 73D + 73E] | 0.01436 |
| Rubber and miscellaneous plastic products [32] | Plastics and synthetic materials [28] | 0.18645 |
| | Industrial and other chemicals [27A] | 0.04564 |
| | Miscellaneous textile goods and floor coverings [17] | 0.01044 |
| | Wholesale and retail trade [69A + 69B] | 0.04989 |
| | Computers, data processing, legal, engineering, accounting and related services; advertising [73A + 73B + 73D + 73E] | 0.02612 |
| Glass and glass products [35] | Nonmetallic minerals, mining [9 + 10] | 0.01144 |
| | Glass and glass products [35] | 0.08137 |
| | Paperboard containers and boxes [25] | 0.04394 |
| | Electric, gas, water, and sanitary services [68A + 68B + 68C] | 0.0679 |
| | Wholesale and retail trade [69A + 69B] | 0.04277 |
| Primary iron and steel manufacturing [37] | Primary iron and steel manufacturing [37] | 0.15029 |
| | Metallic ores, mining [5 + 6] | 0.0289 |
| | Railroads, motor freight, water, air, pipelines, and other related services [65A + 65B + 65C + 65D + 65E] | 0.03615 |
| | Wholesale and retail trade [69A + 69B] | 0.06311 |
| Primary nonferrous metal manufacturing [38] | Primary nonferrous metal manufacturing [38] | 0.30618 |
| | Noncomparable imports [80] | 0.01399 |
| | Nonferrous metal ores, mining [5 + 6] | 0.06975 |
| | Scrap, used, and secondhand goods [81] | 0.04215 |
| | Wholesale and retail trade [69A + 69B] | 0.06108 |

*Source:* Ref. 7. (See also Table 4.10 in Appendix C.)

## 4.3  THE TOTAL-REQUIREMENTS TABLE

Although useful, the direct-requirements table does not enable one to calculate directly the total input requirement for an industry given a change in the final demand. The direct coefficient is the dollar value of of products required "indirectly" via the input from other industries. Returning to the previous example, the $45,200 increase of input required from sector 27A for a change of $10 million in total output to final demand for sector 59A represents only products bought directly from sector 27A. Automobile production, however, requires direct input from rubber and miscellaneous plastic products [32], plastics and synthetic products [28], paints and allied products [30], glass and glass products [35], primary iron and steel manufacturing [37], and so on. These industries, in turn, require input from industrial and other chemicals [27A]. Moreover, these industries will require additional motor vehicles and equipment [59A and 59B], raising the demand for such products even higher than the initial $10 million change. Thus the increase in total demand is equal to

$$\Delta X = \Delta F + (D\Delta F) + (DD\Delta F) + (DDD\Delta F) + \cdots$$

where the first term, $\Delta F$, is the initial change in final demand, the second term, $D\Delta F$, is the first-round increase in production computed by the direct requirement, the third term, $DD\Delta F$, is a new round of increase in production brought about by the need of supplying industries to purchase goods and services to cover the needs of the increase in production, and so on.

To take into account both the direct and indirect requirements of input by one industry needed for a dollar's worth of output to final demand of another industry, we return to Eq. (4.10), where the set of direct-requirement equations is given by in the vector form as follows:

$$X = DX + F$$

If this equation is rearranged

$$X - DX = F$$

$$(I - D)X = F$$

where I is the identity matrix, we can invert the matrix $I - D$ to write:

$$X = (I - D)^{-1}F$$

The matrix $(I - D)^{-1}$ is equal to an infinite series and can be approximated by truncation of term $n$

$$(I - D)^{-1} = I + D + D^2 + D^3 + \cdots + D^n \tag{4.12}$$

The output then becomes

$$X = (I + D + D^2 + \cdots)F \tag{4.13}$$

where $IF$ represents the original final demand, and each subsequent term represents a round of buying of the supplying industries to cover production needs. If the matrix $(I - D)^{-1}$ is defined as the total coefficient matrix $T$, we get

$$X = TF$$

where each element $T_{ij}$ of $T$ is the total input (both direct and indirect) required from industry $i$ for a dollar of output to final demand by industry $j$. If the total coefficients remain constant, changes in output by other industries can be estimated in relation to final demand:

$$\Delta X_i = \sum_{j=1}^{n} T_{ij} \Delta F_j \tag{4.14}$$

Equation (4.14) can be used to estimate the change in output due to a change in the final demand.

A simple example will illustrate how the various relationships are applied and indicate some additional uses of input-output analysis.

Example 4.1.   To illustrate the basic concepts without algebraic complexity, we consider a two-industry economy. We shall assume that some central agency like the federal government has collected statistics on suppliers and supplies needed to develop the transaction table shown in Table 4.12.

Since final demand is known, $F_A = 2$ and $F_B = 7$, a set of two equations can be written to describe the situation:

Industry A:    $10 = 4 + 4 + 2$
Industry B:    $10 = 1 + 2 + 7$

TABLE 4.12   Use Table (two-sector economy)

| Output from | Input to A | Input to B | Intermediate output | Final demand | Total output |
|---|---|---|---|---|---|
| A | 4 | 4 | 8 | 2 | 10 |
| B | 1 | 2 | 3 | 7 | 10 |
| Intermediate input | 5 | 6 | — | — | — |
| Value added | 5 | 4 | — | — | GNP = 9 |
| Total input | 10 | 10 | — | GNP = 9 | — |

The direct-coefficient table is obtained by dividing interindustry transaction by the total output [Eq. (4.8)]:

|   | A | B |
|---|---|---|
| A | 0.4 | 0.4 |
| B | 0.1 | 0.2 |

These equations can be written in symbolic notation, using the direct-coefficient form of Eq. (4.9):

Industry A:

$$X_A = 0.4X_A + 0.4X_B + 2$$

Industry B:

$$X_B = 0.1X_A + 0.2X_B + 7$$
$$X = DX + F$$

In vector form, we have

$$\begin{pmatrix} X_A \\ X_B \end{pmatrix} = \begin{pmatrix} 0.4 & 0.4 \\ 0.1 & 0.2 \end{pmatrix} \begin{pmatrix} X_A \\ X_B \end{pmatrix} + \begin{pmatrix} 2 \\ 7 \end{pmatrix}$$

To use the system of equations for predictive purposes, we must convert to the form shown as Eq. (4.13):

$$X = TF$$
$$T = (I - D)^{-1}$$

$$= \left[ \begin{pmatrix} 1 \\ 0 \end{pmatrix}\begin{pmatrix} 0 \\ 1 \end{pmatrix} - \begin{pmatrix} 0.4 \\ 0.1 \end{pmatrix}\begin{pmatrix} 0.4 \\ 0.2 \end{pmatrix} \right]^{-1}$$

$$= \left[ \begin{pmatrix} 0.6 \\ -0.1 \end{pmatrix}\begin{pmatrix} -0.4 \\ 0.8 \end{pmatrix} \right]^{-1}$$

$$= \frac{1}{\begin{pmatrix} 0.6 \\ 0.1 \end{pmatrix}\begin{pmatrix} -0.4 \\ 0.8 \end{pmatrix}} \begin{pmatrix} 0.8 \\ 0.1 \end{pmatrix}\begin{pmatrix} 0.4 \\ 0.6 \end{pmatrix}$$

$$= \begin{pmatrix} 1.82 \\ 0.23 \end{pmatrix}\begin{pmatrix} 0.91 \\ 1.36 \end{pmatrix}$$

The total coefficient equations now appear as

$$\begin{pmatrix} X_A \\ X_B \end{pmatrix} = \begin{pmatrix} 1.82 \\ 0.23 \end{pmatrix}\begin{pmatrix} 0.91 \\ 1.36 \end{pmatrix}\begin{pmatrix} F_A \\ F_B \end{pmatrix}$$

or

$$X_A = 1.82F_A + 0.91F_B$$

$$X_B = 0.23F_A + 1.36F_B$$

In this form, we can now easily predict changes in any industry output, given changes in final demand. Suppose final demand for industry B changes by one unit and final demand for industry A remains unchanged. Equation (4.14) applies:

$$\Delta X_A = T_{AB}\Delta F_B = (0.91)\,(1) = 0.91$$

$$\Delta X_B = T_{BB}\Delta F_B = (1.36)\,(1) = 1.36$$

(Note: A $2 \times 2$ matrix can be inverted as follows.) If

$$M = \begin{vmatrix} a & b \\ c & d \end{vmatrix}$$

then

$$M^{-1} = \left(\frac{1}{ad-cb}\right)\begin{pmatrix} d \\ -c \end{pmatrix}\begin{pmatrix} -b \\ a \end{pmatrix}$$

A $3 \times 3$ matrix can be inverted as follows. If

$$M = \begin{vmatrix} a & b & c \\ d & e & f \\ g & h & i \end{vmatrix}$$

then

$$M^{-1} = \frac{1}{|M|}\begin{pmatrix} ei-hf & hc-bi & bf-ec \\ gf-di & ai-gc & dc-af \\ dh-ge & gb-ah & ae-bd \end{pmatrix}$$

Computer programs are available for inverting matrices of higher rank.

Thus for a change in final demand of one unit (or 10%) for products from industry B, 0.91 unit of extra output from A is required, and 1.36 extra units of output from B.

Now suppose there is a change in technology that allows products from B to be made more efficiently with less input required from A. Assume that the D matrix changes as follows:

$$\begin{array}{ccc} & A & B \\ A & \begin{pmatrix} 0.40 & 0.30 \\ B & 0.10 & 0.20 \end{pmatrix} \end{array}$$

Although it is not necessary for this simple example, it is instructive to use the series approximation [Eq. (4.12)] to estimate $(\mathbf{I}\text{-}\mathbf{D})^{-1}$:

$$(\mathbf{I} - \mathbf{D})^{-1} = \mathbf{T} = \mathbf{I} + \mathbf{D} + \mathbf{D}^2 + \mathbf{D}^3 + \cdots$$

Using the first two terms of the series produces

$$\mathbf{T} = \begin{vmatrix} 1 & 0 \\ 0 & 1 \end{vmatrix} + \begin{vmatrix} 0.4 & 0.3 \\ 0.1 & 0.2 \end{vmatrix} = \begin{vmatrix} 1.4 & 0.3 \\ 0.1 & 1.2 \end{vmatrix}$$

If the squared term is added, the effect of a second round of purchases to meet final demand is:

$$\mathbf{D}^2 = \begin{vmatrix} 0.19 & 0.18 \\ 0.06 & 0.07 \end{vmatrix} \qquad \mathbf{T} = \begin{vmatrix} 1.59 & 0.48 \\ 0.16 & 1.27 \end{vmatrix}$$

The cubic term takes into account a third round of purchases:

$$\mathbf{D}^3 = \begin{vmatrix} 0.09 & 0.09 \\ 0.03 & 0.03 \end{vmatrix} \qquad \mathbf{T} = \begin{vmatrix} 1.69 & 0.57 \\ 0.19 & 1.30 \end{vmatrix}$$

If the $\mathbf{I} - \mathbf{D}$ matrix is inverted, the same procedure as before yields:

$$\mathbf{T} = \begin{vmatrix} 1.78 & 0.66 \\ 0.22 & 1.33 \end{vmatrix}$$

The total output now is:

$$\mathbf{X} - \mathbf{TF} = \begin{pmatrix} 1.78 & 0.66 \\ 0.22 & 1.33 \end{pmatrix} \begin{pmatrix} 2 \\ 7 \end{pmatrix} = \begin{pmatrix} 8.2 \\ 9.8 \end{pmatrix}$$

There is an 18% drop in output of A merely because B has learned to economize on using supplies from A.

**Note:** Multiplication of two square matrices $[a_{ij}]$ and $[b_{jk}]$, where $a_{ij}$ and $b_{jk}$ are the elements of each matrix, , can be carried out as follows:

$$[a_{ij}][b_{jk}] = \sum_{j=1}^{n} a_{ij} b_{jk}$$

If the transactions table is restructured, the result is Table 4.13. By improving its technology, industry B decreased its total output by 27%, but its value added increased by 22.5% from $4 to $4.90.

Table 4.14 shows the total requirements for the four-sector economy of Table 4.2. The elements of Table 4.14 were obtained by inverting the $\mathbf{I} - \mathbf{D}$ matrix, with D found from Table 4.9. Table 4.15, in Appendix C, shows the total requirements for each dollar of delivery to final demand of the industry of the column. For example, the energy sector requires 40.5 cents of input from the service sector to

**TABLE 4.13**   Use Table (two-sector economy, revised)

| Output from | Input to A | Input to B | Intermediate output | Final demand | Total output |
|---|---|---|---|---|---|
| A | 3.28 | 2.94 | 6.2 | 2 | 8.2 |
| B | 0.82 | 1.96 | 2.8 | 7 | 9.8 |
| Intermediate input | 4.1 | 4.9 | — | — | — |
| Value added | 4.1 | 4.9 | — | — | GNP = 9 |
| Total input | 8.2 | 9.8 | — | GNP = 9 | — |

**TABLE 4.14**   Total Requirements Table for a Four-Sector Economy

|  | Consumer | Basic | Energy | Service |
|---|---|---|---|---|
| Consumer | 1.334 | 0.081 | 0.025 | 0.081 |
| Basic | 0.713 | 1.783 | 0.089 | 0.165 |
| Energy | 0.059 | 0.090 | 1.576 | 0.060 |
| Service | 0.298 | 0.305 | 0.405 | 1.389 |

produce $1 of final demand. This is much more than 18.5 cents shown on the direct-requirements table (Table 4.9).

The total-requirements table for the U.S. economy for 1987 is given as Table 4.15 (see Appendix C). The elements of this table are obtained by inverting the $I - D$ matrix, where $D$ is found from Table 4.10. Each column shows the output required both directly and indirectly from the industry named at the left of each row per dollar of delivery to final demand by the industry named at the head of the column. Returning to our automobile example, a $10 million increase in final demand for the products of industry 59A requires an increase of 0.0.3769 × $10,000,000 = $376,900 in total output in both direct and indirect requirements for products of industry 27A. This is about eight times the direct-requirement increase of $45,200 computed by using the direct requirement only.

Table 4.16 shows the total coefficients as taken from Table 4.15 of food and kindred products [14], apparel [18], furniture and fixtures [22 and 23], drugs [29A], cleaning and toilet preparations [29B], construction [11 and 12], household appliances [54], and motor vehicles [59A] for the input requirements of the CPI and related industries.

You are urged to construct your own tables to illustrate features of the input–output tables that are of personal or professional interest.

TABLE 4.16  Total Coefficients of Several Industries for CPI and Related Industries

| Output from [category] | Input to [category] | | | | | | |
|---|---|---|---|---|---|---|---|
| | Food and kindred product [14] | Apparel [18] | Furniture and fixtures [22 + 23] | Drugs, cleaning, and toilet preparations [29A + 29B] | Maintenance and repair construction [11 + 12] | Household appliances [54] | Motor vehicles, truck and bus bodies, trailers; parts [59A + 59B] |
| Crude petroleum and natural gas [8] | 0.01686 | 0.01402 | 0.01498 | 0.03688 | 0.04335 | 0.16280 | 0.03487 |
| Nonmetallic minerals mining [9 + 10] | 0.00319 | 0.00190 | 0.00238 | 0.00604 | 0.02803 | 0.00279 | 0.00508 |
| Paper and allied products except containers [24] | 0.03828 | 0.01515 | 0.02154 | 0.05846 | 0.02134 | 0.02503 | 0.02991 |
| Chemicals and selected chemical products [27A] | 0.02347 | 0.05419 | 0.36350 | 0.19769 | 0.03662 | 0.04000 | 0.06454 |
| Plastics and synthetic materials [28] | 0.00899 | 0.07537 | 0.02826 | 0.04444 | 0.01656 | 0.03613 | 0.04555 |
| Drugs, cleaning and toilet preparations [29A + 29B] | 0.00845 | 0.00883 | 0.20000 | 1.07908 | 0.00266 | 0.00194 | 0.00342 |
| Paints and allied products [20] | 0.00134 | 0.00112 | 0.01199 | 0.00498 | 0.01847 | 0.00878 | 0.02032 |
| Petroleum refining and related industries [31] | 0.01650 | 0.01369 | 0.01470 | 0.03497 | 0.05821 | 0.01365 | 0.03125 |
| Rubber and miscellaneous plastics products [32] | 0.02800 | 0.01787 | 0.04364 | 0.08928 | 0.03886 | 0.05663 | 0.14042 |
| Glass and glass products [35] | 0.01661 | 0.00286 | 0.00592 | 0.01754 | 0.00826 | 0.01301 | 0.01668 |
| Primary iron and steel manufacturing [37] | 0.01462 | 0.00358 | 0.06510 | 0.01665 | 0.08463 | 0.11100 | 0.19359 |

Source: Ref. 7. (See also Table 4.15 in Appendix C.)

## 4.4  APPLYING INPUT–OUTPUT ANALYSIS

Input–output analysis has been described in this chapter and used to identify the role of the chemical processing industries in the U.S. economy. We have also shown how the industrial responses to changes in demand can be measured. In-put–output analysis can also be applied in a variety of situations by individual corporations. A. D. Little, Inc. conducted a survey among more than 200 compa-nies concerning the use of input–output analysis in some form of corporate plan-ning at one time or another. About one-third of these major firms use input–output analysis on a regular basis. Input–output analysis has been used in the following corporate functions:

*Forecasting.* The major corporate use of input–output analysis has been in providing forecasts of the U.S. economy and forecasts of changes in the coefficients of the direct- and total-requirements tables. Fore-casts have been used in identifying acquisition and diversification opportunities. Studying the effect of changes in final demand for au-tomobiles on the CPI is one such forecasting application of input–output analysis.

*Sensitivity testing.* Input–output analysis can be used to test the conse-quences of "what if" questions. For example, the input–output tables can be used to study the effects on one industry's or company's growth under various scenarios (e.g., different rates of GDP growth or potential changes in technology, as shown by Example 4.1 for a simple two-in-dustry economy).

*Flow and structural analysis.* Input–output analysis can be used effec-tively to determine patterns by which products move through the econ-omy. This exercise can help identify problems like bottlenecks or show how specific markets are coupled. Table 4.7 is an example of tracing in-put–output flows through intermediate markets to final demand.

*Sorting and screening.* Input–output analysis can be used to order markets and industries according to a number of criteria (e.g., size, category, en-ergy intensiveness, etc.). For example, Table 4.10 screens all industries to determine which are the largest suppliers of the CPI.

A variety of additional uses of input–output analysis have been investi-gated, including its use in predicting chemical prices. In spite of its potential, in-put–output analysis has not found widespread use as a business tool. The major barriers to such development appears to be the following:

1.  Lack of detailed and current information for constructing input–out-put tables. The latest table, dated 1987, were not published until 1994. Yet the great advances in technology since 1987 have affected the data in the tables.

2. Lack of knowledge and understanding among people in a position to use input–output analysis.
3. Lack of published case studies of how input–output analysis has been applied by individual companies.

Although input–output analysis has been in existence for several decades, its full potential can be realized only with better-educated managers and better, up-to-date databases.

## 4.5  SUGGESTED READING

A great deal of information concerning definitions and assumptions made in arriving at the U.S. input–output tables can be found in the following fundamental and classical articles, published in the *Survey of Current Business,* published by the Bureau of Economic Analysis of the U.S. Department of Commerce.

The Interindustry Structure of the United States. A report on the 1958 Input–Output Study, November 1964.

The Transaction Table of the 1958 Input–Output Study and Revised Direct and Total Requirements Data, September 1965.

Input–Output Structure of the U.S. Economy, 1963, November 1969.

The Composition of Value Added in the 1963 Input–Output Study, April 1973.

The Input–Output Structure of the U.S. Economy, 1967, February 1974.

Interindustry Transactions in the New Structures and Equipment, 1967, September 1975.

Additional articles on input–output analysis may be of interest:

George A. Gols, "The Use of Input–Output in Industrial Planning," *Business Economics,* May 1975.

David Liebeskind, "A Marketing Tool: Price Forecasting via Input/Output Techniques," *CHEMTECH,* September 1973.

## REFERENCES

1. "Input–Output Analysis: An Appraisal." A report on the National Bureau of Economic Research, Princeton University Press, Princeton, NJ, 1955, pp 215–252.
2. U.S. Department of Commerce, Bureau of Economic Analysis, *Survey of Current Business,* February 1974.
3. U.S. Department of Commerce, Bureau of Economic Analysis, *Survey of Current Business,* September 1975, pp. 9–21.
4. *Business Economics,* May 1975, p. 19.
5. David Liebeskind, "A Marketing Tool: Price Forecasting via Input/Output Techniques," *CHEMTECH,* September 1973, pp. 543–547.

6.  U.S. Department of Commerce, Bureau of Economic Analysis, *Survey of Current Business*, April 1944, pp. 73–115.
7.  U.S. Department of Commerce, Bureau of Economic Analysis, "Benchmark Input–Output Accounts of the United States, 1987." Government Printing Office, Washington, DC, November 1994.

## PROBLEMS

4.1.   a.   Referring to the 1987 U.S. input–output tables in Appendix C, calculate the fraction of total output that is value added for the following industries:

| Input–output sector number | Industry |
|---|---|
| 8 | Crude petroleum and natural gas |
| 9 and 10 | Nonmetallic minerals mining |
| 27A | Industrial and other chemicals |
| 28 | Plastics and synthetic materials |
| 29A | Drugs |
| 29B | Cleaning and toilet preparations |
| 30 | Paints and allied products |
| 31 | Petroleum refining and related products |
| 32 | Rubber and miscellaneous plastics products |

   b.   Most international oil companies are involved in industries 8, 27A, and 31. Can you tell by the value-added figures where these oil companies make most of their money?

4.2.   Does a high value added necessarily mean a high return on investment? Why? Does a low value added necessarily mean a low profit margin?

4.3.   a.   List some of the industries having the highest value added as measured by the fraction of total output. What do these industries have in common?

   b.   List some of the industries having the lowest value added as measured by the fraction of total output.

   c.   For industries 8, 10, 27A, and 32, what percentage of value added goes to employee compensation? How does this compare with the industries given in your answers to part a?

4.4.   The GDP ultimately is distributed into the following categories:

1.   Personal consumption expenditures
2.   Gross private fixed investment
3.   Changes in business inventories
4.   Exports of goods and services
5.   Imports of goods and services

6. Federal government purchases
   a. National defense
   b. Nondefense
7. State and local government purchases
   a. Education
   b. Other

Find the percentages of total final demand (1987) going to each of these categories for industries 27–30 and 32. Can you explain the variations in the fraction amounts going to the different categories from these industries?

4.5. Expand Table 4.7 to include the eighth and ninth largest buyers of industrial and other chemicals [27A]. Determine the six industries buying the most from these two sectors and show the dollar amounts.

4.6. Determine the "direct" effect on output for sector 27A if all the buyers in Table 4.7 increase their final output by $10 million.

4.7. Are the CPI export intensive? Compare the percentage of total final demand going to net exports for the components of the CPI to that of the total economy and to other sectors, such as other agricultural products [2], motor vehicles (passenger cars ant trucks) [59A], scientific and controlling instruments [62], and finance and insurance [70].

4.8. Compare the ratio of property-type income to employee compensation for the various CPI segments. Compared with industries such as finance [70A], apparel [18], motor vehicles [59A], primary metals [37, 38], and food [14], are the CPI more capital or labor intensive?

4.9. New housing starts and new auto sales are two of the most important indicators of business climate. Calculate the impact on the various CPI [27A and 27B] segments for a 10% increase in each of these. List at least 10 different CPI products important in each of these industries.

4.10. Consider all the crude petroleum and natural gas [8], electrical services [68A], and coal mining [7] as energy. What percentage of the GDP is devoted to energy? Which industries are most energy intensive as measured by percentage input by energy per total output? Which industries would be highly affected by an increase in energy costs and which would not? How do the CPI compare?

4.11. If utility services [68A and 68B] were to switch half the present supply of oil into supply from coal, how would the total output of coal mining and oil production be changed? Use Table 4.4a and the following values. A 42-gallon barrel of oil = 5.8 million Btu heating value; 1 ton of coal = 25 million Btu heating value; 1996 energy costs are $30/bbl for fuel oil and $26.50/ton for coal.

4.12. If motor vehicles [59A] reduced the use of steel by 30% but increased the plastics use to make up for the decline in steel use, how would the production of steel and plastics be affected? Hint: Use the following perturbation calculations. Let $T \rightarrow T + e$ be the result of $D \rightarrow D + \Delta$. To estimate e, write

$$(I - D)T = 1 \qquad \text{and} \qquad (I - D - \Delta(T + e) = I$$

so that

$$e - De - \Delta T - \Delta e = 0$$

Assume $\Delta e$ is small. Then

$$(I - D)e = \Delta T \qquad \text{and} \qquad e = T\Delta T$$

**4.13.** An earth reconnaissance satellite reports a drought in Russia. Government experts estimate that Russian wheat production probably will be down by 50 million bushels. What could Russia do to cover this shortage? In which CPI sectors would the shortage have an impact? Estimate the potential impact on the CPI by using input–output tables.

**4.14.** After completing his last assignment and becoming bored with his consulting duties, Professor Sharppencil moved to a small Persian Gulf country where he was promptly hired as head government economist by the ruler Sheik Gushofoil. The country has only three industries A, O (as in oil), and EI (as in export–import). Having read about input–output analysis, Sharppencil decides to impress the sheik by constructing input–output tables. Using a recently installed computerized information retrieval system, Sharppencil finds the following:

1. Industry A has total sales of $100 million annually. In 1 year, industry A buys $10 million worth of goods itself, 5 million worth of goods from O, and $50 million of goods from EI.
2. Industry O has total sales of $100 million annually. In 1 year, industry O buys $20 million worth of goods from A, $50 million worth of goods from O itself, and $50 million worth of goods from EI.
3. Industry EI has a total sales of $200 million. In 1 year, EI buys $50 million worth of goods from A, $80 million worth from O, and none from itself.

Help Sharppencil construct the transactions table, direct-requirements table, and total-requirements table.

**4.15.** Suppose you were asked to determine the input required to produce 10 billion pounds of polyethylene. How would you estimate the dollar value of needed raw materials, energy, and other inputs?

**4.16.** Changes in technology can often be incorporated into an input–output analysis by estimating new direct coefficients for a specific industry or industries. Once a new $D$ matrix has been estimated, the $T$ matrix can be found by inverting $I - D$. However, often it is inconvenient or too expensive to perform the inversion. This problem considers two ways in which changes in the $T$ matrix can be estimated with a new $D$ matrix.

    a. If the new $D$ matrix is written as $D + \Delta$ and the new $T$ matrix is written as $T + e$, the perturbation calculations shown in Problem 4.12 yield $e =$

T$\Delta$T provided $\Delta e$ is small. Using this method, estimate the new **T** matrix for the change in technology given for industry B in Example 4.1. How does this estimation compare with the exact answer?

b.   It is also possible to estimate the new **D** by a series expansion [Eq. (4.12)]:

$$(\mathbf{I} - \mathbf{D})^{-1} = \mathbf{I} + \mathbf{D}^2 + \mathbf{D}^3 + \mathbf{D}^4 + \cdots + \mathbf{D}^n$$

or

$$\mathbf{T} = \mathbf{I} + \mathbf{D} + \cdots + \mathbf{D}^n$$

Estimate the maximum error in **T** that would result from expanding through the $\mathbf{D}^n$ term.

# 5

## Products and Companies of the CPI

The CPI require, as input, many basic raw materials from primary farming, petroleum, and mining industries, as well as a vast array of intermediates from secondary manufacturing industries. Within the CPI the inputs are transformed into chemical products by chemical and physical processes.

Figure 5.1 shows the important raw materials for the manufacturing of some of the intermediates and gives a partial list of chemical products of the chemical industry.

A similar diagram showing some of the chemical transformations for the petrochemical segment of the CPI is given in Figure 5.2. A simplified flow diagram for the petroleum refining industry is shown in Figure 5.3. The processing units that effect the needed chemical and physical transformations are built and operated by more than 10,000 individual companies, all attempting to make a profit by converting raw materials and upgrading intermediates.

The chemical professional needs to know which products and intermediates are of particular importance to the CPI, including the interrelationships between raw materials, intermediates, and products going to final demand. Further, the chemical professional needs to know which companies play important roles in the CPI.

In this chapter, we shall examine the CPI by discussing their products and the companies manufacturing these products. The discussion concentrates on SIC 281, 286, and 291.

---

*Most of the statistics are from the Facts and Figures issues of *Chemical and Engineering News, Statistical Abstracts of the United States,* and from the Fortune 500 issues of *Fortune* magazine. The reader can conveniently update the data by consulting the most recent issues of these readily available publications.

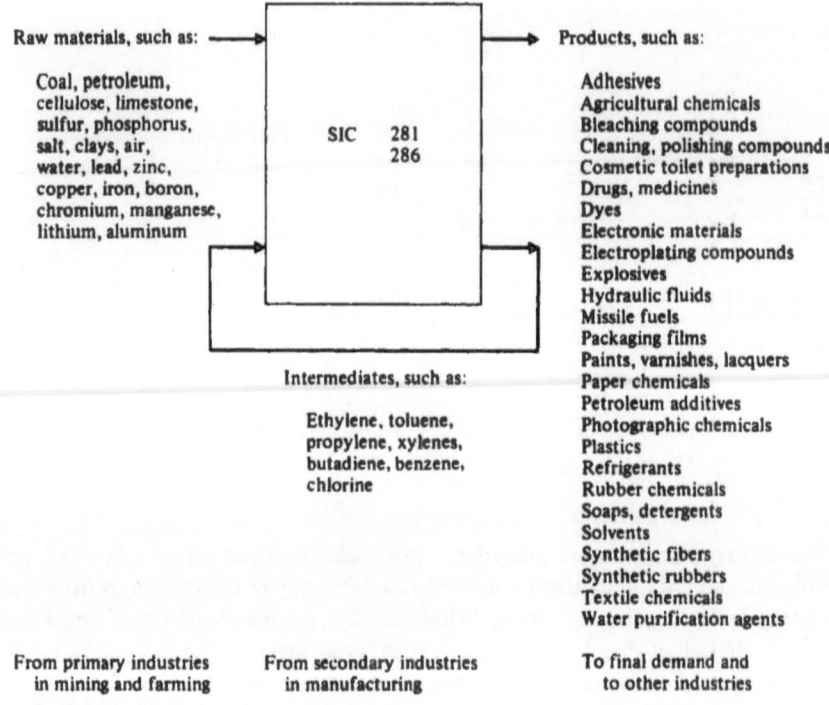

Raw materials, such as: ⟶

Coal, petroleum,
cellulose, limestone,
sulfur, phosphorus,
salt, clays, air,
water, lead, zinc,
copper, iron, boron,
chromium, manganese,
lithium, aluminum

SIC    281
          286

Products, such as:

Adhesives
Agricultural chemicals
Bleaching compounds
Cleaning, polishing compounds
Cosmetic toilet preparations
Drugs, medicines
Dyes
Electronic materials
Electroplating compounds
Explosives
Hydraulic fluids
Missile fuels
Packaging films
Paints, varnishes, lacquers
Paper chemicals
Petroleum additives
Photographic chemicals
Plastics
Refrigerants
Rubber chemicals
Soaps, detergents
Solvents
Synthetic fibers
Synthetic rubbers
Textile chemicals
Water purification agents

Intermediates, such as:

Ethylene, toluene,
propylene, xylenes,
butadiene, benzene,
chlorine

From primary industries        From secondary industries        To final demand and
in mining and farming              in manufacturing                  to other industries

FIGURE 5.1  The chemical industry.

## 5.1.  PRODUCTS OF THE CPI

### 5.1.1  SIC 28: Chemicals and Allied Products

To bring order into any discussion of the immense number of CPI products, some means of classification must be used. The products of the CPI can be classified in several ways.

5.1.1.1  Product and End Use.  Typical end-use groups are shown in Table 5.1. It is evident from our discussion in Chapters 2 and 4 that any reasonable analysis of a CPI operation should include a study of the end use of the product ultimately produced, even though the operation under analysis may not produce the end-use product itself. End-use classifications are generally based upon the industry served by the chemical after it leaves the CPI (e.g., automotive chemicals).

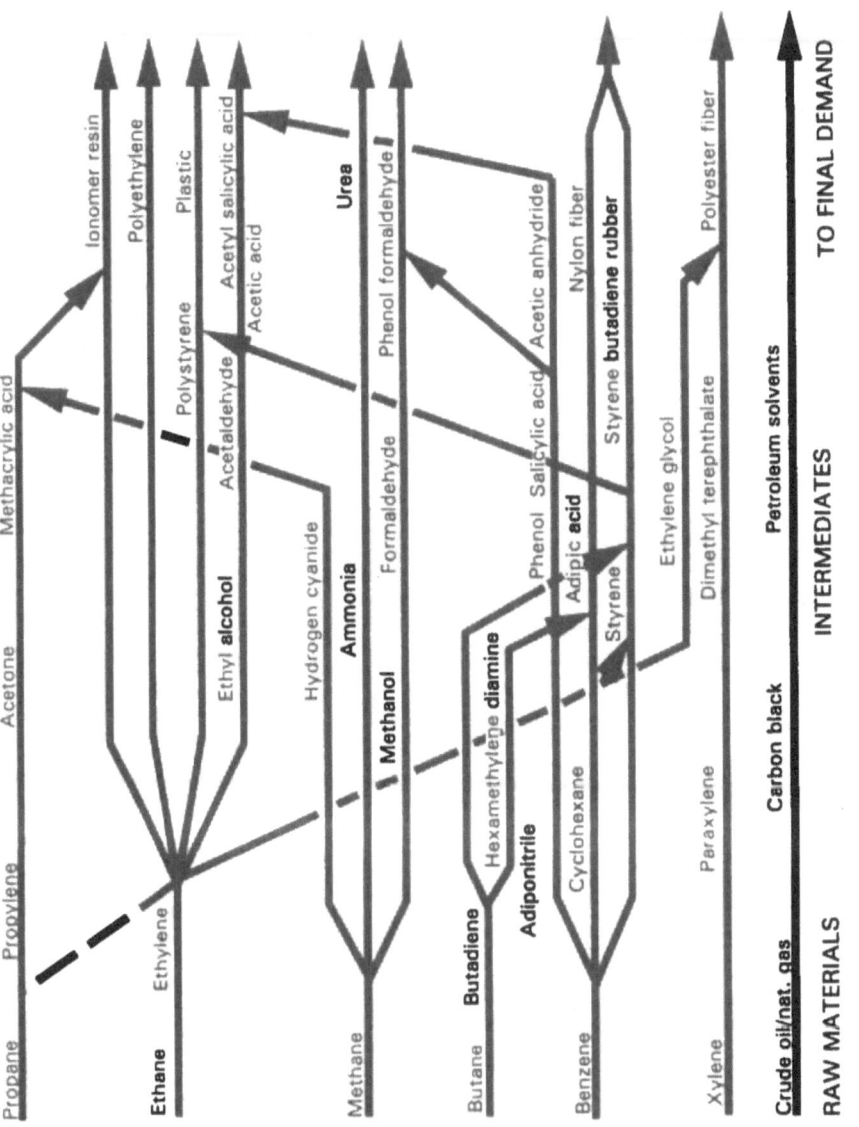

FIGURE 5.2   Raw materials, intermediates, and products of the petrochemical industry.

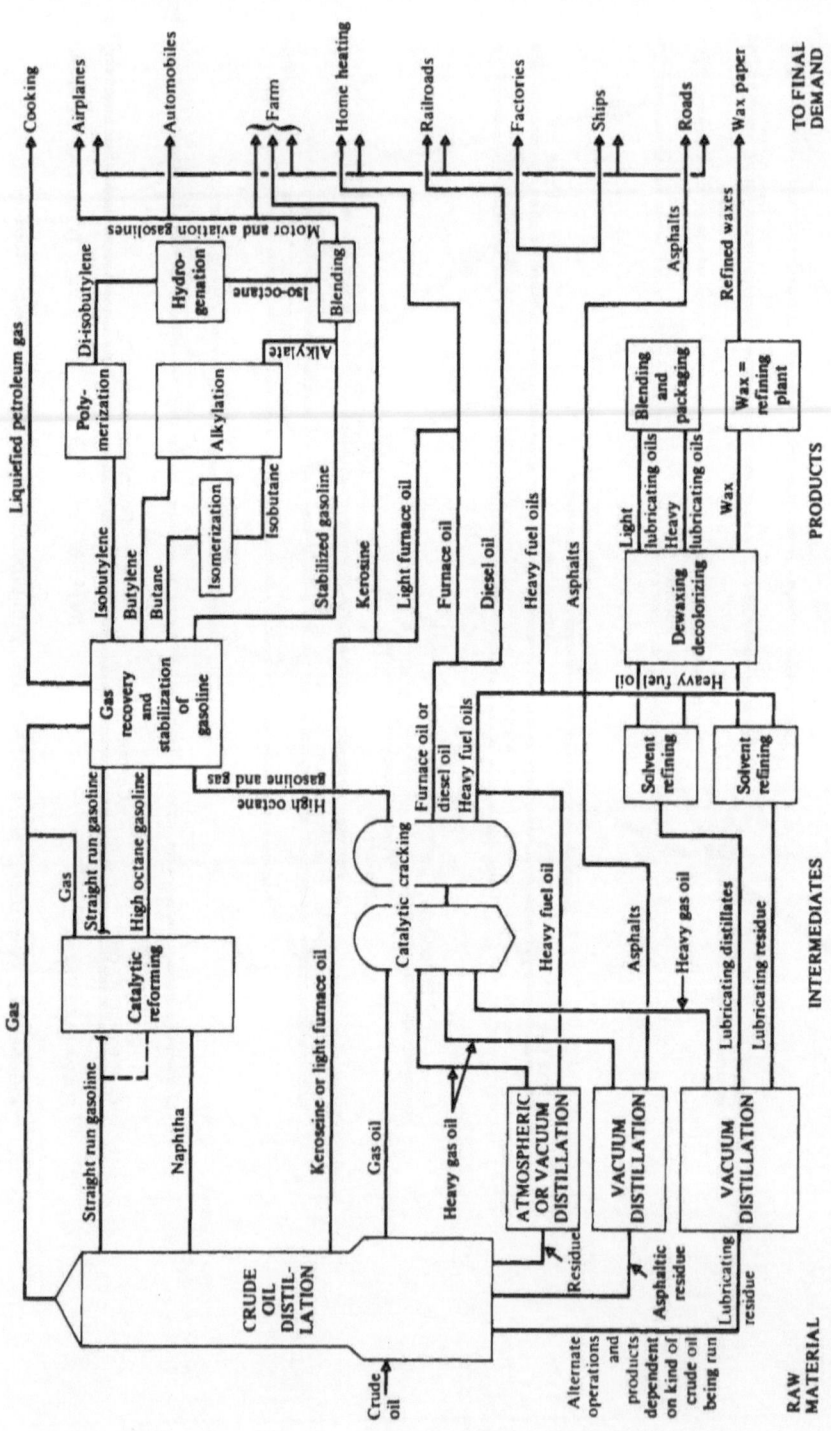

FIGURE 5.3   The petroleum refining industry. (Flow chart courtesy of the American Petroleum Institute.)

**TABLE 5.1** Some End-Use
Products of the CPI

---

Agricultural chemicals
Automotive chemicals
Biochemicals
Diagnostic aids
Drilling-mud additives
Electronic chemicals
Flavors and fragrances
Flotation reagents
Food and food additives
Foundry additives
Industrial cleaning products
Laboratory chemicals
Metal-finishing chemicals
Paint additives
Paper additives
Petroleum additives
Pharmaceuticals
Photographic chemicals
Plastics additives
Rubber-processing chemicals
Textile chemicals
Water treatment chemicals

---

*Source: Chemical and Engineering News,* Facts and Figures issue, June 29, 1998, pp. 44–47.

**5.1.1.2 Differentiated and Undifferentiated Products.** An undifferentiated chemical product is one that has a specific chemical formula and particular physical specifications regardless of who produces it. An example is ethylene glycol defined by the specifications found in Table 5.2. Most of the high-volume chemicals produced by the chemical industry ($H_2SO_4$, $O_2$, $N_2$, ethylene, $NH_3$, benzene, etc.) can be classified as undifferentiated.

A product is classified as differentiated when its producer claims that it has real or imputed difference compared with a similar product made by a competitor, Differentiation can be accomplished by various means, and such products are usually sold on the basis of performance in use.

1. A polymer such as polyethylene or polypropylene can be given different properties by changing molecular weight distribution or crystallinity, or by modifying the physical processing steps.
2. A basic chemical such as ethylene glycol may be sold as a differenti-

**TABLE 5.2** ASTM Specification for Ethylene Glycol (for use in the preparation of surface coatings)

| Apparent specific gravity | |
|---|---|
| At 20/20°C | 1.1151–1.1156 |
| At 25/25°C | 1.1129–1.1134 |
| Color, Pt-Co units, max. | 15 |
| Distillation range, 760 mmHg | |
| Initial boiling point, °C min. | 193 |
| Dry point, °C max. | 204 |
| Water, wt % max. | 0.2 |
| Acidity as acetic acid, wt % max. | 0.005 |
| Diethylene glycol, wt %, max. | 1.0 |
| Iron, ppm, max. | 1.0 |

ated product (antifreeze) by adding compounds that inhibit rust and provide "punctureproofing" to the customer's radiator.

3. Some products that have essentially the same chemical formulation may be classified as differentiated because their method of delivery is unique. Air Products (Chapter 7) did this with industrial gases by providing on-site plants and eliminating railcar or tank-truck delivery. Technical service in the form of providing formulations, process design for the customer's operations, and subsequent troubleshooting has been used by companies like Du Pont to make their pigment business unique.

**5.1.1.3 Annual Production Volume.** This is a common means of ordering and is used by *Chemical and Engineering News* in publishing the top 50 chemicals. A classification into high, intermediate, low and very low volume can be made as follows:

| Volume | U.S. Annual Production ($\times 10^6$ lb/yr) |
|---|---|
| High | >1000 |
| Intermediate | 100–1000 |
| Low | 10–100 |
| Very low | <10 |

**TABLE 5.3** The Most Significant Organic
Chemicals: 1997

| Chemical | Annual production (lb × $10^5$) |
|---|---|
| Ethylene | 51,078 |
| Propylene | 29,335 |
| Ethylene dichloride | 28,366 |
| Benzene | 17,219 |
| Vinyl chloride | 15,988 |
| Urea | 15,530 |
| Methanol | 13,214 |
| Ethylbenzene | 12,691 |
| Styrene | 11,366 |
| Ethylene oxide | 8,241 |
| p-Xylene | 7,789 |
| Cumene | 6,119 |
| 1–3-Butadiene | 4,107 |
| Acrylonitrile | 3,291 |
| Isopropyl alcohol | 1,478 |
| Aniline | 1,339 |
| o-Xylene | 1,091 |
| 2-Ethylhexanol | 768 |

*Sources:* CMAI, *Chemical and Engineering News,* Facts and
Figures issue, June 29, 1998, p. 43.

5.1.1.4 Product Price. A somewhat arbitrary classification of chemicals
into high, intermediate, and low price is as follows:

| Price | U.S. price per pound |
|---|---|
| High | $5 |
| Intermediate | $0.50–$5 |
| Low | <0.50 |

An ordering by gross sales is given in Table 5.3. Figure 5.4 is a generalized
log–log plot of price and production for a range of CPI products. Pharmaceuticals, as well as fine and specialty chemicals, are usually high priced and produced at low volume consistent with market needs. Pseudocommodities are
differentiated intermediate-priced materials produced in millions of pounds per
year. Commodity chemicals like sulfuric acid, oxygen, nitrogen, and ammonia
are undifferentiated, are produced in large tonnages, and sell at low prices.

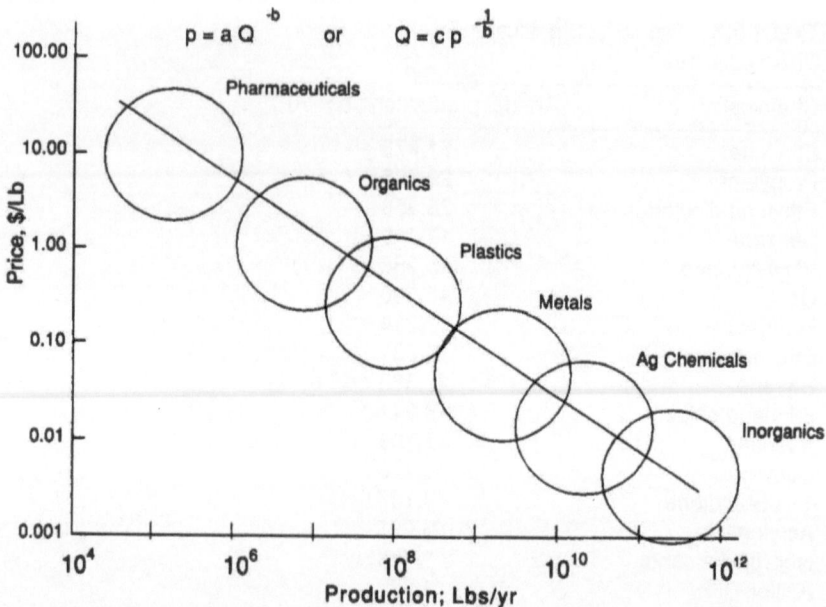

FIGURE 5.4  Price versus production in the United States, 1997.

The following generalized equation relating price and production can be developed by using regression analysis:

$$p = aQ^{-b} \qquad \text{or} \qquad Q = cp^{-1/b}$$

As might be expected, some of the high-volume inorganics (e.g., chlorine and sodium hydroxide) have the lowest prices, and low-volume chemicals like the drug tetracycline hydrate and the rare earth chemical europium in oil, have the high prices of $0.90 and $71.50 per gram, respectively.

5.1.1.5  Chemical Composition.  Chemicals are often separated into two major groups, inorganics and organics. Within the past 5 years, eight additional groups have been designated. They are

| | |
|---|---|
| Aerosols | Pesticides |
| Fertilizers | Plastics |
| Minerals | Synthetic fibers |
| Paints and coatings | Synthetic rubber |

Within each of these eight groups are subgroups classified by further chemical similarities. They are listed in Table 5.4 along with some of the major products

TABLE 5.4 Classification of Chemicals by Composition

| | |
|---|---|
| 1. Inorganics | 4. Plastics |
|   a. Acids |   a. Thermosetting resins |
|     Hydrochloric |     Epoxy |
|     Nitric |     Melamine |
|     Phosphoric |     Phenolic |
|     Sulfuric |     Polyester |
|   b. Chlorine and alkalies |     Urea |
|     Chlorine gas |   b. Thermoplastic resins |
|     Sodium carbonate |     Polyethylene |
|     Sodium hydroxide |     Polypropylene |
|   c. Industrial gases |     Styrene polymers |
|     Carbon dioxide |     Polyamide, nylon type |
|     Hydrogen |     Polyvinyl chloride and co |
|     Nitrogen |     polymers |
|     Oxygen |     Thermoplastic polyester |
|   d. Other | 5. Synthetic rubber |
|     Aluminum sulfate |     Styrene–butadiene rubber (SBR) |
|     Phosphorus |     Polybutadiene |
|     Sodium chlorate |     Ethylene–propylene |
|     Sodium silicate |       Nitrile |
|     Titanium dioxide |     Polychloroprene (Neoprene) |
| 2. Organics |     Other |
|   Acrylonitrile | 6. Synthetic fibers |
|   Aniline |   a. Noncellulosic fibers |
|   Benzene |     Acrylic |
|   1,3-Butadiene |     Nylon |
|   Cumene |     Olefin |
|   Ethylbenzene |     Polyester |
|   Ethylene |   b. Cellulosic fibers |
|   Ethylene dichloride |     Acetate |
|   Ethylene oxide |     Rayon |
|   2-Ethylhexanol | 7. Fertilizers |
|   Isopropyl alcohol |   a. Nitrogen products |
|   Propylene |     Ammonia |
|   Styrene |     Ammonium nitrate |
|   Urea |     Ammonium sulfate |
|   o-Xylene |     Urea |
|   p-Xylene |     Nitrogen solutions |
| 3. Minerals |   b. Phosphate products |
|   Bromine |     Diammonium phosphate |
|   Calcium chloride |     Monoammonium phosphate |
|   Lime |     Concentrated superphosphate |
|   Lithium |     Phosphate rock |
|   Phosphate rock |     Phosphoric acid |
|   Potash |   c. Potash products: |
|   Sodium carbonate |     potassium chloride |
|   Sulfur | 8. Aerosols |
| | 9. Pesticides |
| | 10. Paints and coatings |

Source: Chemical and Engineering News, Facts and Figures issue, June 29, 1998, pp. 44–47.

in each. This classification scheme is used by *Chemical and Engineering News* in their articles on CPI performance.

### 5.1.1.6 Product Function.

Chemicals are often classified by the function they perform. Examples of such functions are presented in Table 5.5.

There are other methods of classifying the products of the CPI. Frequently combinations of classifications are used to give insights into various aspects of the industry. One of the more useful of these is a classification obtained by combining production volume and differentiation to yield the four product classifications shown in Table 5.6. This separation into commodity, pseudocommodity, fine, and specialty chemicals is particularly useful in examining differences in research and development, manufacturing methods, marketing, and finance for

**TABLE 5.5**   Some Functional
Chemical Product Groups

Antioxidants
Biocides
Catalysts
Chelates
Corrosion inhibitors
Flame retardants
Heat transfer fluids
Hydraulic fluids
Octane enhancers
Pigments
Solvents
Thickeners
Ultraviolet absorbers
Viscosity index improvers

**TABLE 5.6**   Chemical Product Classification by a Combination of Volume and Differentiation

| Annual production volume | Undifferentiated | Differentiated |
|---|---|---|
| High | True commodities (e.g., sulfuric acid, benzene, oxygen) | Pseudocommodities (e.g., carbon black, polyethylene, antifreeze) |
| Low | Fine chemicals (e.g., aspirin, vitamin C, food colorings) | Specialty chemicals (e.g. pharmaceuticals, formulated pesticides) |

different chemicals. This separation is used in Chapter 8 to discuss the general characteristics of the CPI.

Several of the classifications are interrelated. For example, price is closely associated with product differentiation and product volume. Commodity chemicals tend to be high in volume and low in price, while specialty chemicals are low-volume, high-priced goods. Classification interrelationships often change the magnitude or meaning of a group within a classification. For example, a high-priced specialty and a high-priced commodity would fall into different price ranges. Differentiated chemicals are those that have particular functions.

Classification and ranking by annual production volume is used in the Facts and Figures issues of *Chemical and Engineering News*. This publication up until 1993 identified the top 50 chemicals by annual production volume. Now the chemicals are listed separately in each of the eight classifications mentioned earlier in this section on chemical composition. Table 5.7, a compilation of the

**TABLE 5.7** The Top 20 Chemicals Produced in the United States Ranked by Production Quantity

| Rank | | | Production (billions of pounds) | | Average annual change (%) | |
|---|---|---|---|---|---|---|
| 1997 | 1996 | Chemical | 1997 | 1996 | 1996–1997 | 1987–1997 |
| 1 | 1 | Sulfuric acid | 95.58 | 95.54 | 0% | 2% |
| 2 | 2 | Nitrogen | 82.90 | 66.29 | 25% | 4% |
| 3 | 3 | Oxygen | 64.89 | 62.04 | 5% | 6% |
| 4 | 4 | Ethylene | 51.08 | 49.10 | 4% | 4% |
| 5 | 5 | Lime | 42.56 | 42.12 | 1% | 3% |
| 6 | 6 | Ammonia | 38.39 | 35.52 | 8% | 2% |
| 7 | 10 | Propylene | 29.34 | 26.05 | 13% | 4% |
| 8 | 7 | Ethylene dichloride | 28.37 | 27.65 | 3% | 8% |
| 9 | 9 | Chlorine | 27.37 | 26.34 | 4% | 2% |
| 10 | 8 | Phosphoric acid | 26.83 | 26.42 | 2% | 2% |
| 11 | 12 | Sodium carbonate | 22.93 | 22.49 | 2% | 3% |
| 12 | 11 | Sodium hydroxide | 22.74 | 23.13 | −2% | 0% |
| 13 | 13 | Nitric acid | 18.15 | 18.41 | −1% | 2% |
| 14 | 16 | Benzene | 17.22 | 15.59 | 10% | 4% |
| 15 | 14 | Ammonium nitrate | 16.52 | 18.05 | −8% | 2% |
| 16 | 17 | Vinyl chloride | 15.99 | 15.29 | 5% | — |
| 17 | 15 | Urea | 15.53 | 17.06 | −9% | 0% |
| 18 | 19 | Methanol | 13.22 | 11.73 | 13% | — |
| 19 | 20 | Ethyl benzene | 12.69 | 10.36 | 22% | 3% |
| 20 | 18 | Styrene | 11.37 | 11.81 | −4% | 4% |

*Sources:* CMAI; *Chemical and Engineering News*, Facts and Figures issue, June 29, 1998.

top 20 inorganic and organic chemicals produced in the United States based on 1996 and 1997 production figures, specifically excludes fibers and plastics. Some initial comments may be made about these top 20 chemicals:

1. The chemicals are all high-volume products, having annual production rates from 11 billion to about 100 billion pounds.
2. All the top 20 are undifferentiated.
3. The chemicals listed as the top 20 are consumed almost exclusively as intermediates in making other chemicals.
4. The top 20 are split almost equally between inorganics and organics, although inorganics dominate the top of the list.

### 5.1.2   SIC 29: Petroleum Refining and Related Industries

SIC 29 includes establishments that are primarily engaged in petroleum refining, in manufacturing paving and roofing materials, as well as in compounding lubricating oils and greases from purchased materials. It does not include establishments that manufacture and distribute gas to customers nor those primarily engaged in producing coke and by-products classified in SIC 33.

The petroleum refining industry is far more homogeneous and coherent in raw material needs and markets served than the chemicals and allied products segment (SIC 28). The major raw material supply for SIC 29 is natural gas and crude oil from wells. Historically, the early main producers in the United States were in the western Pennsylvania oil fields, following Drake's development of the drilled oil well in 1859. The chief petroleum product then was kerosene for illumination, while the unwanted by-product, gasoline, was poured into creeks to the consternation of the local public residents. With the advent of the mass-produced automobiles by Henry Ford at the turn of the century, gasoline became a desired product. Hydrocarbon oil was developed as a lubricant to replace sperm whale oil. As the Pennsylvania fields became depleted, crude oil was obtained from Texas, Oklahoma, Louisiana, California, other parts of the United States, and Canada. On a global basis, the world's petroleum suppliers are in the Persian Gulf, the Far East, Venezuela, Mexico, Africa, and the coastal areas of Norway and Scotland. With the breakup of the Soviet Union, new oil fields are being developed in Kazakhstan, and around the Caspian and Black Seas. The United States imports about 50–60% of its needs in crude oil and petroleum products from abroad. To reduce the dependence on oil and improve our balance of trade, other energy sources have been sought. The most abundant source is coal, but there is a need to develop more economical processes for its conversion. Solar, geothermal, oil shale, tar sands, and the wind have been explored as potential energy sources, but they do not yet contribute substantially to the total energy picture.

The modern refinery produces a large slate of products (see Figure 5.3), which find use in heating fuels, lighting, shipping, lubricating, paving, and many other areas. The value of refinery shipments and volume produced are shown in Table 5.8. The costs are considerably lower than the retail value of the products because taxes, transportation, and marketing costs are not included. In addition to crude petroleum, the United States imports large quantities of re-

TABLE 5.8   U.S. Petroleum Products

| Product | Volume of petroleum products to end users (million gallons/day)[a] | | | | |
|---|---|---|---|---|---|
| | 1983 | 1985 | 1990 | 1995 | 1997 |
| Motor gasoline | 51.1 | 57.5 | 60.3 | 55.9 | 61.4 |
| Aviation gasoline | 0.4 | 0.3 | 0.2 | 0.2 | 0.2 |
| Kerosene-type jet fuel | 30.8 | 34.6 | 39.9 | 45.7 | 49.5 |
| Propane (consumer grade) | 3.1 | 3.7 | 2.7 | 3.2 | 3.2 |
| Kerosene-type jet fuel | 0.2 | 0.3 | 0.2 | 0.6 | 0.3 |
| No. 1 distillate | 0.5 | 0.5 | 0.5 | 0.3 | 0.3 |
| No. 2 distillate | 27.0 | 29.9 | 25.9 | 24.9 | 25.1 |
| No. 4 fuel | 0.7 | 0.5 | 0.8 | 0.5 | 0.3 |
| Residual fuel oil | 28.3 | 25.2 | 25.9 | 11.6 | 13.7 |
| Others[b] | 0.10 | 0.11 | 0.12 | 0.13 | 0.14 |

| Product | Prices (cents/gallon excluding taxes)[c] | | | | |
|---|---|---|---|---|---|
| | 1983 | 1985 | 1990 | 1995 | 1997 |
| Motor gasoline | 95.4 | 91.2 | 88.3 | 76.5 | 83.9 |
| Aviation gasoline | 125.5 | 120.1 | 112.0 | 100.5 | 113.8 |
| Kerosene-type jet fuel | 87.8 | 79.6 | 76.6 | 54.0 | 61.2 |
| Propane (consumer grade) | 70.9 | 71.7 | 74.5 | 49.2 | 55.2 |
| Kerosene-type jet fuel | 96.1 | 103.0 | 92.3 | 58.9 | 74.4 |
| No. 1 distillate | 96.2 | 88.0 | 81.9 | 62.0 | 68.5 |
| No. 2 distillate | 83.9 | 79.9 | 72.6 | 56.0 | 64.1 |
| No. 4 fuel | 76.6 | 77.3 | 62.2 | 50.5 | 56.5 |
| Residual fuel oil | 65.1 | 61.0 | 44.4 | 39.2 | 42.3 |

[a] From *Petroleum Marketing Monthly*, Office of Oil and Gas, U.S. Department of Energy, Washington DC 20585, December 1995.
[b] Includes asphalt, road oil, waxes, and lubricants.
[c] From *Petroleum Marketing Monthly*, Office of Oil and Gas, U.S. Department of Energy, Washington DC 20585, December 1997.

fined products from refineries overseas—for example, residual fuel oil from the Caribbean area.

Most of the products from a petroleum refinery are true commodities, specified by their chemical compositions and physical characteristics rather than by any special effectiveness in use. They are mostly undifferentiated products sold in high volume and at low price. Innovations take place more often in the refinery processes than in the introduction of new products.

Liquefied petroleum gas (LPG) consists principally of propane and butane. It is obtained from condensates in natural gas and in refinery streams. LPG is used mainly as a source of heat in homes and farms as well as a petrochemical fuel stock. It is also used as a fuel in commercial, industrial, and utility heating as well as a fuel for taxis, tractors, and light trucks. It is clean, convenient, and nonpolluting material but must be stored in pressurized containers.

Motor gasoline is the most valuable refinery product, consisting of hydrocarbons with 5–12 carbon atoms per molecule and a boiling point of 120–400°F. It is obtained from distillation of crude oil, from synthesis of smaller molecules by alkylation and polymerization, as well as from breaking larger molecules by catalytic cracking, hydrocracking, and by molecular rearrangement to increase the octane number by reforming. Motor gasoline is cheaper and has a higher heat value, Btu per pound, than any other competitive product (e.g., methanol, ethanol).

Diesel fuel has a boiling point of 375–725°F and is used for a fuel for railroads, trucks, buses, and a few automobiles. It is atomized and burned in high-pressure cylinders without spark ignition; consequently, it must have proper viscosity to prevent leakage and to economize on pumping power. The cetane and diesel indices are designed to specify ease of ignition start as well as reduction of odor and fumes.

Distillate fuel oil, particularly important for home heating, has a boiling point range similar to that of diesel fuel. It is obtained by the distillation of crude oil.

Residual fuel oil, the bottom fraction of the distillation column, contains most of the undesirable sulfur, metals, and carbon deposits from crude oil. It has a very high viscosity and can solidify in cold weather. Its use requires high-pressure nozzles and heated containers, making it unsuited for home heating. It is primarily used in industrial, utility, and marine boilers.

Lubricating oil is a highly refined product used to reduce friction and wear in motors and machinery. It must have a viscosity sufficient to prevent its being squeezed and ejected from between moving parts but not so high that its use consumes excessive power. It must perform over wide temperature ranges, to ensure easy starting in cold weather and sufficient lubrication in hot weather. Resistance to rust and formation of gum and sludge are other important qualities.

Asphalt is a thermoplastic material recovered from distillation bottoms. There is some evidence that it may have been used by Noah to seal the ark. Today, the major uses are in paving for highways and for roofing. Asphalt, depending on its geologic origin, may exhibit Newtonian or non-Newtonian properties.

Petroleum coke, obtained by subjecting the distillation bottoms to pyrolitic polymerization and thermal decomposition in a delayed or fluid coker, is used in the manufacture of calcium and silicon carbides, electrodes for aluminum manufacturing, and graphite. It also finds use as a utility fuel and as a construction material in foundry and blast furnaces.

The specification of petroleum products is controlled by the American Society for Testing and Materials (ASTM). Some of the important physical specifications are API gravity, vapor pressure, and ASTM distillation.

*API gravity.* The American Petroleum Institute (API) defines this by the following equation:

$$\text{degrees API} = \frac{141.5}{\text{specific gravity}} - 131.5$$

The larger the API gravity, the lighter the product. A material with an API gravity of less than 10 will sink in water.

*Vapor pressure.* Reid vapor pressure is measured at 100°F (37.8°C) and gives an indication of a tendency for a fuel to vapor-lock.

*ASTM distillation.* This is expressed by the curve for boiling point versus percentage remaining, including the initial boiling point, the temperature at each 10% incremental distillate, and the end point.

From a consumer's viewpoint, the important performance specifications for petroleum products are as follows:

*Flash and fire points.* These should be sufficiently high to assure safe handling.

*Viscosity and viscosity index.* The viscosity should be such as to prevent excessive loss and sufficiently high to withstand squeezing out within operating and test temperature ranges.

*Cloud and pour point.* These tests measure the cold-temperature handling characteristics of the fuel.

*Octane numbers.* The research method involves performance on a research engine at 600 rpm, simulating city driving, and the motor method at 900 rpm, simulating highway performance of the fuel.

Tests are also performed to determine sulfur, carbon, water, sludge gum, and metal content, as these are important parts of the specification for certain petroleum products.

**TABLE 5.9**  The Top 50 U.S. Chemical Producers, Ranked According to Chemical Sales in 1997

| Rank 1997 | Rank 1996 | Company | Chemical sales 1997 (millions) | Change from 1996 (%) | Chemical sales as percentage of total sales | Industry classification |
|---|---|---|---|---|---|---|
| 1 | 1 | Du Pont | $21,295.0 | 3.8 | 47.2 | Diversified |
| 2 | 2 | Dow Chemical | $19,056.0 | 0.4 | 95.2 | Basic chemicals |
| 3 | 3 | Exxon | $12,195.0 | 6.7 | 8.9 | Petroleum |
| 4 | 6 | General Electric | $6,695.0 | 3.2 | 7.4 | Diversified |
| 5 | 7 | Union Carbide | $6,502.0 | 6.5 | 100.0 | Basic chemicals |
| 6 | 8 | Amoco | $5,941.0 | 4.3 | 16.4 | Petroleum |
| 7 | 11 | Huntsman Corp. | $5,000.0 | 11.1 | 100.0 | Basic chemicals |
| 8 | 5 | Celanese | $4,950.0 | 3.4 | 100.0 | Basic chemicals |
| 9 | 10 | BASF | $4,860.0 | 3.3 | 100.0 | Basic chemicals |
| 10 | 13 | Praxair | $4,735.0 | 6.4 | 100.0 | Basic chemicals |
| 11 | 14 | Shell Oil | $4,725.0 | 9.8 | 16.7 | Petroleum |
| 12 | 9 | Eastman Chemical | $4,678.0 | −2.2 | 100.0 | Basic chemicals |
| 13 | 16 | ICI Americas | $4,615.0 | 15.4 | 100.0 | Specialty chemicals |
| 14 | 12 | Occidental Petroleum | $4,349.0 | −3.0 | 54.3 | Petroleum |
| 15 | 615 | Allied Signal | $4,254.0 | 6.0 | 29.4 | Diversified |
| 16 | 20 | Air Products | $4,122.0 | 12.3 | 88.9 | Basic chemicals |
| 17 | 19 | Ashland | $4,047.0 | 9.5 | 28.5 | Petroleum |
| 18 | 17 | Rohm & Haas | $3,999.0 | 0.4 | 100.0 | Basic chemicals |
| 19 | 18 | Arco Chemical | $3,995.0 | 1.0 | 100.0 | Basic chemicals |
| 20 | 22 | Chevron | $3,632.0 | 6.1 | 8.9 | Petroleum |
| 21 | 23 | Phillips Petroleum | $3,528.0 | 10.8 | 23.2 | Petroleum |
| 22 | 21 | W.R. Grace | $3,312.8 | 0.0 | 100.0 | Specialty chemicals |
| 23 | 25 | Mobil | $3,251.0 | 7.5 | 4.9 | Petroleum |
| 24 | 4 | Monsanto | $3,126.0 | −57.0 | 41.6 | Life sciences |
| 25 | 24 | Millennium | $3,048.0 | 0.3 | 100.0 | Basic chemicals |
| 26 | 26 | IMC Global | $2,988.6 | 0.3 | 100.0 | Agrochemicals |
| 27 | — | Solutia | $5,969.0 | −0.3 | 100.0 | Basic chemicals |
| 28 | 27 | Dow Corning | $2,643.5 | 4.4 | 100.0 | Specialty chemicals |
| 29 | 30 | Akzo Nobel | $2,605.0 | 9.8 | 100.0 | Basic chemicals |
| 30 | 28 | National Starch | $2,455.4 | 0.0 | 100.0 | Specialty chemicals |
| 31 | 29 | Lyondall Petrochemical | $2,346.0 | −2.7 | 100.0 | Basic chemicals |
| 32 | 32 | FMC | $2,254.2 | −1.7 | 52.9 | Machinery |
| 33 | 25 | Solvay America | $2,200.0 | 10.0 | 100.0 | Basic chemicals |
| 34 | 36 | Witco | $2,187.4 | −3.4 | 100.0 | Specialty chemicals |
| 35 | — | American Home Products | $2,119.4 | 6.6 | 14.9 | Pharmaceuticals |
| 36 | 37 | Elf Atochem | $2,000.0 | 11.1 | 100.0 | Basic chemicals |
| 37 | 34 | Hercules | $1,866.0 | −9.4 | 100.0 | Basic chemicals |
| 38 | 60 | Potash Corp. Saskatchewan | $1,826.3 | 81.4 | 100.0 | Basic chemicals |
| 39 | 52 | Henkel | $1,800.0 | 41.9 | 100.0 | Specialty chemicals |
| 40 | 39 | Morton International | $1,680.9 | 4.3 | 71.8 | Specialty chemicals |

| Chemical operating profits (millions) | Change from 1996 (%) | Chemical operating profits/total profits (%) | Operating profit margin (%) | Identifiable chemical sales (millions) | Chemical assets as percentage of total assets | Operating return of chemical sales (%) |
|---|---|---|---|---|---|---|
| $3,554.0 | −6.4 | 66.4 | 16.7 | $19,036.0 | 52.0 | 18.7 |
| $3,031.0 | −9.8 | 111.2 | 15.9 | $17,951.0 | 74.7 | 16.9 |
| NA | NA | NA | NA | NA | NA | NA |
| $1,476.0 | 0.7 | 12.5 | 22.0 | NA | NA | NA |
| $1,045.0 | 13.5 | 100.0 | 16.1 | $6,964.0 | 100.0 | 15.0 |
| $732.0 | −20.8 | 16.2 | 12.3 | $6,351.0 | 20.8 | 11.5 |
| NA | NA | NA | NA | NA | NA | NA |
| $471.0 | −2.7 | 100.0 | 9.5 | $2,743.0 | 100.0 | 17.2 |
| NA | NA | NA | NA | NA | NA | NA |
| $848.0 | 15.8 | 100.0 | 17.9 | $7,810.0 | 100.0 | 10.9 |
| $696.0 | 134.3 | 22.1 | 14.7 | $5,342.0 | 18.0 | 13.0 |
| $568.0 | −14.3 | 100.0 | 12.1 | $5,778.0 | 100.0 | 9.8 |
| NA | NA | NA | NA | NA | NA | NA |
| $497.0 | −27.2 | 94.1 | 11.4 | $5,486.0 | 35.9 | 9.1 |
| $542.0 | −4.4 | 33.1 | 12.7 | $4,100.0 | 29.9 | 13.2 |
| $719.4 | 18.9 | 99.2 | 17.5 | $5,993.1 | 89.6 | 12.0 |
| $160.0 | −5.3 | 29.4 | 4.0 | $1,558.0 | 20.0 | 10.3 |
| $617.0 | −1.6 | 100.0 | 45.4 | $3,900.0 | 100.0 | 15.8 |
| $256.0 | −52.6 | 100.0 | 6.4 | $4,116.0 | 100.0 | 6.2 |
| $362.0 | 21.1 | 6.1 | 10.0 | $3,518.0 | 9.9 | 10.3 |
| $371.0 | 7.8 | 18.1 | 10.5 | $2,638.0 | 19.5 | 14.1 |
| $646.4 | 5.1 | 100.0 | 19.5 | NA | NA | NA |
| $493.0 | 44.2 | 7.0 | 15.2 | $3,111.0 | 7.1 | 15.8 |
| $762.0 | 9.3 | 60.3 | 24.4 | $4,520.0 | 42.0 | 16.9 |
| $449.0 | 58.7 | 100.0 | 14.7 | $4,326.0 | 100.0 | 10.4 |
| $486.7 | −0.3 | 100.0 | 16.3 | $4,673.9 | 100.0 | 10.4 |
| $290.0 | 28.9 | 100.0 | 9.8 | $2,768.0 | 100.0 | 10.5 |
| $380.7 | −3.9 | 100.0 | 14.4 | NA | NA | NA |
| NA | NA | NA | NA | $0.0 | NA | NA |
| $371.3 | 2.9 | 102.9 | 15.1 | NA | NA | NA |
| $534.0 | 58.5 | 100.0 | 22.8 | $1,559.0 | 100.0 | 34.3 |
| $248.0 | −27.3 | 6390.0 | 11.0 | $2,436.0 | 59.2 | 10.2 |
| NA | NA | NA | NA | NA | NA | NA |
| $235.8 | 37.6 | 100.0 | 10.8 | $2,297.7 | 100.0 | 10.3 |
| $429.9 | 27.3 | 15.3 | 20.3 | $4,763.9 | 22.9 | 9.0 |
| NA | NA | NA | NA | NA | NA | NA |
| $228.0 | −48.3 | 100.0 | 12.2 | $2,411.0 | 100.0 | 9.5 |
| $257.0 | 38.0 | 100.0 | 14.1 | $3,019.8 | 100.0 | 8.5 |
| NA | NA | NA | NA | NA | NA | NA |
| $268.6 | 21.0 | 66.1 | 16.0 | $1,540.0 | 54.9 | 17.4 |

**TABLE 5.9**   Continued

| Rank 1997 | Rank 1996 | Company | Chemical sales 1997 (millions) | Change from 1996 (%) | Chemical sales as percentage of total sales | Industry classification |
|-----------|-----------|---------|-------------------------------|----------------------|---------------------------------------------|-------------------------|
| 41 | 40 | Lubrizol | $1,673.8 | 4.8 | 100.0 | Specialty chemicals |
| 42 | 38 | PPG Industries | $1,647.0 | 2.2 | 22.3 | Glass products |
| 43 | — | Ciba Specialty Chemicals | $1,565.0 | NA | 100.0 | Specialty chemicals |
| 44 | 46 | Zeneca | $1,550.0 | 10.7 | 45.6 | Life sciences |
| 45 | 42 | Compton & Knowles | $1,539.5 | 1.3 | 83.2 | Specialty chemicals |
| 46 | 47 | Engelhard | $1,513.1 | 10.7 | 41.7 | Diversified |
| 47 | 50 | Nalco Chemical | $1,433.7 | 10.0 | 100.0 | Specialty chemicals |
| 48 | 43 | CF Industries | $1,431.6 | −2.5 | 100.0 | Agrochemicals |
| 49 | 45 | Cabot | $1,430.3 | −0.3 | 87.7 | Basic chemicals |
| 50 | 44 | International Flavors | $1,426.8 | −0.6 | 100.0 | Specialty chemicals |

*Source: Chemical and Engineering News, Facts and Figures issue, June 29, 1998, p. 50.*

### 5.1.3   Other CPI Sectors

Most chemical professionals are employed in the chemical and petroleum industries, especially in SIC 281, 286, and 29, but other industries in the CPI also employ chemists and chemical engineers and are important customers of the chemical and petroleum industries. Chemical professionals should familiarize themselves with the importance, size, and function of the other CPI sectors, becoming informed about their prosperity and changing needs. The Facts and Figures issues of *Chemical and Engineering News* list other CPI sectors, as well as the products and companies involved in these sectors. Occasionally, *Chemical Week* publishes feature articles on these sectors.

The most important sectors of the CPI are as follows:

Plastics materials and synthetic resins, synthetic rubber, cellulosic, and man-made fibers except glass (SIC 282)

Drugs (SIC 283)

Soap, detergents, and cleaning preparations; perfumes and cosmetics (SIC 284)

Paints, varnishes, lacquers, enamels, and allied products (SIC 285)

Agricultural chemicals (SIC 287)

Miscellaneous chemical products (SIC 289)

Rubber and miscellaneous plastic products (SIC 30)

| Chemical operating profits (millions) | Change from 1996 (%) | Chemical operating profits/total profits (%) | Operating profit margin (%) | Identifiable chemical sales (millions) | Chemical assets as percentage of total assets | Operating return of chemical sales (%) |
|---|---|---|---|---|---|---|
| $232.3 | −84.1 | 100.0 | 13.9 | $1,462.3 | 100.0 | 15.9 |
| $371.0 | −1.3 | 27.8 | 22.5 | $1,166.0 | 17.0 | 31.8 |
| NA | NA | NA | NA | NA | NA | NA |
| NA | NA | NA | NA | NA | NA | NA |
| $238.5 | 10.3 | 93.8 | 15.5 | $1,345.0 | 86.8 | 17.7 |
| $258.7 | 11.8 | 79.9 | 17.1 | $1,539.3 | 59.5 | 16.8 |
| $242.7 | 12.0 | 100.0 | 16.9 | $1,440.9 | 100.0 | 16.8 |
| $155.6 | −42.9 | 100.0 | 10.9 | $1,480.9 | 100.0 | 10.5 |
| $198.5 | −23.8 | 96.8 | 13.9 | $1,442.3 | 79.1 | 13.8 |
| $338.3 | 14.1 | 100.0 | 23.7 | $1,422.3 | 100.0 | 23.8 |

Stone, clay, and glass (SIC 32)
Pulp and paper (SIC 26)
Ferrous metals (SIC 331–3334)
Nonferrous metals SIC 3339–334)
Food processing (SIC 20)

## 5.2 COMPANIES OF THE CPI

### 5.2.1 The United States

*Chemical and Engineering News* has ranked the top chemical producers yearly according to annual chemical sales since 1968. Table 5.9 ranks these firms according to chemical sales in 1997. This list can be used to illustrate certain features of the chemical industry.

1. The companies in this list include a varied group. Only about 36% are classified as basic chemical makers. Sixteen percent are classified as petroleum companies. Diversified companies represent 17% of the list, and the remaining 31% are scattered among specialty chemicals agrochemicals, pharmaceuticals, and life sciences companies. For 30% of the top chemical producers, chemicals make up less than 50% of the sales.
2. The companies include some of the largest Fortune 500 companies in

TABLE 5.10  Top 50 Companies in the United States Ranked by Revenues: Dollar Amounts $\times 10^5$, Except Earnings per Share

| Rank 1997 | Rank 1996 | Company | Revenues Amount | Revenues Percentage change from 1996 | Profits Amount | Profits Rank | Profits Percentage change from 1996 | Assets Amount | Assets Rank | Stockholders' equity Amount | Stockholders' equity Rank | Market value[a] Amount | Market value[a] Rank |
|---|---|---|---|---|---|---|---|---|---|---|---|---|---|
| 1 | 1 | General Motors, Detroit | $178,174.0 | $5.8 | $6,698.0 | 5 | 35.0 | $228,888.0 | 13 | $17,506.0 | 18 | $54,243.8 | 40 |
| 2 | 2 | Ford Motor, Dearborn, MI | $153,627.0 | 4.5 | $6,920.0 | 4 | 55.6 | $279,097.0 | 8 | $30,734.0 | 5 | $73,923.0 | 21 |
| 3 | 3 | Exxon, Irving, TX | $122,379.0 | 2.5 | $8,460.0 | 1 | 12.6 | $96,064.0 | 30 | $43,660.0 | 1 | $158,783.6 | 4 |
| 4 | 4 | Wal-Mart Store, Bentonville, AR | $119,299.0 | 12.4 | $3,526.0 | 14 | 15.4 | $45,525.0 | 64 | $18,502.0 | 17 | $113,730.8 | 8 |
| 5 | 5 | General Electric, Fairfield, CT | $90,840.0 | 14.7 | $8,203.0 | 2 | 12.7 | $304,012.0 | 5 | $34,438.0 | 3 | $260,147.2 | 1 |
| 6 | 6 | IBM, Armonk, NY | $78,508.0 | 3.4 | $6,093.0 | 7 | 12.2 | $81,499.0 | 36 | $19,816.0 | 13 | $98,321.9 | 14 |
| 7 | 9 | Chrysler, Auburn Hills, MI | $61,147.0 | (0.4) | $2,805.0 | 29 | (20.5) | $60,418.0 | 50 | $11,362.0 | 41 | $28,368.8 | 75 |
| 8 | 8 | Mobil, Fairfax, VA | $59,978.0 | (17.0) | $3,272.0 | 19 | 10.4 | $43,559.0 | 67 | $14,461.0 | 15 | $58,409.6 | 36 |
| 9 | 10 | Phillip Morris, New York | $56,114.0 | 2.9 | $6,310.0 | 6 | 0.1 | $55,947.0 | 56 | $14,920.0 | 25 | $102,931.6 | 12 |
| 10 | 7 | AT&T, New York | $53,261.0 | (28.5) | $4,638.0 | 8 | (21.5) | $58,635.0 | 52 | $22,647.0 | 7 | $105,878.7 | 11 |
| 11 | 36 | Boeing, Seattle | $45,800.0 | 101.9 | ($178.0) | 480 | (116.3) | $38,024.0 | 76 | $12,953.0 | 32 | $50,868.1 | 43 |
| 12 | 11 | Texaco, White Plains, NY | $45,187.0 | 104.0 | $2,664.0 | 32 | 32.0 | $29,600.0 | 92 | $12,766.0 | 35 | $31,576.6 | 65 |
| 13 | 12 | State Farm Insurance, Bloomington, IN | $43,957.0 | 2.7 | $3,833.3 | 11 | 49.3 | $103,626.2 | 27 | $37,635.4 | 2 | NA | |
| 14 | 16 | Hewlett-Packard, Palo Alto, CA | $42,895.0 | 11.6 | $3,119.0 | 24 | 20.6 | $31,749.0 | 88 | $16,155.0 | 22 | $65,060.0 | 29 |
| 15 | 14 | E.I. Du Pont, Wilmington, DE | $41,304.0 | 4.1 | $2,405.0 | 35 | (33.6) | $42,942.0 | 68 | $11,270.0 | 43 | $77,018.9 | 19 |
| 16 | 17 | Sears Roebuck, Hoffman Estates, IL | $41,296.0 | 8.0 | $1,188.0 | 68 | (6.5) | $38,700.0 | 75 | $5,862.0 | 98 | $22,573.8 | 89 |
| 17 | 40 | Travelers Group, New York | $37,609.0 | 76.2 | $3,104.0 | 26 | 33.2 | $386,555.0 | 2 | $20,893.0 | 11 | $69,419.7 | 24 |
| 18 | 13 | Prudential Insurance, Newark, NJ | $37,073.0 | — | $610.0 | 144 | — | $259,482.0 | 12 | $19,718.0 | 14 | NA | |
| 19 | 15 | Chevron, San Francisco | $36,376.0 | (6.0) | $3,256.0 | 21 | 24.9 | $35,473.0 | 82 | $17,472.0 | 19 | $54,852.2 | 39 |
| 20 | 18 | Procter & Gamble, Cincinnati, OH | $35,764.0 | 1.4 | $3,415.0 | 16 | 12.1 | $27,544.0 | 101 | $12,046.0 | 38 | $113,634.8 | 9 |
| 21 | 20 | Citicorp, New York | $34,697.0 | 6.4 | $3,591.0 | 13 | (5.2) | $310,897.0 | 4 | $21,196.0 | 10 | $66,105.3 | 27 |
| 22 | 19 | Amoco, Chicago | $32,836.0 | 0.3 | $2,720.0 | 31 | (4.0) | $32,489.0 | 86 | $16,319.0 | 21 | $41,328.7 | 50 |
| 23 | 22 | K-Mart, Troy, MI | $32,183.0 | 2.4 | $249.0 | 291 | — | $13,558.0 | 174 | $5,434.0 | 110 | $8,204.9 | 216 |
| 24 | 30 | Merrill Lynch, New York | $31,731.0 | 26.9 | $1,906.0 | 44 | 17.7 | $292,819.0 | 7 | $8,329.0 | 62 | $29,152.2 | 71 |
| 25 | 33 | J.C. Penney, Plano, TX | $30,546.0 | 29.2 | $566.0 | 151 | 0.2 | $23,493.0 | 116 | $7,357.0 | 77 | $18,605.4 | 108 |

| | | | | | | | | | | | | | |
|---|---|---|---|---|---|---|---|---|---|---|---|---|---|
| 26 | 23 | American International Group, New York | $30,519.5 | 8.2 | $3,332.3 | 17 | 15.0 | $163,970.7 | 17 | $24,001.1 | 6 | $87,964.4 | 15 |
| 27 | 25 | Chase Corp. New York | $30,381.0 | 10.8 | $3,708.0 | 12 | 50.7 | $365,521.0 | 3 | $21,742.0 | 8 | $58,150.6 | 37 |
| 28 | 99 | Bell Atlantic, New York | $30,193.9 | 130.8 | $2,454.9 | 34 | 30.5 | $53,964.1 | 59 | $12,789.1 | 34 | $77,650.0 | 18 |
| 29 | 24 | Motorola, Schaumburg, IL | $29,784.0 | 6.5 | $1,180.0 | 69 | 2.3 | $27,278.0 | 102 | $13,272.0 | 31 | $32,601.3 | 63 |
| 30 | — | TIAA-CREF, New York | $29,348.4 | 18.5 | $1,226.6 | 65 | 31.3 | $214,295.6 | 14 | $5,776.6 | 99 | NA | |
| 31 | 21 | Pepsico, Purchase, NY | $29,292.0 | (7.4) | $2,142.1 | 38 | 86.4 | $20,101.0 | 132 | $6,936.0 | 83 | $63,351.5 | 31 |
| 32 | 26 | Lockheed Martin, Bethseda, MD | $28,069.0 | 4.4 | $1,300.0 | 64 | (3.5) | $28,361.0 | 97 | $5,176.0 | 121 | $22,465.4 | 90 |
| 33 | 29 | Fannie Mae, Washington, DC[b] | $27,776.9 | 10.9 | $3,055.8 | 28 | 12.1 | $391,672.7 | 1 | $13,792.0 | 28 | $67,592.9 | 26 |
| 34 | 27 | Dayton Hudson, Minneapolis | $27,757.0 | 9.4 | $751.0 | 118 | 62.2 | $14,191.0 | 167 | $4,375.0 | 150 | $18,606.7 | 107 |
| 35 | 162 | Morgan Stanley Dean Witter, New York | $27,132.0 | 200.5 | $2,586.0 | 33 | 171.8 | $302,287.0 | 6 | $13,956.0 | 27 | $46,842.4 | 46 |
| 36 | 28 | Kroger, Cincinnati, OH | $26,567.3 | 5.5 | $411.7 | 198 | 17.7 | $6,301.3 | 273 | ($784.8) | 499 | $11,473.7 | 174 |
| 37 | — | Lucent Technologies, Murray Hill, NJ | $26,360.0 | — | $541.0 | 159 | — | $23,811.0 | 115 | $3,387.0 | 180 | $78,366.7 | 17 |
| 38 | 43 | Intel, Santa Clara, CA | $25,070.0 | 20.3 | $6,945.0 | 3 | 34.7 | $28,880.0 | 95 | $19,295.0 | 16 | $125,741.0 | 6 |
| 39 | 32 | Allstate, Northbrook IL | $24,949.0 | 2.7 | $3,105.0 | 25 | 49.6 | $80,918.0 | 37 | $15,610.0 | 23 | $40,657.1 | 52 |
| 40 | 85 | SBC Communications, San Antonio, TX | $24,856.0 | 78.8 | $1,474.0 | 54 | (29.8) | $42,132.0 | 70 | $9,892.0 | 49 | $75,223.7 | 20 |
| 41 | 34 | United Technologies, Hartford, CT | $24,713.0 | 5.1 | $1,072.0 | 77 | 18.3 | $16,719.0 | 149 | $4,073.0 | 160 | $20,988.6 | 99 |
| 42 | 60 | Compaq Computer, Houston, TX | $24,584.0 | 35.8 | $1,855.0 | 47 | 41.3 | $14,631.0 | 164 | $9,429.0 | 54 | $36,052.1 | 57 |
| 43 | 35 | Metropolitan Life Insurance, New York | $24,374.0 | 4.9 | $1,203.0 | 66 | 41.0 | $201,907.0 | 15 | $14,007.0 | 26 | NA | |
| 44 | 50 | Home Depot, Atlanta | $24,155.7 | 23.7 | $1,160.0 | 72 | 23.7 | $11,188.9 | 202 | $7,097.9 | 81 | $50,019.3 | 44 |
| 45 | 31 | Conagra, Omaha | $24,002.1 | (3.3) | $615.0 | 142 | 225.6 | $11,277.1 | 199 | $2,471.7 | 236 | $15,365.5 | 131 |
| 46 | 48 | Merck, Whitehouse Station, NJ | $23,636.9 | 19.2 | $4,614.1 | 9 | 18.9 | $25,811.9 | 105 | $12,613.5 | 36 | $158,964.7 | 5 |
| 47 | 38 | Bankamerica Corp, San Francisco | $23,585.0 | 6.9 | $3,210.0 | 22 | 11.7 | $260,159.0 | 11 | $19,837.0 | 12 | $57,839.8 | 38 |
| 48 | 41 | GTE, Stamford, CT | $23,260.0 | 9.0 | $2,793.6 | 30 | (0.2) | $42,141.7 | 69 | $8,037.6 | 68 | $54,186.9 | 41 |
| 49 | 39 | Johnson & Johnson, New Brunswick, NJ | $22,629.0 | 4.7 | $3,303.0 | 18 | 14.4 | $21,453.0 | 125 | $12,359.0 | 37 | $100,630.6 | 13 |
| 50 | 65 | Safeway, Pleasanton, CA | $22,483.8 | 30.2 | $557.4 | 155 | 21.0 | $8,493.9 | 227 | $2,149.0 | 227 | $16,993.9 | 119 |

a As of March 18, 1998.
b The Federal National Mortgage Association.
Source: Fortune, April 27, 1998, p. F-1.

**TABLE 5.10** Continued

| Rank 1997 | Rank 1996 | | Profits as percentage of — Revenues % | Rank | Assets % | Rank | Stockholders' equity % | Rank | Earnings per share 1997 $ | Percentage change from 1996 | 1987–1997 annual growth rate % | Rank | Total return to investors 1997 % | Rank | 1987–1997 annual rate (%) | Rank | Industry table number | Rank 1997 |
|---|---|---|---|---|---|---|---|---|---|---|---|---|---|---|---|---|---|---|
| 1 | 1 | General Motors, Detroit | 3.8 | 305 | 2.9 | 290 | 38.3 | 24 | $8.62 | 43.2 | 5.5 | 206 | 19.5 | 301 | 12.2 | 291 | 37 | 1 |
| 2 | 2 | Ford Motor, Dearborn, MI | 4.5 | 268 | 2.5 | 314 | 22.5 | 95 | $5.62 | 54.4 | 2.2 | 251 | 57.0 | 88 | 15.3 | 233 | 37 | 2 |
| 3 | 3 | Exxon, Irving, TX | 6.9 | 171 | 8.8 | 79 | 19.4 | 133 | $3.37 | 12.7 | 7.0 | 185 | 28.3 | 246 | 17.4 | 186 | 39 | 3 |
| 4 | 4 | Wal-Mart Store, Bentonville, AR | 3.0 | 334 | 7.7 | 101 | 19.1 | 142 | $1.56 | 17.3 | 18.8 | 45 | 74.8 | 47 | 20.5 | 129 | 25 | 4 |
| 5 | 5 | General Electric, Fairfield, CT | 9.0 | 111 | 2.7 | 305 | 23.8 | 81 | $2.46 | 13.9 | 11.9 | 111 | 51.1 | 117 | 24.3 | 73 | 15 | 5 |
| 6 | 6 | IBM, Armonk, NY | 7.8 | 152 | 7.5 | 111 | 30.7 | 41 | $6.01 | 20.0 | 3.3 | 233 | 39.6 | 180 | 9.5 | 328 | 12 | 6 |
| 7 | 9 | Chrysler, Auburn Hills, MI | 4.6 | 264 | 4.6 | 212 | 24.7 | 74 | $4.09 | (13.7) | 3.3 | 232 | 11.7 | 338 | 17.0 | 195 | 37 | 7 |
| 8 | 8 | Mobil, Fairfax, VA | 5.5 | 223 | 7.5 | 107 | 16.8 | 189 | $4.01 | 10.8 | 10.1 | 129 | 21.7 | 282 | 18.9 | 158 | 39 | 8 |
| 9 | 10 | Phillip Morris, New York | 11.2 | 65 | 11.3 | 36 | 42.3 | 17 | $2.58 | 1.6 | 14.9 | 77 | 25.0 | 269 | 25.3 | 65 | 53 | 9 |
| 10 | 7 | AT&T, New York | 8.7 | 121 | 7.9 | 97 | 20.5 | 119 | $2.84 | (22.4) | 4.2 | 221 | 46.2 | 140 | 16.2 | 216 | 50 | 10 |
| 11 | 36 | Boeing, Seattle | (0.4) | 463 | (0.5) | 462 | (1.4) | 454 | -$0.18 | (109.7) | — | — | (7.1) | 408 | 22.0 | 102 | 2 | 11 |
| 12 | 11 | Texaco, White Plains, NY | 5.9 | 207 | 9.0 | 74 | 20.9 | 115 | $4.87 | 32.3 | — | — | 14.3 | 328 | 18.5 | 161 | 39 | 12 |
| 13 | 12 | State Farm Insurance, Bloomington, IN | 8.7 | 120 | 3.7 | 253 | 10.2 | 332 | NA | | — | — | — | | — | | 31 | 13 |
| 14 | 16 | Hewlett-Packard, Palo Alto, CA | 7.3 | 164 | 9.8 | 58 | 19.3 | 134 | $2.95 | 19.9 | 16.8 | 64 | 25.3 | 265 | 16.8 | 201 | 12 | 14 |
| 15 | 14 | E.I. Du Pont, Wilmington, DE | 5.8 | 212 | 5.6 | 170 | 21.3 | 110 | $2.08 | (34.6) | 5.4 | 209 | 30.3 | 235 | 19.3 | 153 | 7 | 15 |
| 16 | 17 | Sears Roebuck, Hoffman Estates, IL | 2.9 | 338 | 3.1 | 285 | 20.3 | 122 | $2.99 | (4.2) | (3.7) | 302 | 0.2 | 389 | 17.3 | 192 | 25 | 16 |
| 17 | 40 | Travelers Group, New York | 8.3 | 133 | 0.8 | 417 | 14.9 | 230 | $2.54 | 8.9 | 41.3 | 9 | 79.8 | 32 | 32.7 | 19 | 13 | 17 |
| 18 | 13 | Prudential Insurance, Newark, NJ | 1.6 | 400 | 0.2 | 447 | 3.1 | 428 | NA | | — | — | — | | — | | 30 | 18 |
| 19 | 15 | Chevron, San Francisco | 9.0 | 114 | 9.2 | 67 | 18.6 | 152 | $4.95 | 24.4 | 12.9 | 97 | 22.1 | 281 | 19.5 | 148 | 39 | 19 |
| 20 | 18 | Procter & Gamble, Cincinnati, OH | 9.5 | 99 | 12.4 | 27 | 28.3 | 50 | $2.43 | 13.3 | 26.4 | 25 | 50.4 | 118 | 25.2 | 66 | 48 | 20 |
| 21 | 20 | Citicorp, New York | 10.3 | 83 | 1.2 | 391 | 16.9 | 187 | $7.33 | (1.3) | — | — | 24.8 | 270 | 25.2 | 67 | 8 | 21 |
| 22 | 19 | Amoco, Chicago | 8.3 | 131 | 8.4 | 86 | 16.7 | 192 | $5.52 | (2.6) | 7.6 | 176 | 8.9 | 351 | 13.9 | 266 | 39 | 22 |
| 23 | 22 | K-Mart, Troy, MI | 0.8 | 434 | 1.8 | 345 | 4.6 | 416 | $0.51 | — | (11.3) | 321 | 10.8 | 342 | 1.0 | 372 | 25 | 23 |

| | | | | | | | | | | | | | | | | | |
|---|---|---|---|---|---|---|---|---|---|---|---|---|---|---|---|---|---|
| 24 | 30 | Merrill Lynch, New York | 6.0 | 202 | 0.7 | 427 | 22.9 | 91 | $4.83 | 17.5 | 18.4 | 50 | 81.4 | 29 | 32.7 | 20 | 47 | 24 |
| 25 | 33 | J.C. Penney, Plano, TX | 1.9 | 388 | 2.4 | 318 | 7.7 | 385 | $2.10 | (6.7) | 0.2 | 273 | 28.8 | 242 | 15.3 | 234 | 25 | 25 |
| 26 | 23 | American International Group, New York | 10.9 | 73 | 2.0 | 334 | 13.9 | 244 | $4.73 | 15.9 | 12.1 | 108 | 51.1 | 116 | 23.1 | 89 | 32 | 26 |
| 27 | 25 | Chase Corp., New York | 12.2 | 53 | 1.0 | 398 | 17.1 | 183 | $8.03 | 60.0 | — | — | 25.4 | 263 | 24.6 | 71 | 8 | 27 |
| 28 | 99 | Bell Atlantic, New York | 8.1 | 138 | 4.5 | 217 | 19.2 | 138 | $3.13 | (28.2) | 0.0 | 275 | 46.5 | 138 | 16.4 | 213 | 50 | 28 |
| 29 | 24 | Motorola, Schaumburg, IL | 4.0 | 295 | 4.3 | 224 | 8.9 | 359 | $1.94 | 2.1 | 12.5 | 101 | (6.0) | 406 | 17.6 | 178 | 15 | 29 |
| 30 | — | TIAA-CREF, New York | 4.2 | 286 | 0.6 | 433 | 21.2 | 112 | N.A. | — | — | — | — | — | — | — | 29 | 30 |
| 31 | 21 | Pepsico, Purchase, NY | 7.3 | 163 | 10.7 | 42 | 30.9 | 40 | $1.36 | 88.9 | 13.7 | 89 | 36.1 | 196 | 23.6 | 79 | 5 | 31 |
| 32 | 26 | Lockheed Martin, Bethseda, MD | 4.6 | 261 | 4.6 | 216 | 25.1 | 70 | -$3.12 | (151.2) | — | — | 9.0 | 349 | 20.7 | 127 | 2 | 32 |
| 33 | 29 | Fannie Mae, Washington, DC° | 11.0 | 71 | 0.8 | 418 | 22.2 | 96 | $2.83 | 14.1 | 22.0 | 33 | 54.5 | 99 | 39.2 | 9 | 13 | 33 |
| 34 | 27 | Dayton Hudson, Minneapolis | 2.7 | 346 | 5.3 | 180 | 17.2 | 180 | $3.18 | 63.1 | 14.8 | 80 | 74.0 | 50 | 24.6 | 70 | 25 | 34 |
| 35 | 162 | Morgan Stanley Dean Witter, New York | 9.5 | 100 | 0.9 | 415 | 18.5 | 156 | $4.25 | 52.3 | — | — | 81.3 | 30 | — | — | 47 | 35 |
| 36 | 28 | Kroger, Cincinnati, OH | 1.5 | 407 | 6.5 | 135 | (52.5) | 205 | $1.57 | 16.3 | 0.5 | 269 | 58.1 | 84 | 25.9 | 56 | 22 | 36 |
| 37 | — | Lucent Technologies, Murray Hill, NJ | 2.1 | 380 | 2.3 | 329 | 16.0 | 205 | $0.84 | — | — | — | 73.5 | 53 | — | — | 15 | 37 |
| 38 | 43 | Intel, Santa Clara, Ca | 27.7 | 2 | 24.0 | 2 | 36.0 | 27 | $3.87 | 33.4 | 36.5 | 12 | 7.4 | 356 | 35.9 | 12 | 17 | 38 |
| 39 | 32 | Allstate, Northbrook IL | 12.4 | 50 | 3.8 | 250 | 19.9 | 126 | $7.11 | 53.6 | — | — | 58.0 | 85 | — | — | 32 | 39 |
| 40 | 85 | SBC Communications, San Antonio, TX | 5.9 | 205 | 3.5 | 258 | 14.9 | 229 | $0.80 | (53.5) | (0.8) | 284 | 45.6 | 144 | 20.7 | 125 | 50 | 40 |
| 41 | 34 | United Technologies, Hartford, CT | 4.3 | 276 | 6.4 | 142 | 26.3 | 66 | $4.21 | 21.0 | 6.4 | 194 | 11.7 | 339 | 19.3 | 152 | 2 | 41 |
| 42 | 60 | Compaq Computer, Houston, TX | 7.5 | 157 | 12.7 | 25 | 19.7 | 130 | $1.19 | 28.0 | 25.8 | 26 | 89.9 | 20 | 31.4 | 26 | 12 | 42 |
| 43 | 35 | Metropolitan Life Insurance, New York | 4.9 | 249 | 0.6 | 432 | 8.6 | 366 | N.A. | — | — | — | — | — | — | — | 30 | 43 |
| 44 | 50 | Home Depot, Atlanta | 4.8 | 253 | 10.4 | 47 | 16.3 | 197 | $1.55 | 19.8 | 30.1 | 18 | 76.9 | 42 | 41.8 | 5 | 49 | 44 |
| 45 | 31 | Conagra, Omaha | 2.6 | 350 | 5.5 | 172 | 24.9 | 73 | $2.68 | 239.2 | 12.5 | 100 | 35.6 | 198 | 22.0 | 103 | 20 | 45 |
| 46 | 48 | Merck, Whitehouse Station, NJ | 19.5 | 5 | 17.9 | 10 | 36.6 | 26 | $3.74 | 19.9 | 17.6 | 56 | 35.6 | 199 | 22.6 | 95 | 40 | 46 |
| 47 | 38 | Bankamerica Corp, San Francisco | 13.6 | 40 | 1.2 | 381 | 16.2 | 204 | $4.32 | 18.2 | — | — | 49.0 | 130 | 39.4 | 8 | 8 | 47 |
| 48 | 41 | GTE, Stamford, CT | 12.0 | 59 | 6.6 | 130 | 34.8 | 31 | $2.90 | 0.7 | 5.??? | 204 | 20.0 | 296 | 17.4 | 185 | 50 | 48 |
| 49 | 39 | Johnson & Johnson, New Brunswick, NJ | 14.6 | 26 | 15.4 | 19 | 26.7 | 58 | $2.41 | 13.7 | 14.8 | 78 | 34.3 | 211 | 24.0 | 75 | 40 | 49 |
| 50 | 65 | Safeway, Pleasanton, CA | 2.5 | 354 | 6.6 | 133 | 25.9 | 64 | $1.12 | 15.5 | — | — | 48.0 | 132 | — | — | 22 | 50 |

**TABLE 5.11**  Percentage of Total Chemical Sales of the
Top 50 U.S. Chemical Producers Accounted for by Oil
Companies, 1975–1997

| Year | Total chemical sales of top 50 (billions) | Percentage of total chemical sales accounted for by major oil companies |
|---|---|---|
| 1974 | $51.9 | 22.6 |
| 1980 | $100.0 | 29.7 |
| 1985 | $107.7 | 28.0 |
| 1990 | $158.6 | 25.6 |
| 1995 | $198.4 | 23.5 |
| 1997 | $199.1 | 20.9 |

*Source:* Compiled from Facts and Figures issues of *Chemical and Engineering News* over a period of years.

the nation. Table 5.10 indicates that 5 of the top 50 chemical producers are among the top 50 firms in the United States.

3.  Chemical operations are big business, ranging from $1.4 billion to $21.3 billion in annual sales for the 50 companies in the chemical producers list. The combined 1997 chemical sales of the *Chemical and Engineering News* top 50 total $200 billion. For 25 of the top 50, chemicals account for 75% of the total chemical sales.

The *Chemical and Engineering News* top 50 companies list clearly shows the important role of the major oil companies in producing chemicals in the United States. According to Table 5.11, in the top 50 chemical companies in the period from the chemical sales attributed to petroleum companies 1974 to 1997 ranged between 20.9% in 1997 and 29.7% in 1980. Many of the chemical products provided by the oil companies are based on petrochemical feedstocks obtained during the petroleum refining process (see Figures 5.2 and 5.3).

In recent years, there has been some movement at the top of the 10 chemical companies ranked according to chemical sales (Table 5.12). When firms are ranked by total assets, however, it is clear that the companies at the top of the chemical industry have changed a great deal over the past 23 years (see Table 5.13). Some companies have been acquired by or merged with others, disappearing from the lists displayed in Tables 5.12 and 5.13.

From time to time, *Chemical Week* and *Chemical and Engineering News* publish quarterly as well as annual financial data for the basic chemical compa-

**TABLE 5.12** The Top 10 U.S. Chemical Producers Ranked by Chemical Sales: 1940–1996

| 1940 | 1950 | 1960 | 1970 | 1980 | 1985 | 1990 | 1995 | 1996 |
|---|---|---|---|---|---|---|---|---|
| Du Pont | Du Pont | Du Pont | Du Pont | Du Pont | Du Pont | Du Pont | Dow Chemical | Dow Chemical |
| Union Carbide | Union Carbide | Union Carbide | Union Carbide | Dow Chemical | Dow Chemical | Dow Chemical | Du Pont | Du Pont |
| Allied Chemical & Dye | Allied Chemical & Dye | Monsanto | Monsanto | Exxon | Exxon | Exxon | Exxon | Exxon |
| American Cyanamid | American Cyanamid | Dow Chemical | Dow Chemical | Union Carbide | Monsanto | Union Carbide | Hoechst Celanese | Monsanto |
| American Viscose | American Viscose | Allied Chemical | Standard Oil (NJ) | Monsanto | Union Carbide | Monsanto | Monsanto | Hoechst |
| Hercules Powder | Celanese | Olin Mathieson | Celanese | Celanese | Atlantic Richfield | Hoechst Celanese | General Electric | Celanese |
| Monsanto | Monsanto | American Cyanamid | W. R. Grace | Shell Oil | Shell Oil | General Electric | Mobil Union Carbide | General Electric |
| Celanese | Dow Chemical | W. R. Grace | Allied Chemical | W. R. Grace | Celanese | Occidental Petroleum | Union Carbide | Amoco |
| Dow Chemical | Hercules Powder | Hercules | Hercules | Gulf Oil | Amoco | BASF | Amoco | Eastman Chemical |
| Air Reduction | Air Reduction | Celanese | Occidental Petroleum | Occidental Petroleum | W. R. Grace | Amoco | Occidental Petroleum | BASF |

*Source: Compiled from Chemical and Engineering News, January 12, 1998.*

TABLE 5.13  The Top U.S. Chemical Companies Ranked by Total Assets

| 1974 | 1980 | 1985 | 1990 | 1997 |
| --- | --- | --- | --- | --- |
| Du Pont | Dow Chemical | Du Pont | Du Pont | Du Pont |
| Dow Chemical | Union Carbide | Dow Chemical | Dow Chemical | Dow Chemical |
| Union Carbide | Du Pont | Union Carbide | Monsanto | Monsanto |
| Monsanto | Monsanto | Monsanto | Union Carbide | Praxair |
| W. R. Grace | W. R. Grace | W. R. Grace | W. R. Grace | Air Products |
| Allied Chemical | Allied Chemical | American Cyanamid | Hoechst Celanese | Union Carbide |
| Celanese | Celanese | Celanese | American Cyanamid | Eastman Chemical |
| FMC | FMC | FMC | Air Products | IMC |

*Source:* Fortune 500 for respective years.

nies and other CPI-related companies. Table 5.14 presents the top five companies in various industrial groupings ranked by annual revenue. These data were
compiled from a recent Facts and Figure issue of *Chemical and Engineering
News* as well as from the Fortune 500 list of the same year.

The top 50 United States chemical producers are the most visible in the industry. They usually receive the most attention in chemical industry news publications and financial publications as well as from graduating professionals. In
1996 there were a total of 14,400 firms involved in SIC 28. Most of these firms
have total sales less than those of the top 50 companies.

Fifteen of the more profitable small chemical companies with annual sales
in the $50 million to $500 million range are listed in Table 5.15. These small
firms tend to manufacture fine and specialty chemicals, as contrasted with the
top 50, which tend to manufacture commodity and pseudocommodity chemicals.

The small chemical company usually requires less capital but more personnel per pound of product than a large company. The chemical processing is
often batch wise, and the facilities may be multipurpose, being flexible enough
to be used for several different products. Research and development tends to be
oriented toward product applications, compared with the process-oriented research and development of commodity producers. Marketing and product innovation are vital to the success of many small chemical firms. For reasons of
space, this book does not deal with small companies; however, when studying
the CPI and making employment decisions, chemical professionals should keep
in mind what a small company offers in terms of both contribution to the CPI
and professional opportunity. In a small company, an individual with entrepreneurial tendencies may find many opportunities (e.g., having broad responsibili-
ties at an early stage in one's career) not available in large firms. The risks

**TABLE 5.14**  The Top Five Companies of the Various Industrial Groupings: 1997

| Industrial grouping | Annual Revenue (millions)[a] | Industrial grouping | Annual revenue (millions)[a] |
|---|---|---|---|
| Chemicals[b] | | Soaps and Cosmetics | |
| Dow Chemical | $20,018 | Procter & Gamble | $35,764 |
| Occidental Petroleum | $11,061 | Colgate-Palmolive | $9,057 |
| Monsanto | $8,457 | Avon Products | $5,079 |
| PPG Industries | $7,379 | Estee Lauder | $3,382 |
| Union Carbide | $6,502 | Clorox | $2,533 |
| Diversified | | | |
| | | Pharmaceuticals | |
| General Electric | $90,840 | | |
| Du Pont | $41,394 | Merck | $23,637 |
| Allied Signal | $14,472 | Johnson & Johnson | $22,629 |
| Engelhard | $3,631 | Bristol-Myers Squibb | $16,701 |
| B. F. Goodrich | $3,471 | American Home | |
| | | Products | $14,196 |
| | | Pfizer | $12,504 |
| Petroleum refining | | | |
| | | Rubber and plastic products | |
| Exxon | $122,379 | | |
| Mobil | $59,978 | Goodyear Tire | $13,155 |
| Texaco | $45,187 | Rubbermaid | $2,400 |
| Chevron | $36,376 | Mark IV Industries | $2,294 |
| Amoco | $32,836 | M. A. Hanna | $2,200 |
| | | Cooper Tire and Rubber | $1,814 |
| Forest and paper products | | | |
| International Paper | $20,096 | | |
| Georgia-Pacific | $13,094 | | |
| Kimberly-Clark | $12,547 | | |
| Weyerhaeuser | $11,210 | | |
| Fort James | $7,259 | | |

[a] From Fortune 500, April 27, 1998.
[b] From *Chemical and Engineering News*, Facts and Figures issue, June 29, 1998.

TABLE 5.15  Sales and Profitability of 15 Selected Small Chemical Companies
with Sales of $50–$500 Million: 1996

| Rank | Company | Sales (millions) | Net income (millions) | Profit margin (%) | Return on invested capital (%) |
|------|---------|------------------|-----------------------|-------------------|-------------------------------|
| 1 | Forest Laboratories | $446.9 | $104.29 | 23.3 | 12.9 |
| 2 | Mississippi Chemical | $428.9 | $54.18 | 12.6 | 20.8 |
| 3 | Lone Star Industries | $367.7 | $54.16 | 14.7 | 17.4 |
| 4 | Terra Nitrogen | $363.1 | $177.63 | 48.9 | 72.1 |
| 5 | McWhorter Technology | $315.7 | $13.83 | 4.4 | 13.1 |
| 6 | Del Laboratories | $233.0 | $9.78 | 4.2 | 13.4 |
| 7 | Raytech | $217.7 | $15.98 | 7.3 | 23.2 |
| 8 | Learonal | $213.6 | $15.60 | 7.3 | 14.0 |
| 9 | Lawter International | $193.8 | $28.78 | 14.9 | 11.5 |
| 10 | American Filtrona | $193.3 | $12.44 | 6.4 | 13.2 |
| 11 | Sybron Chemicals | $174.3 | $8.51 | 4.9 | 12.1 |
| 12 | WD-4 | $130.9 | $21.30 | 16.3 | 42.3 |
| 13 | Chemfab | $83.9 | $7.71 | 9.2 | 13.0 |
| 14 | Hawkins Chemical | $80.9 | $6.48 | 8.0 | 14.8 |
| 15 | Hitox of America | $11.1 | $0.71 | 6.4 | 12.3 |

Source: CW300, Chemical Week Buyers' Guide, October 1997.

involved in working in a small company may be great, but the rewards for contributing to the growth firm's and prosperity may be correspondingly great.

### 5.2.2  International

Table 5.16 is a list of the top 50 global chemical producers ranked by chemical sales compiled from data in *Chemical and Engineering News.* The contents of this table are discussed more fully in Chapter 9. Note that the United States companies account for only 16 of the top 50 chemical companies and that 3 of the top 5 are located in Germany. This is an indication of the global nature of the chemical business. In Table 5.17 is a list of the top companies outside the United States ranked according to 1997 revenues. Of the top 15 companies, only 3 produce chemicals and allied products. Fourteen percent of the top 50 produce petroleum products, chemicals or allied products. This is a considerable decline from 40% in 1976. Table 5.17 reveals that many of the top companies are involved in the so-called service sector (6 of the top 10) rather than in the manufacturing sector of the global economy. The discussion of the international aspects of the CPI and the global economy is resumed in Chapter 9.

TABLE 5.16 The Top 50 World Chemical Producers, Ranked According to Chemical Sales in 1997

| Rank 1997 | Rank 1996 | Company (country) | Activity (× 10⁵ U.S.$) | | | | |
| --- | --- | --- | --- | --- | --- | --- | --- |
| | | | Chemical sales | Chemical operating profit | Capital spending | R&D spending | Number of employees |
| 1 | 1 | BASF (Germany) | $27,046.9 | $2,619.2 | $2,065.0 | $1213.3 | 104,979 |
| 2 | 5 | Du Pont (U.S.) | $21,295.0 | $3,554.0 | $1,900.0 | NA | 98,396 |
| 3 | 3 | Bayer (Germany) | $19,178.2 | $1,710.4 | $1,384.0 | $914.6 | 144,600 |
| 4 | 4 | Dow Chemical (U.S.) | $19,056.0 | $3,031.0 | $1,198.0 | $785.0 | 42,861 |
| 5 | 2 | Hoechst (Germany) | $16,293.6 | $1,354.0 | $1,258.3 | $526.5 | 118,212 |
| 6 | 6 | Shell (U.K./Netherlands) | $14,251.6 | $1,200.5 | NA | NA | 105,000 |
| 7 | 8 | ICI (U.K.) | $13,349.2 | $795.9 | $1,067.8 | NA | 69,500 |
| 8 | 9 | Exxon (U.S.) | $14,024.0 | $1,368.0 | $1,049.0 | $177.0 | 80,000 |
| 9 | 20 | Alezo Nobel (Netherlands) | $9,997.9 | $806.6 | $538.5 | $322.9 | 68,900 |
| 10 | 10 | Elf Aquitaine (France) | $9,954.3 | $702.5 | $496.6 | $291.3 | 83,700 |
| 11 | 11 | Rhône-Poulenc (France) | $9,868.6 | $737.6 | NA | NA | 68,377 |
| 12 | 17 | Sumitomo Chemical (Japan) | $8,196.9 | $548.5 | $617.5 | NA | NA |
| 13 | 12 | Dainippon Ink & Chemical (Japan) | $7,420.7 | $351.5 | $665.2 | NA | 32,682 |
| 14 | — | Clariant (Switzerland) | $7,017.2 | $709.7 | $325.9 | $240.5 | NA |
| 15 | 18 | Norsk Hydro (Norway) | $6,874.1 | $259.8 | $449.4 | NA | 38,271 |
| 16 | 16 | Huls (Germany) | $6,751.2 | $250.9 | $897.3 | $182.8 | NA |
| 17 | 14 | Mitsubishi Chemical (Japan) | $6,740.0 | $190.9 | $96.6 | $62.0 | 29,185 |
| 18 | 21 | General Electric (U.S.) | $6,695.0 | $1,476.0 | NA | NA | 276,000 |
| 19 | 22 | Union Carbide (U.S.) | $6,502.0 | $1,045.0 | $7,558.0 | $157.0 | NA |
| 20 | 29 | SABIC (Saudi Arabia) | $6,414.7 | $1,228.3 | $2,002.6 | $93.5 | NA |
| 21 | 13 | Toray Industries (Japan) | $6,342.6 | $453.1 | $96.6 | NA | NA |
| 22 | 23 | Henkel (Germany) | $6,186.9 | $873.1 | $344.8 | $150.5 | 54,247 |
| 23 | 26 | DSM (Netherlands) | $3,122.1 | $531.4 | $594.5 | $225.5 | NA |
| 24 | 39 | BOC (U.K.) | $6,017.5 | $840.5 | $1,298.6 | $132.0 | NA |
| 25 | 27 | Amoco (U.S.) | $5,941.0 | $732.0 | NA | NA | NA |

TABLE 5.16 Continued

| Rank 1997 | Rank 1996 | Company (country) | Chemical sales | Chemical operating profit | Capital spending | R&D spending | Number of employees |
|---|---|---|---|---|---|---|---|
| | | | Activity (× 10⁵ U.S.$) | | | | |
| 26 | 25 | Solvay (Belgium) | $5,906.5 | $430.5 | $654.2 | $125.8 | NA |
| 27 | 7 | Novartis (Switzerland) | $5,737.6 | $1,143.1 | $232.2 | $506.4 | 87,239 |
| 28 | 19 | ENI (Italy) | $5,668.2 | $212.6 | $203.2 | $96.9 | 80,178 |
| 29 | 28 | Air Liquide (France) | $5,458.1 | $476.4 | $1,283.1 | NA | NA |
| 30 | — | Ciba Specialties (Switzerland) | $5,389.7 | $588.4 | $360.4 | $208.1 | NA |
| 31 | 30 | British Petroleum (U.K.) | $5,114.6 | $777.9 | NA | NA | 56,450 |
| 32 | 36 | Huntsman Corp. (U.S.) | $5,000.0 | NA | NA | NA | NA |
| 33 | 33 | Total (France) | $4,889.2 | $398.2 | $411.2 | NA | 54,391 |
| 34 | 38 | Praxair (U.S.) | $4,735.0 | $848.0 | $902.0 | $79.0 | NA |
| 35 | 34 | Eastman Chemical (U.S.) | $4,678.0 | $568.0 | $749.0 | $191.0 | NA |
| 36 | 31 | Asahi Chemical (Japan) | $4,590.6 | $153.8 | $351.1 | NA | 27,792 |
| 37 | 37 | Occidental Petroleum (U.S.) | $4,349.0 | $497.0 | NA | NA | 12,380 |
| 38 | 40 | Degussa (Germany) | $4,274.7 | $323.5 | NA | NA | 25,736 |
| 39 | 43 | Allied Signal (U.S.) | $4,254.0 | $542.0 | NA | NA | 70,500 |
| 40 | 49 | Air Products (U.S.) | $4,122.0 | $719.4 | $870.0 | $114.0 | NA |
| 41 | 42 | Zeneca (U.K.) | $4,120.5 | $504.4 | $540.5 | $348.8 | NA |
| 42 | 24 | Formosa Plastics (Taiwan) | $4,049.7 | $443.3 | $1,157.7 | NA | NA |
| 43 | 47 | Ashland (U.S.) | $4,047.0 | $160.0 | NA | NA | 37,200 |
| 44 | 44 | Rohm & Haas (U.S.) | $3,999.0 | $685.0 | $254.0 | $201.0 | NA |
| 45 | 46 | Arco Chemical (U.S.) | $3,995.0 | $256.0 | $263.0 | $82.0 | 24,000 |
| 46 | 32 | Showa Denko (Japan) | $3,979.2 | $226.2 | $351.6 | NA | NA |
| 47 | 45 | Roche (Switzerland) | $3,948.9 | NA | $320.4 | $232.9 | NA |
| 48 | 41 | Sepisui Chemical (Japan) | $3,862.6 | $121.2 | $437.4 | NA | NA |
| 49 | — | Reliance Industries (India) | $3,691.2 | $795.0 | $958.0 | $9.9 | NA |
| 50 | — | Chevron (U.S.) | $3,632.0 | $362.0 | NA | NA | 39,362 |

Source: Abstracted and compiled from Chemical and Engineering News, July 20, 1998, pp. 38–39.

**TABLE 5.17** The 50 Largest Companies Outside the United States Ranked by 1997 Revenues

| Rank | Company | Country | Revenues | Profits | Assets | equity | Employment | Stockholders' Category |
|------|---------|---------|----------|---------|--------|--------|------------|------------------------|
| | | | | | | | Revenues, etc. ($\times 10^5$ U.S.$) | |
| 1 | Mitsui Group | Japan | $142,688 | $269 | $55,071 | $5,272 | 40,000 | Diversified (insurance, trading, etc.) |
| 2 | Mitsubishi Group | Japan | $128,922 | $388 | $71,408 | $7,569 | 38,000 | Diversified (chemicals, electronics, banks) |
| 3 | Royal Dutch/Shell Group | Britain/Neth. | $128,142 | $7,758 | $113,781 | $59,982 | 105,000 | Petroleum refining |
| 4 | Itochu | Japan | $126,632 | ($774) | $56,308 | $2,957 | 6,675 | Trading |
| 5 | Marubeni | Japan | $111,121 | $140 | $55,403 | $3,564 | 64,000 | Trading |
| 6 | Sumitomo Chemical (Japan) | Japan | $106,395 | $210 | $42,866 | $4,319 | 29,500 | Diversified (banks, metals) |
| 7 | Toyota Motor | Japan | $95,137 | $3,701 | $103,894 | $45,158 | 159,035 | Motor vehicles |
| 8 | Nissho Iwai | Japan | $81,893 | $25 | $40,799 | $2,020 | 18,158 | Trading |
| 9 | Nippon Telegraph & Telephone | Japan | $76,984 | $2,361 | $113,410 | $35,990 | 226,000 | Telecommunications |
| 10 | AXA | France | $76,874 | $1,357 | $401,206 | $13,075 | 80,613 | Insurance |
| 11 | Daimler-Benz | Germany | $71,561 | $4,639 | $76,191 | $19,510 | 300,068 | Motor vehicles |
| 12 | Daewoo | South Korea | $71,526 | $527 | $44,861 | $6,325 | 265,044 | Motor vehicles |
| 13 | Nippon Life Insurance | Japan | $71,388 | $2,118 | $316,530 | $5,575 | 75,851 | Insurance |
| 14 | British Petroleum | Britain | $71,194 | $4,046 | $54,099 | $23,221 | 56,450 | Petroleum refining |
| 15 | Hitachi | Japan | $68,567 | $28 | $75,838 | $24,303 | 331,494 | Electronics, electrical equipment |
| 16 | Volkswagen | Germany | $65,238 | $772 | $59,517 | $7,793 | 279,952 | Motor vehicles |

TABLE 5.17 Continued

| Rank | Company | Country | Revenues | Profits | Assets | equity | Employment | Stockholders' Category |
|---|---|---|---|---|---|---|---|---|
| | | | | | Revenues, etc. ($\times 10^5$ U.S.$) | | | |
| 17 | Matsushita Electric Industries | Japan | $64,280 | $763 | $64,218 | $28,272 | 275,962 | Electronics, electrical equipment |
| 18 | Siemens | Germany | $63,754 | $1,427 | $55,546 | $15,109 | 386,000 | Electronics, electrical equipment |
| 19 | Allianz | Germany | $56,785 | $1,172 | $211,442 | $9,588 | 73,290 | Insurance |
| 20 | Sony | Japan | $55,023 | $1,809 | $48,016 | $13,615 | 173,000 | Electronics, electrical equipment |
| 21 | Nissan Motor | Japan | $53,478 | ($114) | $59,121 | $9,617 | 137,201 | Motor vehicles |
| 22 | Fiat | Italy | $52,569 | $1,419 | $69,028 | $14,447 | 239,457 | Motor vehicles |
| 23 | Honda Motor | Japan | $48,876 | $2,123 | $36,110 | $12,058 | 109,400 | Motor vehicles |
| 24 | Unilever | Britain/Neth. | $48,761 | $5,463 | $31,671 | $12,203 | 287,000 | Food |
| 25 | Nestle | Switzerland | $48,254 | $2,761 | $37,489 | $16,736 | 225,808 | Food |
| 26 | Credit Suisse | Switzerland | $48,242 | $274 | $472,768 | $16,212 | 62,412 | Banks; commercial and savings |
| 27 | Dai-Chi Insurance | Japan | $47,442 | $1,492 | $214,994 | $2,247 | 64,598 | Insurance |
| 28 | Toshiba | Japan | $44,467 | $60 | $45,460 | $9,011 | 186,000 | Electronics, electrical equipment |
| 29 | Veba Group | Germany | $43,881 | $1,621 | $44,813 | $12,320 | 129,960 | Trading |
| 30 | Elf Aquitaine | France | $43,572 | $960 | $42,049 | $13,944 | 8,370 | Petroleum refining |
| 31 | Tomen | Japan | $43,400 | ($179) | $17,671 | $640 | 10,920 | Trading |
| 32 | Tokyo Electric Power | Japan | $42,997 | $1,102 | $107,587 | $11,714 | 42,672 | Utilities: gas and electric |

| | Company | Country | | | | | | Industry |
|---|---|---|---|---|---|---|---|---|
| 33 | Sumitomo Life Insurance | Japan | $42,279 | $1,094 | $177,845 | $2,409 | 64,628 | Insurance |
| 34 | Deutschebank | Germany | $40,792 | $552 | $579,992 | $17,843 | 76,141 | Banks; commercial and savings |
| 35 | Fujitsu | Japan | $40,613 | $46 | $38,418 | $8,888 | 180,000 | Computers; office equipment |
| 36 | RWE Group | Germany | $40,233 | $814 | $46,279 | $4,851 | 136,115 | Utilities: gas and electric |
| 37 | NEC | Japan | $39,927 | $337 | $37,299 | $8,030 | 152,450 | Electronics, electrical equipment |
| 38 | Royal Philips Electric | Netherlands | $39,188 | $2,939 | $29,316 | $9,596 | 264,700 | Electronics, electrical equipment |
| 39 | Deutsche Telekom | Germany | $38,969 | $1,905 | $90,543 | $25,967 | 216,006 | Telecommunications |
| 40 | ING Group | Netherlands | $38,674 | $2,104 | $305,984 | $22,745 | 64,162 | Insurance |
| 41 | HSBC Holdings | Britain | $37,474 | $5,496 | $471,256 | $27,055 | 132,969 | Banks; commercial and savings |
| 42 | ENI | Italy | $36,962 | $3,004 | $49,335 | $16,764 | 80,178 | Petroleum refining |
| 43 | Electricité de France | France | $36,673 | $264 | $113,360 | $35,220 | 116,462 | Utilities: gas and electric |
| 44 | Renault | France | $35,624 | $930 | $38,750 | $7,299 | 141,315 | Motor vehicles |
| 45 | PDVSA | Venezuela | $34,801 | $4,773 | $47,148 | $34,555 | 56,692 | Petroleum |
| 46 | Bank of Tokyo—Mitsubishi | Japan | $34,750 | ($4,272) | $690,462 | $18,174 | 18,386 | Banks; commercial and savings |
| 47 | BMW | Germany | $34,692 | $719 | $29,629 | $5,627 | 117,624 | Motor vehicles |
| 48 | Credit Agricole | France | $34,015 | $1,689 | $417,974 | $20,415 | 84,670 | Banks; commercial and savings |
| 49 | SK | South Korea | $33,816 | $125 | $17,930 | $3,095 | 30,595 | Petroleum refining |
| 50 | Metro | Germany | $32,790 | $320 | $14,133 | $2,308 | 117,470 | Food and drug stores |

*Source:* Global 500, *Fortune*, August 3, 1998.

## REFERENCE MATERIALS

*Chemical and Engineering News* has published a Facts and Figures issue annually since 1956. Since 1969 it has ranked the top 50 U.S. chemical producers annually, according to sales. Also found in these issues are data on the significant foreign chemical producers. For the 20-year period 1972–1992, the Facts and Figures issues included the top 50 chemicals in a single list. Since 1993, the Facts and Figures issues have given, instead, the top 10 chemicals in various CPI sectors.

The top 50 U.S. companies were obtained from the 1998 Fortune Directory of the 500 largest industrial corporations ranked according to revenue (*Fortune*, April 27, 1998). The top 50 foreign companies were obtained from the *Fortune* Directory of the 500 world's largest corporations (*Fortune*, August 3, 1998). The data on the top 50 chemical companies in the world were compiled from *Chemical and Engineering News* Facts and Figures issues.

*Chemical Week* also publishes the CW 300, which is a listing of the top 300 CPI companies in the United States. It may be found in each issue of the *Chemical Week Buyers' Guide*.

An article by C. H. Kline, "Maximizing Profits in Chemicals" (*CHEMTECH*, February 1976, p. 110), although dated, contains some helpful insight into the subject and is of historical interest.

## QUESTIONS FOR DISCUSSION

The following questions were developed to enliven classroom discussion or self-directed study. They are intended to help the reader note the significance of tabular data. It is suggested that some of the questions be done as in-class exercises and some as assignments for outside work. All these questions can be answered from data in the chapter; no additional library research is required.

1. List the top 10 organic chemicals in Table 5.7. What is their ranking in Table 5.3? Why are the rankings different? Is the word "chemical" used in the same sense in these two tables?

2. Rank the fibers listed in Chapter 6 in both Tables 5.3 and 5.7.

3. List the top 20 of the top 50 chemicals (Table 5.7) in terms of average annual change in sales volume from 1987 to 1997. Classify as organic and inorganic. Which of your top 20 are listed in Table 5.3?

4. Prepare a list of the top 10 U.S. chemical producers ranked by (a) profit margin and (b) return on stockholders' equity.

5. Prepare a list of companies from Table 5.9 whose chemical sales are less than 20% of total sales. What if anything do these companies have in common?

6. List all the chemical companies in Table 5.9 that appear in Table 5.10.

7. Rank the top 15 companies listed in Table 5.10 with respect to growth rate from 1987 to 1997. Which CPI companies are in the top 15?

8. Compute sales (revenue) per employee for the top 20 companies listed in Table 5.16 and prepare a list ranking them.

9. Which of the companies listed in Table 5.14 also appear in Table 5.10?

10. Compute sales (revenue per employee for the top 20 companies listed in Table 5.16) and prepare a list ranked on this basis. Can you draw any conclusions?

11. Find the top 20 from Table 5.16 in terms of research and development funds spent. Compute the ratio of research and development funds spent to total revenue and rerank. Can you draw any conclusions from this ranking?

12. Prepare a list of the top 10 from Table 5.17 ranked on the basis of revenue per employee.

## PROBLEMS

5.1.  Consider the top 10 inorganic chemicals in Table 5.7. Determine where they would rank in Table 5.3. (Data from Chapter 6 may be used.)

5.2.  Take the top 10 in the list you prepared in Problem 5.1 and determine which important consumer products are made from these chemicals.

5.3.  From Table 5.7, list 10 chemicals whose growth rate is the lowest over the period of 1987–1997. Which firms manufacture these chemicals? List the three most important firms for each of the 10 chemicals.

5.4.  a.  Figure 5.4 is a general plot of $p$ versus $Q$ for the chemical product groups listed in Table 5.4. Using price–production data from Chapter 6, check five specific chemicals. Do they fit in the general areas of Figure 5.4 in accordance with the groupings? If not, why not?

b.  How does Figure 5.4 differ from the $p$–$Q$ curves presented in Chapter 2? Compute an elasticity for the $p$–$Q$ relationship for the chemical you selected in part a. How would you interpret this number?

c.  If you were a corporate planner charged with investigating new business opportunities, how would you use Figure 5.4?

5.5.  Compare Table 5.7 with the latest Facts and Figures issue of *Chemical and Engineering News*. Are there any major changes? If so, can you account for them?

5.6.  Only a few chemical industry products are identified in the text. There are thousands of others. Develop a list of 20 less known chemical products perhaps by perusing the advertisements in *Chemical and Engineering News* and *Chemical Week*. How would you classify the products you found with respect to the categories listed in this section? In your readings, did you find much advertising for the top 50 chemicals?

5.7.  The Buyers' Guide published by *Chemical Week* lists 17,500 chemical products manufactured in the United States and the companies that produce them. It is well worth spending some time to become acquainted with the contents of this publication. You can quickly find, for example, the companies that produce the same products as the company for which you work or may wish to work. Why is it important to know your competition?

5.8.    In view of the uniform specifications for gasoline and all the advertising claims you have heard or seen, would you classify gasoline as a true commodity or a pseudocommodity?

5.9.    Find the latest "refinery-gate" prices for the products listed in Table 5.8. (See the *Oil and Gas Journal* listed in Section 0.1.1.) For which products should a refinery try to maximize production?

5.10.    What vehicles of transportation do not need a fuel derived from oil? How important are they in terms of numbers of people or tons of goods they can carry? Approximately what percentage of the U.S. energy requirement is needed for transportation?

5.11.    Is there any nonpetroleum substitute for gasoline or oil that has suitable properties and convenience and is available in large quantities at a low price? Explain your answer.

5.12.    Where do gasoline and residual fuel oil fit in Figure 5.4?

5.13.    In Problem 5.6 you developed a list of less known chemical products and their manufacturers. Pick five of these products and use the *Chemical Week* Buyers' Guide or some other source to find which other companies manufacture the products. How many of the companies are in the top 50?

5.14.    Find the total sales for each firm which producers one of your products. Are the firms listed on the New York, American, or Over-the-Counter exchanges. (The *Wall Street Journal* is an excellent source.) If not, where would you look for information?

5.15.    Large chemical companies recruit chemical professionals by interviewing at universities. If you were looking for your first job or a different job from one you now have, how would you develop a list of small companies offering promising opportunities? Develop a list of at least three with the statistics you feel are important.

5.16.    Study Table 5.12 and list foreign companies that have U.S. operations. List five products from these firms that compete with those made by the United States firms.

5.17.    Find a chemical company with annual sales of $50–$500 million and analyze its operations. Discuss the products made, the customers served, the manufacturing process used, the sales dollars, the net income, the profit margin, and the return on investment. Several of these companies may be listed as over-the-counter stocks, and information is readily available in the standard references.

# 6

## Specific Chemical Products

The chemical process industries produce an array of products with widely varying properties and uses, but most CPI products, from industrial gases to fibers and plastics, follow a life cycle, identifiable with the seasons of the year (Figure 6.1) (1).

In a traditional society, products like tallow soap and beeswax candles may be continuously in use for centuries without significant change. In our modern dynamic society, most products in the CPI go through complete life cycles in a few decades, from youthful newcomers to tired older items requiring replacement.

In the spring the product is new. There is uncertainty about acceptance in the marketplace. It may be a new artificial sweetener for soft drinks. Will the public find that it has a bitter aftertaste? Will it compare favorably with a new extract from grapefruit peels? Will it induce cancer of the bladder when fed in massive doses to rats? Will one consumer out of a thousand develop a rash after a drink? In the spring, the chemistry is rudimentary and the technology of manufacturing is crude. There is only one manufacturer, who is struggling to finance costly investments in marketing, advertising, manufacturing facilities, inventory buildup, and government certification. The hope of profit lies in the distant future. Many products fail at this phase, leaving the innovative entrepreneur sadder and poorer.

In the summer the product that survives is gaining acceptance in the marketplace. Being the first in the market means a good margin of profit, but it also means that the company is even more cash-hungry than in the spring. The rapid expansion of sales requires even more capital to design and build new plants and to penetrate new markets. The sweet smell of success also draws many imitators and competitors, who try to enter the market despite the presence of any previous, cautiously laid out barriers to entry. Research on product improvements

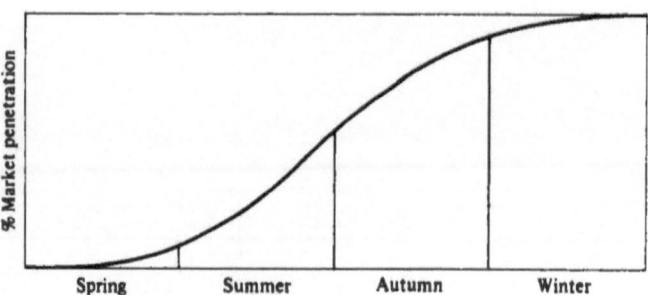

FIGURE 6.1   Product life cycle.

continues to be very important, since it enables the product to penetrate more markets and to compete favorably with new products from competitors.

When autumn arrives, the heady growth period is over, and market saturation is being felt. The profit margin declines from what it was in the summer, and competition between the numerous manufacturers leads to the disappearance of the less efficient producers. Manufacturing costs must be reduced by superior engineering, and larger scale, low-cost manufacturing units need to be designed if a firm is to survive in the period. A less creative but more hard headed management is needed to optimize the manufacturing process. In the autumn the product provides for the first time a large quantity of discretionary cash.

The product is mature and may even turn overripe in winter. Market growth depends strictly on growth in the GNP. There are few new users. The product is now an undifferentiated commodity, produced at a very low profit margin. Manufacturing costs and quantities produced are stable. In fact, the product is in danger of being phased out by more vigorous newcomers in the springtime of their cycle. No new chemistry or technology has been developed in many years. Management may be thought of as caretakers. Occasionally the sparkle of a new idea may occur in deep winter and lead to a new product cycle. But many products in winter are managed by undertakers, waiting for the funeral to begin.

During any particular assignment, chemical professionals in industry usually deal with a limited number of products. It is necessary for them to become expert about those products. To study any product in depth they must answer the types of question posed in Section 0.2.2. This chapter serves two purposes: it illustrates a detailed analysis of one chemical and presents some basic facts for several important organic and inorganic chemicals, plastics, and fibers.

Methanol has been chosen for the detailed analysis in Section 6.1. The facts and written discussion presented for methanol should help readers in preparing their own analyses and should show them how to examine the analyses of others critically.

Section 6.2 gives an overall perspective on a variety of important chemical products. With this foundation, you should be able to expand these data and contrast them with those for other products. The information presented in Section 6.2 will be referred to again in Chapters 8–10. All the products discussed in Section 6.2 are in the autumn or winter phase of their cycles. This is the only time in the life cycle when sufficient data have been accumulated to define methods of manufacture, to allow production and price histories to be established, to identify end-use patterns, and to identify major producers.

## 6.1 A PRODUCT REPORT: METHANOL

### 6.1.1 Method of Manufacture

Methanol ($CH_3OH$) is a toxic, colorless liquid at room temperatures. Its properties are summarized in Table 6.1 and a typical specification is given in Table 6.2. Methanol's identity was established in 1834, and Berthelot first synthesized it in by the *saponification* of methyl chloride. Initially, methanol was commercially

**TABLE 6.1**  Summary of the Physical Properties of Methanol

| Chemical formula | $CH_3OH$ |
|---|---|
| Molecular weight | 32.04 |
| Specific gravity at 20°/20°C | 0.7925 |
| Boiling point at 760 mmHg, °C | 64.6 |
| Freezing point, °C | −97.8 |
| Vapor pressure at 20°C mmHg | 97 |
| Flash point, tag closed cup, °C | 12.2 |
| Refractive index | 1.3286 |
| Solubility in water | Complete |
| Density, lb/gal | 6.59 |
| Vapor density (air = 1) | 1.11 |
| Appearance | Clear, colorless liquid |
| Odor | Alcohol-like |
| Ignition temperature, °C | 470 |
| Explosive limits (vol % in air) | |
|   Lower | 6 |
|   Upper | 36.5 |
| Threshold limit value, ppm | 200 |
| Latent heat of vaporization at boiling point, Btu/lb | 502 |
| Heat of combustion (liquid fuel–liquid $H_2O$), Btu/lb | 9776 |
| Octane number | |
|   Research | 106 |
|   Motor | 92 |

**TABLE 6.2**  ASTM Specification for Methyl Alcohol (Methanol), 99.85 Grade

| | |
|---|---|
| Apparent specific gravity: | |
|   At 20/20°C | 0.7920–0.7930 |
|   At 25/25°C | 0.7883–0.7893 |
| Color, Pt-Co | 5 |
| Distillation range, max | 1.0 (to include 64.6 ≈ 0.1) |
| Nonvolatile matter | 5 mg/100 mL, max |
| Odor | Nonresidual |
| Water, wt % | 0.10 max |
| Acidity (free acid as | 0.003, equivalent to |
|   acetic acid); wt %, max | 0.028 mg KOH per gram |
| Acetone, wt %, max | 0.003 |
| Hydrocarbons | To pass test |
| Sulfuric acid wash test | 50 |
|   (carbonizable impurities) | |
| Permanganate time, min | 50 |

*Source:* ASTM D-1152. American Society for Testing and Materials, Philadelphia, 1987.

produced by wood distillation (hence the common name, wood alcohol), but this process was very inefficient and produced a variety of unwanted by-products. In the early 1920s, the German firm BASF began commercial production by reacting carbon monoxide and hydrogen at approximately 400°C and 200 atm over a metal oxide catalyst.

In the United States both Du Pont and Commercial Solvents Corporation began experimental work on methanol synthesis in 1926. By the end of 1927 both companies had commercial production units operating. Commercial Solvents used a high-pressure reaction in which carbon dioxide and hydrogen react to yield methanol and water:

$$CO_2 + 3H_2 \rightarrow CH_3OH + H_2O$$

This route requires considerable separation equipment to remove the water and by-product higher alcohols formed during the reaction. Virtually all plants using this process have been replaced by newer technology. Du Pont's first methanol plant was based on a by-product carbon monoxide stream from an existing ammonia synthesis plant:

$$CO + 2H_2 \rightarrow CH_3OH$$

As with the Commercial Solvents process, essentially all plants using this process have been replaced by new processes. In fact, in the United States only three processes are used for the manufacture of methanol. Of the total U.S. capacity, 63.3% is based on the Lurgi process; 34.8% is based on the ICI process, and only 1.9% is based on the Haldor Topsoe process. The overall reaction is highly exothermic. As a result, it is desirable to use a relatively low temperature. Reaction is also favored by high pressure. However the copper catalyst is not active at temperatures much below 220°C, and a compromise between reaction kinetics and equilibrium considerations is required. This results in the synthesis reaction being carried out at 240–270°C. These conditions are much milder than the Commercial Solvents and Du Pont processes, which required pressures of 50–350 atm and temperatures of 250–300°C. Energy savings are significant with the newer processes. Plant sizes in the United States vary from 80,000 to 850,000 tons. Plants of 1,200,000 and 1,280,000 tons in Canada and Saudi Arabia, respectively, are the world's largest.

A flowsheet for the ICI process is shown in Figure 6.2.

About 99% of the methanol raw materials currently is obtained by reforming natural gas and refinery light-gas streams. From natural gas the reaction is as follows:

$$3CH_4 + 2H_2O + CO_2 \xrightarrow[\text{800°C}]{\text{Ni cat}} 4CO + 8H_2$$

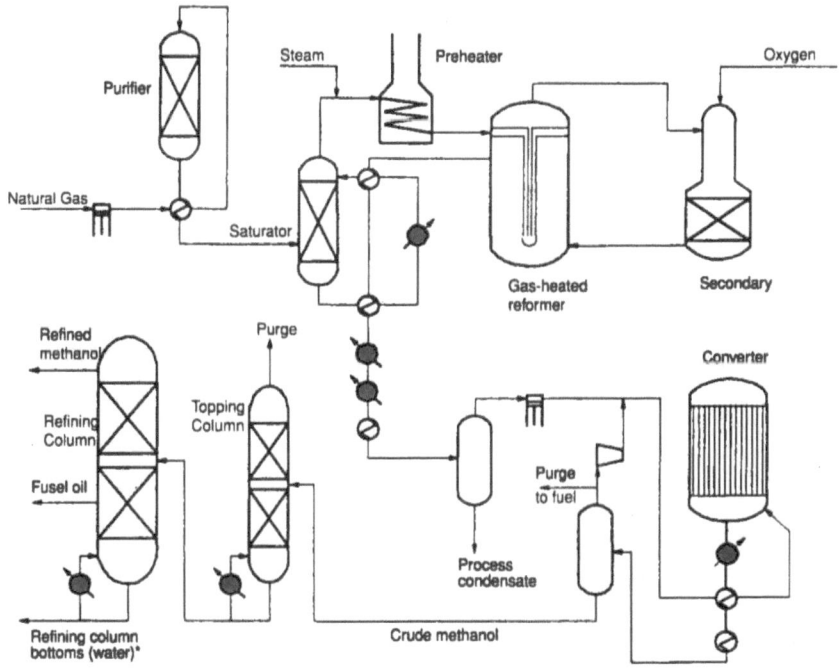

FIGURE 6.2   ICI Methanol process flowsheet. (From Ref. 2.)

Other raw materials such as gas oil and fuel oil have been used, as well as natural gas and coal.

### 6.1.2  Production and Pricing

The production history of synthetic methanol in the United States is shown graphically in Figure 6.3. Summarized in Table 6.3 are production and annual growth rates for 1991–1994.

From 1963 to 1973 average annual growth of U.S. production was 11.8%. The dips in production in 1975 and 1985 were the result of recessions. However, taking that into account it is apparent from Figure 6.3 that production had stabilized between 8.0 and 9.0 billion pounds per year during the 1980s. Demand began to increase again in 1993 as a result of the need to blend oxygenates into gasoline, thereby requiring methanol for the production of methyl tertiary butyl ether (MTBE).

The price history in cents per gallon for synthetic methanol in the United States is shown in Figure 6.4. As production stabilized (1980s), price declined and then remained relatively constant in the range of 40–50 cents per gallon. After demand increased after 1992, prices rose correspondingly. This reflected the gasoline equivalent price for MTBE.

The experience curve for methanol is shown by Figure 6.5. Industry total accumulated production from Figure 6.4 is plotted versus the price in constant

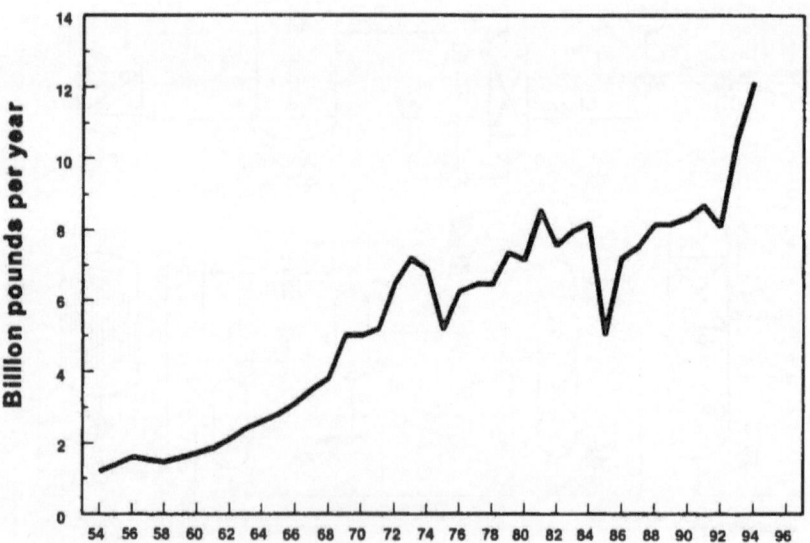

FIGURE 6.3   U.S. methanol production. (Data courtesy of CMAI.)

**TABLE 6.3**  Methanol Production History

| Year | Production (lb × 10⁹) | Change from previous year (%) |
|------|----------------------|-------------------------------|
| 1991 | 8.70 | 4.30 |
| 1992 | 8.08 | −7.14 |
| 1993 | 10.54 | 30.45 |
| 1994 | 12.13 | 15.09 |

*Source:* CMAI.

**Cents per gallon**

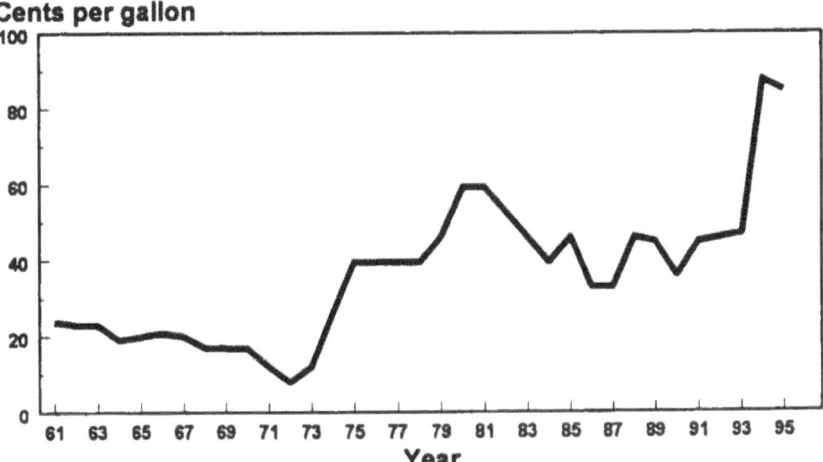

**FIGURE 6.4**  U.S. price history for synthetic methanol. (Data courtesy of CMAI.)

1987 dollars in log–log format. From the beginning of production (in the early 1940s) until 1974, methanol shows a classic experience curve; an early horizontal period is followed by, first, a slightly declining leg and then by a more abrupt declining leg. However, after 1974 the classic curve is interrupted by a different pattern, reflecting the dramatic change in a stable energy market price (crude oil and natural gas) following the oil embargo and later the OPEC price increases in crude oil in 1977/1978.

## 6.1.3  End-Use Consumption Patterns

Methanol is one of the cheapest industrial chemicals, which makes it a very competitive intermediate building block. This is true to a large extent because the raw

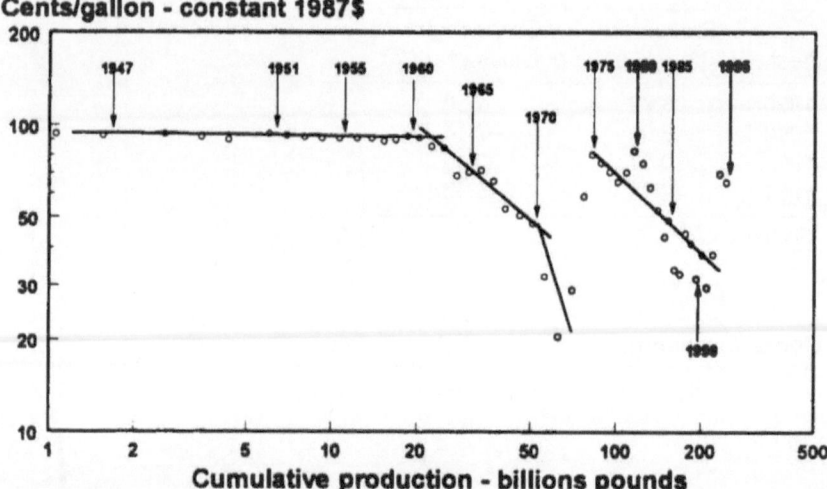

FIGURE 6.5   Experience curve for methanol.

material for methanol manufacture is natural gas or fuel products rather than derived chemicals. Table 6.4 shows U.S. methanol consumption by end use for selected recent years. Major end uses can be categorized into chemical intermediates,
exports, solvents, motor gasoline additives, and other direct uses. Exports are discussed in Section 6.1.5, and each of the other subcategories is discussed below.

Chemical intermediates account for a significant amount of methanol consumption. The more important intermediates and their end uses are summarized
in Tables 6.5 and 6.6. Prior to 1990, the largest single end use for methanol was
for the production of formaldehyde. However, with the congressional action requiring the addition of oxygenates to gasoline to improve combustion and reduce exhaust pipe emissions, the production of MTBE by the reaction of
methanol with isobutylene has become the largest single consumer of methanol.
MTBE has very good octane properties as well as providing the necessary oxygen content for the reformulated gasoline. This paradigm shift has drastically
affected the methanol industry, and while formaldehyde should experience
good growth rates, MTBE probably will continue to grow faster than the demand for formaldehyde. (It should be noted that legislative bodies are considering banning MTBE as an additive to gasoline owing to contamination
considerations. If this occurs, prospects for growth of methanol as a precursor
for MTBE will considerably diminish.

Formaldehyde is used in the production of thermosetting urea–formaldehyde and phenol–formaldehyde resins. These resins are used mostly as adhesives

**TABLE 6.4**   Industry Total Accumulated Production for Methanol from 1947 (see Figure 6.3)

| Year | Accumulated U.S. production quantity (lb × 10⁶) | Price per gallon (constant 1987 dollars) |
|------|------------------------------------------------|------------------------------------------|
| 1947 | 1.6   | 0.93 |
| 1951 | 6.0   | 0.94 |
| 1959 | 16.8  | 0.90 |
| 1963 | 24.8  | 0.85 |
| 1968 | 40.6  | 0.54 |
| 1973 | 69.5  | 0.29 |
| 1978 | 100.7 | 0.66 |
| 1983 | 139.3 | 0.53 |
| 1988 | 175.4 | 0.44 |
| 1993 | 219.2 | 0.47 |
| 1995 | 243.8 | 0.65 |

**TABLE 6.5**   U.S. Synthetic Methanol Consumption (lb × 10⁶) by End Use

|                                    | 1960[a] | 1973[a] | 1993[b] |
|------------------------------------|---------|---------|---------|
| Formaldehyde and inhibitor         | 825     | 2844    | 3,422   |
| Exports                            | 191     | 824     | 525     |
| Solvents                           | 132     | 565     | 562     |
| Dimethyl terephthalate (DMT)       | 66      | 435     | 223     |
| Methyl halides                     | 46      | 435     | 399     |
| Methyl methacrylate                | 86      | 265     | 425     |
| Methyl amines                      | 59      | 232     | 331     |
| Methyl tertiary butyl ether (MTBE) | —       | —       | 4,586   |
| Acetic acid                        | —       | 240     | 1,433   |
| Miscellaneous                      | 555     | 1288    | 1,828   |
| Total                              | 1960    | 7128    | 13,732  |

[a] From Ref. 1.
[b] From CMAI.

in plywood and particleboard manufacture, which is tied to the fortunes of the housing industry. Dimethyl terephthalate (DMT) is used almost exclusively as an intermediate in the production of polyester fibers and films. This use of methanol is dependent upon the fortunes of the textile industry. The decline shown in Table 6.4 for DMT can be attributed to competition from terephthalic acid (TPA) in polyester manufacturing as well as a shift of the polyester production and demand to the Far East, where labor costs for textile workers are low.

TABLE 6.6   Major Derivatives of Methanol

| Derivative and process | End uses of derivative | Major final demand |
|---|---|---|
| Formaldehyde<br>$CH_3OH \rightarrow HCHO + H_2$<br>or<br>$CH_3OH + 1/2\,O_2 \rightarrow HCHO + H_2O$ | Urea–formaldehyde resins<br>Phenol–formaldehyde<br>Acetal resins<br>Pentmerythritol<br>Hexamethylenetetramine<br>Melamine–formaldehyde<br>   resins<br>Miscellaneous and export | Adhesives in plywood<br>   and particleboard<br>   manufacture |
| Dimethyl terephthalate (DMT)<br>   Esterification of terephthalic acid | Polyester fibers<br>Polyester film<br>Export | Textiles |
| Methyl chloride<br>$CH_3OH + HCl \rightarrow CH_3Cl + H_2O$ | Silicones<br>Tetramethyllead<br>Miscellaneous | Gasoline additive |
| Methyl methacrylate<br>   Acetone cyanohydrin process | Acrylic sheet<br>Surface-coating resins<br>Molding and extrusion<br>   powders<br>Oil additives<br>Miscellaneous and export | Lucite, Plexiglas, and<br>   automotive paints |
| Methyl tertiary butyl ether | Oxygenate for gasoline | Gasoline additive |

Tetramethyllead, which once represented a significant demand for methanol, has declined drastically as the U.S. automobile has been equipped with catalytic converters to eliminate exhaust pollution, with the result that tetramethyllead has been eliminated from gasoline except for some commercial vehicles.

Methyl methacrylate, known commercially as Lucite (DuPont/ICI) and Plexiglas (Rohm & Haas), has outstanding optical properties as blocks or sheets. Thus this methanol derivative is very useful in a number of applications that require good transparency and optical qualities for replacement for glass, since it is less susceptible than glass to shattering.

Acetic acid is used mostly in making cellulose acetate and vinyl acetate. The process developed by Monsanto but later exploited by BP Chemicals has provided new and robust demand for methanol as a raw material for acetic acid production.

Industrial solvent use of methanol includes extracting, drying, washing, and crystallizing operations, in which methanol has had the advantage of being available at low cost and high purities. Nevertheless, methanol demand for solvent use is not expected to grow in the future.

Several direct uses of methanol have been significant in the past—for example, use as an automotive antifreeze and as an aircraft fuel additive—but both these uses claim only small amounts of present methanol consumption. Several direct uses have been suggested as potentially large markets for methanol in the future and are discussed in Section 6.1.6.

## 6.1.4  Producers

There are presently 16 synthetic methanol producers in the United States. These are listed in Table 6.7. Except for a small (10,000 ton/yr) plant to be built in West Covina, California, by TeraMeth, there are no expansions planned for the United States.

## 6.1.5  International Aspects

The United States continues to have the largest methanol capacity of any area of the world. However, within the past 20 years, large increments of capacity have been added in various other areas (see Table 6.8). Almost any country (or area) that has a ready supply of cheap natural gas has endeavored to exploit this re-

TABLE 6.7  U.S. Synthetic Methanol Producers

| Company | Process | Feedstock | Capacity (tons × 10³) |
|---|---|---|---|
| Air Products | ICI | Natural gas | 180 |
| Ashland | Lurgi | Natural gas | 460 |
| Beaumont Methanol | Lurgi | Natural gas | 850 |
| Borden | ICI | Natural gas | 600 |
| Coastal | ICI | Natural gas | 80 |
| Enron | Lurgi | Natural gas | 375 |
| Fortier | Lurgi | Natural gas | 570 |
| Georgia Gulf | ICI | Natural gas | 480 |
| Hoechst Celanese | Lurgi | Natural gas | 500 |
| Liquid Carbonic | ICI | Natural gas | 90 |
| Lyondell PC | ICI | Natural gas | 732 |
| Quantum | Lurgi | Residuals | 600 |
| Sand Creek | Lurgi | Natural gas | 90 |
| Tennessee Eastman | Lurgi | Coal | 195 |
| Terra International | Haldor Topsoe | Natural gas | 120 |
| Texaco | Lurgi | Natural gas | 300 |
| Grand total | | | 6222 |

Source: CMAI.

**TABLE 6.8**   Worldwide Methanol Capacity:
1994

| Area | Methanol capacity (metric tons × $10^6$) |
| --- | --- |
| United States | 6.22 |
| Canada | 2.47 |
| South America | 2.78 |
| Western Europe | 2.85 |
| Eastern Europe | 3.35 |
| Africa and Middle East | 3.45 |
| Pacific and Far East | 3.37 |

*Source:* CMAI.

source by upgrading it to a higher value-added product that is easier to transport, namely, methanol. This has been particularly true for Canada, Trinidad, Russia, Libya, Saudi Arabia, Malaysia, and New Zealand. The result is that the United States has shifted being from a net exporter (20 years ago) to being a net importer. In fact, in 1994 imports constituted slightly more than 20% of total U.S. supply.

### 6.1.6   Future Prospects

For the foreseeable future, methyl tertiary butyl ether is expected to provide for the greatest demand for methanol (unless the MTBE is banned from gasoline). Utilization of methanol directly as automotive fuel has not found strong support except in specialized uses such as racing vehicles. However, research is progressing to develop other uses for methanol. Potentially large new markets for methanol may be the following:

> *Agricultural uses:* increased crop yields with smaller requirements for irrigation
> *Single cell protein (SCP) manufacture:* for use as an animal feed supplement
> *Sewage treatment:* denitrification of wastewater

### 6.2   SELECTED CHEMICALS, PLASTICS, AND FIBERS: BASIC DATA

This section provides basic information concerning methods of manufacture, production history, price history, consumption patterns (Table 6.9), and major producers (Table 6.10) for 15 organic chemicals, 3 inorganic chemicals, 5 plastics, and 3

TABLE 6.9   Consumption Patterns: 1975 and 1993

| Chemical | Consumption pattern (%)[a] | 1975 | 1993 | Distribution of product consumption |
|---|---|---|---|---|
| Ethylene | α-Olefins | — | 5 | Plastics; antifreeze; |
| | EB/styrene | 8 | 7 | fibers; solvents |
| | EDC/VCM | 14 | 14 | |
| | Ethanol | 6 | — | |
| | Ethylene oxide | 20 | 13 | |
| | Polyethylene | 40 | — | |
| | LDPE | — | 29 | |
| | HDPE | — | 26 | |
| | Vinyl acetate | — | 2 | |
| | Others | 12 | 4 | |
| Propylene | Acrylonitrile | 16 | 14 | Plastics; fabrics; solvents; |
| | Cumene | 11 | 9 | adhesives; insulation |
| | Isopropanol | 14 | 5 | |
| | Oligomers | — | 4 | |
| | Oxo-aldehydes/alcohols | 8 | 7 | |
| | Polypropylene | 23 | 44 | |
| | Propylene oxide | 13 | 11 | |
| | Others | 15 | 8 | |
| Ethylene oxide | Antifreeze | 60 | 30 | Antifreeze; polyesters; |
| | Polyester fiber | — | 25 | detergents |
| | Polyester film | — | 6 | |
| | PET resins | — | 16 | |
| | Others | 40 | 7 | |
| | Exports | — | 16 | |
| Butadiene | ABS resins | 6 | 5 | Tires; other rubber products |
| | Adiponitrile | 8 | 11 | fibers (nylon); paper coating |
| | Chloroprene | 8 | 6 | |
| | Nitrile rubber | 3 | 3 | |
| | Polybutadiene | 17 | 23 | |
| | SB latex | — | 12 | |
| | SB rubber | 47 | 34 | |
| | Others | 11 | 6 | |
| Toluene | Benzene (Hydrodecalkylation) | 51 | 36 | Plastics; fabrics; tires; solvents |
| | Benzene (TDP) | — | 34 | rubber products; explosives |
| | Phenol | 1 | 1 | |
| | TDI | 5 | 8 | |
| | Solvents | 10 | 11 | |
| | Other | 33 | 4 | |
| | Exports | — | 6 | |
| Benzene | Alkylbenzene | — | 2 | Plastics; fibers; tires; |
| | Chlorobenzene | — | 2 | rubber products; adhesives |

TABLE 6.9   Continued

| Chemical | Consumption pattern (%)[a] | 1975 | 1993 | Distribution of product consumption |
|---|---|---|---|---|
| | Cumene/phenol | 20 | 23 | |
| | Cyclohexane | 17 | 13 | |
| | Ethylbenzene | 48 | 53 | |
| | Maleic anhydride | 3 | 0 | |
| | Nitrobenzene | 5 | 6 | |
| | Other | 7 | 0 | |
| o-Xylene | Phthalic anhydride | 33 | 76 | Fabrics; plastics |
| | Others | 67 | 4 | |
| | Exports | — | 20 | |
| p-Xylene | Dimethylterephthalate/ purified terephthalic acid | 79 | 81 | Fabrics; plastics |
| | Exports | 21 | 19 | |
| Chlorine | VCM/PVC | 59 | 30 | Plastics; solvents; aerosols; |
| | Pulp and paper | 18 | 11 | water and sewage treatment; |
| | Propylene oxide | — | 8 | paper production |
| | EDC export | — | 6 | |
| | Water treatment | 6 | 5 | |
| | Chloroethanes | — | 5 | |
| | Epichlorohydrin | — | 5 | |
| | Chloromethanes | — | 4 | |
| | Other organics | 6 | 15 | |
| | Other inorganics | 11 | 11 | |
| LDPE | Film and sheet | 55 | 50 | Trash bags; industrial liners; |
| | Injection molding | — | 5 | shipping bags; food packaging; |
| | Wire and cable | 8 | 3 | paper; paperboard coatings; |
| | Extrusion coating | 9 | 11 | wire and cable coatings; |
| | Blow molding | — | 1 | tubing |
| | Others | 21 | 15 | |
| | Exports | 7 | 15 | |
| HDPE | Film and sheet | 4 | 14 | Milk containers; pipe and conduit; tubing; |
| | Injection molding | 10 | 16 | housewares |
| | Wire and cable | 3 | 2 | |
| | Blow molding | 40 | 32 | |
| | Pipe and conduit | 12 | 6 | |
| | Others | 23 | 15 | |
| | Exports | 8 | 14 | |
| Polypropylene | Injection molding | 18 | 26 | Appliances; electrical wire and cable; luggage; housewares; |
| | Blow molding | 5 | 2 | |

TABLE 6.9   Continued

| Chemical | Consumption pattern (%)[a] | 1975 | 1993 | Distribution of product consumption |
|---|---|---|---|---|
| | Film and sheet | 15 | 10 | blow-molded bottles; |
| | Fiber and filament | 32 | 27 | closures; food packaging; |
| | Others | 20 | 19 | toys and novelties; auto |
| | Exports | 10 | 16 | fascia and gaskets |
| Ethylbenzene | Styrene | 96 | 99 | Polystyrene; polystyrene |
| | Others | 2 | 0 | copolymers (SBR, ABS resins) SB latex |
| | Exports | 3 | 0 | |
| Cumene | Phenol | — | 94 | Plastics; solvents |
| | Others | — | 2 | |
| | Exports | — | 3 | |
| Phenol | Alkylphenol | — | 7 | Plastics; resins |
| | Aniline | — | 2 | |
| | Bisphenol A | — | 32 | |
| | Cyclohexanol | — | 17 | |
| | Orthoxylenol | — | 3 | |
| | Phenolic resins | — | 28 | |
| | Others | — | 6 | |
| | Exports | — | 4 | |
| Ethylene dichloride | Vinyl chloride | 77 | 91 | Polyvinyl chloride (plastics); solvents |
| | Chlor. solvent | — | 0 | |
| | Others | 15 | 2 | |
| | Exports | 8 | 8 | |
| Vinyl chloride | Polyvinyl chloride | — | 86 | Polyvinyl chloride (plastics) |
| | Others | — | 2 | |
| | Exports | — | 12 | |
| Polyvinyl chloride | Calendering | — | 11 | Pipe and tubing; flexible plastic sheeting; housing fascia |
| | Coatings | — | 3 | |
| | Rigid pipes and tubes | — | 37 | |
| | Other extrusions | — | 23 | |
| | Moldings | — | 5 | |
| | Paste/plastics | — | 2 | |
| | Others | — | 5 | |
| | Exports | — | 13 | |

[a] ABS, acrylonitrile–butadiene–styrene; DMT, dimethyl terephthalate; EB, ethylbenzene; EDC, ethylene dichloride; HDA, hydrodealkylation; HDPE, high-density polyethylene; LDPE, low-density polyethylene; PET, polyethylene terephthalate; PTA, polyethylene terephthalic acid; PVC, polyvinyl chloride; SB, styrene butadiene; TDI, toluene diisocyanate; TDP, thioldiphenol; VCM, vinyl chloride monomer.

**TABLE 6.10** Concentration Ratios for Major Producers in the United States: Top Four Producers in Each Category: 1975 and 1995

| Chemical | 1975 | | 1995 | |
|---|---|---|---|---|
| | Company | Ratio (%) | Company | Ratio (%) |
| Ethylene | Dow Chemical<br>Gulf Oil<br>Shell Chemical<br>Union Carbide | 50–60 | Dow Chemical<br>Shell Chemical<br>Phillips 66<br>Exxon Chemical | 15–20 |
| Propylene | Exxon Chemical<br>Dow Chemical<br>Shell Chemical<br>Union Carbide | 40–50 | Exxon Chemical<br>Lyondell PC<br>Shell Chemical<br>Dow Chemical | 15–20 |
| Ethylene oxide | Dow Chemical<br>Jefferson Chemical<br>PPG Industries<br>Union Carbide | | Union Carbide<br>Shell Chemical<br>Huntsman Chemical<br>PD Glycol | 60–65 |
| Butadiene | Exxon Chemical<br>Neches Butane<br>Petro-Tex Chemical<br>Union Carbide | 50–60 | Texas Petrochemical<br>Shell Chemical<br>Exxon Chemical<br>Huntsman Chemical | 70–75 |
| Toluene | Amoco Chemical<br>Exxon Chemical<br>Phillips 66<br>Sun Oil | 35–45 | Exxon Chemical<br>BP Chemical<br>Amerada Hess<br>Mobil | 45 |
| Benzene | Amoco Chemical<br>Exxon Chemical<br>Corco<br>Phillips 66 | 30–40 | Chevron Chemical<br>Dow Chemical<br>Exxon Chemical<br>Shell Chemical | 35–40 |
| o-Xylene | | | Exxon Chemical<br>Lyondell PC<br>Koch Chemical<br>Phillips 66 | 80–85 |
| p-Xylene | | | Amoco Chemical<br>Koch Chemical<br>Exxon Chemical<br>Phillips 66 | 80–85 |
| Chlorine | Dow Chemical<br>Diamond Shamrock<br>PPG Industries<br>Occidental Petroleum | 50–60 | Dow Chemical<br>Oxychem<br>PPG Industries<br>Olin | 70–75 |

TABLE 6.10 Continued

| Chemical | 1975 | | 1995 | |
|---|---|---|---|---|
| | Company | Ratio (%) | Company | Ratio (%) |
| LDPE | Union Carbide<br>Dow Chemical<br>Du Pont<br>Gulf Oil | 40–50 | Union Carbide<br>Quantum<br>Dow Chemical<br>Exxon Chemical | 55–60 |
| HDPE | | | Quantum<br>Phillips 66<br>Oxychem<br>Solvay | 55–60 |
| Polypropylene | Hercules<br>Amoco Chemical<br>Exxon Chemical<br>Shell Chemical | 70–80 | Himont<br>Amoco Chemical<br>Fina<br>Exxon Chemical | 60–65 |
| Ethylbenzene | Amoco Chemical<br>Dow Chemical<br>Foster Grant<br>Monsanto | 50–60 | ARCO Chemical<br>Sterling Chemical<br>Dow Chemical<br>Chevron Chemical | 60–65 |
| Cumene | | | Georgia Gulf<br>Chevron Chemical<br>Shell Chemical<br>CITGO | 65–70 |
| Phenol | | | Allied Signal<br>Aristech (Mistubishi)<br>Shell Chemical<br>Mt. Vernon Phenol | 60–70 |
| Ethylene<br>dichloride | Conoco<br>Dow Chemical<br>B. F. Goodrich<br>Shell Chemical | 50–60 | Dow Chemical<br>Oxychem<br>Formosa Plastics<br>B. F. Goodrich | 60–70 |
| Vinyl chloride | | | Dow Chemical<br>Formosa Plastics<br>Oxychem<br>B. F. Goodrich | 60–65 |
| Polyvinyl<br>chloride | | | Shintech<br>Oxychem<br>Formosa Plastics<br>B. F. Goodrich | 60–65 |

**TABLE 6.10** Continued

|  | 1975 | | 1995 | |
| --- | --- | --- | --- | --- |
| Chemical | Company | Ratio (%) | Company | Ratio (%) |
| Nylon fibers | Du Pont<br>Fiber Industries<br>Fibers International<br>Monsanto | 85–95 | Du Pont<br>Monsanto<br>BASF<br>Allied Corp. | 96 |
| Polyester fibers | Du Pont<br>Eastman Chemical<br>Fiber Industries<br>Hoechst Fibers | 70–80 | Du Pont<br>Hoechst Celanese<br>Wellman Inc.<br>Nanya | 82 |
| Sulfuric acid | Allied Chemical<br>CF Industries<br>Du Pont<br>Stauffer Chemical | 25–35 |  |  |
| Ammonia | Allied Chemical<br>CF Industries<br>Farmland Industries<br>Socal | 20–30 |  |  |

types of fiber. Ethylene, propylene, butadiene, toluene, benzene, and xylenes were chosen because they are the basic building blocks of most of the organic chemical industry. Other materials (e.g., LDPE/LLDPE, and HDPE) were chosen for discussion to show how one basic building block, ethylene, is connected to intermediates and final demand. Polypropylene, low-density polyethylene, high-density polyethylene and linear low-density polyethylene, which constitute more than 50% of the demand for propylene and ethylene, are widely used throughout the consumer market. The remaining chemicals in the inorganics, plastics, and fibers groups were chosen on the basis of prominence within their respective groups. Figures 6.13–6.25, 6.27, and 6.28 are plots of annual U.S. production and pricing.

## 6.2.1  Method of Manufacture: Organics

Only methods that dominate the industry are discussed.

6.2.1.1  Ethylene. Ethylene is the petrochemical produced in the largest quantity in the United States. Earliest commercial production began about 1936 with the thermal cracking of ethane and/or propane (this was accomplished by companies such as Dow Chemical and Union Carbide, which are primarily "chemical" companies):

$$C_2H_6 \rightarrow C_2H_4 + H_2$$

$$2C_3H_8 \rightarrow C_2H_4 + H_2 + C_3H_6 + CH_4$$

While this was the "U.S. model," the "European model" was for the production of ethylene based on heavy feedstocks such as naphtha and gas oil. In the United States, ethane and propane continued to be the most significant feedstocks for ethylene manufacture until the pyrolysis of naphtha and gas oil began in the mid-1960s. This change of feedstock was brought about by chemical companies that were related to oil companies (e.g., Mobil Chemical, Exxon Chemical, Shell Chemical, Gulf Chemical). These companies had ready uses for the multiplicity of coproducts that resulted from the cracking of naphtha and gas oil inasmuch as many of the resulting components could be blended into the gasoline pool. As a result of the feedstock differences, the prices of coproducts were significantly affected where naphtha/gas oil were the feedstocks of choice. Generally, the coproducts produced from heavy feedstocks surpassed the demand for these coproducts, with the resulting depressive effect on price.

According to 1993 figures about 33% of the ethylene in the United States is made from cracking gas oils and naphtha. This is up from 15% shown for 1970 in the first edition of this book. Product distribution depends on the raw material used and on the severity of the reaction conditions. Typical for various feeds are the figures shown in Table 6.11.

An ethylene production graph plotted on semilog scale and an experience chart (Figures 6.6 and 6.7, respectively) show the same sort of pattern discernible in the methanol experience chart. Thus it seems reasonable to conclude that through 1972 the experience charts provide the expected result. However, with the sharp change in the price of crude oil and other energy sources in 1973, the anticipated patterns are seriously distorted. Another factor affecting the experience chart (particularly in ethylene and other products for which new facilities

**TABLE 6.11**  Ethylene Feedstocks

| | Components (%) | | | |
|---|---|---|---|---|
| Product | Ethane | Propane | Naphthas | Heavy gas oil |
| Ethylene | 76.3 | 42.0 | 31.2 | 23.4 |
| Propylene | 2.9 | 16.2 | 16.1 | 14.3 |
| Butadiene | 1.3 | 3.2 | 4.5 | 4.0 |
| Butylene, butane | 0.6 | 1.4 | 4.5 | 4.4 |
| $C_5$ and greater | 2.8 | 7.2 | 26.5 | 43.9 |
| Off-gas | 16.1 | 30.0 | 17.2 | 10.0 |

**Billion pounds per year**

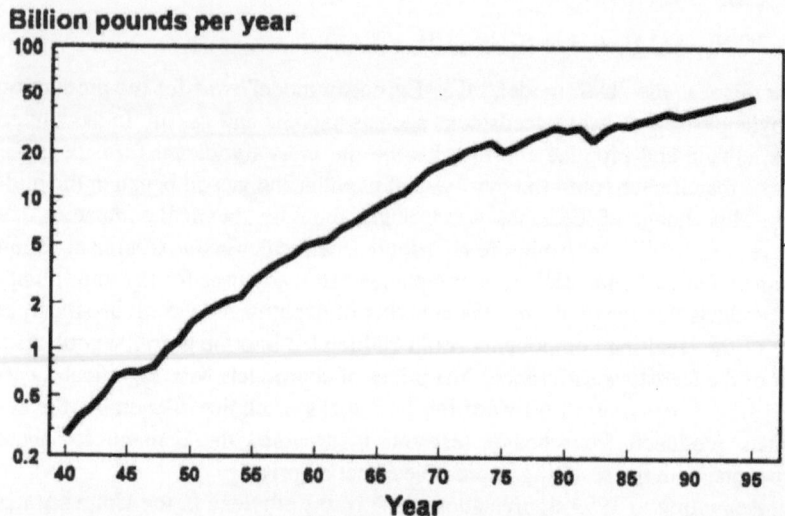

FIGURE 6.6   U.S. production histories for ethylene. (Data courtesy of CMAI.)

**Cents/pound - constant 1987$**

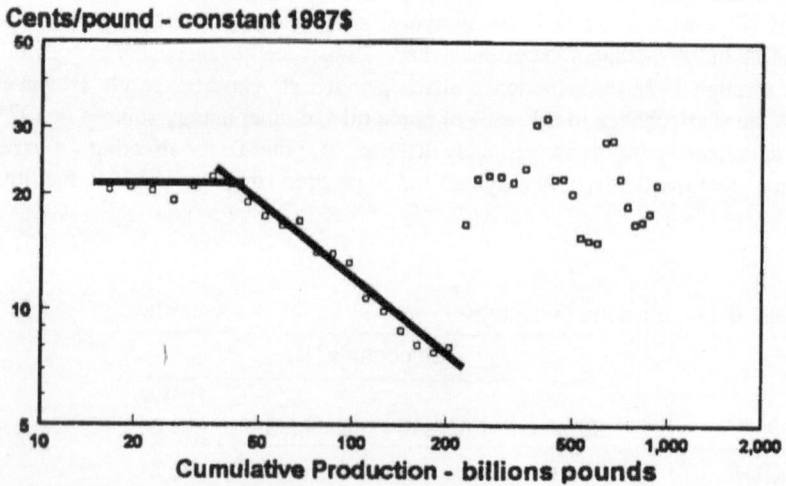

FIGURE 6.7   U.S. experience curve for ethylene.

are very expensive and take 3–5 years to complete) is that the supply–demand balance is seriously displaced.

When demand is strong and profit margins are good, participants often elect to build new plants until production of large quantities of surplus capacity results in severely depressed prices and margins. This fluctuation between strong price/margin and severely depressed price/margin disturbs the smooth results that occurred prior to 1973.

**6.2.1.2 Ethylene oxide (oxidation of ethylene).** Reaction is accomplished at pressures of about 10–30 atm and 270–290°C over a silver catalyst in either air or oxygen. In addition to ethylene oxide, carbon dioxide is formed by the competing reaction:

$$CH_2=CH_2 + 1/2\ O_2\ Ag \rightarrow CH_2 \text{------} CH_2$$

Per-pass conversion of ethylene is typically 30%. Selectivity, defined as the ratio of ethylene converted to ethylene oxide to ethylene reacted, is between 70 and 80%. A variety of reactor systems are used.

**6.2.1.3 Ethylene Dichloride.** Vapor or liquid phase catalytic reaction of ethylene and chlorine:

$$CH_2=CH_2 + Cl_2 \rightarrow ClCH_2CH_2Cl$$

A common catalyst is ethylene dibromide, although a variety of catalysts have been used.

**6.2.1.4 Ethylbenzene.** Over 95% is produced by the alkylation of benzene with ethylene:

$$C_6H_6 + CH_2=CH_2 \rightarrow C_6H_5CH_2CH_3$$

Overall yields of over 99% are achieved. Reaction is either in the liquid phase, using aluminum chloride catalyst, or in the vapor phase, using a boron trifluoride catalyst. Ethylbenzene can also be recovered by fractionation of some gasoline or naphtha fractions or some $C_8$ fractions from aromatics recovery from pyrolysis gasoline (a by-product of ethylene production). Figure 6.8 shows U.S. production figures for both ethylbenzene and ethylene glycol.

**6.2.1.5 Propylene.** Unlike ethylene, propylene production does not represent the requirement for propylene derivatives. With few exceptions, propylene is not made "on purpose" but is obtained as a by-product of other processes. More specifically, large quantities of relatively low purity (40–70%) propylene are produced in refineries as a by-product of gasoline manufacture. Additionally, significant quantities of higher purity propylene originate in olefins plants, where ethylene is the primary product. However, only polymer-grade propylene (>99% pure) can in any way be considered an on-purpose product. To better understand

**Billion pounds per year**

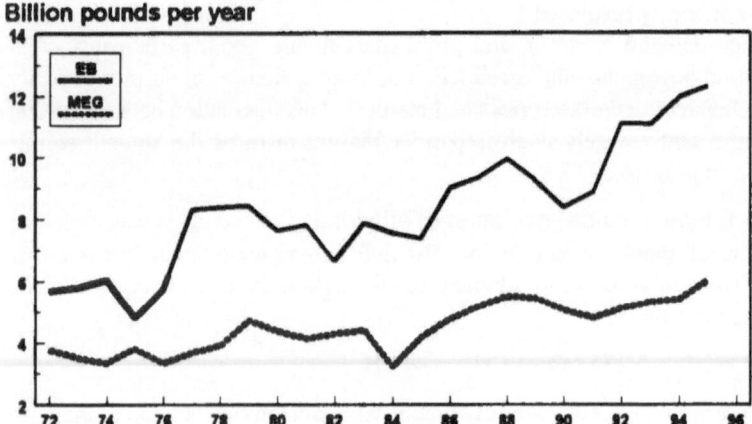

FIGURE 6.8   U.S. production histories of ethylbenzene and ethylene glycol. (Data courtesy of CMAI.)

the situation, the following tabulation shows the demand for various grades of propylene:

> Polymer-grade propylene (>99%): upgraded from either refinery or olefins plant sourced propylene:
>> Polypropylene
>> EP rubber
>> Epichlorohydrin
>
> Chemical-grade propylene (92–96%): recovered from a propylene column in an olefins plant;
>> Acrylonitrile
>> Isopropanol
>> Oxoaldehyde/alcohol
>> Propylene oxide
>
> Refinery-grade propylene (60–75%): recovered from refinery off-gases in gasoline manufacture:
>> Cumene
>> Isopropanol
>> Oligomers (nonene; dodecene)

Figure 6.9 shows price histories for propylene and polypropylene.

It should be noted that at times grades are used other than as shown here. For instance, chemical (and occasionally polymer) grade can be (and is) used for

**Cents per pound**

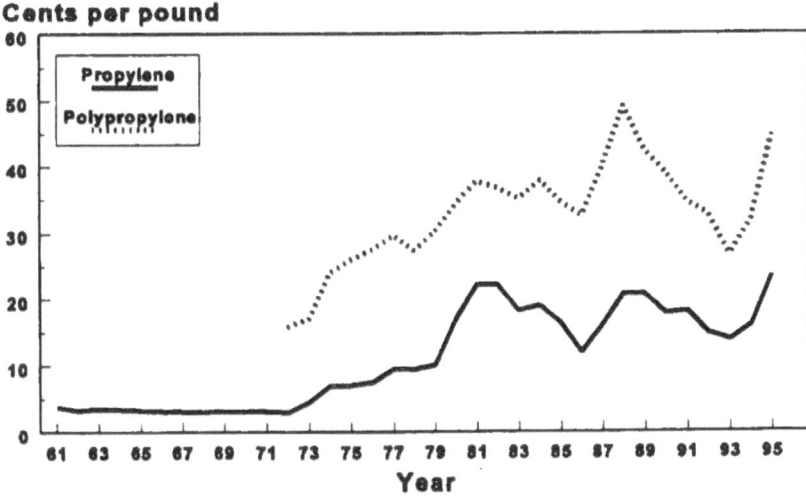

FIGURE 6.9    Price histories for propylene and polypropylene. (Data courtesy of CMAI.)

cumene or isopropanol production, and in some cases polymer grade is used for the production of the products listed for chemical grade. To a large extent, the polymer-grade requirement for the products shown results as much or more from the need to exclude impurities than for considerations of purity. However, the reader should note that economics sometime dictate a preference for polymer grade over chemical grade both from pricing considerations as well as regarding a capacity factor. One propylene oxide producer has chosen to use polymer grade exclusively, since it will provide a 5–8% increase in throughput as a result of the higher purity.

The products in the foregoing tabulation represent only the chemical demand for propylene. Much of the propylene generated in refineries is consumed for gasoline components (polygasoline in an oligomers unit) or is used as fuel in the refinery. The reason is that the combined production of propylene from olefins plants and refineries far exceeds the chemical demand.

The reader should be aware that a few installations have been built during the last 5 years that produce propylene by the dehydrogenation of propane. Plants of this nature are operational or planned for Mexico, Belgium, Russia, Saudi Arabia, South Korea, and Thailand. In most of these cases, the plant is feasible only if there is a large supply of low-valued propane, a significant demand for propylene with an otherwise inadequate supply, and/or a supply of low-priced energy (natural gas). Where one or more of these conditions is not met, propylene from such a unit is generally considered to be uncompetitive with propylene from olefins plants or refinery sources. Locations of these "on-purpose" units are shown in Table 6.12.

6.2.1.6  Butadiene.  During the last 20 years, much of the petrochemical industry has become so internationalized that simply showing U.S. production can be seriously misleading as a reflection of the supply of a product. Butadiene is a particular case in point. Since the early 1970s, butadiene has flowed from Europe to the United States in increasing quantities. As can be seen in the butadiene supply graph (Figure 6.10), butadiene sourced from dehydrogenaton units (dehydro) has declined the point of being an inconsequential factor in the butadiene supply equation. This has resulted from both the increased butadiene production from olefins plants (crackers) and from the European imports. Since virtually all Europe's crackers are based on naphtha/gas oil feedstock, butadiene production far exceeds the demand in Europe. U.S. pricing trends for ethylene and butadiene are shown in Figure 6.11.

Since ethylene demand (and therefore production) has grown at a rate that is greater than the butadiene demand growth, the domestic U.S. deficit has declined gradually. Since much of the U.S. production and the European-sourced material are coproducts of ethylene manufacture, the pricing of this product will be reduced to effectively compete with the on-purpose (dehydro) production. The result is the obvious decline in dehydro-sourced material. One problem with this scenario is that in the mid-1980s when butadiene supply and demand got out of balance, the price "spiked up" until sufficient product could be brought in from Europe. By the latter part of the 1980s, the increasing supply of butadiene produced in the United States resulted in a serious decline in butadiene price. The consequence was that European olefins plant operators began to

TABLE 6.12  Location of Propylene Units from Propane Dehydrogenation

| Country | Status | Capacity (tons × 10³) | Year of initial operation |
|---------|--------|------------------------|----------------------------|
| Thailand | Operational | 115 | 1990 |
| Korea | Operational | 165 | 1991 |
| Belgium | Operational | 250 | 1992 |
| Malaysia | Operational | 80 | 1993 |
| Mexico | Operational | 350 | 1994 |
| Saudi Arabia | Planned | 300 | 2002 |

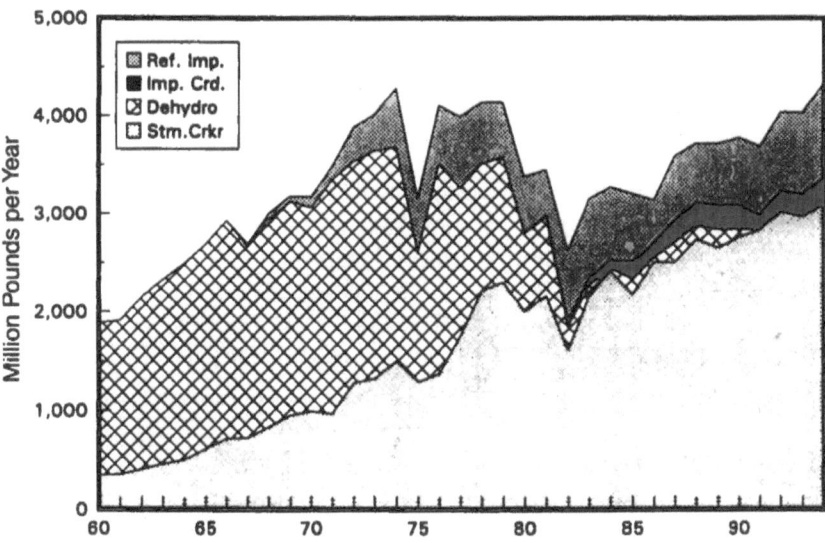

FIGURE 6.10   U.S. butadiene supply. (Data courtesy of CMAI.)

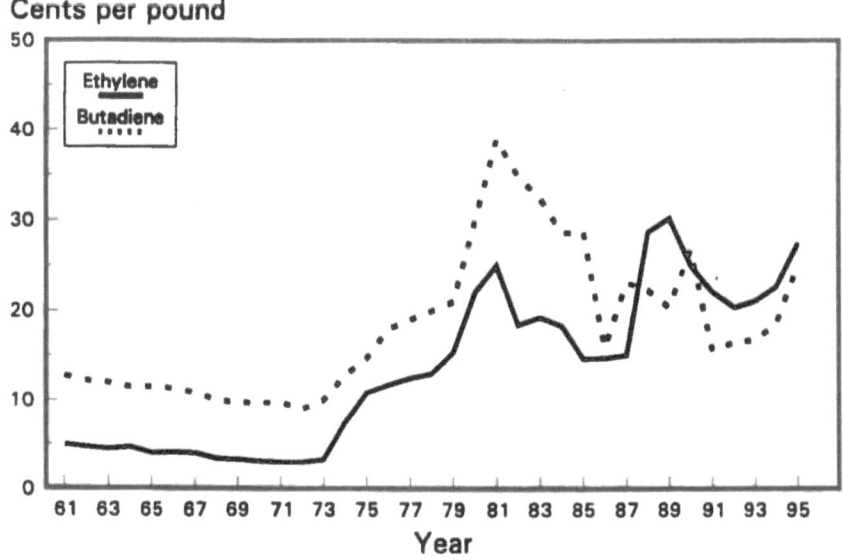

FIGURE 6.11   Price histories for ethylene and butadiene. (Data courtesy of CMAI.)

view the crude butadiene (crude $C_4$ stream) as a viable alternate feedstock stream. Many olefins plant operators began to value crude butadiene at approximately naphtha value (0.8–1.1 times the price of naphtha in dollars per ton). This established a floor for the U.S. butadiene price at the Rotterdam naphtha price, plus extraction cost, plus freight from Europe. This situation prevails to the present (1999).

**6.2.1.7 Toluene.** The sources of toluene lie primarily in the catalytic reforming of selected petroleum fractions rich in naphthenes or in the recovery of toluene contained in aromatic concentrate (pyrolysis gasoline) produced as a by-product of ethylene manufacture—mostly from naphtha/gas oil cracking. U.S. production and pricing for benzene and the aromatics discussed in Sections 6.2.1.8 and 6.2.1.9 are shown in Figures 6.12 and 6.13, respectively.

**6.2.1.8 Benzene.** Benzene is derived from several sources. A small amount is still recovered from coke oven by-product streams. However, the primary sources are from the catalytic reforming of petroleum fractions rich in naphthenes (see Section 6.2.1.7, above), recovery from aromatic concentrate (pyrolysis gasoline) produced as a by-product of ethylene manufacture, hydrodealkylation of

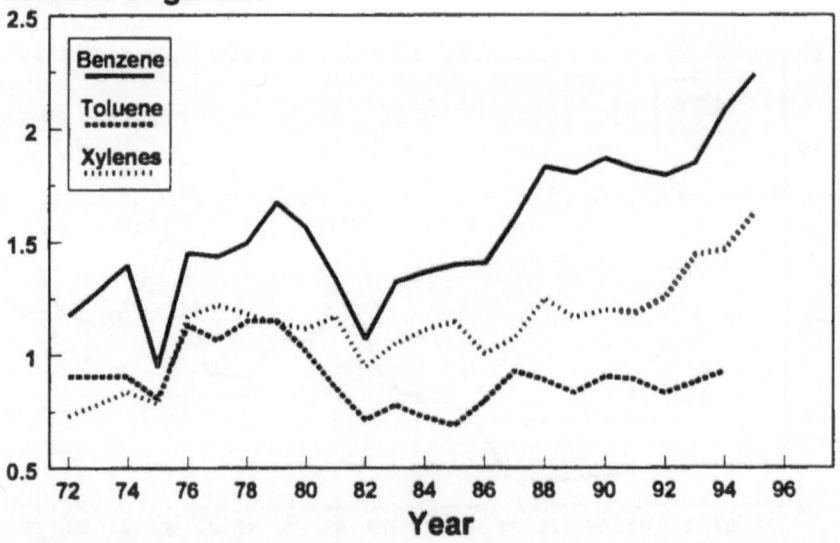

**FIGURE 6.12** Production histories of benzene, toluene, and xylenes. (Data courtesy of CMAI.)

**Cents per pound**

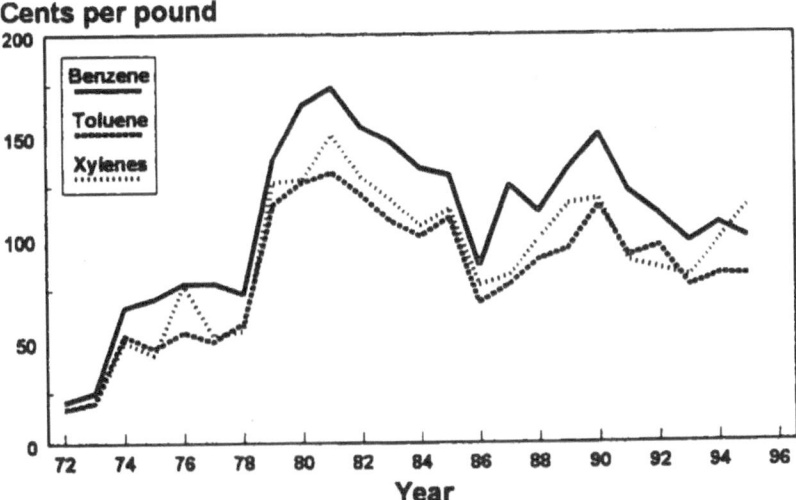

FIGURE **6.13**  U.S. price histories of benzene, toluene, and xylenes. (Data courtesy of CMAI.)

toluene as shown below, or disproportionation of toluene to benzene and xylenes:

$$C_6H_5CH_3 + H_2 \rightarrow C_6H_6 + CH_4$$

**6.2.1.9  Xylenes.**  Mixed xylenes are generally obtained by the catalytic reforming of petroleum fractions (see Section 6.2.1.7, above) or by the recovery of the $C_8$ fraction from an aromatic concentrate (pyrolysis gasoline) stream. An increasingly important source of mixed xylenes is from the disproportionation of toluene. The separation of mixed xylenes into para, meta, and ortho isomers can be accomplished by several methods (e.g., fractional crystallization).

### 6.2.2  Methods of Manufacture: Inorganics

Only methods that dominate the industry are discussed.

**6.2.2.1  Sulfuric Acid Reaction.**  Oxidation of sulfur dioxide by air to sulfur trioxide and adsorption of the $SO_3$ in water (contact process) proceeds as follows:

$$2SO_2 + O_2 \rightarrow 2SO_3$$

$$SO_3 + H_2O \rightarrow H_2SO_4$$

Oxidation takes place over a vanadium pentoxide catalyst at temperatures ranging from 400 to 600°C (temperature varies along the length of the reactor) and atmospheric pressure. Overall sulfur trioxide yield is approximately 97%. The sulfur trioxide is then sent to an oleum tower, and the remaining gas from this tower is sent to a second tower for scrubbing by 97% $H_2SO_4$. Frequently sulfuric acid is produced and sold that is more than 100% sulfuric acid. This is possible by absorbing surplus $SO_3$ in 100% sulfuric acid such that the resulting acid can be diluted with water and still be 100% sulfuric acid. Sulfuric acid plants typically yield from 100 to 5000 tons/day. The sulfur dioxide for reaction is usually obtained by burning elemental sulfur or by roasting iron pyrites. However, waste-gas streams from metallurgical plants, oil refineries, and other sources can also provide feed sulfur dioxide. In recent years, the necessity to remove sulfur and sulfur-containing products from refinery streams has produced surplus quantities of sulfur. Figure 6.14 plots production of sulfuric acid and ammonia between 1955 and 1995.

**6.2.2.2  Ammonia Reaction.**  The following catalytic reaction of nitrogen and hydrogen produces ammonia:

$$N_2 + 3H_2 \rightarrow 2NH_3$$

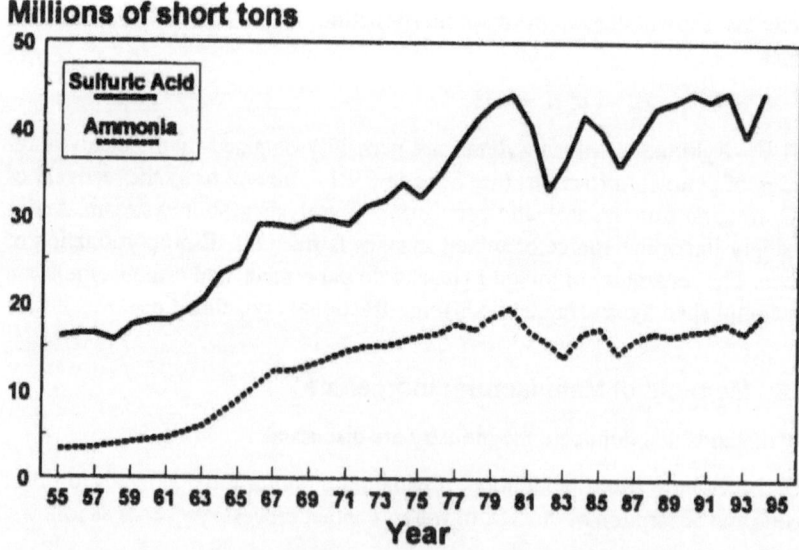

**Millions of short tons**

FIGURE 6.14  U.S. sulfuric acid and ammonia production histories. (Data courtesy of CMAI.)

Reaction takes place over an iron oxide promoted catalyst at high temperatures and pressures (400–600°C; 130–650 atm). Nitrogen is obtained by air liquefaction, by the producer gas reaction, or by removing oxygen from air by burning with hydrogen to leave nitrogen behind. Although hydrogen can be obtained from a variety of ways (e.g., cracking of natural gas or refinery gases deliberately to obtain hydrogen), more often, it is obtained as a by-product of associated processes such as the production of caustic/chlorine or from the production of ethylene and other olefins in a steam cracker from natural gas liquids or naphtha/gas oil.

**6.2.2.3 Chlorine Reaction.** About 98% of all chlorine is produced by the electrolysis of salt:

$$2NaCl + 2H_2O \rightarrow Cl_2 + 2NaOH + H_2$$

Direct current flows through a cell of salt solution to give the chlorine at the anode. Cells of two different kinds, the mercury cathode and the diaphragm, are used. Figure 6.15 gives chlorine production figures in the United States between 1955 and 1995.

Electrolysis of salt also provides the major method of producing sodium hydroxide. The hydrogen by-product may be reacted with the chlorine to make

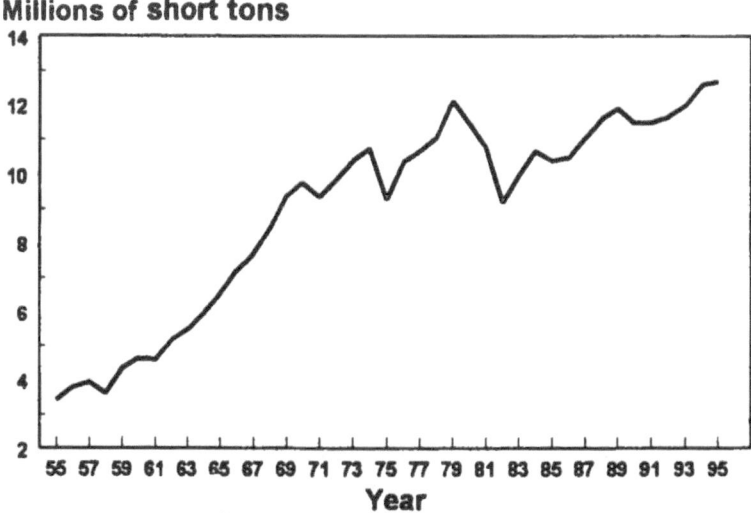

**Millions of short tons**

**FIGURE 6.15**   U.S. chlorine production history. (Data courtesy of CMAI.)

hydrochloric acid, used on site for making other chemicals, or burned. The combination product (i.e., chlorine and sodium hydroxide) is known as an electrochemical unit (ECU).

There are dramatic swings in the price of chlorine (and correspondingly the price of sodium hydroxide also), resulting in times when sodium hydroxide (caustic) is in very strong demand. This situation dictates the operating level of chlorine/caustic units, with the result that chlorine is produced in quantities greater than demand. The result is a very depressed price for chlorine (as Figure 6.16 shows for 1981–1982 and 1991–1992). Conversely, when chlorine in very strong demand and price is high, chlorine/caustic units are operated at levels to supply the chlorine demand, with the result that caustic is in oversupply with a corresponding depression in caustic price. During these drastic swings, chlorine may at times be priced well below the cost of production. Producers then depend on the caustic price to "carry" the chlorine production. Generally producers look at the ECU price rather than only at the chlorine or caustic price for determination of satisfactory economics.

US$ per short ton

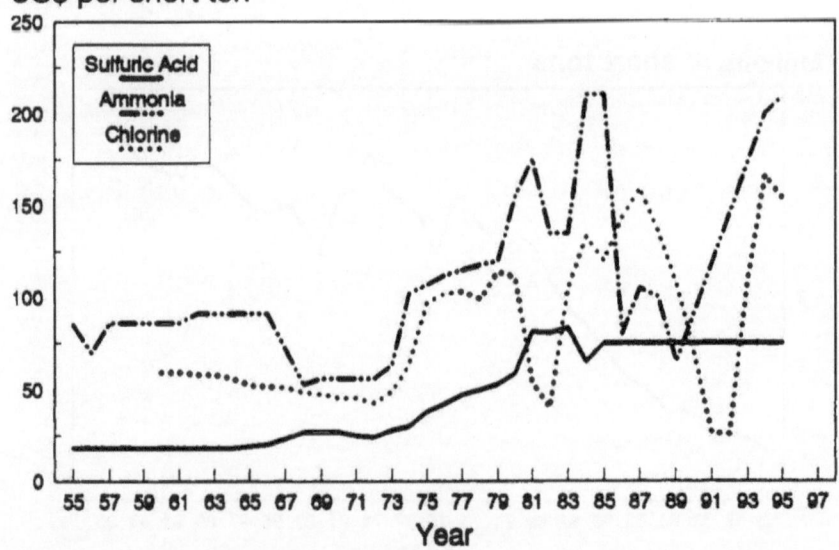

FIGURE 6.16   U.S. prices of sulfuric acid, ammonia, and chlorine. (Data courtesy of CMAI.)

### 6.2.3 Methods of Manufacture: Plastics

Only methods that dominate the industry are discussed.

**6.2.3.1 Polyethylene Reaction.** The addition polymerization of ethylene proceeds as follows:

$$nCH_2 = CH_2 \rightarrow (-CH_2-CH_2-)_n$$

Low-density polyethylene (LDPE: = 0.92 g/cm$^3$) is made continuously by a high-pressure (1200 atm) process at temperatures of about 200°C. Yield per pass through the reactor is about 25%. Unreacted ethylene is recycled to the reactor. Overall yield is approximately 95%. Linear low density polyethylene (LLDPE) is produced in a gas phase reactor at low pressures. LLDPE, as the name implies, does not have the chain branching that characterizes LDPE. High-density polyethylene (HDPE: = 0.97 g/cm$^3$) is produced by a low-pressure, low-temperature suspension polymerization (Ziegler–Natta process), using a triethylaluminum–titanium tetrachloride catalyst. The advent of the production of LLDPE had a double effect on ethylene demand for polyethylene. First, LLDPE has greater tensil strength than conventional LDPE, thereby allowing extruders to "down-gage" (i.e., make the polyethylene film thinner and still achieve the necessary strength for applications such as sandwich or garbage bags). This results in processors requiring less total polyethylene to produce the same number of bags. Additionally, the production of LLDPE requires about 0.92–0.94 pound of ethylene per pound of polyethylene, since an $\alpha$-olefin comonomer such as butene-1 or octene-1 constitutes the remaining olefin (about 0.07–0.08 pound per pound of polyethylene). This compares with the ethylene requirement of about 1.015 pounds of ethylene per pound of conventional (high-pressure) polyethylene.

U.S. price and production data for ethylene, LDPE, and HDPE are plotted in Figures 6.17 and 6.18, respectively. Figure 6.19 gives price data for propylene and polypropylene.

**6.2.3.2 Polypropylene Reaction.** The addition polymerization of propylene proceeds as follows:

$$nCH_2=CH_2-CH_3 \rightarrow [-CH_2-CH_2-]_n$$

The Ziegler process is used. Reaction takes place in the presence of a trialkylaluminum–titanium tetrachloride catalyst at approximately 3 atm pressure and 70°C.

### 6.2.4 Methods of Manufacture: Fibers

Only methods that dominate the industry are discussed.

Cents per pound

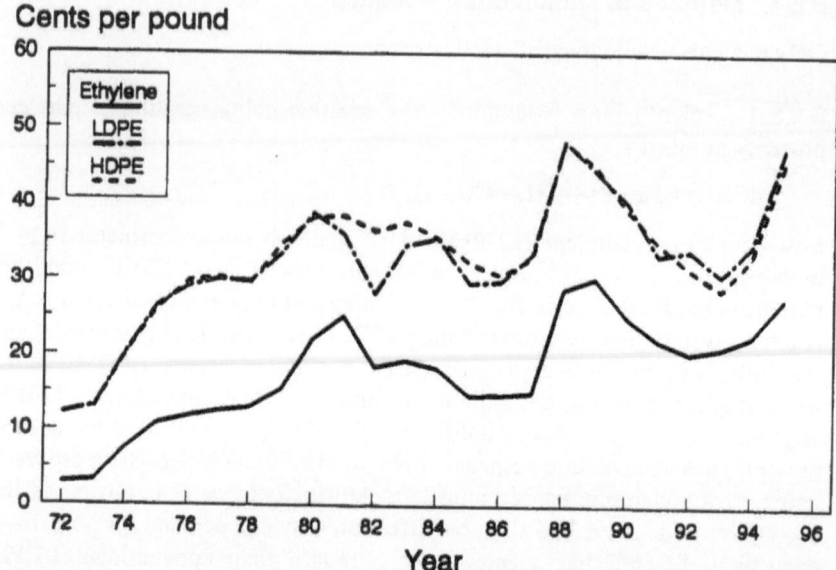

FIGURE 6.17 U.S. ethylene and polyethylene (LDPE, HDPE) price histories. (Data courtesy of CMAI.)

6.2.4.1 Polyester Fibers Reaction. Ester interchange of ethylene glycol with dimethyl terephthalate is followed by condensation polymerization of the intermediate:

Billions pounds per year

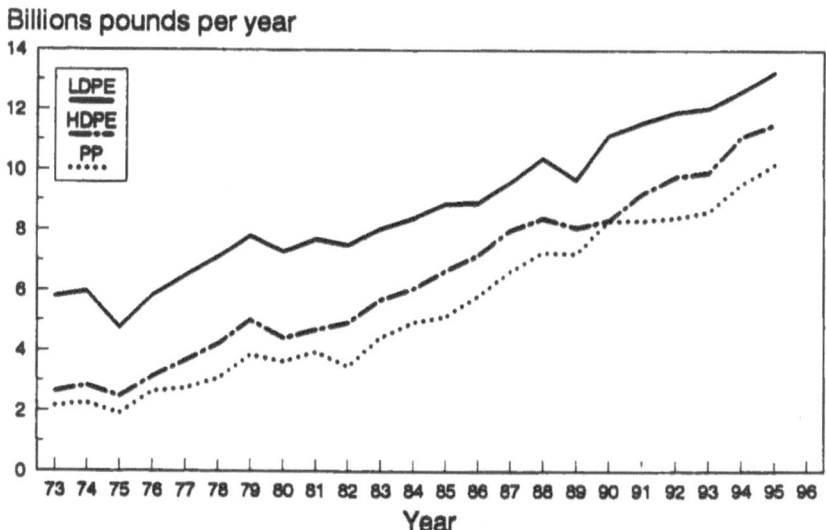

FIGURE 6.18    U.S. production histories of polyolefins. (Data courtesy of CMAI.)

## Cents per pound

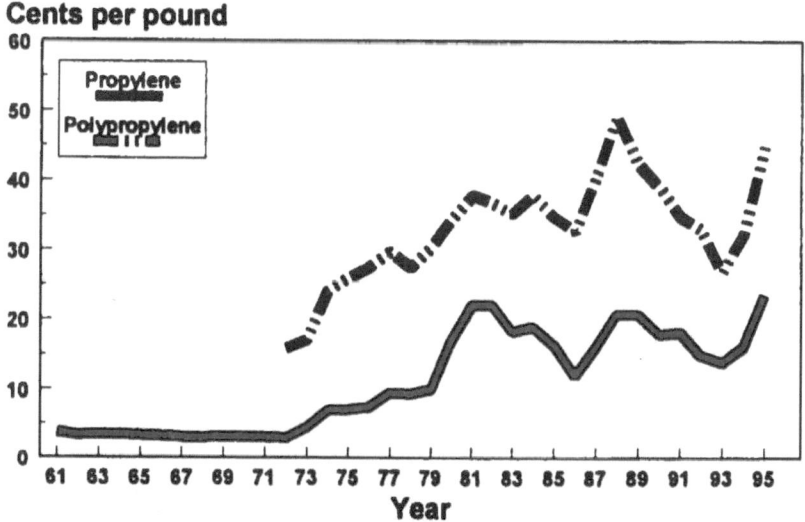

FIGURE 6.19    Price histories for propylene and polypropylene. (Data courtesy of CMAI.)

**Billion pounds per year**

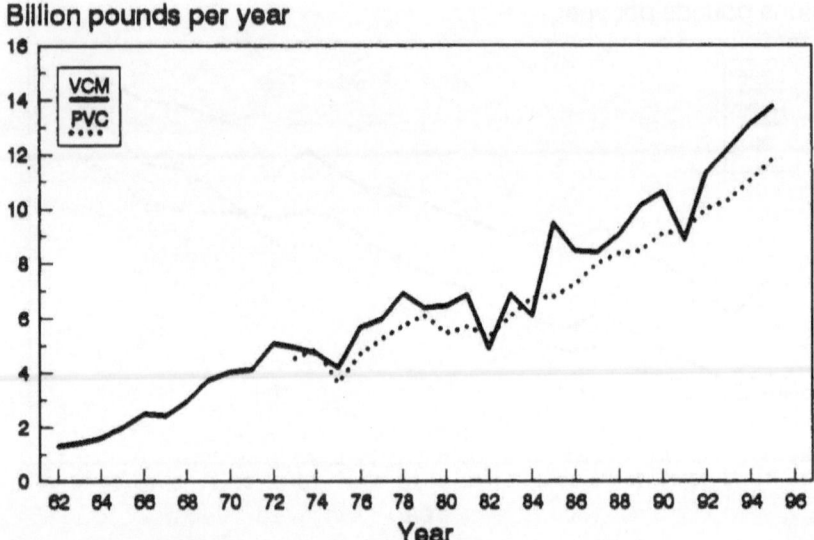

FIGURE 6.20   Industrial routes to nylon.

**Cents per pound**

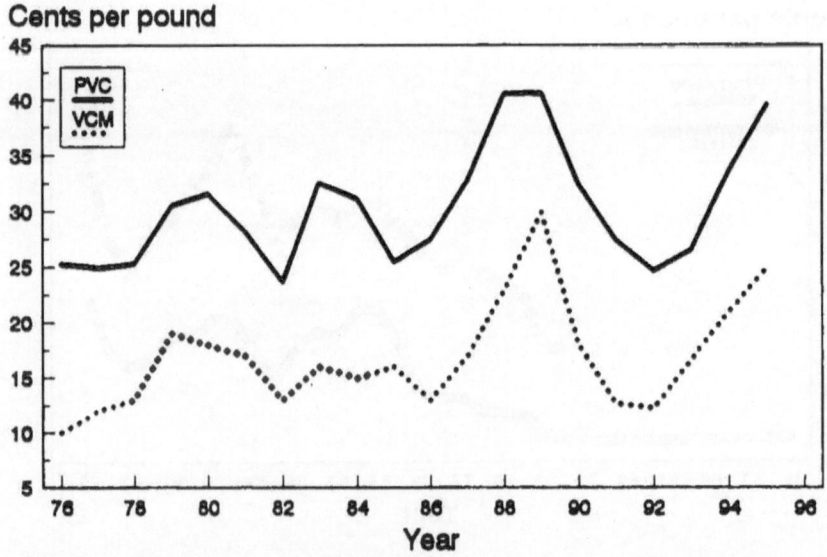

FIGURE 6.21   U.S. VCM and PVC price histories. (Data courtesy of CMAI.)

**Billions of pounds**

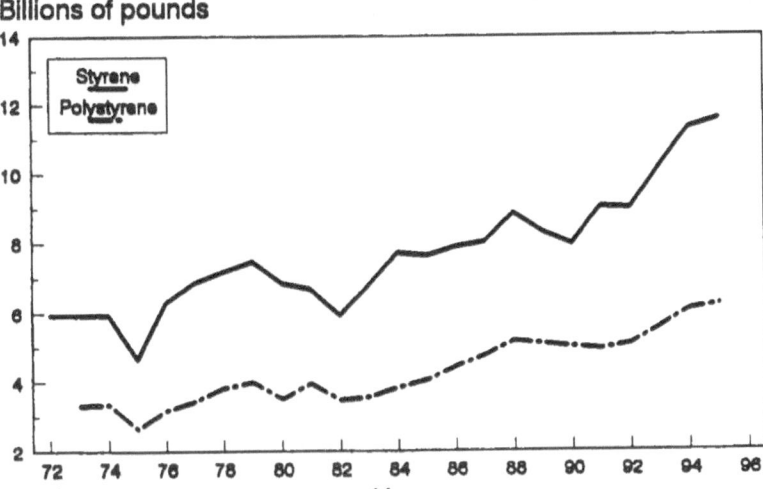

FIGURE 6.22 United States styrene and polystyrene production histories. (Data courtesy of CMAI.)

**Cents per pound**

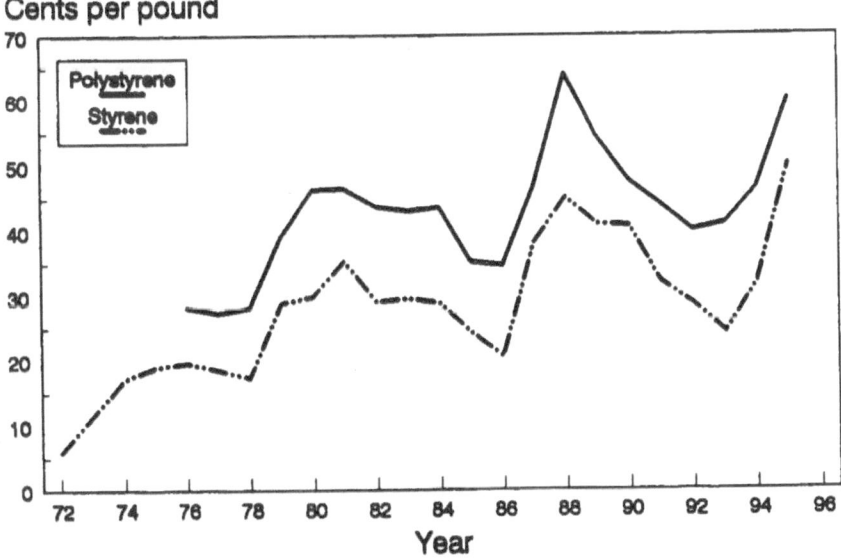

FIGURE 6.23 United States styrene and polystyrene production histories. (Data courtesy of CMAI.)

**Billions of pounds**

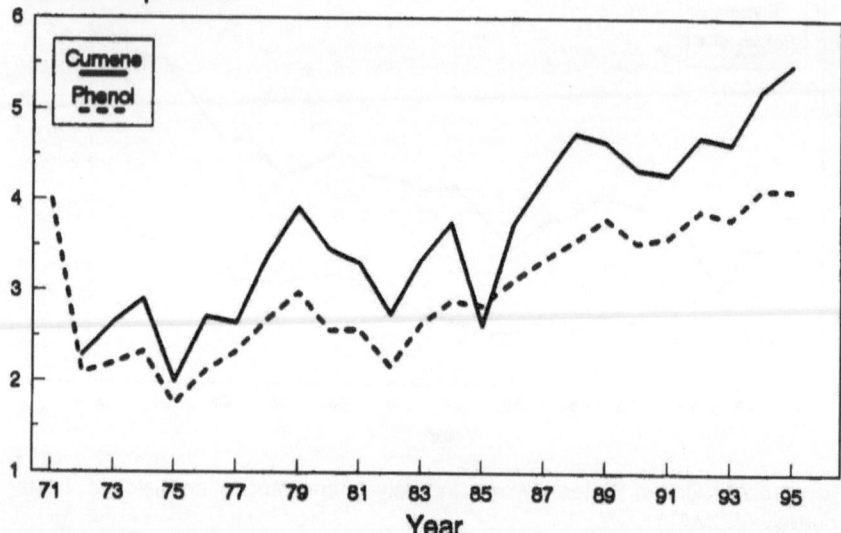

FIGURE 6.24    United States cumene/phenol production histories. (Data courtesy of CMAI.)

**Cents per pound**

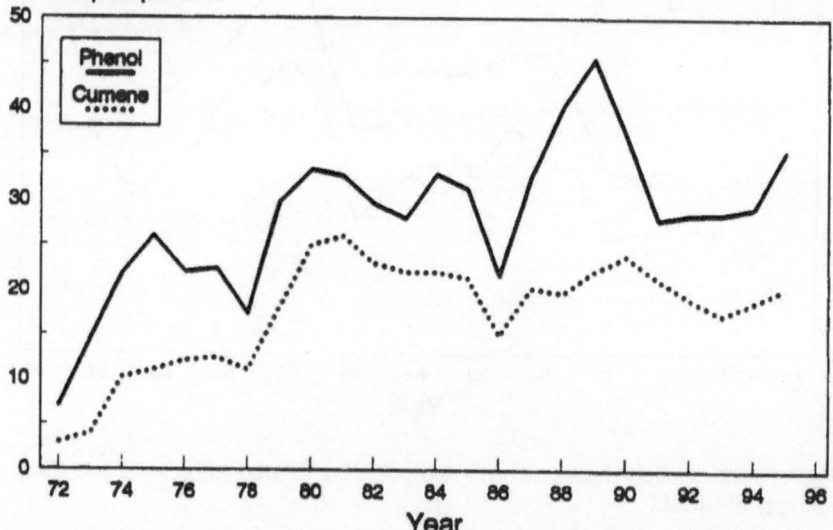

FIGURE 6.25    United States cumene/phenol price histories. (Data courtesy of CMAI.)

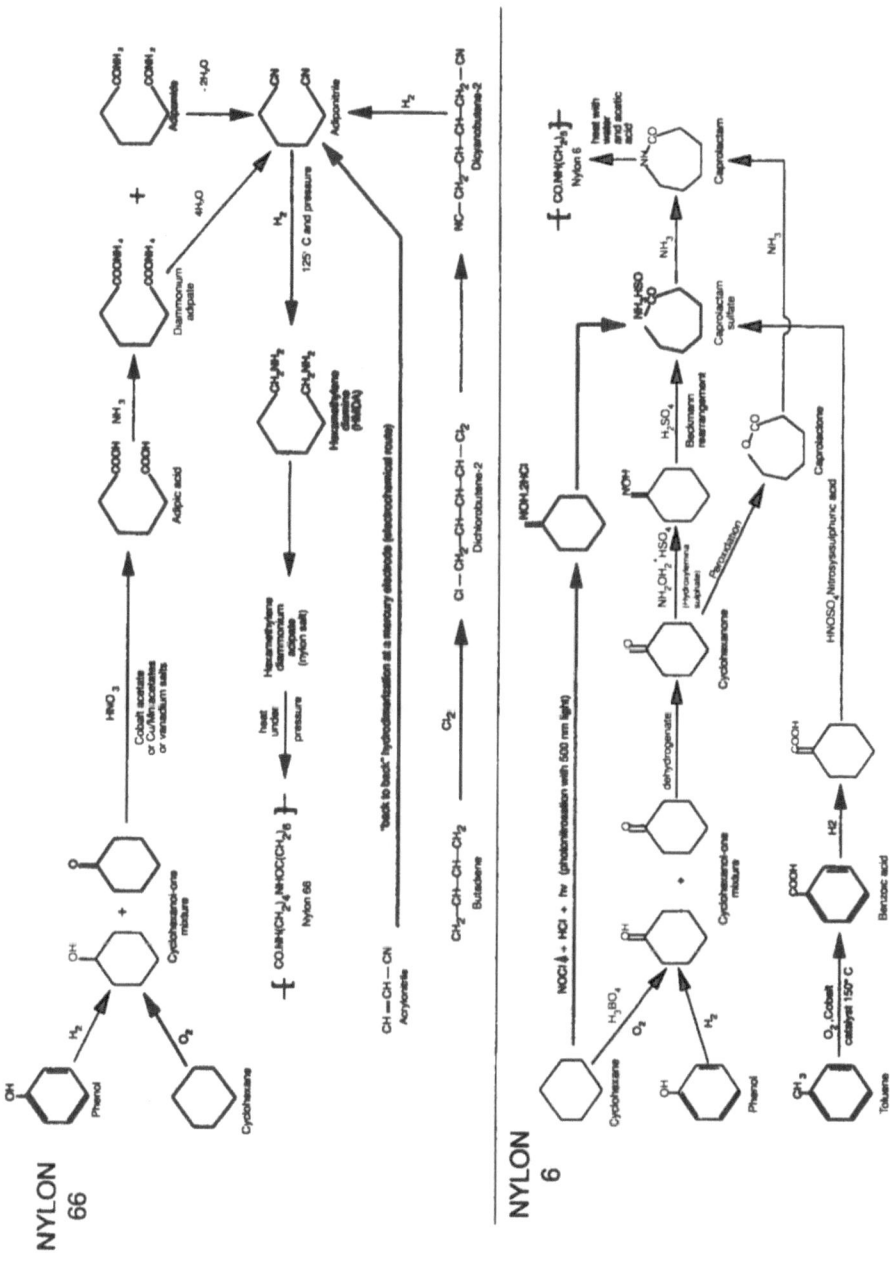

**Figure 6.26**  Industrial routes to nylon.

This route accounts for most polyester manufacture (for production and price data, see above: Figures 6.27 and 6.28, respectively). However, the direct polyesterification of ethylene glycol with terephthalic acid is becoming important:

### 6.2.4.2 Nylon Fibers Reaction.
Industrial routes to nylon are complex, involving several intermediate steps. Nylon 6 and nylon 66, the two types made in the United States, start from phenol, cyclohexane, and toluene. The reaction sequences used for both are shown in Figure 6.26. Molten nylon is transformed into fiber by melt-spinning and stretching operations.

### 6.2.4.3 Acrylic Fibers Reaction.
The free radical or anionic polymerization of acrylonitrile proceeds as follows:

$$n\text{CH}_2=\text{CH}-\text{CN} \rightarrow [\text{CH}_2-\text{CH} \rightarrow -\text{CN}]_n$$

Reaction is usually a suspension or solution–polymerization process. The polymer can be either dry- or wet-spun and subjected to stretching processes to produce fiber.

## REFERENCES

1. J. Wei, T. W. F. Russell and M. W. Schwartzlander, *The Structure of the Chemical Process Industries*, McGraw-Hill, Inc., New York, 1979.
2. Private communication, 1999.

## QUESTIONS FOR DISCUSSION

1. Compare the growth rate of methanol production between 1955 and 1970 with the growth rate of (a) sulfuric acid, (b) benzene and (c) ethylene.
2. What caused the decline in methanol production in 1984–1985?

Millions of pounds

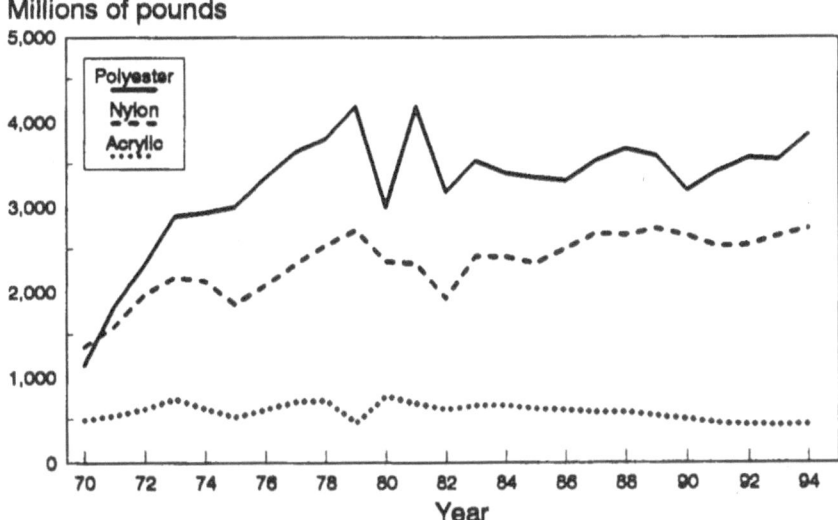

**FIGURE 6.27** U.S. production histories for polyester, nylon, and acrylic. (All fiber production figures are from private sources.)

Cents per pound

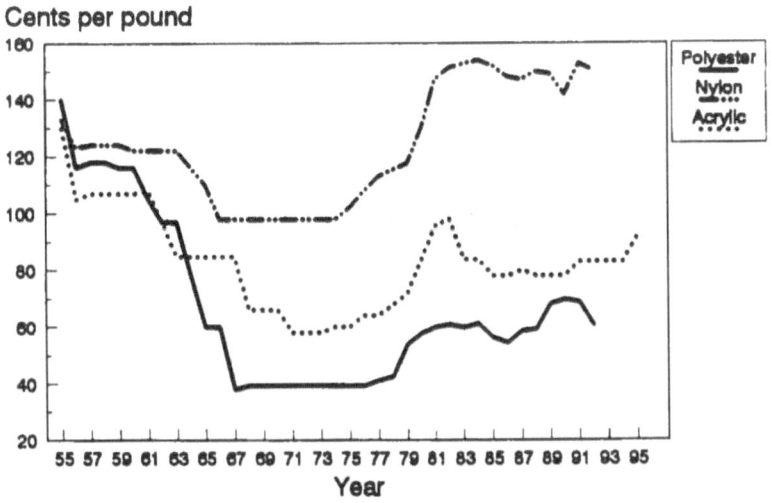

**FIGURE 6.28** U.S. price histories for polyester, nylon, and acrylic. (All fiber prices are from private sources.)

3. What caused the surge in production of methanol beginning in 1992?

4. Which sectors in Table 4.4 would have the greatest effect on methanol production if their final demand is increased by 10%.

5. What general pattern do all production histories presented in this chapter show? Which chemical products seem to show different trends? Suggest reasons for the different behavior.

6. What has happened since 1973 to cause the methanol experience curve to be displaced as shown in Figure 6.5?

7. Which four chemical products in this chapter have shown the highest growth rate from 1980 to 1994? Suggest reasons for this in terms of final consumer demand.

8. What general pattern do price histories show? Which one(s) seems to display unusual trends? Suggest reasons for this behavior.

9. Why are the prices of coproducts depressed where naphtha and gas oil are the primary feedstocks for ethylene production?

10. How can propylene purity affect capacity?

11. What would determine whether refinery-grade propylene is used for chemicals manufacture rather than serving in lower valued fuel uses?

12. Which five chemicals have the highest concentration ratios? What are the reasons for this? Identify chemical products that would have a concentration of 100% and only one company.

## PROBLEMS

6.1. Prepare an experience curve for ethanol using the same format as Figure 6.5. Compare the behavior of methanol and ethanol and write a paragraph of discussion.

6.2. Compare the latent heat, heat of combustion, and octane number of ethanol with the values of methanol in Table 6.1. How does the price of ethanol in dollars per million Btu compare with the price of methanol? What are the important methods of manufacture of ethanol? Who are the five leading producers? How many of these also produce methanol?

6.3. Periodically *Chemical Week* devotes a page to a chemical listing prices, producers, capacity, and end uses. One such chemical in the December 20/27, 1995, issue was MDI (methyl-di-*para*-phenylene isocyanate). Polycarbonates and *o*-xylene were featured in the November 22, 1995, and May 3, 1995, issues. Read two of these articles and summarize your conclusions in one or two paragraphs, citing data to substantiate your conclusions.

6.4. Divide the experience curve for methanol into spring, summer, autumn, and winter periods of its life cycle.

6.5. Identify five chemical products that are in the spring of their life cycles.

Identify five chemicals in the winter of their life cycles. How many of these latter products are likely to be replaced within the next decade?

6.6.   Plot the production history of ethanol since 1930 and compare with the history of synthetic ethanol.

6.7.   In Figure 6.18, what caused the decline in butadiene production in the early 1980s? What conclusions can be drawn from this figure regarding the supply of butadiene from 1980 to the present?

6.8.   In the mid-1970s, methanol consumption was predicted by two groups: How well did these predictions compare with Figure 6.3, assuming that all methanol produced was consumed? If you had an opportunity to interview the people who prepared these estimates, what questions would you ask?

| Use | Worldwide methanol consumption (t/day) | |
| --- | --- | --- |
|  | 1970 | 1980 |
| Chemical use | 25,000 | 50,000 |
| Feedstock for natural gas | 0 | 57,600 |
| Gas turbines | 0 | 53,000 |
| Animal protein production | 0 | 33,000 |
| Boiler fuel | 0 | 42,900 |
| Gasoline substitute | 0 | 13,000 |
| Gasoline additive | 0 | 762 |
| TOTAL | 25,000 | 250,262 |

*Source:* J. Tourtellotte, S. Bangiororno, D. Shah, Eng. Found. Conf., New England College, Henniker, N.H. 1974.

| Use | 1985 World methanol demand (t/day) | |
| --- | --- | --- |
|  | Low estimate | High estimate |
| Steel industry | 223,500 | 287,000 |
| Motor fuel | 30,000 | 241,600 |
| Fuel (turbines, etc.) | 15,100 | 45,000 |
| Ammonia synthesis | 12,100 | 15,100 |
| Protein production | 7,500 | 15,100 |
| TOTAL | 288,200 | 604,100 |

*Source:* Bureau d'Etudes Industrielles et de Cooperation, Paris, 1975.

6.9.   Cumene is a significant raw material for the manufacture of phenol. In Figure 6.10, there was a crossover of the production curves for phenol and cumene. What caused the production decline of cumene in 1984?

6.10.    Polyester, nylon, and acrylic fiber price dropped significantly in the 10-year period from 1957 to 1967. To what is the decline attributed?

6.11.    A product with price/quantity-produced characteristics putting it in the lower left-hand part of Figure 5.4 (to the left of the straight line) can be classified as having unfulfilled potential. (At that selling price, there should be a greater quantity produced.) Try to identify two such products by searching the appropriate sources for $p$ and $q$. What markets must these products penetrate to fulfill their potential?

6.12.    There are a number of products in the upper left-hand side of the line in Figure 5.4. Such products might be termed *vulnerable*, since they seem to be selling at a higher price than might be warranted. Identify five such products and explain why they are selling at this high price. What would happen if a suitable substitute were found?

6.13.    Calculate the concentration ratio (top four) for (a) computers, (b) automobiles, (c) aluminum, and (d) airplanes.

6.14.    Compute an approximate concentration ratio for the chemical industry as a whole. How does it compare with ratios obtained in Problem 6.13?

6.15.    Examine the production and price histories of cumene, LLDPE, HDPE, and acrylic fibers and identify the part of the life cycle of each in the period 1987–1992.

# 7

## Specific Companies

A company's general health, progress, and performance are reviewed by various groups according to the following criteria (1):

*Criterion 1.* Investors asks three sets of questions. The first set is concerned with present performance and how it compares with recent past performance. Is the company making money and paying a dividend? Is the price of the stock rising? Does the company have a good net income as measured by profit per dollar of sales, profit per dollar of investor's equity, and profit per dollar of owner's equity plus long-term debt?

A second set of questions asked by investors concerns the security of the company against unexpected occurrences. Is the company heavily in debt, so that a business downturn would cause it grief? Could it run out of cash? Is it in trouble with the Securities and Exchange Commission (SEC), the Justice Department's Antitrust Division, the Environmental Protection Agency (EPA), the Occupational Safety and Health Agency (OSHA), or the Food and Drug Administration (FDA)? Are its overseas properties in danger of being confiscated by local governments? Is Congress about to pass legislation to curtail its activities or profitability?

The final set of questions concerns future performance. Are the gross sales and net income increasing smoothly and outpacing or at least keeping pace with growth rates of population and GDP? Is there sufficient spending on research and development as well as on new plants and equipment so that the company and its technology will not become obsolete? Is the company innovating, so that it will lead the rest of industry and the economy? Does the firm have the proper portfolio of products to assure that its innovative products are supported by its mature, well-established products?

Each product of the firm may be characterized by two parameters: the market growth potential of the product and the competitive strength of the firm for that product. Each product occupies a point in Figure 7.1, which is divided into four quadrants. A *wildcat* product is characterized by high growth potential but low competitive strength. This is a springlike time in the life cycle and requires management to put a great deal of cash into the project and pay considerable attention to it. If the wildcat product is supported generously enough, it may develop greater competitive strength and turn into a *star*. The star product is characterized by high growth and high competitive strength. Its rapid expansion into the marketplace and the need to build manufacturing facilities demand a heavy flow of capital to sustain its growth. Eventually, the growth of the star slows down to that of the *cash cow* product, characterized by a low growth rate and high competitive strength. During this period there is a large surplus of cash available, which can be used to support other projects in the firm. The *dog* product is characterized by low growth and low competitive strength. It is sometimes a contributor of cash and may be kept; but more often it is a candidate for elimination from the portfolio so that funds can be released to finance other products.

A well-managed company, concerned about the future, should have a suitable mix of products in the various sectors. A company with too many wildcats and stars will have no source of cash to finance the many worthwhile projects that need support and growth. A company with too many cash cows and dogs

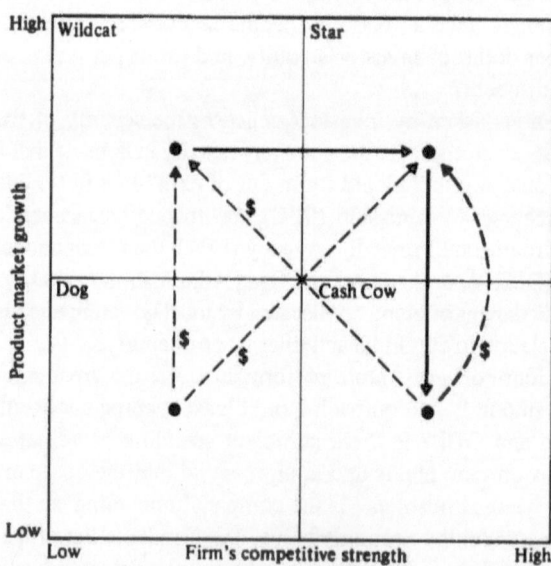

FIGURE 7.1   Classification of products.

will have plenty of cash on hand but an unexciting future. Very often a large and stable company specializes in projects in the cash cow and dog positions and has cash available to acquire smaller venturesome companies that have good ideas but no cash.

*Criterion 2.* Lenders ask the same questions as investors. In addition, lenders are particularly concerned about the company's ability to repay its debt on time—with interest. Lenders are most interested in short-term cash positions, or liquidity, as measured by the ratio of current assets to current liabilities. They are also concerned with the ratio of cash flow (net income after taxes plus depreciation) to interest on debts.

*Criterion 3.* Employees are concerned with the same questions as investors and lenders. They also have special interest in salaries, occupational safety and health, employment stability, and career opportunities. With less mobility than management, they may be opposed to plant relocations and new plant investment abroad. They may be more interested in salary and wage increases than in dividend payout.

*Criterion 4.* Customers are interested in a steady supply of the company's products with increasing quality and decreasing price. They are also interested in new products with superior "performance in use" to satisfy new needs and to replace older products.

*Criterion 5.* The government is concerned with taxes. In addition, the government is interested in making sure that the company conforms to regulations relating to pollution, occupational safety, and employee welfare and that it does not enjoy a monopolistic position. The government is also concerned with the country's general economic welfare— balance of trade with other nations, inflation, unemployment, product liability, and safety of the public. When government agencies review an industry or a specific company, all these factors affect the decisions made.

*Criterion 6.* The public is generally ambivalent to the needs of a large company. The public fears concentrated power that is self-serving and harmful to the general welfare. It is particularly concerned with product availability and price, product liability, encroachment on the rights of individuals, pollution of the environment, safety, and depletion of natural resources.

The major chemical companies have had changes in their focus and in the direction management has led them, in part due to global pressures. Table 5.12 is a display of the top 10 chemical manufacturers from 1940 to 1996. As a group, sales have grown for these 10 companies from $1 billion in sales in 1940 to over $90 billion in 1996. It is interesting to note the relative positions of some of the major firms. Du Pont was number one in sales for about 50 years, but Dow

Chemical was steadily moving up and in 1995 became the leader. Du Pont changed it management focus, becoming a more diversified product manufacturer. For many years, Union Carbide was second in sales with approximately 60,000 employees at its peak. But after the devastating Bhopal incident, a hostile takeover bid by GAF (General Airline and Film), the sale of agricultural chemicals and the consumer products division, as well as corporate restructuring, the company shrunk by about 80%. Monsanto steadily moved up the sales list, maintaining third, fourth or fifth position from 1960 to 1996. In 1997 Monsanto spun off its chemical operations into a new firm called Solutia. Monsanto's current emphasis is in the agrobiochemical business. As a result, Monsanto will fall in its sales position among the top chemical companies.

Other interesting items of note are that American Viscose in the 1940s to mid-1950s was a major marketer of chemical products but disappeared after the mid-1950s when it was sold. Allied Corporation in the early years was a large producer of chemicals and dyestuffs but experienced a steady decline in part because of management problems. It merged with the Signal Companies, Inc., becoming Allied Signal in 1985. Celanese maintained a reasonably strong position until it was beset by financial problems. It was purchased by Hoechst in the 1980s. Exxon has steadily grown in chemical sales from the 1970s and now occupies the third position. Other petroleum companies with sizable but varying chemical sales include Mobil, Amoco, and Occidental. This information in Table 5.12 is a pictorial history of the fortunes of chemical firms in the United States over 57 years. Whenever a company moved from a top position, certain factors caused this to occur, and these are worth pursuing to determine the causes of corporate decline.

To deal with the criteria above adequately, much data must be analyzed. This chapter is designed to give the reader basic data for several selected companies and an idea of how company data can be analyzed. Its format is similar to that in Chapter 6. Section 7.2.1 gives a detailed analysis of a specific company: Air Products and Chemicals, Inc. The analysis answers many of the questions above as well as those posed in Section 0.2.2 in the discussion of company term papers. Air Products has been selected for detailed analysis because it is a stable company with a history of high growth and has excellent future prospects. Sections 7.2.2–7.2.11 give some basic facts for an additional 10 companies.

The companies chosen for background material in Section 7.2 of this chapter include four chemical companies, two oil companies, four medium-sized companies in the agricultural-biotech, food, paper, and pharmaceutical sectors, as well as a major diversified company. They were selected based on their major business lines as examples of the CPI. The data presented were obtained from Standard & Poor's Stock Reports, Moody's Investor Service, company annual reports, and company 10K reports. This material, which is referred to in future chapters, should provide the reader with a basis for understanding some of the similarities and differences of the CPI companies. The short summaries presented in Section 7.2 provide a limited amount of information chosen to illustrate important characteristics of the companies. You are encouraged to broaden your knowledge of these and other CPI firms by analyzing pertinent information from numerous sources including annual reports and 10K reports available on the Internet. A listing of companies in the CPI on the New York, American and Over-the-Counter exchanges is provided in Appendix B.

## 7.1 GENERAL COMMENTS ON COMPANY REPORTS

The information in Section 7.2 is based upon mid-1999 data. It was collected prior to the Exxon–Mobil and Dow–Union Carbide mergers, although mention is made of these impending changes.

## 7.2 SELECTED CPI COMPANIES: BASIC DATA

### 7.2.1 Air Products and Chemicals, Inc.

Air Products and Chemicals has a rich and varied corporate history. The company started as a single business based on a revolutionary idea for selling industrial gases. The company later expanded successfully into the chemicals business, primarily on the strength of shrewd acquisitions. Air Products has entered and exited the engineering design and construction business, as well as the waste management and electrical energy businesses. Because it has such an academically interesting corporate history, Air Products is covered in somewhat more depth and breadth than the other companies cited here.

Air Products was founded by the late Leonard P. Pool in 1940 on the strength of a simple, but then revolutionary, idea: the "on-site" concept of producing and selling industrial gases, primarily oxygen. At the time, most oxygen was sold as a highly compressed gas in cylinders that weighed five times more than the gas product. Air Products built oxygen gas generating facilities adjacent to large-volume gas users, thereby reducing distribution costs. World War II diverted the company's attention to the design and manufacture of mobile generators to produce oxygen for use by the military in high-altitude flights. At the end of the war, the company refocused its sights on commercial markets by setting up operations near Allentown in Pennsylvania's Lehigh Valley, close to the industrial markets of the Northeast. In 1945 Air Products secured a contract with Weirton Steel Company in West Virginia (and later contracts with Bethlehem, U.S. Steel, Acme, and Jones & Laughlin) to lease three generators to produce 6 tons of oxygen per day. The contract confirmed the value of the company's "on-site" marketing and financing strategy. By building and operating its own air separation plants and supplying gas "over the fence" to customers on a take-or-pay basis, Air Products became the first company to put into practice what would later become the marketing norm for the industrial gas industry.

The 1950s brought more civilian business and new government and military contracts. In support of America's emerging missile and space program, Air Products began to design and manufacture plants capable of producing tonnage quantities of liquid oxygen and nitrogen. The experience gained in building these plants—and later liquid hydrogen plants—for the federal government catapulted Air Products into a major position in the commercial industrial gas business. Several independent regional gas distributors were acquired, and the marketing concept known as "piggybacking" was introduced. This concept, which involved adding extra gas liquefaction capacity to the on-site plants, enabled Air Products to serve its tonnage base-load gas customers, such as steelmakers, and produce and deliver additional higher margin liquefied gases economically to other "merchant" gas customers (i.e., users of smaller tanker-load quantities) in the surrounding area.

In 1957 Air Products entered the overseas market for industrial gases through a joint venture (later 100% acquired) with the Butterley Company, a British firm. In Asia, the company took minority positions in industrial gas companies in Korea, Japan, Malaysia, Hong Kong, the People's Republic of China, Thailand, and Taiwan. Further geographic expansion occurred with joint ventures in Spain and Mexico and the acquisition of Inter-City Gas Company in western Canada. In the late 1960s and 1970s, the company expanded its business devoted to the design of cryogenic equipment and other process and related engineering services. Processes for the liquefaction of helium gas and natural gas, and techniques for the manufacturing of giant heat exchangers for such applications, were notable technical achievements. Most of the process designs and heat exchangers for natural gas liquefaction projects around the world have been furnished by the company.

During the 1980s, and beyond, noncryogenic technologies for producing industrial gases were developed and commercialized, and the company embarked on a comprehensive effort aimed at achieving quality throughout its operations in its drive for excellence and future success. In the early 1990s, the company continued to invest globally, expanding into Italy by acquiring a 49% interest in Sapio, a leading Italian industrial gas supplier; forming a 50/50 joint venture in Japan to sell the company's nitrogen trifluoride to semiconductor manufacturers and other industries in that country (thus making its first wholly owned investment in Japan to serve semiconductor manufacturers and chemical customers throughout Asia) establishing a new joint venture in Indonesia; forming a joint venture with a Czechoslovakian manufacturer of cryogenic equipment; and establishing a company in Poland. The company also acquired Permea, Inc., a leading supplier of membrane and adsorption gas separation systems that complement Air Products's internally developed membrane and adsorption technologies. As a result of the acquisition, the product mix that Air Products can now offer in noncryogenic gas supply systems is one of the broadest available in capacity, purity, and cost-effectiveness.

Air Products's latest acquisition (announced in the July and September 1999 issues of *Chemical and Engineering News*) is the purchase of half the $11.2 billion assets of BOC, with the other half going to Air Liquide of France, the world's largest industrial gas producer. This deal will make Air Products the world's second largest industrial gas producer, with assets of $13 billion and yearly sales of $8 billion.

In 1962 Air Products acquired the Houdry Process Company and its subsidiary, the Catalytic Construction Company. Catalytic was well known in the engineering and construction field and had pioneered the contract maintenance concept for the maintenance and repair of chemical and petroleum plants. In 1982 Air Products acquired Stearns-Roger Corporation, a leading engineering and construction services company based in Denver, Colorado. Following several years of profitable operations and a merger of Stearns-Roger with Catalytic, Inc. to form Stearns Catalytic World Corporation, business prospects dimmed, in large part because of to significant declines in energy prices. In November 1986 Air Products divested the majority of its engineering services business by selling the domestic operations of Stearns Catalytic to United Engineers & Constructors, a subsidiary of Raytheon.

In 1961 Air Products began manufacturing chemicals via a joint venture with Tidewater Oil Company to produce oxo-alcohols for use in producing plasticizers. The company's position in chemicals expanded with the 1969 acquisition of Escambia Chemical Corporation, a large chemical complex at Pensacola, Florida. Escambia manufactured industrial chemicals, such as amines and polyurethane intermediates, polyvinyl chloride resins, and fertilizers. The 1980s continued to reflect Air Products' strategy of strengthening its base businesses. In chemicals, plant expansions in both polymer emulsions and polyvinyl alcohol helped the company keep up with its growth in these key product lines. In addition, the acquisition of the industrial chemicals division of Abbott Laboratories in 1985 gave the company increased manufacturing capability for new specialty amines developed in Air Products's research laboratories. During this period the company also completed several acquisitions that extended its polyurethane additives business. A significant expansion of the company's emerging epoxy curing agents business was undertaken in 1988 with the acquisition of a British firm, Anchor Chemical, Plc. The chemicals business was further expanded in the 1970s with acquisition of the chemicals and plastics business of Airco, Inc. This added several chemical products to the company's portfolio, such as polyvinyl acetate emulsions, polyvinyl alcohol, acetylenic chemicals, and fabricated plastics. During this period the company also significantly increased production of DNT and TDA, chemical intermediates in the manufacture of polyurethane foams and in the production of amines, important ingredients in agricultural chemicals.

From the mid-1980s on, Air Products began to explore new businesses to take advantage of the company's proven skills in engineering, owning, and operating large-scale process plants. The company formed an environmental and energy systems division to focus its efforts in the expanding markets for cogeneration, flue gas desulfurization, and energy recovery from solid waste. In partnership with Browning-Ferris Industries, it formed American Ref-Fuel, a joint marketing effort to build, own, and operate waste-to-energy facilities. A general partnership with Mitsubishi Heavy Industries America, Inc. also was formed to participate in markets for flue gas desulfurization systems and services. The company also acquired GSF Energy Inc., a leader in capturing, purifying, and selling the natural gas produced in landfills. However, in 1997 Air Products sold its 50% interest in American Ref-Fuel to concentrate in its core businesses of gases and chemicals.

The current itemization of the Air Products's product lines is as follows:

1. Industrial, medical and specialty gases [59% of sales and 64% of profits in fiscal year (FY) 1998]
   a. Nitrogen, oxygen, argon
   b. Hydrogen, helium, carbon monoxide, carbon dioxide, synthesis gas
   c. Tungsten hexafluoride and nitrogen trifluoride
   d. Equipment for air separation, natural gas liquefaction, hydrogen purification, and air pollution control
2. Chemicals (31% of sales and 28% of profits in FY 1998)
   a. Polymer chemicals
      Acrylic and polyvinyl acetate emulsions
      Vinyl acetate–ethylene copolymer emulsions and emulsions that incorporate vinyl chloride and various acrylates in the polymer
      Polyvinyl aclohol
   b. Performance chemicals
      Specialty additives that are primarily acetylenic alcohols and amines
      Cycloaliphatic, amidoamine, and aliphatic epoxy curing compounds
      Bis(dimethylaminoethyl)ether and other blowing catalysts for polyurethane
      Polyamides, aromatic amines, cylcoaliphatic amines, reactive diluents, and specialty epoxy resins, which are used as performance additives in epoxy formulations
   c. Chemical intermediates
      Amines
      Ammonia, methanol, nitric acid

Dinitrotoluene and toluene diamine

Hydrogen bromide, hydrogen chloride, hydrogen fluoride, and hydrogen sulfide

Silane, sulfur dioxide, and sulfur hexafluoride

3. Equipment and services (10% of sales and 8% of profits in FY 1998)
    a. Cryogenic and process equipment for air separation, gas processing, natural gas liquefaction, hydrogen purification, and nitrogen rejection
    b. Membrane technology for recovering gases

Company financial performance from 1980 through 1998 may be found in Table 7.1 The return on assets (ROA) has been in the range of 6.0–7.5% and the return on equity (ROE) between 15 and 20% except for one year. These returns are typical of the median returns for a chemical company as presented at the end of the chapter in Table 7.23. For comparison purposes, Table 7.24 is a presentation of the returns on assets and equity for the companies in Chapter 7. Table 7.2 summarizes financial information on Air Products for selected years between 1980 and 1998.

## 7.2.2 The Archer-Daniels-Midland (ADM) Company

ADM was founded in 1923 by George Archer and John Daniels from the Midland (Michigan) Linseed Company. ADM now calls itself "supermarket to the world" because it is engaged in the processing and merchandising of many of the most important raw agricultural commodities that are used in the production of

**TABLE 7.1**   Air Products & Chemicals Inc.: Company Financial Performance

| Year | Pretax profit margin (%) | Long-term debt tied to capital expenditures (%) | Return on assets (%) | Return on equity (%) |
|------|------|------|------|------|
| 1998 | 16.8 | 40.3 | 7.4 | 20.6 |
| 1997 | 13.6 | 40.4 | 6.2 | 16.4 |
| 1995 | 14.3 | 29.3 | 6.7 | 15.9 |
| 1990 | 11.6 | 30.8 | 6.3 | 14.6 |
| 1985 | 11.9 | 26.0 | 6.0 | 12.9 |
| 1980 | 13.2 | 32.7 | 7.3 | 18.7 |

*Source:* Standard & Poor's Stock Reports for respective years.

**TABLE 7.2**   Air Products & Chemicals Inc. Information Summary

| | Income and balance sheet statistics[a] | | | | |
|---|---|---|---|---|---|
| | Amounts (millions) | | | | Percentage operating income of net revenue |
| Year | Net revenue | Operating income | Capital expenditures | Long-term debt | |
| 1980 | $1,421 | $302 | $364 | $416 | 21.3 |
| 1985 | $1,830 | $440 | $392 | $625 | 24.0 |
| 1990 | $2,895 | $692 | $468 | $954 | 23.9 |
| 1995 | $3,865 | $958 | $870 | $1,194 | 24.8 |
| 1997 | $4,638 | $1,160 | $870 | $2,292 | 25.0 |
| 1998 | $4,919 | $1,320 | $771 | $2,279 | 26.8 |

| Additional statistics | 1980 | 1985 | 1990 | 1995 | 1997 | 1998 |
|---|---|---|---|---|---|---|
| Total assets, millions[b] | $1,751 | $2,593 | $3,900 | $5,816 | $7,244 | $7,490 |
| Stockholders' equity, millions[b] | $665 | $1,163 | $1,688 | $2,398 | $2,648 | $2,667 |
| Total liabilities, millions[b] | $1,086 | $1,430 | $2,212 | $3,418 | $4,596 | $4,823 |
| R&D spending, ($) millions[c] | $30 | $51 | $72 | $103 | $116 | $113 |
| Current ratio[b] | 1.3 | 1.3 | 1.3 | 1.0 | 1.4 | 1.3 |
| Debt/equity[d] | 63% | 54% | 57% | 50% | 87% | 85% |
| Return on stockholders' equity[b] | 18.7% | 12.9% | 14.6% | 15.9% | 16.4% | 20.6% |
| Number of employees[e] | 15,800 | 18,700 | 14,800 | 14,800 | 16,400 | 16,700 |

[a] From company annual reports.
[b] From Standard & Poor's Stock Reports for respective years.
[c] From *Chemical and Engineering News,* Facts and Figures issues for respective years.
[d] From *Chemical Week,* May 20, 1998 pp. 43–50.
[e] From *Fortune 500* in respective years.

food and beverage products throughout the world. Most of ADM's businesses involve the merchandising and processing of corn, soybeans, cottonseed, sunflower seeds, canola, peanuts, and flaxseed into food and beverage products for human consumption and into animal feeds.

Crude vegetable oils are sold to others or refined and hydrogenated to produce oils for margarine, shortening, salad oils, and other food products. Several products, other than vegetable oil, are produced from the soybean; they include soy flour and grits, lecithin, textured vegetable protein (a meat substitute), and a natural source vitamin E. Corn wet-milling products include starch, liquid and crystalline dextrose, crystalline fructose, high-fructose sweeteners, ethyl alcohol, and animal feeds. Wheat flour is sold principally to large bakeries; durum flour is sold primarily to pasta manufacturers; and bulgur, a gelatinized wheat food, is sold to both domestic and export markets.

ADM uses its strong raw material position for the production of fermentation bioproducts to produce ascorbic acid (vitamin C) monosodium glutamate, sorbitol, citric and lactic acids, xanthan gum, and four amino acids: lysine, methionine, tryptophan, and threonine.

In 1998 the contributions to revenues from the various segments of ADM's businesses were as follows: oilseeds, 63%; corn, 13%; wheat flour, 9%; other, 15%.

In the last few years ADM's financial performance has been hurt by overcapacity in the oilseeds market and by expenses incurred as a result of federal, state, and foreign litigation regarding anticompetitive practices in the marketing of high-fructose corn syrup, lysine, citric acid, and monosodium glutamate. These lawsuits are ongoing, as evidenced by 7 pages devoted to their explanations in the 1999 10K report. However, it is most likely that ADM will "weather the storm" and use its abilities for low-cost processing, competitive size, and vertical integration to recover its former financial health.

The company financial performance for ADM is found in Table 7.3, and Table 7.4 contains further income and balance sheet statistics. The returns on assets and equity are presented in at the end of the chapter (Tables 7.23 and 7.24).

## 7.2.3  The Dow Chemical Company

Dow was formed in 1897 by Herbert H. Dow to extract chemicals from native brine deposits of east-central Michigan. It is now the third largest chemical company (and second largest *domestic* chemical company) in the United States. It has since diversified, and its many activities are classified in the 1998 annual report according to the following operating segments: agricultural chemicals, chemicals, hydrocarbons and energy, performance chemicals, performance plastics, plastics,

**TABLE 7.3**  Archer-Daniels-Midland: Company Financial Performance

| Year | Pretax profit margin (%) | Long-term debt tied to capital expdenditures (%) | Return on assets (%) | Return on equity (%) |
|------|------|------|------|------|
| 1998 | 3.8 | 28.5 | 3.2 | 6.4 |
| 1997 | 4.6 | 26.1 | 3.5 | 6.2 |
| 1995 | 9.3 | 24.5 | 8.7 | 14.8 |
| 1990 | 9.7 | 15.9 | 9.5 | 14.6 |
| 1985 | 5.7 | 21.6 | 5.9 | 9.5 |
| 1980 | 6.6 | 21.0 | 9.2 | 17.3 |

*Source:* Standard & Poor's Stock Reports for respective years.

TABLE 7.4   Archer-Daniels-Midland Information Summary

| | | | | | |
|---|---|---|---|---|---|
| | Income and balance sheet statistics[a] | | | | |
| | Amounts (millions) | | | | Percentage operating income of net revenue |
| Year | Net revenue | Operating income | Capital expenditures | Long-term debt | |
| 1980 | $2,802 | $228 | $186 | $221 | 8.1 |
| 1985 | $4,739 | $357 | $157 | $569 | 7.5 |
| 1990 | $7,751 | $923 | $327 | $751 | 11.9 |
| 1995 | $12,672 | $1,598 | $559 | $2,070 | 12.6 |
| 1997 | $13,853 | $1,071 | $780 | $2,345 | 7.7 |
| 1998 | $16,146 | $ 1,247 | $ 703 | $ 2,847 | 7.7 |

| Additional statistics | 1980 | 1985 | 1990 | 1995 | 1997 | 1998 |
|---|---|---|---|---|---|---|
| Total assets, millions[b] | $1,366 | $2,967 | $5,450 | $9,757 | $11,354 | $13,834 |
| Stockholders' equity, millions[b] | $767 | $1,803 | $3,573 | $5,854 | $6,050 | $6,505 |
| Total liabilities, millions[b] | $599 | $1,164 | $1,877 | $3,903 | $5,304 | $7,329 |
| R&D spending, ($) millions[c] | $30 | $51 | $72 | $103 | $114 | — |
| Current ratio[b] | 2.5 | 4.2 | 3.4 | 3.2 | 1.9 | 1.5 |
| Debt/equity[d] | 0% | 0% | 21% | 35% | 39% | 44% |
| Return on stockholders' equity[b] | 17.3% | 9.5% | 14.6% | 14.8% | 6.2% | 6.4% |
| Number of employees[e] | 5,590 | 9,446 | 11,363 | 14,833 | 17,160 | 23,132 |

[a]From company annual reports.
[b]From Standard & Poor's Stock Reports for respective years.
[c]From *Chemical and Engineering News*, Facts and Figures issues for respective years.
[d]From *Chemical Week*, May 20, 1998 pp. 43–50.
[e]From *Fortune 500* in respective years.

unallocated, and other. Some of the most important products, and product lines, sold by these segments are as follows:

1. Agricultural products: herbicides, insecticides, and seeds
2. Chemicals
   a. Caustic soda, chlorine, and calcium chloride
   b. Chlorinated solvents (carbon tetrachloride, methylene chloride, etc.), ethylene oxide, ethylene glycol, propylene oxide, propylene glycol, and vinyl chloride monomer
3. Hydrocarbons and energy
   a. Butadiene, ethylene, propylene, and styrene
   a. Power and steam
4. Performance chemicals
   a. Glycol ethers, surfactants, antimicrobials, heat transfer fluids, polyethylene, and polypropylene glycols

      b. Superabsorbent polymers, ethylcelloluse resins, amine compounds, and chelating agents

      c. Acrylate latex, polybutadiene rubber, polystyrene and styrene acrylate latex, styrene butadiene (S/B) emulsion rubber, and S/B latex

5. Performance plastics

      a. Engineering plastics: resins (polycarbonate, polyurethane, ABS, SAN, and nylon), polyurethane elastomers, and crystalline polymers

      b Adhesives and sealants: specialty coatings and surface treatments

      b. Epoxy products: epoxy resins and phenolics

      c. Fabricated products: window film, polyethylene and polystyrene foam insulation, adhesive film, and coated metal cable armor

      d. Polyurethanes: liquid catalysts, carpet backing, reaction-moldable products, polyurethane dispersions, toluene and specialty diisocynates, polyether polyols, and copolymers

6. Plastics: polyethylenes, polypropylenes, and polystyrenes

In 1998 Dow sold DowBrands, including such well-known products as Saran Wrap, the environmental services unit, the fabricated metals business, and the advanced composite joint venture. Also, in 1998, Dow exited the magnesium business. In 1999 Dow purchased Angus Chemicals and, on August 4, 1999, Dow agreed to acquire Union Carbide, a proposed acquisition that had to be reviewed by the U.S. government.

Company financial performance may be found in Table 7.5, and additional statistics are shown in Table 7.6 To compare the returns on assets and equity, see Tables 7.23 and 7.24.

TABLE 7.5 Dow Chemical: Company Financial Performance

| Year | Pretax profit margin (%) | Long-term debt tied to capital expdendi-tures (%) | Return on assets (%) | Return on equity (%) |
|------|--------------------------|-----------------------------------|----------------------|----------------------|
| 1998 | 10.9 | 31.6 | 5.5 | 17.3 |
| 1997 | 14.7 | 31.7 | 7.4 | 23.1 |
| 1995 | 17.5 | 31.7 | 8.3 | 26.7 |
| 1990 | 13.0 | 33.6 | 6.0 | 16.5 |
| 1985 | <0.1 | 38.3 | 0.5 | 1.2 |
| 1980 | 11.6 | 39.7 | 7.4 | 19.2 |

Source: Standard & Poor's Stock Reports for respective years.

TABLE 7.6  Dow Chemical Information Summary

| | Income and balance sheet statistics[a] | | | | |
|---|---|---|---|---|---|
| | Amounts (millions) | | | | Percentage operating income of net revenue |
| Year | Net revenue | Operating income | Capital expenditures | Long-term debt | |
| 1980 | $10,626 | $1,940 | $1,184 | $3,438 | 18.3 |
| 1985 | $11,537 | $1,614 | $1,011 | $3,198 | 14.0 |
| 1990 | $19,773 | $4,049 | $2,228 | $5,209 | 20.5 |
| 1995 | $20,200 | $5,333 | $1,417 | $4,705 | 26.4 |
| 1997 | $20,018 | $4,013 | $1,196 | $4,196 | 20.0 |
| 1998 | $18,441 | $3,498 | $1,549 | $4,051 | 19.0 |

| Additional Statistics | 1980 | 1985 | 1990 | 1995 | 1997 | 1998 |
|---|---|---|---|---|---|---|
| Total assets, millions[b] | $11,538 | $11,830 | $23,953 | $23,582 | $24,040 | $23,830 |
| Stockholders' equity, millions[b] | $4,440 | $4,792 | $8,728 | $7,361 | $7,626 | $7,429 |
| Total liabilities, millions[b] | $7,098 | $7,038 | $15,225 | $16,221 | $16,414 | $16,401 |
| R&D spending, millions[c] | $314 | $547 | $1,136 | $808 | $785 | $811 |
| Current ratio[b] | 1.6 | 1.5 | 1.4 | 1.9 | 1.2 | 1.2 |
| Debt/equity[d] | 77% | 67% | 60% | 64% | 55% | 55% |
| Return on stockholders' equity[b] | 19.2% | 1.2% | 16.5% | 26.7% | 23.1% | 19.3% |
| Number of employees[e] | 56,800 | 53,200 | 62,080 | 30,500 | 42,861 | 39,027 |

[a] From company annual reports.
[b] From Standard & Poor's Stock Reports for respective years.
[c] From *Chemical and Engineering News*, Facts and Figures issues for respective years.
[d] From *Chemical Week*, May 20, 1998 pp. 43–50.
[e] From *Fortune 500* in respective years.

### 7.2.4  E. I. Du Pont

In 1802 Eleuthère I. Du Pont founded what is now the world's oldest chemical company and the largest chemical company in the United States. Originally, a gunpowder manufacturer, Du Pont expanded into nitrocellulose plastics and lacquers, dyes, and phenolic resins in the early 1900s. In the 1920s, the company began making ammonia and fibers. Extensive research and development led to the discovery of nylon and chloroprene in the 1930s. Other important fibers and polymeric materials have been developed over the past 60 years. Du Pont has established a reputation as a leader in innovation in the CPI as reflected by its large expenditures for research and development. The company has selectively used acquisitions, joint ventures, and alliances to enter new markets, secure technological advantages, or quickly establish regional and global market presence. At one point in time it owned Conoco but has sold part of its interest to Conoco employees, completing the divestiture in 1999. In the life science areas it has acquired interests in Pioneer Hi-Bred International and Protein Technologies (from Ralston Purina) and formed joint ventures with Merck known as Du Pont Merck Pharmaceutical. It also has a joint venture with Dow in the elastomer area. Worldwide the company has in excess of 120 joint ventures. Du Pont has an established reputation as a leader in innovation in the chemical industry, and management's emphasis on technology is reflected by in the company's research and development spending:

| Year | R & D expenditure (millions) |
|------|------------------------------|
| 1997 | $1116 |
| 1996 | $1032 |
| 1995 | $1067 |
| 1994 | $1047 |
| 1993 | $1132 |

Du Pont has six major product groups and product segments, as follows:

*Chemicals:* a wide range of commodity and specialty products such as titanium dioxide, fluorochemicals, and polymer intermediates used in paper, plastics, chemical processing, refrigeration, textile, and environmental management industries

*Fibers:* a diversified mix of specialty fibers for protective apparel, sportswear, for high-strength composites, carpeting, and industrial applications to serve consumer and industrial markets

*Life sciences:* agricultural products, especially crop protection chemicals, seeds, biotechnology products, and pharmaceuticals

*Polymers:* engineering polymers, elastomers, fluoropolymers, ethylene polymers, finishes, and performance films serving industries such as packaging, construction, chemical processing, electrical, paper, textiles, and automotive

*Petroleum:* develop and produce crude oil and natural gas as well as processing natural gas to recover high-value liquids, intermediates for making petrochemicals, lubricants, and petroleum coke

*Diversified:* films, photopolymers, electronic materials, polyester intermediates, and COMSOL, a coal operation with 50% Du Pont ownership

Du Pont is an example of how various companies in the CPI have changed as a result of mergers, acquisitions, alliances, and divestitures. The sales of Du Pont prior to the partial spin-off of Conoco (see below) can be broken down as follows:

| Group | Percentage of sales |
|-------|---------------------|
| Petroleum | 47 |
| Fibers | 17 |
| Polymers | 15 |
| Chemicals | 10 |
| Diversified | 6 |
| Life sciences | 5 |

The contribution to sales by industry segment in 1998 were as follows:

| Group | Percentage of sales |
|-------|---------------------|
| Agricultural and nutrition | 11 |
| Nylon | 16 |
| Coatings and polymers | 17 |
| Pharmaceuticals | 4 |
| Pigments and chemicals | 13 |
| Polyester | 10 |
| Specialty fibers | 12 |
| Specialty polymers | 15 |
| Other | 2 |

Since 1995, Du Pont has shifted its business mix through acquisitions and divestitures, as well as by the formation of joint ventures and alliances. In 1997–1998 it acquired ICI's titanium dioxide and polyester business, Protein Technologies, Pioneer Hi-Bred Seed, and Herberta Coatings, also becoming 50% owner of Du Pont Merck Pharmaceutical. Du Pont divested itself of In Vitro Diagnostics, Medical Products, Diagnostic Imaging, and Graphic Films and Printing, as well as 30% of Conoco. The Conoco divestiture was completed in August 1999. Du Pont entered into alliances and joint ventures with Du Pont Dow Elastomers, and Optimum Quality Grains (crop protection) with Pioneer and Griffin. It appears that the company strategy is to have parallel paths in materials and life sciences businesses. Growth plans for improving the polyester operations include significant cost and capital reductions, the introduction of new technologies, and participation in joint ventures and alliances. Although diversification brings about stability and hopefully insurance to stockholders, the investors must be patient as traditional chemical companies shift to the life sciences areas.

Historically Du Pont has been a very profitable company; however, the company's financial performance has fluctuated depending on the demand for fibers, polymers, and petroleum products as noted in Tables 7.7 and 7.8. For

TABLE 7.7   Du Pont: Company Financial Performance

| Year | Pretax profit margin (%) | Long-term debt tied to capital expdendi-tures (%) | Return on assets (%) | Return on equity (%) |
|------|------|------|------|------|
| 1998 | 10.5 | 23.3 | 4.4 | 13.3 |
| 1997 | 10.4 | 29.7 | 5.9 | 22.3 |
| 1995 | 12.8 | 35.7 | 8.9 | 31.6 |
| 1990 | 10.5 | 22.2 | 6.4 | 14.7 |
| 1985 | 10.8 | 14.9 | 4.5 | 9.1 |
| 1980 | 8.3 | 14.7 | 7.7 | 13.3 |

*Source:* Standard & Poor's Stock Reports for respective years

many years, from 1940 to 1995, Du Pont ranked first in chemical sales. As it became more diversified, however, chemical sales (SIC 28) declined but were still a major contributor to revenue (see Table 5.12). A comparison of the returns on equity and assets in Table 7.24 with the median values for chemicals and allied products in Table 7.23 indicates that Du Pont exceeded these returns. Revenues declined from 1997 to 1998 in part because of the Conoco and other divestitures, softness in chemical sales, and the new venture in pharmaceuticals, Du Pont Merck Pharmaceuticals.

### 7.2.5   Exxon Corporation

As of 1998, Exxon was the third largest U.S. corporation in terms of revenues, in excess of $122 billion. It was incorporated in the state of New Jersey in 1898 and became a distinct entity upon the dissolution of the Standard Oil Trust in 1911. In 1997 about $14 billion (10.3% of Exxon's total revenue) was attributed to chemicals, making it the third largest U.S. producer of chemicals. The chemical products include basic petrochemicals, including olefins and aromatics, additives for fuels and lubricants, polyethylene and polypropylene plastics, specialty resins, specialty and commodity solvents, and performance chemicals for oil field operations.

Exxon, which is the largest independent nonutility power producer, has four major business groups: (a) exploration and production, (b) refining and marketing, (c) chemicals, and (d) coal, minerals, and power. In recent years, these have been responsible for operating profits as follows:

TABLE 7.8  Du Pont Information Summary

| Year | Amounts (millions) | | | | Percentage operating income of net revenue |
| | Net revenue | Operating income | Capital expenditures | Long-term debt | |
|---|---|---|---|---|---|
| 1980 | $13,652 | $1,893 | $1,279 | $1,102 | 13.9 |
| 1985 | $29,483 | $5,521 | $3,046 | $3,284 | 18.7 |
| 1990 | $39,709 | $7,160 | $5,383 | $5,663 | 18.0 |
| 1995 | $45,163 | $7,675 | $3,240 | $5,678 | 17.0 |
| 1997 | $45,079 | $8,118 | $4,768 | $5,929 | 18.0 |
| 1998 | $24,767 | $5,860 | $2,240 | $4,495 | 23.7 |

Income and balance sheet statistics[a]

| Additional statistics | 1980 | 1985 | 1990 | 1995 | 1996 | 1997 | 1998 | R&D spending as percentage of 1997 sales |
|---|---|---|---|---|---|---|---|---|
| Total assets, millions[b] | $9,560 | $25,140 | $38,128 | $37,312 | $37,870 | $42,942 | $38,536 | |
| Stockholders' equity, millions[b] | $5,453 | $12,422 | $16,181 | $8,199 | $10,593 | $11,033 | $13,954 | |
| Total liabilities, millions[b] | $4,107 | $12,718 | $21,947 | $29,113 | $27,277 | $31,909 | $24,582 | |
| R & D spending, millions[c] | $591 | $1,144 | $1,428 | $1,067 | $1,032 | $1,116 | $1,308 | 2.5 |
| Current ratio[b] | 2.3 | 1.7 | 1.2 | 0.9 | 1.0 | 0.8 | 0.8 | |
| Debt/equity[d] | 20% | 26% | 35% | 69%[d] | 49%[d] | 54%[d] | 32% | |
| Return on stockholders' equity[b] | 13.3% | 9.1% | 14.7% | 31.6% | 38.8% | 22.3% | 25.6% | |
| Number of employees[e] | 135,000 | 146,107 | 143,961 | 105,000 | 97,000 | 98,396 | 135,006 | |

[a] From company annual reports
[b] From Standard & Poor's Stock reports for respective years.
[c] From Chemical Engineering News, Facts and Figures issues for respective years.
[d] From Chemical Week, May 20, 1998, pp. 43–50.
[e] From Fortune 500 in respective years.

| Group | Operating profit (%) | |
| --- | --- | --- |
| | 1998 | 1997 |
| Exploration and production | 40 | 55 |
| Refining and marketing | 36 | 24 |
| Chemicals | 18 | 16 |
| Coal and other | 6 | 5 |

Exxon has a history of a willingness to share its cash with its stockholders, having raised dividends in each of 14 years in the 1980s and 1990s. Exxon's success has been based on the following business strategies:

Maximize profitability of existing oil and gas production.
Identify and pursue all attractive exploration opportunities.
Invest in projects that deliver superior returns.
Capitalize on growing natural gas markets.
Expand potential markets in Asian–Pacific, eastern Europe, and Latin American areas.
Maintain significant focused investment in research and development.
Continually reduce operating expenses and improve productivity.

To broaden its income base and to lessen its dependence on oil and gas in recent years, Exxon has been actively diversifying into other industries like coal and minerals. Although at present it is second only to Royal Dutch Shell in terms of size of its oil and gas reserves, it is likely that in diversifying, Exxon will concentrate on expanding its profitable chemical operations.

In 1998 Exxon and Mobil announced a proposal to merge, having a single management at the top, with each company nevertheless retaining its identity. As of this writing, the merger has not been completed. The emerging company, to be known as Exxon Mobil, would surpass General Motors in annual sales, making it the largest U.S. corporation as well as the world's largest publicly owned oil company. This merger is being scrutinized by the federal government, since both Exxon and Mobil were part of the original oil trust, which was dissolved in 1911.

Among oil companies Exxon has been one of the most profitable companies. In 1998 the return on assets was 6.8% and the return on equity was 14.8% (Table 7.9), compared with the median of the top 20 refiners of 2 and 5%, respectively (see Tables 7.23 and 7.24). Exxon's net revenue in 1998 declined about 15% from 1997 (Table 7.10). This was due in part to restructuring, and softness in the markets, as well as to fluctuation in crude oil prices.

**TABLE 7.9** Exxon Company: Financial Performance

| Year | Pretax profit margin (%) | Long-term debt tied to capital expdendi- tures (%) | Return on assets (%) | Return on equity (%) |
|------|--------------------------|--------------------------|--------------|---------------|
| 1998 | 8.0 | 7.2 | 6.8 | 14.8 |
| 1997 | 9.8 | 10.6 | 8.8 | 19.5 |
| 1995 | 8.6 | 12.4 | 7.2 | 16.6 |
| 1990 | 8.0 | 13.7 | 5.9 | 15.6 |
| 1985 | 11.3 | 10.4 | 7.6 | 17.4 |
| 1980 | 10.9 | 12.5 | 10.7 | 23.7 |

*Source:* Standard & Poor's Stock Reports for respective years.

### 7.2.6 International Paper

International Paper is one of the world's largest paper, paperboard, and packaging producers, serving customers in over 130 countries with 30% of its 1998 sales originating from outside the United States.

The printing papers division makes a wide variety of printing papers for printers, copiers, books, envelopes, and artists. The packaging sector produces container board and corrugated boxes, as well as folding cartons for liquid packaging and food service applications. International Paper also distributes printing, packaging, graphic arts, and industrial supply products, mostly made by others. In the specialty product sector, it produces panels, industrial papers, tissue, and chemicals. The company also explores for oil and gas. The forest products division controls large domestic timberland tracts as well as tracts in New Zealand and Chile.

The distribution of sales by industry sector in 1998 were as follows:

| Sector | Percentage of 1998 sales |
|--------|--------------------------|
| Printing papers division | 24 |
| Packaging sector | 23 |
| Printing, packaging, graphic arts, and industrial supply products | 27 |
| Specialty products | 7 |
| Forest products | 12 |
| Carter Holt Harvey (New Zealand) | 7 |

TABLE 7.10  Exxon Information Summary

|  | Income and balance sheet statistics[a] | | | | |
|---|---|---|---|---|---|
|  | Amounts (millions) | | | | Percentage operating income of net revenue |
| Year | Net revenue | Operating income | Capital expenditures | Long-term debt | |
| 1980 | $103,143 | $12,269 | $6,465 | $4,717 | 11.9 |
| 1985 | $ 86,673 | $14,397 | $8,844 | $4,820 | 16.6 |
| 1990 | $105,519 | $14,099 | $6,474 | $7,687 | 13.4 |
| 1995 | $121,804 | $14,584 | $7,128 | $7,778 | 12.0 |
| 1997 | $135,142 | $16,993 | $7,343 | $7,050 | 12.6 |
| 1998 | $115,417 | $12,326 | $8,359 | $4,530 | 10.7 |

| Additional statistics | 1980 | 1985 | 1990 | 1995 | 1996 | 1997 | 1998 |
|---|---|---|---|---|---|---|---|
| Total assets, millions[b] | $56,577 | $69,160 | $87,797 | $91,296 | $95,527 | $96,064 | $92,630 |
| Stockholder's equity, millions[b] | $25,483 | $29,096 | $33,025 | $39,982 | $43,299 | $43,470 | $43,645 |
| Total liabilities, millions[b] | $31,094 | $40,064 | $54,772 | $51,314 | $52,228 | $52,594 | $48,985 |
| R & D spending, millions[c] | — | — | — | — | $520 | $529 | $549 |
| Current ratio[b] | 1.4 | 0.9 | 0.8 | 0.9 | 2.1 | 1.1 | 0.9 |
| Debt/equity[d] | 19% | 17% | 23% | 19% | 17% | 16% | 10% |
| Return on stockholders' equity[b] | 23.7% | 17.4% | 15.6% | 16.8% | 18.0% | 19.5% | 14.8% |
| Number of employees[e] | 176,615 | 146,000 | 104,000 | 82,000 | 81,000 | 80,000 | 79,000 |

| | Percentage of operating profit[b] | | |
|---|---|---|---|
| Major product groups | 1997 | 1995 | 1998 |
| Exploration/production | 55 | 48 | 40 |
| Refining and marketing | 24 | 18 | 36 |
| Chemical | 16 | 28 | 18 |
| Coal and other | 5 | 6 | 6 |

[a] From company annual reports.
[b] From Standard & Poor's Stock Reports for respective years.
[c] From Chemical and Engineering News, Facts and Figures issues for respective years.
[d] From Chemical Week, May 20, 1998 pp. 43–50.
[e] From Fortune 500 in respective years.

In 1998 the company reorganized into seven more specialized groups: building materials, consumer packaging, distribution, European papers, industrial packaging, printing papers, and specialty businesses. In August 1998 the company sold nonstrategic businesses as part of the restructuring. In 1998 the company acquired a Russian-based paper and pulp business, acquired Zellerbach's distribution business from the Mead Corporation, acquired Weston Paper and Manufacturing Company, and entered into a joint venture to manufacture container board and corrugated boxes with a Turkish company. It substantially expanded its forest products operations through the acquisition of Union Camp in April 1999. The operating results of the mergers and acquisitions are reflected in the earnings statement.

The company financial performance reflected a lower pretax margin as well as lower returns on assets and equity (see Table 7.11). Sales were down from 1997 by about $500 million, resulting in lower operating income as shown in Table 7.12 The median returns (Table 7.23) for 21 forest products companies are low compared with other companies in the CPI (Table 7.24). International Paper's returns are low compared with other forest product companies. International Paper in years past has had a more profitable record, but the restructuring expenses and lower sales in 1998 reflect lower financial performance. Projected price increases for paper, corrugated paper, and printing papers should help sales increase, however.

## 7.2.7  Merck & Company, Inc.

Merck is a global research-driven pharmaceutical company that discovers, develops, manufactures, and markets a broad range of human and animal health products. In addition, the company manufactures specialty chemical and environmental

**TABLE 7.11**  International Paper: Company Financial Performance

| Year | Pretax profit margin (%) | Long-term debt tied to capital expdendi-tures (%) | Return on assets (%) | Return on equity (%) |
|------|------|------|------|------|
| 1998 | 2.0 | 38.0 | 0.9 | 2.7 |
| 1997 | 0.1 | 35.4 | —[a] | —[a] |
| 1995 | 9.5 | 37.8 | 5.5 | 16.1 |
| 1990 | 7.3 | 31.4 | 4.5 | 10.6 |
| 1985 | 3.7 | 23.2 | 2.3 | 3.3 |
| 1980 | 7.9 | 21.4 | 6.2 | 10.7 |

[a]Number missing.
*Source:* Standard & Poor's Stock Reports for respective years.

**TABLE 7.12**   International Paper Information Summary

| | Income and balance sheet statistics[a] | | | | |
|---|---|---|---|---|---|
| | Amounts (millions) | | | | Percentage operating income of net revenue |
| Year | Net revenue | Operating income | Capital expenditures | Long-term debt | |
| 1980 | $ 5,043 | $  446 | $  883 | $  929 | 8.8 |
| 1985 | $ 4,502 | $  605 | $  794 | $1,191 | 13.4 |
| 1990 | $12,960 | $2,088 | $1,409 | $3,096 | 16.1 |
| 1995 | $19,797 | $3,552 | $1,518 | $5,946 | 17.9 |
| 1997 | $20,096 | $2,404 | $1,111 | $7,154 | 12.0 |
| 1998 | $19,541 | $2,202 | $1,049 | $8,212 | 11.3 |

| Additional statistics | 1980 | 1985 | 1990 | 1995 | 1997 | 1998 |
|---|---|---|---|---|---|---|
| Total assets, millions[b] | $5,197 | $6,039 | $13,669 | $23,977 | $26,754 | $26,356 |
| Stockholders' equity, millions[b] | $2,814 | $3,195 | $5,632 | $7,797 | $8,710 | $8,902 |
| Total liabilities, millions[b] | $2,383 | $2,844 | $8,037 | $16,180 | $18,044 | $17,454 |
| Current ratio[b] | 1.8 | 1.5 | 1.2 | 1.2 | 1.2 | 1.7 |
| Debt/equity[c] | 33% | 37% | 55% | 76% | 82% | 92% |
| Return on stockholders' equity[b] | 10.7% | 3.3% | 10.5% | 16.1% | NA | 2.7% |
| Number of employees[d] | 46,048 | 31,900 | 69,000 | 81,500 | 82,000 | 82,000 |

[a] From company annual reports.
[b] From Standard & Poor's Stock Reports for respective years.
[c] From *Chemical Week*, May 20, 1998, pp 43–50.
[d] From *Fortune 500* in respective years.

products. It is the premier United States–based pharmaceutical company that manufactures and markets a wide variety of prescription drugs.

Originally founded in 1887 as a U.S. branch of E. Merck of Germany, the company has grown significantly in subsequent years through aggressive programs of new drug development and acquisition. The company is the undisputed leader in the vast market for high-margin cardiovascular drugs, with five drugs generating aggregate sales of over $9 billion in 1997. Approximately 25% of Merck's sales is derived from foreign operations.

Merck has been actively restructuring its lines of business that were not significant to its financial position, liquidity, or operations. The company's intent was to focus its resources more fully on its core human health and pharmaceutical benefit services. This led to the development of products for the treatment of heart failure, high blood pressure, high cholesterol, depression, osteoporosis, glaucoma, asthma, and ulcers to name a few. The company has also been a leader in over-the-counter products (e.g., antacids).

Merck has formed numerous strategic alliances and joint ventures. In 1982 it entered into a joint venture with Astra AB of Sweden to develop and market Astra's new prescription medicines and others including the antacid Prilosec. A joint venture with Johnson & Johnson in 1989 led to developing, manufacturing and marketing consumer health care products. In 1991 Merck and Du Pont entered into a joint venture to form a worldwide pharmaceutical company to research, manufacture, and market pharmaceutical and imaging agent products. In 1992 the Merck Vaccine Division and Connaught Laboratories, Inc. agreed to collaborate on pediatric vaccines and promote selected vaccines in the United States. Merck has significant joint ventures with Wyeth-Ayerst, Chugai Pharmaceutical, and Rh^one-Poulenc. Merck had about 50 or worldwide alliances or joint ventures as of 1998. Foreign business has been significant, accounting for 25% of sales in 1998.

To focus on more profitable businesses, Merck divested Calgon Destal Laboratories to Bristol-Myers Squibb, Kelso to Monsanto, and Medco Behavioral Care Corporation (MBC) to MBC and Kohlberg Kravis Roberts & Company. In 1997 Merck sold its crop protection business to Novartis.

The pharmaceutical business has always been the highest profitability sector of the CPI (Table 7.23), with 17% return on assets and 36% return on equity in 1998. Table 7.13 lists the Merck's profit margin as well as the returns on equity and assets for selected years between 1980 and 1997. Income and balance sheet statistics may be found in Table 7.14. These are very high compared to the CPI in general (see Table 7.24). The huge success of Merck enabled it to support a $1.8 billion research and development program in 1998.

**TABLE 7.13**   Merck: Company Financial Performance

| Year | Pretax profit margin (%) | Long-term debt tied to capital expdendi-tures (%) | Return on assets (%) | Return on equity (%) |
|------|------|------|------|------|
| 1998 | 30.8 | 15.0 | 18.2 | 41.3 |
| 1997 | 27.9 | 8.1 | 18.4 | 37.5 |
| 1995 | 29.3 | 10.5 | 14.6 | 29.2 |
| 1990 | 35.6 | 2.6 | 24.3 | 48.9 |
| 1985 | 24.8 | 4.7 | 11.5 | 21.1 |
| 1980 | 24.0 | 9.6 | 15.2 | 23.6 |

*Source:* Standard & Poor's Stock Reports for respective years.

TABLE 7.14   Merck Information Summary

| | Income and balance sheet statistics[a] | | | | |
|---|---|---|---|---|---|
| | Amounts (millions) | | | | Percentage |
| Year | Net revenue | Operating income | Capital expenditures | Long-term debt | operating income of net revenue |
| 1980 | $2,734 | $750 | $256 | $211 | 27.4 |
| 1985 | $3,548 | $1,003 | $236 | $171 | 28.3 |
| 1990 | $7,672 | $2,883 | $671 | $124 | 37.6 |
| 1995 | $16,681 | $5,262 | $1,006 | $1,373 | 31.5 |
| 1997 | $23,637 | $6,701 | $1,449 | $1,347 | 28.3 |
| 1998 | $26,898 | $7,655 | $1,973 | $3,221 | 28.5 |

| Additional statistics | 1980 | 1985 | 1990 | 1995 | 1997 | 1998 |
|---|---|---|---|---|---|---|
| Total assets, millions[b] | $2,866 | $4,902 | $8,030 | $23,832 | $25,812 | $31,853 |
| Stockholders' equity, millions[b] | $1,863 | $2,634 | $3,834 | $11,736 | $12,613 | $12,802 |
| Total liabilities, millions[b] | $1,003 | $2,268 | $4,196 | $12,096 | $13,199 | $19,051 |
| R&D spending, millions[c] | $234 | $426 | $854 | $1,331 | $1,684 | $1,800 |
| Current ratio[b] | 2.3 | 1.9 | 1.3 | 1.5 | 1.5 | 1.7 |
| Debt/equity[d] | 11% | 6% | 3% | 12% | 11% | 25% |
| Return on stockholders' equity[b] | 23.6% | 21.1% | 48.9% | 29.2% | 37.5% | 41.3% |
| Number of employees[e] | 31,600 | 30,900 | 36,900 | 45,200 | 53,800 | 57,300 |

[a] From company annual reports.
[b] From Standard & Poor's Stock Reports for respective years.
[c] From *Chemical and Engineering News*, Facts and Figures issues for respective years.
[d] From *Chemical Week*, May 20, 1998, pp. 43–50.
[e] From *Fortune 500* in respective years.

## 7.2.8  Mobil Oil

Like Exxon, Mobil Oil was part of the John D. Rockefeller's Standard Oil Trust until the Supreme Court ordered the breakup of the monopoly in 1911. Mobil is one of the world's largest oil companies. It is involved in all the major segments of the oil industry including exploration, production, refining, and marketing. Through the years, the company grew and expanded into petrochemicals and specialty products including synthetic lubricants, base stocks, and additives for fuels and lubricants.

In the mid-1980s the company underwent significant restructuring, which resulted in write downs and sharp reductions in capital expenditures, assets, and staff. Nonstrategic assets were sold to repay debts incurred in the 1980s.

In terms of proven reserves, Mobil is one of the world's five largest non-state-owned oil companies. In 1997 Mobil's net crude production and natural gas

liquids averaged slightly less than 1 billion barrels a day, 26% of which was obtained from domestic sources. Twenty-four percent of the natural gas produced
by the company was from U.S. production.

Mobil Chemical makes and markets basic petroleum chemicals and is a
leader in polypropylene film, a food packaging product. The company concentrates on high-margin products like synthetic lubricants and premium gasolines.
In 1996 the plastics division was sold to Tenneco. In 1998 Mobil announced an
11% increase in capital and exploration budget over 1997.

Mobil announced a spending cut in 1999 due to weak oil prices globally.
Also, in early 1999, Exxon and Mobil announced that they were considering a
merger in which each company will maintain its own identity, but there will be
one top level management. As of this writing (first quarter 1999), no progress has
been announced.

The pretax profit margin declined about from 9.9% in 1997 to 5.9% in
1998 (see Table 7.15), reflecting a decrease in revenue from sales and services of
about $12 billion, in part because of lower average worldwide prices for crude
oil, natural gas, and petroleum product as well as currency translation effects
(Table 7.16). The returns on assets and equity were about half the values in 1997.
Compared with the median values for 20 petroleum refiners in 1998, Mobil exceeded the returns but did considerably less well than Exxon (see Tables 7.23
and 7.24).

### 7.2.9  Monsanto Company

Monsanto was founded in 1902 to produce saccharin and other specialty chemicals. The company sustained a period of slow growth throughout the 1960s and

**TABLE 7.15**  Mobil Corp.: Company Financial Performance

| Year | Pretax profit margin (%) | Long-term debt tied to capital expdendi-tures (%) | Return on assets (%) | Return on equity (%) |
|------|------|------|------|------|
| 1998 | 5.9 | 14.5 | 3.9 | 9.1 |
| 1997 | 9.9 | 13.6 | 7.3 | 17.3 |
| 1995 | 6.8 | 18.3 | 5.7 | 13.8 |
| 1990 | 7.7 | 16.9 | 4.8 | 11.3 |
| 1985 | 7.7 | 35.8 | 2.5 | 7.5 |
| 1980 | 11.2 | 19.0 | 9.3 | 23.8 |

Source: Standard & Poor's Stock Reports for respective years.

**TABLE 7.16**  Mobil Corp. Information Summary

| | Income and balance sheet statistics[a] | | | | |
|---|---|---|---|---|---|
| | Amounts (millions) | | | | Percentage operating income of net revenue |
| Year | Net revenue | Operating income | Capital expenditures | Long-term debt | |
| 1980 | $59,510 | $7,651 | $3,625 | $3,571 | 12.9 |
| 1985 | $55,960 | $7,732 | $2,725 | $9,745 | 13.8 |
| 1990 | $57,819 | $6,875 | $3,577 | $4,298 | 11.9 |
| 1995 | $64,767 | $6,649 | $4,268 | $4,629 | 10.3 |
| 1997 | $64,327 | $7,768 | $4,689 | $3,760 | 12.1 |
| 1998 | $52,140 | $4,951 | $4,747 | $3,719 | 9.5 |

| Additional statistics | 1980 | 1985 | 1990 | 1995 | 1997 | 1998 |
|---|---|---|---|---|---|---|
| Total assets, millions[b] | $32,705 | $41,752 | $41,665 | $42,138 | $43,559 | $42,754 |
| Stockholders' equity, millions[b] | $13,069 | $14,089 | $17,021 | $17,229 | $18,796 | $17,729 |
| Total liabilities, millions,[b] | $19,636 | $27,663 | $24,644 | $24,909 | $24,763 | $25,025 |
| R&D spending, millions[c] | | | | | | |
| Current ratio[b] | 1.1 | 1.0 | 1.0 | 0.9 | 0.8 | 0.7 |
| Debt/equity[d] | 27% | 69% | 25% | 27% | 20% | 21% |
| Return on stockholders' equity[b] | 23.8% | 7.5% | 11.3% | 13.8% | 17.3% | 9.1% |
| Number of employees[e] | 212,800 | 163,600 | 67,300 | 50,400 | 42,700 | 41,500 |

[a] From company annual reports.
[b] From Standard & Poor's stock reports for respective years.
[c] From *Chemical and Engineering News*, Facts and Figures issues for respective years.
[d] From *Chemical Week*, May 20, 1990, pp. 40-50.
[e] From *Fortune 500* in respective years.

1970s; however, it grew rapidly from 1972 to 1975 when net sales increased an average compounded annual rate of about 15%. Revenues were static in the 1980s up to 1986 at about $6.5 billion. In the 10-year period 1986–1996, revenues increased to $9.3 billion. Through the years Monsanto added many new products and in 1996 was the fourth largest domestic chemical company according to sales. Prior to restructuring in 1997, the company had five product sectors:

*Agricultural products:* fertilizers, herbicides, insecticides, plant-growth regulators, etc.
*Commercial products:* process control and electronics, plastic products, low-density polyethylene film, pollution abatement systems, synthetic turf, engineering construction services, etc.
*Industrial chemicals:* aspirin, pharmaceuticals, rubber-processing chemicals, functional fluids, bleaching compounds, detergent intermediates,

plasticizers, inorganic acids, phthalic and maleic anhydrides, adipic
acid, defoamers, caustic soda, sodium sulfite, etc.

*Polymers and petrochemicals:* styrene, acrylonitrile, phenol, acetic acid,
benzene, naphthalene, *o*-xylene, methanol, butadiene, ethylene, propylene, polystyrene and styrene copolymers, nylon thermoplastics,
polyvinyl butyral sheet, and specialty resins

*Textiles:* nylon 6; acrylic, monoacrylic, and polyester fibers

In 1997 Monsanto divested the $3 billion a year chemicals segment of its
business (now called Solutia). Management shifted its focus to life sciences such
that the company now consists of life sciences and agricultural products, Searle
pharmaceuticals, and food ingredients. Monsanto investors now are likely to be
interested in higher risk and higher returns but this may take time. This is a company in transition.

In the period 1995–1998, Monsanto acquired Kelco, Calgene, Agrocetus,
Asgrow Agronomics, DeKalb Genetics, Holden's Foundation Seeds, Delta and
Pine Land, Plant Breeding International, and Cargill's International Seeds. In
this period it divested the styrenics business, chemicals (Solutia), and the lawn
and garden businesses.

Monsanto also entered into alliances and joint ventures such as Genomics
(with Millenium Pharmaceuticals) and grain and feed products with Cargill.

The business segment contributions to sales in 1997 were as follows:

|                          | Percentage of sales | |
| Segment                  | 1997 | 1998 |
| ------------------------ | ---- | ---- |
| Agricultural products    | 42   | 47   |
| Nutrition and consumer   | 20   | 18   |
| Pharmaceuticals          | 32   | 33   |
| Other                    | 6    | 2    |

International operations accounted for 42% of sales.

In 1998 the company earnings were affected by a 52% increase in research
and development spending, introduction of new products, and other growth ventures as well as an $800 million after-tax charge brought about by acquisitions
and restructuring. As a result, total assets increased significantly and cash declined substantially. The pretax profit margin was negative (Table 7.17), making
any return calculations meaningless. Table 7.18 is a balance sheet and income
statement summary.

In September 1999 a news item in a trade magazine announced that
Monsanto was for sale, partly for reasons of recent setbacks in the development of key drugs and the furor over genetically engineered seeds. The reader

**TABLE 7.17**  Monsanto: Company Financial Performance

| Year | Pretax profit margin (%) | Long-term debt tied to capital expdendi-tures (%) | Return on assets (%) | Return on equity (%) |
|------|------|------|------|------|
| 1998 | −2.8 | 55.7 | —ᵃ | —ᵃ |
| 1997 | 4.9 | 32.0 | 2.7 | 7.3 |
| 1995 | 12.1 | 30.4 | 7.6 | 22.1 |
| 1990 | 9.0 | 25.9 | 6.3 | 13.9 |
| 1985 | 4.4 | 34.3 | —ᵃ | —ᵃ |
| 1980 | 3.1 | 29.7 | 2.6 | 5.3 |

ᵃ Numbers missing.
*Source:* Standard & Poor's Stock Reports for respective years.

may compare Monsanto's performance with other CPI firms by looking at Tables 7.23 and 7.24.

### 7.2.10  Procter & Gamble Company

In the CPI, Procter & Gamble is considered to be a diversified company. It was incorporated in 1905, having been built from a business founded in 1837 by William Procter and James Gamble. Today, the company manufactures a broad range of products and markets them directly to the consumer in many countries throughout the world.

Procter & Gamble has a portfolio of consumer products including laundry/cleaning products, beauty care products, and food and beverage products, as well as paper and health care products. The company creates new products and enhances performance benefits of existing products that are vital to the company's success. For example, the food products group added Olean, a recently developed fat substitute that has been approved by the Food and Drug Administration. Procter & Gamble is focusing on the successful implementation of value pricing as well as the maintenance of key consumer relations in important markets like Mexico, China, Brazil, and Russia. The company manufactures and markets over 300 brands of consumer products marketed in 140 countries. Some of the world-famous brands are Tide, Cascade, Dawn, Bounty, Pampers, Charmin, Always, Pantene, Vidal Sassoon, Oil of Olay, Secret, and Cover Girl. Food and beverage products include Folgers, Jif, Crisco, Olean, and Pringles. Health care products are Scope, Crest, Metamucil, and Vicks labels.

**TABLE 7.18** Monsanto Information Summary

Income and balance sheet statistics[a]

| Year | Amounts (millions) | | | | Percentage operating income of net revenue |
|---|---|---|---|---|---|
| | Net revenue | Operating income | Capital expenditures | Long-term debt | |
| 1980 | 6,574 | 725 | 781 | 1,371 | 11.0 |
| 1985 | 6,747 | 916 | 1,195 | 2,087 | 13.6 |
| 1990 | 8,995 | 1,606 | 750 | 1,652 | 17.9 |
| 1995 | 8,962 | 1,858 | 500 | 1,687 | 20.7 |
| 1997 | 7,514 | 1,670 | 644 | 1,979 | 22.2 |
| 1998 | 8,648 | 1,425 | 854 | 6,259 | 16.5 |

| Additional statistics | 1980 | 1985 | 1990 | 1995 | 1997 | 1998 | R&D spending as percentage of 1997 sales |
|---|---|---|---|---|---|---|---|
| Total assets, millions[b] | $5,796 | $8,877 | $9,236 | $10,611 | $10,774 | $16,724 | |
| Stockholder's equity, millions,[b] | $2,804 | $3,407 | $4,089 | $3,732 | $4,227 | $4,986 | |
| Total liabilities, millions[b] | $2,992 | $5,470 | $5,147 | $6,879 | $6,547 | $11,738 | |
| R&D spending, millions[c] | $208 | $470 | $612 | $658 | $939 | $1,263 | 12.5 (1997) |
| Current ratio[d] | 2.1 | 1.6 | 1.6 | 1.5 | 1.2 | 1.5 | 14.6 (1998) |
| Debt/equity[d] | 49% | 61% | 40% | 45% | 47% | 126% | |
| Return on stockholders' equity[b] | 5.3% | NA | 13.9% | 22.1% | 7.3% | NA | |
| Number of employees[e] | 61,636 | 54,103 | 41,081 | 28,514 | 21,900 | 31,800 | |

[a] From company annual reports.
[b] From Standard & Poor's Stock Reports for respective years.
[c] From Chemical and Engineering News, Facts and Figures issues for respective years.
[d] From Chemical Week, May 20, 1998 pp. 43–50.
[e] From Fortune 500 in respective years.

The product line contributions for 1998 sales were as follows:

| Segment | Percentage of sales |
| --- | --- |
| Laundry/cleaning products | 30 |
| Beauty care products | 19 |
| Food and beverage products | 12 |
| Paper products | 29 |
| Health care products | 8 |
| Corporate and other | 2 |

The contribution of 1998 sales by world geographic areas were as follows:

| Area | Percentage of sales |
| --- | --- |
| North America | 50 |
| Europe, Middle East, and Africa | 32 |
| Asia | 9 |
| Latin America | 7 |
| Corporate | 2 |

Procter & Gamble has announced an intention to restructure according to business lines in FY 2000, reporting its segments by product lines rather than by geographical areas

For many years Procter & Gamble has won the American Management Association's award as the best managed company, a distinction it still enjoys. The company has been a very profitable firm, with a pretax profit margin of 15.4% in 1998 (Table 7.19). Table 7.20 is a summary of balance sheet and income statement statistics. The returns on assets and equity for that year were 12.9 and 35.7%, respectively (Tables 7.19 and 7.24). The company returns are greater than the median values for nine soap and cosmetics firms, as shown in Table 7.23. Sales grew about 4% from 1997 to 1998, financial experts expected them to grow in the low single digits for the next few years.

## 7.2.11 Union Carbide

Union Carbide is now a shadow of its former self, having expended hundreds of millions of dollars in legal settlements due to the Bhopal explosion in 1984. The company was formed in 1898 to produce calcium carbide and from it acetylene. In 1917 the company merged with three other companies, Prest-O-Lite, Linde Air Products Company, and National Carbon Company, to form Union Carbide and Carbon Corporation. From 1940 to 1975, it was second only to Du Pont in chemical sales.

**TABLE 7.19**   Procter & Gamble: Company Financial
Performance

| Year | Pretax profit margin (%) | Long-term debt tied to capital expdendi-tures (%) | Return on assets (%) | Return on equity (%) |
|------|--------------------------|---------------------------------------------------|----------------------|----------------------|
| 1998 | 15.4 | 34.7 | 12.9 | 35.7 |
| 1997 | 14.7 | 27.8 | 12.4 | 33.1 |
| 1995 | 12.0 | 31.7 | 9.9  | 32.7 |
| 1990 | 10.1 | 29.0 | 8.9  | 25.7 |
| 1985 | 7.4  | 12.4 | 6.8  | 12.3 |
| 1980 | 10.0 | 17.1 | 10.8 | 18.8 |

*Source:* Standard & Poor's Stock Reports for respective years.

In the early 1980s, revenues were in the range of $9 billion. In 1986 numerous lawsuits were brought against it seeking damages resulting from the Bhopal incident. In the late 1980s, Carbide began restructuring, selling off parts of the company, reducing employees, and introducing other cost-cutting measures to survive. In 1989 the company paid $425 million to settle the litigation of Bhopal. Revenues steadily dropped, and in the period 1991–1994 were at a low point of $4.6 billion. In the mid-1990s, Union Carbide entered joint ventures with Mitsubishi Corporation (carbon electrodes), Enichem of Italy, and Elf Aquitaine (polyethylene). In 1997 a petrochemical joint venture with Kuwait incurred losses, but when this unit comes on stream, the venture is expected to generate profits. It will make polyethylene, polypropylene, and ethylene glycol primarily for Asian markets.

Segment contributions to sale in 1998 were as follows:

| Segment | Percentage of sales | |
|---------|------|------|
|         | 1997 | 1998 |
| Specialties and intermediates | 65 | 70 |
| Basic chemicals and polymers | 35 | 30 |
| Other | Nil | Nil |

Foreign operations accounted for 29% of sales in 1997 and 1998.

Specialties and intermediates consist of ethylene oxide derivatives (polyethylene glycol, surfactants, amines, deicing fluids, heat transfer fluids, lubricants, and solvents); solvent and emulsion systems (alcohols, acrylics, vinyl acetate, latexes, and modifiers) for use in paints, adhesives, and household

TABLE 7.20 Procter & Gamble Information Summary

### Income and balance sheet statistics[a]

| Year | Net revenue | Operating income | Capital expenditures | Long-term debt | Percentage operating income of net revenue | R&D spending as percentage of 1997 sales |
|------|-------------|------------------|----------------------|----------------|---------------------------------|------------------------------------|
| 1980 | $10,772 | $1,324 | $783 | $835 | 12.3 | |
| 1985 | $13,652 | $1,753 | $1,102 | $2,457 | 12.8 | |
| 1990 | $24,081 | $3,072 | $1,880 | $3,588 | 12.8 | |
| 1995 | $33,434 | $5,432 | $2,146 | $5,161 | 16.2 | |
| 1997 | $35,764 | $6,975 | $2,129 | $4,143 | 19.5 | 12.5 (1997) |
| 1998 | $37,154 | $7,653 | $2,559 | 5,765 | 20.6 | 14.6 (1998) |

| Additional statistics | 1980 | 1985 | 1990 | 1995 | 1997 | 1998 |
|-----------------------|------|------|------|------|------|------|
| Total assets, millions[b] | $6,553 | $9,683 | $18,487 | $28,125 | $27,544 | $30,966 |
| Stockholders' equity, millions[b] | $3,603 | $5,704 | $6,518 | $8,676 | $10,187 | $10,415 |
| Total liabilities, millions[b] | $2,950 | $3,979 | $11,969 | $19,449 | $17,357 | $20,551 |
| R&D spending, millions[c] | — | $400 | 693 | 1,257 | 1,282 | — |
| Current ratio[b] | 1.8 | 1.3 | 1.4 | 1.3 | 1.4 | 1.1 |
| Debt/equity[d] | 23% | 43% | 55% | 59% | 41% | 55% |
| Return on stockholders' equity[d] | 18.8% | 12.8% | 25.7% | 32.7% | 33.1% | 35.7% |
| Number of employees[e] | 61,000 | 62,300 | 88,800 | 56,200 | 106,000 | 110,000 |

[a] From company annual reports.
[b] From Standard & Poor's Stock Reports for respective years.
[c] From Chemical Week, May 20, 1998, pp. 43–50.
[d] From Chemical and Engineering News, Facts and Figures issues for respective years.
[e] From Fortune 500 in respective years.

products; specialty polyolefins; UNIPOL technology licensing for polyethylene and polypropylene; specialty polymers (biocides, water-soluble resins, coating materials, resins, plastics additives and modifiers); and a partnership with Universal Oil Products in catalysts and molecular sieves.

The company is a leading manufacturer of polyethylene, the world's most widely used plastics, for films, bags, bottles, and electrical insulation. Not surprisingly, it is a technology leader in polyethylene and polypropylene.

Union Carbide is continuously transforming itself from a commodity chemical company to a more balanced blend of specialties and commodities. In recent years, the company's profitability has fluctuated as measured by profit margin and stockholders' equity (see Table 7.21). Basic strategies were set in place to increase profitability by the following means:

Strengthening businesses in which it already excels
Withdrawing from business lines with little potential
Shifting the product mix to include a greater number of performance products that offer superior value and performance to the customer
Continued restructuring to reduce the number of employees
A program of stock repurchasing to boost share earnings.

The pretax profit margin declined from 14.9% in 1997 to 11.0% in 1998 as a result of declines in polyethylene and ethylene glycol margins, new ethylene industry capacity, and the Asian economic crisis (Table 7.21). In 1998 the company's returns on assets and equity (Tables 7.21 and 7.24) were about the same as the median values for 37 chemical companies. Table 7.22 is a summary of balance sheet and income statement statistics.

**TABLE 7.21**   Union Carbide: Company Financial Performance

| Year | Pretax profit margin (%) | Long-term debt tied to capital expdendi- tures (%) | Return on assets (%) | Return on equity (%) |
|------|------|------|------|------|
| 1998 | 11.0 | 42.0 | 5.7 | 16.8 |
| 1997 | 14.9 | 38.0 | 10.0 | 29.6 |
| 1995 | 23.2 | 38.3 | 16.4 | 52.0 |
| 1990 | 8.2 | 41.3 | 3.8 | 13.7 |
| 1985 | 11.3 | 27.1 | —[a] | —[a] |
| 1980 | 10.8 | 24.3 | 7.2 | 18.8 |

[a] Numbers missing.
*Source:* Standard & Poor's Stock Reports for respective years.

TABLE 7.22 Union Carbide Information Summary

| | Income and balance sheet statistics[a] | | | | |
| | Amounts (millions) | | | | Percentage |
| Year | Net revenue | Operating income | Capital expenditures | Long-term debt | operating income of net revenue |
|---|---|---|---|---|---|
| 1980 | $9,994 | $1,490 | $1,129 | $1,869 | 14.9 |
| 1985 | $9,003 | $1,187 | $649 | $1,750 | 13.2 |
| 1990 | $7,621 | $1,450 | $744 | $2,340 | 19.0 |
| 1995 | $5,888 | $1,257 | $542 | $1,285 | 21.3 |
| 1997 | $6,502 | $1,215 | $755 | $1,458 | 18.7 |
| 1998 | $5,659 | $918 | $782 | $1,796 | 16.2 |

| Additional statistics | 1980 | 1985 | 1990 | 1995 | 1997 | 1998 | R&D spending as percentage of 1997 sales |
|---|---|---|---|---|---|---|---|
| Total assets, millions[b] | $3,919 | $3,878 | $8,733 | $6,256 | $6,964 | $7,291 | |
| Stockholders' equity, millions[b] | $1,859 | $1,750 | $2,373 | $2,045 | $2,348 | $2,449 | |
| Total liabilities, millions[b] | $2,060 | $2,128 | $6,360 | $4,211 | $4,616 | $4,842 | |
| R&D spending, millions[c] | $166 | $275 | $191 | $144 | $166 | —[f] | 12.5 (1997) |
| Current ratio[b] | 2.2 | 1.3 | 1.2 | 1.6 | 1.2 | 1.3 | 14.6 (1998) |
| Debt/equity[d] | 101% | 100% | 99% | 63% | 62% | 73% | |
| Return on stockholders' equity[b] | 15.1% | —[f] | 13.7% | 52.0% | 29.6% | 16.8% | |
| Number of employees[e] | 116,105 | 91,459 | 37,756 | 11,745 | 11,813 | 11,627 | |

a From company annual reports.
b From Standard & Poor's Stock Reports for respective years.
c From Chemical and Engineering News, Facts and Figures issues for respective years.
d From Chemical Week, May 20, 1998, pp. 43–50.
e From Fortune 500 in respective years.
f Numbers missing.

Dow and Union Carbide are in the discussion stages of a possible merger. The CEO of Dow states that "the objective is to transform Dow into the world's most productive, best 'value-growth' company in the chemical industry." Both companies have faced crises: Dow, part owner of Dow Corning, was a defendant in the silicone breast implant suits, and Union Carbide had the Bhopal incident. The merged company would operate under the Dow Chemical name and would be the world's second largest chemical company, with sales in the range of $20–30 billion per year. The merger with Dow would strengthen Carbide's position in specialties and performance chemicals. In the commodity business, there is great overlap between the companies in ethylene and derivatives, as well as in ethylene glycol and polyolefin technologies.

TABLE 7.23    Median Return for Companies in Seven CPI Segments: 1998

| Number of companies | CPI segment | Return on assets (%) | Return on equity (%) |
|---|---|---|---|
| 37 | Chemical companies | 5 | 15 |
| 21 | Forest products | 2 | 4 |
| 18 | Metals companies | 4 | 12 |
| 20 | Petroleum refineries | 2 | 5 |
| 12 | Pharmaceutical companies | 17 | 36 |
| 10 | Rubber and plastics companies | 4 | 15 |
| 9 | Soap and cosmetics companies | 10 | 29 |

Source: Fortune 500, April 26, 1999.

TABLE 7.24    Returns on Assets and Equity for Companies in Chapter 7: 1998

| Company | ROA (%) | ROE (%) |
|---|---|---|
| Air Products | 7 | 21 |
| ADM | 3 | 6 |
| Dow | 5 | 18 |
| Du Pont | 11 | 32 |
| Exxon | 7 | 15 |
| International Paper | 1 | 2 |
| Merck | 16 | 41 |
| Mobil | 4 | 9 |
| Monsanto | (1) | (5) |
| Procter & Gamble | 12 | 31 |
| Union Carbide | 8 | 14 |

Source: Fortune 500, April 26, 1999.

## 7.2.12 Financial Data for Selected CPI Segments and Companies

Financial data for various CPI segments are presented for comparison purposes in Table 7.23. It is interesting to note that the pharmaceutical industries have had high returns on assets and equity. Not far behind is the soap and cosmetics sectors. The two lead all other groups, and this has been the case for many years.

Table 7.24 gives the return data for the companies presented in this chapter. Merck in the pharmaceutical sector, Procter & Gamble in the soap and cosmetics sector, and Du Pont in the chemicals sector are the leaders with respect to returns on assets and equity.

## REFERENCE

1. J. Wei, T. W. F. Russell and M. W. Schwartzlander, *The Structure of the Chemical Process Industries*, McGraw-Hill, Inc., New York, 1979.

## QUESTIONS FOR DISCUSSION

1. Compare the return on equity and the return on sales for Air Products and each firm in Section 7.2. Prepare two ordered lists.
2. Which of the industrial input–output sectors (Chapter 4) would have the greatest impact on Air Products sales if the final demand changed by 10%?
3. Sketch a new curve postulating how the sales would have looked if Air Products's chemical business had not been acquired.
4. Which firms in Section 7.2 have higher debt-to-equity ratios than Air Products?
5. Compare the firms in this chapter with respect to sales per employee and ratio of long-term debt to sales. Which firms have unusual values in both cases?
6. Compute the percentage change in sales and profits in Europe and in the United States for the gases and equipment groups of Air Products from 1990 to 1998.
7. Compute the percentage change in sales and net income for the top three producers of industrial gases.
8. What percentage of Exxon's and Du Pont's sales were due to chemicals? Compare these companies with other firms listed in Section 7.2 and prepare an ordered list.
9. Prepare an ordered list of all the firms in this chapter with respect to (a) capital expenditures in 1998, (b) total assets in 1998, and (c) research and development expenditures in 1998.

## PROBLEMS

7.1.  a.  At the beginning of this chapter, it was stated that a company is reviewed by various groups according to different criteria. Consult *Moody's Handbook of Common Stock* or *Standard & Poor's Stock Market Encyclopedia* and obtain financial profiles on Dow, Du Pont, Air Products, and Union Carbide.

b. How might the six criteria outlined at the beginning of this chapter evaluate these companies? Support your conclusions with data.

7.2.   Consult *Valueline Investment Survey* and find out how this publication rates the 11 companies in this chapter with respect to investment safety and stock performance. Find two firms with higher performance ratings than those in this chapter and discuss the reasons for their rating. Identify two firms with low performance ratings and explain, using data.

7.3.   Consult *Dun & Bradstreet Million Dollar Directory* and *Who's Who in Finance and Industry* and find the name of the CEO or president of Air Products and Chemicals, Du Pont, Merck, Union Carbide, and Exxon. Give a brief profile of each corporate leader.

7.4.   When we study the history of companies, we see what decisions were made and with hindsight can evaluate such moves. Air Products in 1971 purchased Airco. Which products did Air Products add to their portfolio with this purchase? How did these products do with respect to market growth and cooperative strength characteristics? Find the contribution these operations made to Airco in 1970 and discuss the wisdom of the decision to sell the business.

7.5.   One of the groups interested in the operation of a firm is the public. Generally, the public is ambivalent about the operation of a chemical firm unless it is directly affected. Consult the annual report and 10K report for Exxon and outline the financial costs incurred by the company as a result of the lawsuits brought in connection with the *Exxon Valdez* incident. Who filed suits against the company? What amounts of money were involved? What additional expenditures (if any) were made in public relations efforts? What percentage of 1997 sales was paid out in fines or legal fees?

7.6.   Air Products's gases and equipment group has major European operations. Consult the latest company report and determine the location of these operations and what is being produced. What are the possible threats to the European operation?

7.7.   A great deal of Air Products's growth is attributed to the aggressive expansion of acquired businesses. Consult a source such as *Moody's* or *Valueline* and determine what effect, if any, mergers had on the company's earnings per share and net income. Pick any recent merger for this question.

7.8.   Examine the companies on the American Stock Exchange and the Over-the-Counter Exchange (Appendix B). Are any of these companies listed in the top 50 (Chapter 5)? What are the criteria for being listed on each exchange?

7.9.   Examine the *Chemical Week* 300 list and find 10 companies that are not in the tables of Appendix B. What do these companies have in common?

# 8

## General Characteristics of the CPI

In this chapter we discuss the prevailing and unique characteristics of the CPI and subject them to economic and accounting analysis. The emphasis is on industrial chemicals and petroleum refining, where the predominant economic structure is *oligopolistic;* that is, a handful of large firms shared the manufacturing and market of a product. The large-scale production of chemicals requires sophisticated technologies and heavy capital investments. The industry must have firms that are sufficiently large and stable to carry on these functions. Under these circumstances an oligopoly is inevitable. Whether it is composed of a group of profit-making firms or of government entities (e.g., the Tennessee Valley Authority) it is the only alternative to monopoly.

Firms in the oligopoly environment are motivated and influenced by a set of outside forces that represent carrots and sticks—technological, economic, and sociopolitical. To succeed in this environment, firms must develop patterns of behavior for both short- and long-term survival. We discuss and analyze these forces and the resulting patterns of behavior in this chapter.

### 8.1 MARKETING

Many important industries, such as automobiles and steel, concentrate on producing essentially one product. In contrast, the CPI produce an astounding variety of products by a myriad of processes in many different and widely scattered plants. Most chemical companies produce hundreds of separate products. Moreover, the industry consists of hundreds of separate chemical companies, so that even the largest companies account for a smaller share of the chemical industry

**TABLE 8.1** Concentration Ratios for Selected Industries: 1996

| SIC code | Industry | Percentage of value of shipments accounted for by largest producers | |
|---|---|---|---|
| | | Four largest | Eight largest |
| 2087 | Flavoring, extracts, syrups | 69 | 75 |
| 2023 | Condensed and evaporated milk | 43 | 55 |
| 2082 | Malt beverages | 90 | 98 |
| 2096 | Shortening, table oils, margarine | 70 | 76 |
| 3651 | Radio and TV sets | 39 | 56 |
| 3711 | Motor vehicles and car bodies | 84 | 91 |
| 3721 | Aircraft | 79 | 93 |
| 3724 | Aircraft engines and engine parts | 77 | 84 |
| 3861 | Photographic equipment and supplies | 78 | 83 |
| 3334 | Primary production of aluminum | 59 | 82 |
| 2621 | Paper mills except building paper mills | 29 | 49 |
| 3221 | Glass containers | 84 | 93 |
| 2891 | Printing ink | 25 | 37 |

*Source:* U.S. Bureau of Census, Subject Series *Concentration Ratios in Manufacturing,* 1992 (published in 1997). This document is published approximately every 5 years. The 1997 data will not be available until after 2000.

output than the large companies in other industries, particularly automobiles and aircraft.

The concentration ratios for selected industries are presented in Table 8.1. This ratio measures the percentage of an industry's economic activity that is accounted for by the four and eight largest companies. These data are compiled by the U.S. Bureau of Census and published at 5-year intervals. The data in Tables 8.1 and 8.2 are 1992 information, since the 1997 information was not available at this writing. Although Table 8.1 is based on the value of shipments, one could develop concentration ratios based on value added, employment, net income, or total assets. Of the industries shown in Table 8.2, asphalt paving mixtures has the lowest concentration ratio. It is also obvious to the reader that only four companies supply the market with 84% of the domestic automobiles and eight companies supply 91%.

Although the CPI is not very concentrated in terms of total industry shipments, it does tend to be more concentrated for specific products. Some examples are shown in Table 8.2. Since the number of companies that make plastic foam products (SIC 3086) is large, the concentration ratio is low. However, the number of companies engaged in specific product groups is generally below 10, so that the product concentration ratio tends to be high for the eight largest companies

TABLE 8.2  Concentration Ratios for Specific Sectors in the CPI: 1992

| SIC code | Product | Percentage of value of shipments accounted for by largest producers (%) | |
|---|---|---|---|
| | | Four largest | Eight largest |
| 2812 | Alkalies and chlorine | 75 | 90 |
| 2813 | Industrial gases | 78 | 91 |
| 2816 | Inorganic pigments | 69 | 79 |
| 2823 | Cellulosic man-made fibers | 98 | 100 |
| 2824 | Organic fibers, noncellulosic | 74 | 90 |
| 2835 | Diagnostic substances | 49 | 65 |
| 2836 | Biological products except diagnostic materials | 53 | 71 |
| 2841 | Soap and other detergents | 63 | 77 |
| 2844 | Toilet preparations | 36 | 55 |
| 2851 | Paints and allied product | 29 | 43 |
| 2869 | Industrial organic products, not otherwise classified | 29 | 43 |
| 2879 | Agricultural chemicals, not otherwise classified | 53 | 69 |
| 2892 | Explosives | 57 | 77 |
| 2911 | Petroleum refining | 30 | 49 |
| 2951 | Asphalt paving mixtures | 15 | 24 |
| 2992 | Lubricating oils and greases | 35 | 48 |
| 3011 | Tires and inner tubes | 70 | 91 |
| 3084 | Plastic pipe | 23 | 39 |
| 3085 | Plastic bottles | 39 | 55 |
| 3086 | Plastic foam products | 22 | 32 |

Source: U.S. Bureau of Census, Subject Series Concentration Ratios Manufacturing, 1992 (published in 1997). This document is published approximately every 5 years. The 1997 data will not be available until after 2000.

(e.g., organic fibers, 90%; industrial gases, 91%; explosives, 77%; tires and inner tubes, 91%). Most chemical product markets are classified as oligopolies.

The concentration ratios for selected chemical products are presented in Table 8.3. The four largest producers of butadiene account for 73% of the production, and there are only seven producers in the United States. On the other hand, the eight largest producers of propylene account for 57% of the total propylene.

The classification of products of Chapter 5 designating products as commodities, pseudo-commodities, fine chemicals, and specialty chemicals provides

**TABLE 8.3** Concentration Ratios for Selected U.S. Chemical Products: 1994 Capacity

| Product | Four largest Capacity (lb × 10⁶) | Four largest Percent of total | Eight largest Capacity (lb × 10⁶) | Eight largest Percent of total | Total U.S. capacity (lb × 10⁶) |
|---|---|---|---|---|---|
| Ethylene | 16,940 | 35 | 31,310 | 65 | 48,015 |
| Propylene | 10,785 | 37 | 16,565 | 57 | 28,860 |
| Butadiene[a] | 3,080 | 73 | 4,235 | 100 | 4,235 |
| Ethylene oxide | 4,685 | 59 | 6,971 | 88 | 7,896 |
| Ethylene dichloride | 16,415 | 62 | 23,915 | 91 | 26,295 |
| High-density polyethylene | 6,280 | 50 | 10,450 | 83 | 12,580 |
| Low-density polyethylene | 9,010 | 59 | 13,060 | 86 | 15,155 |
| Polypropylene | 5,590 | 55 | 7,680 | 76 | 10,160 |
| Vinyls | 7,560 | 56 | 11,865 | 88 | 13,450 |

[a]Only seven butadiene producers in the United States.
*Source:* CMAI.

a useful structure for discussing characteristics of the market. Based on dollars of sales, the four categories account for the following approximate percentages of total chemical sales:

| Category | Percentage of sales |
|---|---|
| Commodities | 35 |
| Pseudo commodities | 21 |
| Fine chemicals | 9 |
| Specialty chemicals | 25 |

## 8.1.1 Commodities

A commodity chemical is produced in large volume to meet a reasonably uniform set of specifications. By definition, it is undifferentiated; that is, a commodity chemical produced by Dow is essentially the same as one produced by Union Carbide or Du Pont. Typical commodity chemicals are sulfuric acid, carbon dioxide, ethylene, lime, methanol, urea, and gasoline. These products are also referred to as fungible (i.e., interchangeable) products. For example, specification gasoline of one producer can be introduced into a common carrier pipeline (e.g., to the East Coast) following a gasoline produced by a different company, provided both fuels meet the specifications. However, there are different grades of gasoline (regular, premium, and superunleaded), and these must be separated by some means. There are also some products that because of the way their specifications are established, raise concerns about mixing products from two different producers or even from different lots from the same producer. For instance, some products have a specification that defines an acceptable boiling range of one degree Celsius to include 82.6°C. This can produce problems for different lots, both of which by themselves would pass the specification. Consider that one lot could have a boiling range of 81.8–82.7°C and this would be within the specification limit. A second lot could meet specifications and have a boiling range of 82.5–83.4°C. Both these lots will pass specification and therefore might be considered to be fungible. If the two lots are blended in equal quantities, it is likely that the boiling range of the combined lot will be 81.8–83.4°C, however, such that the boiling range will exceed the specification of one degree Celsius maximum.

A company's share of a commodity market is based mainly on price, reliable delivery, and convenience. After the shortage periods of 1973–1974 and 1979, however, greater emphasis was placed on the reliability of the supplier. Relatively little technical service is required (particularly in comparison to specialty chemicals), although this may depend somewhat upon the product and the company's customers. A typical customer is another industrial concern. Steel companies buy oxygen for steelmaking as well as sulfuric and hydrochloric acids for pickling. Oxygen plants are built adjacent to the producers' steelmaking customers. A chemical company will buy ethylene from a major producer to make ethylene derivatives. Commonly, both the ethylene producer and the consumer will be located on the Gulf Coast ethylene pipeline grid, with deliveries made through that pipeline. Commodity chemical prices tend to be low, hence are not able to "carry much freight." Consuming plants are located near major basic chemical producers/suppliers, so markets tend to be local.

Almost no commodity chemicals are sold directly to the consumer. They satisfy derived demands and provide no pleasure of acquisition from the customer's point of view. Even those few chemicals sold directly to the consumer are adjuncts to the more glamorous products. A person who becomes quite ecstatic over the purchase of a new car may buy gasoline, antifreeze, lubricating oil, and brake fluid only because the car would not move without them. Market research is not product-innovation-oriented but is directed toward prediction of trends in the economy and the economic health of major users. Nevertheless, some commodity chemicals take on increased significance during severe shortage periods such as during the oil embargo of 1973–1974. The price of antifreeze (and therefore ethylene glycol) increased sharply when consumers realized that without that antifreeze their expensive automobiles were in jeopardy. Likewise, the farmer–retailer combinations became aware of the importance of styrene/polystyrene for egg containers, and without this small amount of styrene/polystyrene, egg sales would be severely hampered.

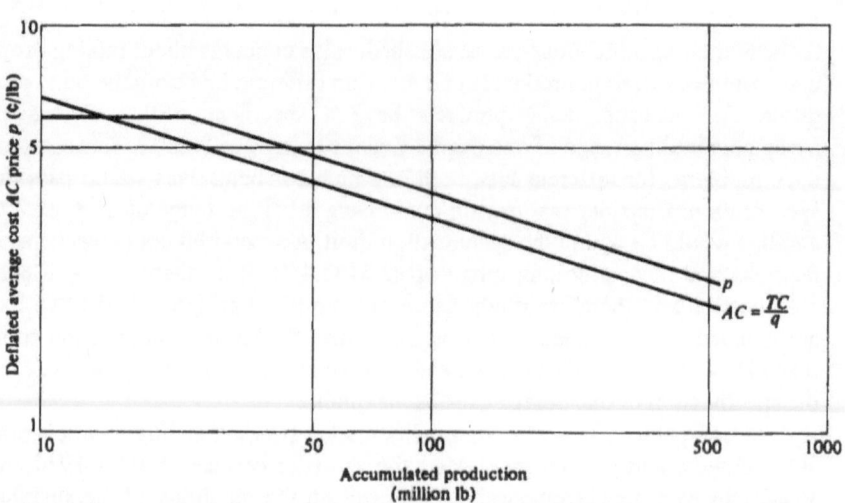

FIGURE 8.1   Typical experience curves showing a stable cost–price pattern.

Commodity prices show distinct patterns with time. Figure 8.1 shows typical curves, the logarithm of deflated cost and price versus the logarithm of accumulated production (see also the experience curves in Chapter 6). The situation shown in Figure 8.1 is typical of a stable cost–price pattern for a commodity chemical. Price frequently starts out below cost, and losses are experienced by the processing firms in the early stages of operation, when the product is being introduced into the market. As experience is gained, cost decreases. As a corporate strategy, prices could be lowered with accumulated volume in the same functional manner as cost, keeping the profit low enough to discourage the entry of new competition and to keep the market share high. Firms that cannot keep their production costs decreasing with time (directly related to accumulated production) must accept lower profits and perhaps eventually lose money and ultimately go out of business. Firms that innovate or expand in size can reduce costs faster than the average producer, make greater profits, and attain larger market shares. However, there is an upper limit to the market share, set by government antimonopoly regulations. When product concentration ratios become too high, the government will sometimes attempt to break up the dominating firms. Some famous examples in the CPI, include the breakup of the Standard Oil Trust in 1911 and the breakup of Du Pont's black-powder business with the formation of Atlas and Hercules.

An unstable cost–price pattern is shown in Figure 8.2. In this case, the producers of the chemical did not lower prices corresponding to decreasing costs (phase A). The high profit margins encourage the building of new processing

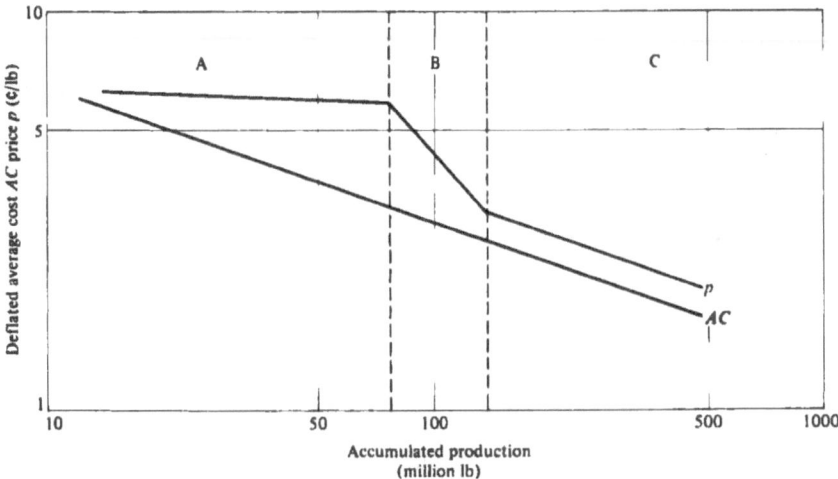

**FIGURE 8.2** Typical experience curves showing unstable cost–price pattern.

units by both existing and new producers. Eventually price cutting must take place (phase B). This may be followed by a stable phase, enjoyed by the surviving firms after marginal producers have been forced out of business (phase C).

The original producers end up with lower market share if many new firms enter the business. Had a different pricing policy been followed, a more stable market situation might have been maintained, with a larger market share for the original producers. This unstable pattern is unfortunately quite typical of many commodity chemicals and is indicative of the intense competition in the CPI. The methanol experience (Figure 6.5) offers a typical example and shows the end result of plant expansion by firms already producing methanol and of new units put on stream by others attracted to the business by potentially high profits. The resultant overcapacity necessitates price-cutting. In the petrochemical industry, this happened so frequently that financial analysts referred to "the death wish of the petrochemical industry." One reason leading to this chain of events is that the only barrier to entry in many commodity chemical businesses is the large amount of capital required. The technology for manufacturing is generally well established, and many engineering firms offer "turnkey" plants for commodity chemicals. Any firm with enough capital can buy technology and enter the commodity chemical market. Of course, the firm must do so with a plant large enough to achieve the economy of scale necessary to keep costs at a profitable level.

In today's market, when most producers are designing plants capable of producing a billion pounds per year, one cannot make a profit by building a comparatively small ethylene plant with a capacity of 100 million pounds per year. Technical expertise is also essential, and it is not possible to remain in the chemical business without a competent staff of chemical professionals. For some commodity chemicals, lack of inexpensive raw materials forms a barrier to entry. This is obviously true in petroleum refining, where crude oil supplies must be available. This problem was graphically demonstrated during the 1973–1974 oil embargo, but since that time it has been possible for anyone with enough money to purchase crude oil supply. Additionally, partnerships such as SABIC (the Saudis) and Texaco or PDVSA (Petroleos de Venezuela S A, the national oil company of Venezuela), and CITGO have helped assure a viable supply of crude oil to a U.S. refiner at a competitive price.

A lack of a local market of sufficient size is another barrier to entry for low-priced commodity chemicals. One must have local customers with sufficient needs to justify the necessary economies of scale for the processing unit. The more mature commodity chemicals have what could be termed a low-profit barrier to entry. A new firm cannot justify a comparatively high-cost processing unit when an older and more experienced firm is operating a similar unit at a much lower cost. Normally, however, the newer units employ the latest technology, and if competitive feedstock is available, the newer units are likely to be the ones with the lower operating cost units. This poses a problem for older and smaller units. Can they justify their continued existence in the face of newer, lower cost units? While it is easy to examine the slate of operating units for a specific commodity chemical and pick out the older, smaller units, one must exercise restraint in assigning them to the scrap heap. In many cases further investigation reveals that these units have other factors that make it possible for management to justify their continued operation. For instance, some older, smaller ethylene plants have continued to operate because they are strategically located next to consuming units that would not have a convenient, economical alternative source of ethylene. However, if an older, smaller unit is dependent on customers that are located on the Gulf Coast ethylene pipeline grid, they are likely to be very vulnerable to being shut down, since other sources of ethylene feed are available to customers also connected to the pipeline grid.

Sometimes price is set by other factors. Firms can attempt to establish price by setting up a cartel. Such an action is illegal in the United States unless it is expressly sanctioned by the government. However, outside the United States, cartels have been in effect for years, with perhaps the most visible one being the Organization of Petroleum Exporting Countries (OPEC). This has cartel attempted with inconsistent success to control the oil price in the world since 1973. The problem is that various countries cheat on their assigned quotas, with the result that too much oil is produced, thereby depressing the price. Similarly, during the period of worldwide ethylene surplus capacity, an ethylene cartel was imposed in Japan by the Ministry of International and Trade Industry (MITI).

This cartel assigned each ethylene producer an "official" capacity that was significantly less than its actual capacity, and the companies were not allowed to exceed that number in their production. While this action had the effect of limiting production and preventing a severe erosion of domestic ethylene price, it is generally acknowledged that there was some overproduction even with MITI monitoring the activity. This cartel was not universally popular with the ethylene producers in Japan and was finally scrapped. Up until 1973, the Texas Railroad Commission regulated the amount of crude oil that could be produced in Texas. All limitations were removed when imports became a major portion of the oil being refined in this country, however, since such limitation would no longer have a pronounced effect on the price of crude oil.

Ultimately, the price of any commodity chemical will be *set by the highest cost producer that is necessary to supply the quantity of product needed by the market.* That producer may just break even (with difficult decisions for company management with respect to whether to continue production or withdraw from the business). Thus, more efficient and lower cost producers will enjoy profit margins determined by the difference in the price required to keep the high-cost producer in business and the cost of production for the more efficient producers. Thus, most of the research and development efforts by commodity–chemical manufacturers are directed toward improvement of existing processes and development of new processes (often using new raw materials). Since the manufacturing units need to be large continuous processes producing large volumes of product, a small change in efficiency can have a marked effect on plant cost. Experienced process engineers capable of innovative design and clever process modifications are essential to the economic health of the producer of commodity chemicals.

The CPI continues to go through pricing and expansion cycles that appear to confirm the "death wish" syndrome. However, consider the situation faced by management of a commodity chemical producer. During strong markets when demand is high and prices increase, profit margins are good, resulting in strong earnings for that company. Management is faced with the question of what to do with the surplus money. The options are to increase retained earnings, pay the money out as extraordinary dividend, purchase another company, or invest in their own new plant and facilities. Each of these options has limitations. Increased retained earnings are scrutinized and limited by the Internal Revenue Service of the federal government. Extraordinary dividends often cause difficulties for both the company and the shareholders Investment in new plant and facilities results very often in overcapacity, which has a depressive effect on future prices and profit margins. However, where a company is dedicated to a particular segment of the petrochemical business and wishes to remain a significant supplier in the U.S. market, this option is one most often adopted.

## 8.1.2 Pseudocommodities

Pseudocommodities are differentiated products produced in large quantities. They are not characterized simply by composition specification, inasmuch as manufacturer promise to meet an in-use performance specification for each customer. Synthetic fibers, resins and plastics, elastomers, and carbon blacks are good examples of pseudocommodities.

A commodity may become a pseudocommodity by the addition of additives or other performance-enhancing chemicals. For example, ethylene glycol—a commodity chemical—perhaps becomes a pseudocommodity such as Prestone™ or Zerex™ antifreeze with the addition of additives. In some of these cases, multiple brands may be packaged in the same facility but with proprietary additives used for each separate brand. There are numerous examples of this sort of activity with gasolines and motor oil. In the past oil companies have claimed that their gasolines were superior because of additives they contained (TCP™, Techron™, etc.).

Market share in a pseudocommodity business is not solely influenced by price, as in a commodity business. The producer of a pseudocommodity can influence market share by stressing differences in product performance by providing technical service and by advertising. The firm that provides superior technical service with the product sold can charge higher prices and may increase its market share. The successful pseudocommodity firm needs chemical professionals who understand the customer's business better than the customer does and have the skill to solve customer problems. The functions of product development, market testing, and customer relations are crucial to success. Such chemical professionals should know about the performance of materials under use (octane number, viscosity index, mechanical strength, thermal conductivity, stability to oxidation, etc.). They should also know the theoretical and empirical relations of these properties to the chemical structure and physical aggregation of the material. They need the ability to design and manufacture these materials by synthesis, use of additives, compounding and formulation, and modification to mechanical operations.

Much research and development effort is directed toward the customer and his needs. It is not uncommon for a single processing unit's output to be custom-made to satisfy a few large customers. Some research and development must be directed toward process changes to satisfy changing customer requirements. Pseudocommodities are produced in large continuously operating plants, and process improvement research is needed to improve operating efficiencies. Companies selling thermoplastic resins can greatly impress the customer by supplying sound engineering solutions to problems in molding and extrusion operations. Du Pont and Phillips Petroleum have laboratories dedicated to assisting customers with injection, extrusion, blow molding, and other processing problems.

Advertising in a monopolistic market is intended to increase sales. Small firms in an atomistic market such as farm cooperatives cannot afford to advertise unless they band together to advertise cooperatively in an effort to increase demand. An example of such jointly sponsored advertising is the current campaign to convince the public that consuming milk is good—the ads show celebrities who have evidently just finished drinking milk. Advertising pays the best dividends in an oligopolistic differentiated product market, like pseudocommodities, where one needs to entice customers from one producer to another. Many CPI firms manufacturing similar products attempt to differentiate by using trade names and extensive advertising (e.g., polyethylene Glad™ bags, Hefty™ bags). Laundry bleaches that are identical in composition are sold under a variety of trade names for a wide variety of prices.

Barriers to entry into the pseudocommodity business include all the barriers for commodities plus the all-important customer know-how. Lack of technical expertise and patent protection can be formidable barriers to potential producers of a pseudocommodity. Du Pont was the sole producer of nylon from 1939 to 1951. Barriers to entry were removed under threat of government antitrust action in 1951, with the licensing of Chemstrand, which later became part of Monsanto. Since then a number of companies have entered the nylon business. Technical know-how and patent protection played a major role in both high- and low-pressure polyethylene manufacture in the years shortly after World War II. ICI developed polyethylene, but Union Carbide had a superior high-pressure process. Ziegler, Du Pont, and Phillips Petroleum all developed low-pressure processes, which they subsequently licensed to other manufacturers. Many pseudocommodities eventually become commodities by the diffusion of technology, standardization of the product, and the entry of many firms into the business.

### 8.1.3  Fine Chemicals

Fine chemicals are undifferentiated products made in lower volume than commodity chemicals. These materials are characterized as very pure substances, produced in low volume, with high manufacturing cost, high selling price, and high profit margin. The production is often under 10 million pounds a year and is carried out in small, flexible batch-processing units. However, in more recent times, some companies have been able to produce large tonnages of fine chemicals rapidly, in flexible multipurpose/multiproduct facilities. These huge state-of-the-art plants are designed for the particular needs of fine chemical manufacturing.

Typical products in this group include pharmaceuticals, food additives, reagent chemicals, and medicinal products isolated from botanical products and herbs. Examples include such varied products as aspirin, tartaric acid, citric acid, vitamins, penicillin, quinine, hormones, antibiotics, biochemicals, and veterinary pharmaceutical products. Many of the products in SIC 283 and 284 can be classified as fine chemicals. These chemicals are produced to a property specification that is generally more stringent than most commodity-chemical specifications, and the product does not vary much from firm to firm.

Market share depends on pricing policy, quality of the product, and advertising to remind buyers of the existence of suppliers. Research and development efforts are directed toward improved product quality. A common process improvement is the conversion of a batch process to a continuous operation if the sales volume is sufficiently high.

There has been a substantial change in the overall fine chemical business in the last four decades. At the end of World War II and into the 1950s, the term *fine chemicals* did not exist. The fine chemical industry started from basic chemicals (ethylene oxide, acetic acid, phenol, etc.) and developed more advanced organic intermediates. Plants were built that were dedicated to the manufacture of, for example, alkylphenols, acetoacetates, and carboxymethylcellulose.

In the late 1960s, specific requirements of these fine chemicals became the impetus for industry growth. Examples of products in this time period were intermediates for pharmaceuticals and agrochemicals such as side chain modifications of penicillins and carbamates for fungicides. Very specific fine chemicals for specific products were developed.

Gradually partnerships were formed between life sciences and fine chemicals industries, the latter having resources to respond to the specific needs in terms of products and technologies. The relationship between life sciences and fine chemicals industries has evolved from a vendor–purchaser approach to a strategic partnership. There has been a much broader sharing of information, be it with respect to cost improvement programs or to commercial outlook for the product(s). Both partners focus their resources on their core competencies.

Not every fine chemical producer will be able to move to *custom* production. It takes more than idle production capacity to enter this field. The fine chemical manufacturer must have the key technology associated with the ability to carry out a wide variety of chemical syntheses. Another key facet is the quality of service. Most suppliers recognize that these projects may have a life of 3–5 years, so they want to do a good job, in the hope of obtaining repeat business. If the project is successful, the customer may consider building a plant. Generally, customer and supplier will share intellectual property.

Custom manufacturing is prevalent in the fine chemical sector and to some extent in specialty chemicals. The partnership formed is usually a long-term arrangement, allowing both parties to obtain the best sustainable commercial advantage. There may be close interfacing of manufacturing, research and development, administrative, operational, and financial functions. There is an equitable sharing of risks and rewards. There are some large custom producers like Lonza (Swiss), Huls (German), and Eastman Chemical (U.S.), but most are small.

Custom manufacturing in 1996 amounted to about $6 billion in total sales: $3 billion in pharmaceuticals, $1.5 billion in agricultural chemicals, and the remainder in food additives, flavors, fragrances, biocides, and so on.

Over the past several years, fine chemical manufacturers have come under increased pressure globally, and this is having a major impact on how the producers do business. They are responding with innovative technologies as well as with the start-up of custom production for major chemical companies. As outsourcing spreads, custom producers are mastering new technologies and creating organizations to manufacture fine chemicals. Technical innovations mean that there are particular chemical reactions such as alkylation, nitration, and halogenation in which a company has the *know-how*.

Most fine chemicals are used to make pesticides, pharmaceuticals, and flavors. The pesticide sales have grown slightly in the past few years, but the pharmaceutical industry, which is the growth area, has grown much more rapidly.

Small firms have lower overhead expenses and they can be more flexible than large firms. Both are faced with the same regulatory requirements, but in a small firm there are fewer layers of management. Outsourcing of intermediates is the most desirable approach for pharmaceutical firms, for it allows them to share some of the risks with fine chemical producers. It also lets pharmaceutical companies focus on what they do best, namely, finding and developing new pharmaceuticals. A product may not live up to its expectations in the marketplace, so the customer limits the loss by not investing in production facilities.

Fine chemical producers are deepening the product and technology mixes that differentiate them from their competitors. The deepening involves developing new technologies based on the company's core technologies. Often *chemical trees* are developed from the core technologies. For example, Lonza Chemical Company has developed *chemical trees* based on diketenes or hydrogen cyanide. Companies with fewer than five core technologies are limited to making *standard* intermediates that have many buyers, or they may offer to process customer's raw materials. Companies with more core technologies are in a better position to carry out multistep syntheses such as in the production of bulk medicinals.

To survive the cost pressures and environmental regulations by means other than developing innovative technologies, fine chemical manufacturers with low sales volume have had to integrate horizontally and vertically through alliances with larger chemical companies.

Barriers to entry include requirements for proprietary manufacturing, a high-technology base, substantial process knowledge, considerable research and development resources, patent protection, specialized marketing ability, proven reliability, and a limited market size.

## 8.1.4  Specialty Chemicals

Specialty chemicals are differentiated end products that are formulated or synthesized in low volume and are designed to solve specific customer needs. Kline (1) developed categories into which specialty chemicals are classified (Table 8.4). Many of these chemicals are highly proprietary items involving trade secrets and protected by patents and are the result of expensive research and development. Chemicals for the microelectronics industry are examples. Over the past two decades, total sales of specialty chemicals have grown from $10 billion to $72 billion.

Since specialty chemicals are often far superior in performance to the next best alternative, their pricing is determined more by a customer's cost savings than by the manufacturer's costs. For example, zeolite-based cracking catalysts enable oil refiners to increase gasoline yield and octane number without modifying their plants. Because of this excellent performance in use, these catalysts are often sold for more than twice the cost of alumina–silica catalysts. Flavors and fragrances are priced not according to raw material and production costs but according to the pleasurable responses they elicit for the customers. Pesticides are selected according to their potency and discrimination between insect species. In the pesticide field, prices reflect the cost of research as well as the extensive laboratory and field testing required to satisfy the regulations of the Food and Drug Administration and the Environmental Protection Agency. Many of the specialty chemicals sell at a premium price and therefore lie in the upper right-hand side of the price–production-volume line in Figure 5.4 (see Chapter 5). They sell at higher prices than average chemical products because of their special characteristics.

In the mid- to late 1970s, new commodity chemical plants were built in the Mideast and Far East. U.S. companies were not interested in building large capital-intensive refineries and commodity-producing plants. Instead, they turned to the specialty chemical field, with the result that there was a flurry of activity in the acquiring or building of specialty chemical facilities. By the mid-1980s, specialty chemicals were in recession. For example, conditions were right for the start of a large number of flavor and fragrance companies, resulting in the proliferation of small company start-ups, mainly in flavors. Ultimately, consolidation in this industry produced a few leading companies with manufacturing and marketing capabilities that spanned the globe and served multinational consumer companies (2).

In the early 1990s, some specialty chemical segments were price sensitive and cyclical, with international producers focusing more on customer service than on research and development. Some producers had not raised prices in years. By the mid-1990s some specialty chemical segments displayed the characteristics of mature commodity sectors. Slowly prices were raised, and some

TABLE 8.4  Sales of Selected Speciality Chemicals in the United States[a]

| | Sales (millions) | | Real growth, 1994–1998 (%)[c] |
|---|---|---|---|
| | 1973[b] | 1994[c] | |
| **Low-volume types of pseudo-commodity** | | | |
| Adhesives | $200 | $4,900 | 4 |
| Elastomers | $185 | $2,100 | 6 |
| Plasticizers | $130 | $620 | 4 |
| Specialty surfactants | $195 | $1,775 | 3 |
| Total | $710 | $9,395 | |
| **Multifunctional compounds** | | | |
| Antioxidants | $180 | $730 | 2 |
| Biocides | $160 | $1,155 | 4 |
| Catalysts | $200 | $1,610 | 4 |
| Chelates | $25 | $165 | 4 |
| Corrosion inhibitors | $100 | $300 | 3 |
| Dyes | $550 | $1,370 | 4 |
| Enzymes | $55 | $315 | 5 |
| Flame retardants | $100 | $665 | 7 |
| Thickeners | $240 | $1,600 | 5 |
| Ultraviolet absorbers | $15 | $160 | 6 |
| Total | $1,625 | $8,070 | |
| **End-use chemicals** | | | |
| Agricultural chemicals | NA | $7,115 | |
| Cosmetic additives | $100 | $995 | |
| Diagnostic aids | $300 | $7,200 | |
| Flavors and fragrances | $225 | $1,755 | 6 |
| Food ingredients | $275 | $3,010 | 6 |
| Foundry chemicals | $75 | $190 | 2 |
| Industrial and institutional cleaning products | $1,200 | $5,335 | 2 |
| Laboratory chemicals | $130 | $1,035 | 6 |
| Paint additives | $95 | $320 | 3 |
| Paper additives | $115 | $735 | 5 |
| Photographic chemicals | $250 | $1,555 | 6 |
| Plastics additives | $255 | $2,920 | 4 |
| Printing chemicals | $50 | $180 | 4 |
| Rubber processing | $220 | $450 | 3 |
| Textile chemicals | $105 | $650 | 3 |
| Water management | $360 | $2,475 | 3 |
| Total | $3,755 | $35,920 | |
| Grand total | $6,090 | $53,385 | |

[a]To compare 1973 sales with 1994, multiply 1973 sales by 1.917 (inflation factor) to bring to 1994 dollars.

[b] C. H. Kline, "Maximizing Profits in Chemicals," CHEMTECH, February 1976.

[c] Kline and Company, Chemical Week, July 26, 1995, p. 53.

TABLE 8.5   Unique Characteristics of Commodity, Fine, and Specialty Chemicals

| Characteristic | Commodity chemicals | Fine chemicals | Specialty chemicals |
|---|---|---|---|
| Product life cycle | Long, >30 years | Moderate, 10–20 years | <10 years |
| Product slate | Narrow | Very broad | Very broad |
| Product volume | >10,000 tons/yr | <10,000 tons/year | Variable |
| Approximate product | <$9/kg | >$10/kg | Variable price |
| Product differentiation | Nonexistent | Low | High |
| Value added | Small | High | Moderate to high |
| R & D focus | Process improvement | Process development | Product development |
| Capacity intensity ratio (assets/sales) | ≥1 | –1 | ≤1 |
| Customer | Not a chemist | Chemist | Not a chemist |
| Pricing | Competitive | Cost plus | Related to value in use |
| Key skill | Cost optimization | Management of production diversity | Customer needs |

Source: Adapted from Chemical and Engineering News, November 25, 1996, p. 46.

markets became highly price competitive. By 1996 it was noted that increased investments were occurring in the custom manufacturing of specialty chemical products. The competition in more mature specialty chemical businesses such as dyes has led to companies outsourcing the production of intermediates usually from companies in the Far East. Frequently, these relationships result in the forming of *alliances*. One danger to this action is that parties may find themselves aligned with potential competitors.

Since unique technologies are incorporated in the manufacture of specialty chemicals, chemical professionals who are innovative and creative in research and development are needed to search for new compounds, new synthesis methods, and new ways to attract customers. Today in some companies, the marketing people are located in close proximity to the research and development groups, forming part of a team to solve customer's requests. The total operation is more chemical professional intensive and less raw material intensive. Specialty chemicals have a wide profit margin, rewarding those companies for being different and first in the field.

Specialty chemical companies concentrate on using their technology to ensure niche positions in the market. Although gross margins have always been high in the specialty chemical sector, the margins are decreasing because of increased raw material prices. Margin pressures are forcing consolidation to reduce costs, boost efficiencies, and increase market share. Large companies eye high value, going for growth in response to the lure of potentially high gross margins. These same chemical companies have made acquiring specialty businesses an important part of their growth strategies.

Barriers to entry include the needs for proprietary technology, highly innovative, expensive research, patent protection, and close customer liaison to become the sole supplier.

Table 8.5 compares some of the unique characteristics of commodity, fine, and specialty chemicals.

## 8.1.5 Oligopoly Marketing and Theory of Games

With few exceptions, all major chemicals are sold in an oligopolistic market. As was noted in Chapter 2, development of a successful marketing strategy in an oligopoly requires the firm to develop insights into the effect of price changes on its market share, $dp/dq$. There is an extensive literature (2–5) devoted to the analysis of oligopoly marketing by means of the theory of games. Some of the more interesting results will be discussed to illustrate the principles of pricing strategy useful to the firm.

Consider a product manufactured by two firms X and Y (a duopoly). These two firms are scheduled to announce the product price for the next quarter. There is some loyalty and there are some long-term contracts, so that if the two firms' announced prices are not identical, the lower priced product will sell more but will not necessarily force the other product completely off the market. Let us assume that the price per unit sold can be set at only three discrete levels and that the two firms announce new prices on the same day without previous knowledge of the other firm's intentions. The resulting market share captured by firm X is displayed in Table 8.6. The market share captured by Y is what is left over after X's price has been subtracted from 100. When the two prices in Table 8.6 are unequal, the lower priced product gains market share at the expense of the higher-priced product. If firm X wishes to maximize its share of the market, it is apparent that the strategy of the 80-cent price is better than any other strategy and constitutes the *dominant strategy*.

**TABLE 8.6** Market Share Payoff Matrix: Market Share of Firm X

| Firm X price | Firm Y price (millions of dollars) | | |
|---|---|---|---|
| | $0.80 | $0.90 | $1.00 |
| $0.80 | 50 | 80 | 100 |
| $0.90 | 20 | 50 | 80 |
| $1.00 | 0 | 20 | 50 |

No matter what the price strategy of firm Y, a firm X price of 80 cents will capture more market share than any other X price. Similarly, firm Y will come to the same conclusion and choose 80 cents. The two firms will share the market equally. This is an example of a *zero-sum* game, where the two firms compete and the gain of one firm is the loss of the other. An underlying assumption is that of constant product demand. The solution of this game results in a *pure strategy* dominated by a price of 80 cents for both firms. This is a stable-equilibrium situation: that is, neither firm can improve its market position by making unilateral moves.

However, if we consider the profits of these two firms, we have to consider that any higher price might lower the total sales revenue but improve the profit margin. Since the sum of the profits of the two firms need not sum to a constant, the profit payoff matrix (Table 8.7) has two numbers in each square; the first is the profit to X and the second is the profit to Y. When the prices are equal and higher both firms enjoy good profits. At some fixed firm X price, say $1.00, a low firm Y price will mean large sales and profits for Y and deficits for X. As the firm Y price increases at a constant firm X price of $1.00, X will capture more market and improve profits. Firm Y will lose market share but increase profits at a Y price of $0.90 because of improved profit margins. When the firm Y price is equal to $1.00, firm Y's profit is equal to firm X's profit. It is apparent that if the two firms want to maximize their profits cooperatively, they should agree on a price of $1.00. Any formal agreement between two firms might result in the corporate executives spending some time at a federal penitentiary; however, this is the prevailing mode of operation for the oil-producing nations and for the bauxite-exporting nations. Having settled on a price of $1.00, firm Y might make secret price concessions to lower the price to $0.90, which would increase its sales volume and profits to $30 million. Firm X will inevitably notice the erosion of its sales orders as well as profits and will lower its price (perhaps all the way to $0.80) to improve on its profit.

In response to this move by firm X, firm Y has no alternative but to lower its price also to $0.80. Notice at that final equilibrium, there is a standoff in market share, but the profits for both firms have declined from $25 million to $5 million. A price of $1.00 for both firms will not be stable, since either firm can improve its own situation by making unilateral moves. This example combines the elements

TABLE 8.7   Profit Payoff Matrix (Millions of Dollars)

| Firm X price | Firm Y price | | |
|---|---|---|---|
|  | $0.80 | $0.90 | $1.00 |
| $0.80 | (5, 5) | (20, 0) | (25, −20 |
| $0.90 | (0, 20) | (15, 15) | (30, 10) |
| $1.00 | (−20, 25) | (5, 30) | (25, 25) |

of both competition and cooperation. If both firms are dominated by the philosophy of profit maximization, both should be reluctant to make any price cuts, since the end result will eventually lower profits from $25 million to $5 million in this example. Some critics claim that such price cartels indeed exist in the United States, if not by formal agreements then by undeclared informal understandings. International cartels have a history of instability and price concessions by new entries and by firms with "weaker hands." OPEC is the classic example.

Consider next a marketing game with no dominant strategy so that the best strategy is a *mixed strategy*. We can develop a payoff matrix (Table 8.8) to illustrate this case by considering a detergent market dominated by firm A. Firm B must decide whether to enter into production and marketing of a similar product. Firm A can announce the price of the detergent and expect that firm B, a new venture with limited production and a higher manufacturing cost, will follow price leadership. Firm A might set the price high to enjoy a good profit, but that would allow firm B to penetrate the market. Firm A might also set the price low enough to force firm B into a discouraging loss position, but this would cause A's own profits to be low. Firm B needs to decide whether there is a reasonable chance of some profit if it enters the market or whether the risk is too great. If firm B stays out, firm A has a monopoly, and its profit will increase with price. If firm B enters, firm A will lose market share. To compensate for lost market share, firm A will have to advertise heavily in hopes of increasing total market volume. The sales of both firms are assumed to increase as a result of such advertising. Lower prices benefit the lower cost producer, firm A, more than firm B, since lower profit margins hurt the higher cost producer, firm B.

There is no dominant strategy for A, since a price of $0.80 maximizes A's profit if B enters, but a price of $1.00 is best if B stays out. Likewise, there is no dominant strategy for B: entry is best if the price is $1.00, but staying out is best if the price is $0.80. There is no stable solution, since in any given situation, each firm can improve its own profit by making a unilateral move. Since no single strategy is best for either firm, the best course of action is to adopt a *mixed strategy*. This is analogous to the childhood game of paper-stone-scissors, where no

TABLE 8.8   Profit Payoff Matrix (Millions of Dollars)

| A profit, B profit | | |
|---|---|---|
| | Firm B strategy | |
| Firm A strategy set price at | Enter | Stay out |
| $1.00 | (10, 6) | (40, 0) |
| $0.90 | (12, 2) | (30, 0) |
| $0.80 | (15, −5) | (20, 0) |

single move is the best and a random choice between three options (plus psych-ing out the opponent) becomes the best strategy. It is impossible to predict the outcome of this game.

In the United States CPI, there are over a thousand duopolies, manufactur-ing mostly specialty chemicals, whose operations can be analyzed with game theory. However, most chemicals are produced by more than two companies in competition, which complicates the game theory models, and the number of strategies is usually higher than three. The most difficult part of game theory analysis is the estimation and assignment of numbers in the payoff matrix. The matrix can be established only by chemical professionals who have had a great deal of experience and possess the necessary intuition. The principles of game theory remains the same, but the analysis becomes more complicated.

## 8.2  MANUFACTURING

The manufacturing of chemical and petroleum products often involves high tem-peratures, high pressures, and corrosive substances, as well as special and expen-sive equipment. While small batches of specialty chemicals can be made in multipurpose equipment, large-volume chemicals are made in special dedicated equipment in continuous and computer-controlled, highly instrumented processes. When the capital cost becomes a significant item in the total cost of the product, the plant operates continuously, except for repair and unexpected difficulties, requiring three shifts of labor over a 24-hour day. Today most chem-ical plants are highly automated, equipped with monitoring instruments and con-trolled by computers.

The CPI is among the nation's heaviest investors in capital equipment. Several of the major chemical companies make annual capital expenditures in the $500 million to $2000 million range. Many oil companies have annual capi-tal expenditures in excess of $1 billion. In terms of capital invested per produc-tion worker, the petroleum industry far outranks all others (Table 8.9) with nearly $775,000 assets per employee in 1996 compared with the all-industry av-erage of $284,000 for all manufacturing industries. The chemical industry ranks fourth, after mining and railroads, with $406,650 assets per employee.

Heavy capital investment in the chemical and petroleum industries is nec-essary to produce large volumes of products efficiently at low prices. If a com-pany produces a commodity or pseudocommodity chemical, it must operate a very large plant to remain competitive in the global market. For example, ethyl-ene is now produced in plants with capacities of 1.0–2.0 billion pounds annu-ally. The current investment for such a plant ranges from $200 million to $1 billion, depending largely on the feedstock and the size of the plant. Also, the complex nature of chemical processing and the high rate of wear due to high temperatures, high pressures, corrosion, and continuous operation dictate the

TABLE 8.9   Assets per Employee

| Industry | 1979[a] | 1985[b] | 1989[c] | 1993[d] | 1996[e] |
|---|---|---|---|---|---|
| Petroleum refining | $234,458 | $469,000 | $576,249 | $667,000 | $774,678 |
| Mining, crude oil production | $246,433 | $334,000 | $533,162 | $419,000 | $479,271 |
| Railroads | NA | NA | NA | NA | $411,340 |
| Chemicals | $68,058 | $115,000 | $192,587 | $283,000 | $406,650 |
| Motor vehicles and parts | NA | NA | NA | NA | $384,684 |
| Pharmaceuticals | $52,034 | $108,000 | $143,275 | $214,306 | $379,500 |
| Beverages | $71,546 | $129,000 | $123,950 | $121,446 | $359,802 |
| Tobacco | NA | NA | NA | NA | $312,568 |
| Forest and paper products | $61,131 | $128,000 | $196,121 | $230,175 | $296,481 |
| Computers, office equipment | $38,248 | $75,000 | $146,999 | $211,000 | $281,218 |
| Electronics, electrical equipment | $30,957 | $61,000 | $156,089 | $300,228 | $280,923 |
| Metals | $40,170 | NA | $189,591 | $241,015 | $241,829 |
| Electronics, semiconductors | NA | NA | NA | NA | $232,765 |
| Soaps and cosmetics | $43,440 | $117,000 | $183,614 | $191,000 | $210,928 |
| Building materials, glass | NA | NA | NA | NA | $196,564 |
| Food | $40,304 | $70,000 | $155,095 | $127,400 | $167,315 |
| Aerospace | $27,990 | NA | $91,656 | $133,052 | $162,553 |
| Rubber and plastics products | NA | NA | NA | NA | $108,790 |
| Apparel | $18,096 | NA | $55,332 | $146,392 | $104,280 |
| Metal products | NA | NA | NA | NA | $103,246 |
| Furniture | NA | $40,000 | $58,211 | $70,021 | NA |
| Textiles | $23,474 | $50,000 | $84,996 | $78,285 | $66,456 |
| All industry average | NA | $79,000 | $182,469 | $231,767 | $284,000 |

[a] *Fortune 500*, May 1980.
[b] *Fortune 500*, April 28, 1986.
[c] *Fortune 500*, April 23, 1990.
[d] *Fortune 500*, April 18, 1994.
[e] *Fortune 500*, April 218, 1997.

use of special materials of construction and intricate equipment. These factors contribute to the capital-intensive nature of the chemical and petroleum industries, as does the increasing need for pollution abatement and energy conservation equipment. Last, the high rate of innovation contributes to the high capital investment in the CPI.

The construction of a plant with a new processing technology carries a significant risk but can lead to significant improvements in efficiency and to the demise of older processes. A heavy rate of plant investment also means that the firm has newer and more efficient processing equipment and is better able to compete by manufacturing at a lower cost.

## 8.2.1  Production Function

The average capital cost, $AC$, of producing a chemical product as we have seen is approximated by

$$AC = \text{constant} \times \text{size}^n \qquad (8.1)$$

where $n$ is usually less than 1. There is a good economy of scale in most chemical processing units, and large-scale plants are more efficient than small ones. In some processes, capacity expansion is achieved only by duplicating process units of a given size. The production of certain biochemicals in many shake flasks is an example of no economy of scale, therefore $n = 1$.

CPI firms are engaged in the transformation of various raw materials and other factors of production into products. The rate of production can be represented in functional form:

$$q = f(K, L, S) \qquad (8.2)$$

where $q$ = quantity of product produced (mass/yr)

$\quad K$ = cost of fixed and working capital per year consisting of depreciation, return on investment (opportunity cost), and annual maintenance and repair (\$/yr)

$\quad S$ = raw material and supply expense (\$/yr)

$\quad L$ = labor expense (\$/yr)

$L$ can be further expressed as follows:

$$L = \sum w_i x_i \qquad (8.3)$$

where $w_i$ is the wage rate and $x_i$ is the number of employees of type $i$—for example, chemical plant operators, needed to produce the product.

To begin any new operation, a firm must design and construct a new plant. In the long-run situation, all the variables above are under the control of the chemical professional who wishes to design the optimum plant. A target production capacity $q_0$, may be suggested by the marketing people, but it would be wise to investigate the profitability of various plant sizes. The cost of raw materials and supplies, $S$, is usually proportional to the quantity of product manufactured, where $p_s$ is the unit cost of the raw materials and supplies:

$$S = p_s q_0 \qquad (8.4)$$

In any design consideration, there are trade-offs between capital and labor. A batch process is more labor intensive than a continuous processing plant for the same amount of chemical product. A continuous unit that is highly instrumented and computer-controlled is more capital intensive than one designed with a minimum amount of instrumentation. A plant designed for minimum maintenance costs more to build but uses less maintenance labor than a plant utilizing equipment that must be repaired or replaced frequently. Although the tactic appears impractical to a process designer used to conditions in the United States today, when capital is unattainable and labor is very cheap, it may be feasible to replace a pump with a bucket brigade. This situation is slowly disappearing, but labor is plentiful and cheap in some developing countries.

An economy of scale can be achieved in many chemical plant designs. The experience curves in Figures 2.9, 6.5, and 8.1 have the slope shown partly as a result of to economy of scale. In functional form, the scale-up process can be approximated by the following relation:

$$q(\alpha K, \alpha L, \alpha S) = \alpha^n q(K, L, S) \tag{8.5}$$

where $\alpha$ is a scale-up factor and $n \geq 1$.

An economic constitutive relationship that has a functional form postulated to take into account capital, labor, and their relationship to process plant size is the Cobb–Douglas equation:

$$q(K, L) = CK^a L^b \tag{8.6}$$

where $C$ is a constant and $a$ and $b$ are constants with values between 0 and 1, such that $a + b \geq 1$.

An alternate way of presenting the equation shows the scale-up aspect more clearly:

$$\frac{q}{q_0} = \left(\frac{K}{K_0}\right)^a \left(\frac{L}{L_0}\right)^b \tag{8.7}$$

where $q_0$, $K_0$, and $L_0$ are the values for a reference plant at a given time and a designated state of technology. The behavior of this equation can best be illustrated with a numerical example:

$$q = 0.25 K^{(0.8)} L^{(0.4)} \tag{8.8}$$

If selected values of $K$ and $L$ in Table 8.10 are substituted into Eq. (8.8), values of $q$ are obtained.

TABLE 8.10   Value of $q$ ($lb \times 10^6$/yr)

| K ($/yr) | L ($/yr) | | |
|---|---|---|---|
| | 100,000 | 200,000 | 300,000 |
| 200,000 | 0.44 | 0.57 | 0.76 |
| 400,000 | 0.76 | 1.00 | 1.32 |
| 800,000 | 1.32 | 1.74 | 2.30 |

A production quota of 0.76 million pounds per year can be attained by two different combinations of $K$ and $L$, one plant being more capital intensive and the other being more labor intensive. In fact any combination of $K$ and $L$ (isoquant) that satisfies the relation

$$K^{(0.8)} L^{(0.4)} = (4) (0.76 \times 10^6)$$

will produce $q = 0.76$ million pounds per year.

In Table 8.8, the effect of diminishing returns is clearly shown. At a fixed value of $K$, each doubling of $L$ results in much less than a doubling of production $q$. As more and more labor is hired, a point is eventually reached where

$$\frac{\partial^2 q}{\partial L^2} < 0 \qquad (8.9)$$

The same diminished-return effect is shown for $K$ when $L$ is constant.. Doubling $K$ shows less than a doubling of production. If the labor force is fixed and more capital is put into the process, eventually we will have

$$\frac{\partial^2 q}{\partial K^2} < 0 \qquad (8.10)$$

Doubling both $K$ and $L$ brings more than a doubling of production. This economy of scale is illustrated by the diagonal elements in the table from upper left to lower right (note that $a + b = 1.2$ for this example).

The optimum ratio of capital to labor must also be a concern of the cost-minimizing professional. This occurs for a given $q$ when the marginal productivity of capital equals the marginal productivity of labor:

$$\frac{\partial q}{\partial K} = \frac{\partial q}{\partial L} \qquad (8.11)$$

If the marginal productivity of capital were greater than the marginal productivity of labor, a dollar added to capital would result in more production than a dollar added to labor. In such a situation, one should withdraw dollars from labor and add to capital, since this would increase productivity. One should keep on doing this until the two marginal productivities are equal, at which point there would be no incentive for transferring funds from capital to labor or vice versa. For the Cobb–Douglas function, Eq. (8.6), we write

$$\frac{\partial q}{\partial K} = caK^{a-1}L^b = \frac{aq}{K}$$

and

$$\frac{\partial q}{\partial L} = cbL^{b-1}K^a = \frac{bq}{L}$$

at the point where the marginal productivity of labor and capital are equal. In our numerical example, the optimum ratio of capital to labor is $a/b = 2$.

At a given production rate $q_0$ an expression for the optimal capital and labor costs can be developed in terms of the three parameters, $a$, $b$, and $c$:

$$\frac{K}{L} = \frac{a}{b} \tag{8.12}$$

$$q_0 = c(K_0)^a(L_0)^b = c(K_0)^a \frac{(bK_0)^b}{a}$$

A similar development yields

$$L_0 = \left[\frac{q_0}{c}\left(\frac{b}{a}\right)^a\right]^{\frac{1}{(a+b)}} \tag{8.13}$$

For our numerical example

$$K_0 = 4.0\, q_0^{0.833} \tag{8.14}$$

Chemical engineers familiar with process economics are used to the "0.6 rule" (6). The cost of process equipment is assumed to follow

$$\text{cost} = \text{constant} \times \text{size}^{0.6}$$

The exponent is not always 0.6, however, since the values may vary from 0.8 or greater for process vessels to 0.5 for specialized equipment like crystallizers (4). The purchased equipment cost is often only one-fourth to one-fifth of the total fixed capital investment. The total capital cost for which $K_0$ is derived includes many items for which the scale-up power is closer to 1 and some items for which $n$ should be closer to 0. The power of $1/(a + b) = 0.833$ used in the numerical example may be considered to be reasonably close to experience.

A word of caution—the equations above and the ones in Section 8.2.2 should be used only for approximate or planning purposes.

## 8.2.2  Long- and Short-Range Average Costs

The average cost of production $AC$ can be computed using the Cobb–Douglas function for both the long-range average cost (LRAC) and the short-range average cost (SRAC). The long-range average cost is

$$\text{LRAC} = \frac{TC}{q}$$

The total cost consists of three major components: the capital cost $K$, the yearly labor expense L, and the raw materials and supplies expense $S$:

$$TC = K + L + S \qquad (8.15)$$

For a unit designed to produce $q_0$, the long-range cost is

$$\text{LRAC} = \frac{K_0 + L_0 + S}{q_0} \qquad (8.16)$$

Equations (8.4), (8.13), and (8.14) can be used to yield an expression for LRAC in terms of the Cobb–Douglas parameters and plant size, $q_0$:

$$\text{LRAC} = p_s + \left[1 + \left(\frac{b}{a}\right)\right] q_0^{\frac{(1+a+b)}{(a+b)}} \left[\left(\frac{1}{c}\right)\left(\frac{a}{b}\right)^b\right]^{\frac{1}{(a+b)}} \qquad (8.17)$$

For the numerical example,

$$LRAC = p_s + 6.0 q^{-0.167} \qquad (8.18)$$

The long-range average cost is an important planning function for capital investment and is shown in Figure 8.3 for $p_s$ = 20 cents per pound. In this example, the LRAC continues to decline as the plant size increases; a larger plant is more efficient. In some cases the LRAC becomes essentially independent of plant size after some critical value has been reached (Figure 8.4, curve $A$); then it is proper to speak of the *minimum effective size*. An inefficient smaller plant can

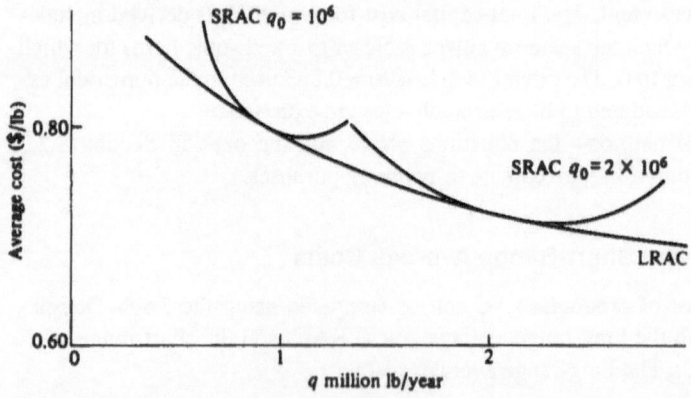

FIGURE 8.3   Long- and short-range average costs.

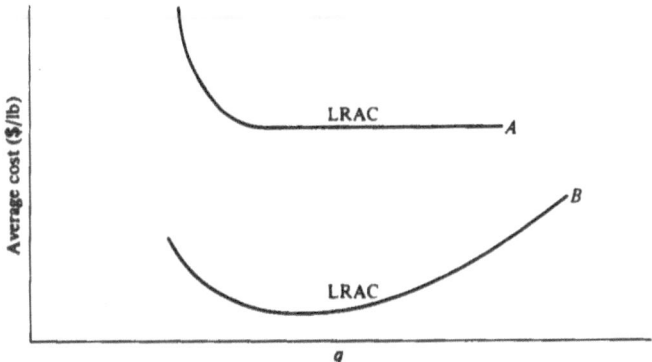

FIGURE 8.4   Long- and short-range average costs.

survive only under special circumstances (e.g., a market sheltered by geographic location, protective tariffs, a favorable raw material supply, or product differentiation). Eventually, many small plants will close before they are physically worn out because of economics.

There are also cases in which LRAC increases beyond some *limiting capacity* that is beyond the control of the corporate planner—for example, a feature of the infrastructure of the community (Figure 8.4, curve *B*) like the water supply, the supply of well-trained labor, or the capacity of waste disposal facilities. For instance, unless there is a national consensus, a company cannot strip-mine coal or bring irrigation water to the arid southwest area of the country for large-scale gasification plants, which were considered in the late 1970s and early 1980s. In the model, $p_s$ becomes a rapidly increasing function of $q$. The prevailing condition in the industrial chemical industry is that of declining LRAC, making larger plants more efficient. This leads to construction of larger and larger plants (e.g., for ethylene and ammonia), to lower costs, in hopes of capturing the expanding market. When multiple firms try to capture the same market, overcapacity and low plant utilization will result. This situation can be very costly as, was the case with methanol in Chapter 6. Such overcapacity is less likely in countries like Japan, where all plant capacity expansions are coordinated by the Ministry of International Trade and Industry (MITI).

After a plant has been built the capital investment is fixed, and the plant manager is concerned with short-run optimization by reducing operating expenses. When the production level changes, the cost of raw materials and supplies is generally proportional to production, if there are no material inefficiencies:

$$S = p_s q$$

$S$ may depend upon $q$ to some power if the raw material efficiency changes with production rate. A function of the following form was implicitly assumed in Chapter 2:

$$S = p_s q^n$$

where $n$ is usually greater than 1. To simplify the mathematics, let us assume $n = 1$. Then the labor needed to satisfy the production requirement is given by

$$q = CK_0^a L^b \tag{8.19}$$

and

$$L = \left[ \left( \frac{q}{C} \right) \left( \frac{1}{K_0^a} \right) \right]^{\frac{1}{b}}$$

If we assume an existing plant with a \$400,000 yearly expense due to capital-related items $K_0$ and use the numerical example, we obtain

$$L = (2 \times 10^{-10}) q^{2.5} \tag{8.20}$$

where $q$ is a production rate that can vary over a limited range of $q_0$, the design capacity.

The SRAC is computed as follows:

$$\text{SRAC} = \frac{K_0 + L + p_s q}{q} = p_s + \frac{K_0}{q} + \left( \frac{1}{cK_0^a} \right)^{\frac{1}{b}} q^{\frac{(1-b)}{b}} \tag{8.21}$$

For our example,

$$\text{SRAC} = \$0.20 + \frac{400,000}{q} + \left(2 \times 10^{-10}\right) q^{1.5} \tag{8.22}$$

This is similar to the average cost expression derived for the Delos production unit [Eq. (2.20)];

$$TC = \frac{AC}{q} = \frac{45L}{q} + \frac{2}{\frac{(1-q)}{B}} \tag{8.23}$$

Equation (2.20) can be rearranged as follows:

$$AC = \frac{45L}{q} + 2 + 2\frac{q}{B} + 2\frac{q^2}{B^2} \tag{8.24}$$

The capital-related terms are easily identified in Eqs. (8.22) and (8.24). The constant terms \$0.20 and 2 arise partly from the raw material expenses. The term $q^{1.5}$ appears in Eq. (8.22) as a result of labor expenses. The terms containing powers of $q$ in Eq. (8.24) appear partly as a result of labor expenses and partly as a result of increased raw material expenses due to poorer conversion at higher throughputs.

The curves of SRAC for two design capacities, $q_0 = 1$ million pounds and $q_0 = 2$ million pounds were presented in Figure 8.3. When the production volume $q$ is less than the design capacity, $q_0$, SRAC increases because the fixed cost $K_0$ is spread over a smaller volume of product; when production is greater than design capacity, there is an increase in SRAC because of the decrease in raw material efficiency, as well as the need for overtime and, possibly, inexperienced labor. A plant of the wrong size is expensive, but if any plant has a lifetime of 20 years, it is seldom of the right size for the entire time period anyway.

Sometimes the plant capacity can be increased by process improvement or "debottlenecking." For example, in the case of an oil pipeline the flow capacity can be increased by installing more pumping stations, say from a station every 50 miles to a station every 25 miles. Computer control can be installed on an older process to gain better raw material efficiencies and increase yield. The LRAC curve forms the envelope of all SRAC curves, representing the minimum average expense for plants of all conceivable sizes. Since the production volume $q$ is tied to the market demand, it cannot be predicted with certainty; therefore, it is important that the design capacity $q_0$ generate an SRAC curve that is the lowest possible within the range of anticipated production requirements. In the design of a plant, there is often a choice between a design that is flexible with regard to varying production volume, raw material supply, and the incorporation of process improvements and another design that is dedicated and narrowly designed for a specific set of conditions. A firm with

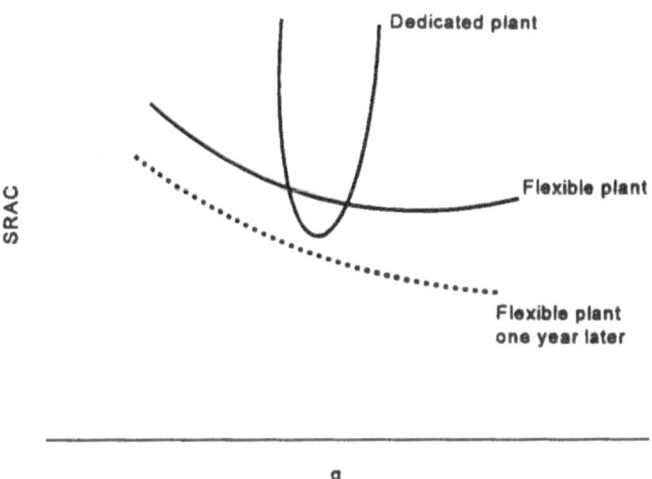

**FIGURE 8.5** Short-range average costs.

several plants may decide to have a mix of flexible and dedicated plants as in Figure 8.5. The design of a flexible plant is more tentative and the equipment employed may have multi-purposes. The operation of a flexible plant requires more imaginative and resourceful operating personnel. There is a significant possibility that experience gained will result in the lowering of the SRAC. The dedicated plant, often a turnkey unit identical to existing plants, is more rigidly specified with specialized equipment. It requires fewer engineers and operators. Further, there is less potential of process improvement and cost reduction.

All the discussion above is concerned with a given process technology. A new technology should increase the value of $c$ and may alter the values of $a$ and $b$ as well. If in the new technology, the value of $a/b$ is increased, it is necessary to increase the design value of $K_0/L_0$. This is an innovation that substitutes capital for labor and is common in the CPI.

Various industries differ greatly in the relative use of capital and labor from the very labor-intensive garment industry to the very capital-intensive petroleum industry (Table 8.9). If we assume that the reported values of capital investment per production worker for the various industries are the optimal ones, the ratio of $a/b$ can be estimated by

$$\frac{K}{L} = \frac{a}{b} = \frac{\text{capital invested}}{\text{production worker}} \times \frac{\text{cost of capital}}{\text{capital invested}} \times \frac{\text{production worker}}{\text{labor wage per year}}$$

Therefore,

$$\frac{K}{L} = \frac{a}{b} = \frac{\text{cost of capital}}{\text{cost of labor}}$$

In the following tabulation of the $a/b$ ratio for various sectors, the petroleum sector should be spending between \$9 and \$12 for capital investment per year for every dollar it spends on labor per year. In the rubber sector, the ratio is between 1 and 3, suggesting the need in this sector for a more even distribution of capital and labor.

| Sector | $a/b$ |
|---|---|
| Petroleum | 9–12 |
| Chemical | 4–6 |
| Paper | 2–4 |
| Rubber | 1–3 |

## 8.3  EMPLOYMENT

### 8.3.1  Size of Labor Force

The CPI has created many jobs through its growth and rapid innovation. In the chemicals and allied products (SIC 28) alone, total employment increased by

**TABLE 8.11**  Total
Employment: Chemical and
Allied Products

| Year | Employees |
|------|-----------|
| 1899 | 134 |
| 1939 | 371 |
| 1949 | 618 |
| 1959 | 809 |
| 1975 | 1,013 |
| 1980 | 1,113 |
| 1985 | 1,042 |
| 1990 | 1,086 |
| 1991 | 1,076 |
| 1992 | 1,084 |
| 1993 | 1,087 |
| 1994 | 1,061 |
| 1995 | 1,045 |
| 1996 | 1,021 |

*Sources:* U.S. Department of Labor
Bulletin 1312-6, 1968 (for years
prior to 1975); then *Chemical and
Engineering News;* Facts and Figures issues, June of year following
one listed.

about 900,000 since nineteenth century to the present (Table 8.11). The peak employment was in 1993 but has declined modestly every year since 1993 to 1.021 million in 1996. As noted in Table 8.12, the total employment in the CPI sector, 1.798 million in 1975, was about 10% of all manufacturing and increased to 2.161 million in 1995, or about 12% of all manufacturing. In the petroleum and coal products sector, chemical engineers accounted for 2.0% and chemists were 1.5% of the total employees.

In a sophisticated manufacturing industry, a high proportion of workers is found in the creative functions of research and development, engineering, and technical service. These people are classified as nonproduction workers in any statistical data. The ratio of nonproduction workers to production workers can be determined from the data in Tables 8.12 and 8.13. In the 20-year period 1975–1995, the ratio of nonproduction to production workers for all manufacturing ranged from 0.404 to 0.490, with 1985 being the peak year. The number of nonproduction workers declined by 10%, while the work force declined by 5% between 1985 and 1995. In the CPI, the ratio of nonproduction to production workers is higher but has steadily declined from 0.570 in 1975 to

TABLE 8.12   Employment in the CPI: All Employees

|  | Employees (thousands) | | | | | |
|---|---|---|---|---|---|---|
|  | 1975 | 1980 | 1985 | 1990 | 1994 | 1995 |
| **All manufacturing** | 18,347 | 20,361 | 19,428 | 19,076 | 18,046 | 18,405 |
| Chemicals and allied products | 1,013 | 1,113 | 1,044 | 1,086 | 1,053 | 1,045 |
| Industrial inorganic chemicals | 324[a] | 166 | 143 | 138 | 131 | 129 |
| Plastic materials and synthetics | 203 | 204 | 172 | 180 | 159 | 158 |
| Drugs | 164 | 199 | 206 | 237 | 264 | 259 |
| Soap, cleaners, and toilet goods | 119 | 138 | 148 | 159 | 152 | 152 |
| Paints and allied products | 65 | 67 | 63 | 61 | 58 | 58 |
| Industrial organic chemicals | NA[a] | 173 | 160 | 155 | 144 | 143 |
| Agricultural chemicals | 55 | 72 | 59 | 56 | 56 | 53 |
| Other chemical products | 83 | 85 | 94 | 100 | 90 | 93 |
| Petroleum and coal products | 197 | 197 | 179 | 157 | 149 | 144 |
| Rubber and plastic products not otherwise classified | 588 | 711 | 818 | 888 | 933 | 972 |
| **Total CPI** | 1,798 | 2,021 | 2,041 | 2,131 | 2,162 | 2,161 |

[a]The 1975 figures were lumped together as industrial chemicals. In 1980 they were classified as industrial inorganic or industrial organic groups.
*Source: Chemical and Engineering News,* Facts & Figures issues: June 7, 1976; June 8, 1981; June 9, 1986; June 24, 1991; June 26, 1996.

TABLE 8.13   Employment in the CPI: Production Workers

|  | Employees (thousands) | | | | | |
|---|---|---|---|---|---|---|
|  | 1975 | 1980 | 1985 | 1990 | 1994 | 1995 |
| **All manufacturing** | 13,070 | 14,277 | 13,084 | 12,947 | 12,433 | 12,727 |
| Chemicals and allied products | 570 | 627 | 577 | 600 | 574 | 582 |
| Industrial inorganic chemicals | 171 | 89 | 72 | 70 | 56 | 57 |
| Plastic materials and synthetics | 133 | 135 | 114 | 116 | 106 | 106 |
| Drugs | 81 | 98 | 95 | 105 | 121 | 127 |
| Soap, cleaners, and toilet goods | 68 | 84 | 93 | 98 | 94 | 94 |
| Paints and allied products | 34 | 34 | 31 | 31 | 30 | 31 |
| Industrial organic chemicals | NA | 88 | 82 | 86 | 79 | 81 |
| Agricultural chemicals | 34 | 45 | 37 | 34 | 32 | 31 |
| Other chemical products | 49 | 54 | 54 | 59 | 53 | 56 |
| Petroleum and coal products | 125 | 124 | 109 | 103 | 96 | 93 |
| Rubber and plastic products not otherwise classified | 450 | 548 | 632 | 689 | 726 | 756 |
| **Total CPI** | 1,145 | 1,299 | 1,318 | 1,389 | 1,417 | 1,431 |

*Source: Chemical and Engineering News,* Facts & Figures, issues: June 7, 1976; June 8, 1981; June 9, 1986; June 24, 1991; June 26, 1996.

0.510 in 1995. In that 20-year period, the total employment in the CPI increased by 363,000, or about 21%, but the production workers increased by 72,000, or about 12%.

In the last 5 years, downsizing resulted in declining employment of non-production workers in all manufacturing as well as in the CPI.

Whenever a recession occurs, there is a downturn in employment. For example, in 1975 the employment in all manufacturing was about 1.7 million lower than in 1974. The CPI employment declined by 134,000, or about 7%, which indicated less distress in these industries in the 1974–1975 period. In the 20-year period 1975–1995, in some sectors of the CPI (e.g., petroleum and coal products, industrial organic and inorganic chemicals, plastics and synthetic materials), there was a decline in employment. In the drug, soap, cleaner, and toilet goods sectors, as well as in rubber and plastics products, there was a slight increase.

Downsizing of companies began to occur in the oil industry in the early 1980s, in the chemical industry in the late 1980s and early 1990s, and in the pharmaceutical industry in 1996. The objective of downsizing was to increase the overall efficiency of company operation and thus improve the "bottom line." Downsizing was achieved by reducing the number of employees (especially middle management positions and redundant positions), and reducing the high cost of benefits programs and operating costs such as raw materials and utilities. In improving efficiency, companies frequently eliminated less profitable product lines or sold them to other manufacturers, thereby realigning the remaining manufacturing areas in keeping with corporate objectives.

Although the pace of downsizing has slowed, it still continues—not because of slow economic growth or recession but because global economic and technological pressures have imposed requirements that the same tasks that existed prior to downsizing be performed by fewer people.

Some companies that used to make raw materials or intermediate products and parts for their captive use are "outsourcing" these steps. A recent example outside the CPI is Chrysler, which is outsourcing parts for use in cars, trucks, and vans. Outsourcing, although practiced to a limited degree prior to downsizing, has increased markedly. To reduce overhead, some CPI firms outsource certain engineering and human relations tasks. Also, in the CPI prior to downsizing, fringe benefits to employees (medical, dental, retirement, thrift, stock options, etc.) accounted for between 40 and 50% of wages or salaries. To be competitive in the global marketplace, these expenses as well as plant operating expenses had to be reduced.

The day when an employee came to work as a young person and remained with the company for 30 years or until retirement is a thing of the past. Not only has company loyalty to the employee suffered, but also loyalty to and trust of the company on the part of the employee. Young professionals may work for several companies before they retire and may even change careers. They will find many varied opportunities with small chemical companies and nontraditional manufacturing suppliers, as well as consulting and entrepreneurial opportunities. Large chemical and petroleum companies will still hire chemicals professionals, but the number will be reduced. Young chemical professionals must be willing to change fields or assignments within a company as well as in the job market at large as the supply and demand for their services change. Chemical professionals must be flexible and attempt to keep up to date in one or more technical areas, and they must learn to work effectively and efficiently in a team environment.

### 8.3.2  Wages and Salaries

**8.3.2.1  Production Worker Wages and Working Conditions.**  The trend since 1947 for average hourly earnings (not corrected for inflation) in the chemical and allied products sector as well as in all manufacturing is presented in Table 8.14. The chemical worker has experienced greater increases in wages than has the average industrial worker. Table 8.15 shows the trend in wages for production and nonsupervisory workers in various sectors of the CPI since 1975. There is a significant difference between the wages of these workers in the petroleum sector and those in rubber products, soap, cleaners, and toilet goods.

Table 8.16 presents the earnings of production workers in other industrial sectors. The highest hourly wage reported is $19.36 per hour in the petroleum industry, followed by followed by the chemical sector at $15.63.

Although the chemical industry as a whole has many potential health and safety hazards, the chemical and petroleum sectors have an excellent safety record. Primary metals, mining, and forestry have the highest number of injuries per thousand workers per year.

Unionization is an indication of worker alienation from management. In the chemical sector union membership has declined from 145,000 in 1978 to 83,000 in 1995, representing less than 0.6% of the total union membership. Nonunionized chemical workers are not at a disadvantage: their wages are better than those of the average manufacturing worker, and their safety record is the best—less than half of the all-manufacturing category (Table 8.16).

The productivity of the worker in the chemical and allied products sector has increased by more than 20% from 1987 to 1995, but all manufacturing productivity has increased nearly 25%, as may be seen in Table 8.17.

The work week has lengthened for all manufacturing from 40.8 hours in 1990 to 41.6 hours in 1995, whereas the work week for the chemical and allied workers has increased from 42.6 to 43.3 over the same time period. Data for other segments of the CPI may be found in Table 8.18. The slight increase in the work week may be beginning to show the incipient effects of downsizing in the various CPI sectors.

**TABLE 8.14** Hourly Earnings Total Employment (Current Dollars, Not Corrected for Inflation)

| Year | Chemical and allied products | All manufacturing |
|---|---|---|
| 1947 | $1.57 | $1.57 |
| 1957 | $2.24 | $2.09 |
| 1967 | $2.82 | $2.43 |
| 1972 | $6.03 | $5.51 |
| 1973 | $4.51 | $4.09 |
| 1974 | $4.85 | $4.41 |
| 1975 | $5.37 | $4.81 |
| 1976 | $5.91 | $5.22 |
| 1977 | $6.43 | $5.68 |
| 1978 | $7.01 | $6.17 |
| 1979 | $7.59 | $6.69 |
| 1980 | $8.29 | $7.27 |
| 1981 | $9.12 | $7.99 |
| 1982 | $9.96 | $8.50 |
| 1983 | $10.59 | $8.84 |
| 1984 | $11.08 | $9.16 |
| 1985 | $11.56 | $9.54 |
| 1986 | $11.98 | $9.73 |
| 1987 | $12.37 | $9.91 |
| 1988 | $12.71 | $10.18 |
| 1989 | $13.09 | $10.47 |
| 1990 | $13.54 | $10.84 |
| 1991 | $14.07 | $11.18 |
| 1992 | $14.51 | $11.46 |
| 1993 | $14.82 | $11.74 |
| 1994 | $15.14 | $12.06 |
| 1995 | $15.66 | $12.36 |

*Sources: Chemical and Engineering News, Facts and Figures issues: June of year following the one listed.*

The employment of women in the work force for the chemical and allied products sector has increased from 27% in 1984 to 32% in 1994, while in all manufacturing the percentage remained essentially constant at about 33% over the same time period. (See Table 8.19 for details in other sectors.)

## 8.3.2.2 Salaries and Employment of Chemical Professionals.

The salaries of chemists are surveyed by the American Chemical Society (ACS) and published in *Chemical and Engineering News* in the June, July, or August issue

TABLE 8.15   Wages ($/h) for Production and Nonsupervisory Workers[a]

|                                | 1975   | 1980    | 1985    | 1990    | 1994    | 1995    |
|--------------------------------|--------|---------|---------|---------|---------|---------|
| **All manufacturing**          | $4.81  | $7.27   | $9.53   | $10.84  | $10.06  | $12.36  |
| Chemicals and allied products  | $5.37  | $8.29   | $11.57  | $13.54  | $15.14  | $15.66  |
| Industrial inorganic chemicals | $5.93  | $9.06   | $12.82  | $14.67  | $16.79  | $17.53  |
| Plastic materials and synthetics | $5.25 | $8.26  | $11.72  | $14.02  | $15.40  | $16.13  |
| Drugs                          | $5.13  | $7.67   | $10.75  | $12.87  | $14.76  | $15.20  |
| Soap, cleaners and toilet goods | $5.21 | $7.56   | $10.26  | $11.76  | $12.69  | $12.79  |
| Paints and allied products     | $4.94  | $7.40   | $10.07  | $12.01  | $12.99  | $13.10  |
| Industrial organic chemicals   | ——     | $9.66   | $13.99  | $15.97  | $18.19  | $19.11  |
| Agricultural chemicals         | $4.73  | $8.11   | $11.02  | $13.58  | $15.34  | $15.74  |
| Other chemical products        | $5.05  | $7.76   | $10.60  | $12.30  | $14.05  | $14.48  |
| Petroleum and coal products    | $6.42  | $10.07  | $14.04  | $16.23  | $19.07  | $19.39  |
| Rubber and plastic products    |        |         |         |         |         |         |
| not otherwise classified       | $4.35  | $6.49   | $8.53   | $9.79   | $10.70  | $10.92  |

[a]Wages reported are the actual wages in the respective year (not corrected for inflation).
*Source: Chemical and Engineering News,* Facts and Figures issues: June 7, 1976; June 8, 1981; June 9, 1986; June 24, 1991; June 24, 1996.

of the following year. The American Institute of Chemical Engineers (AIChE) publishes a salary survey every two years and occasionally publishes yearly figures in *Chemical Engineering Progress* following the close of the calendar year. Often the data reported are a year or two old.

The salaries of chemical professionals have increased in general with the supply and demand. These salaries have been influenced by the number of degrees awarded. Table 8.20 lists the degrees awarded from 1974 to 1994. The peak years for the bachelors in chemistry occurred in 1979, declining to a low in 1990. However, there has been a steady increase up through 1994, whereas the peak for chemical engineers occurred in 1984. The most advanced degrees were awarded to chemists in 1994. Master's degrees awarded to chemical engineers were the greatest in 1985, while doctoral degrees varied, with peaks in 1989, 1991, and 1994. Interestingly, supply and demand for chemical engineers are seldom if ever in phase. Figures 8.6 and 8.7 plot the degrees awarded to chemical professional graduates.

There is a continued gap between the salaries of chemists and chemical engineers, which begins with differences in starting salaries as shown in Table 8.21. There are some reasons for this gap, namely:

1.  Chemical engineers in the sample had longer work experience than chemists.
2.  Chemical engineers have closer identification with industry's needs.

TABLE 8.16 Employment Conditions in Various Sectors

| | Employees in nonagricultural industry (thousands) | Workers injuries per 100 employees per year | Union members (%) | Average hourly wage |
|---|---|---|---|---|
| Total | 117,203 | | 15.5 | NA |
| Private industry | 97,892 | 8.4 | 10.8 | $11.44 |
| Mining | 580 | 6.3 | 13.8 | $15.30 |
| Construction | 5,158 | 11.8 | 17.7 | $15.08 |
| Manufacturing | 18,468 | 12.2 | 17.6 | $l2.37 |
| Stone, clay, glass | 538 | 13.2 | NA | $12.41 |
| Primary metals | 204 | 16.8 | NA | $14.62 |
| Forest | 764 | N/A | NA | $10.12 |
| Paper | 692 | 9.6 | NA | $14.23 |
| Chemicals | 1,035 | 5.7 | NA | $15.63 |
| Petroleum | 144 | 4.7 | NA | $19.36 |
| Rubber | 976 | 14.0 | NA | $10.90 |
| Leather | 106 | 12.0 | NA | $8.17 |
| Transportation | 645 | 9.3 | 27.3 | $14.23 |
| Trade | | | 6.1 | |
| Wholesale | 6,412 | 7.7 | | $12.43 |
| Retail | 21,173 | 7.9 | | $7.69 |
| Finance, insurance, real estate | 6,830 | 22.7 | 2.1 | $12.33 |
| Services | 33,107 | 6.5 | 5.7 | $11.39 |
| Government | 19,310 | NA | 27.8 | NA |

Source: U.S. Bureau of Census, Statistical Abstract of the United States, 1996.

3. A large percentage of chemists worked in the lower paying profession of education.

Starting salaries for chemical professionals are found in Tables 8.21 and 8.22. Although in other industries the starting salaries for women have not kept pace with those of men, in general, average women's salaries were equal to or slightly greater than those of men as chemical professionals. Recent information from schools indicate that women received starting salaries equal to or exceeding those offered men. Data from the latest salary survey by AIChE (Table 8.23) show the median salary based on the year the bachelor's degree was awarded and the effect of advanced degrees earned. The top median salaries for all degrees awarded occurred for those receiving the degrees in the early 1960s. The salaries for professionals who earned MBAs, in many instances, approached

TABLE 8.17 Productivity

| | 1987 | 1988 | 1989 | 1990 | 1991 | 1992 | 1993 | 1994 | 1995 |
|---|---|---|---|---|---|---|---|---|---|
| **Chemical and allied products** | | | | | | | | | |
| Production | 100.0 | 106.0 | 109.2 | 111.8 | 111.1 | 114.4 | 115.4 | 121.3 | 125.1 |
| Work hours | 100.0 | 103.6 | 105.2 | 105.0 | 102.3 | 100.7 | 101.6 | 102.9 | 103.6 |
| Productivity[a] | 100.0 | 102.3 | 103.8 | 106.4 | 108.6 | 113.6 | 113.6 | 117.9 | 120.7 |
| Hourly production wages | 100.0 | 102.7 | 105.8 | 109.5 | 113.5 | 117.3 | 119.8 | 122.4 | 126.7 |
| Unit labor costs[b] | 100.0 | 100.4 | 102.0 | 102.8 | 104.5 | 103.3 | 105.5 | 103.8 | 104.9 |
| **All manufacturing** | | | | | | | | | |
| Production | 100.0 | 104.7 | 106.4 | 106.1 | 103.4 | 108.2 | 112.3 | 119.7 | 124.0 |
| Work hours | 100.0 | 102.1 | 102.1 | 99.4 | 99.4 | 95.0 | 96.4 | 99.9 | 99.8 |
| Productivity[a] | 100.0 | 102.8 | 104.2 | 106.7 | 108.9 | 113.8 | 116.5 | 119.8 | 124.3 |
| Hourly production wages | 100.0 | 102.8 | 105.8 | 109.3 | 112.8 | 115.6 | 118.5 | 121.7 | 124.6 |
| Unit labor costs[b] | 100.0 | 100.3 | 101.5 | 102.4 | 103.6 | 101.6 | 101.6 | 101.6 | 100.2 |

[a]Productivity is output per work hour, calculated by dividing indicates for production by indices for work hour.
[b]Unit labor costs calculated by dividing indices for wages by indices for output per work hour.
Source: Chemical and Engineering News: Facts and Figures issues for the respective years.

TABLE 8.18   Length of Work Week (h)

| | 1990 | 1991 | 1992 | 1993[a] | 1994[b] | 1995[c] |
|---|---|---|---|---|---|---|
| **All manufacturing** | 40.8 | 40.7 | 41.0 | 41.4 | 42.0 | 41.6 |
| Chemicals and allied products | 42.6 | 42.9 | 43.1 | 43.1 | 43.3 | 43.3 |
| Industrial inorganic chemicals | 42.9 | 43.6 | 43.5 | 43.8 | 44.4 | 45.4 |
| Plastic materials and synthetics | 42.5 | 42.6 | 43.8 | 44.1 | 44.2 | 44.2 |
| Drugs | 41.8 | 42.5 | 42.5 | 41.4 | 41.3 | 41.8 |
| Soap, cleaners, and toilet goods | 40.8 | 40.8 | 41.3 | 41.9 | 41.3 | 40.9 |
| Paints and allied products | 41.9 | 41.8 | 41.6 | 42.2 | 43.2 | 42.6 |
| Industrial organic chemicals | 45.2 | 45.6 | 45.6 | 45.4 | 45.5 | 45.1 |
| Agricultural chemicals | 44.1 | 44.5 | 44.7 | 44.7 | 45.4 | 45.0 |
| Other chemical products | 42.4 | 42.3 | 42.6 | 42.9 | 43.4 | 43.5 |
| Petroleum and coal products | 44.6 | 44.1 | 43.8 | 44.2 | 44.5 | 43.8 |
| Rubber and miscellaneous plastic products | 41.1 | 41.1 | 41.7 | 41.8 | 42.2 | 41.5 |

[a]From *Chemical and Engineering News*, Facts & Figures issue, June 1994.
[b]From *Chemical and Engineering News*, Facts & Figures issue, June 26, 1995.
[c]From *Chemical and Engineering News*, Facts & Figures issue, June 24, 1996.

TABLE 8.19   Employment of Women (thousands)

| | 1984 | | 1994 | |
|---|---|---|---|---|
| | Employees | Percentage of total | Employees | Percentage of total |
| All manufacturing | 6,295.0 | 32 | 5,987.0 | 33 |
| Chemical and allied products | 282.5 | 27 | 335.2 | 32 |
| Industrial inorganic chemicals | 24.9 | 17 | 30.3 | 23 |
| Plastic materials and synthetics | 38.5 | 22 | 40.1 | 25 |
| Drugs | 89.3 | 43 | 120.2 | 46 |
| Soap, cleaners, and toilet goods | 61.3 | 42 | 68.3 | 45 |
| Paint and allied products | 11.9 | 19 | 12.0 | 21 |
| Industrial organic chemicals | 26.1 | 16 | 30.6 | 21 |
| Agricultural chemicals | 9.0 | 15 | 10.8 | 20 |
| Miscellaneous chemical products | 21.5 | 23 | 22.8 | 25 |
| Petroleum and coal products | 29.9 | 16 | 24.8 | 17 |
| Rubber and miscellaneous plastic products | 280.4 | 34 | 325.2 | 34 |

*Source: Chemical and Engineering News*, Facts & Figures issue, June 26, 1996.

**TABLE 8.20**   Degrees Awarded[a]

| | Degrees in chemistry | | | Degrees in chemical engineering | | |
|------|------------|----------|--------|------------|----------|--------|
| | Bachelor's | Master's | Ph.D. | Bachelor's | Master's | Ph.D. |
| 1974 | 10,438 | 2,125 | 1,828 | 3,399 | 1,044 | 400 |
| 1975 | 10,549 | 1,986 | 1,822 | 3,070 | 990 | 346 |
| 1976 | 11,022 | 1,783 | 1,621 | 3,140 | 1,031 | 308 |
| 1977 | 11,215 | 1,767 | 1,568 | 3,524 | 1,086 | 291 |
| 1978 | 11,315 | 1,886 | 1,521 | 4,569 | 1,235 | 259 |
| 1979 | 11,509 | 1,757 | 1,516 | 5,588 | 1,149 | 304 |
| 1980 | 11,232 | 1,723 | 1,545 | 6,320 | 1,270 | 284 |
| 1981 | 11,347 | 1,654 | 1,622 | 6,527 | 1,267 | 300 |
| 1982 | 11,025 | 1,618 | 1,722 | 6,740 | 1,285 | 311 |
| 1983 | 10,796 | 1,622 | 1,746 | 7,185 | 1,368 | 319 |
| 1984 | 10,704 | 1,667 | 1,744 | 7,475 | 1,514 | 330 |
| 1985 | 10,482 | 1,719 | 1,789 | 7,146 | 1,544 | 418 |
| 1986 | 10,116 | 1,754 | 1,908 | 5,877 | 1,361 | 446 |
| 1987 | 9,661 | 1,738 | 1,976 | 4,983 | 1,184 | 497 |
| 1988 | 9,052 | 1,708 | 1,995 | 4,917 | 1,088 | 579 |
| 1989 | 8,625 | 1,774 | 2,037 | 3,683 | 1,093 | 602 |
| 1990 | 8,132 | 1,682 | 2,183 | 3,430 | 1,035 | 562 |
| 1991 | 8,321 | 1,665 | 2,238 | 3,444 | 903 | 611 |
| 1992 | 8,641 | 1,780 | 2,280 | 3,754 | 956 | 590 |
| 1993 | 8,917 | 1,842 | 2,261 | 4,459 | 990 | 595 |
| 1994 | 9,425 | 1,999 | 2,353 | 5,163 | 1,032 | 604 |

[a]Data collected from degree-granting institutions.
*Source:* National Center for Education Statistics, *Chemical Engineering News,* August 26, 1996, p. 67.

those of people who obtained doctorates and in some cases exceeded the salaries of the latter group. Experience is also reflected in these salaries.

The decision to obtain an advanced degree must be based on the following considerations:

The type of work one desires to do (e.g., research or technical vs. a business orientation)

The job market at that time for chemical professionals with MBA, MS, and Ph.D. degrees

The amount of salary foregone and the expenses associated with obtaining the advanced degree

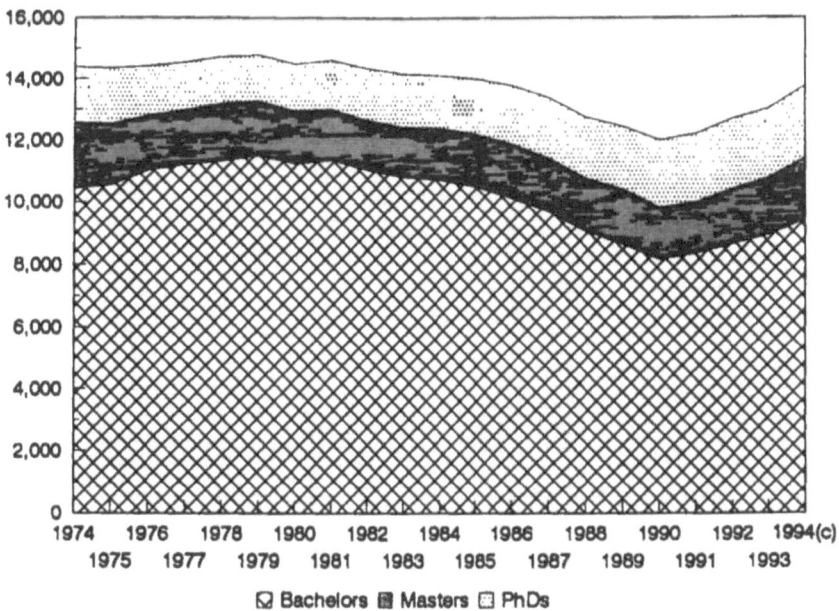

FIGURE 8.6 Degrees awarded in chemistry.

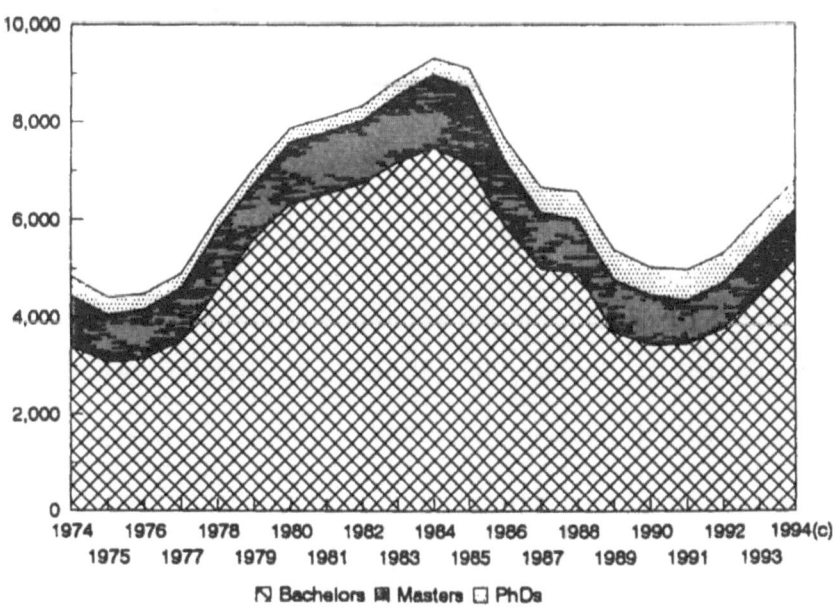

FIGURE 8.7 Degrees awarded in chemical engineering.

TABLE 8.21   Median Starting Salaries (Thousands) for Chemists and Chemical Engineers

| Work experience | Chemists | | | Chemical engineers | | |
|---|---|---|---|---|---|---|
| | Bachelor's | Master's | Ph.D. | Bachelor's | Master's | Ph.D. |
| <12 months | $25.0 | $36.0 | $49.0 | $39.1 | $43.6 | $57.0 |
| 12–36 months | $25.5 | $34.2 | $42.0 | $41.5 | $45.0 | $55.8 |
| >36 months | $32.9 | $42.0 | $50.0 | $40.1 | $46.5 | $56.8 |
| All | $26.0 | $38.0 | $50.0 | $40.0 | $44.4 | $56.0 |

*Source:* Preliminary data from ACS Starting Salary Survey, American Chemical Society, 1995.

TABLE 8.22   Median Starting Salaries (Thousands) in Industry for Men and Women[a]

| Work experience | Chemists | | | Chemical engineers[b] | | |
|---|---|---|---|---|---|---|
| | Bachelor's | Master's | Ph.D. | Bachelor's | Master's | Ph.D. |
| Men | $27.8 | $39.0 | $54.0 | $38.8 | $43.8 | $59.5 |
| Women | $27.0 | $38.3 | $55.0 | $40.8 | NA | NA |
| All | $27.5 | $39.0 | $54.5 | $40.0 | $43.8 | $59.2 |

[a]For inexperienced graduates with less than 12 months of work experience.
[b]NA, not available (sample too small).
*Source:* Preliminary data from ACS Starting Salary Survey, American Chemical Society, 1995.

There appears to be a modest advantage to obtaining a higher degree, but the deciding factor should be a person's desire for additional education. There are also potential disadvantages to obtaining an advanced degree: a person might be overqualified for a given job, and there are the trade-offs of lost salary and the expense of further education. The individual must determine whether the advanced degree is economically feasible. However, an advanced degree indicates a long-term financial gain, as noted in Table 8.23. All these factors must considered, along with an economic analysis, before one decides to pursue an advanced degree.

Salary data for both genders are found in Table 8.24. It is interesting to note that women's salaries in 1976, 1979, 1980 1985, 1987, and 1993 exceeded those reported by men; otherwise their salaries were the same or slightly lower. The national data reflect lower overall salaries for women than for men, but the sample was small, especially prior to 1976. Since 1994, there has been a closing

**TABLE 8.23**   Median Base Salary by Highest Degree Earned and Year of
Bachelor's Degree: Chemical Engineers

| Year of bachelor's degree | Bachelor's | Master's | MBA | Doctorate |
|---|---|---|---|---|
| 1993 | $38,500 | — | — | — |
| 1992 | $40,000 | — | — | — |
| 1991 | $43,100 | $44,500 | — | — |
| 1990 | $44,500 | — | — | — |
| 1989 | $45,800 | $46,800 | — | — |
| 1988 | $47,000 | $44,600 | $46,700 | $45,000 |
| 1987 | $49,000 | $47,750 | $56,500 | $56,600 |
| 1986 | $49,000 | $49,800 | $55,100 | $55,100 |
| 1985 | $52,000 | $52,600 | $52,800 | $58,400 |
| 1984 | $52,900 | $53,350 | $51,300 | $60,000 |
| 1983 | $53,500 | $56,000 | $61,000 | $61,250 |
| 1982 | $56,300 | $54,750 | $60,000 | $62,700 |
| 1981 | $62,000 | $62,100 | $62,400 | $61,650 |
| 1980 | $62,000 | $59,500 | $71,300 | $65,550 |
| 1979 | $65,600 | $65,000 | $70,200 | $66,300 |
| 1978 | $66,800 | $65,500 | $67,350 | $63,000 |
| 1977 | $67,100 | $59,500 | $69,000 | $68,800 |
| 1976 | $65,200 | $67,600 | $74,800 | $76,800 |
| 1975 | $75,000 | $66,750 | $78,000 | $73,000 |
| 1974 | $76,850 | $73,250 | $82,100 | $78,000 |
| 1973 | $74,200 | $71,950 | $70,750 | $78,000 |
| 1972 | $75,650 | $78,100 | $90,000 | $68,000 |
| 1971 | $77,250 | $78,700 | $79,450 | $84,750 |
| 1970 | $79,500 | $83,300 | $95,750 | $77,600 |
| 1969 | $75,100 | $78,150 | $90,000 | $90,000 |
| 1968 | $70,350 | $85,550 | $89,000 | $90,000 |
| 1967 | $89,750 | $90,100 | $86,000 | $82,000 |
| 1966 | $78,900 | $100,000 | — | $85,300 |
| 1965 | $87,100 | $76,950 | $83,400 | $77,150 |
| 1964 | $85,000 | $86,650 | $77,800 | $86,000 |
| 1963 | $81,000 | $80,000 | $110,600 | $92,150 |
| 1962 | $79,800 | $65,000 | $82,100 | $103,500 |
| 1961 | $111,000 | $65,200 | $81,000 | $87,000 |
| 1960 | $79,500 | $76,650 | $103,550 | $86,400 |
| 1959 | $92,700 | $102,000 | — | $99,400 |
| 1958 | $80,700 | $81,000 | — | $103,300 |
| 1957 | $82,900 | $98,200 | — | $81,500 |
| 1956 | $85,800 | $86,700 | — | $79,800 |
| 1955 or earlier | $74,800 | $89,200 | $75,600 | $90,400 |
| Median salary | $56,300 | $66,300 | $73,100 | $73,100 |

Source: AIChE Salary Survey, American Institute of Chemical Engineers, 1994.

TABLE 8.24    Chemical Engineers' Salary by Year of Bachelor's Degree and Gender

| Year of bachelor's degree | Women[a] | | Men | |
|---|---|---|---|---|
| | Number of respondents | Median | Number of respondents | Median |
| 1993 | 34 | $39,600 | 46 | $37,900 |
| 1992 | 35 | $40,000 | 72 | $41,200 |
| 1991 | 29 | $42,700 | 67 | $44,000 |
| 1990 | 22 | $45,000 | 45 | $44,500 |
| 1989 | 37 | $44,000 | 55 | $47,200 |
| 1988 | 22 | $48,000 | 60 | $45,850 |
| 1987 | 16 | $50,650 | 83 | $50,000 |
| 1986 | 32 | $48,550 | 103 | $50,500 |
| 1985 | 28 | $55,050 | 96 | $52,700 |
| 1984 | 29 | $52,700 | 111 | $54,600 |
| 1983 | 19 | $55,600 | 114 | $57,100 |
| 1982 | 14 | $55,100 | 90 | $58,400 |
| 1981 | 17 | $56,800 | 119 | $62,300 |
| 1980 | 11 | $63,300 | 120 | $62,000 |
| 1979 | 15 | $66,600 | 114 | $65,500 |
| 1978 | 12 | $58,450 | 118 | $66,450 |
| 1977 | 12 | $66,450 | 91 | $66,600 |
| 1976 | 7 | $72,000 | 94 | $68,450 |
| 1975 | 4 | | 87 | $73,400 |
| 1974 | 1 | | 97 | $77,100 |
| 1973 | 4 [11] | $65,000 | 67 | $76,300 |
| 1972 | 1 | | 63 | $78,100 |
| 1971 | 1 | | 61 | $79,000 |
| 1970 | 2 | | 92 | $82,950 |
| 1969 | 1 | | 78 | $81,150 |
| 1968 | 2 | | 66 | $78,200 |
| 1967 | 0 | | 63 | $86,400 |
| 1966 | 1 [8] | $80,350 | 61 | $85,300 |
| 1965 | 0 | | 71 | $82,700 |
| 1964 | 0 | | 81 | $84,000 |
| 1963 | 1 | | 61 | $83,000 |
| 1962 | 0 | | 60 | $83,750 |
| 1961 | 0 | | 54 | $84,850 |
| 1960 | 1 | | 56 | $85,800 |
| 1959 | 0 | | 42 | $99,400 |
| 1958 | 0 | | 33 | $81,100 |
| 1957 | 0 | | 35 | $88,300 |
| 1956 | 0 | | 25 | $84,000 |
| 1955 or earlier | 0 | | 107 | $87,800 |
| Overall | 426 | $48,150 | 3064 | $66,000 |

[a]No women who graduated prior to 1960 responded. Because of the low numbers of women respondents who graduated between 1960 and 1975, some years were combined (1975–1971 and 1970–1960) to yield reasonable numbers on which to base medians.

of the salary gap. In part this is because women have moved into upper management in the CPI.

Chemical engineers' median salaries according to job function are given in Table 8.25. Personnel in either technical or general management have the highest median salaries. The larger the company size, the higher the median salary, as shown in Table 8.26. In 1994 chemical engineers' salary exceeded $70,000 in the petrochemicals and petroleum products and glass, ceramics, and refractories sectors, as well as in engineering design, construction, and consulting fields (see Table 8.27).

Finally, Table 8.28 lists base salaries, bonuses, consulting fees, and commissions by salary quartiles and deciles. There is a wide spread in base salaries, from $48,000 to $105,000, depending on a variety of factors including job function, company, location, and employee performance.

**TABLE 8.25** Chemical Engineers' Median Salary by Function

|  | 1990[a] | 1992[b] | 1994[c] |
|---|---|---|---|
| Business; finance; law; licensing | NA | $71,900 | $76,900 |
| Construction | $49,000 | $63,000 | $77,400 |
| Consulting | $54,350 | $60,500 | $56,100 |
| Design | $46,350 | $49,100 | $64,850 |
| Education |  |  |  |
| 12 months | $66,500 | $69,300 | $79,500 |
| 9 months | $61,000 | $58,750 | $63,000 |
| Environmental engineering | $46,000 | $52,300 | $55,050 |
| Maintenance | $54,600 | $55,950 | $56,300 |
| Management |  |  |  |
| General | $82,900 | $87,600 | $96,300 |
| Technical | $71,300 | $72,200 | $81,700 |
| Planning economics | $60,200 | $75,100 | $77,100 |
| Process engineering | $47,550 | $54,800 | $56,400 |
| Product engineering | $41,400 | $46,100 | $54,100 |
| Production | $50,000 | $57,050 | $57,200 |
| Project management | $57,200 | $61,300 | $74,500 |
| Purchasing | $65,000 | $72,000 | $89,400 |
| Process or quality control | $49,000 | $53,600 | $52,900 |
| Research and development | $54,000 | $59,000 | $63,100 |
| Sales and marketing | $56,000 | $60,200 | $66,500 |
| Technical service | $48,750 | $56,400 | $57,700 |
| Testing; analyzers | $46,100 | $47,400 | $48,900 |

[a]From AIChE Salary Survey Report, American Institute of Chemical Engineers, 1990.
[b]From *Chemical Engineering Progress*, May 1992, p. 61.
[c]From *Chemical Engineering Progress*, June 1994, p. 88.

**TABLE 8.26**   Median Chemical Engineering Salaries by Company Size

| Company size (employees) | 1990[a] | 1992[b] | 1994[c] | 1996[d] |
|---|---|---|---|---|
| 1–199 | $48,000 | $55,500 | $56,850 | $61,550 |
| 200–499 | $51,900 | $57,000 | $59,650 | $63,450 |
| 500–999 | $51,750 | $57,600 | $61,000 | $67,000 |
| 1000–4999 | $55,200 | $60,600 | $64,900 | $69,000 |
| 5000 or more | $56,200 | $62,000 | $67,800 | $73,200 |

[a]From AIChE Salary Survey Report, American Institute of Chemical Engineers, 1990.
[b]From *Chemical Engineering Progress*, May 1992, p. 60.
[c]From *Chemical Engineering Progress*, June 1994, p. 88.
[d]From *Chemical Engineering Progress*, August 1996, p. 111.

**TABLE 8.27**   Chemical Engineers' Median Salary by Industry

| | 1990[a] | 1992[b] | 1994[c] |
|---|---|---|---|
| Agricultural chemicals | $46,000 | $63,500 | $65,000 |
| Biotechnology | $54,750 | $67,800 | $57,000 |
| Inorganic chemicals | $56,350 | $62,500 | $61,950 |
| Organic chemicals | $56,600 | $62,900 | $66,850 |
| Chemical specialties | $58,100 | $60,400 | $60,500 |
| Education | $61,000 | $62,000 | $66,900 |
| Electronic materials | $47,350 | $55,500 | $55,000 |
| Engineering; design; construction; consulting | $59,000 | $65,000 | $70,600 |
| Food; beverages | $54,500 | $61,700 | $59,900 |
| Forest products | $50,200 | $56,600 | $60,200 |
| Glass; ceramics; refractories | $55,000 | $47,600 | $75,700 |
| Paint; varnish; lacquer; pigments; inks | $50,300 | $55,300 | $52,000 |
| Petrochemicals and petroleum products | $62,500 | $62,900 | $78,700 |
| Pharmaceuticals; health care | $50,000 | $56,950 | $67,000 |
| Plastics; synthetic resins; composites | $54,100 | $56,300 | $62,900 |
| Rubber; rubber products | $49,200 | $59,500 | $66,650 |
| Soaps; detergents; perfumes; fats; oils: cosmetics | $48,750 | $50,750 | $59,350 |
| Synthetics fibers: textiles flues | $55,850 | $54,650 | $59,800 |
| Waste treatment and disposal | NA | $57,450 | $60,100 |

[a]From AIChE Salary Survey Report, American Institute of Chemical Engineers, 1990.
[b]From *Chemical Engineering Progress*, May 1992, p. 62.
[c]From *Chemical Engineering Progress*, June 1994, p. 89.

**TABLE 8.28** Chemical Engineers' Base Salaries, Bonuses, Fees, and Commissions Received by Full-Time Salaried Employees

| | Base salary | | | Bonuses | | |
|---|---|---|---|---|---|---|
| | 1996 | 1994 | 1992 | 1996 | 1994 | 1992 |
| Lower decile | $48,000 | $42,000 | $40,400 | $1,000 | $700 | $800 |
| Lower quartile | $54,000 | $50,000 | $48,000 | $2,400 | $1,700 | $900 |
| Median | $68,200 | $63,600 | $59,700 | $5,000 | $4,000 | $4,000 |
| Upper quartile | $85,000 | $80,000 | $74,500 | $11,500 | $10,000 | $10,000 |
| Upper decile | $105,000 | $101,000 | $95,000 | $25,000 | $21,600 | $21,000 |
| Number reporting | 3,177 | 3,530 | 3,312 | 1,720 | 1,593 | 1,477 |

| | Consulting fees | | | Commissions | | |
|---|---|---|---|---|---|---|
| | 1996 | 1994 | 1992 | 1996 | 1994 | 1992 |
| Lower decile | $1,000 | $1,000 | $900 | $850 | $1,200 | $640 |
| Lower quartile | $1,800 | $2,000 | $2,000 | $2,900 | $2,500 | $3,700 |
| Median | $5,000 | $5,100 | $5,000 | $11,000 | $8,000 | $10,000 |
| Upper quartile | $15,000 | $13,300 | $12,800 | $20,000 | $19,000 | $22,500 |
| Upper decile | $30,000 | $25,000 | $30,000 | $41,450 | $26,500 | $38,420 |
| Number reporting | 199 | 255 | 239 | 40 | 43 | 50 |

*Sources: Chemical Engineering Progress,* August 1996, p. 111, for 1996 data; June 1994, p. 83, for 1992 and 1994 data.

### 8.3.3 Supply and Demand of Labor

The number of chemical professionals employed is determined by the balance between supply and demand, and the same is true of their salaries.

According to the theory of marginal productivity, a company wishing to maximize profit will try to compare the incremental value gained by hiring another employee with the incremental cost. The profit of a firm is given by

$$\Pi(q) = TR - TC$$

Two expressions have been used for total cost, $TC$, Eqs. (2.10) and (8.15). Since Eq. (8.15) explicitly includes labor, we will use it for the discussion in this section:

$$\Pi(q) = pq - K - S - L$$

An expression for $L$ is given by Eq. (8.3), which we repeat here:

$$L = \sum w_i x_i \tag{8.3}$$

where $x_i$ is the number of people (operators, technicians, chemists, accountants, managers, chemical engineers, etc.) needed to operate the business and $w_i$ is the wage (salary) of each type of employee.

If all else remains constant, the incremental profit of hiring an extra employee of type $i$ is:

$$\frac{\partial \pi}{\partial x_i} = \frac{\partial(pq - K - S)}{\partial x_i} - \frac{\partial L}{\partial x_i}$$

$$\frac{\partial L}{\partial x_i} = \frac{\partial(w_i x_i)}{\partial x_i} = w_i$$

(8.25)

The marginal value of hiring an extra employee, $MV$, is:

$$MV_i = \frac{\partial(pq - K - S)}{\partial x_i} = p\frac{\partial q}{\partial x_i} + q\frac{\partial p}{\partial x_i} - \frac{\partial K}{\partial x_i} - \frac{\partial S}{\partial x_i} = w_i$$

(8.26)

The marginal value of hiring an extra employee may take any one of the forms noted in Eq. (8.26):

1. An increase in reaction yield as a result of creative bench-scale chemistry. For instance, if a chemist could increase the value of the reaction rate constant in the Delos process (Example 2.1) from 0.005 to 0.006 min⁻¹, some $200,000 per year could be saved in total costs. It would be well worthwhile to hire the chemist. This would affect the first term in Eq. (8.26).

2. An increase in sales as a result of hiring more and or better sales personnel and technical service employees. This would affect the second term in Eq. (8.26).

3. A decrease in plant costs as a result of hiring more and better design and construction engineers. This affects $K$ in the third term.

4. A decrease in raw material costs by improved or innovative design. For example, if a chemical engineer could make process changes that increase the concentration of Algol in the feed to the Delos production unit from 0.20 to 0.24 mol/L, production costs would be lowered by some $270,000 annually, affecting the fourth term in Eq. (8.26).

A rational firm should hire an employee whose marginal value is greater than the salary or wage. In a practical sense, the estimation of the marginal value by plant supervision is often obtained intuitively and in an inexact manner.

If all other factors remain constant, the hiring of more and more employees of type $i$ will sooner or later result in a diminished return. For a company engaged in producing and selling commercial catalysts, the presence of a few highly skilled specialists in catalyst preparation is essential. Their departure would bring about a great loss to the firm. However, doubling their numbers will bring the firm much less than double additional value. On the other hand, the marginal value of low-skilled day laborers may have a very flat profile. Both profiles are shown in Figure 8.8. A rational firm should keep adding employees of type $i$ until the $MV_i$ falls to the value of $w_i$. This is also exactly what Eq. (8.26) indicates. The company should also discharge workers of type $j$ when $MV_j$ is less than $w_j$. This process should continue until $MV_j$ equals $w_j$. Therefore, the demand schedule for employees of type $i$ in a firm is the curve of marginal value plotted as a function of the number of employees of type $i$. The industry-wide demand schedule is the sum of all the individual firms.

There are several components to the marginal value of labor to a firm. It has been pointed out that a marketer may increase the sales volume $q$. Technical service personnel can increase the convenience and confidence of customers, thereby increasing $q$. Engineers and chemists concerned with the improved performance, quality, quantity, and usefulness of products increase both $q$ and $p$. Production supervisors and process engineers can decrease the consumption of capital and decrease the raw materials and supplies needed for the manufacture of a pound of product. *An employee is worth his or her keep if the value of his or her contribution is greater than the salary (and benefits).*

The demand schedule for employees is not static. When there is an increase in demand for the products of a company or a price increase, the marginal value of employees increases. An increase in the GDP (or GNP) and the various sectors of final demand affects the employment of chemical professionals by increasing their marginal value to the CPI. The marginal value of chemical professionals may be increased when a new product appears that is much desired by the public or upon the introduction of a more efficient process and/or piece of equipment. In Figure 8.9, the demand schedule for chemical professionals would shift from curve $D_1$ to curve $D_2$.

FIGURE 8.8 Employee demand curve.

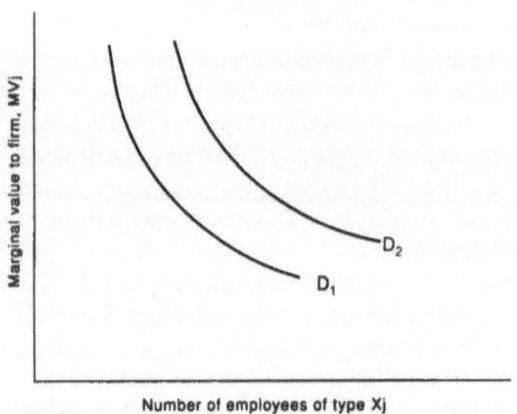

FIGURE 8.9   Employee demand curve.

There are four sources for the supply of chemical professionals:

1. New college graduates with degrees in chemistry and chemical engineering
2. Foreign nationals with permanent resident status with degrees in chemistry and chemical engineering
3. People with degrees in other fields such as mechanical engineering, mathematics, and physics who are willing to change fields
4. People with some college education but without a degree

Eugene Houdry (6), the inventor of cracking catalyst, emigrated from France, as did E. I. Du Pont. Du Pont began manufacturing gunpowder in the early 1800s and ultimately the company diversified into the manufacture of chemicals. Thomas Midgely, inventor of Freon and tetraethyllead, was a mechanical engineer. In any given year, many chemists and chemical engineering graduates seek traditional jobs in the CPI. However, a significant number use the baccalaureate degree as a basis for advanced degrees in medicine, business, or law. With the changing complexion of the chemical business, some chemical professionals are finding employment in industries that service the CPI, such as equipment and instrument manufacturers, as well as computer and computer simulation companies. Still others are finding employment in the food, packaging, and related industries. The supply schedule in Figure 8.10 is the relation between the salary level in the CPI and the number of chemical professionals willing to enter the CPI.

The simple theory predicts that the market clears at the intersection of the supply and demand curves, as in Figure 8.11. Where the demand curves intersect

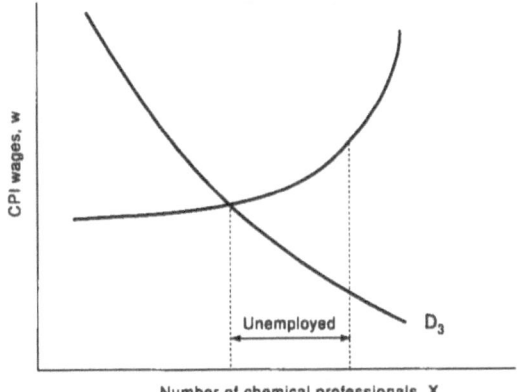

**FIGURE 8.10** Supply and demand of chemical professionals (unemployment).

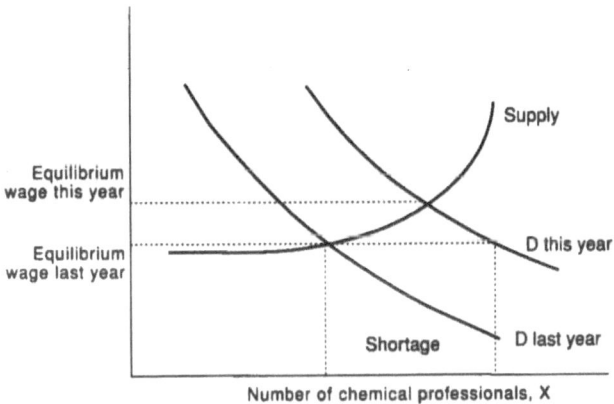

**FIGURE 8.11** Supply and demand of chemical professionals (shortage).

the supply curve, everyone wanting a job in the CPI can have one. If the demand curve expands to curve $D_2$ (see Figure 8.9), a significant number of those employed in jobs peripheral to the CPI may enter employment in the CPI. This group includes people who at first could not find jobs in the CPI and those who were attracted to other fields. With the increased need and better salaries, they are attracted to the CPI field. Such professionals as mechanical engineers may prepare themselves to enter the mainstream CPI through evening classes, short courses, and on-the-job training. If the demand schedule contracts, to $D_3$ in Figure 8.9, a significant number of professionals seeking jobs in the CPI are unable to find them and are unemployed.

To this simple theory, we need to add the speed and the mechanism of market adjustment to new equilibrium conditions. Let us consider the effect of adding a new employee on other employees. Suppose the demand schedule expands but the supply remains essentially the same; then the new equilibrium wage or salary will be higher than the equilibrium wage or salary last year. Firms making offers at last year's rate will encounter a greater number of refusals and will complain of a "shortage" of chemical professionals, as depicted in Figure 8.12. This shortage is interpreted as a shortage of chemical professionals at last year's salary or wage rate. If the firms adjust to the realities of higher compensation, the market is cleared, there is no longer a shortage, and the supply and demand are in equilibrium. What is sometimes termed "shortage" is really a lack of demand or an announced need not backed by sufficient dollars.

Let us examine how the employment market adjusts to a new equilibrium. Information flow in the employment market is slow and uncertain compared to the New York Stock Exchange or the Chicago Mercantile Exchange. At first, a company finds that after making a larger number of offers to new graduates at last year's salary level, the percentage of refusals greatly increased. There are usually other rumors and hints that other companies in the CPI are offering

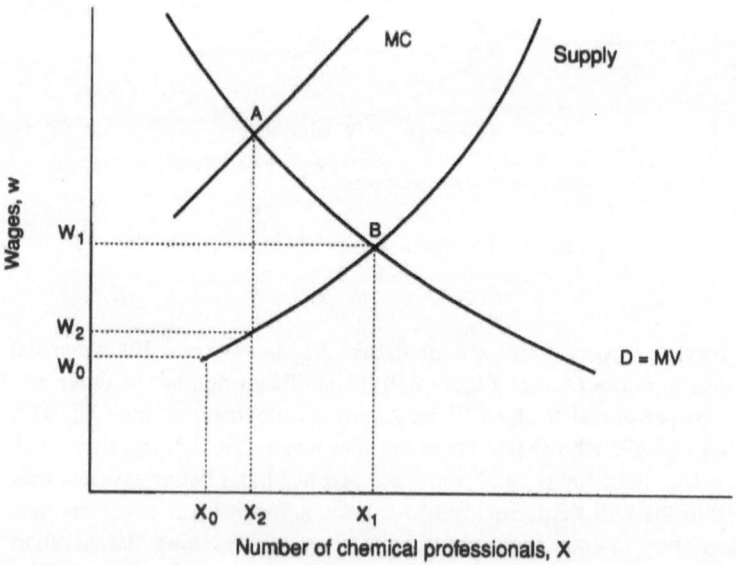

FIGURE 8.12  Supply and demand set by marginal cost of labor.

higher entry-level salaries. College recruiters will often call college placement offices for salary information. Such information usually convinces local company management to propose starting salary increases, but these must be approved at company headquarters. It takes time for this scenario to unfold and for upper level management to become persuaded to execute a new policy. A starting salary adjustment is then communicated to students, some of whom are better informed than others. Some may have accepted offers from other companies. There is an adjustment to the new equilibrium, but it may take months for this to occur.

The hiring of new employees at increased salary levels has an impact on existing employees. To account for any affect on other employees, a total derivative of labor cost with respect to $x_i$ must be considered.

$$\frac{\partial L}{\partial x_i} = \frac{\partial \left( \sum w_j x_j \right)}{\partial x_i}$$

$$\frac{\partial L}{\partial x_i} = \frac{\partial \left( w_i x_i \right)}{\partial x_i} + \frac{\partial \left( \sum w_j x_j \right)}{\partial x_i}$$

here $j$ does not equal $i$

$$\frac{\partial L}{\partial x_i} = w_i + x_i \frac{\partial w_i}{\partial x_i} + \frac{\partial \left( \sum w_j x_j \right)}{\partial x_i} \tag{8.27}$$

again where $j$ does not equal $i$.

The first term on the right-hand side of Eq. (8.27) is the proposed starting salary, $w_i$. The second term, $x_i(\partial w / \partial x_i)$, is the increase in starting salary needed multiplied by the number of employees, $x_i$, of type $i$ already on the payroll. Since an increase in starting salary is proposed to win over the new employee, this news cannot be kept from the older employees, who will certainly expect a similar (perhaps greater) increase to retain parity. If a firm is unwilling to give any increase, older employees have the option of resigning and moving to another company or remaining for nonmonetary reasons. A firm with a very large number of employees of type $i$ already on the payroll would be reluctant to increase $w_i$, since the marginal cost of the new employee is not just $w_i$ but also the much larger $x_i(\partial w_i / \partial x_i)$. An increase in $w_i$ may be inevitable unless all firms agree to suppress wages. This, of course, is illegal. The third term on the right-hand side of Eq. (8.27) needs to be included because there is always a possibility of substitution of one type of employee for another type. For some jobs, a mechanical engineer can substitute for a chemical engineer. An increase $w_i$ could lead to demand for wage increase from all the other type employees, $w_j$.

An example of the other-employee effect is given in Figure 8.12. Suppose a firm employs $x_0$ employees at wage $w_0$. As a result of business expansion, a new $MV$ has been developed, shifting the demand curve to a new position, $D = MV$. The number of employees $x$ should be set at point $A$, where $MC = MV$. The marginal cost of extra labor is:

$$MC = \frac{dL}{dx} = \frac{d(wx)}{dx} = w + x\frac{dw}{dx} \qquad (8.28)$$

If it were possible to keep the salary of the new employees secret, the firm could hire $x_1$-$x_0$ new employees at wage $w_1$ represented by point $B$ in Figure 8.12 and keep all employees at wage $w_0$. This is not an acceptable practice, since all employees should receive comparable pay for comparable jobs. The firm will have to set the employment level at $x_2$, where $MC = MV$, and offer all employees the same wage $w_2$. A firm cannot consider the employment of new personnel in isolation from the needs of existing employees.

The equilibrium theory of supply and demand applies only when there is perfect communication. Any inefficiency in information flow makes adjustment slow and uneven. There is a reluctance of large firms to increase $w_i$, and most firms are unwilling to offer wages lower than last year's to the unemployed. Having employees doing the same job but at different wage rates invites conflict within the company and movement toward unionization. A union or threat of unionization effectively makes the supply schedule horizontal at the low end. It may have the effect of keeping the wage level higher but also may make the employment level lower than it would be otherwise. In Figure 8.13, a union sets a minimum wage $w_u$ and alters the supply schedule. The equilibrium wage, $w_u$, is higher than that without a union $w_n$, but the number of employed, $x_n$, is smaller. It is sometimes argued that unions have a useful role when only one or a few firms are hiring engineers. Such is not the case in the CPI, where thousands of firms seek the services of chemical professionals. There may be a few employers seeking employees with very highly specialized skills, such as Raman spectroscopists. A union of these specialists could develop a "code of ethics" for minimum salary and prosper. Without legal protection of exclusive jurisdiction, chemical professionals have no impregnable barriers against entry by outsiders.

Some chemists and chemical engineers have argued that the way to increase prestige and pay would be to restrict the supply by limiting enrollment in universities and also restricting immigration. The same people point to the income of medical doctors and plumbers for support. The exclusive right to practice medicine and to repair plumbing is specified in law so that the control of supply can be achieved even to the disadvantage of the public. There are no job functions that the law recognizes as the exclusive domain of degreed and registered chemists and chemical engineers. When CPI firms face a shortage of chem-

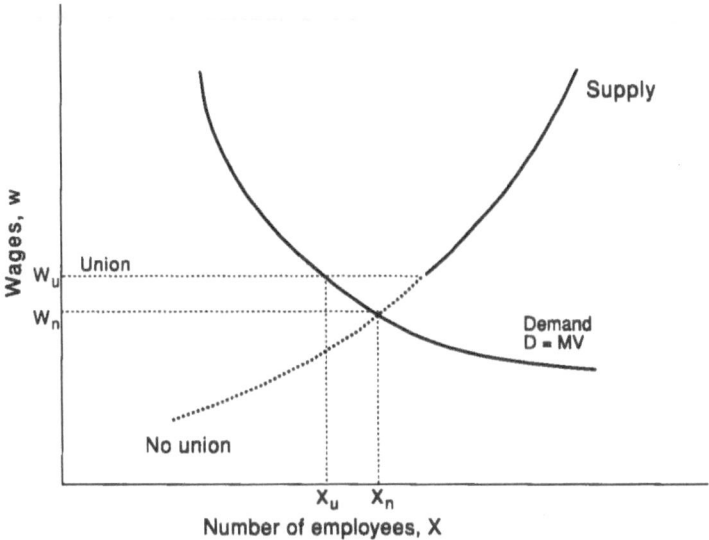

FIGURE 8.13   Union-modified supply–demand equilibrium.

ical professionals, they may establish on-the-job training programs for ambitious young people without such degrees and defeat the purpose of supply contraction.

The salaries of chemical professionals of the same vintage are not uniform. Chemical professionals are a differentiated product not a commodity, some being worth more to a firm than others. Firms in the CPI are also differentiated, and the starting salary is not the only factor a prospective employee considers. Some companies are located in particularly desirable geographic areas with agreeable climate and surroundings, or with outstanding cultural and educational resources. Some companies offer opportunities for rapid advancement, and some companies challenge the entrepreneurial spirit of the graduate, offering participation in creating new business lines.

To some extent, the salary dispersion among employees of the same vintage is a manifestation of ignorance in the market. A college graduate and a company make decisions based on incomplete information. For instance, many graduates might have taken their first jobs with different companies if no recruiters had come to the campus at the right time or if the graduates had been unaware of the other companies' need. The cost of acquiring this rapidly obsolete employment market information can be very high and cannot be supported by an individual alone. A rational professional will continue to search for companies and better salary offers until the expected marginal return is equal to the marginal cost of the search. The expected marginal return is the subjective judgment that the next company would offer more money than the best offer in hand. The expected cost of the search depends on the position of the professional.

The popular market sources are college placement bureaus, professional association meetings, trade magazines, employment brokers, and informal sources referred to as "networking." For recent graduates, the search that was at one time minimal thanks to college placement facilities, bulletin boards, fellow students and so on has come to require considerable resourcefulness. Some positions have come to the notice of the graduates through professors and bulletin boards but mostly by word of mouth from other students and letters or notices sent by companies to the various academic departments. As a result of company downsizing, the student must use both time and ingenuity to obtain a professional job. It does take time to fill out employment applications and to attend on-campus interviews, but this type of search is available at very little expense to the prospective employee. Ignorance of the job market should be minimal because students reside in a designated market. The salary distribution among new graduates is relatively narrow.

For professionals with a few years of industrial experience, the salary distribution is much broader, and as time goes on, large differences can develop between the salaries of two same-age individuals. A large part of the salary dispersion can be explained by growing recognition of a dispersion of knowledge, skill, and drive among employees. Another part of the dispersion may be attributed to employee's ignorance of the market and especially to reluctance to change jobs for noneconomic reasons. In some countries, lifetime employment in one company is the rule and a move to another firm is viewed as disloyalty. This is no longer true in the United States, considering the cost reduction measures (e.g., reduction of benefits) in many corporations. Older professionals who have suffered the downsizing may seek new positions by contacting the placement offices at their former colleges, advertising in journals, or hiring professional placement services. Some companies that have terminated older professionals have established facilities for assisting these erstwhile employees in finding new positions. One of the best ways to become known and to demonstrate one's capabilities is to publish in journals, present papers at national meetings, and become active in technical societies, provided the technical information has been cleared for presentation. Again "networking" has been an effective way to find employment.

## 8.4. FINANCE

Two major financial considerations of importance to any business are profitability and capital structure. As noted in Chapter 5, some companies in the CPI generate a tremendous flow of cash and profit. To see the significance of these companies, examine their financial ratios and compare them with those of various companies in the same product sector (e.g., petroleum, plastics, chemicals). For instance, Exxon generated a net profit of $7.5 billion in 1996, which seems to be exorbitant. Compared to the company's revenues of $119 billion, however, the profit margin was small, namely, 6%. In comparison to assets of $96 billion, Exxon is earning about 8%, on its assets, which may be more risky and not as profitable as investments in other ventures. About 17% of the $96 billion assets is borrowed money, leaving the stockholders with an equity position of $43.6 billion thus the return on equity is 19.0%, and the return on assets is 8.8%, which is better than CD rates (5.6%) and also better than the 16.2% posted by the petroleum refining sector. Exxon out performs the refining sector partly because of its large chemical and petrochemical operations, which generally have higher returns than petroleum refining (see Table 8.29). Globally, the return on assets is about 5.5% and the return on equity is about 14.1% for the petroleum sector (see Table 8.30). Pharmaceuticals have historically posted the strongest performance of any of the CPI with respect to returns on assets and on equity, as noted in Table 8.29. For example, Merck has 17.9% return on assets and 36.6% return on equity, which are the highest performance data for the pharmaceutical sector (6).

The management of a firm is entrusted with capital from lenders and stockholders, who expect that a profit will be produced with reasonable risks. The ultimate sources of capital are from savings of individuals and bank deposits, as

**TABLE 8.29** Return on Equity and Debt-to-Equity Ratio for Several U.S. Industries: 1996

| Industry | Return on equity (%) | Return on assets (%) | Debt-to-equity ratio |
|---|---|---|---|
| Soaps, cosmetics | 33.0 | 9.8 | 0.2 |
| Drugs | 29.1 | 13.8 | 0.1 |
| Petroleum | 16.2 | 5.6 | 0.3 |
| Motor vehicles | 16.1 | 4.3 | 0.5 |
| Railroads | 14.9 | 4.5 | 0.6 |
| Utilities | 11.5 | 3.5 | 1.1 |
| Electronics and electrical equipment | 16.5 | 5.9 | 0.3 |
| Computer and office equipment | 21.1 | 6.5 | 0.3 |
| Food service | 17.7 | 6.7 | 0.5 |
| Chemicals and allied products | 20.5 | 6.8 | 0.5 |
| Beverages | 25.6 | 8.6 | NA |
| Forest and paper products | 14.1 | 3.9 | 0.4 |
| Airlines | 17.9 | 4.6 | 0.6 |
| **All-industry median** | 6.8 | 2.9 | NA |

Source: Fortune 500, April 28, 1997, and compiled from company reports.

**TABLE 8.30**  Global Industry Performance of Selected Industry
Segments 1996

| Segment | Return on assets (%) | Return on stockholders' equity (%) |
|---|---|---|
| Pharmaceuticals | 12.2 | 28.3 |
| Soaps and cosmetics | 8.5 | 25.4 |
| Beverages | 7.6 | 20.1 |
| Metal products | 8.2 | 16.3 |
| Chemicals | 4.8 | 15.0 |
| Food | 4.7 | 14.8 |
| Petroleum refining | 5.5 | 14.1 |
| Rubber and plastics products | 3.3 | 11.3 |
| Forest and paper products | 2.8 | 9.3 |
| All industries | 1.2 | 11.2 |

Source: Fortune. The Global 500, August 4, 1997.

well as insurance and pension plans that may be handled directly by individuals.
Investments may also be made by professional managers of credit unions, pension trust officers, portfolio managers of insurance companies, and investment
houses, as well as by banks. These managers are constantly examining and monitoring alternate investments in terms of yield and potential risk. They may withdraw investments in firms that show weakness and place the money elsewhere.
This action on the part of investors hinders the future ability to raise more capital
in the market when a company wishes to replace obsolete or worn equipment, to
retrofit processes, or to implement new process technologies.

Capital structure is concerned with the ratio of borrowed capital to owner's
equity. When inflation is low and the economy is stable, it is frequently cheaper
to use borrowed money, if a company can secure lenders. However, a highly
leveraged company—that is, one with a high debt-to-equity ratio—faces downside risk when business is bad. Stockholders receive high dividends when profits
are good but bondholders expect the interest and principal to be paid on time,
with the threat to a firm of insolvency and bankruptcy. Section 8.4.3 shows the
effect of debt-to-equity ratio on company operations. Financial officers of companies are concerned with the optimum debt-to-equity ratio.

### 8.4.1  Profitability

Profitability is a measure of how well company management uses the facilities and
how well it assesses the market needs. A company that has a low or negative profit

for a short time period may provide poor service to its customers, its employees, and its investors, as well as to government tax collectors. Such a company cannot invest adequately for future expansion and may ultimately be absorbed by another company that has more efficient management, or it may disappear. Over the past two decades the reporting of profitability data has shifted from net sales, net income, and net income as a percentage of sales to a set of terms more meaningful terms for financial analysis- net income, operating profit margin, and return on equity. As mentioned in Chapter 3, the style of reporting financial data to investors and financial analysts has changed. This is not to imply that terms used in the past are not important; they are, but the emphasis has changed. Table 8.31 contains significant financial data: for example, profit margin as a percentage of sales, and return on assets and equity for sectors in the CPI as well as for all manufacturing. Excise taxes (e.g., gasoline taxes) are not included in the net sales data, so the figures reported in the table would be lower than those found in company reports.

Profit margins may vary widely over time, owing in part to raw material shortages and price increases, as well as other operating expenses. Another factor affecting profit margins is the competitive sales price of the product. The drug sector has consistently led the other sectors in profit margin with double-digit figures. Petroleum companies through the years have had widely varying profit margins dependent upon world oil crises. In general, petroleum and chemical sectors have consistently been two to three percentage points higher than all manufacturing. Rubber and stone, clay and glass sectors frequently lag behind all manufacturing.

TABLE 8.31    Selected Industries, Returns: 1997

| CPI firms | Profit margin (%) | Return on assets (%) | Return on equity (%) |
|---|---|---|---|
| Pharmaceuticals | 18.1 | 14.9 | 29.3 |
| Soaps: cosmetics | 7.4 | 11.5 | 32.9 |
| Beverages | 7.3 | 10.1 | 29.2 |
| Chemicals | 6.1 | 5.9 | 19.8 |
| Petroleum refining | 3.9 | 5.7 | 16.7 |
| Forest products | 0.5 | 0.4 | 2.2 |
| Other | | | |
| Telecommunications | 7.3 | 4.8 | 12.7 |
| Motor vehicles | 3.3 | 3.4 | 17.1 |
| Computers | 7.4 | 6.0 | 19.5 |
| Metal products | 9.0 | 7.4 | 16.9 |
| Food | 3.3 | 6.5 | 20.5 |
| Commercial banks | 13.6 | 1.2 | 16.9 |

Source: Fortune 500, April 17, 1998.

One has to be careful when comparing financial data over long time periods, say 10 years, because the data may not be directly comparable owing to changes in accounting procedures. Also, through mergers, acquisitions, and alliances, accounting procedures may have been modified.

## 8.4.2  Capital Structure

Prior to World War I, most CPI companies were small firms often owned by the founders. In those days, capital expenditures were usually for replacement of obsolete, worn-out equipment or for modest expansion. The funds for these expenditures were obtained from profits generated by the company. Between World War I and World War II, there was significant industrial growth, and mergers and acquisitions occurred, as well. During this period internally generated funds from depreciation and retained earnings were inadequate to meet a company's financial needs, in part because of the large capital expenditures to meet the growth, but also as a result of rampant inflation. External financial sources were required. Many companies that had relied on internal funds were now forced to change their financial policy to meet the capital needs. Eastman and Du Pont were two companies that funded projects internally. In the mid-1960s Du Pont's debt-to-equity ratio was about 0.02. When Du Pont bought Conoco in the early 1980s, it went into enormous debt (in the billions of dollars) for the first time in its history. External sources of funds were banks, insurance companies, and investment houses.

Since World War II, growth has been one of the major corporate management goals. For companies to maintain a regular dividend policy and at the same time grow, external funding was essential. In the last two decades, with corporate mergers, joint ventures, and the high interest on megadolllar projects, external funding was essential, since cash generation within a company could not begin to fund these highly capital-intensive projects. Some chemical companies requiring capital for plant expansions, new plants, environmental equipment, and retrofits have had to seek debt financing. Venture capitalists have been, another source of financing in recent times. These people have access to funds, but they often want a "piece of the action," an arrangement that may not be advantageous to a company, especially for start-up ventures. This is because venture capitalists who lack a thorough understanding of CPI operations may come to have a dominant voice in company management, leaving those who created the idea for the venture with only a minor role in operations.

Table 8.32 presents the sources and applications of funds of the top 12 chemical companies. The data show various problems that beset the chemical companies in the period from 1990 until 1995. In the early 1980s, debt was used to provide the necessary corporate funds. In the mid- to late 1980s, companies began using various sources of funds for corporate restructuring. In the late

TABLE 8.32 Debt/Equity Ratios for Selected Companies[a]

|  | 1990 | 1991 | 1992 | 1993 | 1994 | 1995 |
|---|---|---|---|---|---|---|
| **Exxon Corporation** | | | | | | |
| Long-term debt | 7,687 | 8,582 | 8,637 | 8,506 | 8,831 | 7,778 |
| Stockholders' equity | 33,055 | 34,927 | 33,776 | 34,792 | 37,415 | 40,436 |
| Debt/equity ratio | 0.233 | 0.246 | 0.256 | 0.244 | 0.236 | 0.192 |
| **Du Pont** | | | | | | |
| Long-term debt | — | 6,456 | 7,193 | 6,531 | 6,376 | 5,678 |
| Stockholders' equity | — | 16,739 | 11,765 | 11,230 | 12,822 | 8,436 |
| Debt/equity ratio | — | 0.386 | 0.611 | 0.582 | 0.497 | 0.673 |
| **Dow Chemical** | | | | | | |
| Long-term debt | 5,209 | 6,079 | 6,191 | 5,902 | 5,303 | 4,705 |
| Stockholders' equity | 8,728 | 9,441 | 8,064 | 8,034 | 8,212 | 7,361 |
| Debt/equity ratio | 0.597 | 0.644 | 0.768 | 0.735 | 0.646 | 0.639 |
| **Monsanto** | | | | | | |
| Long-term debt | 1,645 | 1,871 | 1,423 | 1,502 | 1,405 | 1,667 |
| Stockholders' equity | 4,089 | 3,654 | 3,005 | 2,855 | 2,948 | 3,732 |
| Debt/equity ratio | 0.402 | 0.512 | 0.474 | 0.526 | 0.477 | 0.447 |
| **American Home Products** | | | | | | |
| Long-term debt | 105 | 111 | 602 | 859 | 9,973 | 7,809 |
| Stockholders' equity | 2,675 | 3,300 | 3,563 | 3,876 | 4,253 | 5,543 |
| Debt/equity ratio | 0.039 | 0.034 | 0.169 | 0.222 | 2.345 | 1.409 |
| **Procter & Gamble** | | | | | | |
| Long-term debt | 3,588 | 4,111 | 5,223 | 5,174 | 4,980 | 5,161 |
| Stockholders' equity | 7,518 | 7,736 | 9,071 | 7,441 | 8,832 | 10,589 |
| Debt/equity ratio | 0.477 | 0.531 | 0.576 | 0.695 | 0.564 | 0.487 |
| **Bristol Myers** | | | | | | |
| Long-term debt | 231 | 135 | 178 | 588 | 644 | 635 |
| Stockholders' equity | 5,418 | 5,795 | 6,020 | 5,940 | 5,704 | 5,822 |
| Debt/equity ratio | 0.043 | 0.023 | 0.030 | 0.099 | 0.113 | 0.109 |
| **Schering-Plough** | | | | | | |
| Long-term debt | 183 | 754 | 184 | 182 | 186 | 87 |
| Stockholders' equity | 2,081 | 1,346 | 1,597 | 1,582 | 1,574 | 1,623 |
| Debt/equity ratio | 0.088 | 0.560 | 0.115 | 0.115 | 0.118 | 0.054 |
| **Merck** | | | | | | |
| Long-term debt | 124 | 494 | 496 | 1,121 | 1,143 | 1,373 |
| Stockholders' equity | 3,834 | 9,499 | 5,003 | 10,022 | 11,139 | 11,736 |
| Debt/equity ratio | 0.032 | 0.052 | 0.099 | 0.112 | 0.103 | 0.117 |

**TABLE 8.32** Continued

|                      | 1990   | 1991   | 1992   | 1993   | 1994   | 1995   |
|----------------------|--------|--------|--------|--------|--------|--------|
|                      |        |        | **Amoco** |     |        |        |
| Long-term debt       | 5,249  | 4,596  | 5,113  | 4,037  | 4,387  | 3,962  |
| Stockholders' equity | 14,068 | 14,156 | 12,960 | 13,665 | 14,382 | 14,848 |
| Debt/equity ratio    | 0.373  | 0.325  | 0.395  | 0.295  | 0.305  | 0.267  |
|                      |        |        | **Mobil** |     |        |        |
| Long-term debt       | 4,298  | 4,715  | 5,042  | N/A    | 4,714  | 4,629  |
| Stockholders' equity | 17,072 | 17,534 | 16,450 | 17,237 | 17,146 | 17,951 |
| Debt/equity ratio    | 0.252  | 0.269  | 0.307  | —      | 0.275  | 0.258  |
|                      |        |        | **ADM** |       |        |        |
| Long-term debt       | 751    | 980    | 1,562  | 2,039  | 2,021  | 2,070  |
| Stockholders' equity | 3,573  | 3,922  | 4,492  | 4,883  | 5,045  | 5,854  |
| Debt/equity ratio    | 0.210  | 0.250  | 0.348  | 0.418  | 0.401  | 0.354  |

ªLong-term debt and stockholders' equity in millions of dollars.

1980s and into the early 1990s, company profits were down and companies used working capital to finance their needs.

A company that is heavily in debt is said to be highly "leveraged." In general, a company that is highly leveraged may be a candidate for a takeover. So too, companies with high cash ratios are good candidates. Corporate raiders are interested in such companies.

The debt-to-equity ratios for chemical and petroleum firms tend to be more conservative than in overall industry. In the 1970s and 1980s, the CPI debt-to-equity ratio was in the range of 0.2–0.4, compared to overall industry figure exceeding 0.4. In the last decade, there has been an upward trend in the CPI, as noted in Table 8.32. Chemical companies are becoming increasingly dependent on external funding. Some major chemical companies have traditionally had high debt-to-equity ratios, for example, between 1970 and 1973 Dow had a 0.8 ratio, which more recently has come down to 0.64–0.77, but the trend is to reduce it further. To reduce the long-term debt, companies began selling off parts of their operations.

Outside the CPI, utilities, railroads, and airlines tend to have high debt-to-equity ratios, as noted in Table 8.29. In general, companies that are heavily regulated by government also have high debt-to-equity ratios, leaning on the federal government as the ultimate guarantor of their debts. The services these firms provide to the customer are essential, and since there is no alternative, they cannot be permitted to fail. In Japan, where government, banks, and manufacturing companies enjoy a high degree of cooperation, the debt-to-equity ratios are very high, in the range of 3–7.

Another economic indicator, the debt-to-capital ratio, is defined as the long-term debt divided by the total capital. This ratio is an indication of how highly leveraged a company might be. The ratios for a selected industries are found in Table 8.33. The ratios for these companies have been relatively constant in the range of 0.27 to 0.42 over a 10 year period (1984–1994). As an example of how the debt-to-equity ratio affects the return on equity, let us consider two companies.. Company A has a debt-to-equity ratio of 0.35, and company B's ratio is 0.80 (see Table 8.34). Assume that the interest rate on debts is 10% and that each company earns 30 cents per dollar of capital before income tax and interest. The problem is solved by assuming that debt plus equity equals capital assets, which

TABLE 8.33   Debt-to-Capital Ratios for Selected Industries, 1994

| Industry | Return on equity (%) | Return on capital (%) | Debt-to-capital (%) |
|---|---|---|---|
| Food | 14.1 | 10.6 | 38.1 |
| Chemicals | 15.5 | 11.3 | 29.3 |
| Financial services | 16.0 | 12.3 | 33.1 |
| Computers/communications | 13.7 | 10.7 | 19.6 |
| Electric utilities | 11.7 | 6.2 | 37.3 |
| Aerospace/defense | 11.3 | 10.2 | 37.3 |
| Forest products | 5.6 | 4.7 | 41.6 |
| Transportation | 12.8 | 8.2 | 36.2 |
| Energy | 8.9 | 6.6 | 39.6 |
| Metals | 6.7 | 7.3 | 34.0 |
| Entertainment | 14.8 | 13.2 | 28.0 |
| Business services | 14.0 | 11.2 | 30.2 |
| Retailing | 16.1 | 11.5 | 26.5 |

Source: U.S. Bureau of the Census, Statistical Abstract of the United States, 1997, Table 873.

TABLE 8.34   Example: Debt and Equity

| | Company A | Company B |
|---|---|---|
| Debt/equity ratio | 0.35 | 0.80 |
| Debt, % | 30% | 50% |
| Interest payment, costs | $0.30 \times 10\% = 3.0$ | $0.50 \times 10\% = 5.0$ |
| Equity, % | 70% | 50% |
| Gross profit | 30 - 3.0 = 27.0 | 30 - 5 = 25.0 |
| Income tax (35%), cents | 9.45 | 8.75 |
| Net income after taxes, cents | 17.55 | 16.25 |
| Return on equity, % | 17.55/70 = 25.1% | 16.25/50 = 32.5% |

in many cases is a good assumption. Therefore, given the same gross earnings per dollar of capital as company A, company B with a heavy debt structure shows better net earnings on equity after tax.

Now suppose that business takes a downturn and the before-tax gross earnings on capital is 5 cents on the dollar. The results of this example are shown in Table 8.35. Company A has a poor return on equity, namely, 1.9% with a relatively low debt-to-equity ratio. The highly leveraged company B turned the same low gross earnings into zero profit. In fact, it is possible that there could have been a loss, which would result in a negative return. Therefore, using borrowed money is a two-edged sword: it is good during prosperous times but bad during economic downturns.

Funds are generated internally from net income after taxes and depreciation and externally from new debt and selling new stock. The major uses of funds are dividends to stockholders, capital investment, research and development, increase in working capital, and investment in other opportunities. Most companies have an annual capital expenditure program, projecting how much to spend each year and which projects to fund. Scenarios for short-term (perhaps one year), mid-term (say five years), and long-term (10–15 years) expenditures are prepared. There must be projected sales, cash flow, and profit forecasts for each project proposed. The risks and consequences of adverse business conditions on profits are also estimated. Companies rank all projects by their projected return on investment as well as other financial and risk parameters. The projects with the highest merit are funded first, and then the rest in descending order until the total capital available has been committed. Aggressive growth-oriented companies have more worthwhile projects than funds available. Less aggressive companies may have few worthwhile projects to fund with sufficiently high returns or measures of merit. If this be the case, the company might be better off investing in other opportunities, engaging in a stock repurchase program of its own stock, perhaps forming joint ventures, or lending funds to other companies with better ideas on how to invest.

**TABLE 8.35**   Second Example: Debt and Equity

|                          | Company A            | Company B            |
|--------------------------|----------------------|----------------------|
| Debt/equity ratio        | 0.35                 | 0.80                 |
| Interest payment, costs  | $0.30 \times 10\% = 3.0$ | $0.50 \times 10\% = 5.0$ |
| Profit before taxes, cents | $5.0 - 3.0 = 2.0$  | $5.0 - 5.0 = 0$      |
| Income tax (35%), cents  | 0.7                  | 0                    |
| Profit after taxes, cents | 1.3                 | 0                    |
| Return on equity, %      | $1.3/70 = 1.9$       | $0/50 = 0$           |

### 8.4.3   Mergers, Acquisitions, Joint Ventures, and Alliances

In the 1980s, many companies in industrial America were involved in mergers and acquisitions. Toward the end of the 1980s and into the early 1990s, these activities subsided somewhat. Again in the late 1990s, mergers, acquisitions, joint ventures, and alliances were prevalent, especially in the CPI.

Mergers occur when two or more companies find it financially attractive to merge their operations. Usually mergers are based on mutual agreements and are attractive to the companies involved. In acquisitions, on the other hand, the acquirer plays the dominant role and the event may not be friendly. Indeed, the acquired concern may completely disappear from the business line. Joint ventures occur when two or more companies mutually agree that each company can bring either technology or financial benefits to the venture. The companies will often retain their original identity, but a new company name may be introduced.

In the 1980s a fad in industry known as "diversification" allowed companies to become involved in other business lines. Allied Chemical bought Bendix and Signal Oil, Du Pont acquired Conoco, and some companies bought bits and pieces of other companies hoping that through diversification business risks might be spread. Some classical chemical companies have entered the "life science" field, which was touted to be the next big boom. Hoechst, Rhône-Poulenc, Ciba-Geigy, and Sandoz (now called Novartis), and Monsanto, as well as a joint venture between Dow and Du Pont, are a few that have followed this trend. In Monsanto's case, the agrochemicals–life science group retained the Monsanto name, but the industrial chemicals group was spun off as a new company called Solutia.

As mentioned earlier, the merger and acquisitions fever in the CPI accelerated in the 1980s. Chemical companies had become large complex entities, frequently entering fields in which their experience was limited. When this happens, companies often lose their focus and accountability suffers. The profitability declines as does the price of the stock, and the stockholders become dissatisfied. As one might expect, not all mergers, acquisition, joint ventures, and alliances prosper. Companies that do not possess a given marketing and technological ability may soon find that they are in competition with themselves or in a very competitive global market they had not anticipated. In fact, more than one company has entered a technology field for it was ill-equipped. When the anticipated profitability failed to come to fruition, the acquired company was again spun off. Current examples are Hoechst, which wants to divest itself of Celanese, and Du Pont, which has spun off Conoco. In any merger, acquisition, or joint venture, companies must be careful not to be in a monopolistic position or to be perceived as restraining trade, for the federal government will break up such ventures. An interesting article on this subject may be found in *Chemical Week* (8).

One of the business trends in recent years is the formation of alliances. Some of the factors that contributed to this phenomenon have been globalization, escalating research and development expenses, shortened product life cycles, and the convergence of technologies. They are formed to build markets (new and old), to take advantage of technologies, to perform joint research, to leverage material and personnel resources, and to acquire competence. These result in creating new opportunities and in the creation of value for the companies involved (2). Two-firm alliances were the norm, but now multifirm alliances are emerging in which several companies join in a single, large, all-encompassing relationship for a common purpose. Each firm's decision to cooperate depends on and influences the decisions of other firms so that the final joint decision represents the combined decisions among alliance members. Sometimes one partner is foreign, a feature designed to broaden the market in a given product area where each partner shares in the total market.

Because of the complex nature of alliances, game theory (3) provides conceptual backgrounds in exploring and analyzing the interdependent decision making among the firms involved. Conventional two-party strategy games are not adequate for depicting multifirm strategies (3). The most notable examples of multifirm alliances occur in high-tech industries such as aircraft manufacturing, computer hardware and software, and pharmaceuticals.

Fujita sought ICI, Amdahl, and Siemens as partners to build a "critical mass" and acquire a virtual market share and become more competitive with IBM. Chrysler entered into an alliance with Mitsubishi to plug product and component gaps. In the CPI, pharmaceutical companies formed alliances with drug–biotech firms and agricultural–biotech firms to develop genetically engineered products like insulin and herbicide-resistant soybeans (8). The proposed merger of Monsanto and American Home Products is hoped to provide Monsanto with cash to tap synergies between drug and agricultural–biotech firms. Companies that pursue the creation of value by means of alliances turn rival or complementary firms into partners, meld and leverage cospecialized resources, and learn and internalize critical skills.

## 8.5. ROLE IN THE ECONOMY

The qualitative importance of the CPI to the U.S. economy lies mainly in its contribution of synthetic materials and semifinished materials for further processing and manufacturing. It has been exceedingly successful in displacing natural products. Consider the replacement of natural dyes with synthetic dyes; cotton, wool, and silk with synthetic fibers; herbal and natural medicines with antibiotics; smoky coal with cleaner burning oil and clean-burning natural gas, and many other examples. The CPI has been a major contributor to every sector of human need. Table 8.36 is a partial list of the many contributions to society by the CPI.

TABLE 8.36 The CPI's Contribution to Society

| Need | Major CPI contribution |
|---|---|
| Food | Processing additives to prevent spoilage, improve nutritional value, and impart flavor and color<br>Examples: margarine, edible oil, dehydrated foods, vegetable proteins replacing animal proteins<br>Canning, frozen foods, ready-to-serve foods |
| Clothing | Synthetic and nonwoven fabrics<br>Dyes and finishes (permanent press)<br>Additives to decrease flammability and soiling<br>Improved blending of fibers to improve fabric care |
| Shelter | Paints and coatings<br>Air purification and air conditioning<br>Plastics and other materials for buildings, structures, and furniture<br>Adhesives and sealants<br>Heating and power |
| Agriculture | Fertilizers, pesticides, fungicides, biocides<br>Plant growth regulators<br>Nutrients and veterinary medicines<br>Animal feeds<br>Fuels for tractors and drying crops |
| Health | Pharmaceuticals<br>Drug delivery systems<br>Soap and presonal care items |
| Transportation | Automobiles: body (metallic and plastic materials), composites, fuel, lubricants, rubber, fabrics, brake fluids, antifreeze<br>Aircraft: aluminum, composites, plastic materials, fuels, lubricants, deicing fluids |
| Communications | Telephone and TV: plastics, composites, copper wires, semiconductors, fiber optics<br>Computers: electronic chips, copper wires, semiconductors<br>Photography and photocopy: silver chemicals developing fluids, carbon black<br>Publishing: printing inks, paper, paper sizing, paper coating |
| Defense | Gunpowder, shells, rocket fuels, atomic weapons |
| Space exploration | Titanium, high-energy fuels, plastics, composite materials |
| Ecology | Chemicals for air and water pollution control<br>Recycle solid materials |

The importance of the CPI in terms of its contributions and size is evident in Tables 8.37–8.40. Table 8.37 is a summary of the CPI in terms of employees, value added by manufacture, value of shipments, and capital expenditures. A summary of the CPI and related industries is included for comparison purposes in Table 8.38. Tables 8.39 and 8.40 list percentages of all manufacturing for various CPI sectors and for various important industries, respectively.

**TABLE 8.37**  Summary of the Chemical and Allied Products Industries, 1994

| SIC code | Industry group | Employees (thousands) | Value added by manufacture (millions) | Value of shipments (millions) | Capital expenditures (millions) |
|---|---|---|---|---|---|
| 281 | Industrial chemicals | 87 | $15,058 | $24,980 | NA |
| 282 | Plastics materials, synthetics | 128 | $24,497 | $55,314 | $9,852 |
| 283 | Drugs | 206 | $55,528 | $76,238 | $4,140 |
| 284 | Soaps, cleaners, toilet goods | 118 | $27,767 | $46,314 | $1,273 |
| 285 | Paints and allied products | 50 | $8,501 | $17,544 | $280 |
| 286 | Industrial organic chemicals | 115 | $28,942 | $69,596 | $3,524 |
| 287 | Agricultural chemicals | 40 | $10,302 | $20,899 | $633 |
| 289 | Miscellaneous chemical products | 80 | $10,595 | $22,413 | $204 |
| 28 | Chemical and allied products | 824 | $181,189 | $333,259 | $5,455 |
|  | All manufacturing | 18,344 | $1,598,464 | $3,340,223 | $549,912 |

Source: U.S. Bureau of the Census, Statistical Abstract of the United States, 1995, Table 1211.

**TABLE 8.38**  Summary of the CPI and Related Industries, 1994

| SIC code | Industry group | Employees (thousands) | Value added by manufacture (millions) | Value of shipments (millions) | Capital expenditures (millions) |
|---|---|---|---|---|---|
| 20 | Food and kindred products | 1,512 | $171,988 | $430,994 | $10,094 |
| 26 | Paper and allied products | 621 | $63,347 | $143,762 | $7,312 |
| 28 | Chemicals and allied products | 824 | $181,189 | $333,259 | $15,455 |
| 29 | Petroleum and coal products | 112 | $28,594 | $143,150 | $5,215 |
| 30 | Rubber and miscellaneous plastics products | 972 | $69,613 | $134,553 | $4,756 |
| 32 | Stone, clay, glass | 485 | $39,557 | $71,277 | $137 |
| 33 | Primary metal industries | 664 | $61,554 | $160,771 | $3,191 |
|  | All manufacturing | 18,344 | $1,598,464 | $3,340,223 | $549,912 |

Source: U.S. Bureau of the Census, Statistical Abstract of the United States, 1995, Table 1211.

Capital expenditures in the CPI declined in the mid-1990s as companies sold off unprofitable product lines and assessed their goals for the future. In 1995 the first significant capital spending increase in 5 years occurred among 24 major chemical producers, at about $8.3 billion. The highest capital spending in 15 years was in 1990, at $9.7 billion. Between 1990 and 1994 there was a marked decline in capital spending because earnings were under pressure generally owing to overcapacity and the decline in demand for products.

**TABLE 8.39**   Percentage of All Manufacturing Accounted for by Various CPI Sectors

| | | Percentage of all manufacturing, 1994 | | |
|---|---|---|---|---|
| SIC | | Employees (thousands) | Value added by manufacture (millions) | Value of shipments (millions) | Capital expenditures (millions) |
| 20 | Food and kindred products | 8 | 11 | 13 | 2 |
| 26 | Paper and allied products | 3 | 4 | 4 | 1 |
| 28 | Chemicals and allied products | 4 | 11 | 10 | 3 |
| 29 | Petroleum and coal products | 1 | 2 | 4 | 1 |
| 30 | Rubber and miscellaneous plastics products | 5 | 4 | 4 | <1 |
| 32 | Stone, clay, glass | 3 | 2 | 2 | <1 |
| 33 | Primary metal industries | 8 | 4 | 5 | 1 |

*Source:* U.S. Bureau of the Census, *Statistical Abstract of the United States,* 1995, Table 1211.

**TABLE 8.40**   Percentage of All Manufacturing Accounted for by Various Important U. S. Industries, 1995

| | | Percentage of all manufacturing, 1994 | | |
|---|---|---|---|---|
| SIC | | Employees (thousands) | Value added (millions) | Value of shipments (millions) | Capital expenditure (millions) |
| 23 | Apparel and related products | 5 | 2 | 2 | NA |
| 24 | Lumber and wood products | 4 | 3 | 3 | <1 |
| 34 | Fabricated metal products | 8 | 6 | 6 | 1 |
| 35 | Machinery, except electrical | 10 | 10 | 9 | <1 |
| 36 | Electrical and electronic equipment | 8 | 9 | 8 | 1 |
| 37 | Transportation equipment | 8 | 11 | 14 | <1 |

*Source:* U.S. Bureau of the Census, *Statistical Abstract of the United States,* 1995, Table 1211.

## 8.6  RELATION WITH GOVERNMENTS

The starting point of modern economic theory is the free market of Adam Smith, where numerous buyers and sellers meet to transact their business without government interference. Each party takes the action that will maximize his own gain. It is relatively easy to expound theories and to assign problems based on this free-market situation, but a modern firm in the CPI works under an enormous number of supportive and regulative actions of the government. A company's management can make few decisions without considering guidelines and rules established by government to control the activities in question. The larger the company, the greater the influence of government.

An oil company must lease drilling rights for oil and gas. In the case of offshore or public land locations, the company must pay a large bonus to acquire the lease from the government, and yearly fees and royalties on oil recovered. The government decides when the leases will be available for bids, what the terms will be, and whether the eventual bids are high enough. It grants the right-of-way for the building of pipelines and marine terminals for the movement of oil and decides on how these facilities should be built. The government, which can approve or disapprove the refinery construction on a given site, decrees the permissible discharge of waste products into the air and water, regulates plant safety for employees, and may shut down a plant for noncompliance. The government determines a minimum wage and imposes rules relating to the minority composition of each labor force. It decides on the quantity and the supplier of oil for each of many governmental and military uses. It decides on how much to spend on education and government-subsidized research, which produce needed manpower and scientific knowledge that benefit the oil companies. The government may decide that a company is too big and powerful and order it to be broken up into smaller entities (e.g., the breakup of AT&T/Bell Telephone into the numerous smaller local telephone service companies). The government may subsidize a company, lend money to, or guarantee the debts of, a company, and provide protection for domestic operations from foreign competition.

Sometimes government intervention into the affairs of business is thought to be a recent phenomenon. However, for well over 100 years, government has exercised a degree of control over how U.S. businesses conduct themselves. (Businesses in other countries do not always have to conform to the same set of standards imposed by the U.S. government on U.S. businesses—more on that later.) Opinions on this matter can be described by the extreme positions: namely, government should set and prescribe in detail the conditions under which businesses operate versus the opinion that free enterprise, of itself, is ultimately the best regulator of business practices and will eventually correct any abuses that develop in the conduct of business matters. For well over 30 years, the Federal Trade Commission (FTC) has exercised control over business operations. Among other practices, the FTC prohibits competing companies from colluding to establish prices for products. Thus, buyer and seller must negotiate to determine the price at which any transaction will occur without prior agreement among the sellers on the price that will be required. Any violation of this restraint is a serious matter and in a few cases, officers of companies have been sentenced to prison for such violations. While somewhat similar regulations exist in Europe and the Far East, the restrictions in these areas are not nearly as severe as in the United States.

The Federal Drug Administration (FDA) has specified the requirements necessary to bring a new drug to the market. Consider the problems that this sort of control places on a pharmaceutical company. Before approval is given to market a drug, either prescription or generic, the company must test the drug, often for years, to determine the effect on the patients. This process is very costly, and the result is to make drugs expensive when they finally reach the market, to allow the pharmaceutical companies to recover development costs. (It has been said that aspirin would never have reached the market if it had not been produced and distributed before the FDA began regulating this process.) Thus, it is sometimes argued that the FDA should not take so long to evaluate new medications. The other side of this question becomes apparent when one considers that after the drug thalidomide was approved many years ago, it was tragically discovered that pregnant women who took the drug had a high rate of deformed babies. Few FDA officials are willing to chance being the one who approve, "another thalidomide." On occasion, patients suffering from incurable illnesses such as cancer or AIDS, who are willing to risk undesirable side effects from an experimental drug, are not allowed this possibility. Thus, we are confronted with the legal and moral dilemma of exactly what balance to strike between thorough testing to eliminate drugs with bad side effects and providing new medications that may save the lives of patients who have no other alternatives.

Energy supply and price is an issue that is important and pervasive and largely politically determined. Before 1973, oil from the Persian Gulf at $2.00 per barrel was displacing domestic oil, at $3.25 per barrel. Thus, to protect U.S. oil producers, import quotas and tickets to import (worth $1.25 per barrel) were imposed. During this time, U.S. consumers were forced to pay more for their energy requirements than consumers in other countries (notably Europe and Japan), which obtained all their oil from OPEC at $2.00 per barrel.

Prior to 1973, there was a surplus supply of crude oil in the United States. To avoid overproduction with the concomitant depression in the price of crude oil, the Texas Railroad Commission controlled the level of production of oil wells in Texas (other states had similar regulations). This provided some stability in crude oil pricing. However, in 1973, it became apparent to the oil ministers of OPEC countries that oil-importing countries could not do without the Middle

East supply. The price of oil from OPEC went from $2 to about $12 in a short time, and the members of the cartel imposed an oil embargo on the United States because of the country's support for Israel. Since that time, U.S. producing states have operated at full production rates and the Texas Railroad Commission no longer dictates the production levels on the production wells. Thus to a degree, free enterprise returned to the "oil patch," but only to a degree, since OPEC is itself a cartel, which actually eliminates true free enterprise. The OPEC ministers meet regularly to set quotas and prices for the oil production from each of the member countries. In 1979 the OPEC ministers decided to dramatically increase the price of crude oil once again, and the price went from $12 to about $36 per barrel. Not surprisingly this resulted in a "boom" in the U.S. oil fields. Rotary rig (drilling rig) count increased to over 4000. States such as Oklahoma, Texas, and California experienced an economic boom even while other states were enduring a recession. Another effect of this increased oil price included efforts at energy economy (who can forget the 55 mph speed limit); the CAFE regulations, which required U.S. automakers to produce automobiles with increasing fuel economy; and efforts by industry engineers to build plants with very energy-efficient designs. Forecasts were for crude oil to escalate to $100 per barrel by 1990 and for natural gas price to increase to $10–$15 per million Btu.

With these expectations it was easy to justify spending capital money to reduce the consumption and increase the recovery of energy in industrial plants. These crude oil prices also had the effect of encouraging research into development of alternative energy sources (e.g., nuclear, biomass, solar energy, wind energy, oil shale and tar sands recovery, and secondary and tertiary oil recovery). By 1982–1983, these efforts had begun to pay dividends, and the OPEC countries discovered that their price increases had had the result of reducing demand. The OPEC ministers were then faced with the unpleasant task of reducing their production quotas to maintain the higher price structure they had decreed. The reduced quotas were met with production cheating. It eventually became apparent that the higher price level could not be maintained. In due time, the OPEC countries, led by the dominant producer—Saudi Arabia—settled on a crude oil price structure that varied between $15 and $20 per barrel. The reasoning was that with the price at these levels, U.S. consumers would not have adequate economic justification for the strong efforts to save energy, the expense of building very energy-efficient plants, or the cost of research to find and develop alternative energy sources. This situation has more or less prevailed since 1983. With the exception of the OPEC cartel contamination, the oil markets have largely returned to a free enterprise circumstance.

While the FTC and the FDA have been in operation for many years, there is now a proliferation of other agencies that regulate businesses. Much has changed in the arena of business/government relations in the 20 years that have elapsed since the Wei (6) book was published. Perhaps the two most important agencies relative to the CPI are the Occupational Safety and Health Agency (OSHA) and the Environmental Protection Agency (EPA). It would be a mistake, however, to assume that EPA and OSHA concern themselves only with the CPI. Many small private business find themselves having to answer to these two agencies with respect to water or atmospheric emissions (EPA) or the safety of workers (OSHA) from exposure to hazardous conditions to lack of adequate access for the disabled. Even if a company is in complete compliance with all regulations, there is still the burden of completing questionnaires and the occasional inspection by one of these two agencies. Without question, government demands on business have grown ever more costly in terms of the staff and reports needed to satisfy the government that a business is in compliance with standards.

In many cases, business and unions have been adversarial in their relationships. The unions have sometimes demanded work rules that resulted in inefficient operation of a unit or plant. For instance, craftsmen have debated which craft should disconnect a pipe flange from a pump—Should it be a machinist who is assigned to work on the pump or a pipefitter who has responsibility for piping? While this distinction may appear to be unimportant, it has resulted in inefficiency and lost time while the jurisdictional dispute is resolved. However, both businesses and unions have found that they can sometimes gain an advantage by persuading the government to act on their behalf. For instance, during the power days of unions, Congress was persuaded to pass the Jones Act, which specifies that any ship that departs from one U.S. port and docks at another U.S. port must be manned only by U.S. seamen. At one time, a barge that was being used to haul propylene from Ponce, Puerto Rico, to the Gulf Coast (Lake Charles) was disabled. Since there were no U.S. ships manned by U.S. crews that could carry propylene, the company producing and shipping the propylene sought to utilize a foreign ship to carry this material. The unions threatened to strike every port from Brownsville to New York if that was done. Consequently, the propylene in Puerto Rico was burned (flared) until the regular barge could be repaired. Another instance of the problems that arise from the Jones Act is that the cruise ships that sail the inland waterway along the Alaskan coast must not depart and land at U.S. ports, since nearly all these cruise ships are foreign-owned and manned by officers and seamen from various countries. Consequently, Seattle, which would clearly be the port of choice for departure or landing, cannot serve American vacationers in that capacity. Instead, the ships depart from or dock in Vancouver, British Columbia, Canada—an economic boon for Vancouver; a loss for Seattle—solely because of the Jones Act.

Few would argue that it has been (and still is) necessary for government to establish standards of conduct to protect the public from abuses of the less scrupulous operators of businesses. However, government and special interest groups sometimes feel that standards considered too stringent by industry personnel are perfectly appropriate. The effect of this inflexibility is to make American businesses noncompetitive in the world market, where they must compete with goods made in countries that are not saddled with these extreme standards. Certainly no respectable businessman would elect to subject his employees to hazardous conditions, but often federal standards are set by people with little knowledge of science or technology. Consider that some years ago, saccharin was declared a carcinogen as a result of tests conducted on rats in which the rats were force-fed quantities of saccharin that would be almost impossible for a human to consume. Recently, saccharin was taken off of the list of carcinogenic materials.

Perhaps the most controversial aspect of government intervention in business is not the setting of standards by government but that all too often governmental and special interest groups with virtually no training in the sciences choose to dictate the methods of achieving these standards as well. This may, on occasion, be done out of ignorance, but it is often done for purely political reasons. Consider the question of the addition of oxygenates to gasoline to provide for more complete combustion in automotive engines, with the result of reducing contaminants in the exhaust gases of vehicles. While some still question the value of this approach, nearly all industry participants feel that the method of achieving this end should be left to the scientific community. However, a group of politicians from the midwestern corn-producing areas were successful in requiring that a certain amount of these oxygenates be ethanol (produced from the fermentation of corn). Corn-based ethanol is relatively inefficient as an automotive fuel, but the politicians "solved" this problem by providing a subsidy for the blending of ethanol. The end result was an increased burden on the taxpayer.

Another case was the demand by environmental groups that the petroleum refining industry develop a reformulated gasoline (which included the oxygenates mentioned above). This program was estimated to cost the industry $30–$40 billion over a period of about 5 years. It was suggested that the result would be more effective if this amount of money were used to eliminate older, inefficient cars (so-called junkers) from the highways. However, environmental groups did not want to take this approach; they admitted that it was easier and more viable to place this responsibility on the petroleum and automobile industries rather than imposing it on the general public.

It was mentioned earlier that the United States does not operate with the same set of rules that other countries adopt. That is readily apparent to anyone who reads in the newspapers that working conditions in some of our neighboring countries are much worse than those prevailing in the United States. Mexico has been accused of being guilty of pollution to a degree not allowed in the United States. But before we become too self-satisfied, it should be noted that Canada is not happy about the problems associated with "acid rain," which Canadians insists originates in the United States.

Other problems that arise from the differences in standards relate to the cost of producing products if U.S. industry must meet much more rigorous standards in pollution control. Again, no responsible employer wants to subject his workers to dangerous working conditions (i.e., exposure to levels of dangerous pollutants that would be hazardous to their health). The question then becomes, What level of pollutant *is* harmful to a person's health? Some take the position that industry should provide air and water quality standards that are better than the natural, or nonindustrial, state. In the midst of a very serious question, some amusing results develop. One environmentalist, when discussing water quality, said that industry was proposing a pH level that was unacceptable. The environmentalist wanted the pH to be zero.

Another problem that has been addressed by the U.S. government has to do with the rules of conduct of sellers of product. In this country it is illegal to form a cartel or in any way to conspire with competitors to "fix" the price of products being sold. While such collusion does not often occur, officers of some corporations have been sent to prison for this sort of activity. Other countries have similar laws, but degrees of enforcement vary. It seems that there is always a suspicion on the part of the buyers that the sellers are colluding to illegally establish the price at a higher level than a free market would produce. Thus it is common for the public to accuse the oil companies of "setting the price" of gasoline. For the most part, the price of gasoline depends on three factors: (a) the supply of crude oil—and therefore the price of crude oil; (b) the demand for gasoline; and (c) the amount of tax levied by the state and federal government on a gallon of gasoline. While the tax level in the United States is moderate in comparison to countries in Europe and Asia, it constitutes a significant percent of the "pump price" of gasoline. While the federal tax is the same for all states, but state taxes are set by each state and vary somewhat. In Texas, the combined federal and state gasoline tax amounts to about 38.4 cents per gallon, and this constitutes about 37% of the cost of a gallon of gasoline (1999). A "popular" activity is to figure out ways to force the price of gasoline to a lower level. Of the three factors mentioned above, little can be done by the buying public about (a), the price of crude oil, since that is to some degree established by OPEC. Factor (c) is subject to change only by electing officials who feel that the gasoline tax should be reduced (unlikely). Thus, only factor (b), the demand for gasoline, is under any influence of the buying public. However, the public does not, at present, seem inclined to sacrifice anything to achieve better gasmileage and therefore lower gasoline consumption. Although probably an unpopular stand, it could be postulated that the level of gasoline demand is to a significant degree determined by people who drive high gas consumption, four-wheel-drive, 8000-pound vehicles on city streets.

## REFERENCES

1.  C. H. Kline, "Maximizing Profits in Chemicals," *CHEMTECH,* February 1976.
2.  Y. L. Day and G. Hamel, *CHEMTECH,* August 1998, pp. 46–54.
3.  P. Hwang and W. P. Burgers, "The Games Alliances Play," *CHEMTECH,* December 1997, pp. 6–11.
4.  C. E. Ferguson, *Microeconomic Theory,* 3$^{rd}$ ed. Richard D. Irwin, Homewood, IL, 1972.
5.  J. Von Neumann and O. Morgenstern, *Theory of Games and Economic Behavior.* Princeton University Press, Princeton, NJ, 1953.
6.  J. Wei, T. W. F. Russell and M. W. Schwartzlander, *The Structure of the Chemical Process Industries,* McGraw-Hill, Inc., New York, 1979.
7.  Anonymous, *Money,* September 1998, pp 82–93.
8.  Anonymous, "Gripping Tales of M & A," *Chemical Week,* September 10, 1997, pp. 23–40.

## NOTATIONS

| | |
|---|---|
| $AC$ | average cost of producing a chemical ($/lb) |
| $c$ | constant in Eq. (8.6) |
| $K$ | cost of fixed and working capital per year, consisting of depreciation, return on capital (opportunity cost), and maintenance and repair ($/yr) |
| $L$ | cost of labor ($/yr) |
| LRAC | long-range average cost |
| $MC$ | marginal cost = incremental cost of producing product [defined by Eq. (2.12)] |
| $MV_i$ | marginal value of hiring employee of type $i$ |
| $p_s$ | unit cost of supplies per pound of product produced |
| $q$ | quantity of product produced (mass/yr) |
| $S$ | cost of supplies, dollars per year |
| SRAC | short-range average cost |
| $TC$ | total annual production cost |
| $w_i$ | wage of employee $i$ |
| $x_i$ | number of employees of type $i$ needed |

## PROBLEMS

8.1.  Even though the petroleum refining industry is not as concentrated as the auto and aircraft industries, there is public pressure to "break up" (or regulate) the oil companies. What are the reasons behind this movement, and are they justified? Why does the public seem to be less bothered by the much higher concentration in the automotive section?

8.2.   What types of academic training and specialization are particularly appreciated by a firm making a commodity chemical, and why? Answer the same question for firms making pseudocommodity, fine, and specialty chemicals.

8.3.   What are the possible exceptions to the general observations that greater experience leads to lower cost expressed as deflated constant dollars? What are the exceptions to the general observations that greater experience would lead to lower value added expressed as deflated constant dollars?

8.4.   Construct a profit payoff matrix between two firms making Delos (see Section 2.2.3). Firm X has a 30,000-liter reactor and firm Y a 50,000-liter reactor. Let the price be set at three different levels. Estimate the market shares, production levels, and profits of the two firms. Is there a dominant optimal strategy or a mixed optimal strategy?

8.5.   Process engineers are concerned with the development and improvement of manufacturing processes. Product engineers are concerned with the improvement of product properties by blending, introducing additives, and using other techniques. Which courses in the chemistry and chemical engineering curricula would prepare you for work in either of these branches? What elective courses might you take to further your usefulness in either branch of employment?

8.7.   According to the latest statistics, minorities are a small percentage of the CPI work force, and below their demographic proportion of the U.S. population. What might be done to encourage minorities to seek careers as chemical professionals?

8.8.   Suppose a plant's operation can be represented by a Cobb–Douglas production function of

$$q = 0.25K^{0.8} L^{0.4}$$

and suppose that the product demand over a 10-year period is give by

$$q = 0.5 + 0.5t$$

where $t$ takes on values of 1–10.

Determine the optimal value of $K_o$ that will minimize the total cost of manufacturing these products over a 10-year period.

8.9.   A pipeline in the Arctic region is 800 miles long and is serviced by pumping stations every 100 miles. In the initial stages, when the oil flow is less than a million barrels per day, this density of pumping stations conserves capital and pumping expenses. When the oil flow exceeds 2 million barrels per day, the pumping expenses rise steeply, and it would be cheaper to build more pumping stations than to continue to accrue the excess expenses. Sketch quantitatively the cost per barrel delivered versus barrels per day for the SRAC of two different pumping-station densities and the LRAC.

8.10.    Do you believe that the economic rewards and social status of chemical professionals are commensurate with their contributions to society in comparison to doctors, lawyers, and plumbers? Support your answer. If you believe that one profession is unfairly favored over another, what prevents young people from deserting the less favored to enter the more favored profession, to lead to equalization? Do you believe that the salary spread among chemical professionals in the CPI is justified and sustainable?

8.11.    Why is there a larger income spread among self-employed professionals than among those employed in companies?

8.12.    Identify the parameters in Example 2.1 that could be changed and discuss how the chemical professional would alter each to improve profit.

8.13.    Relate the theory of marginal productivity and wages to your own professional life. What type of impact would you strive to make on your company profit, on what time scale, and would it pay your keep? If the answer to the last question is no, what should you do to avoid being laid off? How do you plan to keep up and to expand your usefulness? How do you plan to keep informed about the employment market? How would you find out whether you are underpaid or overpriced?

8.14.    Would a professional union work to your advantage now? Twenty years from now? If a union tends to decrease the income spread between the most and the least accomplished, is it good for the profession and is it good for you? Would a union give you better job protection? Do you believe that a company that cannot reduce forces enough to survive is entitled to shut down altogether, so that all employees lose their jobs?

8.15.    Discuss the relationship between "downsizing" and "outsourcing." What are the pros and cons of each with respect to a company and to an employee? What type of functions do you think can be properly relegated to "outsourcing," and why?

8.16.    The drug sector is consistently more profitable than the stone, clay, and glass sector. Why should investors not shift funds from one to the other until they have equal yields? What are the barriers to entry and exit? Would they persist forever?

8.17.    What would happen if a large company whose policy is to pay out half its income to stockholders suddenly reversed the policy and channeled all profits back into the business? What might be the reaction of common and preferred stockholders? Should this company do this, and if not, why not?

8.18.    Foreign subsidiaries of U.S. companies often have high debt-to-equity ratios. Why? Why are parent firms less concerned about the risk for foreign subsidiaries than for the parent U.S. companies?

8.19.    Do you believe that your friends, relatives, and neighbors understand the contributions to society made by the CPI? How would you respond to comments that the firms in the CPI damage the environment or that a company like Colgate does nothing but put stripes on toothpaste or that PPG makes plate glass? How would you defend the position that firms in the CPI must make reasonable profit to pay for future plants and research?

8.20. Study the contributions made by society by the CPI by developing (a) penicillin, (b) pesticides, (c) biochemically (genetically) produced products, and (d) plastics. Write a short paragraph on each product, discussing the major contributions. Good sources of information are *Chemical Week, Chemical and Engineering News,* and the popular literature.

8.21. Frequently firms in the CPI will select CEOs from outside the CPI sector. What qualifications should such executives have to effectively lead a CPI company? Should they have a degree in chemistry or chemical engineering?

8.22. The salaries, bonuses, and other benefits CEOs, CFOs, and other top executives receive are inordinately high compared to other upper level managerial employees in a firm. In your opinion are such drastic differences justified? Explain your position.

8.23. "Outsourcing" has always been with us. For example, a person's own hair is cut by a hairdresser or barber. What are some of the daily tasks that you outsource?

8.24. At a company, what is the purpose of "outsourcing"? Why are companies doing more "outsourcing"?

# 9

## International Aspects of the CPI

The chemical processing industry is global in scope. Many industrialized nations have a well-developed chemical industry and conduct foreign trade in chemicals, chemical raw materials, and chemical products. A firm in a domestic CPI can regard all foreign nations as potential markets for manufactured chemicals, for technology licensing, and for plant investment. In return, the United States is a market for flow of goods, technology, labor, and investment from foreign countries.

According to the theory of relative advantages, two nations trade because they possess different resources. A nation with large deposits of a natural resource (e.g., crude oil) might wish to trade with a country having a healthy manufacturing industry (e.g. automobiles). Each country has something of benefit to both countries. Relative advantage is a flexible notion, and one finds that nations and people occasionally reorganize their priorities. On a simple level, a professor may hire someone to do his word processing so that he can devote more time to preparing lectures and doing research.

The advantages for the CPI of operating in the United States include an abundant raw material supply derived from farms and mines, a highly skilled and diligent work force, and a large body of technical knowledge in research and engineering, as well as abundant sources of capital and management skills. The United States looks abroad for many things that are insufficient at home, such as crude oil, nickel and tungsten ores, and novel technologies.

Despite the obvious advantages of international commerce, there are various powerful forces to inhibit the movement of goods, technology, capital, and labor. Most nations shelter "inefficient" domestic producers and their employees from superior foreign goods, particularly if these domestic producers qualify as young, start-up industries that require special protection and nurture in their formative years. National security is another impediment to the flow of international commerce. It may be unwise to have a vitally needed raw material supply entirely in the hands of unreliable foreign nations. A nation cannot import more than it exports for an extended period of time without upsetting the balance of payments, which can cause currency devaluation. The free movement of labor may lead to a large immigrant labor force that may disrupt the homogeneity and harmony of the host country, creating unpleasant social and political problems. Such was the case in central Europe, especially in Germany, after World War II. Domination of the economic life of a nation by foreign owners and managers creates difficulties in management style, creates envy, and frequently breeds the threat of nationalization.

The bulk of the world's chemical manufacture is accounted for by a small group of industrialized nations. In 1997, of the world's 50 largest chemical producers, 10 countries, namely, the United States, Germany, Japan, the Netherlands, France, Switzerland, the United Kingdom, Norway, Italy, and Belgium, accounted for 97% of the total world chemical sales (Table 9.1). In 1974 the United States sold about 43% of the chemical market, but this figure has steadily decreased to 28% in 1996. Germany, Japan, The Netherlands, France, and Switzerland have posted strong increased sales in the same time period. In

**TABLE 9.1**   The Leading Countries in Chemical Sales

| Country | Share of market (%)[a] | |
|---|---|---|
| | 1974[b] | 1996[c] |
| United States | 42.7 | 28.3 |
| Germany | 16.5 | 21.1 |
| Japan | 11.4 | 13.5 |
| The Netherlands | 5.2 | 8.0 |
| France | 4.9 | 7.6 |
| Switzerland | 3.1 | 6.8 |
| United Kingdom | 6.5 | 6.7 |
| Norway | NL | 1.7 |
| Italy | 4.2 | 1.7 |
| Belgium | 1.7 | 1.5 |
| | 96.2 | 96.9 |

[a]As measured by percentage of the total sales of the world's 50 largest chemical producers.
[b]From *Chemical Age*, July 22, 1975, p. 53.
[c]From *Chemical and Engineering News*, July 21, 1997, pp. 16–17.

emerging economies, the CPI has been involved in the manufacture of basic products like fertilizers, bulk inorganic chemicals, and in certain instances specialty chemicals for which the national economy enjoys an advantage (e.g., abundant raw materials, low freight for low-priced products).

In this chapter, which is concerned with some of the important features of the world CPI, we consider industries that roughly correspond to the domestic SIC 28 and 29 classifications.

International financial problems created turmoil in global trade in the 1990s. Russia has experienced severe monetary problems since the breakup of the Soviet Union. In the late 1990s, South Korea, Japan, and indeed much of Southeast Asia have had their own financial woes, which have seriously impacted not only U.S. trade but also that of some of the region's trading partners in western Europe. In 1998 Brazil had monetary problems, devalued its currency, and sent trade ripples throughout the Western Hemisphere, upsetting world stock markets.

In reading this chapter, keep in mind that the bulk of the material was written as of a specific date, namely, late 1998. Subsequent global financial problems affect the chapter content. This is another reason for the reader to attempt to keep abreast of current developments in the dynamic world trade using the references cited in Chapter 0.

## 9.1   U.S. FOREIGN TRADE IN CHEMICALS

The chemical industry in colonial North America began a limited export trade in potash and naval stores with the United Kingdom in the early eighteenth century. Today, the United States is the largest exporter in world trade followed closely by Germany, as shown in Table 9.2. However, exports play a small role in the total U.S. economy, amounting to only 7.7% of the GDP. This is in marked contrast to small countries like the Netherlands, Belgium, and Switzerland, where exports amount to 55.4, 52.7, and 50.4% of the GDP, respectively. Small nations like those depend on foreign nations for many of their economic needs. Larger nations, like the United States, have many diverse economic activities and tend to be more self-sufficient. Of particular interest is Japan, which despite its small area is one of the leading exporters in the world. Its exports account for 13.5% of its GDP, and Japan's domestic market is one of the largest in the world.

The U.S. chemical industry competes globally, and its share of international markets is increasingly in unique products in which it has distinct technological advantages. In 1995 the U.S. foreign trade in chemicals amounted to $61.7 billion in exports and $40.4 billion in imports, leaving a positive balance of trade of $21.3 billion in chemicals. This fact is particularly significant in view of the low or negative trade balances in the United States as a whole. Historically, the chemical exports have exceeded the chemical imports. Exports in chemicals accounted for about 10% of all U.S. exports in 1990, 1993, and 1994 (Table 9.3). The U.S.

**TABLE 9.2** World Trade: 1996

| Country | GDP (billions)[a] | Export (billions) | Exports as percentage of GDP |
|---|---|---|---|
| United States | $7,610 | $584.7 | 7.7 |
| Germany | $1,700 | $501.3 | 29.5 |
| France | $1,220 | $275.0 | 22.5 |
| United Kingdom | $1,196 | $240.4 | 20.1 |
| Italy | $1,120 | $250.0 | 22.3 |
| Netherlands | $318 | $176.2 | 55.4 |
| Switzerland | $161 | $81.4 | 50.4 |
| Canada | $721 | $195.4 | 27.1 |
| Japan | $1,850 | $385.0 | 20.8 |
| Russia | $767 | $88.3 | 11.5 |
| China | $3,390 | $151.0 | 4.5 |
| Belgium | $205 | $108.0 | 52.7 |

[a]Estimated.
Source: Central Intelligence Agency, The World Factbook, 1997.

market share of world exports in all manufacturing and chemicals is found in Table 9.4, however, the country's share of world trade has declined in recent years even though chemicals as a group have remained at about 10% of the world trade. In 1996 total exports of chemicals reached $62.9 billion or about 13% of the total world chemical exports of $475 billion. U.S. imports of chemicals also increased to $44.9 billion in 1996, which leaves a positive balance of trade of $21.2 billion. The U.S. chemical industry has traditionally maintained trade surpluses even in the second half of the 1980s, a period when the overall U.S. balance-of-trade deficit was often over $100 billion per year (1).

In comparison with the chemical trade of other nations, the United States accounted for 14.1% of the dollar value of world exports of chemicals in 1995; only Germany exports more chemicals (measured in dollar value) than the United States, or about 16.1% of the world exports, as shown in Table 9.5. Canada represents the largest single market for U.S. chemical exports, followed by Japan. U.S. chemical exports to Mexico, the third largest export market, has grown 280% in the 10 years from 1986 to 1996. Canada is the largest exporter of chemicals to the United States: about $8.2 billion in 1996.

With respect to imports, Germany imports a greater dollar volume than the United States. The dollar value of the U.S. chemical trade with specific countries is presented in Table 9.6. In 1995 the chemical balance of trade with Australia, Latin America, Africa, and Asia (with the exception of Japan and China) was greater than 4:1. With the more highly industrialized nations, like Canada, the

TABLE 9.3 U.S. Trade (millions of dollars)

|  | 1975 | 1980 | 1985 | 1990 | 1993 | 1994 |
|---|---|---|---|---|---|---|
| All commodities |  |  |  |  |  |  |
| Exports | $107,191 | $200,600 | $85,300 | $394,000 | $456,900 | $503,000 |
| Imports | $96,140 | $240,800 | $77,100 | $495,000 | $589,000 | $669,000 |
| Trade balance | $11,051 | ($40,200) | $8,200 | ($101,000) | ($132,100) | ($166,000) |
| Chemicals |  |  |  |  |  |  |
| Exports | $8,705 | $20,740 | $3,988 | $38,983 | $45,066 | $51,600 |
| Imports | $3,696 | $8,594 | $3,997 | $22,468 | $28,502 | $33,925 |

TABLE 9.4 U.S. Market Share of World Exports in All Manufacturing and Chemicals

|  | All manufacturing | | Chemicals | |
|---|---|---|---|---|
| Year | U.S. share (millions) | Percentage of world | U.S. share (millions) | Percentage of world |
| 1974 | $64,559 | 20.3 | $8,891 | 18.5[a] |
| 1980 | $143,000 | 12.8 | $20,740 | 14.7[b] |
| 1985 | $185,000 | 10.7 | $21,759 | 14.2[b] |
| 1990 | $394,000 | 11.3 | $38,983 | 13.2[b] |
| 1995 | $584,700 | NA | $61,701 | 14.1[b] |

[a]From U.S. Bureau of the Census, *Statistical Abstracts 1977*, Table 1303.
[b]From *Chemical and Engineering News*, Facts and Figures issues, June 24, 1991, and June 23, 1007.

countries of western Europe, and Japan, the balance of chemical trade was more nearly 1:1.

U.S. trade of specific product groups is shown in Table 9.7. Four categories, namely, organic chemicals, medicinals and pharmaceuticals, inorganic chemicals, and plastics, account for 69% of the U.S. imports. Organic chemicals, plastics, pharmaceuticals and medicinals, and inorganic chemicals accounted for about 55% of the U.S. chemical exports in 1995.

In 1973, just prior to the Mideast conflict, the cost of a barrel of crude oil was $4.08 and petroleum was 26% of the United States total imports. By 1975 the price had risen to $13.93 per barrel and petroleum was 33% of the total import. In 1997 the price of crude oil was in the range of $19–20 per barrel and imported petroleum is about 55% of the total U.S. imports. In early 1999, owing to the glut of crude oil, the price dropped to $9–11 per barrel.

**TABLE 9.5**  World Chemical Trade (millions of dollars)

|  | 1974 | | 1980 | |
|---|---|---|---|---|
|  | Exports | Imports | Exports | Imports |
| World | $58,000 | $58,000 | $140,748 | $14,748 |
| United States | $8,822 | $3,991 | $20,740 | $8,594 |
| Canada | $1,069 | $1,857 | $3,519 | $3,249 |
| United Kingdom | $5,994 | $4,372 | $12,303 | $7,310 |
| Germany | $11,918 | $5,053 | $24,318 | $13,314 |
| France | $5,137 | $5,009 | $13,249 | $12,169 |
| Italy | $3,793 | $4,070 | $5,505 | $7,993 |
| Netherlands | $4,687 | $2,322 | $11,288 | $6,3934 |
| Switzerland | $2,705 | $1,711 | $5,680 | $3,648 |
| Japan | $4,059 | $2,668 | $6,767 | $6,202 |

|  | 1992 | | 1993 | |
|---|---|---|---|---|
|  | Exports | Imports | Exports | Imports |
| World | $333,245 | $333,245 | $330,291 | $330,291 |
| United States | $43,956 | $27,684 | $45,066 | $28,502 |
| Canada | $4,738 | $6,909 | $5,447 | $8,578 |
| United Kingdom | $23,762 | $18,491 | $27,940 | $20,994 |
| Germany | $60,814 | $39,812 | $57,868 | $34,289 |
| France | $31,343 | $26,312 | $31,795 | $24,540 |
| Italy | $10,661 | $17,275 | $13,192 | $18,590 |
| Netherlands | $26,200 | $16,813 | $24,102 | $16,628 |
| Switzerland | $17,977 | $9,685 | $18,899 | $10,025 |
| Japan | $25,760 | $23,379 | $23,910 | $21,316 |

*Source: Chemical and Engineering News,* Facts and Figures issues for respective years.

Since the Mideast crises, the import of crude oil has had a major impact on the U.S. balance of trade (Tables 9.3 and 9.8), and the 1999 price drop has further exacerbated this problem.

Foreign trade in chemicals is influenced greatly by trade tariffs. A tariff is a charge imposed by a government on imports and occasionally on exports. Free-market economists strongly oppose tariffs, but they are a fact of life and are imposed for the following reasons:

To "keep the money at home" by reducing the ability of foreign products to compete with domestically made products.

To protect domestic wage earners' jobs.

**TABLE 9.5**

| 1985 | | 1990 | |
|---|---|---|---|
| Exports | Imports | Exports | Imports |
| $153,000 | $153,000 | $295,100 | $295,100 |
| $21,759 | $14,533 | $38,983 | $22,468 |
| $4,040 | $4,454 | $6,668 | $8,011 |
| $12,174 | $7,425 | $23,415 | $19,274 |
| $24,253 | $13,858 | $49,563 | $29,944 |
| $13,715 | $10,854 | $28,173 | $24,894 |
| $6,608 | $8,518 | $11,152 | $20,260 |
| $11,324 | $6,825 | $20,164 | $14,224 |
| $5,879 | $3,664 | $13,749 | $7,928 |
| $7,697 | $8,072 | $15,872 | $16,045 |

| 1994 | | 1995 | | Percentage change, 1994–1995 | |
|---|---|---|---|---|---|
| Exports | Imports | Exports | Imports | Exports | Imports |
| $379,835 | $379,835 | $436,810 | $436,810 | 15 | 15 |
| $51,600 | $33,915 | $61,701 | $40,378 | 20 | 19 |
| $7,120 | $10,773 | $8,882 | $12,405 | 25 | 15 |
| $29,847 | $22,590 | $33,585 | $27,343 | 13 | 21 |
| $66,305 | $39,139 | $70,477 | $43,263 | 6 | 11 |
| $36,088 | $28,577 | $40,821 | $32,673 | 13 | 14 |
| $15,225 | $22,340 | $19,575 | $27,553 | 29 | 23 |
| $28,685 | $18,624 | $32,327 | $20,988 | 13 | 13 |
| $19,866 | $10,605 | $20,322 | $10,978 | 2 | 4 |
| $25,239 | $21,220 | $30,077 | $24,548 | 19 | 16 |

To protect special interest groups.

To provide revenue.

To retaliate against a foreign nation.

To protect "essential" industries.

To protect "new" industries until they are established.

To prevent inferior or unsafe foreign products from entering the domestic market, especially those that are potentially environmentally hazardous or could cause product liability problems.

Historically, tariffs have played an essential role in developing the U.S. chemical industry. In the early twentieth century, tariffs on coal tar intermedi-

**TABLE 9.6**  U.S. Chemical Trade by World Area (millions of dollars)

|                  | 1975         |              | 1980         |              |
|------------------|--------------|--------------|--------------|--------------|
|                  | Exports      | Imports      | Exports      | Imports      |
| European Union   | $2,651       | $1,995       | $5,149       | $3,372       |
| Canada           | $1,213       | $760         | $2,134       | $2,494       |
| Latin America    | $2,245       | $220         | $4,846       | $536         |
| Japan            | $766         | $487         | $1,964       | $697         |
| Communist Asia   | NA           | NA           | $386         | $107         |
| Rest of Asia     | $973         | $80          | $2,782       | $163         |
| Australia        | $351         | $195         | $645         | $206         |
| Middle East      | $181         | $16          | $459         | $58          |
| Africa           | $270         | $29          | $721         | $62          |
| Eastern Europe   | $54          | $42          | $88          | $195         |
| Other            | $154         | $167         | $1,566       | $704         |
| Total            | $8,858       | $3,991       | $20,740      | $8,594       |

|                  | 1992         |              | 1993         |              |
|------------------|--------------|--------------|--------------|--------------|
|                  | Exports      | Imports      | Exports      | Imports      |
| European Union   | $12,050      | $12,067      | $11,599      | $12,249      |
| Canada           | $7,285       | $4,950       | $8,224       | $5,494       |
| Latin America    | $7,222       | $1,582       | $7,715       | $1,737       |
| Japan            | $4,708       | $3,175       | $4,966       | $3,572       |
| Communist Asia   | $1,208       | $509         | $843         | $576         |
| Rest of Asia     | $7,054       | $1,589       | $7,431       | $1,676       |
| Australia        | $1,432       | $332         | $1,218       | $177         |
| Middle East      | $766         | $408         | $762         | $506         |
| Africa           | $609         | $108         | $607         | $126         |
| Eastern Europe   | $174         | $270         | $263         | $429         |
| Other            | $1,448       | $2,694       | $1,438       | $1,960       |
| Total            | $43,956      | $27,684      | $45,066      | $28,502      |

[a]Average.
*Source: Chemical and Engineering News,* Facts and Figures issues for respective years.

TABLE 9.6

| 1985 | | 1990 | |
| --- | --- | --- | --- |
| Exports | Imports | Exports | Imports |
| $6,742 | $7,118 | $10,509 | $9,728 |
| $2,698 | $2,913 | $6,050 | $4,305 |
| $3,681 | $1,216 | $5,773 | $1,267 |
| $2,908 | $1,412 | $4,582 | $2,391 |
| $386 | $107 | $1,054 | $337 |
| $3,024 | $457 | $6,566 | $1,267 |
| $704 | $231 | $1,247 | $344 |
| $400 | $218 | $629 | $355 |
| $456 | $203 | $628 | $90 |
| $368 | $380 | $460 | $242 |
| $265 | $222 | $1,485 | $2,122 |
| $21,632 | $14,477 | $38,983 | $22,448 |

| 1994 | | 1995 | | Percentage change, 1994–1995 | |
| --- | --- | --- | --- | --- | --- |
| Exports | Imports | Exports | Imports | Exports | Imports |
| $12,868 | $13,574 | $14,682 | $15,698 | 14 | 16 |
| $9,416 | $6,679 | $10,653 | $8,150 | 13 | 22 |
| $9,337 | $2,290 | $11,465 | $2,859 | 23 | 25 |
| $5,201 | $4,181 | $6,129 | $5,107 | 18 | 22 |
| $1,529 | $741 | $2,027 | $876 | 33 | 18 |
| $8,500 | $1,917 | $11,273 | $2,168 | 33 | 13 |
| $1,270 | $270 | $1,555 | $227 | 22 | −16 |
| $777 | $766 | $900 | $958 | 16 | 25 |
| $577 | $148 | $610 | $112 | 6 | −24 |
| $246 | $802 | $273 | $1,019 | 11 | 27 |
| $1,878 | $2,039 | $2,547 | $3,204 | 14 | 26 |
| $51,599 | $33,407 | $62,114 | $40,378 | 20%[a] | 19%[a] |

**TABLE 9.7**  Dollar Volume (millions) of U.S. Chemical Trade by Product Classification[a]

|  | 1974 | | 1980 | |
| --- | --- | --- | --- | --- |
|  | Exports | Imports | Exports | Imports |
| Organic chemicals | $2,568 | $1,380 | $5,697 | $2,541 |
| Plastics in primary form | $1,618 | $316 | $3,884 | $647 |
| Medicinals and pharmaceuticals | $800 | $212 | $1,932 | $509 |
| Inorganic chemicals | $835 | $742 | $2,939 | $2,314 |
| Plastics in nonprimary form | —[b] | —[b] | —[b] | —[b] |
| Essential oils and perfume materials | $110 | $155 | $806 | $374 |
| Dyeing, tanning, and color materials | $6 | $10 | $490 | $284 |
| Fertilizers, manufactured | $812 | $573 | $2,265 | $1,105 |
| Other | $2,073 | $603 | $2,707 | $820 |
| Total | $8,858 | $3,991 | $20,740 | $8,594 |

|  | 1992 | | 1993 | |
| --- | --- | --- | --- | --- |
|  | Exports | Imports | Exports | Imports |
| Organic chemicals | $10,993 | $9,366 | $11,076 | $9,279 |
| Plastics in primary form | $7,178 | $2,247 | $7,224 | $2,520 |
| Medicinals and pharmaceuticals | $5,357 | $3,812 | $5,750 | $4,135 |
| Inorganic chemicals | $4,119 | $3,300 | $3,809 | $3,284 |
| Plastics in nonprimary form | $3,076 | $2,046 | $3,517 | $2,328 |
| Essential oils and perfume materials | $2,631 | $1,712 | $3,046 | $1,808 |
| Dyeing, tanning, and color materials | $1,867 | $1,626 | $2,013 | $1,700 |
| Fertilizers, manufactured | $2,372 | $952 | $1,797 | $1,136 |
| Other | $6,363 | $2,623 | $6,834 | $2,312 |
| Total | $43,956 | $27,684 | $45,066 | $28,502 |

[a]Data for 1979–1990, *Chemical and Engineering News,* Facts and Figures issue for respective year. Data for 1992–1995, U.S. Department of Commerce.
[b]No such classification at that time.
[c]Average.

**TABLE 9.7**

| 1985 | | 1990 | |
|---|---|---|---|
| Exports | Imports | Exports | Imports |
| $6,012 | $4,576 | $10,400 | $7,392 |
| $3,777 | $1,691 | $6,428 | $1,974 |
| | | | |
| $2,708 | $1,084 | $4,103 | $2,500 |
| $1,967 | $l,984 | $3,816 | $3,234 |
| | | | |
| —[b] | —[b] | $2,633 | $1,774 |
| | | | |
| $478 | $577 | $1,963 | $1,323 |
| | | | |
| $139 | $352 | $1,588 | $1,288 |
| $2,160 | $967 | $2,574 | $955 |
| $4,518 | $3,302 | $5,478 | $2,028 |
| $21,759 | $14,533 | $38,983 | $22,468 |

| 1994 | | 1995 | | Percentage change, 1994–1995 | |
|---|---|---|---|---|---|
| Exports | Imports | Exports | Imports | Exports | Imports |
| $12,826 | $10,799 | $16,395 | $13,325 | 28 | 23 |
| $8,675 | $3,512 | $10,756 | $4,331 | 24 | 23 |
| | | | | | |
| $6,093 | $4,680 | $6,554 | $5,544 | 8 | 18 |
| $4,068 | $4,105 | $4,581 | $4,655 | 13 | 13 |
| | | | | | |
| $3,810 | $2,422 | $1,337 | $2,821 | 14 | 16 |
| | | | | | |
| $3,536 | $1,995 | $3,943 | $2,309 | 12 | 16 |
| | | | | | |
| $2,323 | $1,868 | $2,621 | $2,080 | 13 | 11 |
| | | | | | |
| $2,699 | $1,296 | $3,226 | $1,392 | 20 | 7 |
| $7,570 | $3,238 | $9,288 | $3,921 | 23 | 21 |
| $51,600 | $33,915 | $58,701 | $40,378 | 20[c] | 19[c] |

**TABLE 9.8**  Petroleum Imports by the United States

| Year | Domestic production (bbl × 10³/day) | Total imports (bbl × 10³/day) | Percentage imports | Cost of imports (U.S.$/bbl) | Annual cost of oil imports (U.S.$ × 10⁶) |
|------|------|------|------|------|------|
| 1973 | 9208 | 3244 | 26% | $4.08 | $4,831 |
| 1974 | 8774 | 3477 | 28% | $12.52 | $15,889 |
| 1975 | 8375 | 4105 | 33% | $13.93 | $20,872 |
| 1976 | 8132 | 5287 | 39% | $13.48 | $26,013 |
| 1977 | 8245 | 6615 | 45% | $14.53 | $35,082 |
| 1978 | 8707 | 6356 | 42% | $14.57 | $33,802 |
| 1979 | 8552 | 6519 | 43% | $21.67 | $51,562 |
| 1980 | 8597 | 5263 | 38% | $33.89 | $65,103 |
| 1981 | 8572 | 4396 | 34% | $37.05 | $59,448 |
| 1982 | 8649 | 3488 | 29% | $33.50 | $42,650 |
| 1983 | 8688 | 3329 | 28% | $29.30 | $35,602 |
| 1984 | 8879 | 3426 | 28% | $28.88 | $36,114 |
| 1985 | 8971 | 3201 | 26% | $26.99 | $31,534 |
| 1986 | 8680 | 4178 | 32% | $14.00 | $21,350 |
| 1987 | 8349 | 4674 | 36% | $18.13 | $30,930 |
| 1988 | 8140 | 5107 | 39% | $14.56 | $27,141 |
| 1989 | 7613 | 5843 | 43% | $18.08 | $38,559 |
| 1990 | 7355 | 5894 | 44% | $21.76 | $46,813 |
| 1991 | 7417 | 5782 | 44% | $18.70 | $39,465 |
| 1992 | 7171 | 6083 | 46% | $18.20 | $40,409 |
| 1993 | 6847 | 6787 | 50% | $16.14 | $39,983 |
| 1994 | 6662 | 7063 | 51% | $15.51 | $39,985 |
| 1995 | 6560 | 7230 | 52% | $17.14 | $45,232 |
| 1996 | 6471 | 7482 | 54% | $20.55 | $56,121 |

*Source: Monthly Energy Review,* Energy Information Administration, March 1997.

ates, manufactured dyes, and synthetic medicinals were especially vital in encouraging the growth of the country's fledgling chemical industry. In 1921 Congress enacted the Dye and Chemical Control Act, particularly aimed at the German competition. A 6-month embargo was placed on foreign products for which there was a comparable domestic product available at reasonable terms. In 1934, with the passage of the Trade Agreement Act, tariffs were lowered progressively on chemicals and allied products.

Tariff reductions have been negotiated on an international scale through a series conferences held in the United States, Europe, and Japan by the UN agency known as GATT, the General Agreement on Tariffs and Trade. The impetus for these meetings was the fear of possible trade wars using tariff protection. GATT's intent was to discourage destructive trade protection.

In North America, a free-trade agreement was signed in the mid-1990s by Canada, Mexico, and the United States to eliminate tariffs between these countries. It is too early to assess the effects of this agreement, called after its enabling legislation: the North America Free Trade Act (NAFTA).

## 9.2 MOVEMENT OF CAPITAL INVESTMENTS BETWEEN THE UNITED STATES AND ABROAD

Investments by the CPI abroad through wholly owned subsidiaries, joint ventures, or partnerships with local entrepreneurs has been increasing. The reasons for the increased foreign investments have included the desire (a) to be near important natural resources, (b) to be near important and expanding markets, (c) to minimize or avoid import tariffs and quotas, (d) to participate in local markets, (e) to take advantage of more favorable labor markets, and (f) to avoid shipping costs from the United States.

The book value of U.S. direct investments abroad for 1980, 1990, and 1994 are presented in Table 9.9. Although the figures have varied slightly, ap-

TABLE 9.9   Value of U.S. Investments Abroad (billions)

|  | 1980 | 1990 | 1994 |
|---|---|---|---|
| Total | $215.375 | $430.521 | $612.019 |
| Canada | $42.119 | $69.508 | $72.808 |
| Europe | $96.287 | $214.739 | $300.177 |
|   Germany | $15.415 | $27.609 | $39.886 |
|   United Kingdom | $28.480 | $72.707 | $102.244 |
|   France | $9.347 | $19.164 | $27.894 |
| Latin America | $39.581 | $71.413 | $114.986 |
|   Mexico | $5.986 | $10.313 | $16.375 |
|   Brazil | $7.704 | $14.304 | $18.977 |
|   Argentina | $2.540 | $2.531 | $5.666 |
|   Australia | $7.654 | $15.110 | $20.504 |
| Asia |  |  |  |
|   Japan | $6.225 | $22.599 | $37.027 |
|   South Korea | $0.575 | $2.694 | $3.512 |
|   Singapore | $1.204 | $3.975 | $10.972 |
|   Malaysia | $0.632 | $1.466 | $2.382 |
| Other Asia | $22.963 | $64.716 | $72.218 |

Source: U.S. Bureau of the Census, Statistical Abstract of the United States, 1997, Table 1293.

proximately 70% of the U.S. direct investment in these years has been in Canada, Europe, Japan, and Australia. The investments in petroleum and mining are high in nations that have rich natural resources. The value of direct foreign investments in the United States is significant, as shown in Table 9.10. The largest investments from Canada and Europe are in the manufacturing sector.

In 1994 the U.S. direct foreign investment abroad for chemicals and allied products amounted to $51.6 billion, with 68% of that amount being invested in Europe and Canada (see Table 9.11). Investment in Japan and the Far East is on the increase. Up until the mid-1970s, Japanese law did not permit more than 50% ownership of a company by foreign interests, but this law was relaxed so that in the 1980s and 1990s there was increased foreign investment. In Europe, the concentration of the chemical industry investment has been in the United Kingdom, Germany, France, and Belgium. Nations that were once dominated by the Soviet Union, namely, Poland, Bulgaria, the Czech Republic, Slovakia, Hungary, and Romania, now comprise a new frontier for European investment. The bulk of the U.S. investment in Latin America has been in Mexico, Brazil, and Argentina. Some companies that have invested substantially in Europe are Exxon, Monsanto, Dow, and Du Pont. Most of these companies had plans to invest heavily in the developing areas in the Far East, but many such plans were put on hold in the late 1990s because of that area's financial crises.

Foreign direct investment in chemicals and allied products in the United States has increased more than sixfold from 1980 through 1994, as shown in Table 9.12. New investment outlays by foreign companies amounted to about $80 billion in 1996 but declined to about $70 billion in 1997. European companies in the United Kingdom and Germany are the heaviest investors. Japanese

**TABLE 9.10**  Value of Direct Foreign Investment in the United States 1994 (billions)

| Country or area | Manufacturing | Petroleum | Finance and insurance | Total |
|---|---|---|---|---|
| Canada | $16.911 | $2.585 | $8.496 | $43.223 |
| United Kingdom | $48.190 | $10.906 | $16.343 | $113.504 |
| Netherlands | $19.881 | $12.770 | — | $70.645 |
| Switzerland | $13.111 | — | $6.923 | $25.330 |
| Germany | $21.321 | — | $5.508 | $39.550 |
| Other Europe | $34.301 | $1.804 | $5.398 | $63.847 |
| Japan | — | — | — | $103.120 |
| All areas | $184.484 | $34.048 | $76.040 | $504.401 |

*Source:* U.S. Bureau of the Census, *Statistical Abstract of the United States,* 1997, Table 1288.

TABLE 9.11 U.S. Direct
Investment Abroad in Chemical
and Allied Products (millions: 1994)

| | |
|---|---|
| All countries | $51,638 |
| Canada | $5,856 |
| Europe | $29,239 |
| Belgium | $5,104 |
| France | $4,152 |
| Germany | $2,448 |
| Ireland | $3,870 |
| Italy | $751 |
| Netherlands | $264 |
| Spain | $4,560 |
| Switzerland | $4,657 |
| United Kingdom | $2,628 |
| Latin America | $6,534 |
| Argentina | $561 |
| Brazil | $2,268 |
| Mexico | $2,169 |
| Asia and Pacific | $8,870 |
| Australia | $2,445 |
| China | $186 |
| Japan | $3,634 |
| South Korea | $291 |
| India | $100 |
| Indonesia | $83 |
| Malaysia | $66 |
| Philippines | $417 |
| Singapore | $122 |
| Taiwan | $894 |
| Thailand | $290 |

Source: U.S. Department of Commerce, Survey of Current Business, Vol. 75, No. 8, August 1996, Table 11.3, p. 87.

companies have increased their investments 15-fold in the same time period. Among the largest foreign investors are Hoechst, BASF, Bayer, ICI, Rhône-Poulenc, Pharmacia, and Mitsubishi Chemicals.

U.S. foreign investment from 1980 to 1994 by industry sector is found in Table 9.13. Foreign investment in all U.S. industries has increased over this same period about 2.5 times, from $230 billion to $612 billion. Although the in-

**TABLE 9.12** Direct Foreign Investment in the United States for Chemicals and Allied Products (millions)[a]

| Area or country | 1980[b] | 1985[b] | 1990[b] | 1994[b,c] |
|---|---|---|---|---|
| All countries | $10,439 | $19,502 | $41,678 | $67,303 |
| Canada | $116 | $227 | $508 | $1,018 |
| Europe | $8,163 | $17,150 | $34,604 | $59,668 |
| France | $633 | $2,820 | $4,016 | $7,820 |
| Netherlands | $2,002 | $3,488 | $8,126 | $8,359 |
| United Kingdom | $2,301 | $4,035 | $8,882 | $22,147 |
| Sweden | $67 | D | $589 | $1,104 |
| Switzerland | $1,154 | $2,185 | $2,656 | $6,502 |
| Germany | $1,815 | $3,838 | $8,552 | $11,338 |
| Africa | ($2) | NA | D | NA |
| Latin America | $1,929 | $1,876 | $1,821 | $2,327 |
| Middle East | $1 | ($1) | D | |
| Asia and Pacific (excluding Japan) | $5 | NA | $4,483 | $3,992 |
| Japan | $224 | $267 | $3,918 | $3,456 |

[a]D, data withheld to avoid disclosure of companies; NA, data not reported.
[b]U.S. Bureau of the Census, *Statistical Abstracts of the United States*, 1987, 1990, 1992, 1996, Tables 1316, 1309, 1257, 1226, respectively.
[c]U.S. Department of Commerce, *Survey of Current Business*, Vol. 75, No. 8, August 1996, Table 10.4, p. 62.

**TABLE 9.13** U.S. Direct Investment Abroad by Industry (millions, U.S.)

| Industry | 1980[a] | 1985[b] | 1990[b] | 1994[b] |
|---|---|---|---|---|
| All industries | $215,578 | $229,748 | $430,521 | $612,109 |
| Chemical | $18,888 | $20,372 | $37,988 | $51,638 |
| Petroleum | $47,595 | $58,030 | $52,826 | $65,711 |
| Food and kindred products | $8,277 | $9,266 | $15,570 | $28,796 |
| Paper and allied products | NA | $4,814 | $9,954 | $11,942 |
| Rubber | NA | $2,215 | $3,367 | $4,581 |
| Miscellaneous plastics products | NA | $1,840 | $2,422 | $3,779 |
| Drugs | NA | $5,016 | $9,314 | $15,472 |
| Primary and fabricated metals | $6,322 | $5,214 | $10,520 | $10,974 |

[a]From U.S. Department of Commerce, *Survey of Current Business*, August 1982.
[b]From U.S. Department of Commerce, *Survey of Current Business*, Vol. 75, No. 8, August 1995, Table 18, pp. 115–116.

vestment in the chemical sector increased about the same as the all-industry increase, it is interesting to note that the greatest increases have been in drugs and food products segments of the CPI (by three times each from 1985 to 1994). U.S. investment abroad is expected to increase to take advantage of lower operating expenses and the proximity of new foreign markets.

There are several problems associated with investing in CPI plants overseas. Some major problems are as follows:

Unfamiliar laws and customs

Competition from local companies, including companies that are government-owned and favored

Political instability, anti-American sentiment, unfair treatment, and nationalization of companies

Economic problems (e.g., rampant and chronic inflation, currency fluctuation and/or devaluation, difficulty of withdrawing funds from a country once the funds have been invested)

Lack of necessary stable, skilled labor and management talent; labor unrest

Lack of satisfactory infrastructure (e.g., roads, harbors, telecommunications, adequate supplies of electricity and water, banks and insurance, schools, doctors, and medical facilities), especially in developing nations

In addition to the problems cited above, in the past, U.S. firms investing abroad faced difficulties with the federal government regarding capital outflow, which was considered detrimental to the national economy because it could cause a balance-of-payments deficit. At first, large corporations voluntarily limited the capital outflow but, in 1968, mandatory programs were in place to limit the capital outflow ensure that the practice would continue. As a result, chemical companies were forced to finance foreign operations by selling bonds and/or taking out loans abroad. To raise capital, U.S. companies sold substantial amounts in the foreign bond markets, which led to highly leveraged foreign affiliates. To alleviate some of the problems, U.S. companies began to structure their foreign investments in the form of joint ventures, frequently on a 50/50 ownership basis. However, certain foreign governments (e.g., Spain and Japan) have restrictions on the percentage that a foreign company may own.

A U.S. company in a joint venture often provides a great deal of knowledge, technology, experience, and business skills. A foreign company in such a venture provides local knowledge, facilities, experience, sales organization, and financing. A foreign partner can help with local customs, laws, building codes, and hiring personnel, as well as by establishing good relationships with customers, government, and lenders. Some countries require joint ownership by the government, which may indeed be the required partner.

The U.S. government in more recent years has taken the position that while foreign investments may initially drain capital (i.e., cause capital outflow), in the long run the capital returning in the form of profits and foreign purchases may offset the initial drain. In general, foreign operations of the CPI are characterized by heavy leveraging.

Another difficulty the U.S. chemical firm has in competing with foreign companies is that the federal government and the U.S. firm do not always work together. The governments of Japan and German attempt to cooperate with the overseas operations in financial and legal matters, whereas the U.S. government does not. Although lack of government cooperation may be desirable in some respects, it lessens the amount of protection to be expected from the government. In some instances U.S. firms and the U.S. government are working against each other in overseas operations.

Labor cost and labor productivity are important factors in considerations of overseas operations. In the past labor was less expensive in Canada, the United Kingdom, Australia, and most Asian countries, as shown in Table 9.14. Exceptions are Germany and most recently Japan. Labor productivity is a very important consideration. In most industrialized nations, labor productivity is high; however, some anomalies exist. For example, if labor rates are low in a country, one should look at the productivity. If it is quite low and lags behind that of other nations, then the labor cost per unit of production may in fact be higher

**TABLE 9.14**  Index of Hourly Compensation Costs for Production Workers in Manufacturing: 1974–1994 (United States = 100)

| Country or area | 1974 | 1985 | 1990 | 1996 |
|---|---|---|---|---|
| Canada | 94 | 88 | 106 | 92 |
| Germany | NA | 125 | 147 | 160 |
| Japan | 47 | 56 | 86 | 125 |
| United Kingdom | 55 | 77 | 85 | 80 |
| France | 71 | 91 | 102 | 100 |
| Italy | 73 | 83 | 119 | 95 |
| Australia | 88 | 86 | 88 | 80 |
| South Korea | 5 | 10 | 25 | 37 |
| Singapore | 13 | 15 | 25 | 37 |
| Taiwan | 6 | 10 | 26 | 32 |
| Mexico | 23 | 22 | 11 | 15 |
| Other Asian countries | 8 | 12 | 25 | 34 |

*Source:* U.S. Bureau of Labor Statistics, *Monthly Labor Review,* "Index of Hourly Compensation Costs of Production, 1975–1994," Vol. 118, No. 10. October 1995, p. 6.

than in a country with high labor rates. This was the situation in the United Kingdom in the 1970s and 1980s. The United States generally has a low degree of labor unrest as well as political instability, and the U.S. government is generally favorable to business.

In recent years, the flow of foreign investments in the United States has increased substantially. The major investors have been Japan, Germany, France, and the United Kingdom. The United States is particularly attractive to overseas investors because of the large markets for products, the availability of certain raw materials, and a diligent, well-educated labor pool, which has a high productivity. Further, the United States has political stability. The wealthy nations of OPEC, as well as Germany, the United Kingdom, and Japan, find the United States a favorable place to invest, although the investments have been in the form of debt rather than equity.

Some of the largest investors in the U.S. CPI are as follows:

Germany
  BASF
  Hoechst (in Hoechst Celanese)
  Bayer (in Mobay)
United Kingdom
  British Petroleum (in Standard Oil of Ohio and Amoco and ARCO)
  Unilever (in Lever Brothers)
  ICI
Japan
  Mitsubishi Chemicals
  Mitsui Petrochemical
  Asahi Chemical
France
  Rhône-Poulenc
  Elf Aquitaine
  Air Liquide
The Netherlands
  Royal Dutch Shell (in Shell Oil Company)
  Akzo Nobel (in Akzona)
  DSM
Switzerland
  Ciba-Geigy
  Hoffmann-La Roche
  Novartis
  Nestlé

In some areas of chemical manufacturing, foreign-controlled capacity amounts to a significant percentage of the U.S. capacity (Table 9.15).

**TABLE 9.15**   Foreign-Controlled Capacity: Selected Chemicals

| Chemical | 1996 U.S. capacity ($1b \times 10^6$) | Foreign controlled capacity, (%) | Companies |
|---|---|---|---|
| Polystyrene | 7,413 | 36 | Petrofina, BASF, Nova |
| ABS | 895 | 36 | Bayer |
| Ethylene oxide | 2,995 | 29 | Hoechst, Formosa, Shell |
| PVC | 6,794 | 37 | Shintech, Formosa |
| Polypropylene | 5,683 | 24 | Mitsubishi, PEMEX, Petrofina, Solvay |
| Low-density PE | 3,680 | <10 | — |
| Linear LDPE | 6,800 | 6 | Petrofina, Petromont |
| High-density PE | 5,312 | 25 | Solvay, PEMEX, Petrofina, Formosa |

*Sources; Chemical Week*, Product Focus, various issues, late 1996.

## 9.3   MOVEMENT OF TECHNOLOGY AND LABOR BETWEEN THE UNITED STATES AND ABROAD

Technological innovations in the United States are admired globally. Some of the most important technologies include new polymers and fibers, microchips for computers, antibiotics, and pharmaceuticals, to name a few. These discoveries and inventions have led to technology transfer abroad, earning licensing fees, royalties, and publicity.

The free flow of technologies between nations is a fairly recent phenomenon. In ancient times, the Chinese prevented the movement of silkworms and mulberry seeds abroad, permitting only the export of raw and finished silk. The Venetians protected their fine art of glassblowing by forbidding their artisans to leave Venice to instruct others in the techniques. The current international exchange of technologies is considered to be beneficial to all.

It is important to remember that the flow of technologies is a two-way street. Some important technologies that the United States learned from abroad are as follows:

| Product | Country | Inventor |
|---|---|---|
| Penicillin | United Kingdom | Fleming |
| Silica-alumina cracking catalysts | France | Houdry |
| Oxygen steel furnace | Austria | — |
| Low-density polyethylene | United Kingdom | ICI |
| High-density polyethylene | Italy, Germany | Natta, Ziegler |
| Librium and Valium tranquilizers | Switzerland | Hoffmann, La Roche |

One measure of the technology transfer is the income generated from licenses and royalties. Firms will transfer technology only if they believe the transaction will be to their benefit. Dunning (2) pointed out that "where a firm sells or leases technology to other firms, it is presumed that the terms and conditions of the exchange will fully compensate it for the opportunity costs of supplying that technology." Frequently the technology is released with the stipulation that the licensee will market only in his own country.

The United States is a dominant exporter of technology, as shown in Table 9.16, with receipt-to-payment ratios exceeding 9.0 in 1960 and 12.5 in 1990. Japan and Germany have been moving from the deficit side, and both countries have improved their positions.

There is no generally accepted definition of technology transfer, but broadly it is a system of knowledge, techniques, skills, expertise, and organization used to produce, commercialize, and utilize goods and services that satisfy economic and social needs (3,4). On the international scene, it is a process in which technological inputs are transformed to technological outputs within a country from sources outside that country. It may take the form of the following activities:

Training abroad
Recruitment of key foreign personnel
Reverse engineering
Purchase of equipment
Subcontracting, operating, and maintenance agreements
Licensing
Joint ventures
Strategic partnerships and alliances

Technical assistance and training programs are essential for the effective application of acquired technology. Also, this technology must be blended with the firm's own objectives and development.

As research and innovation become increasingly global, it is essential to think globally with regard to corporate and national (governmental) policies. Scientific and technical achievements of the past will not be sufficient to sustain standards of living in the future. No one company can keep pace with technology, and technology is not the only key to success [5]. The globalization of markets, production, and technology is a defining feature of the worldwide economy.

With the intense competition and the turmoil in the United States due to downsizing and restructuring of companies, intellectual property may walk out the door. What a company knows has become as important as what it produces [6]. Intangible and invisible assets such as knowledge, competence of employees, information systems, and intellectual property (e.g., patent details, trade secrets, customer lists) do not show up on the bottom line but are as valuable as

TABLE 9.16 Incomes from Technology Transfers (millions): Receipt (R), Payment (P), and Receipt/Payment (R/P)

| Country | 1960[a] | | | 1970[a] | | | 1980[b] | | | 1990[b] | | |
|---|---|---|---|---|---|---|---|---|---|---|---|---|
| | R | P | R/P | R | P | R/P | R | P | R/P | R | P | R/P |
| United States (dollars) | 650.0 | 67.0 | 9.70 | 1,805.0 | 194.0 | 9.30 | 4,998.0 | 346.0 | 14.45 | 15,507.0 | 1,238.0 | 12.53 |
| United Kingdom (pounds sterling) | NA | NA | NA | 175.5 | 164.6 | 1.07 | 488.0 | 398.0 | 1.23 | 1,098.0 | 1,074.0 | 1.02 |
| Germany (deutschemarks, DM) | 38.8 | 127.5 | 0.30 | 105.0 | 249.8 | 0.42 | 1,011.0 | 2,079.0 | 0.49 | 2,360.0 | 4,687.0 | 0.50 |
| France (francs) | 48.1 | 90.8 | 0.53 | 195.0 | 230.0 | 0.85 | 2,095.9 | 4,343.9 | 0.48 | 7,311.6 | 11,962.7 | 0.61 |
| Japan (U.S. dollars) | 2.3 | 94.9 | 0.02 | 34.0 | 314.0 | 0.11 | 344.0 | 1,301.0 | 0.26 | 2,490.0 | 6,040.0 | 0.41 |
| Netherlands (guilders) | NA | NA | NA | NA | NA | NA | 831.0 | 1,278.3 | 0.65 | 1,639.3 | 3,399.5 | 0.48 |
| Belgium–Luxembourg (Belgian Francs) | NA | NA | NA | NA | NA | NA | 5,380.0 | 13,334.0 | 0.40 | 27,780.0 | 45,866.0 | 0.61 |
| Canada (Canadian dollars) | NA | NA | NA | NA | NA | NA | 45.0 | 700.0 | 0.06[e] | 120.0 | 1,345.0 | 0.09 |

[a]From J. Wei et al., The Structure of the Chemical Processing Industries, McGraw-Hill, New York, 1979.
[b]From World Investment Directory, Developed Countries, Vol. 3, United Nations, New York, 1993.

financial assets. A hidden value in most companies today is the ability to leverage the firm's knowledge through alliances.

There have been cases of corporate espionage including the theft of trade secrets. One industrial target has been the pharmaceutical industry, in which intellectual property protection is the cornerstone of this highly successful and competitive industry. GATT and NAFTA have set high standards on intellectual property protection. It is important for companies to be on guard and to protect their intellectual capital. In 1996 the U.S. government acknowledged this problem and made this type of theft and the associated espionage a federal offense.

Several excellent publications are available on the subject of technology transfer as well as transfer negotiations and intellectual property protection [7,8,9].

An important factor in the success the United States has enjoyed is its large research manpower in science and technology. Yet science and technology are international endeavors, and there has been considerable movement of chemical professionals between nations. Immigration, in the past, has brought to the United States many notable scientists, engineers, and chemical entrepreneurs like E. I. Du Pont, Leo Baekeland, Vladimir Ipatieff, Eugène Houdry, Herman Mark, and Herman Frasch. Table 9.17 contains the statistics of immigration of scientists and engineers into the United States. The total number of immigrants rose to a peak in 1970, and there has been generally a decline since then.

TABLE 9.17  Immigration of Professionals Into the United States

| Date | Scientists | Engineers | Total scientists and engineers |
|------|-----------|-----------|--------------------------------|
| 1967 | 1,392[a]  | 668[b]    | 12,523 |
| 1970 | 1,495[a]  | 908[b]    | 13,337 |
| 1974 | 592[a]    | 333[b]    | 5,969  |
| 1986 | 3,439     | 1,715     | 5,154  |
| 1990 | 5,102     | 2,666     | 7,768  |
| 1995 | 7,016     | 3,477     | 10,493 |

[a]Chemists only.
[b]Chemical engineers only.
Sources: 1976–1974: J. Wei et al., The Structure of the Chemical Processing Industries, McGraw-Hill, New York, 1979; 1986–1995, National Science Foundation, Division of Science Resources Studies Arlington, VA, 1997.

## 9.4   THE CPI IN OTHER COUNTRIES

The consumption of chemical products is closely related to the GDP of a nation. Also, chemical production correlates with the GDP, as shown in Table 9.18. The relative wealth of nations is measured by the GDP per capita. For the purpose of discussions in this Section, the countries are classified according to the *World Investment Report* published by the United Nations [10] (see Table 9.19). The classification is as follows:

> Developed countries (Section 9.4.1) are nations that have a per-capita GDP greater than $10,000. This group includes the United States, Canada, the United Kingdom, Germany, France, Italy, Spain, Australia and New Zealand, and Japan. Several of these nations have significant and powerful firms in the CPI that compete with United States firms.

> Developing countries (Section 9.4.2) are nations with an increasing GDP in the range of $1000–$10,000 per capita. In the Mediterranean region are Portugal, Greece, and Turkey. The wealthier Latin American nations like Argentina, Brazil, Chile, and Mexico are in this category; in Asia developing economies are found in Taiwan, Singapore, Malaysia, Indonesia, the Philippines, South Korea, and India. Some of these nations have significant and expanding operations in the CPI and are major purchasers of technology and sophisticated equipment.

> Central and eastern European countries (Section 9.4.3) had state-planned economies while under Communist domination. This group includes those countries that were part of the USSR such as Russia, Ukraine, Kazakhstan, and Belarus. Slowly emerging countries (Poland, Romania, Hungary, Slovakia, etc.) are beginning to develop chemical industries. These countries, as well as Poland and Russia, are importers of chemical plants and technology. Some of the nations in this group have large supplies of natural resources.

> OPEC nations (Section 9.4.4) comprise the last group. In general, they have a very high ratio of oil reserves to population. Examples are Saudi Arabia, Kuwait, Qatar, the United Arab Emirates, Libya, and Venezuela. These nations have enormous cash flows as long as the oil reserves are present. Most have invested heavily in chemical plants and refineries as well as modernization of older units.

Nations that have a low GDP rely heavily on agriculture, which is necessary to feed their people. The breakdown of the labor force by industry for selected counties is found in Table 9.20, which gives the percentage of a nation's labor force according to four groups. The Philippines and Mexico have significant forces in agriculture and fishing, for example. To increase a nation's wealth, a shift from low-efficiency agricultural labor to employment in modern manufacturing must take place.

TABLE 9.18 Chemical Product Consumption: Selected Countries

| Country | 1995 GDP per capita (U.S. dollars)[a] | 1996 Sulfuric acid per-capita production (lb/yr)[b] | Consumption of product per capita (lb/yr) | | | | | | |
|---|---|---|---|---|---|---|---|---|---|
| | | | HDPE | All PE combined | LDPE | PVC | PS | PP | Synthetic rubber |
| United States | $27,351 | 359.0 | 46.6 | | 53.3 | 19.8 | 22.8 | 45.2 | 17.7(3)[c] |
| Canada | $24,097 | 333.0 | 59.3 | | 111.4 | 38.7 | 38.9 | | N.A. |
| Japan | $21,347 | 122.0 | | 59.1 | | 44.8 | N.A. | 48.7 | 27.1 |
| United Kingdom | $19,453 | 36.8 | | 29.4 | | 14.6 | 10.7 | 6.7 | 16.2 |
| Italy | $18,939 | 84.3 | | 43.4 | | 24.0 | 20.4 | N.A. | NA |
| France | $18,625 | | | 58.3 | | 44.8 | NA | 42.0 | 22.1 |
| Germany | $17,389 | 88.9 | | 24.2 | | 31.1 | 83.0 | 13.3 | 14.8 |
| Taiwan | $12,100 | 19.2 | | 49.6 | | 115.7 | | 47.1 | 21.9 |
| China | $2,841 | 15.1 | | 9.0[d] | | — | — | — | 0.9 |

[a] From Central Intelligence Agency, The World Factbook, 1995.
[b] Figures derived from data in Chemical and Engineering News, Facts and Figures issue, June 23, 1997.
[c] From Standard & Poor's, Industry Surveys, C41, February 8, 1996.
[d] All plastics combined.

**TABLE 9.19**  Classification of Nations/Economies

| | |
|---|---|
| Developed countries[a] | Oman |
|   Australia | Pakistan |
|   Austria | Paraguay |
|   Belgium and Luxembourg | Peru |
|   Canada | Philippines |
|   Denmark | Republic of Korea |
|   Finland | Saudi Arabia |
|   France | Singapore |
|   Germany | Taiwan |
|   Greece | Uruguay |
|   Iceland | Former Yugoslavia |
|   Ireland | |
|   Italy | Central and eastern Europe[a] |
|   Japan |   Albania |
|   The Netherlands |   Belarus |
|   New Zealand |   Bulgaria |
|   Norway |   Czech and Slovak Federal |
|   Portugal |     Republic |
|   South Africa |   Estonia |
|   Spain |   Hungary |
|   Sweden |   Poland |
|   Switzerland |   Romania |
|   Turkey |   Russian Federation |
|   United Kingdom |   Ukraine |
|   United States | |
| | OPEC members[b] |
| Developing countries[a] |   Algeria |
|   Bolivia |   Indonesia |
|   Brazil |   Iran |
|   China |   Iraq |
|   Colombia |   Kuwait |
|   El Salvador |   Libya |
|   Guatemala |   Nigeria |
|   Hong Kong |   Qatar |
|   India |   Saudi Arabia |
|   Indonesia |   United Arab Emirates |
|   Mexico |   Venezuela |

[a]From *World Investment Report,* United Nations, 1996.
[b]From CMAI.

TABLE 9.20 Breakdown of Labor Force by Industry

| Country (date) | Professional, technical and related workers (%) | Agriculture, animal husbandry forestry, fishing, etc. (%) | Manufacturing, production, and related workers (%) | Service, commerce, and government (%) |
|---|---|---|---|---|
| United Kingdom (1993) | 19.2 | 8.8 | 22.7 | 49.3 |
| United States (1995) | 17.7 | 2.9 | 25.4 | 54.0 |
| Germany (1993) | 9.4 | 2.2 | 35.2 | 53.2 |
| Italy (1994) | 1.1 | 5.1 | 43.5 | 50.3 |
| Japan (1995) | 12.2 | 5.6 | 34.6 | 47.6 |
| Venezuela (1995) | 12.1 | 13.1 | 28.3 | 46.5 |
| Philippines (1995) | 5.6 | 43.6 | 21.7 | 29.1 |
| Mexico (1995) | 9.3 | 24.3 | 23.0 | 43.4 |
| Belgium (1992) | 22.6 | 3.1 | 32.0 | 42.3 |
| The Netherlands (1994) | 24.1 | 4.0 | 21.8 | 50.1 |
| Australia (1993) | 1.5 | 15.7 | 21.8 | 51.2 |
| Canada (1995) | 36.2 | 4.9 | 26.8 | 32.1 |
| Thailand (1991) | 3.5 | 60.4 | 17.4 | 18.7 |

Source: United Nations, Yearbook of Labor Statistics, 1996, United Nations, New York, 1996.

## 9.4.1 The Chemical Industry in Developed Countries

Countries discussed in this section are Germany, the United Kingdom, France, Japan, the Netherlands and Belgium, Italy, and Switzerland. The top chemical producers, ranked by 1996 sales, in these countries are listed in Table 9.21. Also, Chapter 5 gives the 1997 ranking for these companies as reported by *Chemical and Engineering News*.

**TABLE 9.21**　The 20 Largest Chemical Producers by Country[a]

| Country | Company | 1996 Sales (millions) | Ranking |
|---|---|---|---|
| Germany | BASF | $34,801.3 | 1 |
| | Hoechst | $33,951.3 | 2 |
| | Bayer | $32,405.3 | 3 |
| | Degussa | $9,165.3 | 14 |
| United Kingdom | ICI | $16,429.8 | 7 |
| | BOC International | $6,277.5 | 22 |
| | Courtalds | $3,279.1 | — |
| | Laporte | $1,660.8 | — |
| France | Rhône-Poulenc | $16,761.3 | 6 |
| | Air Liquide | $6,725.3 | 18 |
| Netherlands and Belgium | Akzo Nobel | $13,309.2 | 8 |
| | Solvay | $9,106.9 | 15 |
| | DSM | $6,087.5 | 24 |
| Japan | Asahi Chemical | $10,114.0 | 12 |
| | Mitsubishi Chemical's | $9,936.0 | 13 |
| | Sekisui Chemical | $7,494.0 | 16 |
| | Takeda Chemical | $5,858.0 | 25 |
| Others | Norsk Hydro (Norway) | $13,153.9 | 9 |
| | Roche Group (Switzerland) | $12,917.5 | 10 |
| | Enichem (Italy) | $6,628.6 | 20 |

[a]Only the top 30 chemical producers in the world were ranked.
*Source: Chemical and Engineering News,* Facts and Figures issue, June 23, 1997.

For further comparison, data for the top 20 global chemical producers on research and development and capital spending as a percentage of sales, percentage return on assets, and sales per employee are found in Table 9.22.

9.4.1.1　Germany.　Germany has often been thought of as the cradle of the chemical industry. The three top chemical companies, BASF, Hoechst, and Bayer, dominate the German chemical industry. They have had an interesting history. In an attempt to meet world competition following World War I, these

TABLE 9.22   Further Economic Data on the Top 20 Chemical Companies in the World: 1996

| Ranking Company | Net sales (millions) | R&D (%) | Capital spending (%) | Return on assets (%) | Sales per employee (thousands)[a] |
|---|---|---|---|---|---|
| 1 BASF | $34,801 | 4.4 | 7.0 | 6.6 | $336,570 |
| 2 Hoechst | $33,951 | 6.6 | 6.6 | 4.0 | $229,556 |
| 3 Bayer | $32,405 | 7.4 | 7.4 | 5.7 | $227,883 |
| 4 Dow | $18,988 | 3.8 | 6.7 | 7.7 | $520,068 |
| 5 Du Pont | $18,044 | 5.7 | 7.7 | 9.6 | $409,165 |
| 6 Rhône-Poulenc | $16,761 | 9.4 | 8.3 | 2.8 | $222,594 |
| 7 ICI | $16,430 | 1.8 | 9.8 | 2.8 | $256,716 |
| 8 Akzo Nobel | $13,309 | 5.1 | 8.2 | 6.7 | $188,249 |
| 9 Norsk Hydro | $13,154 | 0.9 | 14.3 | 5.9 | $371,579 |
| 10 Roche | $12,198 | 15.3 | 10.2 | 10.7 | $263,622 |
| 11 Exxon[b] | $11,430 | NA | NA | 7.9 | $1,511,823 |
| 12 Asahi Chemical | $10,114 | NA | NA | 2.2 | $648,323 |
| 13 Mitsubishi Chemicals | $9,936 | NA | NA | 1.8 | $770,232 |
| 14 Degussa | $9,165 | 3.4 | 4.2 | 3.7 | $353,281 |
| 15 Solvay | $9,107 | 4.3 | 9.9 | NA | $257,314 |
| 16 Sekisui Chemical | $7,494 | NA | NA | 3.6 | $1,249,000 |
| 17 Monsanto | $7,267 | 8.5 | 7.5 | 3.4 | $330,786 |
| 18 Hoechst Celanese | $6,906 | 6.6 | 7.5 | NA | NA |
| 19 Air Liquide | $6,725 | 2.3 | 19.9 | NA | $241,917 |
| 20 Enichem | $6,629 | 1.2 | 3.0 | NA | $394,560 |

[a]The average revenue per employee among the United States companies is $332,543.
[b]Data are for chemical sales only.
Source: Chemical and Engineering News, Facts and Figures issue, June 23, 1997.

three firms merged in 1925 to form I. G. Farben, which was an important supporter of Hitler from 1932 to the end of World War II. I. G. Farben produced synthetic rubber, synthetic gasoline from $CO-H_2$ mixtures, explosives, methanol, poison gases, tetraethyllead, magnesium, aluminum, and many other chemical products for the war effort. After the war, the firm was split into the three companies again (1953) using the original names; BASF, Hoechst, and Bayer. Each of these companies has had spectacular growth, exceeding that of I. G. Farben. They are the top three in global sales and according to their sales distributions are highly diversified.

The German chemical industry has been the strongest in Europe. Approximately 21% of the total European chemical investment is in Germany. One of the most prominent features of its CPI is the amount of overseas investment.

Many of the German chemical companies have found a fertile field for foreign investment, notably Hoechst and BASF. The industry had some successful years in the early 1990s, but in 1996 and 1997 there was a decline in chemical output with some sectors appearing weak. Germany had a ratio of research and development expenditures to GDP of 2.3% (vs. 2.5% of the United States).

Many of the country's instruments of technology work well, but for it to maintain its international technological competitiveness, a more risk-taking, dynamic spirit is necessary. The cultural environment in Germany is not conducive to entrepreneurial activity. Financial incentives for entrepreneurs are low because there are no tax credits for research and development and the government bureaucracy is cumbersome, especially for start-up companies [11]. Tough regulatory and legal frameworks have been a problem for growth, and there is a long time delay between planning a new project and getting permission from the government to proceed. Another problem for the German chemical industry lies in high labor costs: $27.31 per hour compared to $17.10 average in the United States in 1994 [13]. In early 1999 there were 24 new projects under way [13].

### 9.4.1.2 The United Kingdom.

Imperial Chemicals Industry (ICI) is an order of magnitude in sales greater than the next largest British chemical company, Courtalds. ICI has been active in many segments of the chemical industry, including agricultural chemicals, fibers, paints, petrochemicals, plastics, pharmaceuticals, and metals; but in 1997–1998 ICI spun off the pharmaceuticals, polyester film, and fertilizers segments to increase efficiency. ICI has been an innovator and world leader in the production of polyester and nylon fibers. It is a company with many subsidiaries.

In the 1960s and early 1970s ICI was characterized by low profitability and low efficiency in comparison with German, Japanese, and U.S. counterparts, having been plagued by technical and managerial mistakes. But in the mid-1970s, it was ranked second worldwide among chemical companies in pretax profits. The return on capital in this period was about 23%, which was better than most German companies and comparable to U.S. performance. ICI has increased productivity through increased research and development expenditures and capital restructuring as well as restructuring.

In the past ICI has faced many difficulties not of their own making. The British domestic market for chemicals declined markedly in the mid-1970s. Further, companies had to operate with price controls, a high rate of inflation, and devaluation of the British currency, as well as social and political unrest. In the United Kingdom in early 1999, 10 new projects were in construction [13].

Other major British chemical companies beside Courtalds (now part of Akzo Nobel), are BP Chemicals (which sold its specialty chemicals to Inspec), Shell, and Unilever.

The United Kingdom is the largest foreign investor in the United States. A recent example (1998) is the merger of BP with Amoco.

9.4.1.3 France. The big chemical companies are Rhône-Poulenc, Elf Atochem, L'Oréal, Air Liquide, and Pechiney. Rhône-Poulenc manufactures a broad spectrum of chemical products including pharmaceuticals, plastics, fibers, photographic chemicals, and fertilizers. It has been active in purchasing manufacturing facilities in various countries including the United States. Elf Atochem has a strong base in petroleum, petrochemical products, and the chemicals used in the manufacture of these products. Depending on how companies are classified, L'Oréal, which derives its income from cosmetics, perfumes, and the chemicals used in the firm's products, is second, although Elf Atochem and L'Oréal are often not classified as chemical companies but as petroleum and cosmetics companies, respectively. Air Liquide is the next largest chemical company which markets industrial gases and ranked nineteenth in sales in 1996; see Table 9.22[14]. Pechiney, a widely diversified company, ranked next in France, has been very active in mergers and acquisitions. Nine new CPI facilities were being constructed in 1999 [13].

The hourly wages in the French chemical industry are high compared to other major countries. In fact, the rates are about 10% higher than in the United States, but the labor efficiency is not as high.

9.4.1.4 Japan. The Japanese chemical industry in the mid-1970s was characterized by many small and medium-sized firms. In the 1990s Ashai and Mitsubishi Chemicals are the leaders, with combined sales in excess of $20 billion. The significant chemical products of these companies are agricultural chemicals for the Japanese intensive farming industry as well as petrochemicals, plastics, and synthetic rubber. Japanese firms are characterized by intense innovation, intense domestic competition, and growth in chemical demand. Through the years following the end of World War II, the Japanese received U.S. financial and technical aid; however, this assistance diminished as Japan's chemical industries developed.

On the downside, the Japanese chemical industries have been dependent on obtaining adequate feedstock and fuel for energy. They have been highly dependent on obtaining the needed stocks from the OPEC countries and from the United States. Although the Japanese economy was relatively stable in the recent past, in the late 1990s, currency instability, the practice of making large unsecured loans, rising unemployment, inflation, and some governmental and social unrest have slowed the country's economic progress. Consumer confidence has not fully recovered as of 1998, and a recession drags on. Japanese companies will find progress difficult while they remain small, but mergers are hindered by government regulations. MITI has suggested reducing the number of plants by consolidation, but the government ministry has no power to enforce its advice. Another stumbling block has been the limited set of research and development resources, which has resulted in Japanese firms importing technology rather than developing innovative processes. Also, there is a strong feeling that the Japanese chemical industry cannot flourish in the future if the status quo continues [15]. The industry must overcome the barriers to consolidation and restructuring, which might be the best way to strengthen the industry.

When these problems stabilize, one can look for a strong world chemical industry which will emerge signifying a healthy, growing Japanese CPI. In spite of the financial difficulties, 15 new CPI facilities were listed being under construction in early 1999 [13].

### 9.4.1.5 The Netherlands and Belgium.
The Netherlands is an important contributor to the European chemical industry. In part, this is due to excellent port facilities and the ability to handle large amounts of chemical imports and exports. The chemical industry of this nation is characterized by strong joint ventures with other European nations, for example, Akzo Nobel (Sweden) and its affiliate Shell (United Kingdom). Akzo is the largest chemical producer, with 1996 sales exceeding $13 billion, primarily from synthetic fibers. It is a highly diversified company, however. Five new projects were in construction in early 1999 [13].

Solvay is Belgium's largest chemical company. Soda ash (sodium carbonate) accounts for a large percentage of its sales, however, the firm has also produced agricultural chemicals, halogen compounds, peroxides, and plastics. Solvay is in fifteenth position in world chemical sales at $9.1 billion (Table 9.22). Belgium had seven new CPI projects under construction as of early 1999 [8].

### 9.4.1.6 Switzerland.
The chemical industry of this country is centered about pharmaceuticals, biological chemicals, medicinals, and fine chemicals. Roche is one of the world's largest producers of pharmaceuticals. Other significant CPI firms are Ciba-Geigy and Nestlé. It is difficult to study the performance and profitability of Swiss firms because they are not required to make public financial information, as is required in other nations. Switzerland has a very stable economy and very low inflation. Its labor costs are high, second only to Germany, but the workers are highly efficient and conscientious.

### 9.4.1.7 Italy.
The largest chemical company in Italy is Enichem, with annual sales of $6.6 billion in 1996. In the past, the Italian chemical industry has been subjected to strikes, high inflation, price controls, political instability, lackluster management, inefficient and obsolete plants, and a weak currency. The emergence of the euro may lend financial stability. Recently, there are indications that some stability has been achieved, with the result that the industry is moving forward. The Italian CPI had 15 new or expansion projects under construction in early 1999 [13].

### 9.4.1.8 Spain.
The Spanish chemical industry is a small player in the western European CPI. The labor unit costs are about 60% of those in the United States and very competitive with European rates. Although major problems exist

with respect to finances, there has been little attention paid to the development of the industry by domestic companies or ventures with multinationals. In early 1999 there were 26 new or expansion projects in construction in the petroleum and petrochemical areas [13].

### 9.4.2 The Chemical Industry in Developing Economies

There is a significant CPI in some of the developing economies. This group includes China, Indonesia, Mexico, the Philippines, Malaysia, India, South Korea, Saudi Arabia, Singapore, and Taiwan. These countries produce many essential chemical products as important substitutes to save foreign exchange. The products are primarily simple basic chemicals that have a local raw material or labor advantage and are too low priced to be feasibly transported over long distances.

Mining has produced chemical raw materials for export that are important in world trade, for example:

*Bauxite:* Jamaica, Surinam, Venezuela, Ghana
*Nuclear materials:* Zaire
*Phosphate:* Peru, Morocco, Tunisia, Chile
*Sulfur:* Mexico
*Tin:* Malaysia, Bolivia
*Silver:* Mexico, Peru
*Natural rubber:* Malaysia, Indonesia

In the poorer nations of this group, many chemicals are imported. The governments of Singapore, Malaysia, Indonesia, China, and the Philippines are encouraging foreign investments in the CPI. Besides the fact that the citizens of these countries need the products, there is an element of national prestige in building and operating CPI plants. Further, the production of locally consumed chemical products reduces transportation charges, provides employment for the citizens, and increases the nation's wealth. Foreign investors generally seek political, moral, and economic incentives before giving their support.

A company needs to consider several factors when surveying the potential for developing a chemical industry in countries with emerging economies. Table 9.23 lists some important items and ranges for these factors. A country's potential for developing a CPI must be approached in terms of production infrastructure, market location, and sociopolitical factors. Essential factors to be considered are as follows:

1. Factors of production
   a. Is land available for a plant site?
   b. Is adequate capital available or can it be obtained readily?
   c. Is a trained or trainable work force available?
   d. Is an adequate supply of necessary raw materials available?

2. Infrastructure
   a. Is there other heavy industry in the country?
   b. Are parts, machinery, and equipment available?
   c. Is a plant construction capability availability?
   d. Are adequate ports, airports, and transportation available?
   e. Are communication, medical, educational, banking, and other services available?
3. Markets
   a. Is there a domestic need for chemical products?
   b. Do potential customers have the ability to pay for the products?
   c. Are there nearby markets in other countries?
4. Sociopolitical factors
   a. Are foreign investments and chemical plants welcomed by the government and by the general public?
   b. Does the government offer incentives to encourage CPI investment?
   c. Will the country's people make acceptable trained employees?
   d. Are the people conscientious, dependable, and hardworking?
   e. Is the company relatively safe from foreign military intervention?
   f. Is the country politically and economically stable?

**TABLE 9.23**  Important Factors: Selected Countries

| Factor | Typical value[a] | Example |
|---|---|---|
| Population density, people/km² | | |
| High | 296/km² | India |
| Low | 12.5/km² | Argentina |
| Annual population growth rate | | |
| High | >2.5 | Thailand, Malaysia, Iraq, UAE |
| Low | <0.2 | Italy, Romania, Russia |
| World average | ~1.8 | |
| GDP and size of market (billions) | | |
| High | >2,000[b] | U.S., China, Japan |
| Low | <200 | Singapore, Philippines, Malaysia, Switzerland |
| GDP per capita | | |
| High | >20,000[b] | U.S., Canada, Japan, Belgium |
| Median | 3,000–20,000 | U.K., France, Italy, The Netherlands |
| Low | <3000 | India, China, Philippines, Indonesia |

[a]All values except those referenced in note b from U.S. Bureau of the Census, *Statistical Abstract*, 1997, Table 1325.
[b]From Central Intelligence Agency, *The World Factbook*, 1995.

For many of the countries listed in this group, the answers to these questions are generally positive; however, there are usually several negative answers, as well. For example, India has a large and rapidly growing population and offers large potential for CPI products; on the downside, however, it has limited raw material supplies, limited capital, and many government regulations. Further, the general population cannot afford to buy many of the CPI products. In contrast, several sparsely populated countries in the Middle East that have sufficient raw materials and available capital for a chemical industry may lack adequate infrastructure, political stability, domestic demand, or trained labor for developing a CPI.

Prospects for developing a chemical industry in the near future in underdeveloped countries are slim because of the high level of risk associated with building plants. The GDP per capita is increasing markedly in some less developed countries, however, which may be indicative of future opportunities for the CPI (see Tables 9.24 and 9.25). Brief discussions of the CPI and its potential in selected countries are presented in Sections 9.4.2.1–9.4.2.10.

### 9.4.2.1 Brazil [16–20].

Brazil's chemical industry, after a long stagnation, was growing along with Brazilian economy until the currency devaluation in 1998 hurt local producers with respect to imported raw materials and further limited the building of grassroots plants. Brazil's chemical industry is by far the largest in South America. The growth was the result of changes in past economic policies that involved market protection, government authorization to build new plants, and rampant inflation. There has been a switch to private ownership from government's majority ownership. A number of joint ventures were under consideration and are now on hold. The growth in inorganics, organics, and plastics was slow but steady from 1990 up until 1998. Much of that growth occurred in the petrochemical industry, which assisted the expansion of the plastics segment. Until late 1998 Brazil had a burgeoning automobile industry, which was an outlet for plastic products.

Up until the outbreak of one currency problem, the Brazilian chemical industry was competitive in both quality and prices as a result of the following factors:

Naphtha prices were based on actual prices of oil and naphtha imported rather than on a formula.
Import duties were at international levels.
Domestic prices fluctuated with changes in international markets.
Restructuring and cost reductions were being achieved.
Supply and demand were in relative balance.

As soon as Brazil stabilizes its currency, and it appeared to have begun in early 1999, with banks and lenders agreeing to maintain key credit lines, the

**TABLE 9.24**  GDP, GNP, and Growth Rates of Each for Selected Countries (billions of current dollars)

| Country | 1994[a] | 1995[b] | GNP[c] 1993 | GDP[c] 1994 | Growth rates (%)[c] GDP 1994–1995 | GNP 1993–1994 |
|---|---|---|---|---|---|---|
| United States | $6,738 | $7,248 | $6,348 | $6,727 | 7.6 | 6.0 |
| China | $2,979 | $3,500 | $2,047 | $2,214 | 17.5 | 8.2 |
| Japan | $2,527 | $2,679 | $4,509 | $4,630 | 6.0 | 2.7 |
| Germany | $1,345 | $1,452 | $1,947 | $2,045 | 8.0 | 5.0 |
| India | $1,254 | $1,409 | $267 | $287 | 12.4 | 7.5 |
| France | $1,080 | $1,173 | $1,256 | $1,316 | 8.6 | 4.8 |
| United Kingdom | $1,045 | $1,138 | $970 | $1,027 | 8.9 | 5.9 |
| Italy | $999 | $1,089 | $966 | $1,006 | 9.0 | 4.1 |
| Brazil | $886 | $977 | $547 | $580 | 10.3 | 6.0 |
| Russia | $721 | $796 | $777 | $780 | 10.4 | 0.4 |
| Mexico | $729 | $721 | $344 | $363 | -1.1 | 5.5 |
| Indonesia | $619 | $711 | $153 | $168 | 14.9 | 9.8 |
| Canada | $640 | $694 | $490 | $523 | 8.4 | 6.7 |
| South Korea | $508 | $591 | $322 | $356 | 16.3 | 10.6 |
| Thailand | $355 | NA | $128 | $142 | — | 10.9 |
| Spain | $276 | $565 | $459 | $478 | 104.7 | 4.1 |
| The Netherlands | $276 | NA | $315 | $331 | — | 5.1 |
| Belgium | $182 | NA | $220 | $229 | — | 4.1 |
| Malaysia | $167 | NA | $60 | $67 | — | 11.7 |
| Philippines | $161 | NA | $62 | $66 | — | 6.5 |
| Switzerland | $148 | NA | $256 | $267 | — | 4.3 |
| Singapore | $57 | NA | NA | NA | — | — |

[a]From Central Intelligence Agency, *The World Factbook*, 1995.
[b]From *U.S. News & World Report*, June 23, 1997, p. 36, based on information from the Central Intelligence Agency.
[c]From U.S. Bureau of the Census, *Statistical Abstracts of the United States*, 1997, Table 1334.

country will have potential for strong development of many chemical products [21]. Brazil had 28 new CPI or infrastructure projects under way in early 1999 [13].

**9.4.2.2  Mexico [22–25].**  Mexico has managed to minimize many of the effects of the economic turmoil in Asia and South America, although there has been a supply problem owing to the influx of cheap Asian imports. Although the GDP grew 7.2% in 1997, it dropped to about 5% in 1998, in part as a result of

**TABLE 9.25**   GDP per Capita Growth

| Country | Per-capita GDP[a] (U.S. dollars) | Growth rate of GDP per capita[a] 1985–1994 (%) |
|---|---|---|
| United States | 25,880 | 1.3 |
| Thailand | 2,410 | 8.6 |
| China | 530 | 7.8 |
| South Korea | 8,260 | 7.8 |
| Chile | 3,520 | 6.5 |
| Indonesia | 880 | 6.0 |
| India | 320 | 2.9 |
| Colombia | 1,670 | 2.4 |
| Argentina | 8,110 | 2.0 |
| Kuwait | 19,420 | 1.1 |
| Mexico | 4,180 | 0.9 |
| Venezuela | 2,760 | 0.7 |

[a]From U.S. Bureau of Senses, *Statistical Abstract of the United States*, 1997.

the decline in oil prices. There are positive economic indicators, but chemical producers are faced with a raw material crisis because of lack of investment in petrochemical plants owned by Petroleos Mexicanos (PEMEX), the state oil monopoly.

Indecisiveness on the part of the Mexican government, an unpredictable political climate, as well as a downturn in petrochemicals have caused a lack of interest in purchasing part of the state-owned petrochemicals. Legislation in 1997 allowed private companies to build and operate private enterprises, but such ventures depended on basic petrochemicals produced by PEMEX.

Grupo Idesa has attempted to negotiate buying facilities from PEMEX to integrate ethylene and ethylene oxide. Idesa is negotiating with a multinational on a joint venture to enter new product lines, but multinationals are limited to 24% of the assets. Another company, Alpek, has joint ventures with Du Pont, BASF, and other multinationals, making it cheaper to import some raw materials from the United States. Alpek is making major investments in fibers, assuming that Mexico will produce its own raw materials. As of early 1999, there were 18 new construction projects in the CPI [13]. The Mexican chemical industry should grow if the government climate with respect to foreign investment were liberalized, PEMEX's financial problems ameliorated, and oil prices stabilized.

9.4.2.3   China [26–30].   In recent years, China has seen its role on the world stage change dramatically from a rising star to an experienced performer. It has a massive trade surplus, but chemicals is one sector in which China runs a large deficit. This negative trade balance in chemicals represents a huge opportunity for overseas investors. It is clear that American, European, and Japanese chemical companies are making major capital investment commitments in China. Shell, BASF, Dow Chemical, and British Petroleum all have planned investments in integrated petrochemical facilities, each project exceeding $2 billion. ICI plans a polyurethane complex, Bayer a synthetic rubber plant, Phillips Petroleum an HDPE/LLDPE facility, and Norsk Hydro a PVC facility. The areas in which demand exceeds supply are ethylene, propylene, synthetic resins, synthetic rubber, and synthetic fibers, as well as crude oil and petroleum products. More than a dozen planned or proposed refineries are being considered based on crude oil from Saudi Arabia and the Middle East. China does have significant fertilizer, cement, and paper industries, all of which are being expanded.

Originally barriers to industrial growth in China were as follows:

Lack of available capital
Environmental protection requirements
Energy shortage
Infrastructure limitations
Limited availability of technology
Perceived economic uncertainty
Intellectual property concerns

Certain of the foregoing barriers have been addressed. With the return of Hong Kong to China, some of the concerns about banking and capital availability have been alleviated. New dams and generating facilities are being planned or are under construction to alleviate the energy shortage. Technology availability and intellectual property are being addressed by joint ventures with foreign companies. Environmental requirements, infrastructure limitations (transportation within China), and perceived economic uncertainty are still major concerns.

China is forging ahead in the chemical business, and as soon as the economic situation in the Far East settles, there should be a lucrative market for foreign investors well into the early 2000s. A list of construction projects affecting the CPI, either new facilities, infrastructure or expansion projects, was published in early 1999 [13].

9.4.2.4  Singapore [31,32].   Singapore's chemical industry is the second largest manufacturing sector, just behind the electronics sector. Although China and India are favorites for many U.S. chemical companies, Asia/Pacific investment plans in Singapore, Malaysia, Thailand, South Korea, and Taiwan are not being overlooked despite the region's recent economic upheavals. In Singapore Amoco and ExxonMobil are considering olefins and aromatic complexes. Other companies considering large capital expenditures are Phillips, Hoechst, Du

Pont, and GE Plastics. The chemical growth will largely be in petrochemicals, specialty chemicals, and pharmaceuticals many of which should be on stream in early 2000. Singapore's Economic Development Board (EDB) is optimistic, and chemical industry investment is encouraging in the long run. EDB is encouraging companies to focus on research and development, innovation, and developing new products, processes, or applications. Eastman Chemical, for example, recently opened a technical center devoted to new products and applications for coating, paint, and food ingredients. Shell and a Caltex consortium are building refineries in Southeast Asia. Some projects are on hold until the economic situation clears. Five new CPI projects were under construction in early 1999 [13].

**9.4.2.5  Malaysia [33–36].**  Because of the country's economic problems, the future of the Malaysian chemical industry is uncertain. Investment may be curtailed because of pessimism over the economy. The Malaysian Industrial Development Association (MIDA) has stated that petrochemicals is one of the country's key industries. Joint ventures between Petronas (the state oil and gas company) and Union Carbide, BP Chemicals, and BASF are projected for the early 2000s and beyond. The chemicals of interest are ethylene, ethylene oxide, ethylene glycol, acetic acid, alcohols, and plasticizers. Large multinationals like ExxonMobil, and Shell have preferred Singapore to Malaysia. MIDA has long-term plans (10–15 years) to develop the upstream sector first, and the downstream sectors and specialty chemicals later. Even though economic problems are present, 27 new CPI projects were on hold or under way in early 1999 [13].

**9.4.2.6  Thailand [37–[39].**  Thailand had plans to develop the petrochemical business, but a severe economic crisis in 1997–1998, with currency devaluation, resulted in project delays and cancellations. Monsanto was in joint ventures with Thai companies to produce ABS resins and acrylonitrile styrenes. In the plans were derivative petrochemical facilities as well as base petrochemical plants. Petrochemicals promises to be a major source of Thailand's future economic growth. Thailand has a domestic supply of natural gas but it is limited. In the future the Thais may have to augment the supplies with imports. Of concern is the fact that the petrochemical industry is located in one area and the plant sizes are small.

Thailand is primarily an agrarian economy as evidenced in Table 9.20. The major fertilizer company was considering a large expansion using locally produced ammonia as feedstock. If this occurred, the domestic natural gas supply would have to be apportioned, or imports would be needed, to supply the petrochemical projects. In early 1999 there were 18 projects on hold or under limited construction until the economic situation eases [13].

**9.4.2.7   Indonesia [40, 41].**   Prior to 1997, projects worth billions of dollars were being planned, based on strong forecasts for economic growth. Construction work at most sites had stopped owing to lack of financing, however, even before the antigovernment protests and political unrest that began in 1997. Foreign companies like BP Chemicals and Amoco took measures to evacuate their expatriate employees. Many chemical investors say they are still committed to Indonesia in the long run, but there are possible future complications because a number of these wealthy Indonesians had strong ties to the Suharto family. Executives of foreign chemical companies agree that most of the already announced projects will go ahead, but in a different time frame. A number of western chemical companies are planning to increase their stakes in Indonesian joint ventures when the situation stabilizes. Most experts agree that in the long run (after political unrest, the financial crisis, and inflation have subsided), the chemical industry should grow. Although bank reform in early 1999 was promised in order for the country to receive assistance from the International Monetary Fund, the reform has been delayed as a result of political problems. In early 1999 approximately 18 potential CPI projects were on hold [13].

The chemicals produced in Indonesia included PTA, styrene, propylene, polypropylene, PVC, p-xylene, and EDC.

**9.4.2.8   Philippines [42–43].**   Financial problems, tariff restrictions, and political problems have dealt blows to the building of a major petrochemical complex in the Philippines. In the late 1980s, plans were made, but in 1993 they appeared to be dead. The plans were revived in 1996–1997, however. In the mid-1990s, polymer plants based on imported feedstocks were planned and, ultimately, a steam cracker. In the long range, polystyrene, LLDPE/HDPE, and polypropylene plants are being considered. Seven CPI projects were under way in early 1999 [13].

**9.4.2.9   South Korea [44–46].**   The chemical industry began with foreign capital and technology in the 1970s when full-scale modernization of South Korea began. From the late 1980s, the petrochemical industry grew rapidly as a result of large-scale construction and expansion activities. Because of chronic shortages of synthetic resins, synthetic rubber, and synthetic fiber raw materials, as well as the rapid building of petrochemical facilities, however, the chronic shortage in the 1970s became a surplus in the 1980s. As a result, production/consumption imbalance and high financial costs led to low profitability in the early 1990s. The financial situation seemed to improve in late 1994, but by 1997, the currency weakened, causing another financial crisis in the country. South Korea has been quick to assimilate technology and management techniques developed by more industrialized nations. In early 1999 there were indications that the financial problems were ameliorating and the future of the country's CPI seems to be assured. Investors are keeping faith with South Korea, with 14 mergers and acquisitions being considered and 19 new CPI projects being listed in early 1999 [13].

9.4.2.10  India [45–48].  With the world's second largest population and a corresponding need for CPI products, India has the potential of a large chemical market. Unstable governments, high tariffs, high interest rates, inflation, and overcapacity in certain product lines have resulted in slumps in demand in the late 1990s, however. India has a deficit trade balance in industry as well as in the chemical trade. This deficit increased significantly from 1990 through 1998. Multinational corporations have focused on specialty chemical ventures. Dyestuffs, textiles, and pharmaceuticals are the key attractions for investors. Hoechst, BASF, Dow, and Royal Dutch Shell have invested in some ventures and have exited from joint ventures.

The country has the desire to build its own CPI increases and become self-sufficient, but India has overcapacity in fibers, polymers, and plastics. The fertilizer industry has been heavily dependent on imports, but the availability of cash has been short for the large agricultural industry. There does seem to be a strong demand for intermediates, insecticides, herbicides, and fungicides. A recent tabulation of projects in the CPI shows that over 130 new facilities were under construction as of early 1999 [13]. Some of these were for very small units.

9.4.2.11  Summary of the Far East Situation.  Indications are that for the Asian/Pacific area, the worst may be coming to an end, but 1999 was expected to be a struggle for the chemical industry. Japan is at a turning point, reacting to lost export markets and the financial crisis. The government is attempting to stimulate consumer spending and has recommended restructuring of the banking sector, but progress has been slow. Singapore is striving to recover. Its Economic Development Board is attempting to strengthen links with investors. In Indonesia, there are signs of recovery, however, some chemical assets are for sale and overseas investors have raised the joint venture stakes. Malaysia has encountered the decline in foreign investment enthusiasm owing to economic and political uncertainty. Thailand's petrochemical sector continues to struggle, and producers are attempting to restructure debt and reengineer operations.

It is expected that economies like South Korea, Malaysia, and Singapore will likely emerge from recession. Taiwan, China, and India, on the other hand, have had reasonable growth rates, which may be expected to continue.

### 9.4.3  Central and Eastern Europe [49]

Central and eastern European nations formerly under Communist domination had state-planned economies. This group includes Russia, Ukraine, Belarus, and

Kazakhstan of the former Soviet Union as well as Poland, Romania, Hungary, Bulgaria, Slovakia, and the Czech Republic. Some of the nations in this group have large raw material supplies, especially oil and natural gas. Poland and Russia are major purchasers of western technology. The chemical industries in these countries are emerging slowly, generally owing to financing difficulties, but the markets for their products are building [47].

### 9.4.3.1 Russia.

Russia is different from the other central European countries in that it has access to cheap feedstocks and has a number of companies within the oil and gas industries. In 1998 the chief problem was the financial crisis in the banking industry and in the economy in general, and as of early 1999 there seemed to be little relief in sight. Prior to the economic downturn, Russian banks showed an interest in investing in the CPI. Western companies focus on the exploration for oil, although the risks are high. After year 2000, foreign interest will likely be concerned with industries downstream from the oil industry. The major markets for these products will be in areas around Moscow. Experts think that the Russian petrochemical industry will not develop in accordance with the demand, which suggests that there will be a large export market in the West. In early 1999, despite a financial crunch, 25 new, infrastructure, or retrofit projects were listed as under construction [13].

### 9.4.3.2 Poland.

There are two crackers in Poland: one is relatively, small and old; the other is newer. This latter unit is operated by Petrochemia and receives its feedstock from an adjacent refinery. The petrochemical complex produces ethylene for low-density polyethylene and ethylene oxide. The rest is used elsewhere in Poland to produce vinyl chloride monomer and PVC. Propylene is produced in the newer cracker at Plock, which is used for polypropylene and cumene. The rest goes to oxoalcohol production and to propylene oxide elsewhere in the country. Poland had seven petrochemical or petroleum projects under construction in early 1999 [13].

### 9.4.3.3 Czech Republic.

The major part of the petrochemical industry is controlled by Unipetrol, a holding company. The main petrochemical plant is in Litvinov. In 1995 Shell and Conoco acquired a 49% share at Litvinov and at another site. Ethylene produced at Litvinov is for high-density polyethylene, ethylbenzene, and ethanol. The Czech Republic has excess capacity in ethylene, propylene, and butadiene. Three construction projects in the petroleum and petrochemical areas were listed in early 1999 [13].

### 9.4.3.4 Slovakia.

The Slovakian petrochemical industry is dominated by Slovnaft, which operates two petrochemical sites. The complex at Bratislava has a naphtha cracker, three LDPE units, two propylene plants, and an aromatics unit. Most of the ethylene produced goes to LDPE and ethylene oxide/ethylene glycol production. Some ethylene goes to vinyl chloride monomer production. The cracker at Bratislava produces propylene for polypropylene and cumene. Some propylene is used in propylene oxide manufacturing. There were three petrochemical projects under construction as of early 1999 [13].

### 9.4.3.5  Hungary.

The Hungarian petrochemical industry is in the hands of TVK, a state-owned company that is being prepared for privatization. There is also a state-owned oil company, MOL. Hungary is an importer of ethylene via pipeline from Ukraine. TVK and MOL are producers of propylene, and TVK produces polypropylene. Six CPI construction projects were under way in the petroleum, petrochemical, and environmental cleanup areas in early 1999 [13].

### 9.4.3.6  Romania.

Romania has the second largest ethylene capacity in central and eastern Europe. The country also has five crackers, but most of them have been closed because of the lack of money for importing feedstocks, raw materials, and spare parts. As a result, many derivative plants are out of operation. In early 1999, 14 petrochemical projects were under construction [13].

### 9.4.3.7  Bulgaria.

The Bulgarian chemical industry is dominated by two petrochemical companies, Neftochim and Polinari. Neftochim's petrochemical site is integrated with its refinery on the Black Sea. The location of the facilities is well suited for import/export activities, but the petrochemical industry has suffered from limited feedstock supply, which is caused by the lack of financial resources [13].

### 9.4.3.8  Croatia.

The Croatian petrochemical industry has been greatly affected by the breakup of Yugoslavia. Industrija Nefta owns and operates five petrochemical sites as well as three refineries. There is one small olefins cracker in Zagreb, which supplies ethylene to an adjacent complex for LDPE. The propylene produced is used as a fuel. Three construction projects were under way in the petrochemical area, one being an expansion of an existing facility, as of 1999 [13].

### 9.4.3.9  CIS.

In the Commonwealth of Independent States, there are over 15 ethylene production units. Eleven are located in Russia, two in Ukraine, and one each in Azerbaijan and Belarus.

### 9.4.3.10  Summary of the Central and Eastern European Area.

Most of the existing cracker units in the area were built in the 1970s. There is a need for investments to maintain and update these facilities. The maintenance situation occurs for the olefins and plastics units is similar. Although there is some construction going on in the region, tremendously large investments are needed on top of the need to enhance the infrastructure. At this point, it is an open question whether adequate funds will be found.

### 9.4.4  OPEC Members

From the CPI standpoint, of the OPEC nations Saudi Arabia, Iran, Iraq, the United Arab Emirates (UAE), Kuwait, Indonesia, and Venezuela are the most significant.

The OPEC nations are endowed with large petroleum reserves and therefore have a strong raw material position for the manufacture of chemicals. In establishing oil prices, there are often disagreements among the member nations. In the 1980s, when many U.S. CPI firms were moving from commodity chemicals to specialty and fine chemicals, Saudi Arabia, the UAE and other OPEC nations began building large commodity chemical plants. In the mid-1990s, some U.S. firms recognized their short-sightedness and have returned in a limited way to manufacturing commodities.

To manufacture commodity chemicals requires very large plants with large capital outlays, while the capital requirements for specialties and fine chemicals are lower. The profit margin for the latter products is higher than that for commodity chemicals.

Table 9.26 is a list of the OPEC nations with the reserves as of 1974, 1995, and 1996. OPEC has 76% of the global petroleum reserves, with Saudi Arabia having 25%. The top OPEC producers as of December 1998 are presented in Table 9.27.

#### 9.4.4.1  General Comments About the OPEC Countries.

Through the years since the formation of OPEC, oil prices have provided these nations, particularly those in the Mideast, with adequate capital to fund large and sophisticated petrochemical complexes and downstream facilities. The governments of most of these countries have been active in pursuing such development for a variety of reasons:

> To provide for large domestic demands in countries like Saudi Arabia, UAE, Iran, and Egypt.
> To prolong oil wealth by concentrating on higher value-added products derived from oil.
> To provide non-labor-intensive business, especially in countries like Saudi Arabia, where the population is low.
> To increase national power and prestige in the global arena.

There are many serious and difficult problems faced in building a CPI in the Middle East, although some of these are being overcome:

1. Technical and managerial personnel in many countries have been lacking [50]. Some countries (e.g., Iran and Egypt) have a limited supply of skilled and unskilled labor. In addition, petrochemical complexes require communities with population bases in the have to range of 20,000–50,000 people. Many of these people would be foreigners, with cultural and social backgrounds

TABLE 9.26  OPEC Reserves: 1974 and 1995–1996

| Country | Petroleum reserves (bbl × 10⁹/day) | | | Percentage of world, 1996 | Population (thousands) | | AAGR, 1974–1994 (%)[a] | Reserves (bbl × 10³ per capita) | |
|---|---|---|---|---|---|---|---|---|---|
| | 1974 | 1995 | 1996 | | 1974 | 1994 | | 1974 | 1994–1995 |
| Algeria | 7.6 | 9.2 | 9.2 | 0.89 | 16,275 | 27,400 | 2.64 | 0.47 | 0.34 |
| Indonesia | 10.5 | 5.2 | 5.0 | 0.48 | 127,586 | 190,400 | 2.02 | 0.08 | 0.03 |
| Iran | 60.0 | 88.2 | 93.0 | 8.97 | 32,139 | 62,300 | 3.36 | 1.87 | 1.42 |
| Iraq | 32.0 | 100.0 | 112.0 | 10.80 | 10,765 | 19,700 | 3.07 | 2.97 | 5.08 |
| Kuwait | 64.0 | 96.5 | 96.5 | 9.31 | 929 | 1,600 | 2.76 | 68.89 | 60.31 |
| Libya | 25.6 | 29.5 | 29.5 | 2.85 | 2,346 | 5,200 | 4.06 | 10.91 | 5.67 |
| Nigeria | 19.9 | 20.8 | 15.5 | 1.49 | 61,270 | 108,000 | 2.87 | 0.32 | 0.19 |
| Qatar | 6.5 | 3.7 | 3.7 | 0.36 | 89 | 500 | 9.01 | 73.03 | 7.40 |
| Saudi Arabia | 132.0 | 261.2 | 261.5 | 25.22 | 8,706 | 17,800 | 3.64 | 15.16 | 14.67 |
| United Arab Emirates | 25.5 | 98.1 | 97.8 | 9.43 | 215 | 2,400 | 12.82 | 118.60 | 40.88 |
| Venezuela | 14.2 | 64.5 | 64.9 | 6.26 | 11,632 | 21,200 | 3.05 | 1.22 | 3.04 |
| OPEC total | 397.8 | 776.9 | 788.6 | 76.05 | 271,952 | 456,500 | 2.62 | 1.46 | 1.70 |
| World total | 1016.9 | 1036.9 | | | | | | | |

[a]AAGR, average annual growth rate.
Source: CMAI.

**TABLE 9.27**  Top OPEC Producers, 1998

| Country | Production (bbl $\times 10^9$/day) |
|---|---|
| Saudi Arabia | 7.83 |
| Iran | 3.60 |
| Venezuela | 2.98 |
| Iraq | 2.36 |
| United Arab Emirates | 2.20 |
| Nigeria | 2.04 |
| Kuwait | 1.69 |
| Libya | 1.34 |

*Source:* International Energy Agency. Parts published in *The Wall Street Journal,* February 9, 1999.

different from those in the Mideast countries. Also, infrastructure (housing, transportation, etc.) is limited in many of the OPEC countries.

2. In a number of OPEC countries there is a limited market, although this is rapidly changing. Large quantities of chemicals would have to be exported and sold in foreign markets. Personnel to market and develop markets for their products as well as those to handle the business aspects are being educated and trained abroad.

3. Construction costs are at least 30–40% higher than in the United States. Projects take longer because equipment must be delivered from outside the OPEC countries. Projects take longer owing to the low efficiency of the labor pool. In some cases, there are difficulties in establishing and obtaining financial foundations for large complexes. Therefore, there are severe scheduling problems. Further, when ordering equipment, it is essential that large supplies of spare parts, catalysts, chemicals, and other supplies be on hand, especially during start-up and early operations. The facilities to store these materials must be provided.

4. Certain OPEC countries, like Libya and other Persian Gulf countries, lack large quantities of fresh water for major petrochemical complexes. This problem is slowly being overcome with desalination plants similar to those in Saudi Arabia.

Although there are problems, the United States and many European countries have invested in the region and are planning future ventures. Global chemical plant construction firms are actively pursuing and winning contracts to build huge, modern plants throughout the Middle East. It is expected that this area of the world will become a major area of chemical production in the future.

### 9.4.4.2 Saudi Arabia.

The Middle East's largest chemical producer is Saudi Arabia [51–52]. It has five major core groups:

*Basic chemicals:* olefins, aromatics, and oxygenates

*Fertilizers:* urea, ammonia, sulfuric acid, and compound fertilizers

*Polymers:* polyolefins, vinyls, polystyrene, polyester, melamine, and terephthalic acid

*Intermediates:* ethylene glycol, caustic soda, plasticizers

*Metals:* steel, aluminum; industrial gases are classified here, as well

SABIC, the largest chemical company, has focused on Saudi Arabia, but the company works to develop global markets and appears to be positioned to move aggressively into the world arena.

In Saudi Arabia, 19 large petrochemical plants were under construction as of early 1999 [13].

### 9.4.4.3 Iran.

Iran is the second largest chemical producer in the region [49]. After years of stagnation due to a huge foreign debt as well as sanctions and trade embargoes, the country is still optimistic. Mobil and Atlantic Richfield (ARCO) have notified Iran that they are interested in the country's oil and gas reserves. These companies have paid small sums to receive oil field data. Executives from these two companies want to stay in contact with Iran and have kept the U.S. government informed even though they are barred from doing business there. In the meanwhile, the Iranians have opened 100 billion barrels of oil to multinational companies. Elf Aquitaine of France and ENI of Italy have signed an agreement to produce oil from a large offshore oil field. Attempting to be more attractive to investors, the Iranian government is offering such incentives as feedstocks at preferential prices, exemption from income taxes for 8 years, tariff breaks, and protection of investments from nationalization, and repatriation of profits. On the downside, beside the sanctions, are the double-digit inflation and low oil prices, political instability, and government restrictions.

Iran's petrochemical industry is striving to be a major player in the region. There are eight government-owned petrochemical production companies and three privately owned petrochemical/chemical companies. The major petrochemical products are ethylene and its derivatives, propylene and its derivatives, benzene, toluene, xylenes, pyrolysis gasoline, propane, butane, pentane, ethylbenzene, and styrene, as well as sulfur, ammonia, and its derivatives. In early 1999 there were 21 construction projects in the areas of petrochemicals, petroleum, and environmental cleanup [13].

### 9.4.4.4 Iraq.

Iraq has the second largest petroleum reserves of the OPEC group according to Table 9.26. No reliable project information on chemical, petroleum, and petrochemical industries is available [51].

By the year 2001, 10 more petrochemical units are expected to be in production. Plentiful feedstocks ensure Iran's competitiveness in the world market.

## REFERENCES

1. *Chemical and Engineering News,* Facts and Figures issue, June 1997.
2. J. H. Dunning, *Multinationals, Technology and Competitiveness.* L. Hyman, London, 1998, p. 147.
3. Office of Technology Policy, U.S. Department of Commerce, 1997.
4. *Manual of Technology Transfer.* United Nations Industrial Development Organization, Vienna, 1996, p. 8.
5. *CHEMTECH,* June 1998, pp. 13–21.
6. *Fortune,* November 10, 1997, p. 227.
7. G. J. Mossinghoff and T. Bombelles, *CHEMTECH,* pp. 47–50, May 1997.
8. *CHEMTECH,* June 1998, pp. 13–21.
9. *Chemical Week,* August 18, 1998, pp. 33–53.
10. *World Investment Report,* United Nations, New York, 1996.
11. *CHEMTECH,* January 1998, pp. 14–23.
12. *Chemical Week,* July 31, 1996, p. 31.
13. *HPI Construction Boxscore, Hydrocarbon Processing,* June 1999.
14. *Chemical Week and Engineering News,* June 23, 1997.
15. *Chemistry and Industry,* October 1996, pp. 811–815.
16. *Chemical Engineering,* January 1998, pp. 41–43.
17. *Chemical Engineering,* January 1998, pp. 34–39.
18. *Chemical and Engineering News,* June 16, 1997, pp. 16–21.
19. *Chemical Engineering,* June 1997, p. 47.
20. *Chemical Week,* January 27, 1999, p. 31.
21. *Wall Street Journal,* March 4, 1999, p. A11.
22. *Chemical Week,* January 25, 1997, pp.3–24.
23. *Chemical Week,* June 19, 1996, pp. 1–6.
24. *Chemical Engineering,* August 1997, pp. 41–43.
25. *Chemical Week,* June 24, 1998, pp. 1–12.
26. *CHEMTECH,* January 1996, pp.42–49.
27. *Chemical and Engineering News,* April 7, 1997, pp. 32–38.
28. *Chemical Week,* September 2, 1998, pp.37–42.
29. *Chemical and Engineering News,* November 30, 1998, pp. 16–18.
30. *Hydrocarbon Processing,* March 1998, pp. 47–53.
31. *Chemical Week,* February 4/11, 1998, pp.12–15.
32. *Chemical Week,* February 3/10, 1999, p. 8.
33. *Chemical Week,* February 4/11, 1998, p. 16.
34. *Chemical Week,* February 3/10, 1999, p. 12.
35. *Hydrocarbon Processing,* March 1995, pp. 44–46.
36. *Hydrocarbon Processing,* March 1995, pp. 46–47.
37. *Chemical Week,* February 4/11, 1998, pp. 6–8.

38.  *Chemical Week*, February 3/10, 1999, p. 10.
39.  *Hydrocarbon Processing*, March 1995, pp. 48–50.
40.  *Chemical Week*, June 3, 1998, p. 29.
41.  *Chemical Week*, February 4/11, 1998, p. 11.
42.  *Chemical Week*, June 3, 1998, p. 31.
43.  *Hydrocarbon Processing*, March 1995, p. 48.
44.  *Chemical Week*, February 4/11, 1998, pp. 4–5.
45.  *Chemistry and Industry*, November 18, 1996, pp. 872–875.
46.  *Chemical Week*, November 20, 1996, pp. 31–32.
47.  *Chemical and Engineering News*, August 17, 1998, pp. 20–32.
48.  *Chemical Week*, September 16, 1998, pp. 43–49.
49.  *Chemistry and Industry*, December 1996, pp. 779–782.
50.  Private communication, 1999.
51.  *Chemical and Engineering News*, August, 31, 1998, pp. 15–18.
52.  *Chemical and Engineering News*, August 3, 1998, pp. 22–30.

## 9.5  SUGGESTED READING

*The World Factbook*, Central Intelligence Agency, Washington, DC, 1997.

*Statistical Abstract of the United States*, U.S. Census Bureau, Washington, DC, 1997 (presents not only domestic but international data).

*Survey of Current Business*, U.S. Department of Commerce, Washington DC; current issues.

A. Arora, R. Landau, and N. Rosenberg, *Chemicals and Long-Term Economic Growth*. John Wiley & Sons, New York, 1998.

*The United Nations Yearbook of Labor Statistics*, United Nations, New York, 1998.

*The United Nations Statistical Yearbook*, United Nations, New York, 1998.

*The World Investment Report*, United Nations, New York, 1997.

Articles about the CPI in specific world areas appear frequently in *Chemical and Engineering News, Chemical Week, and Chemical Engineering*. These are very useful as references.

## PROBLEMS

9.1.  Compare Tables 9.1 and 9.2. Table 9.1 lists 10 countries ranked according to chemical sales. Using Table 9.2, rank these countries according to their exports as a percentage of the GDP. Is the ranking the same as in Table 9.1? If so, explain why.

9.2.  Often, biased presentations of economic indicators give misleading impressions of the economic performance of a country or a firm. Illustrate this by using the data from Table 9.4 to draw two graphs, one showing the U.S. income from exports versus time and another depicting U.S. exports as a percentage of world exports versus time.

9.3.   Consult Table 9.22 and note data given for two companies with considerable sales, ICI and Rhône-Poulenc. What are the sales for these companies? What do these companies spend on research and development? What return on assets do these companies enjoy? What are respective sales per employee? Explain any differences you find. Data in Tables 9.17 and 9.20 may be useful in your explanation. If you were seeking a job and both ICI and RhônePoulenc made you an offer, what would you be inclined to do, assuming you have found the answers to the questions above about the two firms? Why?

9.4.   There has been much in the news about the importation of foreign goods. U.S. producers claim they have been damaged by an influx of imported color TV sets and have proposed quotas, but U.S. consumers prefer sets from the Far East because they are cheaper or of better quality. Would the following groups support a quota? Why?

   a.  U.S. producers of color television sets and labor unions
   b.  U.S. producers of picture tubes, transistors, plastics, and other supplies for television sets
   c.  U.S. consumers
   d.  U.S. Department of State
   e.  U.S. Department of Commerce

Is a quota in the national interest? What could Far East nations do to retaliate? Which U.S. sector(s) would suffer?

9.5.   Is the relative decline of the U.S. CPI a national crisis? Is the decline inevitable? Is it caused by increased investing of U.S. firms abroad? Selling technology abroad? Training foreign students in chemistry and chemical engineering? Can we stop these developments? Should we?

9.6.   What is your attitude toward working for a U.S. company abroad or for a foreign company in the United States? What about working for a foreign national company, such as Aramco, in Saudi Arabia? Would you have sufficient skills and career development opportunities for upward movement, job security, and esteem of friends and family?

9.7.   What are the advantages of permitting foreign chemical professionals to emigrate to the United States, and who are the beneficiaries? What are the advantages and disadvantages, and who are the victims? On a whole, should immigration be unrestricted, restricted by quotas, or forbidden?

9.8.   To modernize, less-developed countries must import high-technology products (such as automobiles, chemicals, computers, and airplanes) from the more developed countries. To pay the bills, they must sell things to more developed countries, thus earning foreign exchange; selling raw materials is not enough. Many economists believe that the less-developed countries will sell low-technology products (such as clothing, toys, and shoes) to the more developed countries. Who in the more developed countries would not subscribe to this idea, and why? Should this scenario occur as described?

9.9.    Why are the German and Japanese CPI gaining on the U.S. CPI? Can and should the U.S. government do something to reverse this trend?

9.10.    When a less-developed country starts a plant to produce chemicals locally, usually imports are restricted or shut off. Consumers complain of high prices and low quality. It has been said that in the 1880s, the South would have much preferred consumer products and machinery from England and France to inferior and more expensive products from New England and the rest of the North. Is protected domestic production justified?

9.11.    Multinational corporations are the impetus for change in less-developed countries, introducing modern technology and business organization, a better standard of living, and a social system that is often more democratic than arbitrary local customs. What are the complaints against multinational corporations, and do they outweigh the advantages? Are there other driving forces that offer better advantages?

# 10

## Future Prospects:
## Threats and Opportunities

Even in the midst of prosperity, prudent company management plans not only for continuity but for growth in the firm's business endeavors. We are in an era of the most rapid rate of change in the history of mankind. Toffler [1] in 1970 eloquently described our predicament: "Rip Van Winkle found himself a stranger in his own village after a long sleep; we would feel estrangement after a brief nap. As the Red Queen said to Alice, 'We have to keep on running very hard to stay in the same place.'"

The CPI is one of the most dynamic sectors of the economic life of the United States, responding to both external forces and internal innovations. No chemical company can feel secure for very long with an established technology or market. Examples abound of chemical products and processes that were once important but are now obsolete or dated.

At the founding of the United States, two important products of the colonial chemical industry were spermaceti candles and black powder. The spermaceti candles were replaced by petroleum wax candles, then by kerosene lamps, and finally by the electric lightbulb. An essential ingredient of black powder was potassium nitrate (saltpeter). Congress established a saltpeter works in Philadelphia when the British blockade of the colonies restricted importation of the product. Benjamin Rush, a chemist and signer of the Declaration of Independence, operated the plant. Niter deposits were found on the walls of cellars and stables as incrustations resulting from the decomposition of organic material. Niter beds were constructed of decayed animal and vegetable material, but the production was small; 0.3 pound of saltpeter per cubic foot every 2 years[2]. Eventually, domestic and imported saltpeter was replaced by Chilean saltpeter and subsequently by ammonia synthesis and oxidation. During World War I, blockades prevented the importation of the Chilean product. Finally, black powder was replaced by more modern explosives such as dynamite, trinitroglycerin (TNT), ammonium nitrate and fuel oil, and atomic and hydrogen bombs. Although the Du Pont Company, founded by a French immigrant in 1802, manufactured black powder for the U.S. Army, hunters, and frontiersmen who needed it to clear land during the westward migration, the company sold its last black powder operation in 1973.

A more recent example of the rise and decline of an important product is rayon. As the first mass-produced synthetic fiber, it had phenomenal growth between World War I and II, capturing a large share of the cotton and silk markets. Competition from nylon and polyesters as well as improved cotton products, caused rayon's popularity to wane, resulting in a very small share of the total fiber market. Cellophane is another product that lost its market share—in this case to polyvinyl chloride, polyethylene, and polypropylene wrapping films.

The inorganic raw material supply comes mostly from mining. The shortage or exhaustion of one material requires the substitution of another. Chilean saltpeter replaced niter-bed saltpeter. The exhaustion of naturally occurring bauxite has forced U.S. aluminum companies to seek bauxite ores elsewhere—in Jamaica, Venezuela, and Australia.

Even more dramatic raw material changes have occurred in the supply of organic raw materials. The CPI initially used organic materials derived from farming, fishing, forestry, and animal husbandry sectors. Coal tar and coke oven rose as principal sources for the manufacture of synthetic dyes. In the decades after World War II, coal was displaced as a raw material by natural gas and refined petroleum products because of the lower cost and greater cleanliness of the latter. Early in the twenty-first century, the CPI may have to rely more heavily on vast coal deposits as a resource base, as supplies of oil and natural gas dwindle.

There have been notable shifts in raw materials for the manufacture of maleic anhydride and phenol. Made for many years by the oxidation of benzene, maleic anhydride now is made by a catalytic process from butane. The butane process was found to result in lower costs of operation as well as reduced environmental, safety, and health hazards. Another example is the manufacture of phenol, initially made from benzene or chlorobenzene. Subsequently, however, with large supplies of cumene from the catalytic reaction of benzene and propylene, production came to be dominated by cumene-derived phenol, which, requires a lower capital investment and offers reduced operating expenses as well as reduced environmental and safety problems. A novel process developed by a German company oxidizes benzene catalytically to phenol. It is a direct process with potentially fewer side products but has yet to be commercialized.

CPI companies do not have a guaranteed place in the economy, as evidenced by the disappearance of once-dominant companies through acquisitions or mergers, as noted earlier in this book. For example, Virginia Carolina Chemical Company and American Viscose Company failed to maintain their commanding positions in a rapidly changing world and were absorbed by more adaptable and successful companies. Parts of major companies have been spun off, such as Albermarle from Ethyl Corporation and Solutia from Monsanto, supposedly to let each entity focus on the operation of the line of business.

In addition to the economic threats, companies now must deal with complex sociopolitical factors that may affect their fate. Companies have come under public attack particularly in the areas of environmental problems, product liability, and employee health and safety. Responsible and effective business behavior is required to minimize the threats of governmental regulation and control. When the government becomes concerned about potential monopolistic positions of companies, the breakup of larger companies often results. Du Pont years ago was ordered by the courts to sell Hercules and Atlas (now part of ICI). The Standard Oil Trust was a classic monopoly that was broken into Standard Oil of New Jersey, Mobil, American Oil, Atlantic Richfield, Standard Oil of California, and Standard Oil of Kentucky. Very recently the software giant Microsoft has been the target of government regulators.

With the many changes occurring in the CPI, there are dangers to the continuity of a process, a product, or a chemical company. Table 10.1 lists some of the threats and the measures needed to combat the threats. Also listed are potential opportunities for innovation and growth brought about by the changes and threats.

Chemical professionals used to join a company with the expectation of spending 30–40 years of their life there and were concerned about the company's future rather than immediate prospects. This is no longer the case, as the chemical professional will more likely work for perhaps a half-dozen companies during his or her professional life. In any case, the chemical professional should study the historical growth patterns in the CPI as well as the internal and external forces affecting the CPI and then plan to improve its future.

## 10.1 GROWTH PATTERN

The CPI is characterized by growth as exemplified by the diversity of products and growth rates of different products in Table 10.2. The main reasons for the growth in annual production of a chemical are as follows:

**TABLE 10.1** The CPI: Threats, Needs, and Opportunities

| Part of the CPI | Threats | Needs | Opportunities |
|---|---|---|---|
| Manufacturing plants | Obsolescence, deterioration, inadequate capacity, poor location | Capital for investment in plants | Superior technology<br>Larger and more efficient plants<br>Better locations |
| Technology | Replacement with more efficient processes<br>Dwindling raw material supply | Research and development<br><br>Technical services | New processes<br>New products<br>Alternative or substitute raw materials<br>Solution to safety, health, legal and problems |
| Market | Replacement by substitute or better products<br>Changes in important customers' needs<br>Decline or disappearance of important customers<br>Government intervention | Market research<br>Market surveillance<br>Market development | Penetration of markets not previously serviced with existing products<br>Invasion of existing markets with newly developed or substitute products<br>Competitors entering the market<br>Development of new markets through development of new products |

| | | |
|---|---|---|
| Employees | Obsolesence<br>Low productivity<br>Downsizing<br>Outsourcing<br>Retirements and resignations<br>Low-cost foreign labor<br>Fair and competitive compensation | Recruitment of employees<br>Career development<br>Continuing education<br>Attendance at professional society meetings<br>Provision of employee tools and enhancing of productivity | Upgrading, reeducating, and conferencing education of employees<br>Improvement in productivity |
| Public, corporate, and increased government relations | Opposition to plant site location<br>Inability to obtain needed capital<br>Tax increases<br>Divestiture, confiscation, antitrust action, mergers and acquisitions<br>Restraint of trade action<br>Nationalization<br>Import—export control<br>Price controls<br>Capital flow control<br>Inflation, war, and civil unrest | Lobbying<br>Public relations by employees<br>Public education by corporation | Increased corporate–industry cooperation<br>Increased corporate–community relations<br>Better industry–university understanding<br>Better service to community<br>Greater public acceptance |

**TABLE 10.2** Comparisons of U.S. Chemical Production

| Chemicals | Production ($lb \times 10^6$) | | | | |
|---|---|---|---|---|---|
| | 1974 | 1980 | 1985 | 1990 | 1995 |
| Inorganics | | | | | |
| Carbon dioxide | 2,910 | 7,090 | 8,630 | 10,980 | 10,890 |
| Chlorine | NA | 22,926 | 20,783 | 23,688 | 26,881 |
| Hydrochloric acid | 4,810 | 5,500 | 5,560 | 4,680 | 7,330 |
| Sodium hydroxide | 21,730 | 23,248 | 21,781 | 24,912 | 27,635 |
| Nitric acid | 16,370 | 17,090 | 15,560 | 15,550 | 17,240 |
| Oxygen | 32,120 | 34,350 | 31,370 | 38,990 | 53,480 |
| Sulfuric acid | 64,710 | 80,690 | 79,230 | 88,560 | 95,360 |
| Organics | | | | | |
| Acetic acid | 2,260 | 1,528 | 1,925 | 2,004 | 3,298 |
| Acetone | 2,060 | 2,076 | 1,847 | 2,330 | 2,778 |
| Benzene | 11,070 | 11,506 | 10,198 | 13,949 | 15,761 |
| Ethylene | 23,891 | 28,667 | 30,600 | 37,635 | 46,967 |
| Ethylene dichloride | 7,700 | 10,547 | 13,792 | 18,040 | 24,385 |
| Ethylene oxide | 3,890 | 5,220 | 5,833 | 6,294 | 7,392 |
| Ethylene gylcol | 3,110 | 4,386 | 4,678 | 4,837 | 5,829 |
| Formaldehyde | 5,850 | 5,476 | 6,453 | 6,413 | 8,109 |
| Isopropanol | 1,910 | 1,801 | 1,195 | 1,380 | 1,426 |
| Phenol | 2,320 | 2,568 | 2,787 | 3,593 | 4,242 |
| Propylene | 9,820 | 13,677 | 15,507 | 20,675 | 25,990 |
| Styrene | 5,940 | 6,905 | 7,619 | 7,959 | 11,363 |
| Fibers | | | | | |
| Nylon | 2,124 | 2,358 | 2,343 | 2,662 | 2,703 |
| Polyester | 2,926 | 3,989 | 3,341 | 3,195 | 3,887 |
| Acrylic | 631 | 779 | 631 | 506 | 432 |
| Olefin | 531 | 748 | 1,249 | 1,822 | 2,428 |
| Plastics (resins) | | | | | |
| Low density polyethylene | 5,973 | 7,291 | 8,889 | 11,176 | 12,886 |
| High density polyethylene | 2,837 | 4,405 | 8,871 | 8,334 | 11,211 |
| Polypropylene | 2,249 | 3,648 | 5,139 | 8,318 | 10,890 |
| Polystyrene | 3,364 | 3,521 | 4,054 | 5,012 | 5,656 |
| Epoxy | 249 | 315 | 385 | 499 | 632 |
| ABS | 857 | 761 | 2,038 | 2,335 | 2,908 |
| Polyester | 911 | 847 | 1,223 | 1,221 | 1,577 |
| Melamine | 164 | 167 | 192 | 202 | 290 |
| Phenolic | 1,335 | 1,499 | 2,621 | 2,946 | 3,204 |
| Urea | 835 | 1,165 | 1,210 | 1,496 | 1,816 |

*Source:* Data from various sources.

1. Increasing global population
2. Increasing affluence as measured by real GDP per capita growth (see Chapter 2 and 9)
3. New uses for chemical products (e.g., as in oxygen for steelmaking) and displacement of natural products (e.g., nylon and polyester replacing silk and cotton)
4. Declining chemical prices, which cause chemicals to be used in broader markets (Chapter 2)
5. Economic growth of foreign nations and the expansion of international trade (Chapter 9)
6. Increasing prices caused by inflation or by upgrading a product to higher specifications and performance

The growth rate of a chemical measured in dollars per year can be termed "neutral" (growing at the same rate as the GDP), "aggressive" (growing at a much higher rate than the GDP), or "declining" (accounting for a progressively smaller share of the GDP even though it is still increasing in absolute tonnage).

The growth rates of chemical sales for the 1970–1996 period as measured in terms of the value of shipments (in billions of dollars) and by an industrial production index are given in Table 10.3. There has been a steady growth in manufacturing shipments, namely, 4.8% in the 1986–1996 period, exceeded only by chemicals and allied products, with an annual growth rate of 6.1%. In this latter group, shipments of drugs, soaps, and toiletries led, with an impressive annual growth rate of 7.7%

In Table 10.3, industrial production indices are presented. From 1986 to 1996, the industrial production index increased greater than 29% overall, with an annual growth rate of 2.6%. In the same time period, the chemicals and allied products index increased by 32% with an annual growth rate of 2.8%. The two leading growth areas in the CPI were drugs and medicines, with an overall growth of 54% and plastic materials and resins, at 55% from 1986 to 1996. The annual production index growth rate in that time period was 4.4 and 4.5%, respectively. Industrial inorganic and organic chemicals, man-made fibers, soaps and toiletries, as well as paints and allied products have had annual production growth indices less than the increase in the GDP or GNP for the 1986–1996 period. The low percentage changes over this 10-year period reflect the recession in the early 1990s, company realignments, mergers and acquisitions, and increased global competition.

Over the 1986–1996 period, operating rates, sometimes called capital utilization, has averaged 80.5% for the chemicals and allied products sector, which was more than one percentage point below manufacturing as a whole. In 1991 the manufacturing sector had an operating rate of 78.0%, reflecting the recession and the slow recovery in the 1990s. Although the operating rate for chemicals

TABLE 10.3   Growth of the CPI in Industry Shipments,[a] Industrial Production Indices,[b] and Capacity Utilization[b]

| | Industry shipments (billions) | | | | | | |
|---|---|---|---|---|---|---|---|
| | 1970 | 1980 | 1986 | 1987 | 1988 | 1989 | 1990 |
| Manufacturing | $633.7 | $1,852.7 | $2,335.9 | $2,475.9 | $2,695.4 | $2,840.4 | $2,912.2 |
| Chemicals and allied products | $49.2 | $168.2 | $205.7 | $229.5 | $261.2 | $283.2 | $292.8 |
|   Industrial chemicals, except pigments | $26.8 | $101.7 | $112.2 | $126.2 | $145.8 | $158.2 | $158.4 |
| Drugs, soap, and toiletries | $15.1 | $41.9 | $66.4 | $74.0 | $82.9 | $90.6 | $98.4 |
| Other chemical products | $7.3 | $24.6 | $27.1 | $29.4 | $32.5 | $34.4 | $36.0 |

| | Industrial production indices (1992=100) | | | | | | |
|---|---|---|---|---|---|---|---|
| | 1970 | 1980 | 1986 | 1987 | 1988 | 1989 | 1990 |
| Industry | $58.7 | $79.7 | $89.0 | $93.1 | $97.3 | $99.0 | $98.9 |
| Manufacturing | 54.8 | 75.5 | 88.1 | 92.8 | 97.1 | 99.0 | 98.5 |
| Chemicals and allied products | 48.8 | 75.9 | 82.4 | 87.0 | 92.2 | 95.1 | 97.3 |
| Industrial chemicals | 49.6 | 76.0 | 82.8 | 88.2 | 94.8 | 98.8 | 101.0 |
| Industrial inorganic chemicals | 69.4 | 83.1 | 81.8 | 85.4 | 89.4 | 92.6 | 101.2 |
| Alkalies and chlorine | 96.5 | 87.3 | 94.9 | 99.5 | 101.8 | 101.7 | 98.7 |
| Inorganic pigments | n/a | 79.3 | 84.3 | 88.0 | 97.9 | 100.1 | 95.5 |
| Industrial inorganic chemicals not elsewhere classified | 81.1 | 81.0 | 78.1 | 81.7 | 84.6 | 88.0 | 101.1 |
| Plastics materials and synthetics | 35.9 | 64.0 | 82.9 | 90.7 | 94.5 | 97.5 | 95.9 |

**TABLE** 10.3

| | | | | | | Annual growth rates (%) | | |
|---|---|---|---|---|---|---|---|---|
| | | | | | | 1986– 1996 | 1991– 1996 | 1995– 1996 |
| 1991 | 1992 | 1993 | 1994 | 1995 | 1996 | | | |
| $2,878.2 | $3,004.7 | $3,127.6 | $3,343.8 | $3,566.9 | $3,719.2 | 4.8 | 5.3 | 4.3 |
| $298.5 | $305.4 | $314.9 | $333.3 | $358.5 | 372.4 | 6.1 | 4.5 | 3.9 |
| $155.4 | $157.6 | $158.8 | $168.9 | $186.7 | 184.3 | 5.1 | 3.5 | -1.3 |
| $107.1 | $110.7 | $117.9 | $122.6 | $126.9 | 139.1 | 7.7 | 5.4 | 9.6 |
| $36.1 | $37.1 | $38.2 | $41.8 | $44.9 | 49.0 | 6.1 | 6.3 | 9.1 |

| | | | | | | Annual growth rates (%) | | |
|---|---|---|---|---|---|---|---|---|
| | | | | | | 1986– 1996 | 1991– 1996 | 1995– 1996 |
| 1991 | 1992 | 1993 | 1994 | 1995 | 1996 | | | |
| $96.9 | $100.0 | $103.4 | $108.6 | $112.1 | $115.2 | 2.6 | 3.5 | 2.8 |
| 96.2 | 100.0 | 103.7 | 109.4 | 113.2 | 116.4 | 2.8 | 3.9 | 2.8 |
| 96.4 | 100.0 | 101.0 | 104.1 | 106.5 | 108.8 | 2.8 | 2.4 | 2.2 |
| 97.0 | 100.0 | 97.8 | 101.4 | 103.3 | 103.5 | 2.3 | 1.3 | 0.2 |
| 97.7 | 100.0 | 96.5 | 91.3 | 91.2 | 87.5 | 0.7 | -2.2 | -4.1 |
| 100.0 | 102.5 | 108.2 | 111.5 | 112.6 | | 2.6 | 2.7 | 1.0 |
| 89.1 | 100.0 | 99.3 | 101.3 | 99.0 | 94.4 | 1.1 | 1.2 | -4.6 |
| 97.4 | 100.0 | 93.1 | 86.3 | 86.3 | 81.0 | 0.4 | -3.6 | -6.1 |
| 92.9 | 100.0 | 101.0 | 108.8 | 111.8 | 117.6 | 3.6 | 4.8 | 5.2 |

TABLE 10.3   Continued

|  | Industrial production indices (1992=100) | | | | | | |
|---|---|---|---|---|---|---|---|
|  | 1970 | 1980 | 1986 | 1987 | 1988 | 1989 | 1990 |
| Plastic materials and resins | 27.2 | 57.3 | 78.8 | 89.4 | 92.9 | 94.7 | 95.3 |
| Man-made fibers | 44.9 | 70.6 | 88.5 | 94.4 | 97.7 | 104.2 | 97.0 |
| Drugs and medicines | 36.0 | 62.8 | 75.8 | 78.4 | 82.2 | 85.1 | 88.3 |
| Soap and toiletries | 62.2 | 94.5 | 88.3 | 91.3 | 97.1 | 98.1 | 99.9 |
| Paints and allied products | 87.6 | 93.8 | 105.3 | 107.6 | 111.2 | 106.9 | 104.5 |
| Industrial organic chemicals | n/a | 81.5 | 83.5 | 87.6 | 98.0 | 103.5 | 104.9 |
| Agricultural chemicals | 43.5 | 80.4 | 74.8 | 84.6 | 90.0 | 97.2 | 100.4 |

|  | Capacity utilization (%)[c] | | | | | | |
|---|---|---|---|---|---|---|---|
|  | 1970 | 1980 | 1986 | 1987 | 1988 | 1989 | 1990 |
| Manufacturing | 79.4 | 79.5 | 78.7 | 81.3 | 83.8 | 83.6 | 81.4 |
| Chemicals and allied products | 77.7 | 78.4 | 77.6 | 81.3 | 84.0 | 83.7 | 83.0 |

[a]*Source:* U.S. Bureau of the Census and CMA analysis.
[b]*Source:* The U.S. Federal Reserve Board.
[c]Capacity utilization is also referred to as "operating rate."

and allied products has fallen from its 1988 high of 84.0% to 79.0% in 1996, chemical industry production has risen 18% from 1988 to 1996. There has been no significant change in the operating rate in the period 1993–1996. In the industrial complex, overall operating rate does not appear to be a problem at present.

It is known that growth segments (e.g., drugs and medicines and specialty chemicals) represent the dynamic advances in technology, increases in end uses and heavy investment in research and new manufacturing facilities. The production of noninnovative, established chemicals like sulfuric acid and caustic tends to keep pace with the growth of the GDP and population increases. Certain inorganic chemicals like sodium sulfate and calcium chloride are declining in use. Production of certain organic materials like cellulosic fibers is declining and fail-

TABLE 10.3

| 1991 | 1992 | 1993 | 1994 | 1995 | 1996 | Annual growth rates (%) | | |
|---|---|---|---|---|---|---|---|---|
| | | | | | | 1986– 1996 | 1991– 1996 | 1995– 1996 |
| 90.5 | 100.0 | 98.2 | 111.1 | 114.2 | 122.5 | 4.5 | 6.2 | 7.3 |
| 97.1 | 100.0 | 104.8 | 104.1 | 105.2 | 106.4 | 1.9 | 1.8 | 1.1 |
| 93.1 | 100.0 | 100.5 | 105.9 | 111.2 | 116.9 | 4.4 | 4.7 | 5.1 |
| 98.8 | 100.0 | 107.2 | 106.0 | 104.1 | 104.2 | 1.7 | 1.1 | 0.1 |
| 98.6 | 100.0 | 106.1 | 114.3 | 114.6 | 125.0 | 1.7 | 4.9 | 9.1 |
| 99.9 | 100.0 | 96.1 | 101.5 | 103.9 | 102.1 | 2.0 | 0.4 | −1.7 |
| 97.6 | 100.0 | 100.7 | 99.9 | 103.6 | 103.2 | 3.3 | 1.1 | -0.4 |
| 1991 | 1992 | 1993 | 1994 | 1995 | 1996 | Annual growth rates (%) | | |
| | | | | | | 1986– 1996 | 1991– 1996 | 1995– 1996 |
| $78.0 | $79.5 | $80.8 | $83.1 | $83.1 | $82.1 | 0.4 | 1.0 | −1.2 |
| 80.1 | 80.3 | 78.8 | 79.1 | 79.1 | 78.6 | 0.1 | −0.4 | −0.6 |

ing to keep pace with the GDP. These materials are being replaced by synthetic fibers that have better performance characteristics at lower prices, as noted in Table 10.4. The growth of some chemical products can be attributed to the increase in total use (e.g., synthetic rubber slowly replacing natural rubber and noncellulosic fibers replacing cellulosic man-made fibers). Natural cotton has had an increase in the last 10 years as a result of increased population and better performance of modified cotton fabrics (see Table 10.4).

A chemical may exhibit above-average growth if it serves an industrial customer that has experienced above-average growth or has changed to a new technology requiring the chemical. For example, the change in the manufacture of steel to the use of oxygen-enriched steel mills caused such a boom in oxygen

TABLE 10.4 Rubber and Fiber Use in the United States: 1970–1994

| Year | Rubber (×10³ metric tons) | | | Fiber (×10⁶ lb)[a] | | | | | | |
|---|---|---|---|---|---|---|---|---|---|---|
| | | | | Natural | | | Cellulosic | Man made | | Total fibers |
| | Natural | Synthetic[b] | Total rubber | Cotton | Wool | Total | | Non cellulosic | Total synthetic | |
| 1970 | 559[c] | 1,918 | 2,477 | 3,854 | 177 | 4,031 | 1,414 | 4,086 | 5,500 | 9,531 |
| 1974 | 687[c] | 2,393 | 3,080 | 3,309 | 137 | 3,446 | 1,111 | 6,589 | 7,700 | 11,146 |
| 1980 | 931[a] | NA | — | 3,125 | 167 | 3,292 | 748 | 7,165 | 7,913 | 11,205 |
| 1985 | 840[a] | 1,748 | 2,588 | 3,176 | 194 | 3,370 | 542 | 7,626 | 8,168 | 11,538 |
| 1990 | 1,000[a] | 1,907 | 2,907 | 4,699 | 185 | 4,884 | 599 | 8,528 | 9,127 | 14,011 |
| 1994 | 992[a] | 2,099 | 3,091 | 4,545 | 236 | 4,781 | 594 | 9,691 | 10,285 | 15,066 |

[a]From U.S. Bureau of the Census, *Statistical Abstracts of the United States: 1990*, Table 1322 (for 1970, 1974, 1980, and 1985), and 1997, Table 1229 (for 1990 and 1994). From U.S. Bureau of the Census, *Statistical Abstracts of the United States*, 1997, Table 1099.
[b]From Standard & Poor's Industrial Survey, February 8, 1996, p. 41.
[c]From J. Wei et al., *The Structure of the Chemical Processing Industries*, McGraw-Hill, New York, 1979, p. 343.

consumption that the volume of oxygen is now the third largest of all chemicals produced. Certain segments of the economy grow faster than average rates. An example of this is the vast increase in computer chip manufacture, which is highly dependent on chemicals. Established noninnovative chemicals tend to grow at about the same rate as the growth in customers. In contrast, if the number of these customers dwindles, the demise of that chemical product will follows no matter how innovative the chemical or process happens to be. The measures of growth from 1975 and 1980 to 1995 as well as the growth from 1950 to 1974 are presented in Table 10.5 for selected segments of the CPI.

In the period from 1950 to 1974, there was unprecedented growth in chemicals and in rubber and plastics, 6.16% and 4.82%, respectively, keeping pace or exceeding the GNP. Since then, in the periods 1975–1995 and 1980–1995, the CPI, except for petroleum products, has not kept pace with the GNP or GDP but has exceeded the population growth. In these latter two periods, rubber and plastic products have been the best performers. The increase in the rubber and plastic products segment has been greater than the growth of much of manufacturing toward new-use and replacement of older product roles. These roles are the result of increased research and development as well as marketing efforts. Several key customers of chemicals as well as rubber and plastics have not grown any faster than the industrial production, as noted in Table 10.6

It would be instructive to look at an assortment of growth figures for the United States in terms of personal consumption during selected years. Table 10.7 reflects the growth in durable goods, nondurable goods, and services.

As noted, the U.S. economy is becoming more service oriented. The increased percentage devoted to service is characteristic of advanced societies.

**TABLE 10.5** Measures of U.S. Growth in Value

| | 1995/1975[a] | 1995/1980[b] | 1974/1950[c] |
|---|---|---|---|
| Population | 1.22 | 1.15 | 1.39 |
| GDP, current dollars | 4.71 | 2.60 | 4.91 |
| Production index, industrial | | | 2.78 |
| Chemicals and allied products | 1.81 | 1.42 | 6.16 |
| Petroleum products | 1.18 | 1.09 | 2.58 |
| Rubber and plastics products | 2.94 | 2.26 | 4.82 |
| Paper and products | 1.82 | 1.44 | 3.02 |

[a]*Source:* U.S. Bureau of the Census, *Statistical Abstracts of the United States,* 1985, Table 1433.
[b]*Source:* U.S. Bureau of the Census, *Statistical Abstracts of the United States,* 1996, Table 1218.
[c]J. Wei et al., *The Structure of the Chemical Processing Industries,* McGraw-Hill, New York, 1979.

TABLE 10.6   Growth in Value in Selected Sectors

|                              | 1995/1975 | 1995/1980 |
|------------------------------|-----------|-----------|
| Transportation equipment     | 1.90      | 1.57      |
| Furniture                    | 1.88      | 1.42      |
| Fabricated metal products    | 1.49      | 1.23      |
| Chemicals and allied products| 1.81      | 1.42      |
| Electrical machinery         | 3.88      | 2.39      |
| Primary metals               | 1.11      | 1.08      |
| Stone, clay, and glass products | 1.34   | 1.13      |
| Foods                        | 1.61      | 1.36      |
| Mining                       | 1.02      | 0.91      |
| Utilities                    | 1.45      | 1.27      |

*Source:* U.S. Bureau of the Census, *Statistical Abstracts of the United States,* 1976, Table 1218, and 1997, Table 1224.


TABLE 10.7   Growth in Personal Consumption in the United States

| | Growth (billions) | | | | |
|---|---|---|---|---|---|
| | 1974 | 1980 | 1985 | 1990 | 1995 | 1995/1994 |
|---|---|---|---|---|---|---|
| Personal consumption | — | $1,760.4 | $2,704.8 | $3,839.3 | $4,924.9 | — |
| Durable goods | $127.5 | $213.5 | $361.1 | $476.5 | $606.4 | 4.76 |
| Nondurable goods | $380.2 | $695.5 | $927.6 | $1,245.3 | $1,486.1 | 3.91 |
| Services | $369.0 | $851.4 | $1,416.1 | $2,117.5 | $2,831.8 | 7.67 |
| GDP | — | $2,784 | $4,181 | $5,744 | $7,246 | 2.60 |

*Source:* U.S. Bureau of the Census, *Statistical Abstracts of the United States,* 1996, Table 685.


Table 10.8 is a tabulation of the growth in the contribution to the GDP for specific segments of industry. Between 1980 and 1993, the GDP increased by a factor of 2.34. It is apparent that services are growing faster than durable or nondurable goods: 2.34 versus 1.73 and 2.15, respectively. In the industrial sectors, the strongest growth is in construction, public utilities, and government sectors, while agriculture, mining, as well as petroleum and coal lag behind.

The assortment of spending figures in Table 10.9 indicates that the fastest growing expenditure segments are as follows:

National health
Recreation
Airline revenue

TABLE 10.8  Growth in Contribution to GDP by Industry

| | Growth (billions of current dollars) | | | | 1993/ 1980 |
|---|---|---|---|---|---|
| | 1974 | 1980 | 1990 | 1993 | |
| Industry | | | | | |
| Private industry | — | $2,370.2 | $4,862.1 | $5,559.5 | 2.35 |
| Agriculture | $60.1 | $133.4 | $224.0 | $214.6 | 1.61 |
| Mining | $19.5 | $112.6 | $103.1 | $89.4 | 0.79 |
| Construction | $61.8 | $851.4 | $2,117.5 | $2,831.8 | 3.33 |
| Manufacturing | $325.7 | $588.3 | $1,024.7 | $1,118.3 | 1.90 |
| Chemicals and allied products | — | $47.6 | $103.6 | $117.0 | 2.46 |
| Petroleum and coal | — | $24.3 | $40.1 | $47.7 | 1.96 |
| Rubber and plastics | — | $17.0 | $34.6 | $41.4 | 2.44 |
| Transportation | $50.9 | $102.9 | $176.8 | $207.9 | 2.02 |
| Communications | $30.4 | $68.9 | $146.7 | $162.1 | 2.35 |
| Public utilities | $29.9 | $139.3 | $304.4 | $351.4 | 2.52 |
| Wholesale and retail trade | $218.9 | $436.3 | $878.7 | $1,005.5 | 2.30 |
| Durable goods | — | $348.9 | $563.7 | $603.5 | 1.73 |
| Nondurable goods | — | $239.4 | $461.0 | $514.8 | 2.15 |
| Finance industries | $177.8 | $418.4 | $982.4 | $1,180.6 | 2.82 |
| Services | $148.6 | $377.0 | $1,040.0 | $1,264.8 | 3.35 |
| Government | $167.9 | $324.2 | $676.3 | $781.6 | 2.41 |
| Federal | — | $116.4 | $221.3 | $249.8 | 2.15 |
| State and local | — | $208.8 | $180.1 | $531.8 | 2.55 |
| GDP | — | $2,708.0 | $5,546.2 | $6,343.4 | 2.34 |

Sources: U.S. Bureau of the Census, Statistical Abstracts of the United States: 1975, Table 619 (for 1974), and 1996, Table 686 (for 1980, 1990, and 1993).

Foreign travel
Federal budget
School expenditures

Past behavior is no guarantee of future performance, but the trend is clear: U.S. consumers are continuing to have greater need and utility for services for goods and luxury items that have high income elasticity. There have been large growths in expenditures for government services, although Congress attempted to reduce the budgets and reduce government programs in 1997. If U.S. firms the CPI are to participate in future growth in a better-than-average fashion, they must be cognizant of consumer needs and gear their markets accordingly.

**TABLE 10.9** U.S. Spending Trends: 1974–1995

| Item | Spending (billions) | | | | | 1995/ 1974 | Table[a] |
|---|---|---|---|---|---|---|---|
| | 1974 | 1980 | 1985 | 1990 | 1995 | | |
| National health | $114.3 | $247.2 | $428.2 | $697.5 | $949.4 | 8.3 | 154 |
| School expenditures | $98.8 | $304.0 | $333.8 | $427.9 | $458.0 | 4.6 | 234 |
| Recreation | $52.8 | $159.7 | $215.8 | $291.8 | $369.9 | 7.0 | 401 |
| Foreign travel | $8.4 | $10.4 | $24.6 | $37.3 | $43.1 | 5.1 | 428 |
| Federal budget | $263.2 | $517.1 | $734.1 | $1,031.3 | $1,254.7 | 4.8 | 512 |
| State and local government | $189.0 | $263.4 | $390.7 | $592.2 | $824.0 | 4.4 | 487 |
| Defense | $79.3 | $134.0 | $252.7 | $299.3 | $281.6 | 3.6 | 539 |
| Railroad revenue | $15.2 | $28.0 | $29.0 | $30.0 | $31.0 | 2.0 | 984 |
| Airline revenue | $11.5 | $26.4 | $39.2 | $58.5 | $63.9 | 5.6 | 1039 |
| Car, truck, bus sale | $10.0 | $11.5 | $15.7 | $14.1 | $15.4 | 1.5 | 1261 |
| Food energy (cal/person/day) | 3350 | 3420 | 3360 | 3700 | 3410 | 1.0 | 225 |
| Beef consumption (lb/person/yr) | 83.0 | 72.1 | 74.6 | 64.0 | 63.6 | 0.8 | 226 |
| Pork consumption (lb/person/yr) | 38.7 | 52.1 | 47.7 | 46.4 | 49.5 | 1.3 | 226 |
| College enrollment (millions) | 7.8 | 11.4 | 12.5 | 13.6 | 15.0 | 1.9 | 240 |
| Motor vehicle registration (millions) | 130.8 | 155.8 | 171.7 | 188.8 | 198.0 | 1.5 | 1003 |
| Federal civilian employees (millions) | 2.86 | 2.90 | 3.02 | 3.13 | 2.97 | 1.0 | 531 |
| State and local government employees (millions) | 11.800 | 13.320 | 13.669 | 15.263 | 15.824 | 1.3 | 501 |

[a]In U.S. Bureau of the Census, *Statistical Abstracts of the United States*, 1996.

## 10.2 INNOVATION AND RESEARCH

Technical and marketing innovations are the most important driving forces for economic and social progress in the world today. Some impressive innovations that have had a major impact are the development of high-yield and high-protein agricultural products, the development of effective pharmaceutical and medicinal products for the control and diminution of diseases, the development of commercial-quality membrane technology for the production of high-quality chemical products, the successful commercial application of desalination of brackish and ocean water to produce potable water, and the ever-increasing gain in the productivity of chemical plants.

Until the emergence of the modern research laboratory as a permanent institution, invention was the lonely pursuit of isolated individuals. There is a strong correlation between the expenditures for research and development and the rate of growth of industries. In general, high-technology industries devote a large percentage of their income to research and development. Table 10.10 shows that the chemicals and allied products sector and the professional and scientific instruments sector are among the leaders in their reliance on technological innovation to improve and change their products and processes. Product innovation includes both new products and new uses for existing products. Process innovation has led to lower operating costs and has permitted changes

**TABLE 10.10** Industrial Spending on Research and Development: 1993

| Industry | Spending (millions of 1994 dollars) | | | Company R&D spending as percentage of value added |
| | Total funding[a] | Company funding | Other funding sources | |
| --- | --- | --- | --- | --- |
| Chemicals and allied products | D | $16,559 | — | 8.4 |
| Petroleum refining | $1,950 | $1,939 | $11 | 6.1 |
| Machinery | $8,110 | $8,011 | $99 | 4.6 |
| Electrical equipment | $15,338 | $13,537 | $1,751 | 7.8 |
| Motor vehicles and equipment | D | $11,950 | — | 11.3 |
| Aircraft and missiles | $14,260 | $5,466 | $8,794 | 13.1 |
| Professional and scientific instruments | $11,441 | $8,058 | $3,383 | 8.2 |

[a]D figure withheld to avoid disclosure of information pertaining to an individual or specific organization.
Source: U.S. Bureau of Census, *Statistical Abstracts of the United States*, 1997, Table 971.

to different raw materials. The results usually have been lower product prices. Among the leaders in expenditures of their own funds for research and development, none spend more than chemicals and allied product sector. The government funds much of the research in the aircraft and missile sectors, while in the CPI essentially none of these expenditures are met by government funding. Within the chemicals and allied product sector, the drug group is a very heavy spender in the areas of synthetic organic chemistry, pharmacology, and drug testing. By comparison, research and development expenditures in other CPI sectors are modest.

U.S. expenditures for research and development increased dramatically after the Soviet Union launched *Sputnik* in 1957, reaching an annual compound growth rate of 10–15% in 1968. In that 11-year period, there was the prevalent thought that large expenditures on research and development would cure mankind's ills. In that period, funding increased from 1.5% to 3.0% of the GNP. Subsequent to the recession of 1968, the national research budget shrank in terms of constant dollars and as a percentage of the GNP, partly because of the general disenchantment of the public, which had come to believe that research and development expenditures were unable to deliver promised or implied miracles. For a research idea to move from the laboratory to commercialization requires a minimum of 6–10 years. In the years, since 1968, research and development expenditures have varied in the CPI depending on the economic climate, but they tend to be in the range of 3–5% of a company's income. Research programs in industry and in universities are continually being examined to determine that the research is focused on the stated objectives and will lead to short-term usefulness. Companies want assurance that the research will lead to exclusive patents, potential licensing of processes, and competitive advantages over other companies.

The benefits of research and innovation are widely distributed, with the public benefiting more than the inventors. We live in an era of abundant and relatively cheap technology both domestically and abroad. A nation like Italy and Japan, which spends about 1% of its GDP on research, can have a GDP growth rate equal to or greater than that of nations such as the United States, which spend 3% on research. The growth of high-tech research industries like the electronics and computer industries was in part responsible for the growth in banking and insurance industries, which are heavy users of computers. Strong emphasis of research in agricultural chemicals (fertilizers, pesticides, fungicides, weed control agents, etc.) has resulted in the tremendous growth of productivity on the farms. The revenue and job growth of firms of various types is presented in Table 10.11.

Mature companies like those in the CPI have modest revenue growth and negative job growth as they attempt to reduce costs and increase productivity to

TABLE 10.11 Growth Firms of Various Types

| Industry | Revenue in 1996 (millions) | Revenue growth rate, 1995–1996 (%) | Job growth rate, 1995–1996 (%) |
|---|---|---|---|
| Mature companies | | | |
| Du Pont | $ 39,689 | 6 | –8 |
| Dow | $20,053 | –4 | 2 |
| IBM | $75,947 | 6 | 7 |
| Monsanto | $9,262 | 3 | –2 |
| Exxon | $119,434 | 9 | –4 |
| Innovative companies | | | |
| Lear | $8,249 | 33 | 23 |
| Enron | $13,289 | 45 | 11 |
| ADM | $13,314 | 5 | 0 |
| Hewlett-Packard | $38,420 | 22 | 9 |
| Merck | $19,829 | 19 | 9 |
| Young technology | | | |
| Intel | $20,847 | 29 | 17 |
| Microsoft | $8,671 | 46 | 16 |
| Amgen | $2,240 | 15 | 13 |
| U.S. Robotics (3Com) | $1,978 | 122 | — |
| Micron Technology | $3,654 | 24 | 23 |

*Source: Fortune 500*, April 28, 1997.

meet global competition. Innovative and young technology firms recognize that research and development to produce innovative products is necessary for their survival. They have higher ratios of research and development to sales as well as higher ratios of profit to sales than mature companies. Also, the job growth rate is higher in these industries.

A high rate of expenditure is necessary but not a sufficient condition for growth and profitability. At its inception, an innovation is simply an idea. This idea must be nurtured and developed until it reaches maturity as a product that can be placed on the market to generate a new cash flow. The minimum time for a research idea to move from the laboratory bench to a commercial product is between 6 and 10 years. Research is a risky undertaking, and a reward is not always guaranteed. For example, several oil companies in the 1960s and 1970s invested billions in projects aimed at recovering oil from tar sands in Canada and the United States [3]. The idea behind this pioneering venture was to provide a large, new, and secure source of energy; because of a host of technical, environmental, and political problems, however, these ventures were unprofitable. New and more stringent standards for drug approval by the Food and Drug Administration (FDA) dramatically increased the cost of innovation and consequently lengthened the time needed for new drugs to reach the marketplace.

Research can be classified as "proprietary" or "cooperative"; the former is by a firm attempting to gain a competitive advantage over rival firms, and the latter is performed on behalf of a group of firms or by government and nongovernmental partners to solve common problems.

Proprietary research is carried out only by individuals and firms in an oligopoly, with the aim of developing more desirable products and more efficient processes. The results are protected by trade secrets and patents. In a monopolistic market, proprietary research need not be done, and in an atomistic competition, companies cannot afford it. Proprietary research has been responsible for most of the technical innovations that have enriched and improved our standard of living. Examples are nylons and polyesters for fibers and catalysts for cracking hydrocarbons. This type of research is expensive, and the payoffs are uncertain. Adequate rewards must be probable before such research will be undertaken, and the firm must be able to recover the research cost by exploitation of the market for a limited time period.

When James Watt invented the steam engine, he offered it to mine operators at a royalty equivalent to one-third of the fuel savings. He was plagued by piracy and had to hire detectives to catch cheaters, but eventually lowered the royalty to 15%. It is customary for the inventor to assume full responsibility for the cost and risk associated with the research and innovation. The benefits of his labor accrue, in large part, to the public and to the licensees who have assumed no risk.

Cooperative research is funded jointly by many firms, by nonprofit organizations, and by governments. The results are made available to all sponsors and often to the public in publications. Publicly regulated utilities have banded together to form organizations like the Electric Power Research Institute (EPRI) to address energy supply problems. Farmers, an atomistic sector, have their research done by cooperative investigative efforts in the U.S. Department of Agriculture and by various state experimental farms associated with land-grant universities. Companies in the CPI often form research organizations to solve common problems that may not be proprietary: the Fractionation Research Institute (FRI), the Heat Transfer Research Institute (HTRI), and Fluid Properties Research, Inc. were formed to answer the needs for engineering design data. If the results are not freely published and available to subscribers and nonsubscribers alike, such institutes may be subject to antitrust action because the data were available only to the subscribing firms that support the research, giving them an advantage over nonsubscribers.

## 10.3   CAPITAL INVESTMENT

In the manufacture of chemical products, there is a high degree of automatic process control that results in large capital investments. The high capital costs also reflect large plant capacities required to effect economies of scale, the complex nature of processes and equipment used, the large amount of equipment required, and the high level of technology, all of which are affected by the rapid technological obsolescence and depreciation of process plants [4]. This is reflected in the large investment per employee noted in Chapter 8.

The position of the CPI among the leaders in industrial capital spending is shown in Table 10.12 for selected years. There are various reports of capital spending extant in the literature, but one must recognize how the figures were obtained. Some report the gross capital expenditures, while other report the net expenditures. Table 10.12 is of the latter type. There have been changes in how the Bureau of Economic Activity (BEA) reports data. A revision is under way to "standardize" such reporting, but it was not be available for net investment in plant and equipment at the time of writing. Table 10.13 is the reported capital spending of selected chemical companies at 5 year intervals. For intervening years, the spending figures may be obtained from the appropriate Facts and Figures issues of *Chemical and Engineering News*. Although there was some softening in capital expenditures in 1993, there has been steady growth from 1985 to 1995. Capital expenditures for the years up to the end of the century are expected to decline as a result of the economic problems in the Far East in 1998.

The ratio of capital expenditures to net plant and equipment is the rate of renewal, a ratio presented in Table 10.14 for several industries in 1995. This capital renewal includes new equipment, replacement, "debottlenecking," and ex-

**TABLE 10.12**   Capital Spending in the United States by the CPI: 1975–1994

| Industry | Spending (billions) | | | | |
|---|---|---|---|---|---|
|  | 1975 | 1980 | 1985 | 1990 | 1994 |
| Chemicals | $6.25 | $11.63 | $16.44 | $20.54 | $23.40 |
| Iron and steel | $3.03 | $3.44 | $4.12 | $7.81 | $7.88 |
| Nonferrous metals | $2.28 | $2.13 | $1.68 | $2.88 | $2.36 |
| Paper | $2.95 | $6.52 | $8.59 | $16.72 | $10.87 |
| Petroleum | $10.51 | $19.57 | $26.71 | $34.48 | $27.38 |
| Rubber | $1.00 | $2.03 | $3.86 | $3.48 | $3.82 |
| Stone, clay, and glass | $1.42 | $3.98 | $3.40 | $3.31 | $5.25 |
| All manufacturing | $47.95 | $112.33 | $153.48 | $192.78 | $192.58 |

*Source: Chemical and Engineering News, Facts and Figures issues for respective years.*

TABLE 10.13   Capital Spending of Selected Chemical Companies: 1970–1995

| Company | Spending (billions) | | | | | |
|---|---|---|---|---|---|---|
| | 1970 | 1975 | 1980 | 1985 | 1990 | 1995 |
| Du Pont | $0.499 | $1.000 | $1.297 | $1.176 | $2.608 | $1.476 |
| Dow | $0.348 | $0.940 | $1.184 | $0.806 | $2.119 | $1.417 |
| Monsanto | $0.301 | $0.525 | $0.781 | $0.645 | $0.750 | $0.500 |
| PPG Industries | $0.173 | $0.176 | $0.401 | $0.452 | $0.567 | $0.448 |
| Union Carbide | $0.394 | $0.800 | $1.129 | $0.649 | $0.744 | $0.542 |
| W. R. Grace | $0.116 | $0.290 | $0.658 | $0.348 | $0.514 | $0.538 |
| Air Products & Chemical | $0.051 | $0.151 | $0.364 | $0.392 | $0.468 | $0.870 |
| Rohm & Haas | $0.086 | $0.160 | $0.093 | $0.159 | $0.412 | $0.417 |
| Olin | $0.060 | $0.099 | $0.172 | $0.154 | $0.187 | $0.201 |
| Hercules | $0.091 | $0.160 | $0.229 | $0.234 | $0.273 | $0.117 |
| Total of top 25 chemical companies | $2.711 | $6.049 | $7.364 | $6.469 | $9.688 | $8.333 |

Source: Chemical and Engineering News, Facts and Figures issues for respective years.

pansion of capacity. A 20% capital renewal rate indicates a 5-year turnover of capital for an industry. (Note the petroleum and coal products sector, with a 23.3% renewal rate.) Such turnover rates keep the equipment modern and efficient, in contrast to many sectors (e.g., textiles, rubber, primary metals).

The CPI annual capital expenditures increased by more than 40% from 1985 to 1995, and preliminary unpublished figures indicated another 28% increase from 1994 to 1996. The chemical industry is expected to continue to spend, but at a much lower rate, for the rest of the decade. There were several reasons for the growth noted from 1984 to 1996:

1. In the past when shortages of chemicals have occurred, companies have at times hesitated to add capacity for fear of overcapacity. However, if the need for chemicals cannot be met by U.S. companies, buyers will look to the global markets. To prevent this, U.S. companies have added new investment, but conservatively, remembering that during recessions as in the early 1990s, overcapacity and falling chemical prices resulted in low profit margins. In the late 1990s prices stabilized and increased, motivating companies to expand their facilities.

2. Depreciation allowances are insufficient to cover actual capital replacement needs. In periods of inflation, construction costs in some instances doubled replacement costs. In the mid-1990s to the present, inflation has stabi-

TABLE 10.14    Renewal Rates: Based on Constant Dollars (1994)

| Industry | Capital expenditures (billions) | Net plant and equipment (billions) | Rate of renewal (%) |
|---|---|---|---|
| Chemicals and allied products | $23.3 | $261.5 | 8.9 |
| Petroleum and coal products | $27.4 | $117.2 | 23.3 |
| Rubber | $3.3 | $64.1 | 5.1 |
| Paper and allied products | 10.8 | 132.7 | 8.1 |
| Food and beverage | $21.1 | $183.0 | 11.5 |
| Textile mills | $2.4 | $45.2 | 5.3 |
| Stone, clay, and glass | $5.0 | $56.7 | 8.8 |
| Primary metals | $11.1 | $177.5 | 6.3 |
| All manufacturing | $186.9 | $1,961.9 | 9.5 |

Source: U.S. Bureau of the Census, *Statistical Abstracts of the United States,* 1996, Table 862 and 863.

lized at low levels, and companies are spending for capital replacement, expansion, and new processing facilities.

3.   Major expenditures have been in expensive antipollution equipment required by stringent legislation. For chemicals and allied products in 1996, expenditures for environmental protection exceeded $6.0 billion. These expenditures in the mid-1980s were about 28% of the capital expenditures but declined to 18% as some of the old problems were corrected and newer, more efficient processing minimizing pollutants as well as more efficient environmental protection equipment were developed.

Chemical companies have scrambled to secure needed raw materials, and this also accounts for some of the large capital expenditures. An example was Du Pont's purchase of Conoco for the oil, gas, and coal raw material the latter had. Other companies integrated backward and produced their own raw materials. Still others spent heavily to convert crude oil to petrochemical feedstocks.

There was a period in the 1970s and 1980s when chemical companies spent more for capital expenditures, thereby increasing their long-term debt. When Du Pont bought Conoco, it was the first time in the history of the company that such long-term debt had been assumed. Du Pont had always avoided major indebtedness. CPI companies must maintain the confidence of investors and debtors. This makes it increasingly important for the CPI to earn a good profit to pay the interest on debt and an attractive return to investors.

The two principal uses of economic goods and services are current consumption and capital investment for the future. Modern industry and especially the CPI requires enormous investment in plants and equipment. To ensure continued growth of profits and production, a substantial percentage of industrial profits must be plowed back into the operation via retained earnings. Without the reinvestment of a portion of today's profits, there will be obsolete facilities and lower profits tomorrow.

There is apparently a correlation between fixed investment and growth of the GDP or GNP among nations [6]. The nations in western Europe and Japan have an economic growth rate greater than that of the United States.

## 10.4  PRODUCTIVITY

*Productivity* may be defined as the measure of an industry's efficient use of capital, people, and natural resources. We see dramatic examples of an increase in productivity in the pharmaceutical sector where new drugs are introduced in limited quantities. As production increases, however, the unit cost of the drug drops. Similar examples may be cited in various sectors of the economy.

Measurements of productivity are often given in terms of ratios of outputs to inputs. For example, productivity might be expressed as follows:

1. Dollar output per employee
2. Dollar output per production employee
3. Dollar output per worker-hour
4. Value added per employee
5. Dollar output per dollar assets
6. Dollar output per dollar of raw materials

Measures 1–4 indicate efficiency in the use of people; measure 5 indicates efficiency in the use of capital; and measure 6 indicates efficiency in the use of raw materials. These are partial-productivity measures. To obtain an idea of total productivity, all ratios must be considered rather than the efficiency of one particular input. There are some trade-offs, such as using automatic equipment (automatic controls, robots, etc.) instead of operators. An industry may be very efficient in the use of people but less efficient in the use of capital. Rather than absolute productivity ratios for an industry, the trend in productivity ratios is important. Rising productivity is the key to economic growth, and the increase in output per employee pays for increases in salaries and wages.

The driving forces behind productivity increases are complex and often interrelated. The causes of increased productivity in the United States are as follows:

More people in labor forces
Better educated labor forces
Increase in quantity of plants
Increase in quality of plants
Economy of scale
More efficient uses of automatic processes
More efficient market
Increase in knowledge of business line

For further information on productivity, see Chapter 8.

Table 10.15 is a listing of the value added per employee for SIC 28, SIC 291, and all manufacturing for 1987, 1992, and 1994. Also included is the compound annual growth rate. Only two sectors, soap, cleaners, and toiletries (SIC 284) and plastic materials and synthetics (SIC 282), have fallen below all manufacturing at 3.2 and 3.5%, respectively. In relative terms, SIC 28 and SIC 291 have shown significant growth compared to all manufacturing. Specific industrial groups—namely, drugs (SIC 283), agricultural chemicals (SIC 287), and petroleum refining (SIC 291)—have shown notable differences in growth rate of value added per employee at 6.4, 6.2, and 7.0%, respectively.

Inspection of the values present earlier in Table 10.9, a list of spending trends in the United States for selected years and selected items, indicates that the CPI as a group are suppliers of products for many of the segments, and spending increases contribute to the growth within the CPI. The dollar output in

TABLE 10.15  Value Added per Employee in the U.S. Chemical and Petroleum Refining Industries

| SIC | Industry | Value added per employee | | | Compound annual growth rate (%) |
| --- | --- | --- | --- | --- | --- |
| | | 1987 | 1992 | 1994 | |
| | All manufacturing | $65,794 | $84,058 | $87,138 | 3.6 |
| 28 | Chemicals and allied products | $148,946 | $193,576 | $219,890 | 5.0 |
| 281 | Industrial inorganic chemicals | $118,660 | $162,369 | $173,080 | 4.8 |
| 282 | Plastics materials and synthetics | $145,382 | $162,628 | $191,383 | 3.5 |
| 283 | Drugs | $163,814 | $250,526 | $269,553 | 6.4 |
| 284 | Soaps, cleaners and toilet goods | $181,681 | $212,862 | $235,314 | 3.2 |
| 285 | Paints and allied products | $113,418 | $140,275 | $170,020 | 5.2 |
| 286 | Industrial organic products | $170,968 | $209,056 | $251,670 | 5.0 |
| 287 | Agricultural chemicals | $159,475 | $216,475 | $257,550 | 6.2 |
| 291 | Petroleum refining | $188,373 | $248,093 | $322,667 | 7.0 |

Source: U.S. Bureau of the Census, Statistical Abstracts of the United States, 1996, Table 1211.

**TABLE 10.16**  Sales per Total Assets for Selected Companies

| Basis Company | Sales per assets | | | Revenue per assets: |
|---|---|---|---|---|
| | 1974 | 1985 | 1990 | 1996 |
| Du Pont | 1.17 | 1.17 | 1.07 | 1.14 |
| Dow | 0.98 | 0.91 | 1.00 | 0.88 |
| Monsanto | 1.19 | 1.01 | 1.15 | 1.04 |
| Union Carbide | 1.01 | 0.78 | 0.89 | 0.95 |
| Air Products & Chemical | 0.85 | 0.72 | 0.77 | 0.67 |
| Exxon | 1.34 | 1.25 | 1.21 | 1.25 |
| Mobil | 1.34 | 1.34 | 1.41 | 1.55 |
| Texaco | 1.35 | 1.23 | 1.58 | 1.53 |
| Procter & Gamble | 1.60 | 1.40 | 1.32 | 1.27 |
| PPG Industries | 1.03 | 1.06 | 1.00 | 1.13 |
| American Home Products | 1.65 | 1.46 | 1.27 | 1.05 |
| Bristol Myers | 1.53 | 1.25 | 1.14 | 1.08 |
| Eli Lilly | 0.88 | 0.85 | 0.78 | 0.66 |
| Merck | 1.07 | 0.74 | 0.97 | 0.98 |
| Pfizer | 0.92 | 0.93 | 0.75 | 0.87 |
| Goodyear | 1.24 | 1.42 | 1.28 | 1.35 |

*Sources: Chemical and Engineering News,* Facts and Figures issues, Fortune 500, as well as selected company annual reports for respective years.

terms of sales or revenue per dollar of capital (assets) is not readily available on an industry-wide scale; however, it can be approximated as in Table 10.16 for selected years. It is noted that the ratio ranges from 0.7 to 1.6 for all companies. For the large oil companies, this ratio varies from 1.21 to 1.60.

The productivity of an employee depends heavily on the existence of sophisticated equipment and an up-to-date plant. Industry variations in capital invested per production worker were presented in Chapter 8.

The savings that result from increases in productivity benefit the stockholders by increasing profits and dividends, the employees by increasing salaries and wages, and the consumer by decreasing prices. The experience and learning curves, which are measures of productivity, were discussed in Chapters 2 and 8.

## 10.5  FUTURE PROSPECTS

The future is unpredictable, but prudent men and women plan ahead to exploit new opportunities and to avoid or minimize threats. The CPI is capital intensive and require sophisticated processes; 3–10 years may elapse between the conception of a process and plant start-up. A plant may be expected to be productive for 10–25 years with proper maintenance and minor process upgrading. Thus it is necessary to attempt to predict social and economic conditions that will affect technological operations 10–25 years into the future. Projecting into the future is usually based on a small amount of scientific methodology and a great deal of imagination.

The basic information needed to assess the market in the future consists of the following:

1. The population of the United States, of industrially developed countries in western Europe and Japan, of developing nations in eastern Europe, Asia, and South America, and of the less-developed countries
2. The total buying power or these markets measured as GDP and GDP per capita
3. The changing market demand for specific product owing to product competition, substitution, changes in needs, and changes in lifestyles
4. The social and political climates for industrial activities, the movement of products and capital across national boundaries, and the profitability of chemical processing compared to alternative types of investment

Possible future developments that would necessitate technological changes are as follows:

1. The emergence of completely new sciences and technologies to create possibilities for investment that do not exist presently
2. The exhaustion of traditional raw material sources and the development of new sources that would require new processing technologies
3. The development of superior processing technologies and better products
4. The impact of new government regulations requiring novel solutions

In 1976 the American Institute of Chemical Engineers prepared a study [7] discussing a number of changes and challenges for the decade 1976–1986. These conclusions are quoted below, with the present authors' comment in brackets.

1. The United States and the world will continue to face shortages in energy, food and materials. Due mostly to a lack of political leadership, these problems will be poorly understood by the public, and the solutions will be piecemeal rather than bold steps forward. [These problems are still poorly understood by the public, and attempts to solve some of them have still been piecemeal.]

2. There will be an appreciable shift in the United States population with a decline in the absolute number of 15- to 19-year olds and an increase in the number of 25- to 45-years olds. [This shift is depicted in Table 10.17.]

**TABLE 10.17** Estimated and Projected U.S. Population (millions) by Age Group

| Year | Total | <5 | 5–13 | 14–17 | 18–24 | 25–34 | 35–44 | 45–54 | 55–64 | 65–74 | 75–85 | >85 |
|---|---|---|---|---|---|---|---|---|---|---|---|---|
| 1970 | 204.7 | 17.1 | 36.6 | 15.9 | 24.6 | 25.3 | 23.1 | 23.3 | 18.7 | 20.1 | | |
| 1974 | 211.9 | 16.3 | 34.1 | 16.9 | 26.9 | 29.8 | 22.8 | 23.8 | 19.5 | 21.8 | | |
| 1980 | 222.7 | 17.3 | 30.2 | 15.8 | 29.4 | 36.2 | 25.7 | 22.6 | 21.0 | 24.5 | | |
| 1990 | 249.9 | 18.8 | 31.6 | 13.3 | 26.7 | 42.2 | 37.4 | 25.0 | 21.1 | 18.0 | 10.0 | 3.0 |
| 1995 | 262.8 | 19.6 | 34.4 | 14.8 | 24.9 | 40.9 | 42.5 | 31.1 | 21.1 | 18.8 | 11.1 | 3.6 |
| 2000 | 271.2 | 17.9 | 35.8 | 15.6 | 25.9 | 36.7 | 44.4 | 39.6 | 23.8 | 18.0 | 12.2 | 4.1 |
| 2020 | 288.8 | 17.2 | 31.0 | 14.3 | 26.7 | 40.1 | 36.6 | 35.5 | 39.6 | 29.2 | 13.6 | 5.0 |
| 2050 | 282.5 | 16.3 | 30.1 | 13.9 | 24.8 | 35.6 | 36.5 | 34.9 | 35.7 | 27.7 | 18.6 | 9.6 |

*Sources:* U.S. Bureau of the Census, *Statistical Abstract of the United States: 1975*, Table 3 (for 1970, 1974, and 1980), and 1996, Table 14 (for 1990, 1995, 2000, 2020, and 2050).

3. There will be a shift in manpower from the older industrial Northeast to the Southwest, where energy is cheaper and to foreign energy-rich nations. [The shift in manpower and population has resulted in population growths in the southeastern and southwestern parts of the United States. These areas are attractive to industry owing to strong job markets, lower taxes, and lower costs of living. Energy-rich nations in the Mideast enjoyed growth in the petroleum and commodity petrochemical areas.]

4. The role of government in the CPI, particularly in the oil industries, will continue to increase. Government will impose performance standards, import quotas, production quotas, prices and distribution of products. [While the government has not imposed regulatory power over many of these areas cited, it is true that the government is keeping a closer look at the CPI in general.]

5. Pollution control and environmental impact of products will continue to be important issues. There will be more specific pollution-abatement projects as we learn more about environmental science and engineering. The effects on the CPI will probably intensify as it is the producer of many products having unknown and potentially serious effects. [There is ever-tightening legislation by the states and the federal government to control environmental discharges of the CPI. Product liability cases have grown markedly in part because of lack of information on various products. Continued scrutiny of the CPI and its products by government and the public is expected.]

6. The United States and the developed world will continue in the trend of postindustrial society, in which the growth sectors will not be in the production of non-durable or durable goods but in the providing of services. [It has been noted elsewhere in this text and in this chapter that the United States is moving toward a service economy and away from a manufacturing one.]

7. The demand for energy will slow down but its cost will go up. Construction of conventional oil refineries will be static, but there will be an accelerated growth in petrochemicals. We will also see the construction of massive coal and oil-shale conversion plants to provide clean oil and gas. [There has not been a major grassroots refinery built in the United States in the last three decades. There have been process improvements and upgrading of parts of existing refineries. The only new refineries in the world are in the Mideast and in Southeast Asia. There has been a shift not only in the United States but in the world to the production of petrochemicals. Massive coal and oil-shale conversion units have not come to fruition because massive amounts of capital are required to build them. Stronger environmental regulations as well as processing

problems like the disposal of refuse from the oil-shale plants have also
been deterrents to the building of such units. Further, global oil prices
have declined to a stabilized level, making coal conversion and oil-
shale plants uneconomical.]

Although parts of the predictions of the American Institute of Chemical
Engineers proved to be accurate, some changes have occurred for other reasons-
partly owing to technology improvement and increased government regulations
and partly to economic changes, as noted in the authors' bracketed comments.
The shift in the population mentioned in trend 2 above could be significant to
growth of the CPI. According to input–output analysis, what sectors of the CPI
might prosper or languish because of the age shift? What could this mean to a
chemical professional's career?

The notion of economic growth frequently comes under attack as wasting
limited resources, unnecessary for the "good life," destructive of the environment, and contributing significantly to the uneven distribution of income and
wealth. Such growth is sometimes considered to be the root of anti technology
and anti establishment sentiment. Large corporations in the CPI are often viewed
as having great power, lacking regard for environmental and consumer interests,
and having inadequate concern for employee health and safety. Industry's credi-
bility has frequently been damaged by the handling of such problems as the vinyl
chloride levels in plants and the Bhopal incident, inadequate testing as well as
claims relating to unsafe handling of certain chemicals and pesticides, and the
silicone breast implants.

It cannot be denied that some CPI companies have exhibited less than ex-
emplary conduct; however, in their annual reports, companies devote a section to
how they are addressing potential environmental and safety problems and how
they are becoming better citizens of the community through donation of time and
funds to hospitals, universities, social service organizations, environmental
groups, the arts, and humanities. Even with these efforts there is still public concern about large corporations. It is the duty of such companies to improve their
image as caring and good citizens through positive proof of their motives as
demonstrated by more socially responsible actions.

As the world population grows, a shortage of food is anticipated. Since
1980, the world population has increased by 1.3 billion people, mostly in India
and China. The estimated and projected populations of selected countries are
presented in Table 10.18. Although there have been droughts, floods, tornadoes,
and wars, there is still an increase in the world population, with a projected
growth of 1.4% for the 1990–2000 decade. It is estimated that from 1986 to 2000
there will be an increase of 319 million people worldwide. In undeveloped nations, it was formerly necessary to have five children to make sure that one
reached adulthood. China, India, and some Southeast Asia countries have made

TABLE 10.18 Estimated and Projected World Population (millions) by Selected Countries

| Country | 1980 | 1990 | 1996 | Projected 2000 | Compound growth (%) Actual: 1980–1990 | Compound growth (%) Projected: 1990–2000 |
|---|---|---|---|---|---|---|
| United States | 227.7 | 249.9 | 265.5 | 274.9 | 0.93 | 0.96 |
| Canada | 24.1 | 26.6 | 28.8 | 30.0 | 0.99 | 1.21 |
| Germany | 78.3 | 79.4 | 83.5 | 85.7 | 0.14 | 0.77 |
| France | 53.6 | 56.5 | 58.0 | 58.8 | 0.53 | 0.40 |
| United Kingdom | 56.3 | 57.4 | 58.5 | 58.9 | 0.19 | 0.26 |
| Belgium | 9.8 | 10.0 | 10.4 | 10.5 | 0.20 | 0.49 |
| Netherlands | 14.1 | 15.6 | 15.6 | 15.9 | 1.02 | 0.19 |
| Italy | 56.5 | 57.7 | 57.5 | 57.8 | 0.21 | 0.02 |
| Japan | 116.8 | 123.9 | 125.5 | 126.6 | 0.59 | 0.22 |
| China[a] | 984.7 | 1133.7 | 1210.0 | 1253.4 | 1.42 | 1.01 |
| South Korea | 38.1 | 42.9 | 45.5 | 47.4 | 1.19 | 1.00 |
| India | 692.4 | 855.6 | 952.1 | 1012.9 | 2.14 | 1.70 |
| Brazil | 122.8 | 150.1 | 162.7 | 169.5 | 2.03 | 1.22 |
| Russia | 139.0 | 148.1 | 148.2 | 147.9 | 0.64 | −0.01 |
| Philippines | 51.1 | 65.0 | 74.5 | 81.0 | 2.44 | 2.23 |
| Indonesia | 154.9 | 187.7 | 206.6 | 219.3 | 1.94 | 1.57 |
| Malaysia | 13.8 | 17.5 | 20.0 | 21.6 | 2.40 | 2.13 |
| Singapore | 2.4 | 3.0 | 3.4 | 3.6 | 2.26 | 1.84 |
| Thailand | 47.0 | 55.1 | 58.9 | 61.2 | 1.60 | 1.06 |
| World | 4457.6 | 5281.7 | 5771.9 | 6090.9 | 1.71 | 1.44 |

[a]Includes Taiwan

Source: U.S. Bureau of the Census, *Statistical Abstracts of the United States*, 1976, Table 1325.

efforts in recent years to reduce the population growth through various educational and medical programs. Table 10.18 shows that the projected growth for these nations in the 1990–2000 decade is lower than in the 1980–1990 decade. Russia had a negative growth. Belgium and Canada project a population increase in the current decade.

The United Nations has estimated a daily per-person food requirement of approximately 2,400 calories, including 100 grams of protein. This translates to 243 million metric tons of wheat for every billion additional people in the world. More efficient development and use of fertilizers, pesticides, herbicides, fungicides, and so on will be required to increase food production. As this increase occurs, there will be the environmental problems traditionally associated with these chemicals. The runoff from farms and ranches, for example, pollutes streams and lakes. Therefore, there is a tremendous challenge to develop materials that will increase food production while minimizing environmental damage.

At one time there was a call from government agencies and citizens' groups to have zero discharge of pollutants from any processing unit. Some environmental groups still espouse this goal. Enactment of such regulations would be particularly unfortunate, however, since it is impossible to design technologies for zero discharge or in deed to determine that a new substance will accumulate in the environment some decades hence. If more stringent regulations are imposed, what will happen to CPI companies? Will they be forced out of business, causing increased unemployment and ultimately a lower standard of living? What will such regulations do to the innovation of processes in chemical companies? In a more restrictive regulatory environment, with less innovation, there will be less need for research and development professionals. What skills will be needed for chemical professionals if environmental regulations continue to become more restrictive in the next decade? How do these issues relate to the chemical professional's personal career plans?

When we consider the future growth and threats to the CPI not only in the United States but globally, we need to consider short-term and long-term outlooks. Arbitrarily for this discussion, the short term might be 1–3 years, whereas the long term might be 3–10 years.

In the short term, 1997 appears to be a boom year compared to the early to mid-1990s, despite the currency crises in the Far East and the price decline in basic petrochemicals. The Chemical Manufacturers Association [2] predicted that 1998 also would be a favorable year, with strong growth in profits. Developing countries' consumption of chemicals typically runs two or more times the economic growth rates. Although many unforeseen events might show a downward trend, a good global growth rate should be noted in the increasing demand for chemicals. There has been some discussion about a possible global supply–demand imbalance in certain chemicals, but relatively small differences between supply and demand could create an oversupply, and no one can predict with any certainty whether there will be significant price effects. Overall it appeared that the U.S. chemical industry would show solid growth in sales, income, and profits, while feedstock, energy, and wages would remain nearly the same as in 1996. Research and development expenditures in the U.S. CPI were expected to be slightly higher than in 1996, at about a 5% increase over 1996, then stabilizing in 1998 below the 1997 level. Capital expenditures were expected to post about a 5% increase in 1997 over 1996 and to remain the same in 1998. Industry employment was expected to remain about the same as in 1996, with an occasional decline in some lines of the CPI with continued realignment and productivity increases in the industry[2]

As we consider the long-term outlooks for the CPI, let us look at the big chemical markets:

*Housing.* There is a large amount of chemical products in the average house, approximately $13,000 in 1995. This represents an $18 billion market, although new housing starts have been relatively constant over the past 4–6 years.

*Automobiles.* The average chemical industry content for cars, vans, and pickup trucks is $2160 per unit. This translated to a $32 billion market in 1995. With more plastic materials being used and with the decline in the number of foreign cars in the U.S. market, this should be a strong market. (The number of imported cars as a percentage of the U.S. market declined from 26% in 1982 to 12% in 1997. The key factors that affect this segment of industry are market saturation, slower U.S. population growth, sticker shock/affordability, and longer lasting vehicles.

*Electronics industry.* Semiconductors and circuit boards require the following chemical materials [4]: cleaners, developers, dopants, encapsulators, etchants, photo resins, strippers, and specialty gases. The projected demand for electronic chemicals is expected to expand by 8.5% per year to $4.6 billion in 2001.

*Agricultural chemicals and products.* With the increasing population, there will be an increased need for environmentally compatible fertilizers, herbicides fungicides, biocides, and so on. Roughly, the increase in these products should be in proportion to the population increase in spite of the increased efficiency in the use of these products.

*Drugs and biotech.* Projected drug sales for 1997 were in the range of $20–25 billion and had been growing substantially for the past 5 years. The sales of biotech firms in 1997 were $7.5 billion. These two sectors represent some $30 billion in sales and are expected to increase substantially in the future as new medicines are developed.

Although the U.S. CPI is strong, there are global economic trends as well as government regulations and policies that present potential problems. The long-term outlook for chemical production in the United States is as good as or perhaps better than that in most other manufacturing economies. Although the industry has strengths and competitive advantages, these will probably fade with time, and industry should not be complacent [2].

The long-term strengths of the CPI in the United States are as follows:

World's largest chemical production
A large domestic market
An energy supply that is competitively priced

An excellent internal infrastructure

High levels of research and development as well as capital expenditures

A strong position in international trade

Large, well-established foreign investments

The trends and threats to the U.S. CPI are as follows:

Slow U.S. economic growth

Rapid economic and chemical demands as well as growth in developing economies

Increasing role of foreign trade and foreign investments

Potential global oversupply and strong foreign competition

Increasing regulatory burdens

We have attempted in this chapter to consider some of the points and questions concerning current trends and their effects on the future of the CPI and on the chemical professionals employed by these industries. It is important to realize that many of the issues confronting society and the world have strong implications for the future well-being of the CPI and CPI employees. It is also important to realize that predicting the future is often futile, since trends can change suddenly, and surprises occur.

There is no book that can teach business acumen for anticipating the future. Foresight, however, can be obtained through experience. Throughout this text, we have sought to impart some of the tools and insights necessary for intelligent speculation, pondering, and planning for the future. The future belongs to those who understand the past and present, for they will be those most able to adapt to change and prosper.

## REFERENCES

1.  A. Toffler, *Future Shock*. Random House, New York, 1970.
2.  J. F. Henahan, "200 Years of American Chemicals," *Chemical Week*, February 18, 1976, p. 26.
3.  *Forbes*, April 15, 1976, p. 65.
4   CMAI, *The U.S. Chemical Industry Performance in 1996 and Outlook*. Chemical Manufacturers Association, Arlington, VA, 1996.
5.  CMAI, *U.S. Chemical Industry Statistical Handbook*. Chemical Manufacturers Association, Arlington, VA, 1997.
6.  "Global Forecast '98," *Chemical Week*, January 7, 1998, pp. 28–44.
7.  AIChE, *Dynamic Objectives for Chemical Engineering, 1976–1986*. American Institute of Chemical Engineers, New York.

## PROBLEMS

The problems require the reader to use material from all chapters in the book as well as from outside sources.

10.1.  Would you agree with the following statements? If not, why not?

a. To improve the economic security and job security of chemical engineers and chemists, we should limit enrollment in universities to a minimum and completely exclude immigration.

b. There are about 1.2 billion Chinese, forming the greatest possible market in the world. If they were to buy a stick of chewing gum per person per day, plants to manufacture this gum could be increased in size 100 times.

c. The smart course of action is to let other nations invent new processes and products, letting these others take the risks and have the start-up problems. When the process and product are viable, we can then license them pretty cheaply.

10.2.  In the mid-1970s, there was a high growth of fibers and plastics. Would you expect both these sectors to continue to grow in the future? Did they grow, or have there been downturns at various times? What caused any downturns? Can you extrapolate growth into the future, or must we modify our models for growth for various reasons?

10.3.  Drugs have demonstrated strong growth patterns. What caused this growth? Would you expect the strong growth to continue, or will there be a leveling off? What factors are involved?

10.4.  Will tomorrow's world be more concerned with the basic necessities of food and fuel or with leisure and life enrichment (i.e., luxuries)? Name the CPI products that will support these two divergent sectors of life. What talents should chemical professionals have to satisfy these needs?

10.5.  If better relations between the government and the CPI are of increasing importance to firms in the CPI, what should chemical professionals in industry do to prepare themselves to help better such relations?

10.6.  At various times there appears to be a trend away from product innovations and toward process innovations, and vice versa. Should chemical professionals try to follow this shift in emphasis or should they retain "professional integrity"? How should chemical professionals adapt to these trend changes?

10.7.  It is often said that the CPI will be unable to raise the capital needed for plant expansion in the future unless it can generate more profits. Social critics dismiss such reasoning. Construct a set of cogent arguments to support the critics and to support the CPI.

10.8.  Unions, often fearing that changes directed toward increasing productivity mean fewer jobs, have tended to fight the changes bitterly. Is their fear justified? Do industries with high gains in productivity have falling payrolls and industries with low gains in productivity have stationary-payrolls? Explain your answer.

10.9.   What will the change in age structure of the population of the United States mean to you in your career development?

10.10.   Are the environmentalists, consumer advocates, and other critics of on establishment helpful or harmful to the firms in the CPI? What is their impact on chemical professionals? Who eventually has to pay for the cost of clean air, safer products, and better workplaces? How far should the CPI go to protect the environment, and how is that point determined?

# Appendix A

## SIC Codes for Chemicals and Allied Products

### Major Group 28: CHEMICALS AND ALLIED PRODUCTS

#### The Major Group as a Whole

This major group includes establishments producing basic chemicals, and establishments manufacturing products by predominantly chemical processes. Establishments classified in this major group manufacture three general classes of products: 1) basic chemicals, such as acids, alkalies, salts, and organic chemicals; 2) chemical products to be used in further manufacture, such as synthetic fibers, plastics materials, dry colors, and pigments; and 3) finished chemical products to be used for ultimate consumption, such as drugs, cosmetics, and soaps; or to be used as materials or supplies in other industries, such as paints, fertilizers, and explosives. The mining of natural alkalies and other natural potassium, sodium, and boron compounds, of natural rock salt, and of other natural chemicals and fertilizers are classified in Mining, Industry Group 147. Establishments primarily engaged in manufacturing nonferrous metals and high-percentage ferroalloys are classified in Major Group 33; those manufacturing silicon carbide are classified in Major Group 32; those manufacturing baking powder, other leavening compounds, and starches are classified in Major Group 20; and those manufacturing artists' colors are classified in Major Group 39. Establishments primarily engaged in packaging, repackaging, and bottling of purchased chemical products, but not engaged in manufacturing chemicals and allied products, are classified in Wholesale or Retail Trade industries.

---

Appendix A is excerpted from *Standard Industrial Classification Manual*, U.S. Government Printing Office, Washington, D.C., 1987. The groups listed are the ones pertinent to the chemical process industries.

Industry
Group    Industry
No.        No.

281              INDUSTRIAL INORGANIC CHEMICALS

This industry group includes establishments primarily engaged in manufacturing basic industrial inorganic chemicals. Establishments primarily engaged in manufacturing formulated agricultural pesticides are classified in Industry 2879; those manufacturing medicinal chemicals, drugs, and medicines are classified in Industry Group 283; and those manufacturing soap and cosmetics are classified in Industry Group 284.

2812    *Alkalies and Chlorine*

Establishments primarily engaged in manufacturing alkalies and chlorine. Establishments primarily engaged in mining natural alkalies are classified in Mining, Industry 1474.

| | |
|---|---|
| Alkalies, not produced at mines | Soda ash, not produced at |
| Caustic potash | mines |
| Caustic soda | Sodium bicarbonate, not |
| Chlorine, compressed or | produced at mines |
| liquefied | Sodium carbonate (soda ash), |
| Potassium carbonate | not produced at mines |
| Potassium hydroxide | Sodium hydroxide (caustic soda) |
| Sal soda (washing soda) | Washing soda (sal soda) |

2813    *Industrial Gases*

Establishments primarily engaged in manufacturing industrial gases (including organic) for sale in compressed, liquid, and solid forms. Establishments primarily engaged in manufacturing fluorine and sulfur dioxide are classified in Industry 2819; those manufacturing household ammonia are classified in Industry 2842; those manufacturing other ammonia are classified in Industry 2873; those manufacturing chlorine are classified in Industry 2812; and those manufacturing fluorocarbon gases are classified in Industry 2869. Distributors of industrial gases and establishments primarily engaged in shipping liquid oxygen are classified in Wholesale Trade, Industry 5169.

| | |
|---|---|
| Acetylene | Gases, industrial: compressed, |
| Argon | liquefied, or solid |
| Carbon dioxide | Helium |
| Dry ice (solid carbon dioxide) | Hydrogen |

Industry
Group  Industry
No.    No.

281        INDUSTRIAL INORGANIC CHEMICALS—Cont.

     2813  *Industrial Gases—Cont.*

Neon
Nitrogen
Nitrous oxide

Oxygen, compressed and
liquefied

     2816  *Inorganic Pigments*

Establishments primarily engaged in manufacturing inorganic pigments. Important products of this industry include black pigments, except carbon black, white pigments, and color pigments. Organic color pigments, except animal black and bone black, are classified in Industry 2865, and those manufacturing carbon black are classified in Industry 2895.

Animal black
Barium sulfate, precipitated
  (blanc fine)
Barytes pigments
Black pigments, except carbon
  black
Blanc fixe (barium sulfate,
  precipitated)
Bone black
Chrome pigments: chrome
  green, chrome yellow,
  chrome orange, and zinc
  yellow
Color pigments, inorganic
Ferric oxide pigments
Iron blue pigments
Iron colors
Iron oxide, black
Iron oxide, magnetic
Iron oxide, yellow
Lamp black
Lead oxide pigments
Lead pigments

Litharge
Lithopone
Metallic pigments, inorganic
Mineral colors and
  pigments
Minium (pigments)
Ochers
Paint pigments, inorganic
Pearl essence
Pigments, inorganic
Prussian blue pigments
Red lead pigments
Satin white pigments
Siennas
Titanium pigments
Ultramarine pigments
Umbers
Vermilion pigments
White lead pigments
Whiting
Zinc oxide pigments
Zinc pigments: zinc yellow and
  zinc sulfide

Industry
Group   Industry
No.     No.

281           INDUSTRIAL INORGANIC CHEMICALS—Cont.

2819    *Industrial Inorganic Chemicals, Not Elsewhere Classified*

Establishments primarily engaged in manufacturing industrial inorganic chemicals, not elsewhere classified. Establishments primarily
engaged in mining, milling, or otherwise preparing natural potassium, sodium, or boron compounds (other than common salt) are
classified in Industry 1474. Establishments primarily engaged in
manufacturing household bleaches are classified in Industry 2842;
those manufacturing phosphoric acid are classified in Industry 2874;
and those manufacturing nitric acid, anhydrous ammonia, and other
nitrogenous fertilizer materials are classified in Industry 2873.

Activated carbon and charcoal
Alkali metals
Alumina
Aluminum chloride
Aluminum compounds
Aluminum hydroxide (alumina
   trihydrate)
Aluminum oxide
Aluminum sulfate
Alums
Ammonia alum
Ammonium chloride,
   hydroxide, and molybdate
Ammonium compounds, except
   for fertilizer
Ammonium perchlorate
Ammonium thiosulfate
Barium compounds
Bauxite, refined
Beryllium oxide
Bleach (calcium hypochlorite),
   industrial
Bleach (sodium hypochlorite),
   industrial
Bleaches, industrial
Bleaching powder, industrial
Borax (sodium tetraborate)
Boric acid
Boron compounds, not
   produced at mines

Borosilicate
Brine
Bromine, elemental
Calcium carbide, chloride, and
   hypochlorite
Calcium compounds, inorganic
Calcium metal
Carbide
Catalysts, chemical
Cerium salts
Cerium metal
Charcoal, activated
Chlorosulfonic acid
Chromates and bichromates
Chromic acid
Chromium compounds,
   inorganic
Chromium salts
Cobalt 60 (radioactive)
Cobalt chloride
Cobalt sulfate
Copper chloride
Copper iodide and oxide
Copper sulfate
Cyanides
Desiccants, activated: silica gel
Dichromates
Ferric chloride
Ferric oxides, except
   pigments

281           INDUSTRIAL INORGANIC CHEMICALS—Cont.

  2819   *Industrial Inorganic Chemicals, Not Elsewhere Classified—Cont.*

Ferrocyanides

Fissionable material
  production

Fluorine, elemental

Fuel propellants, solid:
  inorganic

Hydroflouric acid

Hydrogen peroxide

Fuels, high energy: inorganic

Glauber's salt

Heavy water

High purity grade chemicals,
  inorganic: refined from
  technical grades

Hydrated alumina silicate
  powder

Hydrazine

Hydrochloric acid

Hydrocyanic acid

Hydrogen sulfide

Hydrosulfites

Hypophosphites

Indium chloride

Inorganic acids, except nitric or
  phosphoric

Iodides

Iodine, elemental

Iodine, resublimed

Iron sulfate

Isotopes, radioactive

Laboratory chemicals,
  inorganic

Lead oxides, other than
  pigments

Lead silicate

Lime bleaching compounds

Lithium compounds

Lithium metal

Luminous compounds, radium

Magnesium carbonate

Magnesium chloride

Magnesium compounds,
  inorganic

Manganese dioxide powder,
  synthetic

Mercury chlorides (calomel,
  corrosive sublimate), except
  U.S.P.

Mercury compounds, inorganic

Mercury oxides

Mercury, redistilled

Metals, liquid

Mixed acid

Muriate of potash, not produced
  at mines

Nickel ammonium sulfate

Nickel carbonate

Nickel compounds, inorganic

Nickel sulfate

Nuclear cores, inorganic

Nuclear fuel reactor cores,
  inorganic

Nuclear fuel scrap
  reprocessing

Oleum (fuming sulfuric acid)

Oxidation catalyst made from
  porcelain

Porchloric acid

Peroxides, inorganic

Phosphates, except
  defluorinated and
  ammoniated

Phosphorus and phosphorus
  oxychloride

Potash alum

Potassium aluminum sulfate

Potassium bichromate and
  chromate

Potassium bromide

Potassium chlorate

Industry
Group    Industry
No.       No.

281            INDUSTRIAL INORGANIC CHEMICALS—Cont.

   2819    *Industrial Inorganic Chemicals, Not Elsewhere Classified—Cont.*

Potassium chloride

Potassium compounds,
   inorganic: except potassium
   hydroxide and carbonate

Potassium cyanide

Potassium hypochlorate

Potassium iodide

Potassium metal

Potassium nitrate and sulfate

Potassium permanganate

Propellants for missiles, solid:
   inorganic

Radium chloride

Radium luminous compounds

Rare earth metal salts

Reagent grade chemicals,
   inorganic: refined from
   technical grades

Rubidium metal

Salt cake (sodium sulfate)

Salts of rare earth metals

Scandium

Silica gel

Silica, amorphous

Silicofluorides

Silver bromide, chloride, and
   nitrate

Silver compounds, inorganic

Soda alum

Sodium aluminate

Sodium aluminum sulfate

Sodium antimoniate

Sodium arsenite, technical

Sodium bichromate and
   chromate

Sodium borates

Sodium borohydride

Sodium bromide, not produced
   at mines

Sodium chlorate

Sodium compounds, inorganic

Sodium cyanide

Sodium hydrosulfite

Sodium molybdate

Sodium perborate

Sodium peroxide

Sodium phosphate

Sodium polyphosphate

Sodium silicate

Sodium silicoflouride

Sodium stannate

Sodium sulfate-bulk or
   tablets

Sodium tetraborate, not
   produced at mines

Sodium thiosulfate

Sodium tungstate

Sodium uranate

Sodium, metallic

Stannic and stannous chloride

Strontium carbonate,
   precipitated, and oxide

Strontium nitrate

Sublimate, corrosive

Sulfate of potash and potash
   magnesia, not produced at
   mines

Sulfides and sulfites

Sulfocyanides

Sulfur chloride

Sulfur dioxide

Sulfur hexafluoride gas

Sulfur, recovered or refined,
   including from sour natural
   gas

Sulfuric acid

Tanning agents, synthetic
   inorganic

Thiocyanates, inorganic

Tin chloride

281         INDUSTRIAL INORGANIC CHEMICALS—Cont.

2819   *Industrial Inorganic Chemicals, Not Elsewhere Classified—Cont.*

| | |
|---|---|
| Tin compounds, inorganic | by metallurgical |
| Tin oxide | process |
| Tin salts | Uranium slug, radioactive |
| Tungsten carbide powder, | Water glass |
| except abrasives or | Zinc chloride |

282         PLASTICS MATERIALS AND SYNTHETIC RESINS, SYNTHETIC RUBBER, CELLULOSIC AND OTHER MANMADE FIBERS, EXCEPT GLASS

This group includes chemical establishments primarily engaged in manufacturing plastics materials and synthetic resins, synthetic rubbers, and cellulosic and other manmade fibers. Establishments primarily engaged in the manufacture of rubber products, and those primarily engaged in the compounding of purchased resins or the fabrication of plastics sheets, rods, and miscellaneous plastics products, are classified in Major Group 30; and textile mills primarily engaged in throwing, spinning, weaving, or knitting textile products from manufactured fibers are classified in Major Group 22.

2821   *Plastics Materials, Synthetic Resins, and Nonvulcanizable Elastomers*

Establishments primarily engaged in manufacturing synthetic resins, plastics materials, and nonvulcanizable elastomers. Important products of this industry include: cellulose plastics materials; phenolic and other tar acid resins; urea and melamine resins; vinyl resins; styrene resins; alkyd resins; acrylic resins; polyethylene resins; polypropylene resins; rosin-modified resins; coumarone-indene and petroleum polymer resins; miscellaneous resins, including polyamide resins, silicones, polyisobutylenes, polyesters, polycarbonate resins, acetal resins, and fluorohydrocarbon resins; and casein plastics. Establishments primarily engaged in manufacturing fabricated plastics products or plastics film, sheet, rod, non-textile monofilaments and regenerated cellulose products, and vulcanized fiber are classified in Industry Group 308, whether from purchased resins or from resins produced in the same plant.

282          **PLASTICS MATERIALS AND SYNTHETIC RESINS,
SYNTHETIC RUBBER, CELLULOSIC AND OTHER
MANMADE FIBERS, EXCEPT GLASS—Cont.**

2821    *Plastics Materials, Synthetic Resins, and Nonvulcanizable
Elastomers—Cont.*

Establishments primarily engaged in compounding purchased
resins are classified in Industry 3087. Establishments primarily
manufacturing adhesives are classified in Industry 2891.

Acetal resins
Acetate, cellulose (plastics)
Acrylic resins
Acrylonitrile-butadiene-styrene
  resins
Alcohol resins, polyvinyl
Alkyd resins
Allyl resins
Butadiene copolymers,
  containing less than 50
  percent butadione
Carbohydrate plastics
Casein plastics
Cellulose nitrate resins
Cellulose propionate (plastics)
Coal tar resins
Condensation plastics
Coumarone-iodene resins
Cresol resins
Cresol-furfural resins
Dicyandiamine resins
Diisocyanate resins
Elastomers, nonvulcanizable
  (plastics)
Epichlorohydrin hisphenol
Epichlorohydrin diphenol
Epoxy resins
Enter gum
Ethyl cellulose plastics
Ethylene-vinyl acetate resins
Flourohydrocarbon resins
Ion exchange resins

Ionomer resins
Isobutylene polymers
Lignin plastics
Melamine resins
Methyl acrylate resins
Methyl cellulose plastics
Methyl methacrylate resins
Molding compounds, plastics
Nitrocellulose plastics
  (pyroxylin)
Nylon resins
Petroleum polymer resins
Phenol-furfural resins
Phenolic resins
Phenoxy resins
Phthalic alkyd resins
Phthalic anhydride resins
Polyacrylonitrile resins
Polyamide resins
Polycarbonate resins
Polyesters
Polyethylene resins
Polyhexamothylemediamine
  adipamide resins
Polyisobutylenes
Polymerization plastics, except
  fibers
Polypropylene resins
Polyetyrene resins
Polyurethane resins
Polyvinyl chloride resins
Polyvinyl halide resins

Industry
Group   Industry
No.     No.

282         PLASTICS MATERIALS AND SYNTHETIC RESINS,
            SYNTHETIC RUBBER, CELLULOSIC AND OTHER
            MANMADE FIBERS, EXCEPT GLASS—Cont.

    2821    *Plastics Materials, Synthetic Resins, and Nonvulcanizable*
            *Elastomers—Cont.*

| | |
|---|---|
| Polyvinyl resins | Silicone resins |
| Protein plastics | Soybean plastics |
| Pyroxylin | Styrene resins |
| Rosins, synthetic | Styrene-acrylonitrile resins |
| Rosin modified resins | Tar acid resins |
| Silicone fluid solution (fluid for | Urea resins |
| sonar transducers) | Vinyl resins |

    2822    *Synthetic Rubber (Vulcanizable Elastomers)*

    Establishments primarily engaged in manufacturing synthetic
    rubber by polymerization or copolymerization. An elastomer for
    the purpose of this classification is a rubber-like material capable of
    vulcanization, such as copolymers of butadiene and styrene, or bu-
    tadiene and acrylonitrile, polybutadienes, chloroprene rubbers,
    and isobutylene-isoprene copolymers. Butadiene copolymers con-
    taining less than 50 percent butadiene are classified in Industry
    2821. Natural chlorinated rubbers and cyclized rubbers are consid-
    ered as semifinished products and are classified in Industry 3069.

| | |
|---|---|
| Acrylate-type rubbers | Elastomers, vulcanizable |
| Acrylate-butadiene rubbers | (synthetic rubber) |
| Acrylic rubbers | Epichlorohydrin elastomers |
| Butadiene rubbers | Estane |
| Butadiene acrylonitrile | Ethylene-propylene rubbers |
| copolymers (more than 50 | Fluoro rubbers |
| percent butadiene) | Fluorocarbon derivative rubbers |
| Butadiene-styrene copolymers | Isobutylene-isoprene rubbers |
| (more than 50 percent | Isocyanate type rubber |
| butadiene) | Isoprene rubbers, synthetic |
| Butyl rubber | N-type rubber |
| Chlorinated rubbers, synthetic | Neoprene |
| Chlorophene type rubbers | Nitrile type rubber |
| Chlorosulfonated polyethylenes | Nitrile-butadiene rubbers |
| Cyclo rubbers, synthetic | Polybutadienes |
| EPDM polymers | Polyethylenes, chlorosulfonated |

Industry
Group   Industry
No.     No.

282         **PLASTICS MATERIALS AND SYNTHETIC RESINS,**
            **SYNTHETIC RUBBER, CELLULOSIC AND OTHER**
            **MANMADE FIBERS, EXCEPT GLASS—Cont.**

2822        *Synthetic Rubber (Vulcanizable Elastomers)—Cont.*

Polyisobutylene (synthetic          Silicone rubbers
  rubber)                           Stereo regular elastomers
Polyisobutylene-isoprene            Styrene-butadiene rubbers (50
  elastomers                          percent or less styrene
Polymethylene rubbers                 content)
Polysulfides                        Styrene-chloroprene rubbers
Pyridine-butadiene copolymers       Styrene-isoprene rubbers
Pyridine-butadiene rubbers          Thiol rubbers
Rubber, synthetic                   Urethane rubbers
S-type rubber                       Vulcanized oils

2823        *Cellulosic Manmade Fibers*

Establishments primarily engaged in manufacturing cellulosic
fibers (including cellulose acetate and regenerated cellulose
such as rayon by the viscose or cuprammonium process) in the
form of monofilament, yarn, staple, or tow suitable for further
manufacturing on spindles, looms, knitting machines, or other
textile processing equipment.

Acetate fibers                      Rayon yarn, made in chemical
Cellulose acetate monofilament,       plants
  yarn, staple, or tow             Regenerated cellulose fibers
Cellulose fibers, manmade           Textured yarns and fibers,
Cigarette tow, cellulosic fiber       cellulosic made in chemical
Cuprammonium fibers                   plants
Fibers, rayon                       Triacetate fibers
Horeshair, artificial: rayon        Viscose fibers, bonds, strips,
Nitrocellulose fibers                 and yarn
Rayon primary products: fibers,     Yarn, cellulosic: made in
  straw, strips, and yarn             chemical plants

2824        *Manmade Organic Fibers, Except Cellulosic*

Establishments primarily engaged in manufacturing manmade organic fibers, except cellulosic (including those of regenerated pro-
teins, and of polymers or copolymers of such components as vinyl
chloride, vinylidene chloride, linear esters, vinyl alcohols, acry-

282      PLASTICS MATERIALS AND SYNTHETIC RESINS,
         SYNTHETIC RUBBER, CELLULOSIC AND OTHER
         MANMADE FIBERS, EXCEPT GLASS—Cont.

   2824   *Manmade Organic Fibers, Except Cellulosic—Cont.*
          lonitrile, ethylenes, amides, and related polymeric materials), in
          the form of monofilament, yarn, staple, or tow suitable for further
          manufacturing on spindles, looms, knitting machines, or other tex-
          tile processing equipment. Establishments primarily engaged in
          manufacturing textile glass fibers are classified in Industry 3229.

| | |
|---|---|
| Acrylic fibers | Polyester fibers |
| Acrylonitrile fibers | Polyvinyl outer fibers |
| Anidex fibers | Polyvinylidene chloride |
| Casein fibers | fibers |
| Elastomeric fibers | Protein fibers |
| Fibers, manmade: except | Saran fibers |
| cellulosic | Soybean fibers (manmade |
| Fluorocarbon fibers | textile materials) |
| Horsehair, artificial: nylon | Textured fibers and yarns, |
| Linear esters fibers | noncellulosic: made in |
| Modacrylic fibers | chemical plants |
| Nylon fibers and bristles | Vinyl fibers |
| Olefin fibers | Vinylidene chloride |
| Organic fibers, synthetic: | fibers |
| except cellulosic | Zein fibers |

283      DRUGS

         This group includes establishments primarily engaged in manu-
         facturing, fabricating, or processing medicinal chemicals and
         pharmaceutical products. Also included in this group are estab-
         lishments primarily engaged in the grading, grinding, and milling
         of botanicals.

   2833   *Medicinal Chemicals and Botanical Products*
          Establishments primarily engaged in: (1) manufacturing bulk or-
          ganic and inorganic medicinal chemicals and their derivatives
          and (2) processing (grading, grinding, and milling) bulk botanical
          drugs and herbs. Included in this industry are establishments pri-
          marily engaged in manufacturing agar-agar and similar products
          of natural origin, endocrine products, manufacturing or isolating

Industry
Group   Industry
No.     No.

283              DRUGS—Cont.

2833    *Medicinal Chemicals and Botanical Products—Cont.*

basic vitamins, and isolating active medicinal principals such as
alkaloids from botanical drugs and herbs.

Adrenal derivative: bulk,
  uncompounded
Agar-agar (ground)
Alkaloids and salts
Anesthetics, in bulk form
Antibiotics: bulk
  uncompounded
Atropine and derivatives
Barbituric acid and derivatives:
  bulk, uncompounded
Botanical products, medicinal:
  ground, graded, and milled
Brucine and derivatives
Caffeine and derivatives
Chemicals, medicinal: organic
  and inorganic—bulk,
  uncompounded
Cinchona and derivatives
Cocaine and derivatives
Codeine and derivatives
Digitoxin
Drug grading, grinding, and
  milling
Endocrine products
Ephedrine and derivatives
Ergot alkaloids
Fish liver oils, refined and
  concentrated for medicinal use
Gland derivatives: bulk,
  uncompounded
Glycosides
Herb grinding, grading, and
  milling

Hormones and derivatives
Insulin: bulk, uncompounded
Kelp plants
Mercury chlorides, U.S.P.
Mercury compounds,
  medicinal: organic and
  inorganic
Morphins and derivatives
N-methylpiperazine
Oils, vegetable and animal:
  medicinal grade—refined and
  concentrated
Opium derivatives
Ox bile salts and derivatives:
  bulk, uncompounded
Penicillin: bulk, uncompounded
Physostigmine and derivatives
Pituitary gland derivatives:
  bulk, uncompounded
Procaine and derivatives: bulk,
  uncompounded
Quinine and derivatives
Reserpines
Salicylic acid derivatives,
  medicinal grade
Strychnine and derivatives
Sulfa drugs: bulk,
  uncompounded
Sulfonamides
Theobromine
Vegetable gelatin (agar-agar)
Vitamins, natural and synthetic:
  bulk, uncompounded

2834    *Pharmaceutical Preparations*

Establishments primarily engaged in manufacturing, fabricating,
or processing drugs in pharmaceutical preparations for human or

283          DRUGS—Cont.

2834    *Pharmaceutical Preparations—Cont.*

veterinary use. The greater part of the products of these establishments are finished in the form intended for final consumption, such as ampoules, tablets, capsules, vials, ointments, medicinal powders, solutions, and suspensions. Products of this industry consist of two important lines, namely: (1) pharmaceutical preparations promoted primarily to the dental, medical, or veterinary professions, and (2) pharmaceutical preparations promoted primarily to the public.

Adrenal pharmaceutical
   preparations
Analgesics
Anesthetics, packaged
Antacids
Anthelmintics
Antibiotics, packaged
Antihistamine preparations
Antipyretics
Antiseptics, medicinal
Astringents, medicinal
Barbituric and pharmaceutical
   preparations
Belladonna pharmaceutical
   preparations
Botanical extracts: powdered,
   pilular, solid, and fluid,
   except diagnostics
Chlorination tablets and kits
   (water purification)
Cold remedies
Cough medicines
Cyclopropane for anesthetic use
   (U.S.P. par N.F.), packaged
Dermatological preparations
Dextrose and sodium chloride
   injection, mixed
Dextrose injection
Digitalis pharmaceutical
   preparations

Diuretics
Effervescent salts
Emulsifiers, fluorescent,
   inspection
Emulsions, pharmaceutical
Fever remedies
Galenical preparations
Hormone preparations, except
   diagnostics
Insulin preparations
Intravenous solutions
Iodine, tincture of
Laxatives
Liniments
Lip balms
Lozenges, pharmaceutical
Medicines, capsuled or
   ampuled
Nitrofuran preparations
Ointments
Parenteral solutions
Penicillin preparations
Pharmaceuticals
Pills, pharmaceutical
Pituitary gland pharmaceutical
   preparations
Poultry and animal remedies
Powders, pharmaceutical
Procaine pharmaceutical
   preparations

Industry
Group   Industry
No.     No.
283            DRUGS—Cont.

2834   *Pharmaceutical Preparations—Cont.*

Proprietary drug products
Remedies, human and animal
Sodium chloride solution for
  injection, U.S.P.
Sodium salicylate tablets
Solutions, pharmaceutical
Spirits, pharmaceutical
Suppositories
Syrups, pharmaceutical
Tablets, pharmaceutical
Thyroid preparations

Tinctures, pharmaceutical
Tranquilizers and mental drug
  preparations
Vermifuges
Veterinary pharmaceutical
  preparations
Vitamin preparations
Water decontamination or
  purification tablets
Water, sterile: for injections
Zinc ointment

2835   *In Vitro and In Vivo Diagnostic Substances*

Establishments primarily engaged in manufacturing in vitro and
in vivo diagnostic substances, whether or not packaged for retail
sale. These materials are chemical, biological, or radioactive substances used in diagnosing or monitoring the state of human or
veterinary health by identifying and measuring normal or abnor-
mal constituents of body fluids or tissues.

Angiourographic diagnostic
  agents
Barium diagnostic agents
Blood derivative diagnostic
  reagents
Clinical chemistry reagents
  (including toxicology)
Clinical chemistry standards
  and controls (including
  toxicology)
Coagulation diagnostic reagents
Cold kits for labeling with
  technetium
Contrast media diagnostic
  products (e.g., iodine and
  barium)
Cytology and histology
  diagnostic products

Diagnostic agents, biological
Electrolyte diagnostic reagents
Enzyme and isoenzyme
  diagnostic reagents
Hematology diagnostic reagents
In vitro diagnostics
In vivo diagnostics
In vivo radioactive reagents
Iodinated diagnostic agents
Metabolite diagnostic
  reagents
Microbiology, virology, and
  serology diagnostic products
Pregnancy test kits
Radioactive diagnostic
  substances
Technetium products
Viral test diagnostic reagents

Industry
Group Industry
No. No.

283 DRUGS—Cont.

2836 *Biological Products, Except Diagnostic Substances*

Establishments primarily engaged in the production of bacterial and virus vaccines, toxoids, and analogous products (such as allergenic extracts), serums, plasmas, and other blood derivatives for human or veterinary use, other than in vitro and in vivo diagnostic substances. Included in this industry are establishments primarily engaged in the production of microbiological products for other uses. Establishments primarily engaged in manufacturing in vitro and in vivo diagnostic substances are classified in Industry 2835.

Agar culture media, except in vitro and in vivo
Aggressions, except in vitro and in vivo
Allergenic extracts, except in vitro and in vivo
Allergens
Anti-hog-cholera serums
Antigums
Antiserums
Antitoxins
Antivenin
Bacterial vaccines
Bacterins, except in vitro and in vivo
Bacteriological media, except in vitro and in vivo
Biological and allied products: antitoxins, bacterins, vaccines, viruses, except in vitro and in vivo

Blood derivatives, for human or veterinary use, except in vitro and in vivo
Coagulation products
Culture media or concentrates, except in vitro and in vivo
Diphtheria toxin
Hematology products, except in vitro and in vivo
Plasmas
Pollen extracts, except in vitro and in vivo
Serobacterins
Serums, except in vitro and in vivo
Toxins
Toxoids except in vitro and in vivo
Tuberculins
Vaccines
Venoms
Viruses

284 SOAP, DETERGENTS, AND CLEANING PREPARATIONS; PERFUMES, COSMETICS, AND OTHER TOILET PREPARATIONS

This industry group includes establishments primarily engaged in manufacturing soap and other detergents and in producing glycerin from vegetable and animal fats and oils; specialty cleaning,

284                 SOAP, DETERGENTS, AND CLEANING
                    PREPARATIONS; PERFUMES, COSMETICS, AND
                    OTHER TOILET PREPARATIONS—Cont.

polishing, and sanitation preparations; and surface active preparations used as emulsifiers, wetting agents, and finishing agents, including sulfonated oils; and perfumes, cosmetics, and other toilet
preparations.

2841    *Soap and Other Detergents, Except Specialty Cleaners*

Establishments primarily engaged in manufacturing soap, synthetic organic detergents, inorganic alkaline detergents, or any
combination thereof, and establishments producing crude and refined glycerin from vegetable and animal fats and oils. Establishments primarily engaged in manufacturing shampoos or shaving
products, whether from soap or synthetic detergents, are classified in Industry 2844; and those manufacturing synthetic glycerin
are classified in Industry 2869.

| | |
|---|---|
| Detergents, synthetic organic | Presoaks |
| and inorganic alkaline | Scouring compounds |
| Dishwashing compounds | Soap: granulated, liquid, cake, |
| Dye-removing cream, soap base | flaked, and chip |
| Glycerin, crude and refined: | Textile soap |
| from fats—except synthetic | Washing compounds |
| Mechanic paste | |

2842    *Specialty Cleaning, Polishing, and Sanitation Preparations*

Establishments primarily engaged in manufacturing furniture,
metal, and other polishes; waxes and dressings for fabricated
leather and other materials; household, institutional, and industrial
plant disinfectants; nonpersonal deodorants; drycleaning preparations; household bleaches; and other sanitation preparations. Establishments primarily engaged in manufacturing industrial
bleaches are classified in Industry 2819, and those manufacturing
household pesticidal preparations are classified in Industry 2879.

| | |
|---|---|
| Ammonia, household | Blackings |
| Aqua ammonia, household | Burnishing ink |
| Beeswax, processing of | Chlorine bleaching compounds, |
| Belt dressing | household: liquid or dry |

284　　　　SOAP, DETERGENTS, AND CLEANING
PREPARATIONS; PERFUMES, COSMETICS, AND
OTHER TOILET PREPARATIONS—Cont.

2842　*Specialty Cleaning, Polishing, and Sanitation Preparations—Cont.*

Cleaning and polishing
preparations
Cloths, dusting and polishing:
chemically treated
Degreasing solvent
Deodorants, nonpersonal
Disinfectants, household and
industrial plant
Drain pipe solvents and
cleansers
Dressings for fabricated leather
and other materials
Drycleaning preparations
Dust mats, gelatin
Dusting cloths, chemically
treated
Dye removing cream,
petroleum base
Fabric softeners
Floor wax emulsion
Floor waxes
Furniture polish and war
Glass window cleaning
preparations
Harness dressing
Household bleaches, dry or
liquid
Industrial plant disinfectants
and deodorants
Ink eradicators

Ink, burnishing
Leather dressings and finishes
Dye, household
Paint and wallpaper cleaners
Polishes: furniture,
automobile, metal, shoe,
and stove
Re-refining drycleaning
fluid
Rug, upholstery, and
drycleaning detergents and
spotters
Rust removers
Saddle soap
Sanitation preparations
Shoe cleaners and polishes
Sodium hypochlorite
(household bleach)
Stain removers
Starch preparations, laundry
Starches, plastics
Sweeping compounds, oil and
water absorbent, clay or
sawdust
Wallpaper cleaners
Wax removers
Waxes for wood, fabricated
leather, and other materials
Window cleaning
preparations

2843　*Surface Active Agents, Finishing Agents, Sulfonated Oils, and
Assistants*

Establishments primarily engaged in producing surface active
preparations for use as wetting agents, emulsifiers, and penetrants. Establishments engaged in producing sulfonated oils and
fats and related products are also included.

Industry
Group  Industry
No.    No.

284          SOAP, DETERGENTS, AND CLEANING
             PREPARATIONS; PERFUMES, COSMETICS, AND
             OTHER TOILET PREPARATIONS—Cont.

       2843  *Surface Active Agents, Finishing Agents, Sulfonated Oils, and*
             *Assistants—Cont.*

| | |
|---|---|
| Assistants, textile and leather processing | Penetrants |
| Calcium salts of sulfonated oils, fats, or greases | Sodium salts of sulfonated oils, fats, or greases |
| Cod oil, sulfonated | Softeners (textile assistants) |
| Emulsifiers, except food and pharmaceutical | Soluble oils and greases |
| Finishing agents, textile and leather | Sulfonated oils, fats, and greases |
| Mordants | Surface active agents |
| Oil, turkey red | Textile processing assistants |
| Oils, soluble (textile assistants) | Textile scouring compounds and wetting agents |
| | Thin water (admixture) |

Softeners (textile assistants)

       2844  *Perfumes, Cosmetics, and Other Toilet Preparations*

Establishments primarily engaged in manufacturing perfumes (natural and synthetic), cosmetics, and other toilet preparations. This industry also includes establishments primarily engaged in blending and compounding perfume bases; and those manufacturing shampoos and shaving products, whether from soap or synthetic detergents. Establishments primarily engaged in manufacturing synthetic perfume and flavoring materials are classified in Industry 2869, and those manufacturing essential oils are classified in Industry 2899.

| | |
|---|---|
| Bath salts | Dressings, cosmetic |
| Bay rum | Face creams and lotions |
| Body powder | Face powders |
| Colognes | Hair coloring preparations |
| Concentrates, perfume | Hair preparations: dressings, rinses, tonics, and scalp conditioners |
| Cosmetic creams | |
| Cosmetic lotions and oils | Home permanent kits |
| Cosmetics | Lipsticks |
| Dentifrices | Manicure preparations |
| Denture cleaners | Mouthwashes |
| Deodorants, personal | |
| Depilatories, cosmetic | |

284          SOAP, DETERGENTS, AND CLEANING
             PREPARATIONS; PERFUMES, COSMETICS, AND
             OTHER TOILET PREPARATIONS—Cont.

     2844    *Perfumes, Cosmetics, and Other Toilet Preparations—Cont.*

Perfume bases, blending and
    compounding
Perfumes, natural and
    synthetic
Rouge, cosmetic
Sachet
Shampoos, hair
Shaving preparations (e.g.,
    cakes, creams, lotions,
    powders, tablets)

Soap-impregnated papers and
    paper washcloths
Suntan lotions and oils
Talcum powders
Toilet creams, powders, and
    waters
Toilet preparations
Toothpastes and powders
Towelettes, premoistened
Washes, cosmetic

285          PAINTS, VARNISHES, LACQUERS, ENAMELS, AND
             ALLIED PRODUCTS

     2851    *Paints, Varnishes, Lacquers, Enamels, and Allied Products*

Establishments primarily engaged in manufacturing paints (in paste and ready-mixed form); varnishes; lacquers; enamels and shellac; putties, wood fillers, and sealers; paint and varnish removers; paintbrush cleaners; and allied paint products. Establishments primarily engaged in manufacturing carbon black are classified in Industry 2895; those manufacturing bone black, lamp black, and inorganic color pigments are classified in Industry 2816; those manufacturing organic color pigments are classified in Industry 2865; those manufacturing plastics materials are classified in Industry 2821; those manufacturing printing ink are classified in Industry 2893; those manufacturing caulking compounds and sealants are classified in Industry 2891; those manufacturing artists' paints are classified in Industry 3952; and those manufacturing turpentine are classified in Industry 2861.

Calcimines, dry and paste
Coating, air curing
Colors in oil, except artists'
Dispersions, thermoplastics and
    colloidal: paint
Dopes, paint

Driers, paint
Enamels, except dental and
    china painting
Epoxy coatings, made from
    purchased resin
Intaglio ink vehicle

284            SOAP, DETERGENTS, AND CLEANING
               PREPARATIONS; PERFUMES, COSMETICS, AND
               OTHER TOILET PREPARATIONS—Cont.

   2851    *Paints, Varnishes, Lacquers, Enamels, and Allied Products—
            Cont.*

Japans, baking and drying
Kalsomines, dry or paste
Lacquer bases and dopes
Lacquer thinner
Lacquer, clear and
   pigmented
Lacquers, plastics
Lead-in-oil paints
Linoleates, paint driers
Lithographic varnishes
Marine paints
Naphthanate driers
Oleate driers
Paint driers
Paint removers
Paintbrush cleaners
Paints, asphalt and
   bituminous
Paints, plastics texture: paste
   and dry
Paints, waterproof
Paints: oil and alkyd vehicle,
   and water thinned

Phenol formaldehyde coatings,
   baking and air curing
Plastics base paints and
   varnishes
Plastisol coating compound
Polyurethane coatings
Primers, paint
Putty
Resinate driers
Shellac, protective coating
Soyate driers
Stains: varnish, oil, and wax
Tallate driers
Thinners, paint: prepared
Undercoatings, paint
Varnish removers
Varnishes
Vinyl coatings, strippable
Vinyl plastisol
Water paints
Wood fillers and sealers
Wood stains
Zinc oxide in oil, paint

286            INDUSTRIAL ORGANIC CHEMICALS

Establishments primarily engaged in manufacturing industrial
organic chemicals. Important products of this group include: 1)
noncyclic organic chemicals, such as acetic, chloroacetic, adipic,
formic, oxalic, and tartaric acids, and their metallic salts; chloral,
formaldehyde, and methylamine; 2) solvents such as amyl, butyl,
and ethyl alcohols; methanol; amyl, butyl and ethyl acetates;
ethyl ether, ethylene glycol ether, and diethylene glycol ether;
acetone, carbon disulfide, and chlorinated solvents, such as carbon tetrachloride, perchloroethylene, and trichloroethylene; 3)
polyhydric alcohols, such as ethylene glycol, sorbitol, pentaery-

286        INDUSTRIAL ORGANIC CHEMICALS—Cont.

thritol, synthetic glycerin; 4) synthetic perfume and flavoring materials, such as coumarin, methyl salicylate, saccharin, citral, citronellal, synthetic geraniol, ionone, terpineol, and synthetic vanillin; 5) rubber-processing chemicals, such as accelerators and antioxidants, both cyclic and acyclic; 6) plasticizers, both cyclic and acyclic, such as esters of phosphoric acid, phthalic anhydride, adipic acid, lauric acid, oleic acid, sebacic acid, and stearic acid; 7) synthetic tanning agents, such as naphthalene sulfonic acid condensates; 8) chemical warfare gases; 9) esters, amines, etc., of polyhydric alcohols and fatty and other acids; 10) cyclic crudes and intermediates; 11) cyclic dyes and organic pigments; and 12) natural gum and wood chemicals. Establishments primarily engaged in manufacturing plastics materials and nonvulcanizable elastomers are classified in Industry 2821; those manufacturing synthetic rubber are classified in Industry 2822; those manufacturing essential oils are classified in Industry 2899; those manufacturing rayon and other manmade fibers are classified in Industries 2823 and 2824; those manufacturing specialty cleaning, polishing, and sanitation preparations are classified in Industry 2842; those manufacturing paints are classified in Industry 2851; and those manufacturing inorganic pigments are classified in Industry 2816. Distillers engaged i the manufacture of grain alcohol for beverage purposes are classified in Industry 2085.

2861   Gum and Wood Chemicals

Establishments primarily engaged in manufacturing hardwood and softwood distillation products, wood and gum naval stores, charcoal, natural dyestuffs, and natural tanning materials. Establishments primarily engaged in manufacturing synthetic organic tanning materials are classified in Industry 2869, and those manufacturing synthetic organic dyes are classified in Industry 2865.

Acetate of lime, natural
Acetone, natural
Annato extract
Brazilwood extract
Brewers' pitch, product of
    softwood distillation

Calcium acetate, product
    of hardwood
    distillation
Charcoal, except activated
Chestnut extract
Dragon's blood

Industry
Group    Industry
No.      No.

286              INDUSTRIAL ORGANIC CHEMICALS—Cont.

2861     *Gum and Wood Chemicals—Cont.*

Dyeing and extract materials, natural
Dyestuffs, natural
Ethyl acetate, natural
Fustic wood extract
Gambier extract
Gum naval stores, processing but not gathering or warehousing
Hardwood distillates
Hemlock extract
Logwood extract
Mangrove extract
Methanol, natural (wood alcohol)
Methyl acetone
Methyl alcohol, natural (wood alcohol)
Myrobalans extract
Naval stores, wood
Oak extract
Oils, wood: product of hardwood distillation
Pine oil, produced by distillation of pine gum or pine wood
Pit charcoal
Pitch, wood
Pyroligneous acid
Quebracho extract
Quercitron extract
Rosin, produced by distillation of pine gum or pine wood
Softwood distillates
Sumac extract
Tall oil, except skimmings
Tanning extracts and materials, natural
Tar and tar oils, products of wood distillation
Turpentine, produced by distillation of pine gum or pine wood
Valonia extract
Wattle extract
Wood alcohol, natural
Wood creosote
Wood distillates

2865     *Cyclic Organic Crudes and Intermediates, and Organic Dyes and Pigments*

Establishments primarily engaged in manufacturing cyclic organic crudes and intermediates, and organic dyes and pigments. Important products of this industry include: 1) aromatic chemicals, such as benzene, toluene, mixed xylenes naphthalene; 2) synthetic organic dyes; and 3) synthetic organic pigments. Establishments primarily engaged in manufacturing coal tar crudes in chemical recovery ovens are classified in Industry 3312, and petroleum refineries which produce such products as byproducts of petroleum refining are classified in Industry 2911.

Acid dyes, synthetic
Acids, coal tar: derived from coal tar distillation
Alkylated diphenylamines, mixed
Alkylated phenol, mixed

Industry
Group   Industry
No.       No.

286        INDUSTRIAL ORGANIC CHEMICALS—Cont.

2865   *Cyclic Organic Crudes and Intermediates, and Organic Dyes*
       *and Pigments—Cont.*

Aminoanthraquinone
Aminoazobenzene
Aminoazotoluene
Aminophenol
Aniline
Aniline oil
Anthracene
Anthraquinone dyer
Azine dyes
Azo dyes
Azobenzene
Azoic dyes
Benzaldehyde
Benzene hexachloride (BHC)
Benzene, made in chemical
   plants
Benzoic acid
Biological stains
Chemical indicators
Chlorobenzene
Chloronapthalene
Chlorophenol
Chlorotoluene
Coal tar crudes, derived from
   coal tar distillation
Coal tar distillates
Coal tar intermediates
Color lakes and toners
Color pigments, organic: except
   animal black and bone
   black
Colors, dry: lakes, toners, or
   full strength organic colors
Colors, extended (color
   lakes)
Cosmetic dyes, synthetic
Creasote oil, made in chemical
   plants

Cresols, made in chemical
   plants
Cresylic acid, made in chemical
   plants
Cyclic crudes, coal tar: product
   of coal tar distillation
Cyclic intermediates, made in
   chemical plants
Cyclohexane
Diphenylamine
Drug dyes, synthetic
Dye (cyclic) intermediates
Dyes, food: synthetic
Dyes, synthetic organic
Eosine toners
Ethylbenzene
Hydroquinone
Isocyanates
Lake red C toners
Leather dyes and stains,
   synthetic
Lithol rubine lakes and toners
Maleic anhydride
Methyl violet toners
Naphtha, solvent: made in
   chemical plants
Naphthalene chips and flakes
Naphthalene, made in chemical
   plants
Naphthol, alpha and beta
Nitro dyes
Nitroaniline
Nitrobenzene
Nitrophenol
Nitroso dyes
Oils: light, medium, and
   heavy: made in chemical
   plants

286          INDUSTRIAL ORGANIC CHEMICALS—Cont.

   2865   *Cyclic Organic Crudes and Intermediates, and Organic Dyes
          and Pigments—Cont.*

| | |
|---|---|
| Organic pigments (lakes and toners) | Pitch, product of coal tar distillation |
| Orthodichlorobenzene | Pulp colors, organic |
| Paint pigments, organic | Quinoline dyes |
| Peacock blue lake | Resorcinol |
| Pentachlorophenol | Scarlet 2 R lake |
| Persian orange lake | Stilbone dyes |
| Phenol | Styrene |
| Phlorine toners | Styrene monomer |
| Phosphomolybdic acid lakes and toners | Tar, product of coal tar distillation |
| Phosphotungstic acid lakes and toners | Toluene, made in chemical plants |
| Phthalic anhydride | Toluidines |
| Phthalocyanine toners | Vat dyes, synthetic |
| Pigment scarlet lake | Xylene, made in chemical plants |

   2869   *Industrial Organic Chemicals, Not Elsewhere Classified*

   Establishments primarily engaged in manufacturing industrial
   organic chemicals, not elsewhere classified. Important products
   of this industry include: 1) aliphatic and other acylic organic
   chemicals, such as ethylene, butylene, and butadiene; acetic,
   chloroacetic, adipic, formic, oxalic, and tartaric acids and their
   metallic salts; chloral, formaldehyde, and methylamine; 2) solvents, such as amyl, butyl, and ethyl alcohols; methanol; amyl,
   butyl, and ethyl acetates; ethyl ether, ethylene glycol ether, and
   diethylene glycol ether; acetone, carbon disulfide, and chlorinated
   solvents, such as carbon tetrachloride, perchloroethylene, and -
   trichloroethylene; 3) polyhydric alcohols, such as ethylene glycol, sorbitol, pentaerythritol, synthetic glycerin; 4) synthetic perfume and flavoring materials, such as coumarin, methyl
   salicylate, saccharin, citral, citronellal; synthetic geraniol, ionone,
   terpineol, and synthetic vanillin; 5) rubber-processing chemicals,
   such as accelerators and antioxidants, both cyclic and acyclic; 6)
   plasticizers, both cyclic and acyclic, such as esters of phosphoric
   acid, phthalic anhydride, adipic acid, lauric acid, oleic acid, sebacic acid, and stearic acid; 7) synthetic tanning agents, such as

Industry
Group    Industry
No.      No.

## 286            INDUSTRIAL ORGANIC CHEMICALS—Cont.

2869   *Industrial Organic Chemicals, Not Elsewhere Classified—Cont.*
naphthalene sulfonic acid condensates; 8) chemical warfare
gases; and 9) esters, amines, etc., of polyhydric alcohols and fatty
and other acids. Establishments primarily engaged in manufacturing plastics materials and nonvulcanizable elastomers are classified in Industry 2821; those manufacturing synthetic rubber are
classified in Industry 2822; those manufacturing essential oils are
classified in Industry 2899; those manufacturing wood distillation
products, naval stores and natural dyeing and tanning materials
are classified in Industry 2861; those manufacturing manmade
textile fibers are classified in Industries 2823 and 2824; those
manufacturing specialty cleaning, polishing, and sanitation
preparations are classified in Industry 2842; those manufacturing
paints are classified in Industry 2851 those manufacturing urea
are classified in Industry 2873; those manufacturing organic pigments are classified in Industry 2865; those manufacturing inorganic pigments are classified in Industry 2816 and those
manufacturing aliphatics and aromatics as byproducts of petroleum refining are classified in 2911. Distilleries engaged in the
manufacture of grain alcohol for beverage purposes are classified
in Industry 2085.

Acetaldehyde
Acetates, except natural acetate
  of lime
Acetic acid, synthetic
Acetic anhydride
Acetin
Acetone, synthetic
Acid esters and amines
Acids, organic
Acrolein
Acrylonitrile
Adipic acid
Adipic acid esters
Adiponitrile
Alcohol, aromatic
Alcohol, fatty: powdered

Alcohol, methyl: synthetic
  (methanol)
Alcohols, industrial: denatured
  (nonbeverage)
Algin products
Amyl acetate and alcohol
Aspartame
Bromochloromethane
Butadiene, made in chemical
  plants
Butyl acetate, alcohol, and
  propionate
Butyl ester solution of 2,4-D
Butylene, made in chemical
  plants
Calcium oxalate

Industry
Group    Industry
No.       No.

286            INDUSTRIAL ORGANIC CHEMICALS—Cont.

2869    *Industrial Organic Chemicals, Not Elsewhere Classified—Cont.*

Camphor, synthetic
Caprolactam
Carbon bisulfide (disulfide)
Carbon tetrachloride
Casing fluids for curing fruits,
  spices, and tobacco
Cellulose acetate,
  unplasticized
Chemical warfare gases
Chloral
Chlorinated solvents
Chloroacetic acid and metallic
  salts
Chloroform
Chloropicrin
Citral
Citrates
Citric acid
Citronellal
Coumarin
Cream of tartar
Cyclopropane
DDT, technical
Decahydronaphthalene
Dichlorodifluoromethane
Diethylcyclobezane (mixed
  isomera)
Diethylene glycol ether
Dimethyl divinyl acetylene
  (disopropenyl acetylene)
Dimethylhydrazine,
  unsymmetrical
Enzymes, except diagnostic
  substances
Esters of phosphoric, adipic,
  lauric, oleic, sebacic, and
  stearic acids
Esters of phthalic anhydride
Ethanol, industrial
Ether

Ethyl acetate, synthetic
Ethyl alcohol, industrial
  (nonbeverage)
Ethyl butyrate
Ethyl cellulose, unplasticized
Ethyl chloride
Ethyl ether
Ethyl formate
Ethyl nitrite
Ethyl perhydrophenanthrene
Ethylene glycol
Ethylene glycol ether
Ethylene glycol, inhibited
Ethylene oxide
Ethylene, made in chemical
  plants
Fatty acid esters and amines
Ferric ammonium oxalate
Flavors and flavoring materials,
  synthetic
Fluorinated hydrocarbon gases
Formaldehyde (formalin)
Formic acid and metallic salts
Fuel propellants, solid: organic
Fuels, high energy: organic
Geraniol, synthetic
Glycerin, except from fats
  (synthetic)
Grain alcohol, industrial
  (nonbeverage)
Hexamethylenediamine
Hexamethylenetetramine
High purity grade chemicals,
  organic: refined from
  technical grades
Hydraulic fluids, synthetic base
Industrial organic cyclic
  compounds
Ionone
Isopropyl alcohol

286            INDUSTRIAL ORGANIC CHEMICALS—Cont.

  2869   *Industrial Organic Chemicals, Not Elsewhere Classified—Cont.*

Ketone, methyl ethyl
Ketone, methyl isobutyl
Laboratory chemicals, organic
Lauric acid esters
Lime citrate
Malanonitrile, technical grade
Metallic salts of acyclic organic
  chemicals
Metallic stearate
Methanol, synthetic (methyl
  alcohol)
Methyl chloride
Methyl perhydrofluorine
Methyl salicylate
Methylamine
Methylene chloride
Monochlorodifluoremethane
Monomethylparaminophenol
  sulfate
Monosodium glutamate
Mustard gas
Naphthalene sulfonic acid
  condensates
Naphthonic acid soaps
Normal hexyl decalin
Nuclear fuels, organic
Oleic acid esters
Organic acid esters
Organic chemicals, acyclic
Oxalates
Oxalic acid and metallic salts
Perchloroethylene
Perfume materials, synthetic
Phosgene
Phthalates
Plasticizers, organic: cyclic and
  acyclic
Polyhydric alcohol esters and
  amines
Polyhydric alcohols

Potassium bitartrate
Propellants for missiles, solid:
  organic
Propylene glycol
Propylene, made in chemical
  plants
Quinuclidinol ester of benzylic
  acid
Reagent grade chemicals,
  organic: refined from
  technical grades, except
  diagnostic and substances
Rocket engine fuel, organic
Rubber processing chemicals,
  organic: accelerators and
  antioxidants
Saccharin
Sebacic acid
Silicones
Sodium acetate
Sodium alginate
Sodium benzoate
Sodium glutamate
Sodium pentachlorophenate
Sodium sulfoxalate
  formaldehyde
Solvents, organic
Sorbitol
Stearic acid salts
Sulfonated naphthalene
Sweeteners, synthetic
Tackifiers, organic
Tannic acid
Tanning agents, synthetic
  organic
Tartaric acid and metallic
  salts
Tartrates
Tear gas
Terpineol

Industry
Group   Industry
No.     No.

286              INDUSTRIAL ORGANIC CHEMICALS—Cont.

        2869   *Industrial Organic Chemicals, Not Elsewhere Classified—Cont.*

*tert*-Butylated bis(*p*-                tetrachlorodifluoroethane
  phenoxyphenyl) ether fluid             isopropyl alcohol
Tetrachloroethylene                      Tricresyl phosphate
Tetraethyl lead                          Tridecyl alcohol
Thioglycolic acid, for                   Trimethyltrithiophosphite
  permanent wave lotions                   (rocket propellants)
Trichloroethylene                        Triphenyl phosphate
Trichlorophenoxyacetic acid              Vanillin, synthetic
Trichlorotrifluoroethane                 Vinyl acetate


287              AGRICULTURAL CHEMICALS

This group includes establishments primarily engaged in manufacturing nitrogenous and phosphatic basic fertilizers, mixed
fertilizers, pesticides, and other agricultural chemicals. Establishments primarily engaged in manufacturing basic chemicals,
which require further processsing or formulation before use as
agricultural pest control agents, are classified in Industry
Groups 281 or 286.

        2873   *Nitrogenous Fertilizers*

Establishments primarily engaged in manufacturing nitrogenous fertilizer materials or mixed fertilizers from nitrogenous
materials produced in the same establishment. Included are ammonia fertilizer compounds and anhydrous ammonia, nitric
acid, ammonium nitrate, ammonium sulfate and nitrogen solutions, urea, and natural organic fertilizers (except compost) and
mixtures.

Ammonia liquor                           Fertilizers: natural (organic),
Ammonium nitrate and sulfate               except compost
Anhydrous ammonia                        Nitric acid
Aqua ammonia, made in                    Nitrogen solutions (fertilizer)
  ammonia plants                         Plant foods, mixed: made in
Fertilizers, mixed: made in                plants producing nitrogenous
  plants producing nitrogenous             fertilizer materials
  fertilizer materials                   Urea

287              AGRICULTURAL CHEMICALS—Cont.

*2874   Phosphatic Fertilizers*

Establishments primarily engaged in manufacturing phosphatic fertilizer materials, or mixed fertilizers from phosphatic materials produced in the same establishment. Included are phosphoric acid; normal, enriched, and concentrated superphosphates; ammonium phosphates; nitrophosphates; and calcium metaphosphates.

| | |
|---|---|
| Ammonium phosphates | Phosphoric acid |
| Calcium metaphosphates | Plant foods, mixed: made in |
| Defluorinated phosphates | plants producing phosphatic |
| Diammonium phosphates | fertilizer materials |
| Fertilizers, mixed: made in | Superphosphates, |
| plants producing phosphatic | ammoniated and |
| fertilizer materials | not ammoniated |

*2875   Fertilizers, Mixing Only*

Establishments primarily engaged in mixing fertilizers from purchased fertilizer materials.

| | |
|---|---|
| Compost | Potting soil, mixed |
| Fertilizers, mixed: made in | |
| plants not manufacturing | |
| fertilizer materials | |

*2879   Pesticides and Agricultural Chemicals, Not Elsewhere Classified*

Establishments primarily engaged in the formulation and preparation of ready-to-use-agricultural and household pest control chemicals, including insecticides, fungicides, and herbicides, from technical chemicals or concentrates; and the production of concentrates which require further processing before use as agricultural pesticides. This industry also includes establishments primarily engaged in manufacturing or formulating agricultural chemicals, not elsewhere classified, such as minor or trace elements and soil conditioners. Establishments primarily engaged in manufacturing basic or technical agricultural pest control chemicals are classified in Industry Group 281 if the chemicals are inorganic and in Industry Group 286 if they are organic. Establishments primarily engaged in manufacturing agricultural lime products are classified in Major Group 32.

287          AGRICULTURAL CHEMICALS—Cont.

    2879    *Pesticides and Agricultural Chemicals, Not Elsewhere*
           *Classified—Cont.*

| | |
|---|---|
| Agricultural disinfectants | Lindane, formulated |
| Agricultural pesticides | Moth repellants |
| Arsenates: calcium, copper, and lead-formulated | Nicotine and salts |
| | Nicotine-bearing insecticides |
| Arsenites, formulated | Paris green (insecticide) |
| Bordeaux mixture | Pesticides, household |
| Calcium arsenate and arsenite, formulated | Phytoactin |
| | Plant hormones |
| Cattle dips | Poison: ant, rat, roach, and rodent (household) |
| Copper arsenate, formulated | |
| DDT (insecticide), formulated | Pyrethrin-bearing preparations |
| Defoliants | |
| Elements, minor or trace (agricultural chemicals) | Pyrethrin concentrates |
| | Rodenticides |
| Exterminating products, for household and industrial use | Rotenone-bearing preparations |
| | Rotenone concentrates |
| Fly sprays | Sheep dips, chemical |
| Fungicides | Sodium arsenite (formulated) |
| Growth regulants, agricultural | Soil conditioners |
| Herbicides | Sulfur dust (insecticide) |
| Household insecticides | Thiocyanates, organic (formulated) |
| Insect powder, household | |
| Insecticides, agricultural | Trace elements (agricultural chemicals) |
| Lead arsenate, formulated | |
| Lime-sulfur, dry and polution | Xanthone (formulated) |

289          MISCELLANEOUS CHEMICAL PRODUCTS

    2891    *Adhesives and Sealants*

Establishments primarily engaged in manufacturing industrial and
household adhesives, glues, caulking compounds, sealants, and
linoleum, tile, and rubber cements from vegetable, animal, or synthetic plastics materials, purchased or produced in the same establishment. Establishments primarily engaged in manufacturing
gelatin and sizes are classified in Industry 2899, and those manufacturing vegetable gelatin or agar-agar are classified in Industry 2833.

| | |
|---|---|
| Adhesives | Cement (cellulose nitrate base) |
| Adhesives, plastics | Cement, linoleum |
| Caulking compounds | Cement, mending |

289          MISCELLANEOUS CHEMICAL PRODUCTS—Cont.

2891    *Adhesives and Sealants—Cont.*

| | |
|---|---|
| Epoxy adhesives | Paste, adhesive |
| Glue, except dental: animal, | Porcelain cement, household |
| vegetable, fish, casein, and | Rubber cement |
| synthetic resin | Sealing compounds for pipe |
| Iron cement, household | threads and joints |
| Joint compounds | Sealing compounds, synthetic |
| Laminating compounds | rubber and plastics |
| Mucilage | Wax, sealing |

2892    *Explosives*

Establishments primarily engaged in manufacturing explosives. Establishments primarily engaged in manufacturing ammunition for small arms are classified in Industry 3482, and those manufacturing fireworks are classified in Industry 2899.

| | |
|---|---|
| Ametol (explosives) | High explosives |
| Axides (explosives) | Lead acids (explosives) |
| Blasting powder and blasting | Mercury oxide (explosives) |
| caps | Nitrocellulose powder |
| Carbohydrates, altrated | (explosives) |
| (explosives) | Nitroglycerin (explosives) |
| Cordeau détonant (explosives) | Nitromannitol (explosives) |
| Cordite (explosives) | Nitrostarch (explosives) |
| Detonating caps for safety fuses | Nitrosugars (explosives) |
| Detonators (explosive | Pentolite (explosives) |
| compounds) | Permissible explosives |
| Dynamite | Picric acid (explosives) |
| Explosive cartridges for | Powder, explosive: pellet, |
| concussion forming of metal | smokeless, and sporting |
| Explosive compounds | RDX (explosives) |
| Explosives | Squibbs, electric |
| Fulminate of mercury | Styphnic acid |
| (explosive compounds) | TNT (trinitrotoluene) |
| Fuse powder | Tetryl (explosives) |
| Fuses, safety | Well shooting torpedoes |
| Gunpowder | (explosives) |

2893    *Printing Ink*

Establishments primarily engaged in manufacturing printing ink, including gravure ink, screen process ink, and lithographic ink.

Industry
Group    Industry
No.       No.

289              MISCELLANEOUS CHEMICAL PRODUCTS—Cont.

2893    *Printing Ink—Cont.*

Establishments primarily engaged in manufacturing writing ink
and fluids are classified in Industry 2899, and those manufacturing drawing ink are classified in Industry 3952.

| | |
|---|---|
| Bronze ink | Letterpress ink |
| Flexographic ink | Lithographic ink |
| Gold ink | Offset ink |
| Gravure ink | Printing ink: base or finished |
| Ink, duplicating | Screen process ink |

2895    *Carbon Black*

Establishments primarily engaged in manufacturing carbon black
(channel and furnace black). Establishments primarily engaged in
manufacturing bone and lamp black are classified in Industry
2816.

| | |
|---|---|
| Carbon black | Furnace black |
| Channel black | |

2899    *Chemicals and Chemical Preparations, Not Elsewhere Classified*

Establishments primarily engaged in manufacturing miscellaneous chemical preparations, not elsewhere classified, such as
fatty acids, essential oils, gelatin (except vegetable), sizes, bluing,
laundry sours, writing and stamp pad ink, industrial compounds,
such as boiler and heat insulating compounds, metal, oil, and water treating compounds, waterproofing compounds, and chemical
supplies for foundries. Establishments primarily engaged in manufacturing vegetable gelatin (agar-agar) are classified in Industry
2833; those manufacturing dessert preparations based on gelatin
are classified in Industry 2099; those manufacturing printing ink
are classified in Industry 2893; and those manufacturing drawing
ink are classified in Industry 3952.

| | |
|---|---|
| Acid resist for etching | Bay oil |
| Acid, battery | Binders (chemical foundry |
| Anise oil | supplies) |
| Antifreeze compounds, except | Bluing |
| industrial alcohol | Boiler compounds, antiscaling |

289         MISCELLANEOUS CHEMICAL PRODUCTS—Cont.

2899    *Chemicals and Chemical Preparations, Not Elsewhere
Classified—Cont.*

Bombs, flashlight

Caps, for toy pistols

Carbon removing solvent

Chemical cotton (processed cotton limiters)

Chemical supplies for foundries

Citronella oil

Concrete curing compounds (blends of pigments, waxes, and resins)

Concrete hardening compounds

Core oil and binders

Core wash

Core wax

Correction fluid

Corrosion preventive lubricant, synthetic base: for jet engines

Deicing fluid

Desalter kits, sea water

Dextrine sizes

Drilling mud

Dyes, household

Essential oils

Ethylene glycol antifreeze preparations

Eucalyptus oil

Exothermics for metal industries

Facings (chemical foundry supplies)

Fatty acids: margaric, eleic, and stearic

Fire extinguisher charges

Fire retardant chemical preparations

Fireworks

Flares

Fluidifier (retarder) for concrete

Fluorescent inspection oil

Fluxes: bracing, soldering, galvanizing, and welding

Foam charge mixtures

Food contamination testing and screening kits

Foundry supplies, chemical preparations

Frit

Fuel tank and engine cleaning chemicals, automotive and aircraft

Fusses: highway, marine, and railroad

Gelatin capsules, empty

Gelatin: edible, technical, photographic, and pharmaceutical

Glue size

Grapefruit oil

Grouting material (concrete mending compound)

Gum sizes

Gun slushing compounds

Heat insulating compounds

Heat treating salts

Hydrofluoric acid compound, for etching and polishing glass

Igniter grains, boron potassium nitrate

Incense

Industrial sizes

Insulating compounds

Jet fuel igniters

Laundry sours

Lemon oil

Lighter fluid

Industry
Group  Industry
No.    No.

289        MISCELLANEOUS CHEMICAL PRODUCTS—Cont.

2899   *Chemicals and Chemical Preparations, Not Elsewhere*
       *Classified—Cont.*

Magnetic inspection oil and
    powder
Margaric acid
Metal drawing compound
    lubricants
Metal treating compounds
Military pyrotechnics
Napalm
Oil treating compounds
Oleic acid (red oil)
Orange oil
Orris oil
Ossein
Oxidizers, inorganic
Packers' salt
Parting compounds (chemical
    foundry supplies)
Patching plaster, household
Penetrants, inspection
Peppermint oil
Plating compounds
Pyrotechnic ammunition: flares,
    signals, flashlight bombs, and
    rockets
Railroad torpedoes

Red oil (oleic acid)
Rifle bore cleaning compounds
Rosin sizes
Rubber processing preparations
Rust resisting compounds
Salt
Signal flares, marine
Sizes: animal, vegetable, and
    synthetic plastics materials
Sodium chloride, refined
Soil testing kits
Spearmint oil
Spirit duplicating fluid
Stearic acid
Stencil correction compounds
Tints and dyes, household
Torches (fireworks)
Vegetable oils, vulcanized or
    sulfurized
Water treating compounds
Water, distilled
Waterproofing compounds
Wintergreen oil
Wood, plastic
Writing ink and fluids

## Major Group 29: PETROLEUM REFINING AND RELATED INDUSTRIES

### The Major Group as a Whole

This major group includes establishments primarily engaged in petroleum refining, manufacturing paving and roofing materials, and compounding lubricating oils and greases from purchased materials. Establishments manufacturing and distributing gas to consumers are classified in public utilities industries, and those primarily engaged in producing coke and byproducts are classified in Major Group 33.

Industry
Group    Industry
No.       No.

291              PETROLEUM REFINING

2911     *Petroleum Refining*

Establishments primarily engaged in producing gasoline, kerosene, distillate fuel oils, residual fuel oils, and lubricants, through fractionation or straight distillation of crude oil, redistillation of unfinished petroleum derivatives, cracking or other processes. Establishments of this industry also produce aliphatic and aromatic chemicals as byproducts. Establishments primarily engaged in producing natural gasoline from natural gas are classified in mining industries. Those manufacturing lubricating oils and greases by blending and compounding purchased materials are included in Industry 2992. Establishments primarily re-refining used lubricating oils are classified in Industry 2992. Establishments primarily engaged in manufacturing cyclic and acyclic organic chemicals are classified in Major Group 28.

Acid oil, produced in petroleum refineries

Alkylates, produced in petroleum refineries

Aromatic chemicals, made in petroleum refineries

Asphalt and asphaltic materials: liquid and solid—produced in petroleum refineries

Benzene, produced in petroleum refineries

Butadiene, produced in petroleum refineries

Butylene, produced in petroleum refineries

Coke, petroleum: produced in petroleum refineries

Ethylene, produced in petroleum refineries

Industry
Group   Industry
No.     No.

291          PETROLEUM REFINING—Cont.

   2911    *Petroleum Refining—Cont.*

Fractionation products of crude
petroleum, produced in
petroleum refineries

Gas, refinery or still oil:
produced in petroleum
refineries

Gases, liquefied petroleum:
produced in petroleum
refineries

Gasoline blending plants

Gasoline, except natural
gasoline

Greases, lubricating: produced
in petroleum
refineries

Hydrocarbon fluid, produced
in petroleum refineries

Jet fuels

Kerosene

Mineral jelly, produced in
petroleum refineries

Mineral oils, natural: produced
in petroleum refineries

Mineral waxes, natural:
produced in petroleum
refineries

Naphtha, produced in
petroleum refineries

Naphthanic acids, produced in
petroleum refineries

Oils, partly refined: sold for
rerunning—produced in
petroleum refineries

Oils: fuel, lubricating, and
illuminating—produced in
petroleum refineries

Paraffin wax, produced in
petroleum refineries

Petroleum, produced in
petroleum refineries

Petroleum refining

Propylene, produced in
petroleum refineries

Road materials, bituminous:
produced in petroleum
refineries

Road oils, produced in
petroleum refineries

Solvents, produced in
petroleum refineries

Tar or residuum, produced in
petroleum refineries

295          ASPHALT PAVING AND ROOFING MATERIALS

   2951    *Asphalt Paving Mixtures and Blocks*

Establishments primarily engaged in manufacturing asphalt and
tar paving mixtures; and paving blocks made of asphalt and various compositions of asphalt or tar with other materials. Establishments primarily engaged in manufacturing brick, concrete,
granite, and stone paving blocks are classified in Major Group 32.

Asphalt and asphaltic mixtures
for paving, not made in
refineries

Asphalt paving blocks, not
made in petroleum refineries

Asphaltic concrete, not made in
petroleum refineries

Industry
Group   Industry
No.     No.

295          ASPHALT PAVING AND ROOFING MATERIALS—Cont.

2951   *Asphalt Paving Mixtures and Blocks—Cont.*

Coal tar paving materials, not made in petroleum refineries
Composition blocks for paving
Concrete, bituminous
Mastic floor composition, hot and cold
Road materials, bituminous:
not made in petroleum refineries
Tar and asphalt mixtures for paving, not made in petroleum refineries

2952   *Asphalt Felts and Coatings*

Establishments primarily engaged in manufacturing, from purchased materials [mfpm], asphalt, and other saturated felts in roll or shingle form, either smooth or faced with grit, and in manufacturing roofing cements and coatings. Establishments primarily engaged in manufacturing paint are classified in Industry 2851, and those manufacturing linoleum and tile cement are classified in Industry 2891.

Asphalt roof cement—mfpm
Asphalt saturated board—mfpm
Brick siding, asphalt—mfpm
Coating compounds, tar—mfpm
Fabrics, roofing: asphalt or tar saturated—mfpm
Insulating siding, impregnated—mfpm
Mastic roofing composition—mfpm
Pitch, roofing—mfpm
Roof cement: asphalt, fibrous, and plastics—mfpm
Roof coatings and cements: liquid and plastics—mfpm
Roofing felts, cements, and
coatings: asphalt, tar, and composition—mfpm
Roofing, asphalt or tar saturated felt: built-up, roll, and shingle—mfpm
Sheathing, asphalt saturated—mfpm
Shingles, asphalt or tar saturated felt: strip and individual—mfpm
Siding, insulating: impregnated—mfpm
Tar paper, roofing—mfpm

299          MISCELLANEOUS PRODUCTS OF PETROLEUM AND COAL

2992   *Lubricating Oils and Greases*

Establishments primarily engaged in blending, compounding, and re-refining lubricating oils and greases from purchased mineral,

299            MISCELLANEOUS PRODUCTS OF PETROLEUM AND
               COAL—Cont.

   2992    *Lubricating Oils and Greases—Cont.*
           animal, and vegetable materials. Petroleum refineries engaged in
           the production of lubricating oils and greases are classified in In-
           dustry 2911.

| | |
|---|---|
| Brake fluid, hydraulic—mfpm | Re-refining lubricating oils |
| Cutting oils, blending and | and greases— |
| compounding from | mfpm |
| purchased material | Rust arresting compounds, |
| Greases, lubricating—mfpm | animal and vegetable |
| Hydraulic fluids—mfpm | oil base— |
| Lubricating greases and oils— | mfpm |
| mfpm | Transmission fluid—mfpm |

   2999    *Products of Petroleum and Coal, Not Elsewhere Classified*
           Establishments primarily engaged in manufacturing packaged
           fuel, powdered fuel, and other products of petroleum and coal,
           not elsewhere classified.

| | |
|---|---|
| Calcined petroleum coke— | Fuel briquettes or boulets, made |
| mfpm | with petroleum binder |
| Coke, petroleum: not produced | Waxes, petroleum: not |
| in petroleum refineries | produced in petroleum |
| Fireplace logs, made from coal | refineries |

## Major Group 30: RUBBER AND MISCELLANEOUS PLASTICS PRODUCTS

### The Major Group as a Whole

This major group includes establishments manufacturing products, not elsewhere classified, from plastics resins and from natural, synthetic, or reclaimed rubber, gutta percha, balata, or gutta siak. Numerous products made from these materials are included in other major groups, such as boats in Major Group 37, and toys, buckles, and buttons in Major Group 39. This group includes establishments primarily manufacturing tires, but establishments primarily recapping and retreading automobile tires are classified in Services, Industry 7534. Establishments primarily engaged in manufacturing synthetic rubber and synthetic plastics resins are classified in Industry Group 282.

Industry
Group   Industry
No.     No.

### 301          TIRES AND INNER TUBES

3011   *Tires and Inner Tubes*

Establishments primarily engaged in manufacturing pneumatic casings, inner tubes, and solid and cushion tires for all types of vehicles, airplanes, farm equipment, and children's vehicles; tiring; camelback; and tire repair and retreading materials. Establishments primarily engaged in retreading tires are classified in Services, Industry 7534.

| | |
|---|---|
| Camelback for tire retreading | Tire sundries and tire repair |
| Inner tubes: airplane, | materials, rubber |
| automobile, bicycle, | Tires, cushion or solid rubber |
| motorcycle, and tractor | Tiring, continuous lengths: |
| Pneumatic casings (rubber | rubber, with or without metal |
| tires) | core |

### 302          RUBBER AND PLASTICS FOOTWEAR

3021   *Rubber and Plastics Footwear*

Establishments primarily engaged in manufacturing fabric upper footwear having rubber or plastics soles vulcanized, injection molded, or cemented to the uppers, and rubber and plastics protective footwear. Establishments primarily engaged in manufacturing

Industry
Group   Industry
No.     No.

302           RUBBER AND PLASTICS FOOTWEAR—Cont.

_3021_   _Rubber and Plastics Footwear—Cont._

rubber, composition, and fiber heels, soles, soling strips, and related shoe making and repairing materials are classified in Industry 3069; those manufacturing plastics soles and soling strips are
classified in Industry 3089; and those manufacturing other
footwear of rubber or plastics are classified in Industry Group 314.

| | |
|---|---|
| Arctics, rubber or rubber soled fabric | Overshoes, plastics |
| Boots, plastics | Overshoes, rubber or rubber soled fabric |
| Boots, rubber or rubber soled fabric | Pacs, rubber or rubber soled fabric |
| Canvas shoes, rubber soled | Sandals, rubber |
| Footholds, rubber | Shoes, plastics soles molded to fabric uppers |
| Footwear, rubber or rubber soled fabric | Shoes, rubber or rubber soled fabric uppers |
| Gaiters, rubber or rubber soled fabric | Shower sandals or slippers, rubber |
| Galoshes, plastics | |
| Galoshes, rubber or rubber soled fabric | |

305           GASKETS, PACKING, AND SEALING DEVICES AND
              RUBBER AND PLASTICS HOSE AND BELTING

_3052_   _Rubber and Plastics Hose and Belting_

Establishments primarily engaged in manufacturing rubber and
plastics hose and belting, including garden hose. Establishments
primarily engaged in manufacturing rubber tubing are classified
in Industry Group 306; those manufacturing plastics tubing are
classified in Industry 3082; and those manufacturing flexible
metallic hose are classified in Industry 3599.

| | |
|---|---|
| Air brake and air line hose, rubber or rubberized fabric | Hose, plastics or rubber |
| Automobile hose, plastics or rubber | Hose: cotton fabric, rubber lined |
| Belting, rubber (e.g., conveyor, elevator, transmission) | Pneumatic hose, rubber or rubberized (e.g., air brake and air line) |
| Firehose, rubber | V-belts, rubber or plastics |
| Garden hose, plastics or rubber | Vacuum cleaner hose, plastics or rubber |
| Hester hose, plastics or rubber | |

Industry
Group   Industry
No.        No.

305          GASKETS, PACKING, AND SEALING DEVICES
             AND RUBBER AND PLASTICS HOSE AND
             BELTING—Cont.

       3053  *Gaskets, Packing, and Sealing Devices*
             Establishments primarily engaged in manufacturing gaskets, gas-
             keting materials, compression packings, mold packings, oil seals,
             and mechanical seals. Included are gaskets, packing, and sealing
             devices made of leather, rubber, metal, asbestos, and plastics.

|  |  |
|---|---|
| Gaskets, regardless of material | Packing for steam engines, pipe joints, air compressors, etc. |
| Grease retainers, leather | Packing, metallic |
| Grease seals, asbestos | Packing, rubber |
| Oil seals, asbestos | Packing: cup, U-valve, etc.— leather |
| Oil seals, leather | Steam and other packing |
| Oil seals, rubber | Washers, leather |

306          FABRICATED RUBBER PRODUCTS, NOT ELSEWHERE
             CLASSIFIED

       3061  *Molded, Extruded, and Lathe-Cut Mechanical Rubber Goods*
             Establishments primarily engaged in manufacturing molded, ex-
             truded, and lathe-cut mechanical rubber goods. The products are
             generally parts for machinery and equipment. Establishments pri-
             marily engaged in manufacturing other industrial rubber goods,
             rubberized fabric, and miscellaneous rubber specialties and sun-
             dries are classified in Industry 3069.

|  |  |
|---|---|
| Appliance mechanical rubber goods: molded, extruded, and lathe-cut | goods: molded, extruded, and lathe-cut |
| Automotive mechanical rubber goods: molded, extruded, and lathe-cut | Oil and gas field machinery and equipment mechanical rubber goods: molded, extruded, and lathe-cut |
| Mechanical rubber goods: molded, extruded, and lathe-cut | Rubber goods, mechanical: molded, extruded, and lathe-cut |
| Off-highway machinery and equipment mechanical rubber | Surgical and medical tubing: extruded and lathe-cut |

Industry
Group    Industry
No.       No.

306              FABRICATED RUBBER PRODUCTS, NOT ELSEWHERE
                 CLASSIFIED—Cont.

         3069    *Fabricated Rubber Products, Not Elsewhere Classified*

                 Establishments primarily engaged in manufacturing industrial rub-
                 ber goods, rubberized fabrics, and vulcanized rubber clothing, and
                 miscellaneous rubber specialties and sundries, not elsewhere classi-
                 fied. Included in this industry are establishments primarily engaged
                 in reclaiming rubber and rubber articles. Establishments primarily
                 engaged in the wholesale distribution of scrap rubber are classified
                 in Wholesale Trade, Industry 5093. Establishments primarily en-
                 gaged in rebuilding and retreading tires are classified in Services,
                 Industry 7534; those manufacturing rubberized clothing from pur-
                 chased materials are classified in Industry 2385; and those manu-
                 facturing gaskets and packing are classified in Industry 3053.

Acid bottles, rubber
Air-supported rubber structures
Aprons, vulcanized rubber and
    rubberized fabric—mitse
Bags, rubber or rubberized
    fabric
Balloons, advertising and toy:
    rubber
Balloons, metal foil laminated
    with rubber
Balls, rubber: except athletic
    equipment
Bath sprays, rubber
Bathing caps and suits,
    rubber—mitse
Battery boxes, jars, and parts:
    hard rubber
Bibs, vulcanized rubber and
    rubberized fabric—mitse
Bottles, rubber
Boxes, hard rubber
Brake lining, rubber
Brushes, rubber
Bulbs for medicine droppers,
    syringes, atomizers, and
    sprays: rubber

Bushings, rubber
Capes, vulcanized rubber and
    rubberized fabric—mitse
Caps, rubber—mitse
Castings, rubber
Chlorinated rubbers, natural
Cloaks, vulcanized rubber and
    rubberized fabric—
    mitse
Clothing, vulcanized rubber
    and rubberized fabric—
    mitse
Combs, hard rubber
Culture cups, rubber
Custom compounding of rubber
    materials
Cyclo rubbers, natural
Diaphragms, rubber: separate
    and in kits
Dress shields, vulcanized
    rubber and rubberized
    fabric—mitse
Druggists' sundries, rubber
Erasers: rubber, or rubber and
    abrasive combined
Fabrics, rubberized

Industry
Group   Industry
No.       No.

306        FABRICATED RUBBER PRODUCTS, NOT ELSEWHERE
             CLASSIFIED—Cont.

   3069   *Fabricated Rubber Products, Not Elsewhere Classified—Cont.*

Film, rubber
Finger cots, rubber
Flooring, rubber: tile or sheet
Foam rubber
Fountain syringes, rubber
Friction tape, rubber
Fuel coils, rubber
Fuel tanks, collapsible:
   rubberized fabric
Funnels, rubber
Gloves: e.g., surgeons',
   electricians', household—
   rubber
Grips and handles, rubber
Grommets, rubber
Gutta, percha compounds
Hair curlers, rubber
Hairpins, rubber
Handles, rubber
Hard rubber products
Hard surface floor coverings:
   rubber
Heels, boot and shoe: rubber,
   composition, and fiber
Jar rings, rubber
Laboratory sundries: e.g., cases,
   covers, funnels, cups,
   bottles—rubber
Latex, foamed
Lifejackets: inflatable
   rubberized fabric
Liferafts, rubber
Liner strips, rubber
Linings, vulcanizable
   elastomeric: rubber
Mallets, rubber
Mats and matting: e.g., both,
   door—rubber

Mattress protectors, rubber
Mattresses, pneumatic: fabric
   coated with rubber
Medical sundries, rubber
Mittens, rubber
Mouthpieces for pipes and
   cigarette holders, rubber
Nipples, rubber
Orthopedic sundries, molded
   rubber
Pacifiers, rubber
Pads, kneeling: rubber
Pants, baby: vulcanized rubber
   and rubberized fabric—mitse
Pillows, sponge rubber
Pipe stems and bits, tobacco:
   hard rubber
Platens, except printers': solid
   or covered rubber
Plumbers' rubber goods
Pontoons, rubber
Printers' blankets, rubber
Printers' rolls, rubber
Prophylactics, rubber
Pump sleeves, rubber
Reclaimed rubber (reworked
   by manufacturing
   processes)
Rods, hard rubber
Roll coverings: rubber for
   papermill; industrial,
   steelmills, printers'
Roller covers, printers':
   rubber
Rolls, solid or covered
   rubber
Roofing, single ply membrane:
   rubber

306          **FABRICATED RUBBER PRODUCTS, NOT ELSEWHERE CLASSIFIED—Cont.**

    3069    *Fabricated Rubber Products, Not Elsewhere Classified—Cont.*

Rubber heels, soles, and soling
strips
Rubber-covered meter
mounting rings (rubber
bonded)
Rubberbands
Rug bucking compounds, latex
Separators, battery: rubber
Sheeting, rubber or rubberized
fabric
Sheets, hard rubber
Sleeves, pump: rubber
Soles, boot and shoe: rubber,
composition, and fiber
Soling strips, boot and shoe:
rubber, composition, and
fiber
Spatulas, rubber
Sponge rubber and sponge
rubber products
Stair treads, rubber
Stationers' sundries, rubber
Stoppers, rubber

Tape, pressure sensitive
(including friction), rubber
Teething rings, rubber
Thermometer cases, rubber
Thread, rubber: except fabric
covered
Tile, rubber
Top lift sheets, rubber
Top roll covering, for textile
mill machinery: rubber
Toys, rubber: except dolls
Trays, rubber
Tubing, rubber: except extruded
and lothe-cut
Type, rubber
Urinals, rubber
Valves, hard rubber
Wainscoting, rubber
Wallcoverings, rubber
Weather strip, sponge rubber
Water bottles, rubber
Wet suits, rubber

308          **MISCELLANEOUS PLASTICS PRODUCTS**

    3081    *Unsupported Plastics Film and Sheet*

Establishments primarily engaged in manufacturing unsupported plastics film and sheet, from purchased resins or from resins produced in the same plant. Establishments primarily engaged in manufacturing plastics film and sheet for blister and bubble-formed packaging are classified in Industry 3089.

Cellulosic plastics film and
sheet, unsupported
Film, plastics: unsupported
Photographic, micrographic,
and X-ray plastics, sheet, and
film: unsupported

Polyester film and sheet,
unsupported
Polyethylene film and sheet,
unsupported
Polypropylene film and sheet,
unsupported

308        MISCELLANEOUS PLASTICS PRODUCTS—Cont.

*3081*   *Unsupported Plastics Film and Sheet—Cont.*

| | |
|---|---|
| Polyvinyl film and sheet, unsupported | Vinyl and vinyl copolymer film and sheet, unsupported |
| Sheet, plastics: unsupported | |

*3082*   *Unsupported Plastics Profile Shapes*

Establishments primarily engaged in manufacturing unsupported plastics profiles, rods, tubes, and other shapes. Establishments primarily engaged in manufacturing plastics hose are classified in Industry 3052.

| | |
|---|---|
| Profiles, unsupported plastics | Tubes, unsupported plastics |
| Rods, unsupported plastics | |

*3083*   *Laminated Plastics Plate, Sheet, and Profile Shapes*

Establishments primarily engaged in manufacturing laminated plastics plate, sheet, profiles, rods, and tubes. Establishments primarily engaged in manufacturing laminated flexible packaging are classified in Industry Group 267.

| | |
|---|---|
| Plastics, laminated: plate, rods, tubes, profiles and sheet, except flexible packaging | except flexible packaging |
| | Thermosetting laminates: rods, tubes, plates, and sheet, |
| Thermoplastics laminates: rods, tubes, plates, and sheet, | except flexible packaging |

*3084*   *Plastics Pipe*

Establishments primarily engaged in manufacturing plastics pipe. Establishments primarily engaged in manufacturing plastics pipe fittings are classified in Industry 3089.

Pipe, plastics

*3085*   *Plastics Bottles*

Establishments primarily engaged in manufacturing plastics bottles.

Bottles, plastics

Industry
Group   Industry
No.     No.

308              MISCELLANEOUS PLASTICS PRODUCTS—Cont.

     3086    *Plastics Foam Products*

             Establishments primarily engaged in manufacturing plastics foam
             products.

|  |  |
|---|---|
| Cups, foamed plastics | Insulation and cushioning: |
| Cushions, carpet and rug: | foamed plastics |
| plastics foam | Packaging foamed plastics |
| Foamed plastics products | Plates, foamed plastics |
| Ice chests or coolers, portable: | Shipping pads, plastics |
| foamed plastics | foam |

     3087    *Custom Compounding of Purchased Plastics Resins*

             Establishments primarily engaged in custom compounding of
             purchased plastics resins.

             Custom compounding of
                 purchased resins

     3088    *Plastics Plumbing Fixtures*

             Establishments primarily engaged in manufacturing plastics
             plumbing fixtures. Establishments primarily engaged in assem-
             bling plastics plumbing fixture fittings are classified in Industry
             3432. Establishments primarily engaged in manufacturing plas-
             tics plumbing fixture components are classified in Industry 3089.

|  |  |
|---|---|
| Bathroom fixtures, plastics | Portable chemical toilets, |
| Drinking fountains, except | plastics |
| mechanically refrigerated: | Shower stalls, plastics |
| plastics | Sinks, plastics |
| Flush tanks, plastics | Toilet fixtures, plastics |
| Hot tubs, plastics or fiberglass | Tubes, plastics: bath, shower, |
| Laundry tubs, plastics | and laundry |
| Lavatories, plastics | Urinals, plastics |
| Plumbing fixtures, plastics | Water closets, plastics |

     3089    *Plastics Products, Not Elsewhere Classified*

             Establishments primarily engaged in manufacturing plastics
             products, not elsewhere classified. Establishments primarily en-
             gaged in manufacturing artificial leather are classified in Industry
             2295.

Industry
Group Industry
No. No.

308         MISCELLANEOUS PLASTICS PRODUCTS—Cont.

3089   *Plastics Products, Not Elsewhere Classified—Cont.*

Air mattresses, plastics
Aquarium accessories,
    plastics
Awnings, fiberglass and plastics
    combination
Bands, plastics
Bathware, plastics: except
    plumbing fixtures
Battery cases, plastics
Bearings, plastics
Billfold inserts, plastics
Blister packaging, plastics
Boats, nonrigid: plastics
Bolts, plastics
Bowl covers, plastics
Boxes, plastics
Brush handles, plastics
Bubble formed packaging,
    plastics
Buckets, plastics
Buoys and floats, plastics
Caps, plastics
Carafes, plastics
Casein products, molded for the
    trade
Cases, plastics
Casting of plastics for the trade,
    except foam plastics
Ceiling tile, unsupported
    plastics
Celluloid products, molded for
    the trade
Closures, plastics
Clothes hangers, plastics
Clothespins, plastics
Combs, plastics
Composition stone, plastics
Containers, plastics: except
    foam, bottles, and bags
Corrugulated panels, plastics

Cotter pins, plastics
Counter coverings, plastics
Cups, plastics: except foam
Dinnerware, plastics: except
    foam
Dishes, plastics: except foam
Doors, folding: plastics or
    plastics coated fabric
Downspouts, plastics
Drums, plastics (containers)
Engraving of plastics
Fascia, plastics (siding)
Fittings for pipe, plastics
Fittings, plastics
Flat panels, plastics
Floor coverings, plastics
Flower pots, plastics
Food casings, plastics
Garbage containers, plastics
Gate hooks, plastics
Glazing panels, plastics
Gloves and mittens, plastics
Grower pots, plastics
Gutters, plastics: glass fiber
    reinforced
Hardware, plastics
Heels, boot and shoe: plastics
Holders, plastics: paper towel,
    grocery bag, dust mon, and
    broom
Hospitalware, plastics: except
    foam
Ice buckets, plastics: except
    foam
Ice chests or coolers, portable,
    plastics: except insulated or
    foam plastics
Jars, plastics
Kitchenware, plastics: except
    foam

308              MISCELLANEOUS PLASTICS PRODUCTS—Cont.

   3089    *Plastics Products, Not Elsewhere Classified—Cont.*

Laboratoryware, plastics
Ladders, plastics
Lamp bases, plastics
Lamp shades, plastics
Lenses, plastics: except
   ophthalmic or optical
Lifejackets, plastics
Liferafts, nonrigid: plastics
Lock washers, plastics
Machine nuts, plastics
Microwaveware, plastics
Molding of plastics for the
   trade, except foam
Monofilaments, plastics: not
   suited for textile use
Netting, plastics
Nuts, plastics
Organizers for closets, drawers,
   and shelves: plastics
Ovenware, plastics
Pails, plastics
Picnic jugs, plastics
Planters, plastics
Pontoons, nonrigid: plastics
Printer acoustic covers, plastics
Rivets, plastics
Saucers, plastics: except foam
Screw eyes, plastics
Scrubbing pads, plastics
Septic tanks, plastics
Shutters, plastics
Siding, plastics
Sinkware, plastics
Skirts, plastics (siding)
Soffit, plastics (siding)
Soles, boot and shoe: plastics
Soling strips, boot and shoe:
   plastics

Sponges, plastics
Spouting, plastics: glass fiber
   reinforced
Spring pins, plastics
Spring washers, plastics
Suitcase shells, plastics
Swimming pool covers and
   blankets: plastics
Tableware, plastics: except
   foam
Tires, plastics
Tissue dispensers, plastics
Toggle bolts, plastics
Tool handles, plastics
Tops, plastics (e.g., dispenser,
   shaker)
Trash containers, plastics
Trays, plastics: except foam
Tube, plastics (containers)
Tumblers, plastics: except
   foam
Unions, plastics
Utility containers, plastics
Vials, plastics
Vulcanized fiber plate, sheet,
   rods, and tubes
Wall coverings, plastics
Warmers, bottle: plastics,
   except foam
Washers, plastics
Watering pots, plastics
Window frames and sash,
   plastics
Window screening, plastics
Windows, louver: plastics
Windows, storm: plastics
Windshields, plastics
Work gloves, plastics

# Appendix B

## Chemical Companies Listed on the New York, American, and Over-the-Counter (NASDAQ) Exchanges

Data in Appendix B is based on the first quarter of 1999.

TABLE B.1  Selected Chemical Companies Listed on the New York Stock Exchange

| Name of firm | Ticker symbol | Principal business |
|---|---|---|
| Abbott Laboratories | ABT | Diversified health care products |
| Air Products & Chemicals Inc. | APD | Industrial gases, equipment, and chemicals |
| Albemarle Corporation | ALB | Manufacturing industrial chemicals—split off from Ethyl |
| Alberto-Culver | ACV | Hair care, health, beauty aids |
| Alco Standard | ASN | Distributor of paper and office equipment |
| Alcoa[a] | AA | Leading U.S. aluminum products |
| AlliedSignal Inc.[a] | ALD | Aerospace, automotive, fibers |
| Alumax Inc. | AMX | Produce aluminum products |

Note: The listings of companies in Table B.1–B.5 are not to be construed as a definitive listing of all chemical processing companies.

TABLE B.1   Continued

| Name of firm | Ticker symbol | Principal business |
|---|---|---|
| American Home Products | AHP | Drugs, food, household/ware |
| Amoco Corporation | AN | International oil company |
| Archer-Daniels-Midland | ADM | Drugs, food, household/ware |
| Arco Chemical | RCM | Major commodity chemical manufacturer |
| Armstrong World Industries | ACK | Manufacturer of floor covering/building products |
| ASARCO | AR | Producer nonferrous metals |
| Ashland Inc. | ASH | Petroleum/refining/chemicals/ construction |
| Asia Pacific Resources International | ARH | Pulp, paper/rayon fiber |
| Asia Pulp & Paper ADS | PAP | Manufacturer of pulp and paper products |
| Atlantic Richfield Company | ARC | Integrated oil enterprise |
| Avon Products | AVP | Cosmetics; jewelry; gift products |
| Baxter International | BAX | Manufacturer, distributor of hospital/ laboratory products |
| Betz Laboratories | BIL | Water treatment chemicals |
| Borden Chem/Plastics L.P. | BCU | Produces chemicals and PVC resins |
| Bristol-Meyers Squibb | BMY | Pharmaceutical, medical products |
| British Petroleum ADS | BP | Major integrated world oil |
| British Steel ADS | BST | Largest steel producer in U.K. |
| Broken Hill Prop ADR | BHP | Petroleum, minerals, steel |
| Buckeye Cellulose | BKI | Manufacturer and marketer of cellulose pulp |
| Cabot Corporation | CBT | Carbon black, oil and gas |
| Carlisle Corporation | CSL | Rubber, plastics, and metal products |
| Carlisle Plastics | CPA | Manufacturer of plastic bags and containers |
| Carter Wallace, Inc. | CAR | Drug and family products |
| Champion International | CHA | Building materials; paper products |
| Chemed Corporation | CHE | Specialty chemicals/health care products |
| Chesapeake Corporation | CSK | Kraft paper, board containers |
| Chevron[a] | CHV | Major integrated international oil company |
| Chris-Craft Industries | CCN | TV stations, video, chemicals |
| Clorox Company | CLX | Household products/specialty foods |
| Coastal Corporation | CGP | Refining/marketing oil and gas/chemicals |
| Colgate Palmolive Company | CL | Household and personal care |
| Cooper Companies | COO | Manufacturers of health care products |
| CPC International | CPC | International food processor |
| Crompton & Knowles | CNK | Specialty chemicals/industrial machinery |

TABLE B.1    Continued

| Name of firm | Ticker symbol | Principal business |
|---|---|---|
| Desoto Inc. | DSO | Detergent production and manufacturing |
| Dexter Corporation | DEX | Specialty chemicals/fibers |
| Diagnostic Products | DP | Medical immunodiagnostic kits |
| Diamond Shamrock | DRM | Oil and gas refining/marketing |
| Domtar, Inc. | DTC | Pulp/paper, construction materials/chemicals |
| Dow Chemical | DOW | Chemical/metals/plastics/packaging |
| Dresser Industries | DI | Energy and natural resources marketing |
| Du Pont, E. I., de Nemours & Company[a] | DD | Large chemical company/plastics, filters |
| Eastman Chemical | EMN | Manufacturer of chemicals, plastics, fibers |
| Eastman Kodak | EK | Photographic apparatus/chemicals |
| Englehard Corporation | EC | Specialty chemicals |
| ENI S.p.A | E | Oil and gas/chemicals |
| Enron | ENE | Major gas pipeline/oil and gas products |
| Ensearch Corporation | EEX | Oil and natural gas; exploration, development, distribution |
| Ethyl Corporation | EY | Manufacturer of fuel additives |
| Exxon[a] | XON | World's leading oil company |
| Ferro Corporation | FOE | Chemical specialties |
| First Brands Corporation | FBR | Manufacturer of home/auto consumer products |
| First Mississippi Corporation | FRM | Industrial and agricultural chemicals |
| FMC Corporation | FMC | Chemicals/machinery |
| Forest Labs | FRX | Pharmaceutical manufacturer and distributor |
| GenCorp | GY | Aerospace, auto, polymer products |
| General Chemical Group | GCG | Producer of inorganic chemicals |
| General Electric[a] | GE | Consumer/industrial products |
| General Mills | GIS | Consumer foods/restaurants |
| Geon Company | GON | Manufacturer of polyvinyl chloride resins |
| Georgia Gulf | GGC | Integrated chemical producer |
| Gillette Company | G | Shaving, personal care |
| Goodrich, B. F. | GR | Aerospace/specialty chemicals |
| Goodyear Tire & Rubber[a] | GT | Manufacturer of tires and rubber products |
| Grace, W. R. & Company | GRA | Specialty chemical products |
| Great Lakes Chemical | GLK | Specialty chemical products |
| Hanna (M.A.) Company | MAH | Polymers, specialty chemicals |
| Helmerich & Payne | HP | Contract driller/oil and gas products |

TABLE B.1   Continued

| Name of firm | Ticker symbol | Principal business |
|---|---|---|
| Hercules Inc. | HPC | Specialty chemicals/food products |
| Hexcel Corporation | HXL | Honeycomb cores; plastics |
| Howell Corporation | HWL | Oil and gas exploration and development of products |
| ICN Pharmaceuticals | ICN | Manufacturer of specialty/generic products |
| IMC Global | IGL | Manufacturers of chemical fertilizer |
| Imperial Chemical Industries (ICI) | ICI | Leading industrial chemical company |
| Inco Ltd. | N | Producer of nickel and copper |
| International Flavors & Fragrances | IFF | Development and manufacturing of flavor and fragrance products |
| International Paper[a] | IP | Manufacturer of paper, pulp, and wood products |
| James River Corporation | JR | Chemically treated secondary paper products |
| Jilin Chemical Ind. ADS | JCC | Chemical products; China |
| Johnson & Johnson | JNJ | Health care products |
| Kerr-McGee Corporation | KMG | Petroleum, chemicals, coal |
| KN Energy | KNE | Natural gas/mineral resources |
| Lilly, Eli, & Company | LLY | Ethical drugs; agricultural chemicals |
| Loctite Corporation | LOC | Chemical sealants, adhesives |
| Louisiana Pacific | LPX | Lumber, plywood, and pulp |
| Lubrizol Corporation | LZ | Specialty chemicals |
| Lyondell Petrochemical | LYO | Petrochemical refining; marketing |
| Mallinckrodt Group | MKG | Health care products; specialty chemicals |
| Merck[a] | MRK | Ethical drugs/specialty chemicals |
| Minnesota Mining & Manufacturing[a] | MMM | Scotch tapes; coated abrasives |
| Mississippi Chemical | MIS | Production/supply nitrogen fertilizers |
| Mitchell Energy/Dev. | MND | Oil and gas; pipeline; real estate |
| Mobil Corporation | MOB | International oil/gas/chemicals |
| Monsanto Company | MTC | Major chemical producer |
| Montedison | MNT | International chemical concern |
| Murphy Oil Corporation | MUR | International integrated oil |
| Mylan Laboratories | MLY | Pharmaceutical products |
| Nalco Chemical Company | NLC | Specialty chemicals |
| NL Industries Inc. | NL | Manufacturer of commodity chemicals |
| NOVA Corporation (Canada) | NVA | Petrochemical; gas PL; petroleum |
| Occidental Petroleum Corporation | OXY | Major diversified energy company |

TABLE B.1 Continued

| Name of firm | Ticker symbol | Principal business |
|---|---|---|
| Olin Corporation | OLN | Chemical; metals and defense products |
| P.T. Tri Polyta Indonesia ADS | TPI | Polypropylene resins, plastics |
| Parker & Parsley | PDP | Oil and gas exploration, development, production |
| Pfizer Corporation | PFE | Health care; consumer products; specialty chemicals |
| Phillip Morris[a] | MO | Cigarettes, food products, brewing |
| Phillips Petroleum Company | P | Domestic integrated oil; chemicals |
| PPG Industries | PPG | Coatings and chemicals/flat glass |
| Praxair Inc. | PX | Industrial gases/specialty coatings |
| Procter & Gamble[a] | PG | Household, personal care, food products |
| Publicker Industries | PUL | Industrial products and services |
| Quaker Oats | OAT | Human and pet food products |
| Quaker State Corporation | KSF | Motor oil, lubricants, coal |
| Revlon, Inc. | REV | Cosmetic/personal care products |
| Rexene Corporation | RXN | Thermoplastic/petrochemical products |
| Reynolds Metals | RLM | Aluminum manufacturing, finished products |
| Rhône-Poulenc, ADR | RP | Chemicals medical, agricultural, consumer products |
| Rohm & Haas | ROH | Manufacturer of specialty chemicals and plastics |
| Royal Dutch Petroleum Company | RD | Owns 64% Royal Dutch Shell |
| Rubbermaid, Inc. | RBD | Manufacturer of plastic/rubber housewares |
| Schering-Plough Corporation | SGP | Pharmaceutical/consumer products |
| Schlumberger Ltd. | SLB | Oil field services; electronics |
| Seagull Energy | SGO | Natural gas exploration, production |
| Shell Transport & Trading Company | SC | Owns 36% Royal Dutch Shell |
| Sherwin-Williams | SHW | Large paint and varnish manufacturing |
| SmithKline Beecham | SBH | Ethical drugs/health care products |
| Sonoco Products | SON | Manufacturer of paper packaging products |
| Sterling Chemical | STX | Manufacturer of petrochemicals |
| Sun Company | SUN | Petroleum refining; marketing |
| Sybron Corporation | SYB | Manufacturer of orthonodontic/dental lab products |
| Terra Industries | TRA | Produce fertilizers/agricultural services |
| Tesoro Petroleum Corporation | TSO | Integrated oil company |
| Texaco, Inc.[a] | TX | Major international oil company |

TABLE B.1   Continued

| Name of firm | Ticker symbol | Principal business |
|---|---|---|
| Texfi Industries | TXF | Fabrics for apparel industry |
| Textron, Inc. | TXT | Aerospace/commercial products, financial services |
| Thiokol Corporation | TKC | Aerospace propulsion systems |
| U.S. Filter | USF | Wastewater treatment equipment |
| Unilever N.V. | UN | Controls vast international enterprise |
| Union Camp | UCC | Paper, packaging, chemicals, and building products distributor |
| Union Carbide[a] | UK | Chemicals and plastics |
| Union Texas Petroleum | UTH | Oil and gas exploration, production |
| Univar Corporation | UVX | Distributor of industrial chemicals |
| Unocal Corporation | UCL | Oil and gas exploration, production; development |
| USX—Marathon Group | MRO | Oil and gas exploration, production development |
| USX—U.S. Steel Group | X | Integrated steel producer |
| Vulcan Materials | VMC | Construction materials/chemicals |
| Warner-Lambert Company | WLA | Drugs; toiletries; food; gum |
| Wheelabrator Technologies | WTI | Refuse to energy |
| Whittaker Corporation | WKR | Aerospace; fluid control |
| Williams Companies | WMB | Gas pipeline; petroleum production |
| Witco Chemical Corporation | WIT | Petroleum and specialty chemicals |

[a]Included in 30 stocks comprising the Dow Jones Industrial Averages.

TABLE B.2 Selected Chemical Companies Listed on the American Stock Exchange

| Name of firm | Ticker symbol | Principal business |
|---|---|---|
| American Israeli Paper Organization | AIP | Paper manufacturer from imported pulp |
| AT Plastics | ATJ | Polymers, films, and packaging |
| Balchem Corporation | BCP | Manufacturer of specialty chemicals |
| Barr Laboratories | BRL | Manufacturer of generic pharmaceuticals |
| Canadian Occidental Petroleum | CXY | Oil, gas, chemicals |
| Cominco, Ltd. | CLT | Lead, zinc mines; fertilizers |
| Courtaulds PLC ADR | COU | Rayon fibers, British Isles |
| Crown Central | CNP | Independent producer; refiner |
| Del Laboratories | DU | Drugs and cosmetics |
| Fina Inc. | FI | Integrated oil; petrochemical |
| Halsey Drug | HDG | Manufacturers of generic drug products |
| Health-Chem | HCH | Industrial and health care products |
| Imperial Oil, Ltd. | IMO | Canadian oil: Exxon owns 70% |
| Ionics, Inc. | ION | Water purifying equipment: water and chemicals |
| K-V Pharmaceutical | KVB | Manufacturer of drugs for major firms |
| Kinark Corporation | KIN | Chemical packaging/distribution/ marketing |
| Park Electrochemical | PKE | Printed circuit material |
| Stepan Company | SCL | Basic/intermediate chemicals |
| Synalloy Corporation | SYO | Metal fabricator: chemicals |
| Tipperary Corporation | TPY | Oil and gas exploration; chemical production |
| Total Petroleum, Ltd. | TPN | Petroleum refining/marketing |
| Valspar Corporation | VAL | Manufacturer of paints and coatings |

TABLE B.3   Selected Chemical Companies Listed on the NASDAQ (Over-the-Counter) Exchange

| Name of firm | Ticker symbol | Principal business |
|---|---|---|
| Aceto Corporation | ACET | Manufacturer and distributor of chemicals |
| AgriBio Tech Inc. | ABTX | Manufacture of chemical agricultural products |
| Agrium Inc. | AGMF | Produces nitrate/fertilizer products |
| Akzo Nobel N.V. ADS | AKZOY | Chemical, fiber, health care products |
| Alliance Pharmaceutical | ALLP | Development, manufacturing of medical/pharmaceutical products |
| Amgen Inc. | AMGN | Res/development of biological products |
| BeautiControl Cosmetics | BUTI | Manufacturing/marketing cosmetics, toiletries |
| Bio-Rad Laboratories | BRLS | Research chemicals/medical test kits |
| Carrington Laboratories | CARN | Manufacturer of pharmaceuticals/drugs |
| Chattem Drug & Chemical | CHTT | Consumer products/specialty chemicals |
| ChiRex Inc. | CHRX | Development/manufacturing of chemical and generic drugs |
| Copley Pharmaceuticals | CPLY | Manufacturer of off-patent pharmaceutical products |
| Fuller, H.B., Company | FULL | Industrial adhesives |
| MacDermid, Inc. | MACD | Industrial metal finishing chemicals |
| Melamine Chemicals | MTWO | Specialty chemicals melamine |
| OM Group | OMGI | Produce specialty chemicals |
| Petrolite Corporation | PLIT | Specialty chemicals; process wax products |
| Pride Petroleum Services | PRDE | Oil and gas well services in U.S. |
| Quaker Chemical | QCHM | Specialty chemical products |
| RPM, Inc. | RPOW | Protective coatings; fabrics |
| Sigma-Aldrich | SIAL | Specialty chemical production |
| Uniroyal Chemical | UCHM | Specialty chemicals/elastomers |


TABLE B.4   Other Category Stocks—Not Traded

| | |
|---|---|
| Formosa Plastics | Taiwan company—olefins/plastics |
| Hoechst Celanese | Owned by Hoechst (German) |
| Huntsman Chemical | Privately owned—integrated chemicals |
| Solvay | Belgium company—plastics/chemicals |
| Vista Chemical | Owned by German company—vinyls/chemicals |
| Westlake Chemical | Privately owned Taiwan company—olefins/plastics |

**TABLE B.5**   Stocks Removed from the New York Stock Exchange

| | |
|---|---|
| Amerace Corporation | |
| American Cyanamid Corporation | Bought by American Home Products |
| American Hospital Supply | |
| Ansul Company | |
| Basic Inc. | |
| Beatrice Foods | |
| Beker Industries Corporation | |
| Big Three Industries Inc. | |
| Borg-Warner | |
| Celanese | Divested from Hoechst (German) in 1999 |
| Central Soya | |
| Chelsea Industries | |
| Chesebrough-Ponds | |
| Cities Service Company | |
| Commonwealth Oil Refining Company | |
| Consolidated Foods | |
| Continental Oil Company (Conoco) | Divested from Du Pont in 1999 |
| Dart Industries | |
| Dayco Corporation | |
| Diversified Industries | |
| Eagle-Picher Industries | |
| El Paso Company | Became Rexene |
| Emery Industries | |
| Esmark, Inc. | |
| Fabergé, Inc. | |
| Filtrol Corporation | |
| Foremost-McKesson | |
| Freeport Minerals | |
| GAF Corporation | |
| General American Oil Company of Texas | |
| Gulf & Western | |
| Gulf Resources & Chemical Corporation | |
| Gulf States Utilities | |
| Helene Curtis Industries | |
| Hoover Ball & Bearing | |
| Houston Natural Gas | |
| Hunt, Philip A., Chemical | |
| Inmont Corporation | |
| International Minerals & Chemicals | |
| International Telephone & Telegraph | |
| Ipco Hospital Supply Corporation | |
| Koppers Company, Inc. | |
| Kraftco Corporation | |
| Marathon Oil Company | Bought by USX |

TABLE B.5   Continued

| | |
|---|---|
| Marion Laboratories, Inc. | |
| Martin Marietta | |
| Miles Laboratories | |
| Molycorp., Inc. | |
| Morton-Norwich Products Inc. | |
| National Distillers & Chemical Corporation | |
| North American Philips | |
| Northern Natural Gas | |
| Northwest Industries | |
| Oakite Products | |
| Pennwalt Corporation | |
| Products Research & Chemical Corporation | |
| Purex Corporation | |
| Reichhold Chemicals | Bought by Japanese company |
| Richardson Company | |
| Richardson-Merrell | |
| Robins, A.H., Company, Inc. | |
| Rorer-Amchem, Inc. | |
| SCM Corporation | |
| Seagrave Corporation | |
| Searle, G. D., & Company | |
| Shell Oil Company | Merged into Royal Dutch Shell |
| Skelly Oil Company | |
| Squibb Corporation | |
| Staley, A.E., Manufacturing | |
| Standard Brands Paint | |
| Standard Oil of California | |
| Standard Oil of Indiana | |
| Standard Oil of Ohio | |
| Stauffer Chemical Company | |
| Sterling Drug, Inc. | Bought by Du Pont |
| Tenneco, Inc. | |
| Union Oil Company of California | |
| United Merchants & Manufacturers | |
| UOP, Inc. | |
| Upjohn Company | |

# Appendix C

## Input–Output Tables

The data in Appendix C is from *Benchmark Input–Output Accounts of the United States*, 1987 published by U.S. Department of Commerce, Bureau of Economic Analysis, U.S. Government Printing Office, Washington, DC 20402, November 1994.

TABLE 4.4A　The Use of Commodities by Industry (millions of dollars at producers' prices)

| Commodity number | For the distribution of output of a commodity, read the row for that commodity<br><br>For the composition of inputs to an industry, read the column for that industry<br><br>Industry number | Livestock and livestock products<br><br>1 | Other agricultural products<br><br>2 |
|---|---|---|---|
| 1 | Livestock and livestock products | 16,818 | 1,584 |
| 2 | Other agricultural products | 23,778 | 3,855 |
| 3 | Forestry and fishery products | | |
| 4 | Agricultural, forestry, and fishery services | 4,003 | 6,542 |
| 5+6 | Metallic ores mining | | |
| 7 | Coal mining | | |
| 8 | Crude petroleum and natural gas | | |
| 9+10 | Nonmetallic minerals mining | 6 | 254 |
| 11 | New construction | | |
| 12 | Maintenance and repair construction | 458 | 710 |
| 13 | Ordnance and accessories | | |
| 14 | Food and kindred products | 11,566 | |
| 15 | Tobacco products | | |
| 16 | Broad and narrow fabrics, yarn and thread mills | | 44 |
| 17 | Miscellaneous textile goods and floor coverings | 25 | 26 |
| 18 | Apparel | | |
| 19 | Miscellaneous fabricated textile products | | 88 |
| 20+21 | Lumber and wood products | 36 | 295 |
| 22+23 | Furniture and fixtures | | |
| 24 | Paper and allied products, except containers | 110 | 139 |
| 25 | Paperboard containers and boxes | 5 | 333 |
| 26A | Newspapers and periodicals | 9 | 10 |
| 26B | Other printing and publishing | 9 | 10 |
| 27A | Industrial and other chemicals | 102 | 64 |
| 27B | Agricultural fertilizers and chemicals | 142 | 4,607 |
| 28 | Plastics and synthetic materials | | |
| 29A | Drugs | 196 | |
| 29B | Cleaning and toilet preparations | 54 | |
| 30 | Paints and allied products | | |
| 31 | Petroleum refining and related products | 335 | 1,175 |
| 32 | Rubber and miscellaneous plastics products | 162 | 357 |
| 33+34 | Footwear, leather, and leather products | 25 | (*) |
| 35 | Glass and glass products | 6 | |
| 36 | Stone and clay products | | 100 |
| 37 | Primary iron and steel manufacturing | 14 | 15 |
| 38 | Primary nonferrous metals manufacturing | | |
| 39 | Metal containers | | |
| 40 | Heating, plumbing, and fabricated structural metal products | 17 | 19 |
| 41 | Screw machine products and stampings | 27 | |
| 42 | Other fabricated metal products | 68 | 153 |
| 43 | Engines and turbines | | |
| 44+45 | Farm, construction, and mining machinery | 249 | 663 |
| 46 | Materials handling machinery and equipment | | |
| 47 | Metalworking machinery and equipment | 83 | 92 |
| 48 | Special industry machinery and equipment | | |
| 49 | General industrial machinery and equipment | 27 | 50 |
| 50 | Miscellaneous machinery, except electrical | 47 | 123 |

Table 4.4a

| Forestry and fishery products | Agri-cultural, forestry, and fishery services | Metallic ores mining | Coal mining | Crude petroleum and natural gas | Non-metallic minerals mining | Con-struction | Ordnance and acces-sories |
|---|---|---|---|---|---|---|---|
| 3 | 4 | 5+6 | 7 | 8 | 9+10 | 11+12 | 13 |
| 27 | 1,251 | | | | | | |
| | 2,089 | | | | | 241 | |
| 168 | 32 | | | | | | |
| 1,288 | 8 | (*) | 1 | 1 | 2 | 3,250 | 2 |
| | | 519 | | | | | |
| | | 11 | 2,730 | | 61 | 1 | 4 |
| | | | | 3,149 | | | |
| | 2 | 7 | 32 | | 452 | 4,834 | |
| | | | | | | 44 | |
| 83 | 286 | 88 | 195 | 1,844 | 116 | 338 | 208 |
| 29 | | | | | | 13 | 899 |
| 305 | 33 | 1 | (*) | 3 | 2 | (*) | (*) |
| | | 1 | 18 | | 4 | | 4 |
| 72 | 114 | | | | | 1,760 | 1 |
| | | (*) | 8 | 4 | 4 | 120 | 12 |
| 31 | 79 | | | (*) | | 223 | (*) |
| | | 38 | 62 | (*) | 3 | 33,521 | 27 |
| | | | | | | 1,271 | |
| 1 | 10 | 1 | 8 | 8 | 36 | 1,184 | 7 |
| 1 | 185 | (*) | | 2 | 4 | 45 | 22 |
| 1 | 3 | (*) | 1 | 2 | 24 | 52 | 3 |
| 40 | 24 | 3 | 12 | 32 | 8 | 204 | 22 |
| 14 | 7 | 204 | 138 | 837 | 220 | 1,310 | 127 |
| 40 | 2,972 | 1 | 2 | | (*) | 8 | 6 |
| | | | | | | | 31 |
| | 1 | | | | | | |
| | | | | 7 | 2 | 10 | 2 |
| 4 | | | | 6 | | 4,688 | 4 |
| 294 | 215 | 127 | 390 | 289 | 194 | 11,220 | 31 |
| 2 | 36 | 82 | 247 | 26 | 122 | 6,677 | 198 |
| | 3 | | (*) | 1 | | 29 | 1 |
| 2 | 10 | 1 | (*) | 8 | 4 | 993 | 3 |
| | 12 | 23 | 80 | 265 | 2 | 31,054 | 50 |
| | | 134 | 19 | 260 | 43 | 10,023 | 473 |
| | | 15 | 13 | | 70 | 6,194 | 562 |
| 18 | | | | | | | |
| | | 43 | 66 | 46 | 61 | 31,335 | |
| | | 35 | 146 | | 33 | 280 | 189 |
| 53 | 26 | 9 | 82 | 383 | 37 | 8,090 | 462 |
| 14 | 44 | 37 | 111 | 22 | 57 | | 28 |
| 21 | 78 | 129 | 1,084 | 239 | 253 | 1,419 | |
| | | 35 | 95 | | 142 | 1,416 | |
| (*) | 2 | 8 | 8 | 77 | 8 | 200 | 87 |
| 8 | 4 | 78 | 428 | 157 | 164 | 1,513 | 235 |
| 3 | 4 | 12 | 78 | 36 | 19 | 126 | 119 |

**TABLE 4.4A** Continued

| | | | |
|---|---|---:|---:|
| 51 | Computer and office equipment | .................. | .................. |
| 52 | Service industry machinery | .................. | .................. |
| 53 | Electrical industrial equipment and apparatus | 10 | 27 |
| 54 | Household appliances | .................. | .................. |
| 55 | Electric lighting and wiring equipment | 19 | 43 |
| 56 | Audio, video, and communication equipment | .................. | .................. |
| 57 | Electronic components and accessories | .................. | .................. |
| 58 | Miscellaneous electrical machinery and supplies | 152 | 404 |
| 59A | Motor vehicles (passenger cars and trucks) | .................. | .................. |
| 59B | Truck and bus bodies, trailers, and motor vehicles parts | 90 | 217 |
| 60 | Aircraft and parts | .................. | .................. |
| 61 | Other transportation equipment | .................. | .................. |
| 62 | Scientific and controlling instruments | .................. | .................. |
| 63 | Ophthalmic and photographic equipment | .................. | .................. |
| 64 | Miscellaneous manufacturing | 15 | 16 |
| 65A | Railroads and related services; passenger ground transportation | 824 | 249 |
| 65B | Motor freight transportation and warehousing | 1,914 | 1,290 |
| 65C | Water transportation | 107 | 46 |
| 65D | Air transportation | 20 | 100 |
| 65E | Pipelines, freight forwarders, and related services | 4 | 15 |
| 66 | Communications, except radio and TV | 222 | 246 |
| 67 | Radio and TV broadcasting | .................. | .................. |
| 68A | Electric services (utilities) | 972 | 515 |
| 68B | Gas production and distribution (utilities) | .................. | 164 |
| 68C | Water and sanitary services | 111 | 368 |
| 69A | Wholesale trade | 3,861 | 3,567 |
| 69B | Retail trade | 75 | 228 |
| 70A | Finance | 796 | 849 |
| 70B | Insurance | 438 | 1,797 |
| 71A | Owner-occupied dwellings | .................. | .................. |
| 71B | Real estate and royalties | 3,000 | 7,097 |
| 72A | Hotels and lodging places | 49 | 54 |
| 72B | Personal and repair services (except auto) | 26 | 70 |
| 73A | Computer and data processing services | .................. | .................. |
| 73B | Legal, engineering, accounting, and related services | 108 | 121 |
| 73C | Other business and professional services, except medical | 362 | 954 |
| 73D | Advertising | 19 | 21 |
| 74 | Eating and drinking places | 16 | 17 |
| 75 | Automotive repair and services | 50 | 179 |
| 76 | Amusements | .................. | .................. |
| 77A | Health services | 744 | .................. |
| 77B | Educational and social services, and membership organizations | .................. | .................. |
| 78 | Federal Government enterprises | 12 | 13 |
| 79 | State and local government enterprises | 15 | 32 |
| 80 | Noncomparable imports | .................. | 17 |
| 81 | Scrap, used and secondhand goods | .................. | .................. |
| 82 | General government industry | .................. | .................. |
| 83 | Rest-of-the-world adjustment to final uses | .................. | .................. |
| 84 | Household industry | .................. | .................. |
| 85 | Inventory valuation adjustment | .................. | .................. |
| I | Total intermediate inputs | 72,410 | 40,021 |
| VA | Value added | 15,074 | 46,721 |
| T | Total Industry output | 87,484 | 86,742 |

* Less than $500,000.

TABLE 4.4A

| | | | | | | | |
|---|---|---|---|---|---|---|---|
| | 2 | | | | | | |
| | | | | | | 6,638 | |
| | | 27 | 90 | 166 | 64 | 2,743 | 25 |
| | | | | | | 1,505 | |
| 1 | 8 | 4 | 28 | 19 | 8 | 9,894 | 2 |
| | (*) | | (*) | 1 | | 2,011 | 874 |
| | | | | | | | 870 |
| 1 | 10 | 6 | 2 | 2 | 5 | 845 | 19 |
| 5 | 40 | 13 | 9 | 7 | 3 | 414 | 1 |
| | 1 | | | | | | 3,195 |
| 156 | 12 | 5 | | | | 2 | |
| 15 | 1 | 5 | 3 | 3 | 1 | 1,421 | 398 |
| (*) | 8 | 1 | 3 | 10 | 2 | 116 | 19 |
| 1 | 14 | 2 | 4 | 5 | 6 | 878 | 7 |
| 7 | 92 | 20 | 652 | 45 | 36 | 1,557 | 18 |
| 24 | 208 | 52 | 159 | 154 | 182 | 8,274 | 153 |
| 24 | 33 | 7 | 54 | 103 | 12 | 302 | 2 |
| 7 | 418 | 25 | 19 | 86 | 48 | 853 | 145 |
| 2 | 3 | 2 | 6 | 4 | 3 | 37 | 1 |
| 4 | 1 | 9 | 21 | 120 | 22 | 2,315 | 145 |
| 3 | 59 | 647 | 611 | 1,292 | 651 | 1,047 | 260 |
| 1 | 2 | 74 | 11 | 556 | 218 | 322 | 89 |
| 6 | | 15 | 60 | 131 | 135 | 252 | 31 |
| 186 | 1,452 | 133 | 720 | 523 | 260 | 26,466 | 585 |
| 8 | 67 | 8 | 9 | 6 | 3 | 24,114 | 4 |
| 77 | 103 | 69 | 169 | 223 | 231 | 7,098 | 58 |
| 102 | 80 | 23 | 31 | 18 | 6 | 1,966 | 32 |
| | 179 | 85 | 622 | 14,541 | 154 | 2,711 | 264 |
| 14 | 24 | 12 | 5 | 27 | 106 | 707 | 27 |
| 7 | 182 | 2 | 4 | 10 | 20 | 139 | 12 |
| 16 | 90 | 33 | 3 | 5 | 115 | 6 | 15 |
| 252 | 243 | 113 | 207 | 801 | 96 | 36,297 | 325 |
| 132 | 206 | 38 | 82 | 144 | 112 | 14,948 | 346 |
| 3 | 92 | 5 | 46 | 853 | 19 | 244 | 615 |
| 11 | 80 | 8 | 19 | 69 | 23 | 1,440 | 88 |
| 131 | 588 | 187 | 197 | 191 | 17 | 6,117 | 28 |
| | 282 | | 1 | 2 | 3 | 42 | 2 |
| 21 | 40 | 14 | 42 | 29 | 15 | 25 | 9 |
| 6 | 50 | 10 | 3 | 2 | 17 | 243 | 7 |
| 5 | 17 | 6 | 4 | 3 | 3 | 121 | 6 |
| | 37 | 31 | 18 | 792 | 9 | 5 | 18 |
| | | | | | | 14 | |
| 3,748 | 12,253 | 3,331 | 9,964 | 28,744 | 4,751 | 327,813 | 12,510 |
| 3,708 | 9,948 | 3,476 | 15,488 | 55,484 | 8,213 | 291,000 | 18,928 |
| 7,456 | 22,201 | 6,807 | 25,452 | 84,228 | 12,964 | 618,813 | 31,438 |

TABLE 4.4A   Continued

| Food and kindred products | Tobacco products | Broad and narrow fabrics, yarn and thread mills | Miscellaneous textile goods and floor coverings | Apparel | Miscellaneous fabricated textile products | Lumber and wood products | Furniture and fixtures |
|---|---|---|---|---|---|---|---|
| 14 | 15 | 16 | 17 | 18 | 19 | 20+21 | 22+23 |
| 60,821 | | 262 | | 13 | | | |
| 22,262 | 1,707 | 3,192 | 34 | 31 | | | |
| 2,033 | | | | 295 | | 5,874 | |
| 8 | (*) | 3 | 1 | | | 22 | 1 |
| 105 | 15 | 28 | 10 | 4 | 4 | 21 | 7 |
| 8 | | | | | | | |
| 810 | 39 | 179 | 48 | 129 | 42 | 423 | 289 |
| | 1 | (*) | | (*) | | 1 | |
| 54,695 | | 1 | 22 | 1 | 4 | 7 | 15 |
| | 3,664 | | | | | | |
| | | 9,897 | 3,621 | 13,040 | 3,875 | 5 | 1,236 |
| 13 | | 346 | 683 | 19 | 1,264 | 136 | 973 |
| 11 | 2 | 2 | 11 | 12,117 | 161 | 12 | 13 |
| 72 | | | 17 | 1,506 | 552 | 17 | 58 |
| 61 | (*) | 3 | 1 | | 52 | 20,956 | 3,229 |
| | | | | | | 71 | 146 |
| 3,142 | 131 | 24 | 94 | 111 | 39 | 64 | 51 |
| 5,856 | 952 | 131 | 99 | 200 | 200 | 279 | 555 |
| 10 | 2 | 2 | 1 | 4 | 2 | 7 | 4 |
| 1,880 | 377 | 11 | 7 | 54 | 30 | 38 | 33 |
| 1,505 | 51 | 852 | 594 | 78 | 103 | 848 | 215 |
| 187 | | | | | | 186 | |
| 121 | | 5,251 | 3,825 | 1,526 | 509 | 329 | 114 |
| 890 | | | | | | | |
| 154 | 18 | 88 | 78 | 294 | | 1 | |
| | | 1 | 1 | (*) | | 386 | 376 |
| 369 | 35 | 104 | 31 | 112 | 13 | 446 | 84 |
| 5,261 | 71 | 198 | 117 | 302 | 496 | 690 | 1,197 |
| 1 | | 2 | | 313 | 230 | 9 | 104 |
| 3,923 | 1 | 207 | 6 | | | 211 | 137 |
| 19 | 1 | 4 | 2 | 1 | 1 | 409 | 131 |
| 5 | 1 | 2 | 4 | (*) | 1 | 31 | 1,528 |
| | | | | | | 53 | 469 |
| 8,683 | 2 | | | | | | |
| | | | | | | 479 | |
| 662 | | | | | | 1,007 | 372 |
| 1,020 | 185 | 2 | 1 | 7 | 1 | 1,505 | 1,378 |
| | | | | | | | |
| (*) | 1 | 12 | | 1 | | 39 | |
| 38 | 6 | 14 | 5 | 9 | 6 | 147 | 79 |
| 90 | | 109 | 224 | 160 | 18 | 83 | 13 |

TABLE 4.4A

| Paper and allied products, except containers | Paperboard containers and boxes | News-papers and peri-odicals | Other printing and publishing | Industrial and other chemicals | Agri-cultural fertilizers and chemicals | Plastics and synthetic materials | Commodity number |
|---|---|---|---|---|---|---|---|
| 24 | 25 | 26A | 26B | 27A | 27B | 28 | |
| .......... | .......... | .......... | .......... | .......... | .......... | .......... | 1 |
| .......... | .......... | .......... | .......... | 86 | .......... | .......... | 2 |
| 108 | .......... | .......... | .......... | 68 | .......... | .......... | 3 |
| 6 | (*) | 1 | 2 | 2 | 1 | 1 | 4 |
| 15 | .......... | .......... | .......... | 636 | .......... | 6 | 5+6 |
| 428 | 2 | 1 | 13 | 273 | 5 | 132 | 7 |
| .......... | .......... | .......... | .......... | 1,043 | 552 | 4 | 8 |
| 303 | .......... | .......... | .......... | 576 | 926 | .......... | 9+10 |
| .......... | .......... | .......... | .......... | .......... | .......... | .......... | 11 |
| 397 | 88 | 187 | 238 | 575 | 55 | 158 | 12 |
| .......... | .......... | .......... | .......... | .......... | .......... | .......... | 13 |
| 344 | 3 | 3 | 9 | 348 | 92 | 53 | 14 |
| .......... | .......... | .......... | .......... | .......... | .......... | .......... | 15 |
| 677 | .......... | .......... | 94 | .......... | .......... | 123 | 16 |
| 356 | .......... | 5 | 54 | .......... | .......... | .......... | 17 |
| 11 | 1 | 1 | 5 | (*) | (*) | 5 | 18 |
| 1 | .......... | 1 | 1 | 23 | 1 | (*) | 19 |
| 4,804 | .......... | 2 | .......... | 48 | .......... | .......... | 20+21 |
| 1 | .......... | .......... | .......... | .......... | .......... | .......... | 22+23 |
| 13,139 | 11,548 | 6,485 | 14,631 | 721 | 48 | 463 | 24 |
| 1,088 | 53 | 12 | 336 | 409 | 55 | 269 | 25 |
| 4 | 1 | 443 | 206 | 7 | 1 | 2 | 26A |
| 110 | 13 | 3,358 | 7,054 | 52 | 190 | 15 | 26B |
| 3,444 | 491 | 379 | 2,389 | 18,226 | 1,371 | 13,509 | 27A |
| 205 | .......... | .......... | .......... | 366 | 2,533 | 95 | 27B |
| 2,005 | 557 | .......... | 129 | 757 | .......... | 1,470 | 28 |
| .......... | .......... | .......... | .......... | 136 | .......... | .......... | 29A |
| 386 | .......... | .......... | 39 | 104 | 25 | 230 | 29B |
| 19 | .......... | .......... | .......... | 408 | .......... | 89 | 30 |
| 565 | 182 | 81 | 238 | 1,012 | 69 | 103 | 31 |
| 1,999 | 73 | 33 | 1,816 | 994 | 115 | 1,685 | 32 |
| 1 | 3 | 2 | 13 | 2 | .......... | (*) | 33+34 |
| 5 | (*) | .......... | 1 | 133 | 39 | 26 | 35 |
| 90 | 1 | 1 | 9 | 68 | 26 | 14 | 36 |
| 2 | 68 | 1 | 7 | 177 | (*) | (*) | 37 |
| 47 | 54 | .......... | 178 | 59 | .......... | 2 | 38 |
| .......... | .......... | .......... | .......... | 462 | 55 | 3 | 39 |
| .......... | .......... | .......... | .......... | .......... | .......... | .......... | 40 |
| .......... | .......... | .......... | 1 | 10 | 1 | .......... | 41 |
| 502 | 123 | 1 | 53 | 650 | 40 | 13 | 42 |
| .......... | .......... | .......... | .......... | .......... | .......... | .......... | 43 |
| .......... | .......... | .......... | .......... | .......... | .......... | .......... | 44+45 |
| 1 | .......... | .......... | .......... | .......... | .......... | .......... | 46 |
| 44 | 18 | 7 | 23 | 35 | 2 | 21 | 47 |
| 368 | 100 | 77 | 359 | 410 | .......... | 9 | 48 |

**TABLE 4.4A**   Continued

| | | | | | | | |
|---|---|---|---|---|---|---|---|
| 7 | | | | | | 22 | |
| | | | | | | 44 | 59 |
| (*) | | (*) | | 1 | (*) | 147 | |
| 7 | (*) | 1 | (*) | 1 | 1 | 114 | 6 |
| 1 | (*) | (*) | (*) | | (*) | 1 | 1 |
| | | 2 | | | | | |
| 4 | 1 | 1 | | (*) | | 12 | 1 |
| 19 | 4 | 4 | (*) | 5 | | 191 | 9 |
| | | | | | | 6 | |
| 31 | 5 | 4 | 1 | 4 | 3 | 12 | 14 |
| 19 | 2 | 5 | 2 | 6 | 2 | 14 | 10 |
| 18 | 1 | 3 | 2 | 305 | 105 | 43 | 62 |
| 1,842 | 10 | 75 | 43 | 18 | 15 | 663 | 160 |
| 5,132 | 110 | 277 | 238 | 280 | 123 | 1,026 | 369 |
| 498 | 3 | 9 | 27 | 7 | 7 | 111 | 15 |
| 484 | 29 | 29 | 9 | 114 | 21 | 81 | 83 |
| 3 | (*) | 1 | 1 | 1 | | 5 | 1 |
| 447 | 28 | 48 | 28 | 95 | 29 | 125 | 89 |
| 2,750 | 87 | 1,088 | 216 | 519 | 105 | 1,044 | 344 |
| 1,336 | 20 | 152 | 112 | 96 | 40 | 344 | 82 |
| 442 | 9 | 114 | 26 | 34 | 52 | 182 | 66 |
| 16,850 | 536 | 1,807 | 433 | 2,171 | 951 | 3,806 | 2,039 |
| 53 | 8 | 11 | 2 | 19 | 1 | 73 | 16 |
| 937 | 131 | 106 | 60 | 281 | 129 | 446 | 379 |
| 355 | 37 | 41 | 24 | 76 | 26 | 196 | 58 |
| 866 | 78 | 71 | 57 | 385 | 102 | 305 | 257 |
| 209 | 8 | 3 | 2 | 16 | 16 | 22 | 19 |
| 192 | 7 | 191 | 52 | 144 | 69 | 32 | 27 |
| 123 | 11 | 30 | 10 | 13 | 4 | 22 | 16 |
| 866 | 61 | 91 | 36 | 175 | 46 | 230 | 421 |
| 1,802 | 97 | 466 | 99 | 342 | 136 | 546 | 749 |
| 8,657 | 783 | 227 | 343 | 992 | 202 | 987 | 730 |
| 358 | 19 | 76 | 32 | 191 | 51 | 218 | 160 |
| 477 | 101 | 113 | 20 | 292 | 4 | 502 | 178 |
| 5 | 2 | 1 | 1 | 2 | (*) | 5 | 15 |
| 80 | 12 | 22 | 3 | 25 | 97 | 175 | 63 |
| 235 | 54 | 38 | 24 | 149 | 38 | 71 | 63 |
| 288 | 8 | 18 | 7 | 12 | | 25 | 14 |
| 5,056 | 47 | 25 | 166 | 59 | 88 | 13 | 32 |
| | | | 5 | | | | |
| 225,473 | 9,588 | 26,104 | 11,628 | 37,181 | 10,072 | 46,952 | 19,518 |
| 100,498 | 16,795 | 12,140 | 4,354 | 27,003 | 6,915 | 25,923 | 17,259 |
| 325,972 | 26,383 | 38,244 | 15,982 | 64,184 | 16,987 | 72,875 | 36,777 |

TABLE 4.4A

| | | 7 | 47 | | | | |
|---|---|---|---|---|---|---|---|
| | | | | 21 | | | 51 |
| | | | | | | | 52 |
| | | | (*) | 41 | | | 53 |
| | | | | | | | 54 |
| 5 | 1 | 1 | 3 | 3 | (*) | 1 | 55 |
| 1 | (*) | 1 | 3 | 1 | | (*) | 56 |
| | | | | | | | 57 |
| 4 | 1 | 38 | 29 | (*) | (*) | | 58 |
| | | | | | | | 59A |
| 30 | 8 | 13 | 35 | 4 | 1 | 1 | 59B |
| | | | | | | | 60 |
| | | | | | | | 61 |
| 38 | 7 | 43 | 172 | 84 | 6 | 5 | 62 |
| 12 | 3 | 121 | 527 | 20 | 1 | 5 | 63 |
| 9 | 3 | 13 | 141 | 5 | (*) | 3 | 64 |
| 828 | 287 | 174 | 430 | 908 | 281 | 475 | 65A |
| 1,627 | 570 | 327 | 1,005 | 1,494 | 781 | 570 | 65B |
| 107 | 17 | 5 | 30 | 156 | 49 | 131 | 65C |
| 387 | 48 | 940 | 368 | 232 | 15 | 59 | 65D |
| 9 | 1 | 2 | 4 | 42 | 1 | 12 | 65E |
| 152 | 72 | 206 | 241 | 207 | 39 | 91 | 66 |
| | | | | | | | 67 |
| 2,292 | 277 | 246 | 725 | 2,850 | 304 | 964 | 68A |
| 994 | 80 | 21 | 95 | 2,389 | 466 | 771 | 68B |
| 1,074 | 29 | 21 | 77 | 318 | 113 | 149 | 68C |
| 3,637 | 999 | 782 | 3,080 | 3,141 | 747 | 1,743 | 69A |
| 80 | 16 | 27 | 69 | 37 | 3 | 10 | 69B |
| 303 | 41 | 256 | 449 | 274 | 226 | 104 | 70A |
| 216 | 47 | 70 | 191 | 159 | 26 | 77 | 70B |
| | | | | | | | 71A |
| 221 | 118 | 722 | 973 | 441 | 37 | 118 | 71B |
| 167 | 2 | 86 | 347 | 41 | 3 | 30 | 72A |
| 127 | 11 | 53 | 48 | 38 | 4 | 81 | 72B |
| 184 | 29 | 451 | 266 | 66 | 6 | 17 | 73A |
| 205 | 69 | 897 | 616 | 1,701 | 83 | 1,038 | 73B |
| 450 | 184 | 1,945 | 1,064 | 1,164 | 335 | 358 | 73C |
| 857 | 40 | 869 | 1,061 | 1,135 | 194 | 479 | 73D |
| 132 | 53 | 201 | 533 | 220 | 17 | 57 | 74 |
| 732 | 184 | 327 | 879 | 116 | 30 | 91 | 75 |
| 3 | 1 | 5 | 16 | 5 | (*) | 1 | 76 |
| | | | | | | | 77A |
| 62 | 4 | 78 | 124 | 25 | 10 | 53 | 77B |
| 72 | 22 | 586 | 594 | 85 | 5 | 19 | 78 |
| 153 | 8 | 8 | 21 | 67 | 7 | 19 | 79 |
| 84 | 2 | 55 | 234 | 636 | 101 | 147 | 80 |
| 828 | | | | | 33 | | 81 |
| | | | | | | | 82 |
| | | | | | | | 83 |
| | | | | | | | 84 |
| | | | | | | | 85 |
| 47,704 | 16,705 | 20,689 | 42,232 | 47,098 | 10,148 | 26,308 | I |
| 34,278 | 8,806 | 29,037 | 45,145 | 37,277 | 3,364 | 14,365 | VA |
| 81,982 | 25,511 | 49,727 | 87,378 | 84,375 | 13,512 | 40,672 | T |

TABLE 4.4A   Continued

| Commodity number | For the distribution of output of a commodity, read the row for that commodity / For the composition of inputs to an industry, read the column for that industry | Drugs | Cleaning and toilet preparations |
|---|---|---|---|
| | Industry number | 29A | 29B |
| 1 | Livestock and livestock products | 86 | 15 |
| 2 | Other agricultural products | 27 | .......... |
| 3 | Forestry and fishery products | 6 | .......... |
| 4 | Agricultural, forestry, and fishery services | 1 | (*) |
| 5+6 | Metallic ores mining | .......... | .......... |
| 7 | Coal mining | 9 | 5 |
| 8 | Crude petroleum and natural gas | .......... | .......... |
| 9+10 | Nonmetallic minerals mining | .......... | 9 |
| 11 | New construction | .......... | .......... |
| 12 | Maintenance and repair construction | 131 | 75 |
| 13 | Ordnance and accessories | .......... | .......... |
| 14 | Food and kindred products | 61 | 558 |
| 15 | Tobacco products | .......... | .......... |
| 16 | Broad and narrow fabrics, yarn and thread mills | .......... | .......... |
| 17 | Miscellaneous textile goods and floor coverings | .......... | 18 |
| 18 | Apparel | (*) | 1 |
| 19 | Miscellaneous fabricated textile products | (*) | 2 |
| 20+21 | Lumber and wood products | .......... | 6 |
| 22+23 | Furniture and fixtures | .......... | .......... |
| 24 | Paper and allied products, except containers | 161 | 61 |
| 25 | Paperboard containers and boxes | 314 | 1,146 |
| 26A | Newspapers and periodicals | 1 | 2 |
| 26B | Other printing and publishing | 182 | 345 |
| 27A | Industrial and other chemicals | 697 | 2,971 |
| 27B | Agricultural fertilizers and chemicals | 54 | .......... |
| 28 | Plastics and synthetic materials | .......... | 468 |
| 29A | Drugs | 3,758 | .......... |
| 29B | Cleaning and toilet preparations | 21 | 1,535 |
| 30 | Paints and allied products | .......... | 63 |
| 31 | Petroleum refining and related products | 39 | 329 |
| 32 | Rubber and miscellaneous plastics products | 749 | 1,713 |
| 33+34 | Footwear, leather, and leather products | (*) | .......... |
| 35 | Glass and glass products | 233 | 231 |
| 36 | Stone and clay products | 1 | 3 |
| 37 | Primary iron and steel manufacturing | .......... | 2 |
| 38 | Primary nonferrous metals manufacturing | 1 | .......... |
| 39 | Metal containers | 138 | 433 |
| 40 | Heating, plumbing, and fabricated structural metal products | 12 | 16 |
| 41 | Screw machine products and stampings | 64 | 196 |
| 42 | Other fabricated metal products | 65 | 390 |
| 43 | Engines and turbines | .......... | .......... |
| 44+45 | Farm, construction, and mining machinery | .......... | .......... |
| 46 | Materials handling machinery and equipment | .......... | .......... |
| 47 | Metalworking machinery and equipment | 3 | 6 |
| 48 | Special industry machinery and equipment | .......... | .......... |
| 49 | General industrial machinery and equipment | 17 | 59 |

TABLE 4.4A

| Paints and allied products | Petroleum refining and related products | Rubber and miscellaneous plastics products | Footwear, leather, and leather products | Glass and glass products | Stone and clay products | Primary iron and steel manufacturing | Primary nonferrous metals manufacturing |
|---|---|---|---|---|---|---|---|
| 30 | 31 | 32 | 33+34 | 35 | 36 | 37 | 38 |
|  |  |  |  |  |  |  |  |
| 15 |  |  |  |  |  |  |  |
|  | 2 | 3 |  | 1 | 2 | 2 | 2 |
| 43 |  |  |  |  | 33 | 1,969 | 3,933 |
|  | 21 | 25 | 1 | 2 | 399 | 1,449 | 34 |
| 13 | 75,971 | 75 |  |  |  | 10 |  |
| 17 | 490 | 35 | 2 | 184 | 3,513 | 231 | 18 |
| 46 | 952 | 392 | 28 | 130 | 328 | 1,294 | 273 |
|  |  | 1 |  |  | 57 | 20 |  |
| 160 | 42 | 17 | 893 | 1 | 25 | 6 | 6 |
|  |  | 812 | 239 |  | 126 |  | 43 |
|  | 39 | 894 | 197 |  | 2 |  | 1 |
| (*) | 1 | 12 | 4 | 3 | 6 | 8 | 4 |
|  | (*) | 24 |  |  | 2 | 1 | (*) |
|  | 59 | 189 | 24 | 237 | 95 | 136 | 151 |
|  |  |  |  |  |  | 28 |  |
| 2 | 9 | 789 | 6 | 15 | 551 | 15 | 17 |
|  | 191 | 987 | 66 | 707 | 142 | 79 | 103 |
| (*) | 2 | 9 | 1 | 1 | 5 | 3 | 3 |
| 26 | 17 | 99 | 5 | 25 | 25 | 35 | 25 |
| 2,334 | 1,758 | 3,905 | 271 | 982 | 1,269 | 1,723 | 715 |
| 1,441 | 60 | 15,955 | 109 |  | 189 |  | 784 |
| 5 | 417 | 29 | 33 |  | 96 | 1 | 1 |
| 252 | 6 | 65 |  | 20 | 70 | 23 | 31 |
| 80 | 9,933 | 235 | 10 | 45 | 238 | 332 | 364 |
| 1 | 620 | 4,074 | 250 | 173 | 125 | 139 | 647 |
|  | 3 | 5 | 1,601 | 2 | 1 | 2 |  |
| 8 | 277 | 450 |  | 1,309 | 71 | 5 | 48 |
| 132 | 51 | 247 | 7 | 320 | 5,111 | 996 | 263 |
| 16 | 42 | 300 | 1 | 6 | 277 | 10,233 | 515 |
| 7 |  | 115 |  | 1 | 23 | 1,470 | 17,261 |
| 534 | 167 |  |  |  | (*) | 1 | (*) |
|  |  | 44 |  |  | 7 |  |  |
| 6 |  | 439 | 9 | 35 | 100 | 275 | 90 |
| 21 | 397 | 860 | 43 | 2 | 219 | 772 | 528 |
|  |  | 15 |  |  | 2 | 11 |  |
|  |  |  |  |  | 13 |  |  |
|  |  | 5 | (*) |  | 2 |  |  |
| 1 | 19 | 161 | 4 | 87 | 25 | 17 | 10 |
|  |  | 233 | 1 | 11 |  | 524 | 488 |
| (*) | 1 | 29 | (*) | 4 | 32 | 868 | 556 |

TABLE 4.4A   Continued

| | | | |
|---|---|---:|---:|
| 50 | Miscellaneous machinery, except electrical | 10 | 25 |
| 51 | Computer and office equipment | 8 | |
| 52 | Service industry machinery | | |
| 53 | Electrical industrial equipment and apparatus | | |
| 54 | Household appliances | | |
| 55 | Electric lighting and wiring equipment | 1 | (*) |
| 56 | Audio, video, and communication equipment | (*) | (*) |
| 57 | Electronic components and accessories | | |
| 58 | Miscellaneous electrical machinery and supplies | 1 | (*) |
| 59A | Motor vehicles (passenger cars and trucks) | | |
| 59B | Truck and bus bodies, trailers, and motor vehicles parts | 5 | 3 |
| 60 | Aircraft and parts | | |
| 61 | Other transportation equipment | | |
| 62 | Scientific and controlling instruments | 28 | 6 |
| 63 | Ophthalmic and photographic equipment | 5 | 5 |
| 64 | Miscellaneous manufacturing | 3 | 45 |
| 65A | Railroads and related services; passenger ground transportation | 27 | 112 |
| 65B | Motor freight transportation and warehousing | 119 | 369 |
| 65C | Water transportation | 12 | 38 |
| 65D | Air transportation | 56 | 80 |
| 65E | Pipelines, freight forwarders, and related services | 1 | 2 |
| 66 | Communications, except radio and TV | 126 | 79 |
| 67 | Radio and TV broadcasting | | |
| 68A | Electric services (utilities) | 288 | 189 |
| 68B | Gas production and distribution (utilities) | 167 | 167 |
| 68C | Water and sanitary services | 34 | 48 |
| 69A | Wholesale trade | 1,439 | 1,583 |
| 69B | Retail trade | 9 | 5 |
| 70A | Finance | 186 | 88 |
| 70B | Insurance | 32 | 52 |
| 71A | Owner-occupied dwellings | | |
| 71B | Real estate and royalties | 211 | 150 |
| 72A | Hotels and lodging places | 13 | 25 |
| 72B | Personal and repair services (except auto) | 48 | 21 |
| 73A | Computer and data processing services | 5 | 7 |
| 73B | Legal, engineering, accounting, and related services | 1,568 | 169 |
| 73C | Other business and professional services, except medical | 1,071 | 395 |
| 73D | Advertising | 366 | 853 |
| 74 | Eating and drinking places | 66 | 61 |
| 75 | Automotive repair and services | 111 | 55 |
| 76 | Amusements | 2 | 2 |
| 77A | Health services | | |
| 77B | Educational and social services, and membership organizations | 172 | 38 |
| 78 | Federal Government enterprises | 26 | 51 |
| 79 | State and local government enterprises | 17 | 19 |
| 80 | Noncomparable imports | 777 | 187 |
| 81 | Scrap, used and secondhand goods | | |
| 82 | General government industry | | |
| 83 | Rest-of-the-world adjustment to final uses | | |
| 84 | Household industry | | |
| 85 | Inventory valuation adjustment | | |
| I | Total intermediate inputs | 13,840 | 15,583 |
| VA | Value added | 22,172 | 17,646 |
| T | Total industry output | 36,012 | 33,229 |

* Less than $500,000.

## TABLE 4.4A

| | | | | | | | |
|---:|---:|---:|---:|---:|---:|---:|---:|
| 2 | 41 | 255 | 9 | 49 | 48 | 207 | 114 |
| | | | | | | 4 | |
| | | | | | | 4 | |
| | | 13 | | 34 | 12 | 434 | 329 |
| | | (*) | 1 | 5 | | (*) | |
| (*) | 13 | 120 | (*) | 3 | 32 | 10 | 2 |
| | (*) | 1 | | (*) | (*) | (*) | (*) |
| | | 55 | | | 1 | | |
| | 1 | 23 | | 1 | 1 | 4 | 1 |
| (*) | 54 | 13 | (*) | 4 | 9 | 7 | 12 |
| | | | | | | 4 | |
| 2 | 24 | 43 | 1 | 16 | 16 | 21 | 14 |
| 1 | 6 | 19 | 1 | 3 | 10 | 11 | 8 |
| 1 | 3 | 19 | 83 | 2 | 40 | 11 | 5 |
| 158 | 153 | 603 | 18 | 185 | 644 | 1,149 | 336 |
| 280 | 563 | 2,331 | 88 | 222 | 2,527 | 916 | 1,707 |
| 19 | 889 | 115 | 3 | 12 | 154 | 319 | 57 |
| 11 | 67 | 127 | 24 | 67 | 61 | 78 | 102 |
| 2 | 5,485 | 6 | (*) | 2 | 3 | 5 | 2 |
| 40 | 187 | 249 | 22 | 122 | 295 | 150 | 111 |
| 74 | 1,653 | 1,829 | 68 | 478 | 1,159 | 2,813 | 2,501 |
| 43 | 1,260 | 511 | 19 | 580 | 789 | 1,898 | 807 |
| 8 | 209 | 202 | 10 | 34 | 120 | 489 | 95 |
| 343 | 6,367 | 4,238 | 369 | 682 | 1,269 | 4,274 | 3,417 |
| (*) | 19 | 32 | (*) | 6 | 28 | 23 | 27 |
| 15 | 1,232 | 386 | 50 | 69 | 286 | 247 | 267 |
| 6 | 361 | 181 | 10 | 26 | 91 | 138 | 112 |
| 58 | 614 | 538 | 45 | 86 | 218 | 157 | 184 |
| 7 | 27 | 23 | 43 | 15 | 17 | 10 | 36 |
| 1 | 54 | 118 | 12 | 12 | 55 | 76 | 85 |
| 1 | 126 | 265 | 5 | 23 | 89 | 108 | 35 |
| 54 | 376 | 580 | 30 | 53 | 168 | 191 | 140 |
| 62 | 1,156 | 685 | 90 | 93 | 408 | 1,246 | 561 |
| 55 | 321 | 704 | 130 | 375 | 864 | 2,010 | 537 |
| 18 | 51 | 246 | 28 | 37 | 122 | 119 | 91 |
| 7 | 128 | 318 | 11 | 88 | 223 | 103 | 285 |
| (*) | 3 | 7 | (*) | 1 | 2 | 2 | 2 |
| 4 | 145 | 297 | 2 | 8 | 39 | 13 | 37 |
| 8 | 72 | 86 | 31 | 19 | 49 | 87 | 46 |
| 1 | 16 | 31 | 8 | 7 | 11 | 43 | 25 |
| 65 | 392 | 1,155 | 2 | 87 | 130 | 84 | 789 |
| | | | | 72 | | 2,583 | 2,376 |
| 6,504 | 113,613 | 47,948 | 5,018 | 7,875 | 23,169 | 42,721 | 42,163 |
| 5,568 | 24,258 | 37,624 | 3,681 | 8,210 | 20,563 | 25,370 | 14,213 |
| 12,072 | 137,871 | 85,572 | 8,700 | 16,085 | 43,732 | 68,091 | 56,376 |

TABLE 4.4A   Continued

| Metal containers | Heating, plumbing, and fabricated structural metal products | Screw machine products and stampings | Other fabricated metal products | Engines and turbines | Farm, construction, and mining machinery | Materials handling machinery and equipment | Metalworking machinery and equipment |
|---|---|---|---|---|---|---|---|
| 39 | 40 | 41 | 42 | 43 | 44+45 | 46 | 47 |
| .......... | .......... | .......... | .......... | .......... | .......... | .......... | .......... |
| .......... | .......... | .......... | .......... | .......... | .......... | .......... | .......... |
| (*) | 1 | (*) | 1 | 1 | 2 | (*) | 1 |
| .......... | 13 | .......... | .......... | .......... | .......... | .......... | .......... |
| 1 | 5 | 12 | 21 | 4 | 12 | (*) | 6 |
| .......... | 5 | .......... | 8 | .......... | .......... | .......... | .......... |
| 33 | 518 | 324 | 552 | 92 | 223 | 74 | 127 |
| (*) | 6 | 2 | 4 | 1 | 3 | 2 | 4 |
| .......... | .......... | .......... | .......... | .......... | .......... | .......... | .......... |
| .......... | .......... | .......... | 1 | .......... | .......... | .......... | .......... |
| 1 | 2 | 2 | 10 | 2 | 1 | (*) | 2 |
| .......... | (*) | 103 | .......... | .......... | .......... | .......... | (*) |
| 11 | 139 | 32 | 177 | .......... | 46 | 11 | 26 |
| .......... | .......... | .......... | .......... | .......... | (*) | 1 | .......... |
| 4 | 33 | 31 | 10 | 7 | 7 | 2 | 6 |
| 28 | 231 | 132 | 329 | 24 | 64 | 5 | 83 |
| (*) | 5 | 2 | 4 | 1 | 2 | 1 | 3 |
| 220 | 27 | 15 | 29 | 7 | 18 | 34 | 20 |
| 61 | 119 | 242 | 742 | 4 | 35 | 10 | 168 |
| 17 | 33 | 49 | 169 | .......... | .......... | .......... | .......... |
| .......... | .......... | .......... | .......... | .......... | .......... | .......... | .......... |
| 6 | 15 | 15 | 17 | .......... | .......... | .......... | .......... |
| 196 | 243 | 79 | 342 | 5 | 73 | 7 | 23 |
| 17 | 104 | 40 | 118 | 13 | 43 | 18 | 66 |
| 19 | 425 | 97 | 859 | 139 | 727 | 101 | 151 |
| .......... | 1 | .......... | .......... | .......... | 3 | .......... | (*) |
| (*) | 325 | 55 | 114 | (*) | (*) | .......... | (*) |
| 13 | 125 | 76 | 171 | 48 | 57 | 13 | 235 |
| 2,571 | 8,294 | 7,905 | 5,190 | 1,938 | 2,988 | 607 | 1,467 |
| 3,008 | 3,194 | 980 | 2,049 | 652 | 225 | 95 | 427 |
| 199 | .......... | 4 | 8 | .......... | .......... | .......... | .......... |
| .......... | 996 | .......... | 4 | 175 | 997 | 193 | 307 |
| 15 | 1,209 | 476 | 729 | 233 | 394 | 134 | 179 |
| 159 | 1,475 | 585 | 2,453 | 213 | 440 | 260 | 170 |
| .......... | 1 | .......... | 33 | 1,306 | 1,249 | 66 | .......... |
| .......... | .......... | .......... | .......... | .......... | 971 | .......... | .......... |
| .......... | .......... | .......... | .......... | .......... | 22 | 383 | .......... |
| 17 | 401 | 962 | 256 | 91 | 122 | 37 | 847 |
| .......... | .......... | 2 | .......... | .......... | .......... | .......... | .......... |

TABLE 4.4A

| Special industry machinery and equipment | General industrial machinery and equipment | Miscellaneous machinery, except electrical | Computer and office equipment | Service industry machinery | Electrical industrial equipment and apparatus | Household appliances | Commodity number |
|---|---|---|---|---|---|---|---|
| 48 | 49 | 50 | 51 | 52 | 53 | 54 | |
| | | | | | | | 1 |
| | | | | | | | 2 |
| | | | | | | | 3 |
| (*) | 1 | 1 | 1 | (*) | 1 | (*) | 4 |
| | | | | | | | 5+6 |
| (*) | 1 | 1 | | 2 | 6 | 4 | 7 |
| | | | | | | | 8 |
| | | | | 18 | | | 9+10 |
| | | | | | | | 11 |
| 123 | 156 | 192 | 434 | 136 | 185 | 73 | 12 |
| | | | | | | | 13 |
| 3 | 2 | 1 | 1 | 2 | 3 | 2 | 14 |
| | | | | | | | 15 |
| | | | | | | 34 | 16 |
| | 136 | 17 | | | | 4 | 17 |
| 1 | 3 | 1 | | 2 | 2 | 2 | 18 |
| | | (*) | (*) | (*) | | | 19 |
| 27 | 25 | 5 | 1 | 107 | 29 | 98 | 20+21 |
| | | | (*) | | | 17 | 22+23 |
| 4 | 26 | 7 | 78 | 11 | 120 | 35 | 24 |
| 29 | 83 | 64 | 90 | 130 | 132 | 308 | 25 |
| 2 | 2 | 2 | 6 | 2 | 2 | 1 | 26A |
| 12 | 17 | 15 | 71 | 12 | 19 | 12 | 26B |
| 193 | 18 | 19 | 13 | 86 | 43 | 96 | 27A |
| | | | | | | | 27B |
| | 41 | 36 | 56 | 96 | 89 | 366 | 28 |
| | | | | | | | 29A |
| | | | | | | | 29B |
| | 4 | 1 | 3 | 66 | 69 | 112 | 30 |
| 33 | 44 | 34 | 57 | 24 | 147 | 15 | 31 |
| 257 | 325 | 85 | 1,045 | 435 | 468 | 642 | 32 |
| | | (*) | 2 | (*) | (*) | | 33+34 |
| 42 | 1 | 1 | 3 | 20 | 1 | 166 | 35 |
| 46 | 99 | 147 | 13 | 98 | 222 | 70 | 36 |
| 1,017 | 2,135 | 1,338 | 335 | 1,229 | 1,117 | 1,144 | 37 |
| 365 | 639 | 910 | 709 | 1,140 | 1,214 | 312 | 38 |
| | | | | | | | 39 |
| 247 | 157 | 214 | 174 | 192 | 110 | | 40 |
| 103 | 189 | 189 | 194 | 486 | 400 | 243 | 41 |
| 216 | 215 | 253 | 302 | 389 | 220 | 567 | 42 |
| 60 | 83 | 12 | | 30 | 104 | | 43 |
| | | | | | | | 44+45 |
| 6 | | | | | | | 46 |
| 150 | 146 | 283 | 35 | 122 | 61 | 37 | 47 |
| 372 | | | | | | | 48 |

TABLE 4.4A   Continued

| | | | | | | | |
|---|---|---|---|---|---|---|---|
| 3 | 202 | 63 | 30 | 188 | 709 | 225 | 130 |
| 20 | 115 | 370 | 187 | 451 | 682 | 165 | 612 |
| | 34 | | | | | | |
| 1 | 110 | 50 | 216 | 476 | 117 | 287 | 601 |
| | | 1 | | | | | |
| (*) | 4 | 13 | 3 | 3 | 7 | 1 | 6 |
| (*) | 1 | (*) | (*) | (*) | (*) | | (*) |
| | | | 11 | | | | 12 |
| (*) | 1 | | 16 | 155 | 129 | 48 | 10 |
| | 9 | | | | 9 | | |
| 2 | 11 | 21 | 11 | 22 | 72 | 1 | 9 |
| 1 | 96 | 5 | 13 | 2 | 5 | 1 | 4 |
| 1 | 11 | 5 | 10 | 3 | 7 | 2 | 7 |
| 1 | 22 | 4 | 15 | 1 | 11 | 4 | 7 |
| 41 | 137 | 107 | 107 | 8 | 38 | 7 | 27 |
| 190 | 680 | 433 | 551 | 105 | 306 | 58 | 151 |
| 3 | 13 | 13 | 21 | 3 | 15 | 2 | 5 |
| 50 | 128 | 27 | 156 | 30 | 120 | 20 | 91 |
| (*) | 2 | (*) | 2 | | 1 | (*) | 1 |
| 11 | 117 | 85 | 173 | 39 | 80 | 29 | 86 |
| 139 | 378 | 423 | 685 | 139 | 267 | 54 | 269 |
| 70 | 159 | 165 | 262 | 44 | 116 | 20 | 57 |
| 26 | 34 | 79 | 54 | 14 | 29 | 11 | 16 |
| 796 | 2,479 | 1,639 | 2,302 | 707 | 1,742 | 460 | 784 |
| 3 | 30 | 3 | 21 | 2 | 8 | 3 | 15 |
| 38 | 186 | 205 | 306 | 50 | 104 | 32 | 108 |
| 20 | 84 | 57 | 85 | 22 | 58 | 14 | 46 |
| 54 | 285 | 139 | 221 | 41 | 79 | 38 | 177 |
| 10 | 207 | 11 | 30 | 1 | 25 | 5 | 40 |
| 5 | 20 | 40 | 49 | 21 | 34 | 9 | 22 |
| 10 | 23 | 81 | 104 | 52 | 80 | 13 | 41 |
| 23 | 179 | 717 | 231 | 45 | 171 | 44 | 135 |
| 47 | 409 | 225 | 481 | 90 | 198 | 64 | 205 |
| 241 | 278 | 265 | 821 | 104 | 612 | 149 | 355 |
| 14 | 139 | 69 | 134 | 22 | 74 | 24 | 91 |
| 37 | 246 | 93 | 227 | 26 | 35 | 12 | 184 |
| (*) | 5 | 1 | 2 | 2 | 1 | 2 | 1 |
| 9 | 38 | 97 | 40 | 5 | 6 | 2 | 19 |
| 4 | 46 | 36 | 52 | 11 | 43 | 12 | 22 |
| 6 | 14 | 11 | 29 | 5 | 7 | (*) | 7 |
| 7 | 20 | 20 | 65 | 26 | 168 | 12 | 76 |
| | 41 | | 34 | | 12 | | 11 |
| 8,483 | 24,930 | 17,787 | 22,155 | 7,870 | 14,902 | 3,884 | 8,756 |
| 3,421 | 19,001 | 14,187 | 22,269 | 6,226 | 11,852 | 3,309 | 12,470 |
| 11,904 | 43,930 | 31,973 | 44,424 | 14,096 | 26,753 | 7,194 | 21,227 |

TABLE 4.4A

| | | | | | | | |
|---|---|---|---|---|---|---|---|
| 380 | 1,559 | 210 | 27 | 398 | 89 | 101 | 49 |
| 427 | 382 | 1,076 | 42 | 242 | 127 | 23 | 50 |
| 3 | 8 | 1 | 11,591 | | 3 | | 51 |
| 2 | 1 | | | 1,322 | | 320 | 52 |
| 979 | 976 | 80 | 1,398 | 1,456 | 843 | 592 | 53 |
| 1 | | | | | | 80 | 54 |
| 1 | 1 | 17 | 206 | 142 | 64 | 190 | 55 |
| (*) | 1 | (*) | 913 | (*) | (*) | (*) | 56 |
| 30 | 44 | 46 | 4,063 | 6 | 584 | 137 | 57 |
| 11 | 16 | 9 | 63 | | 20 | 2 | 58 |
| | | | | | | | 59A |
| 2 | 4 | 2 | 9 | 1 | 4 | | 59B |
| | | 22 | | | 27 | | 60 |
| | | | | | | | 61 |
| 3 | 11 | 3 | 17 | 396 | 40 | 303 | 62 |
| 5 | 7 | 7 | 15 | 4 | 6 | 3 | 63 |
| 2 | 4 | 2 | 24 | 32 | 17 | 108 | 64 |
| 21 | 22 | 22 | 26 | 30 | 61 | 31 | 65A |
| 84 | 132 | 100 | 109 | 156 | 156 | 166 | 65B |
| 3 | 6 | 4 | 6 | 8 | 5 | 3 | 65C |
| 91 | 123 | 64 | 740 | 120 | 220 | 78 | 65D |
| 1 | 1 | (*) | 2 | (*) | 1 | | 65E |
| 76 | 101 | 72 | 192 | 76 | 126 | 36 | 66 |
| | | | | | | | 67 |
| 154 | 290 | 253 | 338 | 186 | 289 | 135 | 68A |
| 40 | 81 | 43 | 30 | 51 | 81 | 69 | 68B |
| 16 | 18 | 8 | 45 | 37 | 20 | 18 | 68C |
| 929 | 1,153 | 565 | 4,147 | 1,526 | 1,407 | 1,114 | 69A |
| 6 | 8 | 7 | 18 | 4 | 8 | 1 | 69B |
| 106 | 110 | 166 | 361 | 59 | 246 | 110 | 70A |
| 28 | 40 | 28 | 76 | 31 | 41 | 23 | 70B |
| | | | | | | | 71A |
| 127 | 130 | 188 | 521 | 84 | 120 | 41 | 71B |
| 29 | 38 | 20 | 337 | 36 | 128 | 33 | 72A |
| 15 | 29 | 22 | 22 | 14 | 25 | 17 | 72B |
| 22 | 58 | 56 | 52 | 27 | 42 | 20 | 73A |
| 112 | 142 | 150 | 414 | 110 | 128 | 42 | 73B |
| 150 | 222 | 485 | 473 | 132 | 221 | 118 | 73C |
| 272 | 291 | 318 | 151 | 346 | 775 | 318 | 73D |
| 57 | 76 | 84 | 199 | 56 | 77 | 35 | 74 |
| 63 | 78 | 114 | 229 | 26 | 101 | 16 | 75 |
| 1 | 5 | 2 | 5 | 1 | 1 | (*) | 76 |
| | | | | | | | 77A |
| 6 | 19 | 20 | 29 | 20 | 9 | 25 | 77B |
| 29 | 33 | 16 | 22 | 8 | 26 | 50 | 78 |
| 2 | 6 | 4 | 9 | 5 | 7 | 8 | 79 |
| 67 | 58 | 43 | 1,066 | 8 | 142 | 52 | 80 |
| | 11 | 7 | | | | | 81 |
| | | | | | | | 82 |
| | | | | | | | 83 |
| | | | | | | | 84 |
| | | | | | | | 85 |
| 7,659 | 10,836 | 8,164 | 31,625 | 11,987 | 11,056 | 8,767 | I |
| 8,595 | 12,400 | 11,839 | 24,195 | 10,422 | 11,609 | 6,594 | VA |
| 16,254 | 23,236 | 20,003 | 55,819 | 22,409 | 22,665 | 15,361 | T |

TABLE 4.4A    Continued

| Commodity number | For the distribution of output of a commodity, read the row for that commodity<br><br>For the composition of inputs to an industry, read the column for that industry | Electric lighting and wiring equipment | Audio, video, and communication equipment |
|---|---|---|---|
| | Industry number | 55 | 56 |
| 1 | Livestock and livestock products | .......... | .......... |
| 2 | Other agricultural products | .......... | .......... |
| 3 | Forestry and fishery products | .......... | .......... |
| 4 | Agricultural, forestry, and fishery services | 1 | 2 |
| 5+6 | Metallic ores mining | .......... | .......... |
| 7 | Coal mining | 2 | 3 |
| 8 | Crude petroleum and natural gas | .......... | .......... |
| 9+10 | Nonmetallic minerals mining | .......... | .......... |
| 11 | New construction | .......... | .......... |
| 12 | Maintenance and repair construction | 107 | 153 |
| 13 | Ordnance and accessories | .......... | .......... |
| 14 | Food and kindred products | 2 | 1 |
| 15 | Tobacco products | .......... | .......... |
| 16 | Broad and narrow fabrics, yarn and thread mills | 11 | .......... |
| 17 | Miscellaneous textile goods and floor coverings | .......... | .......... |
| 18 | Apparel | 1 | 10 |
| 19 | Miscellaneous fabricated textile products | .......... | (*) |
| 20+21 | Lumber and wood products | 29 | 27 |
| 22+23 | Furniture and fixtures | .......... | 446 |
| 24 | Paper and allied products, except containers | 5 | 101 |
| 25 | Paperboard containers and boxes | 239 | 137 |
| 26A | Newspapers and periodicals | 2 | 3 |
| 26B | Other printing and publishing | 15 | 102 |
| 27A | Industrial and other chemicals | 99 | 107 |
| 27B | Agricultural fertilizers and chemicals | .......... | .......... |
| 28 | Plastics and synthetic materials | 409 | 105 |
| 29A | Drugs | .......... | .......... |
| 29B | Cleaning and toilet preparations | .......... | .......... |
| 30 | Paints and allied products | 30 | 28 |
| 31 | Petroleum refining and related products | 30 | 24 |
| 32 | Rubber and miscellaneous plastics products | 372 | 1,418 |
| 33+34 | Footwear, leather, and leather products | .......... | 1 |
| 35 | Glass and glass products | 641 | 23 |
| 36 | Stone and clay products | 24 | 14 |
| 37 | Primary iron and steel manufacturing | 747 | 117 |
| 38 | Primary nonferrous metals manufacturing | 806 | 420 |
| 39 | Metal containers | .......... | .......... |
| 40 | Heating, plumbing, and fabricated structural metal products | .......... | 81 |
| 41 | Screw machine products and stampings | 558 | 324 |
| 42 | Other fabricated metal products | 278 | 411 |
| 43 | Engines and turbines | .......... | .......... |
| 44+45 | Farm, construction, and mining machinery | .......... | .......... |
| 46 | Materials handling machinery and equipment | .......... | .......... |
| 47 | Metalworking machinery and equipment | 53 | 36 |
| 48 | Special industry machinery and equipment | .......... | .......... |
| 49 | General industrial machinery and equipment | .......... | 36 |
| 50 | Miscellaneous machinery, except electrical | (*) 42 | 42 |

TABLE 4.4A

| Electronic components and accessories | Miscellaneous electrical machinery and supplies | Motor vehicles (passenger cars and trucks) | Truck and bus bodies, trailers, and motor vehicles parts | Aircraft and parts | Other transportation equipment | Scientific and controlling instruments | Ophthalmic and photographic equipment |
|---|---|---|---|---|---|---|---|
| 57 | 58 | 59A | 59B | 60 | 61 | 62 | 63 |
| 2 | 1 | 4 | 3 | 3 | (*) | 4 | 1 |
|  | 89 |  |  |  |  | 13 |  |
| 1 | 2 | 63 | 18 | 8 | 3 | 6 | 20 |
|  |  |  | 1 |  |  | (*) |  |
| 493 | 111 | 431 | 469 | 495 | 471 | 407 | 91 |
|  |  | 1 | 1 | 38 | (*) | (*) |  |
| 5 | 1 | (*) | 2 | 5 | 2 | 77 | 1 |
|  |  | 117 | 2 | 84 | 11 | 312 |  |
|  | 8 | 328 | 112 | 96 | 122 | 394 | 18 |
| 12 | 3 | 14 | 3 | 6 | 3 | 32 | 1 |
| (*) | (*) | 3,520 | 48 | 163 | 182 | (*) | (*) |
|  | 1 | 4 | 199 | 31 | 471 | 173 |  |
| 11 | 2 | 1,678 | 4 | 25 | 60 | 62 |  |
| 34 | 13 | 99 | 28 | 13 | 9 | 262 | 1,197 |
| 80 | 140 | 57 | 115 | 7 | 7 | 343 | 158 |
| 4 | 4 | 5 | 3 | 4 | 1 | 11 | 2 |
| 32 | 25 | 43 | 32 | 53 | 9 | 145 | 18 |
| 789 | 390 | 606 | 216 | 36 | 55 | 298 | 470 |
| 169 | 111 | 55 | 365 | 96 | 184 | 574 | 123 |
|  |  |  | 8 |  |  | 3 |  |
| 1 | 1 | 1,615 | 222 | 140 | 144 | 45 | 1 |
| 42 | 47 | 236 | 145 | 114 | 47 | 126 | 31 |
| 2,539 | 898 | 8,393 | 2,158 | 703 | 289 | 1,444 | 402 |
| 1 | (*) | 5 | 1 | 1 | (*) | 4 | (*) |
| 403 | 1 | 1,291 | 85 | 15 | 236 | 194 | 99 |
| 53 | 19 | 247 | 366 | 203 | 58 | 117 | 9 |
| 130 | 244 | 717 | 4,421 | 1,349 | 934 | 1,291 | 29 |
| 1,956 | 1,205 | 85 | 3,046 | 3,539 | 437 | 1,357 | 125 |
|  |  |  |  |  |  | 20 |  |
| 187 | 86 | 2 | 1,278 | 183 | 853 | 439 |  |
| 443 | 213 | 9,934 | 2,280 | 957 | 181 | 1,032 | 72 |
| 1,743 | 475 | 1,645 | 1,384 | 928 | 478 | 1,312 | 235 |
|  |  | 2,371 | 58 |  | 1,057 |  |  |
|  |  |  |  |  | 110 |  |  |
|  |  | 13 | 8 |  | 1 |  |  |
| 88 | 48 | 1,105 | 209 | 1,145 | 50 | 178 | 23 |
| 126 |  |  |  |  |  |  |  |
| 12 | 47 | 68 | 1,411 | 164 | 503 | 155 | 19 |
| 108 | 59 | 863 | 2,658 | 772 | 71 | 191 | 37 |

**TABLE 4.4A** Continued

| | | | |
|---|---|---:|---:|
| 51 | Computer and office equipment | .............. | 108 |
| 52 | Service industry machinery | .............. | 5 |
| 53 | Electrical industrial equipment and apparatus | 413 | 237 |
| 54 | Household appliances | .............. | .............. |
| 55 | Electric lighting and wiring equipment | 485 | 272 |
| 56 | Audio, video, and communication equipment | (*) | 1,472 |
| 57 | Electronic components and accessories | 98 | 8,193 |
| 58 | Miscellaneous electrical machinery and supplies | 16 | 156 |
| 59A | Motor vehicles (passenger cars and trucks) | .............. | .............. |
| 59B | Truck and bus bodies, trailers, and motor vehicles parts | 3 | 2 |
| 60 | Aircraft and parts | .............. | .............. |
| 61 | Other transportation equipment | .............. | .............. |
| 62 | Scientific and controlling instruments | 3 | 54 |
| 63 | Ophthalmic and photographic equipment | 6 | 10 |
| 64 | Miscellaneous manufacturing | 37 | 13 |
| 65A | Railroads and related services; passenger ground transportation | 31 | 21 |
| 65B | Motor freight transportation and warehousing | 146 | 138 |
| 65C | Water transportation | 2 | 3 |
| 65D | Air transportation | 148 | 220 |
| 65E | Pipelines, freight forwarders, and related services | 1 | (*) |
| 66 | Communications, except radio and TV | 65 | 159 |
| 67 | Radio and TV broadcasting | .............. | .............. |
| 68A | Electric services (utilities) | 185 | 247 |
| 68B | Gas production and distribution (utilities) | 54 | 52 |
| 68C | Water and sanitary services | 18 | 21 |
| 69A | Wholesale trade | 1,138 | 1,970 |
| 69B | Retail trade | 5 | 5 |
| 70A | Finance | 163 | 163 |
| 70B | Insurance | 27 | 59 |
| 71A | Owner-occupied dwellings | .............. | .............. |
| 71B | Real estate and royalties | 105 | 326 |
| 72A | Hotels and lodging places | 87 | 30 |
| 72B | Personal and repair services (except auto) | 18 | 178 |
| 73A | Computer and data processing services | 30 | 96 |
| 73B | Legal, engineering, accounting, and related services | 109 | 235 |
| 73C | Other business and professional services, except medical | 166 | 449 |
| 73D | Advertising | 268 | 782 |
| 74 | Eating and drinking places | 66 | 93 |
| 75 | Automotive repair and services | 69 | 50 |
| 76 | Amusements | 2 | 4 |
| 77A | Health services | .............. | .............. |
| 77B | Educational and social services, and membership organizations | 8 | 49 |
| 78 | Federal Government enterprises | 21 | 79 |
| 79 | State and local government enterprises | 6 | 8 |
| 80 | Noncomparable imports | 32 | 233 |
| 81 | Scrap, used and secondhand goods | 2 | .............. |
| 82 | General government industry | .............. | .............. |
| 83 | Rest-of-the-world adjustment to final uses | .............. | .............. |
| 84 | Household industry | .............. | .............. |
| 85 | Inventory valuation adjustment | .............. | .............. |
| I | Total intermediate inputs | 8,532 | 20,363 |
| VA | Value added | 9,083 | 20,337 |
| T | Total industry output | 17,615 | 40,700 |

* Less than $500,000.

TABLE 4.4A

| 170 | 21 | | | 41 | | 757 | 10 |
|---|---|---|---|---|---|---|---|
| | | 2,773 | 176 | | 80 | | |
| 155 | 203 | 212 | 90 | 93 | 402 | 1,295 | 66 |
| | | (*) | (*) | | 148 | | |
| 78 | 163 | 495 | 28 | 1 | 90 | 206 | 32 |
| 36 | 13 | 1,347 | 14 | 962 | 13 | 20 | (*) |
| 4,625 | 1,304 | 856 | 381 | 1,211 | 7 | 7,877 | 1,573 |
| 26 | 1,025 | 3,740 | 871 | 77 | 129 | 217 | 28 |
| | | 1,548 | 184 | | 602 | | |
| 3 | 6 | 40,005 | 6,259 | 5 | 313 | 16 | 2 |
| | | 51 | | 15,912 | 40 | | |
| | | 21 | 14 | | 626 | | |
| 195 | 9 | 1,234 | 27 | 2,217 | 70 | 2,295 | 186 |
| 13 | 9 | 17 | 14 | 18 | 3 | 29 | 269 |
| 11 | 4 | 59 | 11 | 15 | 5 | 46 | 5 |
| 46 | 36 | 572 | 173 | 39 | 50 | 85 | 37 |
| 214 | 151 | 2,168 | 936 | 220 | 202 | 362 | 115 |
| 10 | 7 | 43 | 26 | 12 | 8 | 16 | 16 |
| 297 | 241 | 685 | 194 | 1,182 | 62 | 338 | 101 |
| 2 | 1 | 3 | 2 | 3 | (*) | 2 | 1 |
| 187 | 82 | 159 | 310 | 359 | 92 | 432 | 42 |
| 789 | 231 | 491 | 731 | 660 | 191 | 758 | 122 |
| 99 | 51 | 342 | 261 | 144 | 26 | 132 | 40 |
| 34 | 41 | 113 | 59 | 47 | 28 | 98 | 35 |
| 2,125 | 1,379 | 10,545 | 3,688 | 1,552 | 1,328 | 3,212 | 720 |
| 7 | 11 | 124 | 43 | 27 | 10 | 30 | 6 |
| 450 | 222 | 493 | 159 | 883 | 50 | 452 | 208 |
| 70 | 45 | 345 | 142 | 113 | 29 | 158 | 50 |
| 389 | 144 | 117 | 159 | 459 | 346 | 774 | 87 |
| 150 | 134 | 79 | 24 | 766 | 9 | 72 | 38 |
| 39 | 20 | 103 | 80 | 46 | 14 | 90 | 9 |
| 86 | 37 | 106 | 92 | 103 | 14 | 100 | 19 |
| 317 | 189 | 220 | 212 | 508 | 90 | 803 | 117 |
| 586 | 244 | 543 | 484 | 1,016 | 194 | 1,030 | 273 |
| 484 | 602 | 2,434 | 1,807 | 1,929 | 183 | 2,115 | 489 |
| 157 | 118 | 193 | 150 | 148 | 37 | 308 | 59 |
| 71 | 137 | 482 | 2,196 | 101 | 100 | 487 | 72 |
| 23 | 24 | 23 | 4 | 21 | 12 | 28 | 12 |
| 123 | 16 | 254 | 81 | 122 | 8 | 155 | 111 |
| 34 | 20 | 148 | 172 | 120 | 15 | 175 | 18 |
| 19 | 10 | 64 | 42 | 17 | 4 | 24 | 4 |
| 195 | 74 | 291 | 918 | 99 | 16 | 340 | 306 |
| | 108 | | 155 | | | | |
| 21,758 | 11,371 | 109,111 | 42,721 | 42,620 | 12,687 | 36,349 | 8,462 |
| 26,895 | 9,452 | 25,004 | 26,270 | 39,508 | 11,396 | 49,114 | 11,264 |
| 48,654 | 20,823 | 134,115 | 68,991 | 82,128 | 24,082 | 85,463 | 19,725 |

TABLE 4.4A  Continued

| Miscellaneous manufacturing | Railroads and related services; passenger ground transportation | Motor freight transportation and warehousing | Water transportation | Air transportation | Pipelines, freight forwarders, and related services | Communications, except radio and TV |
|---|---|---|---|---|---|---|
| 64 | 65A | 65B | 65C | 65D | 65E | 66 |
| 2 | | | 1 | | | |
| 20 | | | 3 | | | |
| | | | 3 | | | |
| 3 | 3 | 5 | 3 | 6 | 2 | 4 |
| 3 | (*) | | 9 | | | |
| 17 | | | | | | |
| | | | | | 93 | |
| 189 | 4,213 | 438 | 128 | 470 | 441 | 6,678 |
| | (*) | (*) | (*) | 1 | 4 | 1 |
| 65 | 9 | 1 | 65 | 279 | 8 | 2 |
| 324 | | | | 4 | | |
| 43 | 1 | 5 | 63 | (*) | | |
| 18 | 15 | 7 | 95 | 14 | | 73 |
| 126 | (*) | 13 | 74 | 9 | 116 | 1 |
| 724 | | 15 | | | 35 | |
| 9 | | | | | | |
| 289 | 19 | 52 | 7 | 56 | 35 | 80 |
| 434 | 7 | 67 | 13 | 1 | 29 | 43 |
| 7 | 7 | 25 | 3 | 14 | 8 | 18 |
| 88 | 115 | 268 | 48 | 149 | 154 | 689 |
| 337 | 86 | 80 | 18 | 17 | 8 | 93 |
| | 33 | | 21 | | | |
| 692 | | | | | | |
| | | | | | | 52 |
| 1 | 1 | 3 | | 4 | | |
| 143 | 2 | 17 | 42 | 4 | | 109 |
| 90 | 2,132 | 5,794 | 731 | 8,583 | 89 | 117 |
| 947 | 144 | 1,117 | 76 | 19 | 66 | 394 |
| 101 | 1 | 7 | 1 | 2 | 6 | 5 |
| 28 | 37 | 5 | | (*) | 8 | 15 |
| 144 | 8 | 9 | 8 | 3 | 2 | 1 |
| 548 | 83 | 1 | (*) | 2 | | 1 |
| 1,839 | 7 | | 195 | | | 44 |
| | | | | | | |
| 1 | | | | | | |
| 131 | 9 | | 2 | 48 | | 329 |
| 288 | 267 | 263 | 499 | 55 | 36 | 181 |
| 3 | 44 | 42 | 163 | | 24 | 242 |
| | | | | | | |
| | | 3 | | | | |
| 76 | 72 | 16 | 178 | 20 | 1 | 4 |
| 25 | 291 | 76 | 573 | 4 | 116 | 301 |

## TABLE 4.4A

| Radio and TV broadcasting | Electric services (utilities) | Gas production and distribution (utilities) | Water and sanitary services | Wholesale trade | Retail trade | Finance | Commodity number |
|---|---|---|---|---|---|---|---|
| 67 | 68A | 68B | 68C | 69A | 69B | 70A | |
| | | | | | | | 1 |
| | | | | | | | 2 |
| | | | | | | | 3 |
| 4 | 20 | 2 | 6 | 145 | 82 | 28 | 4 |
| | | | | | | | 5+6 |
| | 13,284 | | 5 | 24 | 9 | | 7 |
| | | 16,417 | | | | | 8 |
| | | | | | | | 9+10 |
| | | | | | | | 11 |
| 285 | 8,764 | 650 | 878 | 2,816 | 5,741 | 1,966 | 12 |
| | (*) | | | 15 | 1 | | 13 |
| 10 | (*) | (*) | 4 | 69 | 949 | 3 | 14 |
| | | | | | | | 15 |
| | | | | 1 | | | 16 |
| | | | | 58 | 26 | | 17 |
| (*) | 4 | (*) | 1 | 63 | 23 | | 18 |
| (*) | (*) | (*) | | 163 | 48 | 128 | 19 |
| | | | 142 | 1,335 | 56 | | 20+21 |
| | | | | 5 | 4 | | 22+23 |
| 14 | 36 | 5 | 68 | 1,746 | 2,948 | 832 | 24 |
| 1 | 12 | | | 4,237 | 526 | 3 | 25 |
| 6 | 6 | 2 | 2 | 92 | 86 | 304 | 26A |
| 53 | 122 | 17 | 20 | 3,792 | 181 | 3,667 | 26B |
| 30 | 423 | 4 | 509 | 70 | 27 | 11 | 27A |
| | | | 2 | | | | 27B |
| | | | | | 30 | | 28 |
| | | | | | | | 29A |
| | | 5 | | 36 | 29 | 29 | 29B |
| | | | | | | | 30 |
| 9 | 3,078 | 338 | 260 | 2,759 | 2,112 | 418 | 31 |
| 1 | 118 | 6 | 253 | 1,174 | 421 | 80 | 32 |
| 2 | 1 | 1 | (*) | 61 | 22 | 20 | 33+34 |
| (*) | 11 | 1 | 35 | 164 | 8 | 43 | 35 |
| | 9 | (*) | 12 | 109 | 18 | 10 | 36 |
| | 2 | 1 | | 32 | 7 | 1 | 37 |
| | 110 | | | | | | 38 |
| | 15 | | | 631 | | | 39 |
| | | | | 39 | | | 40 |
| | 98 | | | | | | 41 |
| 1 | 20 | 2 | 312 | 236 | 596 | 14 | 42 |
| | 625 | | | | | | 43 |
| | | | | 27 | | | 44+45 |
| | 12 | | | 262 | 2 | | 46 |
| 1 | 81 | 2 | 2 | 98 | 33 | 3 | 47 |
| | | | | 4 | 9 | | 48 |
| | 73 | 5 | 5 | 9 | 12 | 64 | 49 |

**TABLE 4.4A** Continued

| | | | | | | |
|---:|---:|---:|---:|---:|---:|---:|
| 124 | 155 | 26 | 157 | 43 | 18 | 7 |
| 5 | 1 | 8 | 1 | 2 | 12 | 58 |
| 26 | 1 | 20 | 1 | 2 | | |
| 90 | 148 | 98 | 12 | 10 | 27 | 264 |
| | | (*) | 13 | (*) | | (*) |
| 22 | 22 | 48 | 36 | 9 | 1 | 55 |
| 4 | 1 | 4 | (*) | 7 | 7 | 2,179 |
| 397 | 11 | 6 | | 90 | 8 | 1,683 |
| 12 | 55 | 69 | 33 | 25 | 45 | 206 |
| | | 19 | | | | |
| 10 | 132 | 397 | 1 | | 12 | 14 |
| | | | | 13 | | |
| | | | | 3,336 | | |
| 4 | 856 | 5 | 364 | | 11 | 14 |
| 8 | 2 | 4 | 32 | 30 | 1 | 9 |
| 14 | 10 | 45 | 5 | 14 | 16 | 68 |
| 1,218 | 14 | 31 | 49 | 37 | 24 | 94 |
| 83 | 2,322 | 260 | 15 | 65 | 78 | 72 |
| 494 | 122 | 18,968 | 74 | 174 | 96 | 142 |
| 23 | 38 | 91 | 2,472 | 155 | 31 | 15 |
| 77 | 82 | 242 | 69 | 4,158 | 579 | 234 |
| 1 | 720 | 3,543 | 1,425 | 4,821 | 1,086 | 2 |
| 71 | 47 | 1,425 | 113 | 926 | 477 | 33,184 |
| 293 | 99 | 964 | 570 | 244 | 973 | 480 |
| 64 | 2 | 43 | 4 | 14 | 37 | 22 |
| 109 | 33 | 42 | 100 | 36 | 130 | 1,136 |
| 1,731 | 788 | 1,789 | 565 | 2,266 | 131 | 883 |
| 19 | 200 | 3,230 | 4 | 67 | 26 | 41 |
| 260 | 336 | 489 | 1,303 | 2,110 | 1,026 | 1,898 |
| 71 | 189 | 957 | 8 | 371 | 531 | 65 |
| 242 | 301 | 2,165 | 794 | 1,005 | 936 | 2,806 |
| 24 | 69 | 166 | 14 | 38 | 86 | 81 |
| 31 | 36 | 91 | 23 | 421 | 50 | 403 |
| 14 | 349 | 342 | 270 | 717 | 452 | 1,647 |
| 332 | 172 | 547 | 73 | 351 | 440 | 710 |
| 362 | 617 | 1,353 | 1,501 | 1,073 | 329 | 1,058 |
| 1,683 | 34 | 238 | 246 | 1,741 | 644 | 1,176 |
| 236 | 179 | 555 | 56 | 2,288 | 476 | 503 |
| 240 | 589 | 4,401 | 34 | 286 | 271 | 523 |
| 11 | 4 | 15 | 1 | 43 | 11 | 699 |
| 124 | 133 | 56 | 62 | 7 | 115 | 231 |
| 91 | 58 | 147 | 7 | 7 | 58 | 251 |
| 8 | 62 | 135 | (*) | 22 | 7 | 34 |
| 968 | 180 | 10 | 2,880 | 4,262 | 61 | 3,758 |
| | | | | | | |
| | | | | | | |
| | | | | | | |
| | | | | | | |
| 17,347 | 16,774 | 51,373 | 16,406 | 41,048 | 10,599 | 66,178 |
| 15,742 | 26,684 | 64,722 | 7,647 | 35,205 | 15,309 | 94,949 |
| 33,089 | 43,458 | 116,095 | 24,053 | 76,253 | 25,908 | 161,127 |

TABLE 4.4A

| | | | | | | | |
|---:|---:|---:|---:|---:|---:|---:|:---|
| 2 | 47 | 5 | 5 | 217 | 71 | 6 | 50 |
| 20 | 4 | 47 | ......... | 49 | 38 | 48 | 51 |
| ......... | ......... | ......... | 3 | 121 | 87 | ......... | 52 |
| 22 | 198 | (*) | ......... | 1 | (*) | 9 | 53 |
| ......... | (*) | ......... | ......... | 1 | 8 | 1 | 54 |
| (*) | 130 | 9 | 24 | 36 | 37 | 41 | 55 |
| 51 | 1 | (*) | (*) | 21 | 33 | 37 | 56 |
| 778 | ......... | 3 | 1 | 22 | 4 | 211 | 57 |
| 46 | 213 | 100 | 18 | 214 | 130 | 368 | 58 |
| | | | | | | | 59A |
| 2 | 13 | 5 | 726 | 348 | 238 | 50 | 59B |
| ......... | ......... | ......... | ......... | ......... | ......... | ......... | 60 |
| 2 | 8 | 2 | ......... | ......... | ......... | 66 | 61 |
| 22 | 93 | 17 | 398 | 66 | 23 | 8 | 62 |
| 191 | 26 | 6 | 5 | 190 | 116 | 1,047 | 63 |
| 11 | 22 | 6 | 10 | 387 | 413 | 410 | 64 |
| 34 | 3,550 | 10 | 36 | 498 | 724 | 561 | 65A |
| 62 | 408 | 16 | 73 | 967 | 1,150 | 5,272 | 65B |
| 2 | 415 | 25 | 14 | 110 | 56 | 23 | 65C |
| 210 | 336 | 108 | 28 | 4,336 | 777 | 2,087 | 65D |
| 1 | 15 | 290 | 4 | 396 | 33 | 10 | 65E |
| 190 | 191 | 25 | 74 | 5,573 | 4,574 | 5,192 | 66 |
| 471 | ......... | ......... | ......... | ......... | ......... | ......... | 67 |
| 24 | 27 | 165 | 37 | 3,277 | 9,939 | 1,702 | 68A |
| 3 | 5,812 | 20,956 | 498 | 1,831 | 1,011 | 104 | 68B |
| 33 | 202 | 24 | 610 | 311 | ·687 | 293 | 68C |
| 154 | 1,342 | 291 | 360 | 9,954 | 1,696 | 1,246 | 69A |
| 4 | 48 | 20 | 58 | 764 | 902 | 135 | 69B |
| 293 | 1,545 | 445 | 159 | 5,778 | 3,679 | 48,309 | 70A |
| 4 | 952 | 51 | 1,068 | 494 | 532 | 3,280 | 70B |
| | | | | | | | 71A |
| 1,568 | 507 | 140 | 80 | 9,905 | 24,309 | 9,486 | 71B |
| 33 | 42 | 8 | 38 | 2,978 | 599 | 804 | 72A |
| 346 | 268 | 37 | 142 | 1,761 | 1,982 | 427 | 72B |
| 123 | 909 | 239 | 89 | 399 | 1,961 | 12,499 | 73A |
| 401 | 508 | 98 | 137 | 4,900 | 14,163 | 7,237 | 73B |
| 470 | 519 | 171 | 134 | 20,686 | 7,477 | 13,020 | 73C |
| 82 | 79 | 20 | 18 | 8,482 | 20,898 | 3,548 | 73D |
| 231 | 135 | 36 | 54 | 5,786 | 5,504 | 3,357 | 74 |
| 39 | 474 | 218 | 4 | 9,026 | 6,870 | 1,554 | 75 |
| 9,383 | 5 | 1 | 1 | 1,033 | 199 | 119 | 76 |
| | | | | | | | 77A |
| 130 | 346 | 23 | 10 | 658 | 424 | 774 | 77B |
| 22 | 325 | 129 | 43 | 871 | 1,455 | 7,115 | 78 |
| 26 | 19 | 8 | 1 | 379 | 427 | 42 | 79 |
| 35 | 7 | 14 | 1 | 2,710 | 134 | 3,898 | 80 |
| ......... | ......... | ......... | ......... | ......... | ......... | ......... | 81 |
| ......... | ......... | ......... | ......... | ......... | ......... | ......... | 82 |
| ......... | ......... | ......... | ......... | ......... | ......... | ......... | 83 |
| ......... | ......... | ......... | ......... | ......... | ......... | ......... | 84 |
| | | | | | | | 85 |
| 15,936 | 46,665 | 41,224 | 7,477 | 125,804 | 127,371 | 142,016 | I |
| 13,460 | 85,706 | 26,325 | 3,786 | 297,947 | 293,322 | 144,596 | VA |
| 29,396 | 132,371 | 67,549 | 11,262 | 423,751 | 420,694 | 286,613 | T |

TABLE **4.4A**   Continued

| Commodity number | For the distribution of output of a commodity, read the row for that commodity / For the composition of inputs to an industry, read the column for that industry | Insurance | Owner-occupied dwellings |
|---|---|---|---|
| | Industry number | 70B | 71A |
| 1 | Livestock and livestock products | | |
| 2 | Other agricultural products | | |
| 3 | Forestry and fishery products | | |
| 4 | Agricultural, forestry, and fishery services | 6 | 2,584 |
| 5+6 | Metallic ores mining | | |
| 7 | Coal mining | | |
| 8 | Crude petroleum and natural gas | | |
| 9+10 | Nonmetallic minerals mining | | |
| 11 | New construction | | |
| 12 | Maintenance and repair construction | 517 | 15,921 |
| 13 | Ordnance and accessories | (*) | |
| 14 | Food and kindred products | 2 | |
| 15 | Tobacco products | | |
| 16 | Broad and narrow fabrics, yarn and thread mills | | |
| 17 | Miscellaneous textile goods and floor coverings | | |
| 18 | Apparel | | |
| 19 | Miscellaneous fabricated textile products | 11 | |
| 20+21 | Lumber and wood products | | |
| 22+23 | Furniture and fixtures | | |
| 24 | Paper and allied products, except containers | 143 | |
| 25 | Paperboard containers and boxes | 1 | |
| 26A | Newspapers and periodicals | 24 | |
| 26B | Other printing and publishing | 1,389 | |
| 27A | Industrial and other chemicals | 5 | 1 |
| 27B | Agricultural fertilizers and chemicals | | 269 |
| 28 | Plastics and synthetic materials | | |
| 29A | Drugs | | |
| 29B | Cleaning and toilet preparations | | |
| 30 | Paints and allied products | | |
| 31 | Petroleum refining and related products | 98 | |
| 32 | Rubber and miscellaneous plastics products | 21 | 73 |
| 33+34 | Footwear, leather, and leather products | 16 | |
| 35 | Glass and glass products | 1 | |
| 36 | Stone and clay products | 1 | |
| 37 | Primary iron and steel manufacturing | | |
| 38 | Primary nonferrous metals manufacturing | | |
| 39 | Metal containers | | |
| 40 | Heating, plumbing, and fabricated structural metal products | | 100 |
| 41 | Screw machine products and stampings | | |
| 42 | Other fabricated metal products | 2 | |
| 43 | Engines and turbines | | |
| 44+45 | Farm, construction, and mining machinery | | 195 |
| 46 | Materials handling machinery and equipment | | |
| 47 | Metalworking machinery and equipment | 1 | |
| 48 | Special industry machinery and equipment | | |

TABLE 4.4A

| Real estate and royalties | Hotels and lodging places | Personal and repair services (except auto) | Computer and data processing services | Legal, engineering, accounting, and related services | Other business and professional services, except medical | Advertising |
|---|---|---|---|---|---|---|
| 71B | 72A | 72B | 73A | 73B | 73C | 73D |
| | 2 | | | | 27 | |
| 19 | 3 | | | | | |
| | 1 | | | | | |
| 2,002 | 217 | 10 | 4 | 11 | 57 | 1 |
| 1 | 7 | 4 | | | 4 | |
| | | | | | | |
| | | | | | | |
| 21,677 | 1,483 | 646 | 165 | 313 | 1,466 | 89 |
| (*) | (*) | | | 7 | 205 | |
| 9 | 104 | 20 | 1 | 9 | 140 | (*) |
| | 48 | 189 | | | 1 | |
| (*) | 8 | 15 | (*) | 1 | 7 | |
| 1 | 97 | 339 | | | 77 | (*) |
| 5 | 531 | 359 | 1 | 1 | 10 | 18 |
| 34 | 5 | 25 | 9 | 31 | 28 | 1 |
| 2 | (*) | 1 | 1 | 3 | 3 | (*) |
| 452 | 295 | 304 | 365 | 639 | 898 | 52 |
| 22 | 6 | 22 | 11 | 36 | 34 | 9 |
| 16 | 34 | 22 | 17 | 31 | 78 | 3 |
| 1,284 | 332 | 689 | 2,067 | 1,518 | 2,955 | 688 |
| 24 | 5 | 234 | 3 | 14 | 774 | 3 |
| 111 | 212 | | | | 291 | |
| | | 34 | | | | |
| | | | | | 47 | |
| 36 | 242 | 898 | 9 | 37 | 717 | 1 |
| | | 1 | | | 137 | 1 |
| 389 | 242 | 310 | 121 | 458 | 690 | 46 |
| 540 | 772 | 1,107 | 991 | 282 | 1,753 | 19 |
| 10 | 9 | 691 | 4 | 37 | 20 | 1 |
| 9 | 479 | | 3 | 34 | 181 | |
| 31 | 30 | 246 | 1 | 1 | 60 | (*) |
| 2 | 1 | 4 | 1 | 4 | 18 | 1 |
| | | 29 | | | | |
| | | | | | 246 | |
| 61 | | | | | | |
| | | | | | 152 | |
| 71 | 26 | 61 | 26 | 106 | 214 | 4 |
| | | 184 | | | 101 | |
| 4 | | | | | 510 | |
| | | | | | 258 | |
| 1 | 2 | 13 | 2 | 7 | 584 | 1 |
| 2 | (*) | 1 | 1 | 3 | 274 | (*) |

TABLE 4.4A   Continued

| | | | |
|---|---|---|---|
| 49 | General industrial machinery and equipment | ............ | ............ |
| 50 | Miscellaneous machinery, except electrical | 1 | ............ |
| 51 | Computer and office equipment | 10 | ............ |
| 52 | Service industry machinery | ............ | ............ |
| 53 | Electrical industrial equipment and apparatus | (*) | ............ |
| 54 | Household appliances | (*) | ............ |
| 55 | Electric lighting and wiring equipment | 6 | ............ |
| 56 | Audio, video, and communication equipment | 24 | ............ |
| 57 | Electronic components and accessories | 8 | ............ |
| 58 | Miscellaneous electrical machinery and supplies | 170 | ............ |
| 59A | Motor vehicles (passenger cars and trucks) | ............ | ............ |
| 59B | Truck and bus bodies, trailers, and motor vehicles parts | 42 | ............ |
| 60 | Aircraft and parts | ............ | ............ |
| 61 | Other transportation equipment | 2 | ............ |
| 62 | Scientific and controlling instruments | 5 | ............ |
| 63 | Ophthalmic and photographic equipment | 254 | ............ |
| 64 | Miscellaneous manufacturing | 141 | ............ |
| 65A | Railroads and related services; passenger ground transportation | 325 | 8 |
| 65B | Motor freight transportation and warehousing | 406 | 12 |
| 65C | Water transportation | 7 | 3 |
| 65D | Air transportation | 502 | 1 |
| 65E | Pipelines, freight forwarders, and related services | 69 | ............ |
| 66 | Communications, except radio and TV | 2,405 | ............ |
| 67 | Radio and TV broadcasting | ............ | ............ |
| 68A | Electric services (utilities) | 98 | ............ |
| 68B | Gas production and distribution (utilities) | 4 | ............ |
| 68C | Water and sanitary services | 298 | ............ |
| 69A | Wholesale trade | 277 | 144 |
| 69B | Retail trade | 46 | 256 |
| 70A | Finance | 5,832 | 1,619 |
| 70B | Insurance | 54,111 | 9,705 |
| 71A | Owner-occupied dwellings | ............ | ............ |
| 71B | Real estate and royalties | 4,830 | 12,218 |
| 72A | Hotels and lodging places | 1,047 | ............ |
| 72B | Personal and repair services (except auto) | 192 | ............ |
| 73A | Computer and data processing services | 1,153 | ............ |
| 73B | Legal, engineering, accounting, and related services | 2,823 | 1,657 |
| 73C | Other business and professional services, except medical | 1,963 | 1,347 |
| 73D | Advertising | 2,025 | ............ |
| 74 | Eating and drinking places | 2,856 | ............ |
| 75 | Automotive repair and services | 799 | ............ |
| 76 | Amusements | 66 | ............ |
| 77A | Health services | ............ | ............ |
| 77B | Educational and social services, and membership organizations | 63 | ............ |
| 78 | Federal Government enterprises | 688 | ............ |
| 79 | State and local government enterprises | 25 | ............ |
| 80 | Noncomparable imports | 618 | ............ |
| 81 | Scrap, used and secondhand goods | ............ | ............ |
| 82 | General government industry | ............ | ............ |
| 83 | Rest-of-the-world adjustment to final uses | ............ | ............ |
| 84 | Household industry | ............ | ............ |
| 85 | Inventory valuation adjustment | ............ | ............ |
| I | Total intermediate inputs | 86,428 | 46,111 |
| VA | Value added | 86,422 | 279,033 |
| T | Total industry output | 172,850 | 325,144 |

* Less than $500,000.

TABLE 4.4A

| | | | | | | |
|---|---|---|---|---|---|---|
| 24 | … | … | … | … | 724 | … |
| 3 | 3 | 53 | 5 | 44 | 262 | 1 |
| 21 | 6 | 464 | 972 | 259 | 354 | 9 |
| … | … | 157 | … | … | 253 | … |
| … | … | 39 | 88 | … | 559 | … |
| 59 | 7 | 658 | … | … | 29 | … |
| 64 | 123 | 27 | 6 | 66 | 180 | 1 |
| 22 | 4 | 24 | 11 | 17 | 95 | 1 |
| … | … | 1,608 | 1,746 | 15 | 3,335 | … |
| 138 | 3 | 28 | 441 | 339 | 636 | 17 |
| 48 | 16 | 31 | 15 | 178 | 113 | 10 |
| … | 6 | … | … | … | 264 | … |
| 3 | 34 | 26 | 37 | 6 | 136 | 2 |
| 91 | 18 | 600 | 29 | 718 | 1,397 | 10 |
| 134 | 116 | 1,412 | 25 | 378 | 353 | 20 |
| 445 | 106 | 78 | 62 | 130 | 366 | 28 |
| 595 | 173 | 303 | 113 | 224 | 683 | 37 |
| 14 | 10 | 21 | 4 | 22 | 126 | 2 |
| 424 | 46 | 187 | 551 | 1,673 | 3,276 | 43 |
| 115 | 49 | 4 | 3 | 10 | 15 | 1 |
| 1,539 | 587 | 981 | 1,571 | 2,204 | 3,403 | 210 |
| … | … | … | … | … | … | 453 |
| 308 | 1,278 | 1,192 | 295 | 467 | 1,274 | 43 |
| 60 | 604 | 476 | 39 | 85 | 418 | 10 |
| 75 | 372 | 144 | 14 | 169 | 84 | 3 |
| 493 | 558 | 1,517 | 904 | 1,199 | 3,037 | 70 |
| 229 | 50 | 83 | 41 | 223 | 258 | 18 |
| 6,294 | 1,931 | 844 | 543 | 1,412 | 2,512 | 154 |
| 6,322 | 47 | 72 | 36 | 316 | 748 | 19 |
| 32,320 | 1,523 | 3,767 | 2,051 | 8,826 | 6,374 | 1,069 |
| 910 | 28 | 84 | 87 | 3,128 | 1,050 | 191 |
| 763 | 684 | 1,436 | 56 | 557 | 277 | 46 |
| 74 | 18 | 356 | 7,058 | 8,284 | 6,035 | 21 |
| 2,190 | 630 | 2,765 | 491 | 15,381 | 5,285 | 200 |
| 7,363 | 2,690 | 2,246 | 1,674 | 16,011 | 14,020 | 437 |
| 6,263 | 658 | 1,492 | 533 | 493 | 2,320 | 92 |
| 3,338 | 50 | 462 | 473 | 1,142 | 2,537 | 199 |
| 1,387 | 541 | 870 | 464 | 2,717 | 2,412 | 318 |
| 72 | 17 | 19 | 11 | 121 | 152 | 123 |
| 194 | 130 | 898 | 344 | 921 | 1,339 | 37 |
| 467 | 189 | 305 | 259 | 1,357 | 1,559 | 63 |
| 112 | 233 | 122 | 20 | 114 | 118 | 23 |
| 56 | 60 | 10 | 179 | 465 | 928 | 24 |
| … | … | … | … | … | … | … |
| … | … | … | … | … | … | … |
| … | … | … | … | … | … | … |
| … | … | … | … | … | … | … |
| 99,839 | 18,787 | 32,319 | 25,051 | 73,250 | 82,309 | 4,941 |
| 280,436 | 22,211 | 33,983 | 35,770 | 104,682 | 138,418 | 10,942 |
| 380,275 | 40,997 | 66,302 | 60,821 | 177,931 | 220,728 | 15,884 |

TABLE 4.4A  Continued

| Eating and drinking places | Automotive repair and services | Amuse- ments | Health services | Educational and social services, and mem- bership or- ganizations | Federal Govern- ment enter- prises | State and local government enterprises |
|---|---|---|---|---|---|---|
| 74 | 75 | 76 | 77A | 77B | 78 | 79 |
| 391 | | 20 | 62 | 83 | 4 | |
| 1,446 | | 8 | 121 | 95 | 14 | 1 |
| 1,673 | | 4 | 26 | 31 | 16 | |
| 9 | 16 | 571 | 158 | 485 | 3 | 125 |
| | | | | | | |
| 5 | 6 | 2 | 6 | 8 | 1,061 | 1,062 |
| | | | | | | |
| | | | | 3 | | 25 |
| | | | | | | |
| 1,836 | 654 | 1,017 | 2,742 | 7,154 | 571 | 16,093 |
| | | (*) | 2 | 5 | (*) | 1 |
| 47,037 | 7 | 655 | 3,035 | 2,970 | 393 | 4 |
| | | | | | | |
| | | 69 | | 46 | 14 | |
| 25 | 16 | 9 | 16 | 10 | 4 | 10 |
| | 169 | 138 | 374 | 243 | 1 | 30 |
| 17 | (*) | 91 | 776 | 69 | 78 | 6 |
| 27 | 9 | 209 | 46 | 200 | | |
| (*) | 1 | 10 | 4 | 1 | | |
| 735 | 189 | 152 | 1,542 | 1,529 | 32 | 50 |
| 869 | 16 | 6 | 124 | 120 | 15 | (*) |
| 6 | 43 | 28 | 180 | 578 | 3 | 6 |
| 294 | 80 | 387 | 2,619 | 9,745 | 383 | 162 |
| 62 | 58 | 167 | 6,516 | 226 | 8 | 1,137 |
| | | 9 | | 62 | | 119 |
| | | | | 11 | | |
| | | 11 | 6,464 | 127 | (*) | 3 |
| 117 | 10 | | 839 | 189 | 29 | 52 |
| | 785 | 3 | | 39 | | 1 |
| 128 | 3,663 | 104 | 922 | 984 | 641 | 4,249 |
| 1,461 | 1,331 | 386 | 7,834 | 1,035 | 49 | 115 |
| 66 | 5 | 53 | 12 | 36 | 14 | 2 |
| 339 | 847 | 3 | 789 | 252 | 3 | 9 |
| 37 | 37 | 2 | 289 | 5 | | 142 |
| 1 | (*) | 1 | 5 | 18 | | (*) |
| 21 | | 28 | | (*) | 1 | |
| | | | | | | |
| | | | | 4 | 1 | 2 |
| 9 | 3,331 | 30 | 244 | 167 | 90 | 3 |
| 52 | 4,815 | 59 | 176 | 289 | 8 | 31 |
| | 245 | | | | | 61 |
| | | 1 | | (*) | | 211 |
| | | | | | 4 | |
| 1 | 18 | 18 | 8 | 11 | 6 | 15 |
| 53 | 13 | 1 | 3 | 1 | | |
| 92 | 53 | 1 | 1 | | 17 | 13 |

TABLE 4.4A

| General government industry | Household industry | Inventory valuation adjustment | Total inter-mediate use | Personal consump-tion expend-itures | Gross private fixed investment | Change in business inventories | Commodity number |
|---|---|---|---|---|---|---|---|
| 82 | 84 | 85 | | 91 | 92 | 93 | |
| .......... | .......... | .......... | 81,465 | 3,090 | .......... | -719 | 1 |
| .......... | .......... | .......... | 59,031 | 15,682 | .......... | -4,261 | 2 |
| .......... | .......... | .......... | 10,351 | 3,763 | .......... | 101 | 3 |
| .......... | .......... | .......... | 21,754 | 647 | .......... | .......... | 4 |
| .......... | .......... | .......... | 7,268 | .......... | 446 | 19 | 5+6 |
| .......... | .......... | .......... | 21,498 | 138 | .......... | 1,100 | 7 |
| .......... | .......... | .......... | 97,326 | .......... | 84 | -1,758 | 8 |
| .......... | .......... | .......... | 11,974 | 36 | .......... | -8 | 9+10 |
| .......... | .......... | .......... | 44 | .......... | 358,627 | .......... | 11 |
| .......... | .......... | .......... | 124,100 | .......... | 17,300 | .......... | 12 |
| .......... | .......... | .......... | 1,304 | 1,099 | 198 | 457 | 13 |
| .......... | .......... | .......... | 125,260 | 201,153 | .......... | 1,771 | 14 |
| .......... | .......... | .......... | 3,664 | -20,774 | .......... | 242 | 15 |
| .......... | .......... | .......... | 35,123 | 1,047 | .......... | 599 | 16 |
| .......... | .......... | .......... | 8,549 | 4,992 | 2,369 | 412 | 17 |
| .......... | .......... | .......... | 14,438 | 71,153 | .......... | 1,446 | 18 |
| .......... | .......... | .......... | 9,369 | 10,088 | .......... | 333 | 19 |
| .......... | .......... | .......... | 68,635 | 1,820 | 3,920 | 1,157 | 20+21 |
| .......... | .......... | .......... | 3,867 | 19,469 | 15,467 | 596 | 22+23 |
| .......... | .......... | .......... | 69,529 | 11,902 | .......... | 916 | 24 |
| .......... | .......... | .......... | 24,501 | 292 | .......... | 127 | 25 |
| .......... | .......... | .......... | 2,547 | 11,741 | .......... | 449 | 26A |
| .......... | .......... | .......... | 49,322 | 10,923 | .......... | 1,188 | 26B |
| .......... | .......... | .......... | 79,565 | 978 | 795 | 515 | 27A |
| .......... | .......... | .......... | 12,543 | 784 | .......... | 138 | 27B |
| .......... | .......... | .......... | 39,534 | .......... | .......... | 502 | 28 |
| .......... | .......... | .......... | 11,676 | 23,958 | .......... | 1,199 | 29A |
| .......... | .......... | .......... | 6,952 | 25,019 | .......... | 558 | 29B |
| .......... | .......... | .......... | 11,365 | 194 | .......... | 197 | 30 |
| .......... | .......... | .......... | 70,488 | 60,189 | .......... | 3,001 | 31 |
| .......... | .......... | .......... | 78,219 | 11,669 | 155 | 1,292 | 32 |
| .......... | .......... | .......... | 3,578 | 13,619 | .......... | 467 | 33+34 |
| .......... | .......... | .......... | 15,366 | 1,518 | .......... | 179 | 35 |
| .......... | .......... | .......... | 43,245 | 2,705 | .......... | 606 | 36 |
| .......... | .......... | .......... | 74,182 | 11 | 13 | 1,204 | 37 |
| .......... | .......... | .......... | 57,761 | 72 | 36 | 864 | 38 |
| .......... | .......... | .......... | 11,619 | .......... | 21 | 24 | 39 |
| .......... | .......... | .......... | 39,226 | 525 | 2,811 | 557 | 40 |
| .......... | .......... | .......... | 29,847 | 1,464 | .......... | 237 | 41 |
| .......... | .......... | .......... | 43,780 | 3,600 | 1,945 | 604 | 42 |
| .......... | .......... | .......... | 8,316 | 461 | 2,302 | 208 | 43 |
| .......... | .......... | .......... | 6,177 | 248 | 16,909 | 333 | 44+45 |
| .......... | .......... | .......... | 2,749 | .......... | 5,032 | 42 | 46 |
| .......... | .......... | .......... | 10,100 | 583 | 13,439 | 50 | 47 |
| .......... | .......... | .......... | 3,126 | 176 | 15,053 | 198 | 48 |
| .......... | .......... | .......... | 14,038 | .......... | 11,072 | 153 | 49 |

**TABLE 4.4A**   Continued

| | | | | | | |
|---|---|---|---|---|---|---|
| 115 | 1,252 | 14 | 19 | 32 | 14 | 429 |
| 3 | 11 | 7 | 43 | 107 | 3 | 2 |
| 15 | 675 | 55 | 9 | 11 | 22 | 19 |
| ......... | 173 | (*) | 1 | 8 | 2 | 504 |
| 1 | ......... | 1 | 8 | 13 | 3 | 50 |
| 87 | 955 | 66 | 308 | 202 | 26 | 105 |
| 7 | 181 | 4 | 7 | 122 | 6 | 1 |
| ......... | ......... | 7 | 22 | 190 | 6 | 39 |
| 4 | 532 | 36 | 310 | 167 | 38 | 113 |
| 23 | 9,976 | 10 | 108 | 119 | 267 | 194 |
| ......... | 34 | 103 | 5 | 10 | 33 | 129 |
| ......... | 36 | 13 | 6,698 | 124 | 4 | 17 |
| 7 | 41 | 186 | 683 | 498 | 13 | 15 |
| 293 | 47 | 183 | 261 | 767 | 53 | 43 |
| 385 | 294 | 93 | 337 | 237 | 724 | 440 |
| 1,541 | 802 | 188 | 842 | 791 | 1,616 | 309 |
| 62 | 112 | 41 | 84 | 41 | 77 | 200 |
| 171 | 486 | 234 | 955 | 2,803 | 926 | 98 |
| 2 | 92 | 2 | 20 | 22 | 10 | 75 |
| 804 | 1,021 | 658 | 2,774 | 1,762 | 146 | 234 |
| 4,445 | 1,006 | 1,100 | 3,172 | 2,038 | 281 | 4,483 |
| 364 | 520 | 220 | 1,499 | 964 | 83 | 3,082 |
| 329 | 82 | 125 | 289 | 255 | 83 | 303 |
| 9,302 | 4,700 | 498 | 6,618 | 2,462 | 376 | 1,566 |
| 27 | 4,855 | 45 | 288 | 131 | 10 | 23 |
| 2,236 | 4,201 | 718 | 1,449 | 2,085 | 45 | 403 |
| 33 | 2,263 | 77 | 1,223 | 715 | 43 | 492 |
| 8,896 | 4,643 | 4,145 | 21,791 | 16,283 | 529 | 686 |
| 28 | 153 | 230 | 466 | 1,052 | 26 | 51 |
| 569 | 1,790 | 412 | 986 | 409 | 9 | 17 |
| 216 | 18 | 327 | 4,764 | 1,402 | 30 | 260 |
| 2,226 | 764 | 2,138 | 2,169 | 2,319 | 79 | 2,202 |
| 4,140 | 4,027 | 3,869 | 9,636 | 5,121 | 451 | 856 |
| 4,629 | 1,517 | 2,605 | 912 | 3,256 | 3 | 104 |
| 790 | 865 | 578 | 1,939 | 1,163 | 77 | 113 |
| 404 | 2,503 | 476 | 3,374 | 1,300 | 520 | 180 |
| 965 | 18 | 15,940 | 87 | 969 | 38 | 4 |
| | | | 6,761 | | | |
| 299 | 168 | 379 | 718 | 355 | 13 | 81 |
| 118 | 528 | 365 | 1,866 | 2,088 | 381 | 70 |
| 187 | 374 | 85 | 226 | 227 | 25 | 5 |
| 73 | 12 | 154 | 45 | 855 | 1,085 | ......... |
| ......... | 169 | ......... | ......... | ......... | ......... | ......... |
| ......... | ......... | ......... | ......... | ......... | ......... | ......... |
| ......... | ......... | ......... | ......... | ......... | ......... | ......... |
| ......... | ......... | ......... | ......... | ......... | ......... | ......... |
| 100,603 | 68,309 | 40,640 | 119,710 | 80,088 | 11,636 | 41,734 |
| 108,791 | 62,395 | 37,552 | 218,801 | 72,590 | 33,760 | 27,750 |
| 209,394 | 130,704 | 78,192 | 338,511 | 152,678 | 45,396 | 69,484 |

## TABLE 4.4A

| | | | | | | | |
|---|---|---|---|---|---|---|---|
| | | | 14,927 | 117 | 747 | 101 | 50 |
| | | | 15,346 | 3,290 | 33,476 | 331 | 51 |
| | | | 12,881 | 883 | 7,186 | 306 | 52 |
| | | | 17,719 | 161 | 5,878 | 110 | 53 |
| | | | 2,743 | 11,997 | 2,657 | 3 | 54 |
| | | | 15,549 | 2,278 | 435 | 608 | 55 |
| | | | 10,561 | 18,387 | 21,728 | 446 | 56 |
| | | | 42,174 | 263 | | 787 | 57 |
| | | | 12,838 | 5,277 | 2,755 | 361 | 58 |
| | | | 2,370 | 101,875 | 62,933 | 8,115 | 59A |
| | | | 61,127 | 3,133 | 6,591 | 1,745 | 59B |
| | | | 22,583 | 316 | 8,843 | 2,132 | 60 |
| | | | 2,763 | 11,043 | 3,183 | 1,070 | 61 |
| | | | 17,485 | 4,456 | 33,814 | 1,285 | 62 |
| | | | 7,770 | 4,625 | 5,653 | 398 | 63 |
| | | | 9,277 | 27,179 | 3,876 | 2,181 | 64 |
| | | | 27,231 | 13,080 | 827 | 553 | 65A |
| | | | 80,137 | 20,258 | 2,343 | 755 | 65B |
| | | | 8,029 | 4,177 | 167 | 49 | 65C |
| | | | 36,314 | 31,439 | 819 | 97 | 65D |
| | | | 18,525 | 2,553 | | 39 | 65E |
| | | | 81,923 | 61,963 | 4,389 | | 66 |
| | | | 924 | 1,326 | | | 67 |
| | | | 79,596 | 63,318 | | | 68A |
| | | | 55,987 | 25,544 | | | 68B |
| | | | 12,350 | 14,864 | | | 68C |
| | | | 210,780 | 111,741 | 39,161 | 4,929 | 69A |
| | | | 37,597 | 373,725 | 11,178 | | 69B |
| | | | 121,959 | 135,789 | | | 70A |
| | | | 93,582 | 81,638 | | | 70B |
| | | | | 325,144 | | | 71A |
| | | | 225,105 | 122,178 | 23,701 | | 71B |
| | | | 18,052 | 20,180 | | | 72A |
| | | | 17,186 | 48,030 | | | 72B |
| | | | 53,578 | 855 | 10 | 39 | 73A |
| | | | 127,255 | 31,456 | 7,509 | | 73B |
| | | | 100,951 | 12,802 | | 59 | 73C |
| | | | 107,841 | 661 | | | 73D |
| | | | 43,381 | 169,638 | | | 74 |
| | | | 61,098 | 67,684 | | 7 | 75 |
| | | | 30,722 | 47,411 | | | 76 |
| | | | 7,505 | 363,015 | | | 77A |
| | | | 12,086 | 148,974 | | | 77B |
| | | | 24,980 | 6,430 | | | 78 |
| | | | 4,424 | 14,152 | | | 79 |
| | | | 39,151 | 29,295 | | 85 | 80 |
| | | | 6,460 | 13,705 | -24,960 | 1,969 | 81 |
| | | | | | | | 82 |
| | | | | -31,136 | | | 83 |
| | | | | 7,709 | | | 84 |
| | | | | | | -17,817 | 85 |
| | | | 3,602,186 | | | | I |
| 466,785 | 7,709 | -17,817 | | | | | VA |
| 466,785 | 7,709 | -17,817 | | 3,072,252 | 732,891 | 28,037 | T |

TABLE 4.4A   Continued

| Commodity number | For the distribution of output of a commodity, read the row for that commodity / For the composition of inputs to an industry, read the column for that industry | Exports of goods and services | Imports of goods and services |
|---|---|---|---|
| | Industry number | 94 | 95 |
| 1 | Livestock and livestock products | 485 | −808 |
| 2 | Other agricultural products | 12,747 | −2,353 |
| 3 | Forestry and fishery products | 544 | −3,747 |
| 4 | Agricultural, forestry, and fishery services | 122 | −16 |
| 5+6 | Metallic ores mining | 559 | −1,349 |
| 7 | Coal mining | 2,663 | −65 |
| 8 | Crude petroleum and natural gas | 1,494 | −28,965 |
| 9+10 | Nonmetallic minerals mining | 633 | −734 |
| 11 | New construction | 15 | ................ |
| 12 | Maintenance and repair construction | 81 | ................ |
| 13 | Ordnance and accessories | 2,725 | −467 |
| 14 | Food and kindred products | 12,111 | −18,538 |
| 15 | Tobacco products | 2,591 | −879 |
| 16 | Broad and narrow fabrics, yarn and thread mills | 1,407 | −3,601 |
| 17 | Miscellaneous textile goods and floor coverings | 782 | −919 |
| 18 | Apparel | 1,197 | −25,395 |
| 19 | Miscellaneous fabricated textile products | 362 | −1,772 |
| 20+21 | Lumber and wood products | 3,645 | −6,399 |
| 22+23 | Furniture and fixtures | 684 | −5,287 |
| 24 | Paper and allied products, except containers | 5,922 | −9,914 |
| 25 | Paperboard containers and boxes | 262 | −126 |
| 26A | Newspapers and periodicals | 555 | −226 |
| 26B | Other printing and publishing | 1,062 | −1,335 |
| 27A | Industrial and other chemicals | 14,630 | −10,727 |
| 27B | Agricultural fertilizers and chemicals | 542 | −990 |
| 28 | Plastics and synthetic materials | 5,364 | −2,009 |
| 29A | Drugs | 2,959 | −7,590 |
| 29B | Cleaning and toilet preparations | 983 | −1,281 |
| 30 | Paints and allied products | 342 | −214 |
| 31 | Petroleum refining and related products | 6,128 | −13,332 |
| 32 | Rubber and miscellaneous plastics products | 3,233 | −9,702 |
| 33+34 | Footwear, leather, and leather products | 666 | −9,700 |
| 35 | Glass and glass products | 777 | −1,837 |
| 36 | Stone and clay products | 1,019 | −4,513 |
| 37 | Primary iron and steel manufacturing | 1,407 | −10,824 |
| 38 | Primary nonferrous metals manufacturing | 3,303 | −6,992 |
| 39 | Metal containers | 166 | −155 |
| 40 | Heating, plumbing, and fabricated structural metal products | 869 | −961 |
| 41 | Screw machine products and stampings | 2,123 | −2,261 |
| 42 | Other fabricated metal products | 2,634 | −6,573 |
| 43 | Engines and turbines | 2,899 | −2,102 |
| 44+45 | Farm, construction, and mining machinery | 6,063 | −5,402 |
| 46 | Materials handling machinery and equipment | 540 | −1,321 |
| 47 | Metalworking machinery and equipment | 2,335 | −4,911 |
| 48 | Special industry machinery and equipment | 2,696 | −4,993 |
| 49 | General industrial machinery and equipment | 4,182 | −6,947 |

TABLE 4.4A

| Federal Government purchases | | | State and local government purchases | | | GDP | Total commodity output |
|---|---|---|---|---|---|---|---|
| Total | National defense | Non-defense | Total | Education | Other | | |
| | 96 | 97 | | 98 | 99 | | |
| 12 | 2 | 10 | 84 | 30 | 54 | 2,144 | 83,609 |
| 750 | ............. | 750 | 587 | 220 | 368 | 23,152 | 82,183 |
| -1,112 | ............. | -1,112 | -413 | 6 | -419 | -864 | 9,488 |
| 121 | 38 | 84 | 1,040 | 284 | 756 | 1,914 | 23,668 |
| -141 | -142 | 1 | ............. | ............. | ............. | -466 | 6,802 |
| 86 | 56 | 29 | 32 | 21 | 11 | 3,953 | 25,451 |
| -173 | 5 | -177 | ............. | ............. | ............. | -29,318 | 68,008 |
| 2 | -2 | 4 | -19 | ............. | -19 | -90 | 11,884 |
| 15,550 | 7,495 | 8,055 | 71,111 | 10,091 | 61,020 | 445,303 | 445,347 |
| 6,258 | 4,358 | 1,900 | 25,728 | 5,912 | 19,816 | 49,367 | 173,466 |
| 22,745 | 20,365 | 2,380 | 117 | 1 | 116 | 26,873 | 28,177 |
| 2,025 | 189 | 1,836 | 5,854 | 3,739 | 2,114 | 204,376 | 329,636 |
| ............. | ............. | ............. | -11 | -1 | -10 | 22,717 | 26,381 |
| 115 | 105 | 10 | 142 | 58 | 84 | -291 | 34,832 |
| 30 | 3 | 27 | 51 | 8 | 43 | 7,717 | 16,266 |
| 567 | 566 | 1 | 853 | 15 | 838 | 49,821 | 64,259 |
| 185 | 140 | 45 | 441 | 95 | 347 | 9,637 | 19,006 |
| 45 | 32 | 13 | 113 | 59 | 55 | 4,301 | 72,936 |
| 129 | 39 | 90 | 1,775 | 992 | 784 | 32,833 | 36,700 |
| 366 | 124 | 243 | 2,240 | 1,183 | 1,057 | 11,432 | 80,961 |
| 74 | 42 | 31 | 158 | 32 | 125 | 786 | 25,288 |
| 153 | 15 | 138 | 456 | 302 | 154 | 13,128 | 15,674 |
| 1,097 | 428 | 669 | 4,870 | 3,403 | 1,466 | 17,804 | 67,126 |
| 1,893 | 1,793 | 100 | 2,203 | 567 | 1,635 | 10,286 | 89,852 |
| 25 | 9 | 15 | 324 | 111 | 213 | 823 | 13,365 |
| 13 | 13 | 1 | 2 | 1 | 1 | 3,872 | 43,407 |
| 795 | 472 | 324 | 2,005 | 88 | 2,778 | 24,186 | 35,862 |
| 202 | 160 | 42 | 439 | 119 | 321 | 25,920 | 32,872 |
| 6 | 1 | 5 | 294 | 228 | 66 | 818 | 12,183 |
| 3,193 | 2,649 | 545 | 7,931 | 3,789 | 4,142 | 67,111 | 137,599 |
| 636 | 480 | 157 | 1,348 | 88 | 1,260 | 8,631 | 86,851 |
| 50 | 47 | 4 | 106 | (*) | 105 | 5,209 | 8,787 |
| 59 | 22 | 37 | 273 | 66 | 207 | 968 | 16,335 |
| 108 | 51 | 57 | 170 | 66 | 104 | 95 | 43,340 |
| 151 | 78 | 72 | 57 | 6 | 51 | -7,982 | 66,201 |
| 644 | 395 | 250 | 58 | 4 | 54 | -2,014 | 55,746 |
| 57 | 57 | ............. | 7 | 4 | 3 | 120 | 11,739 |
| 658 | 491 | 168 | 1 | ............. | 1 | 4,460 | 43,686 |
| 138 | 110 | 28 | 277 | 216 | 61 | 1,978 | 31,826 |
| 481 | 407 | 75 | 550 | 177 | 373 | 3,241 | 47,022 |
| 2,045 | 1,879 | 167 | 265 | ............. | 265 | 6,078 | 14,394 |
| 321 | 303 | 18 | 1,358 | 84 | 1,274 | 19,829 | 26,005 |
| 321 | 312 | 9 | 13 | 4 | 9 | 4,627 | 7,376 |
| 220 | 180 | 40 | 236 | 134 | 102 | 11,951 | 22,051 |
| 82 | 76 | 6 | 74 | 69 | 5 | 13,285 | 16,411 |
| 560 | 542 | 18 | 162 | (*) | 162 | 9,183 | 23,221 |

**TABLE 4.4A** Continued

| | | | |
|---|---|---|---:|
| 50 | Miscellaneous machinery, except electrical | 1,660 | −604 |
| 51 | Computer and office equipment | 13,167 | −17,329 |
| 52 | Service industry machinery | 1,217 | −1,504 |
| 53 | Electrical industrial equipment and apparatus | 1,847 | −3,346 |
| 54 | Household appliances | 943 | −2,950 |
| 55 | Electric lighting and wiring equipment | 1,358 | −3,341 |
| 56 | Audio, video, and communication equipment | 4,137 | −20,190 |
| 57 | Electronic components and accessories | 12,596 | −13,704 |
| 58 | Miscellaneous electrical machinery and supplies | 2,404 | −4,511 |
| 59A | Motor vehicles (passenger cars and trucks) | 12,918 | −61,157 |
| 59B | Truck and bus bodies, trailers, and motor vehicles parts | 10,874 | −16,950 |
| 60 | Aircraft and parts | 22,891 | −6,875 |
| 61 | Other transportation equipment | 1,278 | −2,937 |
| 62 | Scientific and controlling instruments | 10,311 | −9,990 |
| 63 | Ophthalmic and photographic equipment | 2,224 | −5,696 |
| 64 | Miscellaneous manufacturing | 2,831 | −15,769 |
| 65A | Railroads and related services; passenger ground transportation | 3,377 | −135 |
| 65B | Motor freight transportation and warehousing | 4,606 | |
| 65C | Water transportation | 7,512 | 3,264 |
| 65D | Air transportation | 11,216 | −5,711 |
| 65E | Pipelines, freight forwarders, and related services | 1,958 | |
| 66 | Communications, except radio and TV | 2,496 | |
| 67 | Radio and TV broadcasting | | |
| 68A | Electric services (utilities) | 134 | −986 |
| 68B | Gas production and distribution (utilities) | 161 | −1,763 |
| 68C | Water and sanitary services | 37 | |
| 69A | Wholesale trade | 26,294 | 15,533 |
| 69B | Retail trade | 85 | |
| 70A | Finance | 12,598 | −161 |
| 70B | Insurance | 2,906 | −3,078 |
| 71A | Owner-occupied dwellings | | |
| 71B | Real estate and royalties | 10,830 | |
| 72A | Hotels and lodging places | 49 | |
| 72B | Personal and repair services (except auto) | 31 | |
| 73A | Computer and data processing services | 928 | −104 |
| 73B | Legal, engineering, accounting, and related services | 2,398 | −391 |
| 73C | Other business and professional services, except medical | 1,546 | −740 |
| 73D | Advertising | 475 | −253 |
| 74 | Eating and drinking places | 271 | |
| 75 | Automotive repair and services | 31 | |
| 76 | Amusements | 1,222 | −64 |
| 77A | Health services | 16 | |
| 77B | Educational and social services, and membership organizations | 144 | −9 |
| 78 | Federal Government enterprises | 169 | |
| 79 | State and local government enterprises | | |
| 80 | Noncomparable imports | | −78,696 |
| 81 | Scrap, used and secondhand goods | 4,267 | −2,068 |
| 82 | General government industry | | |
| 83 | Rest-of-the-world adjustment to final uses | 31,653 | |
| 84 | Household industry | | |
| 85 | Inventory valuation adjustment | | |
| I | Total intermediate inputs | | |
| VA | Value added | | |
| T | **Total Industry output** | **348,572** | **−490,442** |

* Less than $500,000.

*Source:* U.S. Department of Commerce, Bureau of Economic Analysis, "Benchmark Input–Output Accounts of the United States, 1987." Washington, D.C.: U.S. Government Printing Office, November 1994.

## TABLE 4.4A

| | | | | | | | |
|---|---|---|---|---|---|---|---|
| 2,657 | 2,573 | 84 | 251 | 116 | 135 | 4,929 | 19,855 |
| 4,168 | 3,493 | 675 | 1,982 | 1,196 | 786 | 39,085 | 54,431 |
| 123 | 101 | 21 | 655 | 476 | 179 | 8,865 | 21,746 |
| 636 | 467 | 168 | 273 | 140 | 132 | 5,557 | 23,277 |
| 42 | 38 | 4 | 235 | 101 | 134 | 12,927 | 15,670 |
| 62 | 41 | 21 | 472 | 284 | 189 | 1,871 | 17,421 |
| 5,265 | 4,964 | 301 | 703 | 335 | 368 | 30,476 | 41,037 |
| 5,912 | 5,884 | 28 | 174 | 81 | 94 | 6,029 | 48,203 |
| 1,671 | 1,577 | 95 | 191 | 71 | 120 | 8,149 | 20,987 |
| 740 | 609 | 131 | 5,714 | 1,288 | 4,426 | 131,139 | 133,509 |
| 1,093 | 1,050 | 43 | 716 | 294 | 422 | 7,200 | 68,327 |
| 34,512 | 33,306 | 1,206 | 20 | ............... | 20 | 61,838 | 84,421 |
| 7,160 | 6,754 | 405 | 515 | 103 | 411 | 21,311 | 24,074 |
| 25,249 | 23,710 | 1,540 | 3,249 | 517 | 2,732 | 68,373 | 85,858 |
| 1,129 | 704 | 425 | 2,604 | 1,015 | 1,589 | 10,937 | 18,707 |
| -424 | 123 | -547 | 1,932 | 1,184 | 747 | 21,805 | 31,083 |
| 810 | 320 | 490 | 2,653 | 2,256 | 397 | 21,164 | 48,394 |
| 5,150 | 2,729 | 2,421 | 1,944 | 897 | 1,048 | 35,056 | 115,194 |
| 834 | 697 | 137 | 167 | 35 | 133 | 16,169 | 24,198 |
| 2,837 | 2,159 | 678 | 2,048 | 956 | 1,093 | 42,745 | 79,060 |
| 80 | 42 | 38 | 146 | 40 | 106 | 4,776 | 23,301 |
| 3,853 | 1,954 | 1,899 | 5,540 | 2,519 | 3,021 | 78,241 | 160,164 |
| ............... | ............... | ............... | ............... | ............... | ............... | 1,326 | 2,250 |
| 2,671 | 1,734 | 937 | 11,720 | 4,541 | 7,180 | 76,857 | 156,453 |
| 579 | 465 | 114 | 1,672 | 685 | 987 | 26,192 | 82,180 |
| 236 | 184 | 53 | 983 | 1,018 | -35 | 16,120 | 28,469 |
| 5,860 | 5,039 | 821 | 9,454 | 3,940 | 5,514 | 212,971 | 423,751 |
| 114 | 92 | 22 | 263 | -41 | 304 | 385,364 | 422,960 |
| 1,400 | ............... | 1,400 | 9,366 | ............... | 9,366 | 158,991 | 280,950 |
| 1,746 | 36 | 1,710 | 827 | 666 | 161 | 84,039 | 177,621 |
| ............... | ............... | ............... | ............... | ............... | ............... | 325,144 | 325,144 |
| 1,243 | 547 | 696 | 6,563 | 713 | 5,850 | 164,515 | 389,620 |
| 947 | 692 | 254 | 836 | -1,237 | 2,073 | 22,012 | 40,064 |
| 125 | 65 | 60 | 898 | 301 | 597 | 49,085 | 66,271 |
| 4,133 | 2,833 | 1,300 | 5,248 | 1,402 | 3,845 | 11,110 | 64,687 |
| 8,099 | 7,561 | 538 | 1,656 | 1,729 | -73 | 50,727 | 177,982 |
| 20,907 | 15,944 | 4,963 | 11,433 | 4,136 | 7,298 | 45,807 | 211,758 |
| 85 | 79 | 5 | 597 | 416 | 181 | 1,564 | 109,406 |
| 1,139 | 371 | 768 | -3,409 | -5,546 | 2,138 | 167,639 | 211,021 |
| 185 | 94 | 92 | 2,223 | 823 | 1,400 | 70,129 | 131,228 |
| 1,001 | 823 | 179 | -1,452 | 189 | -1,641 | 48,119 | 78,841 |
| 613 | -352 | 965 | -32,757 | -10 | -32,747 | 330,888 | 338,393 |
| 7,326 | 1,127 | 6,199 | -16,252 | -15,934 | -318 | 140,184 | 152,270 |
| 409 | 312 | 97 | 1,482 | 128 | 1,354 | 8,490 | 33,469 |
| 111 | 80 | 31 | 359 | 171 | 187 | 14,621 | 19,045 |
| 10,116 | 8,673 | 1,443 | 49 | 43 | 6 | -39,151 | ............... |
| 675 | -104 | 778 | 2,272 | 683 | 1,589 | -4,139 | 2,321 |
| 150,627 | 108,244 | 42,383 | 316,158 | 173,286 | 142,873 | 466,785 | 466,785 |
| -517 | -161 | -356 | ............... | ............... | ............... | ............... | ............... |
| ............... | ............... | ............... | ............... | ............... | ............... | 7,709 | 7,709 |
| ............... | ............... | ............... | ............... | ............... | ............... | -17,817 | -17,817 |
| ............... | ............... | ............... | ............... | ............... | ............... | 4,572,829 | ............... |
| 384,927 | 292,052 | 92,875 | 496,592 | 218,272 | 278,320 | ............... | 8,175,016 |

**TABLE 4.4B** Input Components of Total Industry Output (millions of dollars at producers' prices)

| Industry number | | Value added |
|---|---|---|
| | | Total |
| 1 | Livestock and livestock products | 15,074 |
| 2 | Other agricultural products | 46,721 |
| 3 | Forestry and fishery products | 3,708 |
| 4 | Agricultural, forestry, and fishery services | 9,948 |
| 5+6 | Metallic ores mining | 3,476 |
| 7 | Coal mining | 15,488 |
| 8 | Crude petroleum and natural gas | 55,484 |
| 9+10 | Nonmetallic minerals mining | 8,213 |
| 11+12 | Construction | 291,000 |
| 13 | Ordnance and accessories | 18,928 |
| 14 | Food and kindred products | 100,498 |
| 15 | Tobacco products | 16,795 |
| 16 | Broad and narrow fabrics, yarn and thread mills | 12,140 |
| 17 | Miscellaneous textile goods and floor coverings | 4,354 |
| 18 | Apparel | 27,003 |
| 19 | Miscellaneous fabricated textile products | 6,915 |
| 20+21 | Lumber and wood products | 25,923 |
| 22+23 | Furniture and fixtures | 17,259 |
| 24 | Paper and allied products, except containers | 34,278 |
| 25 | Paperboard containers and boxes | 8,806 |
| 26A | Newspapers and periodicals | 29,037 |
| 26B | Other printing and publishing | 45,145 |
| 27A | Industrial and other chemicals | 37,277 |
| 27B | Agricultural fertilizers and chemicals | 3,364 |
| 28 | Plastics and synthetic materials | 14,365 |
| 29A | Drugs | 22,172 |
| 29B | Cleaning and toilet preparations | 17,646 |
| 30 | Paints and allied products | 5,568 |
| 31 | Petroleum refining and related products | 24,258 |
| 32 | Rubber and miscellaneous plastics products | 37,624 |
| 33+34 | Footwear, leather, and leather products | 3,681 |
| 35 | Glass and glass products | 8,210 |
| 36 | Stone and clay products | 20,563 |
| 37 | Primary iron and steel manufacturing | 25,370 |
| 38 | Primary nonferrous metals manufacturing | 14,213 |
| 39 | Metal containers | 3,421 |
| 40 | Heating, plumbing, and fabricated structural metal products | 19,001 |
| 41 | Screw machine products and stampings | 14,187 |
| 42 | Other fabricated metal products | 22,269 |
| 43 | Engines and turbines | 6,226 |
| 44+45 | Farm, construction, and mining machinery | 11,852 |
| 46 | Materials handling machinery and equipment | 3,309 |
| 47 | Metalworking machinery and equipment | 12,470 |
| 48 | Special industry machinery and equipment | 8,595 |
| 49 | General industrial machinery and equipment | 12,400 |
| 50 | Miscellaneous machinery, except electrical | 11,839 |
| 51 | Computer and office equipment | 24,195 |

TABLE 4.4B

| Value added | | | Total inter-<br>mediate inputs | Total industry<br>output | Industry<br>number |
|---|---|---|---|---|---|
| Compensation<br>of employees | Indirect<br>business tax<br>and nontax<br>liability | Other value<br>added | | | |
| 3,284 | 1,091 | 10,700 | 72,410 | 87,484 | 1 |
| 5,619 | 2,536 | 38,566 | 40,021 | 86,742 | 2 |
| 779 | 158 | 2,771 | 3,748 | 7,456 | 3 |
| 9,941 | .................. | 7 | 12,253 | 22,201 | 4 |
| 1,836 | 501 | 1,139 | 3,331 | 6,807 | 5+6 |
| 8,383 | 2,033 | 5,072 | 9,964 | 25,452 | 7 |
| 11,699 | 3,939 | 39,847 | 28,744 | 84,228 | 8 |
| 4,008 | 688 | 3,518 | 4,751 | 12,964 | 9+10 |
| 189,998 | 4,487 | 96,515 | 327,813 | 618,813 | 11+12 |
| 12,370 | 235 | 6,323 | 12,510 | 31,438 | 13 |
| 43,805 | 7,225 | 49,468 | 225,473 | 325,972 | 14 |
| 2,853 | 4,701 | 9,242 | 9,588 | 26,383 | 15 |
| 8,413 | 235 | 3,491 | 26,104 | 38,244 | 16 |
| 2,729 | 102 | 1,523 | 11,628 | 15,982 | 17 |
| 17,503 | 239 | 9,262 | 37,181 | 64,184 | 18 |
| 4,048 | 83 | 2,784 | 10,072 | 16,987 | 19 |
| 16,168 | 1,251 | 8,503 | 46,952 | 72,875 | 20+21 |
| 11,412 | 230 | 5,617 | 19,518 | 36,777 | 22+23 |
| 16,521 | 1,345 | 16,412 | 47,704 | 81,982 | 24 |
| 6,370 | 205 | 2,231 | 16,705 | 25,511 | 25 |
| 15,391 | 255 | 13,392 | 20,689 | 49,727 | 26A |
| 27,499 | 957 | 16,689 | 42,232 | 87,378 | 26B |
| 15,582 | 1,824 | 19,871 | 47,098 | 84,375 | 27A |
| 1,950 | 226 | 1,188 | 10,148 | 13,512 | 27B |
| 6,560 | 766 | 7,040 | 26,308 | 40,672 | 28 |
| 8,292 | 152 | 13,728 | 13,840 | 36,012 | 29A |
| 5,308 | 184 | 12,155 | 15,583 | 33,229 | 29B |
| 2,505 | 32 | 0,031 | 6,504 | 12,072 | 30 |
| 6,857 | 10,590 | 6,812 | 113,613 | 137,871 | 31 |
| 23,433 | 1,672 | 12,519 | 47,948 | 85,572 | 32 |
| 2,362 | 28 | 1,292 | 5,018 | 8,700 | 33+34 |
| 4,875 | 218 | 3,117 | 7,875 | 16,085 | 35 |
| 11,952 | 803 | 7,807 | 23,169 | 43,732 | 36 |
| 17,894 | 1,183 | 6,293 | 42,721 | 68,091 | 37 |
| 10,442 | 590 | 3,182 | 42,163 | 56,376 | 38 |
| 2,019 | 85 | 1,318 | 8,483 | 11,904 | 39 |
| 12,772 | 492 | 5,737 | 24,930 | 43,930 | 40 |
| 11,245 | 512 | 2,430 | 17,787 | 31,973 | 41 |
| 14,716 | 501 | 7,053 | 22,155 | 44,424 | 42 |
| 3,973 | 175 | 2,077 | 7,870 | 14,096 | 43 |
| 7,478 | 449 | 3,925 | 14,902 | 26,753 | 44+45 |
| 2,409 | 66 | 835 | 3,884 | 7,194 | 46 |
| 9,843 | 275 | 2,353 | 8,756 | 21,227 | 47 |
| 6,147 | 163 | 2,285 | 7,659 | 16,254 | 48 |
| 8,544 | 262 | 3,595 | 10,836 | 23,236 | 49 |
| 9,391 | 250 | 2,198 | 8,164 | 20,003 | 50 |
| 13,585 | 440 | 10,170 | 31,625 | 55,819 | 51 |

**TABLE 4.4B**   Continued

| | | |
|---|---|---:|
| 52 | Service industry machinery | 10,422 |
| 53 | Electrical industrial equipment and apparatus | 11,609 |
| 54 | Household appliances | 6,594 |
| 55 | Electric lighting and wiring equipment | 9,083 |
| 56 | Audio, video, and communication equipment | 20,337 |
| 57 | Electronic components and accessories | 26,895 |
| 58 | Miscellaneous electrical machinery and supplies | 9,452 |
| 59A | Motor vehicles (passenger cars and trucks) | 25,004 |
| 59B | Truck and bus bodies, trailers, and motor vehicles parts | 26,270 |
| 60 | Aircraft and parts | 39,508 |
| 61 | Other transportation equipment | 11,396 |
| 62 | Scientific and controlling instruments | 49,114 |
| 63 | Ophthalmic and photographic equipment | 11,264 |
| 64 | Miscellaneous manufacturing | 15,742 |
| 65A | Railroads and related services; passenger ground transportation | 26,684 |
| 65B | Motor freight transportation and warehousing | 64,722 |
| 65C | Water transportation | 7,647 |
| 65D | Air transportation | 35,205 |
| 65E | Pipelines, freight forwarders, and related services | 15,309 |
| 66 | Communications, except radio and TV | 94,949 |
| 67 | Radio and TV broadcasting | 13,460 |
| 68A | Electric services (utilities) | 85,706 |
| 68B | Gas production and distribution (utilities) | 26,325 |
| 68C | Water and sanitary services | 3,786 |
| 69A | Wholesale trade | 297,947 |
| 69B | Retail trade | 293,322 |
| 70A | Finance | 144,596 |
| 70B | Insurance | 86,422 |
| 71A | Owner-occupied dwellings | 279,033 |
| 71B | Real estate and royalties | 280,436 |
| 72A | Hotels and lodging places | 22,211 |
| 72B | Personal and repair services (except auto) | 33,983 |
| 73A | Computer and data processing services | 35,770 |
| 73B | Legal, engineering, accounting, and related services | 104,682 |
| 73C | Other business and professional services, except medical | 138,418 |
| 73D | Advertising | 10,942 |
| 74 | Eating and drinking places | 108,791 |
| 75 | Automotive repair and services | 62,395 |
| 76 | Amusements | 37,552 |
| 77A | Health services | 218,801 |
| 77B | Educational and social services, and membership organizations | 72,590 |
| 78 | Federal Government enterprises | 33,760 |
| 79 | State and local government enterprises | 27,750 |
| 82 | General government industry | 466,785 |
| 84 | Household industry | 7,709 |
| 85 | Inventory valuation adjustment | −17,817 |
| T | **Total** | **4,572,829** |

* Less than $500,000.

TABLE 4.4B

| | | | | | |
|---:|---:|---:|---:|---:|---:|
| 6,580 | 151 | 3,691 | 11,987 | 22,409 | 52 |
| 7,919 | 243 | 3,447 | 11,056 | 22,665 | 53 |
| 3,660 | 127 | 2,807 | 8,767 | 15,361 | 54 |
| 5,249 | 158 | 3,675 | 8,532 | 17,615 | 55 |
| 11,383 | 358 | 8,596 | 20,363 | 40,700 | 56 |
| 18,527 | 852 | 7,517 | 21,758 | 48,654 | 57 |
| 6,579 | 235 | 2,639 | 11,371 | 20,823 | 58 |
| 15,227 | 2,108 | 7,669 | 109,111 | 134,115 | 59A |
| 19,067 | 1,597 | 5,607 | 42,721 | 68,991 | 59B |
| 30,002 | 614 | 8,892 | 42,620 | 82,128 | 60 |
| 8,713 | 117 | 2,566 | 12,687 | 24,082 | 61 |
| 33,494 | 955 | 14,665 | 36,349 | 85,463 | 62 |
| 4,276 | 220 | 6,768 | 8,462 | 19,725 | 63 |
| 8,637 | 309 | 6,796 | 17,347 | 33,089 | 64 |
| 18,648 | 1,581 | 6,456 | 16,774 | 43,458 | 65A |
| 40,701 | 3,083 | 20,938 | 51,373 | 116,095 | 65B |
| 5,732 | 687 | 1,229 | 16,406 | 24,053 | 65C |
| 23,231 | 5,749 | 6,225 | 41,048 | 76,253 | 65D |
| 7,945 | 642 | 6,722 | 10,599 | 25,908 | 65E |
| 36,761 | 11,910 | 46,278 | 66,178 | 161,127 | 66 |
| 9,886 | 600 | 2,975 | 15,936 | 29,396 | 67 |
| 19,453 | 9,242 | 57,012 | 46,665 | 132,371 | 68A |
| 8,626 | 3,151 | 14,548 | 41,224 | 67,549 | 68B |
| 3,510 | 584 | −309 | 7,477 | 11,262 | 68C |
| 174,697 | 57,724 | 65,525 | 125,804 | 423,751 | 69A |
| 187,889 | 53,073 | 52,360 | 127,371 | 420,694 | 69B |
| 109,452 | 8,317 | 26,827 | 142,016 | 286,613 | 70A |
| 62,328 | 12,429 | 11,666 | 86,428 | 172,850 | 70B |
| ..................... | 50,971 | 228,062 | 46,111 | 325,144 | 71A |
| 27,230 | 53,227 | 199,979 | 99,839 | 380,275 | 71B |
| 10,663 | 3,698 | 7,850 | 18,787 | 40,997 | 72A |
| 21,130 | 1,187 | 11,666 | 32,319 | 66,302 | 72B |
| 25,443 | 655 | 9,673 | 25,051 | 60,821 | 73A |
| 79,014 | 818 | 24,850 | 73,250 | 177,931 | 73B |
| 92,121 | 3,952 | 42,345 | 82,309 | 220,728 | 73C |
| 7,404 | 126 | 3,412 | 4,941 | 15,884 | 73D |
| 81,909 | 9,606 | 17,276 | 100,603 | 209,394 | 74 |
| 28,995 | 4,301 | 29,099 | 68,309 | 130,704 | 75 |
| 24,710 | 2,857 | 9,986 | 40,640 | 78,192 | 76 |
| 178,143 | 1,901 | 38,757 | 119,710 | 338,511 | 77A |
| 68,100 | 418 | 4,072 | 80,088 | 152,678 | 77B |
| 31,077 | ..................... | 2,683 | 11,636 | 45,396 | 78 |
| 19,296 | 26 | 8,428 | 41,734 | 69,484 | 79 |
| 466,785 | ..................... | ..................... | ..................... | 466,785 | 82 |
| 7,709 | ..................... | ..................... | ..................... | 7,709 | 84 |
| ..................... | ..................... | −17,817 | ..................... | −17,817 | 85 |
| **2,698,657** | **364,986** | **1,509,186** | **3,602,186** | **8,175,016** | T |

TABLE 4.10  Commodity-by-Industry Direct Requiremenets (direct requirements per dollar of industry output at producers' prices)

| Commodity number | For the composition of inputs to an industry, read the column for that industry | Livestock and livestock products | Other agricultural products |
|---|---|---|---|
| | Industry number | 1 | 2 |
| 1 | Livestock and livestock products | 0.19224 | 0.01826 |
| 2 | Other agricultural products | .27179 | .04444 |
| 3 | Forestry and fishery products | | |
| 4 | Agricultural, forestry, and fishery services | .04576 | .07542 |
| 5+6 | Metallic ores mining | | |
| 7 | Coal mining | | |
| 8 | Crude petroleum and natural gas | | |
| 9+10 | Nonmetallic minerals mining | .00007 | .00292 |
| 11 | New construction | | |
| 12 | Maintenance and repair construction | .00524 | .00819 |
| 13 | Ordnance and accessories | | |
| 14 | Food and kindred products | .13220 | |
| 15 | Tobacco products | | |
| 16 | Broad and narrow fabrics, yarn and thread mills | | .00051 |
| 17 | Miscellaneous textile goods and floor coverings | .00029 | .00030 |
| 18 | Apparel | | |
| 19 | Miscellaneous fabricated textile products | | .00102 |
| 20+21 | Lumber and wood products | .00041 | .00340 |
| 22+23 | Furniture and fixtures | | |
| 24 | Paper and allied products, except containers | .00125 | .00160 |
| 25 | Paperboard containers and boxes | .00006 | .00384 |
| 26A | Newspapers and periodicals | .00010 | .00012 |
| 26B | Other printing and publishing | .00011 | .00012 |
| 27A | Industrial and other chemicals | .00117 | .00074 |
| 27B | Agricultural fertilizers and chemicals | .00162 | .05311 |
| 28 | Plastics and synthetic materials | | |
| 29A | Drugs | .00227 | |
| 29B | Cleaning and toilet preparations | .00062 | |
| 30 | Paints and allied products | | |
| 31 | Petroleum refining and related products | .00383 | .01355 |
| 32 | Rubber and miscellaneous plastics products | .00186 | .00412 |
| 33+34 | Footwear, leather, and leather products | .00029 | (*) |
| 35 | Glass and glass products | .00007 | |
| 36 | Stone and clay products | | .00115 |
| 37 | Primary iron and steel manufacturing | .00015 | .00018 |
| 38 | Primary nonferrous metals manufacturing | | |
| 39 | Metal containers | | |
| 40 | Heating, plumbing, and fabricated structural metal products | .00019 | .00022 |
| 41 | Screw machine products and stampings | .00031 | |
| 42 | Other fabricated metal products | .00078 | .00176 |
| 43 | Engines and turbines | | |
| 44+45 | Farm, construction, and mining machinery | .00285 | .00764 |
| 46 | Materials handling machinery and equipment | | |
| 47 | Metalworking machinery and equipment | .00095 | .00106 |
| 48 | Special industry machinery and equipment | | |
| 49 | General industrial machinery and equipment | .00031 | .00058 |

TABLE 4.10

| Forestry and fishery products | Agricultural, forestry, and fishery services | Metallic ores mining | Coal mining | Crude petroleum and natural gas | Non-metallic minerals mining | Construction | Ordnance and accessories | Food and kindred products |
|---|---|---|---|---|---|---|---|---|
| 3 | 4 | 5+6 | 7 | 8 | 9+10 | 11+12 | 13 | 14 |
| 0.00355 | 0.05633 | | | | | | | 0.18658 |
| | .09409 | | | | | .00039 | | .06829 |
| .02250 | .00144 | | | | | | | .00624 |
| .17281 | .00037 | .00006 | .00002 | .00002 | .00014 | .00525 | .00005 | .00002 |
| | | .07622 | | | | | | |
| | | .00154 | .10724 | | .00473 | (*) | .00014 | .00032 |
| | | | | .03738 | | | | |
| | .00010 | .00106 | .00127 | | .03484 | .00781 | | .00003 |
| | | | | | | .00007 | | |
| .01111 | .01299 | .01285 | .00766 | .02189 | .00895 | .00055 | .00660 | .00249 |
| .00394 | | | | | | .00002 | .02860 | |
| .04090 | .00150 | .00018 | .00002 | .00004 | .00015 | | .00001 | .16779 |
| | | .00009 | .00070 | | .00027 | | .00013 | |
| .00960 | .00515 | | | | | .00284 | .00004 | .00004 |
| | | .00006 | .00031 | .00005 | .00027 | .00019 | .00039 | .00003 |
| .00414 | .00355 | | | (*) | | .00036 | .00001 | .00022 |
| | | .00557 | .00242 | (*) | .00019 | .05417 | .00085 | .00019 |
| | | | | | | .00205 | | |
| .00013 | .00047 | .00012 | .00033 | .00009 | .00292 | .00191 | .00023 | .00964 |
| .00017 | .00834 | .00003 | | .00002 | .00031 | .00007 | .00068 | .01797 |
| .00007 | .00014 | .00004 | .00004 | .00003 | .00184 | .00008 | .00011 | .00003 |
| .00539 | .00109 | .00047 | .00046 | .00038 | .00059 | .00033 | .00069 | .00577 |
| .00181 | .00033 | .02993 | .00540 | .00994 | .01700 | .00212 | .00402 | .00462 |
| .00535 | .13389 | .00013 | .00008 | | .00003 | .00001 | .00018 | .00058 |
| | | | | | | | .00097 | .00037 |
| | .00005 | | | | | | | .00273 |
| | | | | .00008 | .00012 | .00002 | .00005 | .00047 |
| | | | | .00007 | | .00758 | .00013 | |
| .00056 | | | | | | | | |
| .03948 | .00968 | .01861 | .01534 | .00343 | .01495 | .01813 | .00099 | .00113 |
| .00026 | .00160 | .01205 | .00969 | .00031 | .00942 | .01079 | .00631 | .01614 |
| | .00011 | | .00001 | .00001 | | .00005 | .00002 | (*) |
| .00024 | .00045 | .00021 | .00001 | .00010 | .00031 | .00161 | .00008 | .01204 |
| | .00055 | .00339 | .00316 | .00315 | .00014 | .05018 | .00160 | .00006 |
| | | .01972 | .00075 | .00309 | .00328 | .01620 | .01504 | .00001 |
| | | .00220 | .00052 | | .00536 | .01001 | .01787 | |
| .00236 | | | | | | | | .02664 |
| | | .00633 | .00259 | .00054 | .00470 | .05064 | | |
| | | .00510 | .00572 | | .00253 | .00045 | .00601 | .00203 |
| .00712 | .00118 | .00125 | .00324 | .00455 | .00288 | .01307 | .01468 | .00313 |
| .00192 | .00196 | .00548 | .00435 | .00026 | .00439 | | .00088 | |
| .00276 | .00352 | .01892 | .04259 | .00283 | .01954 | .00229 | | |
| | | .00507 | .00371 | | .01093 | .00229 | | (*) |
| .00003 | .00007 | .00113 | .00030 | .00091 | .00064 | .00032 | .00276 | .00012 |
| | | | | | | | | .00028 |
| .00109 | .00016 | .01152 | .01680 | .00186 | .01263 | .00245 | .00747 | .00068 |
| .00036 | .00020 | .00181 | .00308 | .00043 | .00143 | .00020 | .00380 | .00028 |

TABLE 4.10   Continued

| | | | |
|---|---|---|---|
| 50 | Miscellaneous machinery, except electrical | .00053 | .00142 |
| 51 | Computer and office equipment | .......... | .......... |
| 52 | Service industry machinery | .......... | .......... |
| 53 | Electrical industrial equipment and apparatus | .00012 | .00031 |
| 54 | Household appliances | .......... | .......... |
| 55 | Electric lighting and wiring equipment | .00022 | .00050 |
| 56 | Audio, video, and communication equipment | .......... | .......... |
| 57 | Electronic components and accessories | .......... | .......... |
| 58 | Miscellaneous electrical machinery and supplies | .00173 | .00466 |
| 59A | Motor vehicles (passenger cars and trucks) | .......... | .......... |
| 59B | Truck and bus bodies, trailers, and motor vehicles parts | .00103 | .00251 |
| 60 | Aircraft and parts | .......... | .......... |
| 61 | Other transportation equipment | .......... | .......... |
| 62 | Scientific and controlling instruments | .......... | .......... |
| 63 | Ophthalmic and photographic equipment | .......... | .......... |
| 64 | Miscellaneous manufacturing | .00017 | .00019 |
| 65A | Railroads and related services; passenger ground transportation | .00942 | .00287 |
| 65B | Motor freight transportation and warehousing | .02188 | .01487 |
| 65C | Water transportation | .00123 | .00053 |
| 65D | Air transportation | .00023 | .00115 |
| 65E | Pipelines, freight forwarders, and related services | .00004 | .00017 |
| 66 | Communications, except radio and TV | .00253 | .00284 |
| 67 | Radio and TV broadcasting | .......... | .......... |
| 68A | Electric services (utilities) | .01111 | .00594 |
| 68B | Gas production and distribution (utilities) | .......... | .00189 |
| 68C | Water and sanitary services | .00127 | .00425 |
| 69A | Wholesale trade | .04413 | .04112 |
| 69B | Retail trade | .00086 | .00262 |
| 70A | Finance | .00910 | .00979 |
| 70B | Insurance | .00501 | .02071 |
| 71A | Owner-occupied dwellings | .......... | .......... |
| 71B | Real estate and royalties | .03429 | .08181 |
| 72A | Hotels and lodging places | .00056 | .00082 |
| 72B | Personal and repair services (except auto) | .00030 | .00081 |
| 73A | Computer and data processing services | .......... | .......... |
| 73B | Legal, engineering, accounting, and related services | .00124 | .00140 |
| 73C | Other business and professional services, except medical | .00414 | .01100 |
| 73D | Advertising | .00021 | .00024 |
| 74 | Eating and drinking places | .00018 | .00020 |
| 75 | Automotive repair and services | .00058 | .00206 |
| 76 | Amusements | .......... | .......... |
| 77A | Health services | .00850 | .......... |
| 77B | Educational and social services, and membership organizations | | |
| 78 | Federal Government enterprises | .00013 | .00015 |
| 79 | State and local government enterprises | .00017 | .00036 |
| 80 | Noncomparable imports | .......... | .00019 |
| 81 | Scrap, used and secondhand goods | .......... | .......... |
| 82 | General government industry | .......... | .......... |
| 83 | Rest-of-the-world adjustment to final uses | .......... | .......... |
| 84 | Household industry | .......... | .......... |
| 85 | Inventory valuation adjustment | .......... | .......... |
| I | Total intermediate inputs | .82769 | .46138 |
| VA | Value added | .17231 | .53862 |
| T | Total | 1.00000 | 1.00000 |

*Less than .000005.

## TABLE 4.10

| | | | | | | | | |
|---|---|---|---|---|---|---|---|---|
| .00036 | .00020 | .00181 | .00308 | .00043 | .00143 | .00020 | .00380 | .00028 |
|  | .00007 |  |  |  |  |  |  |  |
|  |  |  |  |  |  | .01073 |  | .00002 |
|  |  | .00395 | .00352 | .00197 | .00491 | .00443 | .00078 |  |
|  |  |  |  |  |  | .00243 |  |  |
| .00016 | .00034 | .00057 | .00110 | .00022 | .00063 | .01599 | .00005 | .00002 |
|  | .00001 |  | (*) | .00001 |  | .00325 | .02780 | (*) |
|  |  |  |  |  |  |  | .02769 |  |
| .00008 | .00046 | .00081 | .00008 | .00003 | .00042 | .00137 | .00059 | .00001 |
| .00067 | .00179 | .00185 | .00034 | .00008 | .00027 | .00067 | .00004 | .00006 |
|  | .00002 |  |  |  |  |  | .10162 |  |
| .02096 | .00052 | .00071 |  |  |  | (*) |  |  |
| .00207 | .00003 | .00074 | .00011 | .00003 | .00008 | .00230 | .01264 | .00010 |
| .00005 | .00034 | .00013 | .00010 | .00012 | .00013 | .00019 | .00061 | .00006 |
| .00012 | .00064 | .00031 | .00015 | .00006 | .00047 | .00142 | .00021 | .00006 |
| .00098 | .00414 | .00297 | .02561 | .00053 | .00278 | .00252 | .00058 | .00565 |
| .00319 | .00935 | .00768 | .00625 | .00183 | .01402 | .01337 | .00488 | .01574 |
| .00326 | .00146 | .00106 | .00211 | .00122 | .00090 | .00049 | .00006 | .00153 |
| .00090 | .01881 | .00361 | .00073 | .00103 | .00373 | .00138 | .00461 | .00148 |
| .00028 | .00015 | .00024 | .00024 | .00005 | .00020 | .00006 | .00003 | .00001 |
| .00047 | .00002 | .00134 | .00083 | .00143 | .00171 | .00374 | .00460 | .00137 |
| .00046 | .00266 | .09506 | .02401 | .01534 | .05023 | .00169 | .00828 | .00844 |
| .00017 | .00007 | .01087 | .00043 | .00660 | .01683 | .00052 | .00284 | .00410 |
| .00082 |  | .00222 | .00236 | .00156 | .01039 | .00041 | .00098 | .00136 |
| .02497 | .06541 | .01955 | .02830 | .00621 | .02007 | .04277 | .01862 | .05169 |
| .00107 | .00301 | .00121 | .00037 | .00007 | .00024 | .03897 | .00013 | .00016 |
| .01037 | .00464 | .01012 | .00663 | .00265 | .01783 | .01147 | .00185 | .00288 |
| .01369 | .00361 | .00342 | .00122 | .00022 | .00049 | .00318 | .00102 | .00109 |
|  | .00806 | .01249 | .02444 | .17382 | .01185 | .00438 | .00840 | .00266 |
| .00182 | .00108 | .00179 | .00020 | .00032 | .00817 | .00114 | .00087 | .00064 |
| .00093 | .00819 | .00029 | .00017 | .00012 | .00156 | .00022 | .00039 | .00059 |
| .00220 | .00404 | .00488 | .00011 | .00006 | .00889 | .00001 | .00047 | .00038 |
| .03376 | .01096 | .01663 | .00813 | .00951 | .00737 | .05866 | .01034 | .00266 |
| .01776 | .00927 | .00558 | .00322 | .00171 | .00867 | .02416 | .01102 | .00553 |
| .00043 | .00414 | .00075 | .00179 | .01012 | .00144 | .00039 | .01958 | .02656 |
| .00146 | .00362 | .00118 | .00073 | .00082 | .00174 | .00233 | .00280 | .00110 |
| .01761 | .02348 | .02743 | .00775 | .00226 | .00134 | .00989 | .00088 | .00146 |
|  | .01269 |  | .00002 | .00003 | .00026 | .00007 | .00006 | .00002 |
| .00286 | .00182 | .00206 | .00166 | .00034 | .00113 | .00004 | .00029 | .00025 |
| .00078 | .00226 | .00143 | .00011 | .00003 | .00131 | .00039 | .00021 | .00072 |
| .00070 | .00078 | .00085 | .00015 | .00004 | .00025 | .00020 | .00018 | .00088 |
|  | .00167 | .00451 | .00072 | .00940 | .00071 | .00001 | .00059 | .01551 |
|  |  |  |  |  |  | .00092 |  |  |
| .50271 | .55191 | .48939 | .39148 | .34126 | .36645 | .52974 | .39794 | .69170 |
| .49729 | .44809 | .51061 | .60852 | .65874 | .63355 | .47026 | .60207 | .30830 |
| 1.00000 | 1.00000 | 1.00000 | 1.00000 | 1.00000 | 1.00000 | 1.00000 | 1.00000 | 1.00000 |

TABLE 4.10  Continued

| Tobacco products | Broad and narrow fabrics, yarn and thread mills | Miscellaneous textile goods and floor coverings | Apparel | Miscellaneous fabricated textile products | Lumber and wood products | Furniture and fixtures | Paper and allied products, except containers | Paperboard containers and boxes |
|---|---|---|---|---|---|---|---|---|
| 15 | 16 | 17 | 18 | 19 | 20+21 | 22+23 | 24 | 25 |
|  | 0.00686 |  | 0.00020 |  |  |  |  |  |
| .06472 | .08346 | .00210 | .00048 |  |  |  |  |  |
|  |  |  | .00460 |  | .08060 |  | .00132 |  |
| .00002 | .00007 | .00003 |  |  | .00030 | .00002 | .00007 | .00001 |
|  |  |  |  |  |  |  | .00018 |  |
| .00056 | .00072 | .00061 | .00006 | .00025 | .00029 | .00019 | .00522 | .00009 |
|  |  |  |  |  |  |  |  |  |
|  |  |  |  |  |  |  | .00370 |  |
| .00146 | .00469 | .00297 | .00201 | .00244 | .00580 | .00785 | .00484 | .00343 |
|  | .00001 |  | (*) |  | .00001 |  |  |  |
| .00004 | .00002 | .00140 | .00002 | .00024 | .00009 | .00040 | .00419 | .00011 |
| .13889 |  |  |  |  |  |  |  |  |
|  | .25878 | .22654 | .20317 | .22810 | .00007 | .03361 | .00826 |  |
|  | .00905 | .04274 | .00030 | .07441 | .00187 | .02645 | .00434 |  |
| .00007 | .00004 | .00066 | .18879 | .00945 | .00017 | .00035 | .00014 | .00005 |
|  |  | .00106 | .02346 | .03248 | .00023 | .00159 | .00001 |  |
| (*) | .00008 | .00008 |  | .00306 | .28756 | .08781 | .05859 |  |
|  |  |  |  |  | .00098 | .00398 | .00001 |  |
| .00498 | .00063 | .00591 | .00174 | .00228 | .00087 | .00138 | .16027 | .45268 |
| .03608 | .00342 | .00617 | .00312 | .01177 | .00383 | .01509 | .01327 | .00209 |
| .00007 | .00005 | .00006 | .00006 | .00009 | .00010 | .00011 | .00005 | .00005 |
| .01429 | .00029 | .00046 | .00084 | .00175 | .00052 | .00089 | .00134 | .00052 |
| .00193 | .02227 | .03719 | .00122 | .00606 | .01164 | .00584 | .04201 | .01924 |
|  |  |  |  |  | .00255 |  | .00250 |  |
|  | .13730 | .23932 | .02378 | .02999 | .00451 | .00309 | .02446 | .02183 |
|  |  |  |  |  |  |  |  |  |
| .00067 | .00230 | .00487 | .00458 |  | .00001 |  | .00471 |  |
|  | .00004 | .00007 | .00001 |  | .00529 | .01022 | .00024 | .00257 |
| .00132 | .00272 | .00196 | .00174 | .00078 | .00612 | .00229 | .00689 | .00713 |
| .00269 | .00519 | .00733 | .00471 | .02922 | .00946 | .03256 | .02438 | .00285 |
|  | .00004 |  | .00487 | .01351 | .00012 | .00282 | .00001 | .00012 |
| .00002 | .00542 | .00036 |  |  | .00289 | .00371 | .00006 | .00001 |
| .00002 | .00010 | .00009 | .00001 | .00007 | .00561 | .00355 | .00110 | .00003 |
| .00003 | .00005 | .00027 | (*) | .00006 | .00042 | .04155 | .00003 | .00265 |
|  |  |  |  |  | .00073 | .01274 | .00057 | .00210 |
| .00009 |  |  |  |  |  |  |  |  |
|  |  |  |  |  | .00657 |  |  |  |
|  |  |  |  |  | .01382 | .01012 |  |  |
| .00702 | .00004 | .00004 | .00012 | .00008 | .02065 | .03748 | .00612 | .00483 |
|  |  |  |  |  |  |  |  |  |
| .00004 | .00032 |  | .00002 |  | .00054 |  | .00001 |  |
| .00023 | .00036 | .00029 | .00014 | .00035 | .00202 | .00215 | .00053 | .00072 |
|  | .00286 | .01399 | .00249 | .00107 | .00113 | .00036 | .00449 | .00391 |

TABLE 4.10

| Newspapers and periodicals | Other printing and publishing | Industrial and other chemicals | Agricultural fertilizers and chemicals | Plastics and synthetic materials | Drugs | Cleaning and toilet preparations | Paints and allied products | Commodity number |
|---|---|---|---|---|---|---|---|---|
| 26A | 26B | 27A | 27B | 28 | 29A | 29B | 30 | |
| | | | | | 0.00237 | 0.00046 | | 1 |
| | | .00102 | | | .00076 | | | 2 |
| | | .00080 | | | .00016 | | .00128 | 3 |
| .00001 | .00002 | .00003 | .00004 | .00002 | .00004 | .00001 | | 4 |
| | | .00754 | | .00015 | | | .00358 | 5+6 |
| .00002 | .00015 | .00323 | .00034 | .00325 | .00024 | .00016 | | 7 |
| | | .01236 | .04084 | .00010 | | | .00104 | 8 |
| | | .00682 | .06855 | | | .00027 | .00141 | 9+10 |
| | | | | | | | | 11 |
| .00375 | .00273 | .00682 | .00409 | .00388 | .00363 | .00227 | .00381 | 12 |
| | | | | | | | | 13 |
| .00006 | .00011 | .00412 | .00684 | .00130 | .00168 | .01680 | .01327 | 14 |
| | | | | | | | | 15 |
| | .00107 | | | .00301 | | | | 16 |
| .00009 | .00061 | | | | | .00054 | | 17 |
| .00002 | .00006 | (*) | .00002 | .00013 | .00001 | .00002 | .00002 | 18 |
| .00001 | .00001 | .00028 | .00007 | .00001 | .00001 | .00006 | | 19 |
| .00004 | | .00057 | | | | .00019 | | 20+21 |
| | | | | | | | | 22+23 |
| .13042 | .16745 | .00855 | .00356 | .01138 | .00448 | .00183 | .00015 | 24 |
| .00024 | .00385 | .00485 | .00410 | .00660 | .00871 | .03448 | | 25 |
| .00890 | .00236 | .00008 | .00004 | .00005 | .00003 | .00005 | .00003 | 26A |
| .06752 | .08073 | .00062 | .01405 | .00036 | .00505 | .01038 | .00216 | 26B |
| .00763 | .02734 | .21601 | .10145 | .33215 | .01937 | .08941 | .19337 | 27A |
| | | .00433 | .18749 | .00232 | .00150 | | | 27B |
| | .00147 | .00898 | | .03615 | | .01409 | .11933 | 28 |
| | | .00161 | | | .10434 | | | 29A |
| | | .00124 | .00187 | .00565 | .00059 | .04618 | .00040 | 29B |
| | .00044 | .00484 | | .00219 | | .00189 | .02083 | 30 |
| .00163 | .00272 | .01199 | .00509 | .00254 | .00107 | .00989 | .00659 | 31 |
| .00066 | .01849 | .01178 | .00853 | .04143 | .02081 | .05155 | .00009 | 32 |
| .00003 | .00015 | .00003 | | (*) | .00001 | | | 33+34 |
| | .00001 | .00157 | .00289 | .00065 | .00648 | .00696 | .00062 | 35 |
| .00002 | .00010 | .00080 | .00192 | .00034 | .00003 | .00010 | .01091 | 36 |
| .00003 | .00007 | .00210 | .00001 | .00001 | | .00005 | .00132 | 37 |
| | .00204 | .00069 | | .00005 | .00003 | | .00058 | 38 |
| | | .00547 | .00409 | .00007 | .00384 | .01304 | .04422 | 39 |
| | | | | | .00033 | .00047 | | 40 |
| | .00001 | .00012 | .00010 | | .00179 | .00591 | .00053 | 41 |
| .00001 | .00061 | .00770 | .00294 | .00032 | .00181 | .01173 | .00175 | 42 |
| | | | | | | | | 43 |
| | | | | | | | | 44+45 |
| | | | | | | | | 46 |
| .00013 | .00026 | .00041 | .00013 | .00051 | .00009 | .00019 | .00007 | 47 |
| .00155 | .00411 | .00486 | | .00021 | | | | 48 |

TABLE 4.10   Continued

| | | | | | | | | |
|---|---|---|---|---|---|---|---|---|
| .00027 | | | .00011 | | .00155 | .00193 | .00068 | |
| .00053 | .00081 | .00061 | .00032 | .00045 | .00261 | .00103 | .00116 | .00105 |
| | | | | | .00031 | | | |
| | | | | | .00061 | .00160 | | |
| | .00001 | | .00001 | .00001 | .00202 | | | |
| .00001 | .00003 | .00001 | .00002 | .00004 | .00157 | .00015 | .00006 | .00002 |
| (*) | .00001 | .00001 | | .00001 | .00001 | .00002 | .00001 | (*) |
| | .00006 | | | | | | | |
| .00002 | .00002 | | (*) | | .00017 | .00002 | .00005 | .00003 |
| .00014 | .00012 | .00003 | .00008 | | .00262 | .00024 | .00037 | .00031 |
| | | | | | .00008 | | | |
| .00017 | .00010 | .00006 | .00006 | .00015 | .00017 | .00039 | .00047 | .00027 |
| .00006 | .00012 | .00013 | .00009 | .00014 | .00019 | .00028 | .00014 | .00011 |
| .00005 | .00008 | .00012 | .00475 | .00620 | .00058 | .00168 | .00011 | .00013 |
| .00039 | .00196 | .00271 | .00028 | .00087 | .00910 | .00434 | .01010 | .01123 |
| .00415 | .00724 | .01488 | .00436 | .00724 | .01408 | .01002 | .01984 | .02236 |
| .00011 | .00023 | .00170 | .00011 | .00042 | .00152 | .00040 | .00131 | .00067 |
| .00110 | .00075 | .00056 | .00177 | .00124 | .00111 | .00227 | .00472 | .00189 |
| .00001 | .00003 | .00003 | .00001 | | .00007 | .00002 | .00011 | .00004 |
| .00107 | .00125 | .00177 | .00148 | .00171 | .00172 | .00242 | .00186 | .00282 |
| .00329 | .02844 | .01354 | .00809 | .00618 | .01433 | .00936 | .02796 | .01087 |
| .00074 | .00397 | .00700 | .00149 | .00234 | .00472 | .00223 | .01212 | .00314 |
| .00032 | .00297 | .00165 | .00052 | .00308 | .00250 | .00178 | .01310 | .00115 |
| .02032 | .04726 | .02708 | .03383 | .05600 | .05223 | .05543 | .04437 | .03917 |
| .00031 | .00028 | .00013 | .00029 | .00004 | .00100 | .00045 | .00097 | .00063 |
| .00497 | .00278 | .00377 | .00438 | .00761 | .00612 | .01032 | .00370 | .00162 |
| .00138 | .00108 | .00149 | .00118 | .00153 | .00269 | .00157 | .00263 | .00183 |
| .00296 | .00185 | .00356 | .00600 | .00603 | .00419 | .00699 | .00270 | .00464 |
| .00031 | .00007 | .00014 | .00024 | .00092 | .00030 | .00051 | .00204 | .00006 |
| .00025 | .00499 | .00327 | .00224 | .00406 | .00045 | .00074 | .00155 | .00044 |
| .00042 | .00080 | .00064 | .00200 | .00222 | .00030 | .00042 | .00224 | .00113 |
| .00230 | .00238 | .00228 | .00273 | .00268 | .00316 | .01146 | .00250 | .00269 |
| .00368 | .01218 | .00619 | .00533 | .00800 | .00749 | .02036 | .00549 | .00723 |
| .02968 | .00593 | .02146 | .01545 | .01187 | .01354 | .01986 | .01045 | .00156 |
| .00072 | .00198 | .00202 | .00297 | .00297 | .00299 | .00436 | .00161 | .00209 |
| .00384 | .00296 | .00124 | .00455 | .00025 | .00689 | .00483 | .00893 | .00721 |
| .00006 | .00002 | .00004 | .00004 | .00002 | .00007 | .00042 | .00003 | .00003 |
| .00044 | .00057 | .00020 | .00040 | .00572 | .00240 | .00170 | .00075 | .00014 |
| .00205 | .00099 | .00150 | .00232 | .00225 | .00097 | .00170 | .00087 | .00085 |
| .00029 | .00047 | .00046 | .00018 | | .00035 | .00037 | .00187 | .00030 |
| .00178 | .00065 | .01039 | .00092 | .00518 | .00018 | .00087 | .00103 | .00007 |
| | | .00033 | | | | | .01011 | |
| | | | | | | | | |
| | | | | | | | | |
| | | | | | | | | |
| .36341 | .68257 | .72757 | .57928 | .59293 | .64428 | .53071 | .58188 | .65481 |
| .63659 | .31743 | .27244 | .42072 | .40707 | .35572 | .46930 | .41812 | .34519 |
| 1.00000 | 1.00000 | 1.00000 | 1.00000 | 1.00000 | 1.00000 | 1.00000 | 1.00000 | 1.00000 |

Table **4.10**

| | | | | | | | | |
|---|---|---|---|---|---|---|---|---|
| .00004 | .00002 | .00025 | .00121 | .00205 | .00047 | .00176 | .00003 | 49 |
| .00026 | .00046 | .00091 | .00039 | .00118 | .00027 | .00076 | .00012 | 50 |
| .00015 | .00053 | .............. | .............. | .............. | .00023 | .............. | .............. | 51 |
| .............. | .............. | .00025 | .............. | .............. | .............. | .............. | .............. | 52 |
| .............. | (*) | .00049 | .............. | .............. | .............. | .............. | .............. | 53 |
| .............. | .............. | .............. | .............. | .............. | .............. | .............. | .............. | 54 |
| .00003 | .00003 | .00004 | .00002 | .00001 | .00002 | .00001 | .00001 | 55 |
| .00002 | .00003 | .00001 | .............. | .00001 | .00001 | .00001 | .............. | 56 |
| .............. | .............. | .............. | .............. | .............. | .............. | .............. | .............. | 57 |
| .00076 | .00033 | .00001 | .00001 | .............. | .00002 | .00001 | .............. | 58 |
| .............. | .............. | .............. | .............. | .............. | .............. | .............. | .............. | 59A |
| .00026 | .00040 | .00004 | .00010 | .00002 | .00013 | .00008 | .00003 | 59B |
| .............. | .............. | .............. | .............. | .............. | .............. | .............. | .............. | 60 |
| .............. | .............. | .............. | .............. | .............. | .............. | .............. | .............. | 61 |
| .00087 | .00197 | .00099 | .00045 | .00013 | .00078 | .00017 | .00019 | 62 |
| .00243 | .00604 | .00024 | .00007 | .00013 | .00014 | .00016 | .00007 | 63 |
| .00027 | .00161 | .00006 | .00003 | .00006 | .00008 | .00134 | .00011 | 64 |
| .00350 | .00493 | .01076 | .02081 | .01168 | .00074 | .00337 | .01310 | 65A |
| .00658 | .01150 | .01771 | .05778 | .01401 | .00330 | .01112 | .02321 | 65B |
| .00009 | .00035 | .00185 | .00363 | .00322 | .00034 | .00115 | .00157 | 65C |
| .01891 | .00422 | .00275 | .00110 | .00145 | .00155 | .00241 | .00088 | 65D |
| .00003 | .00005 | .00049 | .00009 | .00029 | .00003 | .00006 | .00015 | 65E |
| .00415 | .00275 | .00246 | .00291 | .00224 | .00350 | .00239 | .00332 | 66 |
| .............. | .............. | .............. | .............. | .............. | .............. | .............. | .............. | 67 |
| .00495 | .00830 | .03378 | .02246 | .02370 | .00801 | .00568 | .00609 | 68A |
| .00043 | .00109 | .02831 | .03452 | .01895 | .00463 | .00503 | .00354 | 68B |
| .00042 | .00088 | .00377 | .00833 | .00366 | .00095 | .00143 | .00062 | 68C |
| .01572 | .03525 | .03723 | .05528 | .04286 | .03996 | .04764 | .02840 | 69A |
| .00054 | .00079 | .00044 | .00019 | .00024 | .00026 | .00015 | .00002 | 69B |
| .00515 | .00513 | .00324 | .01671 | .00255 | .00517 | .00265 | .00133 | 70A |
| .00141 | .00219 | .00189 | .00192 | .00189 | .00088 | .00155 | .00050 | 70B |
| .............. | .............. | .............. | .............. | .............. | .............. | .............. | .............. | 71A |
| .01452 | .01113 | .00522 | .00275 | .00291 | .00585 | .00451 | .00480 | 71B |
| .00174 | .00397 | .00048 | .00023 | .00075 | .00037 | .00076 | .00056 | 72A |
| .00106 | .00055 | .00045 | .00030 | .00200 | .00132 | .00063 | .00006 | 72B |
| .00907 | .00304 | .00079 | .00041 | .00041 | .00014 | .00020 | .00008 | 73A |
| .01805 | .00705 | .02016 | .00615 | .02551 | .04353 | .00507 | .00447 | 73B |
| .03910 | .01217 | .01379 | .02476 | .00880 | .02974 | .01188 | .00513 | 73C |
| .01747 | .01214 | .01345 | .01435 | .01178 | .01017 | .02566 | .00459 | 73D |
| .00403 | .00610 | .00261 | .00124 | .00139 | .00183 | .00182 | .00145 | 74 |
| .00657 | .01006 | .00137 | .00223 | .00225 | .00307 | .00165 | .00057 | 75 |
| .00011 | .00018 | .00006 | .00002 | .00003 | .00006 | .00005 | .00003 | 76 |
| .............. | .............. | .............. | .............. | .............. | .............. | .............. | .............. | 77A |
| .00156 | .00142 | .00029 | .00076 | .00130 | .00478 | .00114 | .00036 | 77B |
| .01179 | .00680 | .00101 | .00040 | .00047 | .00071 | .00154 | .00066 | 78 |
| .00017 | .00024 | .00080 | .00048 | .00047 | .00047 | .00056 | .00007 | 79 |
| .00111 | .00268 | .00753 | .00748 | .00361 | .02158 | .00562 | .00536 | 80 |
| .............. | .............. | .............. | .00247 | .............. | .............. | .............. | .............. | 81 |
| .............. | .............. | .............. | .............. | .............. | .............. | .............. | .............. | 82 |
| .............. | .............. | .............. | .............. | .............. | .............. | .............. | .............. | 83 |
| .............. | .............. | .............. | .............. | .............. | .............. | .............. | .............. | 84 |
| .............. | .............. | .............. | .............. | .............. | .............. | .............. | .............. | 85 |
| .41606 | .48333 | .55820 | .75105 | .64682 | .38431 | .46895 | .53877 | I |
| .58394 | .51667 | .44180 | .24895 | .35319 | .61569 | .53105 | .46123 | VA |
| 1.00000 | 1.00000 | 1.00000 | 1.00000 | 1.00000 | 1.00000 | 1.00000 | 1.00000 | T |

**TABLE 4.10**   Continued

| Commodity number | For the composition of inputs to an industry, read the column for that industry | Petro-leum refining and re-lated products | Rubber and miscel-laneous plastics products |
|---|---|---|---|
| | Industry number | 31 | 32 |
| 1 | Livestock and livestock products | | |
| 2 | Other agricultural products | | |
| 3 | Forestry and fishery products | | |
| 4 | Agricultural, forestry, and fishery services | .00001 | .00003 |
| 5+6 | Metallic ores mining | | |
| 7 | Coal mining | .00015 | .00029 |
| 8 | Crude petroleum and natural gas | .55103 | .00088 |
| 9+10 | Nonmetallic minerals mining | .00355 | .00041 |
| 11 | New construction | | |
| 12 | Maintenance and repair construction | .00691 | .00458 |
| 13 | Ordnance and accessories | | .00001 |
| 14 | Food and kindred products | .00031 | .00020 |
| 15 | Tobacco products | | |
| 16 | Broad and narrow fabrics, yarn and thread mills | | .00949 |
| 17 | Miscellaneous textile goods and floor coverings | .00029 | .01044 |
| 18 | Apparel | .00001 | .00014 |
| 19 | Miscellaneous fabricated textile products | (*) | .00028 |
| 20+21 | Lumber and wood products | .00043 | .00221 |
| 22+23 | Furniture and fixtures | | |
| 24 | Paper and allied products, except containers | .00007 | .00922 |
| 25 | Paperboard containers and boxes | .00138 | .01154 |
| 26A | Newspapers and periodicals | .00001 | .00010 |
| 26B | Other printing and publishing | .00013 | .00116 |
| 27A | Industrial and other chemicals | .01275 | .04564 |
| 27B | Agricultural fertilizers and chemicals | | |
| 28 | Plastics and synthetic materials | .00044 | .18645 |
| 29A | Drugs | | |
| 29B | Cleaning and toilet preparations | .00303 | .00034 |
| 30 | Paints and allied products | .00005 | .00076 |
| 31 | Petroleum refining and related products | .07204 | .00275 |
| 32 | Rubber and miscellaneous plastics products | .00449 | .04761 |
| 33+34 | Footwear, leather, and leather products | .00002 | .00005 |
| 35 | Glass and glass products | .00201 | .00526 |
| 36 | Stone and clay products | .00037 | .00289 |
| 37 | Primary iron and steel manufacturing | .00031 | .00351 |
| 38 | Primary nonferrous metals manufacturing | | .00134 |
| 39 | Metal containers | .00121 | |
| 40 | Heating, plumbing, and fabricated structural metal products | | .00052 |
| 41 | Screw machine products and stampings | | .00513 |
| 42 | Other fabricated metal products | .00288 | .01005 |
| 43 | Engines and turbines | | .00017 |
| 44+45 | Farm, construction, and mining machinery | | |
| 46 | Materials handling machinery and equipment | | .00006 |
| 47 | Metalworking machinery and equipment | .00014 | .00188 |

TABLE 4.10

| Foot-wear, leather, and leather products | Glass and glass products | Stone and clay products | Primary iron and steel manu-facturing | Primary non-ferrous metals manu-facturing | Metal contain-ers | Heating, plumbing, and fabricated structural metal products | Screw machine products and stamp-ings | Other fabricated metal products |
|---|---|---|---|---|---|---|---|---|
| 33+34 | 35 | 36 | 37 | 38 | 39 | 40 | 41 | 42 |
| ............ | ............ | ............ | ............ | ............ | ............ | ............ | ............ | ............ |
| ............ | ............ | ............ | ............ | ............ | ............ | ............ | ............ | ............ |
| ............ | ............ | ............ | ............ | ............ | ............ | ............ | ............ | ............ |
| ............ | .00003 | .00006 | .00002 | .00003 | .00002 | .00001 | .00001 | .00002 |
| ............ | ............ | .00074 | .02892 | .06975 | ............ | .00029 | ............ | ............ |
| .00013 | .00012 | .00912 | .02128 | .00060 | .00005 | .00012 | .00037 | .00048 |
| ............ | ............ | ............ | .00014 | ............ | ............ | ............ | ............ | ............ |
| .00017 | .01144 | .08032 | .00339 | .00031 | ............ | .00011 | ............ | .00018 |
| .00324 | .00806 | .00751 | .01901 | .00484 | .00276 | .01179 | .01013 | .01243 |
| ............ | ............ | .00131 | .00030 | ............ | ............ | ............ | ............ | ............ |
| .10266 | .00008 | .00057 | .00009 | .00010 | .00003 | .00014 | .00008 | .00010 |
| .02742 | ............ | .00288 | ............ | .00076 | ............ | ............ | ............ | ............ |
| .02265 | ............ | .00003 | ............ | .00002 | ............ | ............ | ............ | .00002 |
| .00043 | .00017 | .00014 | .00012 | .00006 | .00007 | .00003 | .00008 | .00022 |
| ............ | .00002 | .00005 | .00002 | (*) | ............ | (*) | .00323 | ............ |
| .00270 | .01473 | .00218 | .00200 | .00267 | .00096 | .00314 | .00100 | .00398 |
| ............ | ............ | ............ | .00041 | ............ | ............ | ............ | ............ | ............ |
| .00074 | .00095 | .01260 | .00022 | .00030 | .00034 | .00075 | .00098 | .00022 |
| .00762 | .04394 | .00324 | .00117 | .00184 | .00233 | .00525 | .00414 | .00741 |
| .00007 | .00008 | .00010 | .00005 | .00005 | .00003 | .00010 | .00006 | .00008 |
| .00055 | .00153 | .00057 | .00051 | .00044 | .01851 | .00062 | .00048 | .00064 |
| .03114 | .06106 | .02902 | .02530 | .01268 | .00516 | .00270 | .00756 | .01671 |
| .01251 | ............ | .00432 | ............ | .01390 | .00143 | .00075 | .00153 | .00380 |
| .00382 | ............ | .00223 | .00002 | .00001 | .00054 | .00035 | .00047 | .00037 |
| ............ | .00121 | .00159 | .00034 | .00054 | .01646 | .00554 | .00247 | .00770 |
| .00109 | .00281 | .00544 | .00487 | .00646 | .00145 | .00237 | .00126 | .00265 |
| .02878 | .01077 | .00286 | .00204 | .01147 | .00164 | .00967 | .00302 | .01933 |
| .18407 | .00012 | .00001 | .00003 | ............ | ............ | .00002 | ............ | ............ |
| ............ | .08137 | .00162 | .00008 | .00085 | .00002 | .00739 | .00173 | .00257 |
| .00081 | .01987 | .11687 | .01462 | .00466 | .00105 | .00283 | .00238 | .00384 |
| .00008 | .00035 | .00632 | .15029 | .00913 | .21596 | .18879 | .24725 | .11683 |
| ............ | .00008 | .00054 | .02159 | .30618 | .25264 | .07272 | .03065 | .04613 |
| ............ | ............ | .00001 | .00002 | (*) | .01675 | ............ | .00011 | .00018 |
| ............ | ............ | .00016 | ............ | ............ | ............ | .02268 | ............ | .00009 |
| .00108 | .00219 | .00229 | .00403 | .00160 | .00128 | .02753 | .01488 | .01641 |
| .00495 | .00012 | .00502 | .01134 | .00936 | .01338 | .03357 | .01828 | .05522 |
| ............ | ............ | .00004 | .00016 | ............ | ............ | .00003 | ............ | .00073 |
| ............ | ............ | .00029 | ............ | ............ | ............ | ............ | ............ | ............ |
| .00002 | ............ | .00003 | .00024 | .00018 | ............ | ............ | ............ | ............ |
| .00051 | .00541 | .00057 | .00769 | .00865 | .00146 | .00913 | .02978 | .00576 |

**TABLE 4.10**   Continued

| | | | |
|---|---|---:|---:|
| 48 | Special industry machinery and equipment | ............. | .00273 |
| 49 | General industrial machinery and equipment | .00001 | .00034 |
| 50 | Miscellaneous machinery, except electrical | .00030 | .00298 |
| 51 | Computer and office equipment | ............. | ............. |
| 52 | Service industry machinery | ............. | ............. |
| 53 | Electrical industrial equipment and apparatus | ............. | .00015 |
| 54 | Household appliances | ............. | (*) |
| 55 | Electric lighting and wiring equipment | .00009 | .00140 |
| 56 | Audio, video, and communication equipment | (*) | .00001 |
| 57 | Electronic components and accessories | ............. | .00065 |
| 58 | Miscellaneous electrical machinery and supplies | .00001 | .00027 |
| 59A | Motor vehicles (passenger cars and trucks) | ............. | ............. |
| 59B | Truck and bus bodies, trailers, and motor vehicles parts | .00040 | .00015 |
| 60 | Aircraft and parts | ............. | ............. |
| 61 | Other transportation equipment | ............. | ............. |
| 62 | Scientific and controlling instruments | .00017 | .00050 |
| 63 | Ophthalmic and photographic equipment | .00004 | .00022 |
| 64 | Miscellaneous manufacturing | .00002 | .00022 |
| 65A | Railroads and related services; passenger ground transportation | .00111 | .00705 |
| 65B | Motor freight transportation and warehousing | .00408 | .02724 |
| 65C | Water transportation | .00645 | .00134 |
| 65D | Air transportation | .00049 | .00148 |
| 65E | Pipelines, freight forwarders, and related services | .03978 | .00007 |
| 66 | Communications, except radio and TV | .00135 | .00291 |
| 67 | Radio and TV broadcasting | ............. | ............. |
| 68A | Electric services (utilities) | .01199 | .02137 |
| 68B | Gas production and distribution (utilities) | .00914 | .00597 |
| 68C | Water and sanitary services | .00152 | .00236 |
| 69A | Wholesale trade | .04618 | .04952 |
| 69B | Retail trade | .00014 | .00037 |
| 70A | Finance | .00894 | .00451 |
| 70B | Insurance | .00262 | .00212 |
| 71A | Owner-occupied dwellings | ............. | ............. |
| 71B | Real estate and royalties | .00446 | .00629 |
| 72A | Hotels and lodging places | .00019 | .00027 |
| 72B | Personal and repair services (except auto) | .00039 | .00138 |
| 73A | Computer and data processing services | .00092 | .00310 |
| 73B | Legal, engineering, accounting, and related services | .00273 | .00678 |
| 73C | Other business and professional services, except medical | .00838 | .00801 |
| 73D | Advertising | .00233 | .00823 |
| 74 | Eating and drinking places | .00037 | .00287 |
| 75 | Automotive repair and services | .00093 | .00372 |
| 76 | Amusements | .00002 | .00008 |
| 77A | Health services | ............. | ............. |
| 77B | Educational and social services, and membership organizations | .00105 | .00347 |
| 78 | Federal Government enterprises | .00052 | .00101 |
| 79 | State and local government enterprises | .00012 | .00036 |
| 80 | Noncomparable imports | .00284 | .01350 |
| 81 | Scrap, used and secondhand goods | ............. | ............. |
| 82 | General government industry | ............. | ............. |
| 83 | Rest-of-the-world adjustment to final uses | ............. | ............. |
| 84 | Household industry | ............. | ............. |
| 85 | Inventory valuation adjustment | ............. | ............. |
| I | Total intermediate inputs | .82405 | .56032 |
| VA | Value added | .17595 | .43968 |
| T | Total | 1.00000 | 1.00000 |

*Less than .000005.

## TABLE 4.10

| | | | | | | | | |
|---|---|---|---|---|---|---|---|---|
| .00014 | .00067 | | | | | | .00005 | |
| .00002 | .00027 | .00072 | .01275 | .00986 | .00023 | .00460 | .00197 | .00068 |
| .00100 | .00305 | .00110 | .00304 | .00202 | .00170 | .00262 | .01157 | .00421 |
| | | | .00006 | | | | | |
| | | | .00006 | | | .00076 | | |
| | .00208 | .00027 | .00638 | .00583 | .00008 | .00250 | .00155 | .00487 |
| .00008 | .00033 | (*) | (*) | | | | .00002 | |
| .00003 | .00017 | .00073 | .00015 | .00004 | .00001 | .00008 | .00041 | .00006 |
| | .00001 | .00001 | (*) | (*) | .00001 | .00002 | .00001 | .00001 |
| | | .00002 | | | | | | .00024 |
| | .00003 | .00002 | .00005 | .00002 | .00002 | .00003 | | .00037 |
| | | | | | | .00021 | | |
| .00002 | .00022 | .00022 | .00010 | .00021 | .00016 | .00025 | .00065 | .00024 |
| | | | .00006 | | | | | |
| .00013 | .00098 | .00036 | .00031 | .00025 | .00009 | .00217 | .00017 | .00029 |
| .00016 | .00021 | .00024 | .00017 | .00015 | .00010 | .00026 | .00017 | .00023 |
| .00948 | .00010 | .00091 | .00016 | .00009 | .00007 | .00051 | .00013 | .00033 |
| .00212 | .01151 | .01472 | .01687 | .00595 | .00346 | .00311 | .00334 | .00241 |
| .01015 | .01378 | .05778 | .01345 | .03027 | .01593 | .01548 | .01353 | .01241 |
| .00039 | .00073 | .00352 | .00469 | .00101 | .00024 | .00030 | .00040 | .00047 |
| .00275 | .00415 | .00139 | .00114 | .00181 | .00421 | .00292 | .00085 | .00352 |
| .00002 | .00009 | .00006 | .00007 | .00004 | .00002 | .00004 | .00001 | .00004 |
| .00258 | .00760 | .00674 | .00220 | .00196 | .00094 | .00266 | .00266 | .00389 |
| .00782 | .02974 | .02651 | .04131 | .04436 | .01164 | .00861 | .01323 | .01542 |
| .00214 | .03606 | .01804 | .02787 | .01431 | .00587 | .00362 | .00517 | .00590 |
| .00109 | .00210 | .00275 | .00718 | .00169 | .00219 | .00078 | .00247 | .00123 |
| .04246 | .04238 | .02902 | .06277 | .06061 | .06690 | .05642 | .05126 | .05182 |
| .00005 | .00039 | .00065 | .00034 | .00047 | .00029 | .00067 | .00010 | .00046 |
| .00570 | .00429 | .00655 | .00363 | .00473 | .00321 | .00423 | .00642 | .00688 |
| .00109 | .00159 | .00208 | .00202 | .00198 | .00171 | .00191 | .00179 | .00192 |
| .00514 | .00536 | .00497 | .00230 | .00326 | .00449 | .00648 | .00435 | .00497 |
| .00495 | .00090 | .00039 | .00015 | .00064 | .00087 | .00471 | .00034 | .00068 |
| .00138 | .00074 | .00127 | .00111 | .00151 | .00042 | .00045 | .00126 | .00111 |
| .00052 | .00142 | .00203 | .00158 | .00063 | .00087 | .00051 | .00254 | .00235 |
| .00346 | .00328 | .00385 | .00280 | .00248 | .00194 | .00407 | .02243 | .00520 |
| .01032 | .00576 | .00933 | .01830 | .00996 | .00396 | .00931 | .00704 | .01082 |
| .01493 | .02331 | .01977 | .02952 | .00953 | .02021 | .00634 | .00829 | .01849 |
| .00318 | .00232 | .00278 | .00174 | .00162 | .00119 | .00317 | .00215 | .00301 |
| .00131 | .00547 | .00510 | .00152 | .00506 | .00310 | .00559 | .00290 | .00512 |
| .00002 | .00004 | .00004 | .00003 | .00004 | .00002 | .00011 | .00004 | .00005 |
| .00026 | .00052 | .00090 | .00019 | .00066 | .00076 | .00086 | .00305 | .00091 |
| .00358 | .00119 | .00113 | .00127 | .00082 | .00029 | .00106 | .00114 | .00118 |
| .00093 | .00043 | .00025 | .00063 | .00044 | .00050 | .00031 | .00034 | .00064 |
| .00021 | .00538 | .00298 | .00124 | .01399 | .00062 | .00045 | .00064 | .00146 |
| | .00445 | | .03794 | .04215 | | .00093 | | .00076 |
| | | | | | | | | |
| | | | | | | | | |
| | | | | | | | | |
| .57683 | .48960 | .52980 | .62741 | .74788 | .71260 | .56748 | .55629 | .49871 |
| .42317 | .51040 | .47020 | .37259 | .25212 | .28740 | .43252 | .44371 | .50129 |
| 1.00000 | 1.00000 | 1.00000 | 1.00000 | 1.00000 | 1.00000 | 1.00000 | 1.00000 | 1.00000 |

**TABLE 4.10**  Continued

| Engines and turbines | Farm, construction, and mining machinery | Materials handling machinery and equipment | Metalworking machinery and equipment | Special industry machinery and equipment | General industrial machinery and equipment | Miscellaneous machinery, except electrical | Computer and office equipment | Service industry machinery |
|---|---|---|---|---|---|---|---|---|
| 43 | 44+45 | 46 | 47 | 48 | 49 | 50 | 51 | 52 |
| | | | | | | | | |
| | | | | | | | | |
| .00004 | .00006 | .00001 | .00003 | .00002 | .00003 | .00004 | .00002 | .00002 |
| .00028 | .00046 | .00004 | .00027 | .00001 | .00005 | .00006 | | .00010 |
| | | | | | | | | .00080 |
| | | | | | | | | |
| .00655 | .00834 | .01031 | .00598 | .00756 | .00669 | .00961 | .00777 | .00607 |
| .00006 | .00010 | .00028 | .00017 | .00017 | .00010 | .00005 | .00001 | .00008 |
| | | | | | | | | |
| | | | | | .00585 | .00086 | | |
| .00016 | .00005 | .00006 | .00011 | .00006 | .00015 | .00006 | .00003 | .00008 |
| | | | .00001 | | (*) | .00001 | .00001 | (*) |
| | .00173 | .00156 | .00123 | .00168 | .00109 | .00026 | .00002 | .00479 |
| | (*) | .00008 | | | | | .00001 | |
| .00052 | .00026 | .00028 | .00029 | .00026 | .00111 | .00037 | .00141 | .00048 |
| .00169 | .00240 | .00063 | .00390 | .00179 | .00358 | .00320 | .00161 | .00578 |
| .00006 | .00008 | .00011 | .00012 | .00009 | .00010 | .00011 | .00011 | .00007 |
| .00048 | .00066 | .00475 | .00092 | .00076 | .00074 | .00074 | .00127 | .00054 |
| .00027 | .00132 | .00133 | .00789 | .01188 | .00079 | .00097 | .00023 | .00382 |
| | | | | | .00175 | .00180 | .00100 | .00428 |
| | | | | | | | | |
| | | | | | | .00004 | | |
| .00034 | .00273 | .00103 | .00110 | | .00016 | .00014 | .00037 | .00293 |
| .00095 | .00162 | .00252 | .00313 | .00201 | .00191 | .00170 | .00103 | .00109 |
| .00988 | .02718 | .01403 | .00711 | .01579 | .01398 | .00423 | .01873 | .01942 |
| | .00009 | | .00001 | | | .00002 | .00003 | .00001 |
| .00001 | .00002 | | .00001 | .00260 | .00005 | .00004 | .00006 | .00089 |
| .00341 | .00214 | .00175 | .01105 | .00282 | .00426 | .00734 | .00022 | .00436 |
| .13748 | .11169 | .08434 | .06912 | .06255 | .09187 | .06689 | .00601 | .05485 |
| .04628 | .00842 | .01323 | .02012 | .02247 | .02749 | .04547 | .01269 | .05086 |
| | | | | | | | | |
| .01239 | .03727 | .02679 | .01445 | .01522 | .00674 | .01071 | .00312 | .00856 |
| .01655 | .01474 | .01868 | .00841 | .00634 | .00812 | .00942 | .00347 | .02168 |
| .01514 | .01646 | .03618 | .00800 | .01330 | .00925 | .01266 | .00542 | .01735 |
| .09263 | .04669 | .00922 | | .00367 | .00355 | .00060 | | .00133 |
| | .03629 | | | | | | | |
| | .00084 | .05318 | | .00038 | | | | |
| .00648 | .00456 | .00516 | .03991 | .00924 | .00628 | .01416 | .00063 | .00545 |

## TABLE 4.10

| Electrical industrial equipment and apparatus | Household appliances | Electric lighting and wiring equipment | Audio, video, and communication equipment | Electronic components and accessories | Miscellaneous electrical machinery and supplies | Motor vehicles (passenger cars and trucks) | Truck and bus bodies, trailers, and motor vehicles parts | Commodity number |
|---|---|---|---|---|---|---|---|---|
| 53 | 54 | 55 | 56 | 57 | 58 | 59A | 59B | |
| ............ | ............ | ............ | ............ | ............ | ............ | ............ | ............ | 1 |
| ............ | ............ | ............ | ............ | ............ | ............ | ............ | ............ | 2 |
| ............ | ............ | ............ | ............ | ............ | ............ | ............ | ............ | 3 |
| .00005 | .00001 | .00003 | .00004 | .00004 | .00004 | .00003 | .00004 | 4 |
| ............ | ............ | ............ | ............ | ............ | .00428 | ............ | ............ | 5+6 |
| .00024 | .00029 | .00011 | .00008 | .00002 | .00011 | .00047 | .00025 | 7 |
| ............ | ............ | ............ | ............ | ............ | ............ | ............ | ............ | 8 |
| ............ | ............ | ............ | ............ | ............ | ............ | ............ | .00001 | 9+10 |
| ............ | ............ | ............ | ............ | ............ | ............ | ............ | ............ | 11 |
| .00818 | .00472 | .00605 | .00376 | .01014 | . .00534 | .00321 | .00680 | 12 |
| ............ | ............ | ............ | ............ | ............ | ............ | .00001 | .00001 | 13 |
| .00011 | .00012 | .00010 | .00002 | .00010 | .00006 | (*) | .00002 | 14 |
| ............ | ............ | ............ | ............ | ............ | ............ | ............ | ............ | 15 |
| ............ | .00222 | .00060 | ............ | ............ | ............ | .00087 | .00003 | 16 |
| .00017 | ............ | ............ | ............ | ............ | .00038 | .00245 | .00162 | 17 |
| .00009 | .00010 | .00007 | .00024 | .00026 | .00014 | .00010 | .00004 | 18 |
| ............ | ............ | ............ | .00001 | .00001 | .00001 | .02625 | .00070 | 19 |
| .00127 | .00638 | .00164 | .00066 | ............ | .00005 | .00003 | .00288 | 20+21 |
| ............ | .00108 | ............ | .01097 | .00022 | .00011 | .01251 | .00005 | 22+23 |
| .00529 | .00227 | .00028 | .00248 | .00070 | .00061 | .00074 | .00040 | 24 |
| .00583 | .02003 | .01354 | .00336 | .00164 | .00672 | .00042 | .00167 | 25 |
| .00010 | .00008 | .00012 | .00008 | .00009 | .00017 | .00003 | .00005 | 26A |
| .00085 | .00077 | .00086 | .00251 | .00067 | .00122 | .00032 | .00047 | 26B |
| .00188 | .00625 | .00564 | .00262 | .01622 | .01874 | .00452 | .00312 | 27A |
| ............ | ............ | ............ | ............ | ............ | ............ | ............ | ............ | 27B |
| .00394 | .02000 | .02320 | .00257 | .00348 | .00531 | .00041 | .00529 | 28 |
| ............ | ............ | ............ | ............ | ............ | ............ | ............ | ............ | 29A |
| ............ | ............ | ............ | ............ | ............ | ............ | ............ | .00012 | 29B |
| .00303 | .00729 | .00171 | .00064 | .00002 | .00006 | .01204 | .00321 | 30 |
| .00650 | .00098 | .00171 | .00058 | .00087 | .00225 | .00176 | .00210 | 31 |
| .02064 | .04178 | .02112 | .03483 | .05218 | .04310 | .06258 | .03128 | 32 |
| (*) | ............ | ............ | .00002 | .00003 | .00001 | .00004 | .00001 | 33+34 |
| .00005 | .01078 | .03636 | .00057 | .00829 | .00004 | .00963 | .00123 | 35 |
| .00977 | .00453 | .00134 | .00034 | .00109 | .00091 | .00184 | .00531 | 36 |
| .04928 | .07446 | .04239 | .00288 | .00267 | .01170 | .00535 | .06408 | 37 |
| .05356 | .02031 | .04578 | .01031 | .04019 | .05788 | .00064 | .04415 | 38 |
| ............ | ............ | ............ | ............ | ............ | ............ | ............ | ............ | 39 |
| .00485 | ............ | ............ | .00199 | .00384 | .00411 | .00002 | .01853 | 40 |
| .01764 | .01579 | .03167 | .00795 | .00910 | .01022 | .07407 | .03305 | 41 |
| .00972 | .03691 | .01577 | .01010 | .03583 | .02279 | .01227 | .02006 | 42 |
| .00459 | ............ | ............ | ............ | ............ | ............ | .01768 | .00083 | 43 |
| ............ | ............ | ............ | ............ | ............ | ............ | ............ | ............ | 44+45 |
| ............ | ............ | ............ | ............ | ............ | ............ | .00010 | .00012 | 46 |
| .00269 | .00240 | .00300 | .00093 | .00181 | .00232 | .00824 | .00302 | 47 |

**TABLE 4.10**   Continued

|  |  |  |  |  |  |  |  |  |
|---|---|---|---|---|---|---|---|---|
|  |  |  |  | .02287 |  |  |  |  |
| .01330 | .02648 | .03126 | .00614 | .02336 | .06707 | .01052 | .00049 | .01777 |
| .03198 | .02547 | .02298 | .02883 | .02627 | .01645 | .05379 | .00074 | .01078 |
|  |  |  |  | .00020 | .00036 | .00004 | .20764 |  |
|  |  |  |  | .00012 | .00003 |  |  | .05897 |
| .03379 | .00438 | .03991 | .02832 | .06023 | .04200 | .00399 | .02505 | .06498 |
|  |  |  |  | .00004 |  |  |  |  |
| .00023 | .00028 | .00008 | .00027 | .00004 | .00003 | .00085 | .00370 | .00634 |
| .00001 | .00001 |  | .00001 | .00001 | .00002 | .00002 | .01636 | .00001 |
|  |  |  | .00058 | .00183 | .00189 | .00228 | .07278 | .00028 |
| .01103 | .00481 | .00670 | .00045 | .00068 | .00068 | .00047 | .00112 |  |
|  | .00032 |  |  |  |  |  |  |  |
| .00159 | .00268 | .00015 | .00043 | .00015 | .00017 | .00012 | .00016 | .00005 |
|  |  |  |  |  |  | .00110 |  |  |
| .00011 | .00017 | .00018 | .00021 | .00017 | .00046 | .00015 | .00030 | .01769 |
| .00020 | .00026 | .00029 | .00033 | .00028 | .00028 | .00033 | .00027 | .00018 |
| .00010 | .00043 | .00049 | .00032 | .00012 | .00015 | .00011 | .00042 | .00141 |
| .00054 | .00143 | .00100 | .00127 | .00130 | .00093 | .00108 | .00047 | .00135 |
| .00746 | .01142 | .00812 | .00710 | .00516 | .00567 | .00498 | .00196 | .00694 |
| .00023 | .00056 | .00024 | .00025 | .00017 | .00025 | .00020 | .00012 | .00036 |
| .00215 | .00448 | .00281 | .00431 | .00557 | .00529 | .00322 | .01325 | .00536 |
|  | .00002 | .00001 | .00004 | .00003 | .00002 | .00001 | .00003 | .00002 |
| .00274 | .00301 | .00409 | .00403 | .00466 | .00433 | .00360 | .00344 | .00337 |
| .00987 | .00998 | .00745 | .01267 | .00946 | .01249 | .01263 | .00605 | .00828 |
| .00309 | .00432 | .00275 | .00268 | .00244 | .00346 | .00213 | .00054 | .00227 |
| .00097 | .00108 | .00150 | .00077 | .00098 | .00076 | .00039 | .00081 | .00167 |
| .05013 | .06511 | .06394 | .03692 | .05717 | .04961 | .02824 | .07429 | .06810 |
| .00013 | .00029 | .00035 | .00071 | .00035 | .00034 | .00034 | .00032 | .00016 |
| .00355 | .00389 | .00445 | .00509 | .00653 | .00473 | .00830 | .00646 | .00262 |
| .00155 | .00218 | .00195 | .00219 | .00174 | .00173 | .00139 | .00136 | .00138 |
| .00292 | .00296 | .00530 | .00832 | .00784 | .00559 | .00938 | .00933 | .00373 |
| .00005 | .00092 | .00072 | .00189 | .00176 | .00165 | .00100 | .00603 | .00159 |
| .00150 | .00128 | .00125 | .00102 | .00095 | .00125 | .00108 | .00039 | .00060 |
| .00367 | .00299 | .00184 | .00194 | .00137 | .00249 | .00278 | .00093 | .00122 |
| .00318 | .00638 | .00614 | .00638 | .00690 | .00609 | .00749 | .00741 | .00489 |
| .00635 | .00742 | .00894 | .00965 | .00924 | .00955 | .02424 | .00848 | .00588 |
| .00740 | .02286 | .02070 | .01674 | .01675 | .01250 | .01588 | .00270 | .01542 |
| .00158 | .00277 | .00329 | .00427 | .00348 | .00328 | .00421 | .00356 | .00250 |
| .00186 | .00130 | .00165 | .00866 | .00389 | .00337 | .00571 | .00411 | .00117 |
| .00011 | .00005 | .00022 | .00007 | .00005 | .00023 | .00011 | .00008 | .00004 |
| .00033 | .00023 | .00029 | .00089 | .00038 | .00082 | .00101 | .00051 | .00088 |
| .00080 | .00160 | .00163 | .00102 | .00176 | .00141 | .00079 | .00039 | .00038 |
| .00037 | .00027 | .00006 | .00033 | .00012 | .00024 | .00022 | .00015 | .00022 |
| .00186 | .00627 | .00188 | .00356 | .00413 | .00251 | .00216 | .01909 | .00036 |
|  | .00046 |  | .00054 |  | .00047 | .00034 |  |  |
| .55833 | .55700 | .53996 | .41252 | .47119 | .46633 | .40816 | .56655 | .53492 |
| .44167 | .44300 | .46004 | .58748 | .52881 | .53367 | .59184 | .43345 | .46508 |
| 1.00000 | 1.00000 | 1.00000 | 1.00000 | 1.00000 | 1.00000 | 1.00000 | 1.00000 | 1.00000 |

## TABLE 4.10

| | | | | | | | | |
|---|---|---|---|---|---|---|---|---|
| | | | | .00258 | | | | 48 |
| .00394 | .00655 | .00001 | .00093 | .00025 | .00228 | .00051 | .02045 | 49 |
| .00560 | .00148 | .00240 | .00103 | .00221 | .00281 | .00644 | .03853 | 50 |
| .00014 | | | .00266 | .00350 | .00098 | | | 51 |
| | .02084 | | .00012 | | | .02068 | .00256 | 52 |
| .03719 | .03853 | .02346 | .00583 | .00319 | .00974 | .00158 | .00131 | 53 |
| | .00518 | | | | | (*) | (*) | 54 |
| .00282 | .01236 | .02753 | .00669 | .00160 | .00782 | .00369 | .00040 | 55 |
| .00002 | .00001 | .00002 | .03618 | .00074 | .00060 | .01005 | .00020 | 56 |
| .02576 | .00889 | .00554 | .20131 | .09505 | .06263 | .00638 | .00552 | 57 |
| .00089 | .00015 | .00090 | .00384 | .00054 | .04922 | .02788 | .01263 | 58 |
| | | | | | | .01154 | .00266 | 59A |
| .00019 | | .00017 | .00005 | .00007 | .00027 | .29829 | .09072 | 59B |
| .00120 | | | | | | .00038 | | 60 |
| | | | | | | .00016 | .00020 | 61 |
| .00176 | .01971 | .00017 | .00132 | .00400 | .00043 | .00920 | .00039 | 62 |
| .00028 | .00018 | .00031 | .00024 | .00028 | .00043 | .00013 | .00020 | 63 |
| .00074 | .00700 | .00212 | .00032 | .00022 | .00019 | .00044 | .00017 | 64 |
| .00270 | .00202 | .00178 | .00050 | .00094 | .00172 | .00426 | .00251 | 65A |
| .00689 | .01083 | .00826 | .00340 | .00440 | .00724 | .01616 | .01357 | 65B |
| .00023 | .00020 | .00011 | .00008 | .00021 | .00034 | .00032 | .00038 | 65C |
| .00972 | .00509 | .00837 | .00540 | .00611 | .01156 | .00511 | .00281 | 65D |
| .00003 | | .00006 | .00001 | .00004 | .00005 | .00002 | .00003 | 65E |
| .00556 | .00233 | .00370 | .00391 | .00384 | .00391 | .00119 | .00449 | 66 |
| | | | | | | | | 67 |
| .01274 | .00880 | .01053 | .00606 | .01622 | .01110 | .00366 | .01059 | 68A |
| .00359 | .00447 | .00305 | .00127 | .00204 | .00247 | .00255 | .00378 | 68B |
| .00088 | .00119 | .00099 | .00050 | .00070 | .00199 | .00084 | .00085 | 68C |
| .06207 | .07255 | .06463 | .04840 | .04367 | .06621 | .07863 | .05346 | 69A |
| .00037 | .00005 | .00030 | .00012 | .00015 | .00054 | .00092 | .00062 | 69B |
| .01087 | .00715 | .00926 | .00400 | .00925 | .01064 | .00368 | .00231 | 70A |
| .00180 | .00147 | .00155 | .00146 | .00144 | .00214 | .00257 | .00206 | 70B |
| | | | | | | | | 71A |
| .00530 | .00266 | .00593 | .00800 | .00799 | .00693 | .00087 | .00231 | 71B |
| .00563 | .00212 | .00493 | .00074 | .00309 | .00643 | .00059 | .00035 | 72A |
| .00110 | .00107 | .00104 | .00437 | .00080 | .00098 | .00077 | .00115 | 72B |
| .00185 | .00128 | .00170 | .00237 | .00176 | .00177 | .00079 | .00133 | 73A |
| .00566 | .00276 | .00620 | .00578 | .00652 | .00907 | .00164 | .00307 | 73B |
| .00975 | .00766 | .00941 | .01104 | .01205 | .01174 | .00405 | .00702 | 73C |
| .03419 | .02068 | .01523 | .01921 | .00953 | .02889 | .01815 | .02619 | 73D |
| .00338 | .00229 | .00376 | .00228 | .00323 | .00567 | .00144 | .00218 | 74 |
| .00446 | .00102 | .00390 | .00122 | .00146 | .00659 | .00359 | .03183 | 75 |
| .00004 | .00002 | .00010 | .00010 | .00047 | .00115 | .00017 | .00005 | 76 |
| | | | | | | | | 77A |
| .00040 | .00165 | .00048 | .00120 | .00252 | .00074 | .00189 | .00117 | 77B |
| .00114 | .00324 | .00117 | .00194 | .00070 | .00095 | .00110 | .00249 | 78 |
| .00031 | .00049 | .00032 | .00020 | .00040 | .00047 | .00048 | .00060 | 79 |
| .00627 | .00341 | .00179 | .00573 | .00400 | .00354 | .00217 | .01330 | 80 |
| | | .00009 | | | .00519 | | .00225 | 81 |
| | | | | | | | | 82 |
| | | | | | | | | 83 |
| | | | | | | | | 84 |
| | | | | | | | | 85 |
| .48779 | .57072 | .48436 | .50032 | .44721 | .54607 | .81356 | .61922 | I |
| .51221 | .42928 | .51564 | .49968 | .55279 | .45393 | .18644 | .38078 | VA |
| 1.00000 | 1.00000 | 1.00000 | 1.00000 | 1.00000 | 1.00000 | 1.00000 | 1.00000 | T |

**TABLE 4.10**   Continued

| Commodity number | For the composition of inputs to an industry, read the column for that industry | Aircraft and parts | Other transportation equipment |
|---|---|---|---|
| | Industry number | 60 | 61 |
| 1 | Livestock and livestock products | .......... | .......... |
| 2 | Other agricultural products | .......... | .......... |
| 3 | Forestry and fishery products | .......... | .......... |
| 4 | Agricultural, forestry, and fishery services | .00003 | (*) |
| 5+6 | Metallic ores mining | .......... | .......... |
| 7 | Coal mining | .00010 | .00013 |
| 8 | Crude petroleum and natural gas | .......... | .......... |
| 9+10 | Nonmetallic minerals mining | .......... | .......... |
| 11 | New construction | .......... | .......... |
| 12 | Maintenance and repair construction | .00602 | .01954 |
| 13 | Ordnance and accessories | .00046 | .00001 |
| 14 | Food and kindred products | .00007 | .00009 |
| 15 | Tobacco products | .......... | .......... |
| 16 | Broad and narrow fabrics, yarn and thread mills | .00103 | .00044 |
| 17 | Miscellaneous textile goods and floor coverings | .00117 | .00508 |
| 18 | Apparel | .00007 | .00012 |
| 19 | Miscellaneous fabricated textile products | .00199 | .00755 |
| 20+21 | Lumber and wood products | .00038 | .01957 |
| 22+23 | Furniture and fixtures | .00031 | .00250 |
| 24 | Paper and allied products, except containers | .00016 | .00038 |
| 25 | Paperboard containers and boxes | .00009 | .00028 |
| 26A | Newspapers and periodicals | .00005 | .00006 |
| 26B | Other printing and publishing | .00064 | .00038 |
| 27A | Industrial and other chemicals | .00044 | .00227 |
| 27B | Agricultural fertilizers and chemicals | | |
| 28 | Plastics and synthetic materials | .00117 | .00766 |
| 29A | Drugs | | |
| 29B | Cleaning and toilet preparations | .......... | .......... |
| 30 | Paints and allied products | .00170 | .00599 |
| 31 | Petroleum refining and related products | .00139 | .00194 |
| 32 | Rubber and miscellaneous plastics products | .00856 | .01200 |
| 33+34 | Footwear, leather, and leather products | .00001 | .00001 |
| 35 | Glass and glass products | .00018 | .00980 |
| 36 | Stone and clay products | .00248 | .00242 |
| 37 | Primary iron and steel manufacturing | .01643 | .03878 |
| 38 | Primary nonferrous metals manufacturing | .04310 | .01815 |
| 39 | Metal containers | | |
| 40 | Heating, plumbing, and fabricated structural metal products | .00223 | .03541 |
| 41 | Screw machine products and stampings | .01166 | .00753 |
| 42 | Other fabricated metal products | .01130 | .01987 |
| 43 | Engines and turbines | .......... | .04390 |
| 44+45 | Farm, construction, and mining machinery | .......... | .00458 |
| 46 | Materials handling machinery and equipment | .......... | .00005 |
| 47 | Metalworking machinery and equipment | .01394 | .00206 |
| 48 | Special industry machinery and equipment | .......... | .......... |

## TABLE 4.10

| Scientific and controlling instruments | Ophthalmic and photographic equipment | Miscellaneous manufacturing | Railroads and related services; passenger ground transportation | Motor freight transportation and warehousing | Water transportation | Air transportation | Pipelines, freight forwarders, and related services |
|---|---|---|---|---|---|---|---|
| 62 | 63 | 64 | 65A | 65B | 65C | 65D | 65E |
|  |  | 0.00005 |  |  | 0.00005 |  |  |
|  |  | .00060 |  |  | .00012 |  |  |
|  |  |  |  |  | .00012 |  |  |
| .00004 | .00007 | .00008 | .00006 | .00005 | .00013 | .00008 | .00007 |
| .00016 |  |  |  |  |  |  |  |
| .00007 | .00103 | .00010 | (*) |  | .00036 |  |  |
|  |  |  |  |  |  |  | .00360 |
| (*) |  | .00051 |  |  |  |  |  |
|  |  |  |  |  |  |  |  |
| .00477 | .00459 | .00570 | .09693 | .00377 | .00532 | .00616 | .01704 |
| .00001 |  |  | .00001 | (*) | .00002 | .00001 | .00016 |
| .00090 | .00005 | .00198 | .00021 | .00001 | .00272 | .00366 | .00031 |
| .00365 |  | .00980 |  |  |  | .00006 |  |
| .00461 | .00092 | .00130 | .00002 | .00004 | .00280 | (*) |  |
| .00038 | .00003 | .00054 | .00035 | .00006 | .00394 | .00019 | .00002 |
| (*) | .00001 | .00361 | .00001 | .00011 | .00309 | .00012 | .00447 |
| .00203 |  | .02187 |  | .00013 |  |  | .00137 |
| .00073 |  | .00026 |  |  |  |  |  |
| .00306 | .06070 | .00873 | .00044 | .00045 | .00031 | .00073 | .00134 |
| .00401 | .00803 | .01312 | .00016 | .00058 | .00054 | .00001 | .00111 |
| .00012 | .00011 | .00021 | .00015 | .00021 | .00011 | .00019 | .00032 |
| .00169 | .00092 | .00265 | .00264 | .00231 | .00199 | .00196 | .00596 |
| .00348 | .02383 | .01018 | .00197 | .00069 | .00073 | .00022 | .00032 |
|  |  |  | .00076 |  | .00087 |  |  |
| .00671 | .00622 | .02092 |  |  |  |  |  |
|  |  |  |  |  |  |  |  |
| .00003 |  | .00002 | .00002 | .00003 |  | .00005 |  |
| .00052 | .00005 | .00433 | .00005 | .00015 | .00173 | .00005 |  |
| .00148 | .00159 | .00271 | .04905 | .04991 | .03040 | .11256 | .00342 |
| .01690 | .02040 | .02863 | .00330 | .00963 | .00315 | .00024 | .00253 |
| .00005 | .00002 | .00306 | .00003 | .00006 | .00003 | .00002 | .00022 |
| .00227 | .00503 | .00085 | .00085 | .00005 |  | .00001 | .00030 |
| .00136 | .00046 | .00434 | .00019 | .00008 | .00034 | .00004 | .00006 |
| .01510 | .00148 | .01657 | .00190 | .00001 | .00001 | .00003 |  |
| .01587 | .00633 | .05557 | .00016 |  | .00810 |  |  |
| .00023 |  |  |  |  |  |  |  |
| .00514 |  | .00002 |  |  |  | .00063 |  |
| .01207 | .00365 | .00395 | .00021 |  | .00008 |  | .00138 |
| .01535 | .01192 | .00872 | .00615 | .00226 | .02075 | .00073 | .00092 |
|  |  | .00009 | .00102 | .00036 | .00678 |  |  |
|  |  |  |  | .00003 |  |  |  |
| .00208 | .00117 | .00228 | .00165 | .00013 | .00738 | .00027 | .00005 |

**TABLE 4.10**  Continued

| | | | |
|---|---|---|---|
| 49 | General Industrial machinery and equipment | .00200 | .02087 |
| 50 | Miscellaneous machinery, except electrical | .00940 | .00296 |
| 51 | Computer and office equipment | .00049 | ............... |
| 52 | Service industry machinery | ............... | .00331 |
| 53 | Electrical industrial equipment and apparatus | .00113 | .01667 |
| 54 | Household appliances | ............... | .00615 |
| 55 | Electric lighting and wiring equipment | .00001 | .00375 |
| 56 | Audio, video, and communication equipment | .01171 | .00055 |
| 57 | Electronic components and accessories | .01474 | .00030 |
| 58 | Miscellaneous electrical machinery and supplies | .00094 | .00536 |
| 59A | Motor vehicles (passenger cars and trucks) | ............... | .02501 |
| 59B | Truck and bus bodies, trailers, and motor vehicles parts | .00006 | .01298 |
| 60 | Aircraft and parts | .19375 | .00166 |
| 61 | Other transportation equipment | ............... | .02599 |
| 62 | Scientific and controlling instruments | .02699 | .00292 |
| 63 | Ophthalmic and photographic equipment | .00022 | .00014 |
| 64 | Miscellaneous manufacturing | .00018 | .00020 |
| 65A | Railroads and related services; passenger ground transportation | .00048 | .00206 |
| 65B | Motor freight transportation and warehousing | .00268 | .00838 |
| 65C | Water transportation | .00014 | .00034 |
| 65D | Air transportation | .01439 | .00256 |
| 65E | Pipelines, freight forwarders, and related services | .00004 | .00001 |
| 66 | Communications, except radio and TV | .00437 | .00380 |
| 67 | Radio and TV broadcasting | ............... | ............... |
| 68A | Electric services (utilities) | .00804 | .00794 |
| 68B | Gas production and distribution (utilities) | .00175 | .00110 |
| 68C | Water and sanitary services | .00058 | .00115 |
| 69A | Wholesale trade | .01889 | .05513 |
| 69B | Retail trade | .00033 | .00043 |
| 70A | Finance | .01075 | .00206 |
| 70B | Insurance | .00137 | .00120 |
| 71A | Owner-occupied dwellings | ............... | ............... |
| 71B | Real estate and royalties | .00559 | .01437 |
| 72A | Hotels and lodging places | .00933 | .00037 |
| 72B | Personal and repair services (except auto) | .00056 | .00056 |
| 73A | Computer and data processing services | .00125 | .00059 |
| 73B | Legal, engineering, accounting, and related services | .00619 | .00373 |
| 73C | Other business and professional services, except medical | .01237 | .00806 |
| 73D | Advertising | .02349 | .00760 |
| 74 | Eating and drinking places | .00180 | .00155 |
| 75 | Automotive repair and services | .00123 | .00417 |
| 76 | Amusements | .00025 | .00050 |
| 77A | Health services | ............... | ............... |
| 77B | Educational and social services, and membership organizations | .00149 | .00032 |
| 78 | Federal Government enterprises | .00146 | .00064 |
| 79 | State and local government enterprises | .00021 | .00016 |
| 80 | Noncomparable imports | .00120 | .00068 |
| 81 | Scrap, used and secondhand goods | ............... | ............... |
| 82 | General government industry | ............... | ............... |
| 83 | Rest-of-the-world adjustment to final uses | ............... | ............... |
| 84 | Household industry | ............... | ............... |
| 85 | Inventory valuation adjustment | ............... | ............... |
| I | Total intermediate inputs | .51895 | .52680 |
| VA | Value added | .48105 | .47320 |
| T | Total | 1.00000 | 1.00000 |

*Less than .000005.

## TABLE 4.10

| | | | | | | | |
|---|---|---|---|---|---|---|---|
| .00181 | .00096 | .00077 | .00670 | .00066 | .02381 | .00005 | .00449 |
| .00224 | .00190 | .00376 | .00357 | .00022 | .00652 | .00057 | .00071 |
| .00886 | .00051 | .00015 | .00002 | .00007 | .00005 | .00003 | .00044 |
| | | .00080 | .00003 | .00018 | .00003 | .00002 | |
| .01515 | .00337 | .00273 | .00341 | .00084 | .00049 | .00013 | .00105 |
| | | | | (*) | .00055 | (*) | |
| .00241 | .00163 | .00065 | .00051 | .00042 | .00150 | .00012 | .00004 |
| .00023 | .00002 | .00011 | .00001 | .00003 | .00001 | .00009 | .00029 |
| .09217 | .07973 | .01200 | .00025 | .00005 | | .00118 | .00029 |
| .00254 | .00144 | .00036 | .00127 | .00060 | .00138 | .00033 | .00173 |
| | | | | .00016 | | | |
| | | | | .00342 | .00006 | .00017 | .00047 |
| .00019 | .00011 | .00030 | .00303 | | | .04374 | |
| | | .00012 | .01970 | .00005 | .01514 | | .00043 |
| .02685 | .00942 | .00025 | .00005 | .00003 | .00135 | .00039 | .00005 |
| .00034 | .01365 | .00043 | .00024 | .00039 | .00020 | .00019 | .00060 |
| .00054 | .00023 | .03681 | .00032 | .00026 | .00202 | .00048 | .00092 |
| .00100 | .00188 | .00251 | .05344 | .00224 | .00063 | .00086 | .00302 |
| .00424 | .00583 | .01492 | .00280 | .16339 | .00309 | .00228 | .00369 |
| .00019 | .00079 | .00071 | .00087 | .00079 | .10279 | .00203 | .00119 |
| .00396 | .00513 | .00233 | .00189 | .00209 | .00285 | .05453 | .02236 |
| .00002 | .00004 | .00002 | .01656 | .03052 | .05922 | .06322 | .04193 |
| .00505 | .00213 | .00214 | .00109 | .01227 | .00471 | .01215 | .01840 |
| .00886 | .00618 | .00885 | .00228 | .00831 | .02369 | .00320 | .03755 |
| .00155 | .00204 | .00192 | .00005 | .00037 | .00015 | .00019 | .00144 |
| .00114 | .00176 | .00329 | .00077 | .00036 | .00415 | .00047 | .00501 |
| .03759 | .03851 | .05230 | .01814 | .01541 | .02347 | .02972 | .00506 |
| .00035 | .00030 | .00058 | .00461 | .02782 | .00017 | .00087 | .00102 |
| .00529 | .01053 | .00786 | .00773 | .00421 | .05419 | .02767 | .03960 |
| .00185 | .00255 | .00214 | .00436 | .00824 | .00032 | .00486 | .02048 |
| .00905 | .00439 | .00731 | .00693 | .01865 | .03301 | .01318 | .03611 |
| .00084 | .00191 | .00072 | .00158 | .00143 | .00058 | .00050 | .00332 |
| .00105 | .00044 | .00094 | .00083 | .00078 | .00096 | .00552 | .00194 |
| .00116 | .00097 | .00042 | .00804 | .00294 | .01122 | .00941 | .01743 |
| .00939 | .00591 | .01004 | .00395 | .00471 | .00303 | .00460 | .01700 |
| .01205 | .01385 | .01094 | .01419 | .01165 | .06238 | .01408 | .01269 |
| .02475 | .02477 | .05086 | .00079 | .00205 | .01021 | .02283 | .02487 |
| .00360 | .00299 | .00713 | .00412 | .00478 | .00235 | .03001 | .01838 |
| .00570 | .00365 | .00725 | .01354 | .03791 | .00143 | .00375 | .01045 |
| .00033 | .00062 | .00033 | .00010 | .00013 | .00005 | .00056 | .00041 |
| .00182 | .00565 | .00376 | .00306 | .00048 | .00256 | .00010 | .00444 |
| .00204 | .00093 | .00274 | .00133 | .00127 | .00029 | .00009 | .00223 |
| .00028 | .00020 | .00025 | .00143 | .00116 | .00002 | .00029 | .00029 |
| .00398 | .01553 | .02925 | .00413 | .00008 | .11975 | .05589 | .00236 |
| | | | | | | | |
| | | | | | | | |
| | | | | | | | |
| | | | | | | | |
| .42532 | .42898 | .52426 | .38598 | .44251 | .68209 | .53832 | .40911 |
| .57468 | .57103 | .47574 | .61402 | .55749 | .31791 | .46168 | .59089 |
| 1.00000 | 1.00000 | 1.00000 | 1.00000 | 1.00000 | 1.00000 | 1.00000 | 1.00000 |

TABLE 4.10 Continued

| Communications, except radio and TV | Radio and TV broadcasting | Electric services (utilities) | Gas production and distribution (utilities) | Water and sanitary services | Wholesale trade | Retail trade | Finance | Insurance |
|---|---|---|---|---|---|---|---|---|
| 66 | 67 | 68A | 68B | 68C | 69A | 69B | 70A | 70B |
| | | | | | | | | |
| | | | | | | | | |
| .00002 | .00013 | .00015 | .00003 | .00052 | .00034 | .00020 | .00010 | .00004 |
| | | .10035 | | .00044 | .00006 | .00002 | | |
| | | | .24304 | | | | | |
| | | | | | | | | |
| .04145 | .00971 | .06621 | .00962 | .07798 | .00665 | .01365 | .00686 | .00299 |
| (*) | | (*) | | | .00004 | | (*) | (*) |
| .00001 | .00035 | (*) | (*) | .00037 | .00016 | .00226 | .00001 | .00001 |
| | | | | | (*) | | | |
| .00045 | .00001 | .00003 | | .00011 | .00014 | .00006 | | |
| .00001 | .00001 | (*) | (*) | | .00015 | .00005 | | |
| | | | | | .00039 | .00011 | .00045 | .00007 |
| | | | | .01258 | .00315 | .00013 | | |
| | | | | | .00001 | .00001 | | |
| .00049 | .00046 | .00027 | .00007 | .00605 | .00412 | .00701 | .00290 | .00083 |
| .00027 | .00003 | .00009 | | | .01000 | .00125 | .00001 | .00001 |
| .00011 | .00022 | .00005 | .00002 | .00020 | .00022 | .00021 | .00106 | .00014 |
| .00427 | .00180 | .00092 | .00026 | .00177 | .00895 | .00043 | .01279 | .00803 |
| .00058 | .00100 | .00319 | .00006 | .04520 | .00016 | .00006 | .00004 | .00003 |
| | | | | .00020 | | | | |
| | | | | | | .00009 | | |
| .00032 | | | | | | | | |
| | | | .00008 | | .00009 | .00007 | .00010 | |
| .00068 | | | | | | | | |
| .00073 | .00032 | .02325 | .00500 | .02308 | .00651 | .00502 | .00146 | .00057 |
| .00244 | .00003 | .00089 | .00008 | .02250 | .00277 | .00100 | .00028 | .00012 |
| .00003 | .00008 | .00001 | .00001 | .00004 | .00015 | .00005 | .00007 | .00010 |
| .00009 | .00001 | .00009 | .00001 | .00306 | .00039 | .00002 | .00015 | .00001 |
| .00001 | | .00007 | (*) | .00107 | .00026 | .00004 | .00004 | .00001 |
| .00001 | | .00001 | .00001 | | .00008 | .00002 | (*) | |
| .00027 | | .00083 | | | | | | |
| | | .00011 | | | .00149 | | | |
| | | | | | .00009 | | | |
| .00204 | | .00074 | | | | | | |
| .00113 | .00002 | .00015 | .00003 | .02770 | .00056 | .00142 | .00005 | .00001 |
| .00150 | | .00472 | | | | | | |
| | | | | | .00006 | | | |
| | | .00009 | | | .00062 | (*) | | |
| .00002 | .00004 | .00061 | .00003 | .00020 | .00023 | .00008 | .00001 | (*) |
| | | | | | .00001 | .00002 | | |

TABLE 4.10

| Owner-occupied dwellings | Real estate and royalties | Hotels and lodging places | Personal and repair services (exc. auto) | Computer and data processing services | Legal, engineering, accounting, and related services | Other business and professional services, except medical | Advertising | Commodity number |
|---|---|---|---|---|---|---|---|---|
| 71A | 71B | 72A | 72B | 73A | 73B | 73C | 73D | |
| | | 0.00004 | | | | 0.00012 | | 1 |
| | .00005 | .00008 | | | | | | 2 |
| | | .00003 | | | | | | 3 |
| .00795 | .00527 | .00529 | .00015 | .00006 | .00006 | .00026 | .00004 | 4 |
| | | | | | | | | 5+6 |
| | (*) | .00018 | .00006 | | | .00002 | | 7 |
| | | | | | | | | 8 |
| | | | | | | | | 9+10 |
| | | | | | | | | 11 |
| .04897 | .05700 | .03618 | .00975 | .00271 | .00176 | .00664 | .00558 | 12 |
| | (*) | .00001 | | | .00004 | .00093 | | 13 |
| | .00003 | .00254 | .00031 | .00001 | .00005 | .00064 | .00001 | 14 |
| | | | | | | | | 15 |
| | | .00116 | .00286 | | | (*) | | 16 |
| | (*) | .00019 | .00023 | (*) | .00001 | .00003 | | 17 |
| | (*) | .00235 | .00512 | | | .00035 | .00001 | 18 |
| | .00001 | .01294 | .00542 | .00001 | (*) | .00005 | .00111 | 19 |
| | .00009 | .00012 | .00038 | .00015 | .00018 | .00013 | .00007 | 20+21 |
| | .00001 | .00001 | .00002 | .00002 | .00002 | .00002 | .00001 | 22+23 |
| | .00119 | .00719 | .00458 | .00600 | .00359 | .00407 | .00329 | 24 |
| | .00006 | .00014 | .00033 | .00019 | .00020 | .00016 | .00054 | 25 |
| | .00004 | .00082 | .00033 | .00028 | .00018 | .00035 | .00021 | 26A |
| | .00338 | .00809 | .01039 | .03398 | .00853 | .01339 | .04330 | 26B |
| | .00006 | .00012 | .00353 | .00006 | .00008 | .00351 | .00020 | 27A |
| .00083 | .00029 | .00518 | | | | .00132 | | 27B |
| | | | .00052 | | | | | 28 |
| | | | | | | .00021 | | 29A |
| | .00010 | .00589 | .01354 | .00014 | .00021 | .00325 | .00007 | 29B |
| | | | .00002 | | | .00062 | .00006 | 30 |
| | .00102 | .00589 | .00468 | .00199 | .00257 | .00312 | .00290 | 31 |
| .00023 | .00142 | .01883 | .01669 | .01630 | .00159 | .00794 | .00118 | 32 |
| | .00003 | .00021 | .01042 | .00007 | .00021 | .00009 | .00007 | 33+34 |
| | .00002 | .01168 | | .00005 | .00019 | .00082 | | 35 |
| | .00008 | .00073 | .00372 | .00001 | .00001 | .00027 | .00001 | 36 |
| | (*) | .00002 | .00006 | .00001 | .00002 | .00008 | .00004 | 37 |
| | | | .00044 | | | | | 38 |
| | | | | | | .00112 | | 39 |
| .00031 | .00016 | | | | | .00069 | | 40 |
| | | | .00092 | | | .00069 | | 41 |
| | .00019 | .00062 | .00277 | .00043 | .00060 | .00097 | .00027 | 42 |
| | | | | | | .00046 | | 43 |
| .00060 | .00001 | | | | | .00231 | | 44+45 |
| | | | | | | .00117 | | 46 |
| | (*) | .00004 | .00020 | .00004 | .00004 | .00265 | .00003 | 47 |
| | (*) | .00001 | .00001 | .00002 | .00002 | .00124 | .00001 | 48 |

**TABLE 4.10**  Continued

|  |  |  |  |  |  |  |  |  |
|---|---|---|---|---|---|---|---|---|
| .00187 |  | .00055 | .00007 | .00045 | .00002 | .00003 | .00022 |  |
| .00004 | .00008 | .00035 | .00008 | .00044 | .00051 | .00017 | .00002 | .00001 |
| .00036 | .00069 | .00003 | .00070 |  | .00012 | .00009 | .00017 | .00006 |
|  |  |  |  | .00028 | .00029 | .00021 |  |  |
| .00164 | .00074 | .00150 | (*) |  | (*) |  | .00003 | (*) |
| (*) |  | (*) |  |  | .00002 |  | (*) | (*) |
| .00034 | .00001 | .00098 | .00014 | .00212 | .00009 | .00009 | .00014 | .00003 |
| .01352 | .00172 | .00001 | (*) | .00004 | .00005 | .00008 | .00013 | .00014 |
| .01044 | .02646 |  | .00004 | .00012 | .00005 | .00001 | .00074 | .00005 |
| .00128 | .00156 | .00161 | .00147 | .00162 | .00050 | .00031 | .00128 | .00098 |
|  |  |  |  |  |  |  |  |  |
| .00009 | .00008 | .00010 | .00008 | .06445 | .00082 | .00057 | .00017 | .00024 |
|  |  |  |  |  |  |  |  |  |
| .00009 | .00007 | .00006 | .00003 |  | .00016 | .00005 | .00023 | .00001 |
| .00006 | .00074 | .00071 | .00025 | .03534 | .00045 | .00003 | .00003 | .00003 |
| .00042 | .00648 | .00020 | .00009 | .00046 | .00045 | .00028 | .00365 | .00147 |
| .00058 | .00036 | .00017 | .00008 | .00087 | .00091 | .00098 | .00143 | .00082 |
| .00045 | .00116 | .02682 | .00014 | .00317 | .00118 | .00172 | .00196 | .00188 |
| .00088 | .00210 | .00308 | .00023 | .00646 | .00228 | .00273 | .01839 | .00235 |
| .00009 | .00005 | .00314 | .00037 | .00123 | .00026 | .00013 | .00008 | .00004 |
| .00145 | .00716 | .00254 | .00160 | .00248 | .01023 | .00185 | .00728 | .00290 |
| .00001 | .00002 | .00011 | .00430 | .00039 | .00093 | .00008 | .00003 | .00040 |
| .20595 | .00646 | .00144 | .00037 | .00654 | .01315 | .01087 | .01812 | .01391 |
|  | .01601 |  |  |  |  |  |  |  |
| .00298 | .00083 | .00020 | .00245 | .00329 | .00773 | .02363 | .00594 | .00057 |
| .00014 | .00011 | .04391 | .31024 | .04425 | .00432 | .00240 | .00036 | .00003 |
| .00705 | .00112 | .00152 | .00036 | .05416 | .00073 | .00163 | .00102 | .00172 |
| .00548 | .00523 | .01014 | .00431 | .03193 | .02349 | .00403 | .00435 | .00160 |
| .00025 | .00013 | .00036 | .00029 | .00513 | .00180 | .00214 | .00047 | .00027 |
| .01178 | .00998 | .01167 | .00658 | .01410 | .01364 | .00874 | .16855 | .03374 |
| .00040 | .00012 | .00719 | .00075 | .09486 | .00117 | .00127 | .01144 | .31305 |
|  |  |  |  |  |  |  |  |  |
| .01741 | .05335 | .00383 | .00207 | .00710 | .02338 | .05778 | .03310 | .02795 |
| .00050 | .00113 | .00031 | .00011 | .00333 | .00703 | .00142 | .00281 | .00606 |
| .00250 | .01178 | .00203 | .00054 | .01262 | .00416 | .00471 | .00149 | .00111 |
| .01022 | .00418 | .00686 | .00354 | .00786 | .00094 | .00466 | .04361 | .00667 |
| .00441 | .01365 | .00384 | .00146 | .01217 | .01156 | .03367 | .02525 | .01633 |
| .00657 | .01598 | .00392 | .00253 | .01188 | .04882 | .01777 | .04543 | .01136 |
| .00730 | .00278 | .00059 | .00029 | .00160 | .02002 | .04968 | .01238 | .01171 |
| .00312 | .00786 | .00102 | .00053 | .00476 | .01365 | .01308 | .01171 | .01652 |
| .00325 | .00134 | .00358 | .00323 | .00032 | .02130 | .01633 | .00542 | .00462 |
| .00434 | .31920 | .00004 | .00002 | .00012 | .00244 | .00047 | .00042 | .00038 |
|  |  |  |  |  |  |  |  |  |
| .00143 | .00441 | .00262 | .00034 | .00089 | .00155 | .00101 | .00270 | .00036 |
| .00156 | .00075 | .00246 | .00191 | .00385 | .00206 | .00346 | .02482 | .00398 |
| .00021 | .00089 | .00015 | .00012 | .00008 | .00089 | .00102 | .00015 | .00014 |
| .02332 | .00118 | .00005 | .00021 | .00007 | .00640 | .00032 | .01360 | .00358 |
|  |  |  |  |  |  |  |  |  |
|  |  |  |  |  |  |  |  |  |
|  |  |  |  |  |  |  |  |  |
|  |  |  |  |  |  |  |  |  |
| .41072 | .54212 | .35253 | .61028 | .66387 | .29688 | .30277 | .49550 | .50002 |
| .58928 | .45788 | .64747 | .38972 | .33613 | .70312 | .69724 | .50450 | .49998 |
| 1.00000 | 1.00000 | 1.00000 | 1.00000 | 1.00000 | 1.00000 | 1.00000 | 1.00000 | 1.00000 |

## TABLE 4.10

| | | | | | | | | |
|---|---|---|---|---|---|---|---|---|
| | .00006 | | | | | | .00328 | 49 |
| | .00001 | .00008 | .00081 | .00009 | .00024 | .00119 | .00006 | 50 |
| | .00006 | .00014 | .00700 | .01598 | .00146 | .00160 | .00059 | 51 |
| | | | .00237 | | | .00115 | | 52 |
| | | | .00059 | .00145 | (*) | .00253 | | 53 |
| | .00016 | .00018 | .00992 | | | .00013 | | 54 |
| | .00017 | .00299 | .00040 | .00011 | .00037 | .00082 | .00006 | 55 |
| | .00006 | .00010 | .00036 | .00018 | .00009 | .00043 | .00007 | 56 |
| | | | .02425 | .02871 | .00008 | .01511 | | 57 |
| | .00036 | .00008 | .00042 | .00725 | .00191 | .00288 | .00106 | 58 |
| | | | | | | | | 59A |
| | .00013 | .00040 | .00047 | .00024 | .00100 | .00051 | .00061 | 59B |
| | | | | | | | | 60 |
| | | .00015 | | | | .00119 | | 61 |
| | .00001 | .00082 | .00039 | .00061 | .00003 | .00062 | .00011 | 62 |
| | .00024 | .00044 | .00905 | .00048 | .00404 | .00633 | .00060 | 63 |
| | .00035 | .00283 | .02130 | .00040 | .00213 | .00160 | .00127 | 64 |
| .00003 | .00117 | .00257 | .00118 | .00102 | .00073 | .00166 | .00176 | 65A |
| .00004 | .00157 | .00422 | .00457 | .00186 | .00126 | .00309 | .00231 | 65B |
| .00001 | .00004 | .00025 | .00031 | .00006 | .00013 | .00057 | .00015 | 65C |
| (*) | .00112 | .00112 | .00282 | .00906 | .00940 | .01484 | .00268 | 65D |
| | .00030 | .00119 | .00007 | .00004 | .00005 | .00007 | .00004 | 65E |
| | .00405 | .01433 | .01479 | .02582 | .01239 | .01542 | .01324 | 66 |
| | | | | | | | .02854 | 67 |
| | .00081 | .03116 | .01797 | .00486 | .00262 | .00577 | .00271 | 68A |
| | .00016 | .01474 | .00718 | .00063 | .00048 | .00190 | .00062 | 68B |
| | .00020 | .00906 | .00218 | .00023 | .00095 | .00038 | .00020 | 68C |
| .00044 | .00130 | .01361 | .02289 | .01486 | .00674 | .01376 | .00441 | 69A |
| .00079 | .00060 | .00123 | .00125 | .00067 | .00125 | .00117 | .00112 | 69B |
| .00498 | .01655 | .04711 | .01273 | .00893 | .00794 | .01138 | .00969 | 70A |
| .02985 | .01663 | .00115 | .00109 | .00059 | .00177 | .00339 | .00120 | 70B |
| | | | | | | | | 71A |
| .03758 | .08499 | .03716 | .05681 | .03372 | .04960 | .02888 | .06731 | 71B |
| | .00239 | .00069 | .00126 | .00143 | .01758 | .00476 | .01201 | 72A |
| | .00201 | .01668 | .02165 | .00092 | .00313 | .00125 | .00287 | 72B |
| | .00020 | .00043 | .00537 | .11605 | .04656 | .02734 | .00133 | 73A |
| .00510 | .00576 | .01537 | .04170 | .00807 | .08644 | .02394 | .01257 | 73B |
| .00414 | .01936 | .06561 | .03387 | .02752 | .08999 | .06352 | .02751 | 73C |
| | .01647 | .01604 | .02250 | .00876 | .00277 | .01051 | .00579 | 73D |
| | .00878 | .00123 | .00697 | .00778 | .00642 | .01149 | .01255 | 74 |
| | .00363 | .01319 | .01312 | .00763 | .01527 | .01093 | .01999 | 75 |
| | .00019 | .00040 | .00028 | .00018 | .00068 | .00069 | .00774 | 76 |
| | | | | | | | | 77A |
| | .00051 | .00317 | .01355 | .00565 | .00518 | .00607 | .00236 | 77B |
| | .00123 | .00461 | .00460 | .00426 | .00763 | .00706 | .00399 | 78 |
| | .00030 | .00567 | .00184 | .00032 | .00064 | .00054 | .00147 | 79 |
| | .00015 | .00146 | .00015 | .00295 | .00261 | .00420 | .00153 | 80 |
| | | | | | | | | 81 |
| | | | | | | | | 82 |
| | | | | | | | | 83 |
| | | | | | | | | 84 |
| | | | | | | | | 85 |
| .14182 | .26254 | .45824 | .48745 | .41188 | .41167 | .37290 | .31110 | I |
| .85818 | .73746 | .54176 | .51255 | .58812 | .58833 | .62710 | .68890 | VA |
| 1.00000 | 1.00000 | 1.00000 | 1.00000 | 1.00000 | 1.00000 | 1.00000 | 1.00000 | T |

TABLE 4.10  Continued

| Commodity number | For the composition of inputs to an industry, read the column for that industry | Eating and drinking places | Automotive repair and services |
|---|---|---|---|
| | Industry number | 74 | 75 |
| 1 | Livestock and livestock products | 0.00187 | ................ |
| 2 | Other agricultural products | .00691 | ................ |
| 3 | Forestry and fishery products | .00799 | ................ |
| 4 | Agricultural, forestry, and fishery services | .00005 | .00013 |
| 5+6 | Metallic ores mining | ................ | ................ |
| 7 | Coal mining | .00002 | .00005 |
| 8 | Crude petroleum and natural gas | ................ | ................ |
| 9+10 | Nonmetallic minerals mining | ................ | ................ |
| 11 | New construction | ................ | ................ |
| 12 | Maintenance and repair construction | .00877 | .00501 |
| 13 | Ordnance and accessories | ................ | ................ |
| 14 | Food and kindred products | .22463 | .00005 |
| 15 | Tobacco products | ................ | ................ |
| 16 | Broad and narrow fabrics, yarn and thread mills | ................ | ................ |
| 17 | Miscellaneous textile goods and floor coverings | .00012 | .00012 |
| 18 | Apparel | ................ | .00129 |
| 19 | Miscellaneous fabricated textile products | .00008 | (*) |
| 20+21 | Lumber and wood products | .00013 | .00007 |
| 22+23 | Furniture and fixtures | (*) | (*) |
| 24 | Paper and allied products, except containers | .00351 | .00145 |
| 25 | Paperboard containers and boxes | .00415 | .00012 |
| 26A | Newspapers and periodicals | .00003 | .00033 |
| 26B | Other printing and publishing | .00141 | .00061 |
| 27A | Industrial and other chemicals | .00030 | .00045 |
| 27B | Agricultural fertilizers and chemicals | ................ | ................ |
| 28 | Plastics and synthetic materials | ................ | ................ |
| 29A | Drugs | ................ | ................ |
| 29B | Cleaning and toilet preparations | .00056 | .00008 |
| 30 | Paints and allied products | ................ | .00600 |
| 31 | Petroleum refining and related products | .00061 | .02802 |
| 32 | Rubber and miscellaneous plastics products | .00698 | .01018 |
| 33+34 | Footwear, leather, and leather products | .00032 | .00004 |
| 35 | Glass and glass products | .00162 | .00648 |
| 36 | Stone and clay products | .00018 | .00028 |
| 37 | Primary iron and steel manufacturing | (*) | (*) |
| 38 | Primary nonferrous metals manufacturing | .00010 | ................ |
| 39 | Metal containers | ................ | ................ |
| 40 | Heating, plumbing, and fabricated structural metal products | ................ | ................ |
| 41 | Screw machine products and stampings | .00004 | .02549 |
| 42 | Other fabricated metal products | .00025 | .03684 |
| 43 | Engines and turbines | ................ | .00187 |
| 44+45 | Farm, construction, and mining machinery | ................ | ................ |
| 46 | Materials handling machinery and equipment | ................ | ................ |
| 47 | Metalworking machinery and equipment | (*) | .00014 |
| 48 | Special industry machinery and equipment | .00025 | .00010 |
| 49 | General industrial machinery and equipment | .00044 | .00041 |

TABLE 4.10

| Amusements | Health services | Educational and social services, and membership organizations | Federal Government enterprises | State and local government enterprises | General government industry | Household industry | Inventory valuation adjustment |
|---|---|---|---|---|---|---|---|
| 76 | 77A | 77B | 78 | 79 | 82 | 84 | 85 |
| 0.00026 | 0.00018 | 0.00054 | 0.00009 | ........ | ........ | ........ | ........ |
| .00011 | .00036 | .00062 | .00032 | .00001 | ........ | ........ | ........ |
| .00005 | .00008 | .00020 | .00034 | ........ | ........ | ........ | ........ |
| .00730 | .00047 | .00317 | .00006 | .00180 | ........ | ........ | ........ |
| ........ | ........ | ........ | ........ | ........ | ........ | ........ | ........ |
| .00002 | .00002 | .00005 | .02336 | .01528 | ........ | ........ | ........ |
| ........ | ........ | ........ | ........ | ........ | ........ | ........ | ........ |
| ........ | ........ | .00002 | ........ | .00036 | ........ | ........ | ........ |
| ........ | ........ | ........ | ........ | ........ | ........ | ........ | ........ |
| .01301 | .00810 | .04686 | .01259 | .23160 | ........ | ........ | ........ |
| (*) | .00001 | .00003 | .00001 | .00001 | ........ | ........ | ........ |
| .00837 | .00897 | .01945 | .00865 | .00006 | ........ | ........ | ........ |
| ........ | ........ | ........ | ........ | ........ | ........ | ........ | ........ |
| .00088 | ........ | .00030 | .00032 | ........ | ........ | ........ | ........ |
| .00011 | .00005 | .00006 | .00010 | .00014 | ........ | ........ | ........ |
| .00177 | .00111 | .00159 | .00002 | .00044 | ........ | ........ | ........ |
| .00116 | .00229 | .00045 | .00171 | .00008 | ........ | ........ | ........ |
| .00267 | .00014 | .00131 | ........ | ........ | ........ | ........ | ........ |
| .00012 | .00001 | .00001 | ........ | ........ | ........ | ........ | ........ |
| .00195 | .00456 | .01001 | .00070 | .00073 | ........ | ........ | ........ |
| .00008 | .00037 | .00079 | .00032 | .00001 | ........ | ........ | ........ |
| .00036 | .00053 | .00378 | .00007 | .00009 | ........ | ........ | ........ |
| .00495 | .00774 | .06383 | .00843 | .00233 | ........ | ........ | ........ |
| .00214 | .01925 | .00148 | .00017 | .01637 | ........ | ........ | ........ |
| .00012 | .00003 | .00040 | ........ | .00171 | ........ | ........ | ........ |
| ........ | ........ | ........ | ........ | ........ | ........ | ........ | ........ |
| ........ | .01910 | .00083 | .00001 | .00005 | ........ | ........ | ........ |
| .00014 | .00248 | .00124 | .00064 | .00075 | ........ | ........ | ........ |
| .00003 | ........ | .00025 | ........ | .00001 | ........ | ........ | ........ |
| .00133 | .00272 | .00644 | .01412 | .06116 | ........ | ........ | ........ |
| .00494 | .02314 | .00678 | .00108 | .00165 | ........ | ........ | ........ |
| .00068 | .00004 | .00024 | .00031 | .00002 | ........ | ........ | ........ |
| .00004 | .00233 | .00165 | .00007 | .00013 | ........ | ........ | ........ |
| .00003 | .00085 | .00003 | ........ | .00205 | ........ | ........ | ........ |
| .00001 | .00002 | .00012 | ........ | (*) | ........ | ........ | ........ |
| .00035 | ........ | (*) | .00002 | ........ | ........ | ........ | ........ |
| ........ | ........ | ........ | ........ | ........ | ........ | ........ | ........ |
| ........ | ........ | .00003 | .00002 | .00003 | ........ | ........ | ........ |
| .00038 | .00072 | .00110 | .00198 | .00005 | ........ | ........ | ........ |
| .00075 | .00052 | .00189 | .00018 | .00044 | ........ | ........ | ........ |
| ........ | ........ | ........ | ........ | .00088 | ........ | ........ | ........ |
| .00002 | ........ | (*) | ........ | .00304 | ........ | ........ | ........ |
| ........ | ........ | ........ | .00009 | ........ | ........ | ........ | ........ |
| .00023 | .00002 | .00007 | .00012 | .00022 | ........ | ........ | ........ |
| .00002 | .00001 | .00001 | ........ | ........ | ........ | ........ | ........ |
| .00002 | (*) | ........ | .00038 | .00018 | ........ | ........ | ........ |

**TABLE 4.10**   Continued

| | | | |
|---|---|---|---|
| 50 | Miscellaneous machinery, except electrical | .00055 | .00958 |
| 51 | Computer and office equipment | .00002 | .00008 |
| 52 | Service industry machinery | .00007 | .00516 |
| 53 | Electrical industrial equipment and apparatus | ............... | .00133 |
| 54 | Household appliances | (*) | ............... |
| 55 | Electric lighting and wiring equipment | .00042 | .00730 |
| 56 | Audio, video, and communication equipment | .00003 | .00138 |
| 57 | Electronic components and accessories | ............... | ............... |
| 58 | Miscellaneous electrical machinery and supplies | .00002 | .00407 |
| 59A | Motor vehicles (passenger cars and trucks) | ............... | ............... |
| 59B | Truck and bus bodies, trailers, and motor vehicles parts | .00011 | .07633 |
| 60 | Aircraft and parts | ............... | ............... |
| 61 | Other transportation equipment | ............... | .00026 |
| 62 | Scientific and controlling instruments | ............... | .00028 |
| 63 | Ophthalmic and photographic equipment | .00003 | .00031 |
| 64 | Miscellaneous manufacturing | .00140 | .00036 |
| 65A | Railroads and related services; passenger ground transportation | .00184 | .00225 |
| 65B | Motor freight transportation and warehousing | .00736 | .00614 |
| 65C | Water transportation | .00030 | .00086 |
| 65D | Air transportation | .00082 | .00371 |
| 65E | Pipelines, freight forwarders, and related services | .00001 | .00070 |
| 66 | Communications, except radio and TV | .00384 | .00781 |
| 67 | Radio and TV broadcasting | ............... | ............... |
| 68A | Electric services (utilities) | .02123 | .00770 |
| 68B | Gas production and distribution (utilities) | .00174 | .00398 |
| 68C | Water and sanitary services | .00157 | .00063 |
| 69A | Wholesale trade | .04442 | .03596 |
| 69B | Retail trade | .00013 | .03714 |
| 70A | Finance | .01068 | .03214 |
| 70B | Insurance | .00016 | .01731 |
| 71A | Owner-occupied dwellings | ............... | ............... |
| 71B | Real estate and royalties | .04250 | .03553 |
| 72A | Hotels and lodging places | .00013 | .00117 |
| 72B | Personal and repair services (except auto) | .00272 | .01369 |
| 73A | Computer and data processing services | .00103 | .00013 |
| 73B | Legal, engineering, accounting, and related services | .01063 | .00585 |
| 73C | Other business and professional services, except medical | .01977 | .03081 |
| 73D | Advertising | .02211 | .01161 |
| 74 | Eating and drinking places | .00377 | .00662 |
| 75 | Automotive repair and services | .00193 | .01915 |
| 76 | Amusements | .00461 | .00014 |
| 77A | Health services | ............... | ............... |
| 77B | Educational and social services, and membership organizations | .00143 | .00129 |
| 78 | Federal Government enterprises | .00057 | .00404 |
| 79 | State and local government enterprises | .00069 | .00286 |
| 80 | Noncomparable imports | .00035 | .00009 |
| 81 | Scrap, used and secondhand goods | ............... | .00129 |
| 82 | General government industry | ............... | ............... |
| 83 | Rest-of-the-world adjustment to final uses | ............... | ............... |
| 84 | Household industry | ............... | ............... |
| 85 | Inventory valuation adjustment | ............... | ............... |
| I | Total intermediate inputs | .48045 | .52263 |
| VA | Value added | .51955 | .47738 |
| T | Total | 1.00000 | 1.00000 |

*Less than .000005.

Table 4.10

| | | | | | | | |
|---|---|---|---|---|---|---|---|
| .00018 | .00006 | .00021 | .00030 | .00618 | | | |
| .00009 | .00013 | .00070 | .00006 | .00003 | | | |
| .00070 | .00003 | .00007 | .00049 | .00028 | | | |
| (*) | (*) | .00005 | .00005 | .00726 | | | |
| .00002 | .00002 | .00009 | .00007 | .00072 | | | |
| .00085 | .00091 | .00132 | .00058 | .00151 | | | |
| .00005 | .00002 | .00080 | .00012 | .00001 | | | |
| .00009 | .00007 | .00124 | .00014 | .00057 | | | |
| .00046 | .00092 | .00110 | .00083 | .00163 | | | |
| .00013 | .00032 | .00078 | .00589 | .00279 | | | |
| .00131 | .00001 | .00006 | .00074 | .00186 | | | |
| .00017 | .01979 | .00081 | .00009 | .00024 | | | |
| .00238 | .00202 | .00326 | .00029 | .00021 | | | |
| .00234 | .00077 | .00502 | .00116 | .00062 | | | |
| .00118 | .00099 | .00155 | .01594 | .00633 | | | |
| .00241 | .00249 | .00518 | .03561 | .00445 | | | |
| .00052 | .00025 | .00027 | .00169 | .00288 | | | |
| .00300 | .00282 | .01705 | .02041 | .00141 | | | |
| .00002 | .00006 | .00015 | .00023 | .00108 | | | |
| .00841 | .00819 | .01154 | .00322 | .00336 | | | |
| .01407 | .00937 | .01335 | .00619 | .06452 | | | |
| .00282 | .00443 | .00632 | .00182 | .04435 | | | |
| .00160 | .00086 | .00167 | .00182 | .00437 | | | |
| .00637 | .01955 | .01612 | .00829 | .02254 | | | |
| .00058 | .00085 | .00086 | .00023 | .00032 | | | |
| .00918 | .00428 | .01366 | .00099 | .00580 | | | |
| .00099 | .00361 | .00468 | .00096 | .00708 | | | |
| .05301 | .06437 | .10665 | .01166 | .00987 | | | |
| .00294 | .00138 | .00689 | .00057 | .00073 | | | |
| .00527 | .00291 | .00268 | .00020 | .00025 | | | |
| .00418 | .01407 | .00918 | .00066 | .00374 | | | |
| .02735 | .00641 | .01519 | .00174 | .03169 | | | |
| .04947 | .02847 | .03354 | .00994 | .01232 | | | |
| .03332 | .00269 | .02133 | .00006 | .00150 | | | |
| .00740 | .00573 | .00762 | .00170 | .00163 | | | |
| .00608 | .00997 | .00851 | .01145 | .00259 | | | |
| .20386 | .00026 | .00635 | .00085 | .00005 | | | |
| | .01997 | | | | | | |
| .00485 | .00212 | .00233 | .00028 | .00117 | | | |
| .00467 | .00551 | .01368 | .00839 | .00101 | | | |
| .00109 | .00067 | .00148 | .00054 | .00008 | | | |
| .00197 | .00013 | .00560 | .02389 | | | | |
| | | | | | | | |
| | | | | | | | |
| | | | | | | | |
| .51974 | .35364 | .52456 | .25633 | .60063 | | | |
| .48026 | .64636 | .47545 | .74367 | .39937 | 1.00000 | 1.00000 | 1.00000 |
| 1.00000 | 1.00000 | 1.00000 | 1.00000 | 1.00000 | 1.00000 | 1.00000 | 1.00000 |

**TABLE 4.15**   Industry-by-Commodity Total Requirements (total requirements, direct and indirect, per dollar of delivery to final demand)

| Industry number | Each entry represents the output required, directly and indirectly, of the industry named at the beginning of the row for each dollar of delivery to final demand of the commodity named at the head of the column | Livestock and livestock products | Other agri-cultural products |
|---|---|---|---|
| | Commodity number | 1 | 2 |
| 1 | Livestock and livestock products | 1.31515 | 0.03468 |
| 2 | Other agricultural products | .39555 | 1.07001 |
| 3 | Forestry and fishery products | .00167 | .00072 |
| 4 | Agricultural, forestry, and fishery services | .08565 | .07862 |
| 5+6 | Metallic ores mining | .00089 | .00069 |
| 7 | Coal mining | .00362 | .00220 |
| 8 | Crude petroleum and natural gas | .01953 | .02247 |
| 9+10 | Nonmetallic minerals mining | .00514 | .00964 |
| 11+12 | Construction | .02848 | .02485 |
| 13 | Ordnance and accessories | .00011 | .00010 |
| 14 | Food and kindred products | .20984 | .00819 |
| 15 | Tobacco products | (*) | (*) |
| 16 | Broad and narrow fabrics, yarn and thread mills | .00193 | .00234 |
| 17 | Miscellaneous textile goods and floor coverings | .00147 | .00122 |
| 18 | Apparel | .00024 | .00019 |
| 19 | Miscellaneous fabricated textile products | .00095 | .00142 |
| 20+21 | Lumber and wood products | .00703 | .00822 |
| 22+23 | Furniture and fixtures | .00016 | .00013 |
| 24 | Paper and allied products, except containers | .01522 | .01009 |
| 25 | Paperboard containers and boxes | .00865 | .00687 |
| 26A | Newspapers and periodicals | .00549 | .00340 |
| 26B | Other printing and publishing | .01000 | .00672 |
| 27A | Industrial and other chemicals | .01820 | .02144 |
| 27B | Agricultural fertilizers and chemicals | .03862 | .07457 |
| 28 | Plastics and synthetic materials | .00523 | .00492 |
| 29A | Drugs | .00461 | .00066 |
| 29B | Cleaning and toilet preparations | .00294 | .00311 |
| 30 | Paints and allied products | .00081 | .00066 |
| 31 | Petroleum refining and related products | .02288 | .02432 |
| 32 | Rubber and miscellaneous plastics products | .01369 | .00997 |
| 33+34 | Footwear, leather, and leather products | .00060 | .00014 |
| 35 | Glass and glass products | .00366 | .00096 |
| 36 | Stone and clay products | .00342 | .00419 |
| 37 | Primary iron and steel manufacturing | .00788 | .00595 |
| 38 | Primary nonferrous metals manufacturing | .00603 | .00365 |
| 39 | Metal containers | .00634 | .00092 |
| 40 | Heating, plumbing, and fabricated structural metal products | .00266 | .00241 |
| 41 | Screw machine products and stampings | .00238 | .00148 |
| 42 | Other fabricated metal products | .00574 | .00493 |
| 43 | Engines and turbines | .00097 | .00091 |
| 44+45 | Farm, construction, and mining machinery | .00780 | .00933 |
| 46 | Materials handling machinery and equipment | .00030 | .00030 |
| 47 | Metalworking machinery and equipment | .00243 | .00181 |
| 48 | Special industry machinery and equipment | .00050 | .00039 |

TABLE 4.15

| Forestry and fishery products | Agri-cultural, forestry, and fishery services | Metallic ores mining | Coal mining | Crude petroleum and natural gas | Non-metallic minerals mining | New con-struction | Mainte-nance and repair con-struction |
|---|---|---|---|---|---|---|---|
| 3 | 4 | 5+6 | 7 | 8 | 9+10 | 11 | 12 |
| 0.05975 | 0.10349 | 0.00074 | 0.00049 | 0.00056 | 0.00066 | 0.00181 | 0.00184 |
| .23398 | .16985 | .00120 | .00084 | .00074 | .00099 | .00412 | .00420 |
| .80076 | .00164 | .00088 | .00048 | .00029 | .00032 | .00523 | .00533 |
| .13957 | .95618 | .00074 | .00057 | .00138 | .00063 | .00639 | .00651 |
| .00079 | .00089 | 1.08467 | .00136 | .00061 | .00238 | .00378 | .00376 |
| .00151 | .00257 | .01578 | 1.12441 | .00257 | .01181 | .00385 | .00383 |
| .02964 | .02818 | .03279 | .01847 | 1.04824 | .02851 | .02152 | .02163 |
| .00318 | .01276 | .00392 | .00287 | .00133 | .99280 | .01391 | .01412 |
| .02041 | .02717 | .03185 | .02190 | .03962 | .02313 | 1.01325 | 1.01322 |
| .00313 | .00014 | .00014 | .00011 | .00005 | .00017 | .00031 | .00031 |
| .05311 | .02258 | .00264 | .00159 | .00162 | .00232 | .00358 | .00365 |
| (*) | (*) | (*) | (*) | (*) | (*) | (*) | (*) |
| .00544 | .00448 | .00120 | .00200 | .00038 | .00128 | .00284 | .00290 |
| .00901 | .00565 | .00062 | .00055 | .00025 | .00052 | .00360 | .00367 |
| .00031 | .00040 | .00029 | .00059 | .00015 | .00052 | .00048 | .00049 |
| .00385 | .00357 | .00025 | .00019 | .00011 | .00028 | .00066 | .00068 |
| .00371 | .00513 | .01234 | .00674 | .00375 | .00386 | .07898 | .08064 |
| .00018 | .00015 | .00022 | .00016 | .00013 | .00016 | .00256 | .00261 |
| .00800 | .01415 | .00553 | .00457 | .00392 | .00919 | .01059 | .01075 |
| .00415 | .01202 | .00201 | .00169 | .00074 | .00216 | .00364 | .00369 |
| .00347 | .00554 | .00330 | .00307 | .00530 | .00489 | .00483 | .00485 |
| .01097 | .01100 | .00588 | .00497 | .00624 | .00597 | .00780 | .00788 |
| .01514 | .03796 | .04277 | .01275 | .01388 | .02610 | .01799 | .01863 |
| .02956 | .14947 | .00143 | .00064 | .00069 | .00100 | .00196 | .00199 |
| .00591 | .00826 | .00650 | .00432 | .00108 | .00493 | .00819 | .00837 |
| .00002 | .00144 | .00026 | .00010 | .00009 | .00018 | .00017 | .00018 |
| .00178 | .00598 | .00128 | .00057 | .00057 | .00126 | .00112 | .00119 |
| .00127 | .00097 | .00119 | .00078 | .00067 | .00078 | .00914 | .00933 |
| .04257 | .02590 | .03575 | .02615 | .00861 | .04223 | .02903 | .02918 |
| .00734 | .01026 | .01931 | .01610 | .00306 | .01544 | .01921 | .01965 |
| .00046 | .00044 | .00009 | .00008 | .00006 | .00010 | .00021 | .00021 |
| .00194 | .00205 | .00133 | .00074 | .00052 | .00113 | .00409 | .00417 |
| .00236 | .00419 | .00764 | .00655 | .00638 | .04435 | .05925 | .05981 |
| .00783 | .00641 | .04054 | .01963 | .00826 | .01622 | .04325 | .04138 |
| .00532 | .00425 | .01175 | .00778 | .00285 | .01368 | .02840 | .02884 |
| .00386 | .00183 | .00056 | .00031 | .00022 | .00042 | .00089 | .00090 |
| .00240 | .00227 | .01052 | .00701 | .00301 | .00827 | .05228 | .05288 |
| .00197 | .00217 | .00891 | .00903 | .00089 | .00454 | .00623 | .00632 |
| .00934 | .00582 | .00732 | .00806 | .00628 | .00696 | .02059 | .02109 |
| .00306 | .00255 | .00793 | .00775 | .00066 | .00563 | .00079 | .00077 |
| .00477 | .00578 | .02218 | .04889 | .00337 | .02065 | .00374 | .00308 |
| .00022 | .00036 | .00575 | .00443 | .00018 | .01138 | .00272 | .00277 |
| .00111 | .00115 | .00302 | .00222 | .00136 | .00210 | .00271 | .00274 |
| .00049 | .00061 | .00068 | .00057 | .00022 | .00063 | .00076 | .00077 |

TABLE 4.15   Continued

| | | | |
|---|---|---|---|
| 49 | General industrial machinery and equipment | .00234 | .00208 |
| 50 | Miscellaneous machinery, except electrical | .00278 | .00286 |
| 51 | Computer and office equipment | .00072 | .00061 |
| 52 | Service industry machinery | .00064 | .00055 |
| 53 | Electrical industrial equipment and apparatus | .00163 | .00142 |
| 54 | Household appliances | .00017 | .00016 |
| 55 | Electric lighting and wiring equipment | .00147 | .00137 |
| 56 | Audio, video, and communication equipment | .00045 | .00038 |
| 57 | Electronic components and accessories | .00197 | .00174 |
| 58 | Miscellaneous electrical machinery and supplies | .00471 | .00544 |
| 59A | Motor vehicles (passenger cars and trucks) | .00031 | .00029 |
| 59B | Truck and bus bodies, trailers, and motor vehicles parts | .00467 | .00468 |
| 60 | Aircraft and parts | .00048 | .00043 |
| 61 | Other transportation equipment | .00064 | .00036 |
| 62 | Scientific and controlling instruments | .00096 | .00066 |
| 63 | Ophthalmic and photographic equipment | .00094 | .00077 |
| 64 | Miscellaneous manufacturing | .00136 | .00102 |
| 65A | Railroads and related services; passenger ground transportation | .01749 | .00722 |
| 65B | Motor freight transportation and warehousing | .05475 | .03191 |
| 65C | Water transportation | .00341 | .00178 |
| 65D | Air transportation | .00647 | .00573 |
| 65E | Pipelines, freight forwarders, and related services | .00469 | .00346 |
| 66 | Communications, except radio and TV | .01363 | .01019 |
| 67 | Radio and TV broadcasting | .00417 | .00253 |
| 68A | Electric services (utilities) | .02348 | .01338 |
| 68B | Gas production and distribution (utilities) | .00846 | .00946 |
| 68C | Water and sanitary services | .00208 | .00252 |
| 69A | Wholesale trade | .11019 | .06977 |
| 69B | Retail trade | .00616 | .00580 |
| 70A | Finance | .03226 | .02546 |
| 70B | Insurance | .02790 | .03771 |
| 71A | Owner-occupied dwellings | ............ | ............ |
| 71B | Real estate and royalties | .10169 | .11091 |
| 72A | Hotels and lodging places | .00374 | .00295 |
| 72B | Personal and repair services (except auto) | .00330 | .00294 |
| 73A | Computer and data processing services | .00556 | .00445 |
| 73B | Legal, engineering, accounting, and related services | .01448 | .01183 |
| 73C | Other business and professional services, except medical | .03303 | .03027 |
| 73D | Advertising | .00236 | .00144 |
| 74 | Eating and drinking places | .00631 | .00516 |
| 75 | Automotive repair and services | .01250 | .01038 |
| 76 | Amusements | .00366 | .00272 |
| 77A | Health services | .01158 | .00032 |
| 77B | Educational and social services, and membership organizations | .00156 | .00120 |
| 78 | Federal Government enterprises | .00470 | .00352 |
| 79 | State and local government enterprises | .01146 | .00956 |
| 82 | General government industry | ............ | ............ |
| 83 | Rest-of-the-world adjustment to final uses | ............ | ............ |
| 84 | Household industry | ............ | ............ |
| 85 | Inventory valuation adjustment | ............ | ............ |
| | Total industry output multiplier | 2.83175 | 1.91916 |

* Less than .000005.

## TABLE 4.15

| | | | | | | | |
|---|---|---|---|---|---|---|---|
| .00267 | .00196 | .01561 | .02212 | .00272 | .01535 | .00562 | .00555 |
| .00200 | .00215 | .00500 | .00666 | .00116 | .00380 | .00276 | .00277 |
| .00085 | .00105 | .00106 | .00050 | .00035 | .00119 | .00113 | .00115 |
| .00063 | .00077 | .00095 | .00068 | .00058 | .00061 | .01183 | .01206 |
| .00142 | .00140 | .00701 | .00678 | .00278 | .00729 | .00773 | .00784 |
| .00023 | .00022 | .00020 | .00015 | .00017 | .00015 | .00274 | .00280 |
| .00111 | .00145 | .00203 | .00207 | .00112 | .00152 | .01663 | .01697 |
| .00044 | .00043 | .00042 | .00029 | .00029 | .00034 | .00372 | .00379 |
| .00196 | .00212 | .00203 | .00134 | .00093 | .00194 | .00374 | .00380 |
| .00176 | .00223 | .00207 | .00103 | .00048 | .00134 | .00253 | .00258 |
| .00070 | .00038 | .00050 | .00033 | .00007 | .00020 | .00030 | .00031 |
| .00443 | .00590 | .00642 | .00324 | .00097 | .00236 | .00371 | .00377 |
| .00108 | .00137 | .00093 | .00069 | .00023 | .00083 | .00067 | .00069 |
| .01722 | .00097 | .00128 | .00102 | .00016 | .00040 | .00055 | .00055 |
| .00237 | .00078 | .00162 | .00076 | .00041 | .00092 | .00355 | .00361 |
| .00092 | .00131 | .00096 | .00065 | .00062 | .00082 | .00138 | .00141 |
| .00110 | .00188 | .00116 | .00082 | .00077 | .00122 | .00278 | .00283 |
| .00453 | .01085 | .00887 | .02954 | .00226 | .00625 | .00724 | .00720 |
| .01690 | .03405 | .01851 | .01447 | .00621 | .02709 | .03013 | .03045 |
| .00417 | .00328 | .00271 | .00340 | .00181 | .00216 | .00192 | .00193 |
| .00690 | .02250 | .00732 | .00340 | .00274 | .00700 | .00613 | .00635 |
| .00383 | .00528 | .00380 | .00333 | .00123 | .00365 | .00347 | .00351 |
| .00732 | .00895 | .00764 | .00546 | .00548 | .00761 | .01327 | .01342 |
| .00251 | .00420 | .00254 | .00240 | .00432 | .00242 | .00373 | .00374 |
| .00825 | .01490 | .09611 | .02892 | .01684 | .04938 | .01424 | .01435 |
| .00542 | .01200 | .02403 | .00563 | .01172 | .02258 | .00864 | .00865 |
| .00089 | .00139 | .00159 | .00143 | .00088 | .00449 | .00098 | .00098 |
| .05566 | .10251 | .04581 | .05087 | .01532 | .04080 | .07547 | .07607 |
| .00415 | .00724 | .00484 | .00262 | .00230 | .00258 | .04140 | .04219 |
| .02341 | .02241 | .02356 | .01588 | .01118 | .02608 | .02483 | .02526 |
| .02331 | .01587 | .01039 | .00563 | .00646 | .00495 | .00923 | .00938 |
| .03050 | .04747 | .03227 | .04050 | .19925 | .02750 | .02700 | .02682 |
| .00386 | .00362 | .00381 | .00176 | .00215 | .00991 | .00454 | .00461 |
| .00352 | .01025 | .00211 | .00136 | .00113 | .00305 | .00235 | .00200 |
| .00771 | .00919 | .01000 | .00370 | .00260 | .01405 | .00821 | .00835 |
| .03943 | .02344 | .02929 | .01667 | .01768 | .01636 | .07443 | .07587 |
| .03496 | .03567 | .02314 | .01656 | .01450 | .02255 | .04965 | .05043 |
| .00142 | .00238 | .00144 | .00136 | .00245 | .00137 | .00211 | .00212 |
| .00552 | .00899 | .00503 | .00384 | .00410 | .00500 | .00782 | .00793 |
| .02340 | .03368 | .03570 | .01329 | .00558 | .00661 | .01903 | .01937 |
| .00364 | .01737 | .00142 | .00134 | .00203 | .00164 | .00228 | .00230 |
| .00050 | .00093 | .00002 | .00001 | .00001 | .00002 | .00004 | .00004 |
| .00373 | .00332 | .00362 | .00282 | .00105 | .00227 | .00198 | .00201 |
| .00377 | .00566 | .00922 | .00337 | .00266 | .00597 | .00438 | .00444 |
| .00505 | .00856 | .01880 | .01062 | .00840 | .01531 | .00599 | .00602 |
| 1.91932 | 2.16999 | 1.92761 | 1.72137 | 1.54762 | 1.69939 | 2.02832 | 2.03943 |

TABLE 4.15 Continued

| Industry number | Each entry represents the output required, directly and indirectly, of the industry named at the beginning of the row for each dollar of delivery to final demand of the commodity named at the head of the column | Ordnance and accessories | Food and kindred products |
|---|---|---|---|
| | Commodity number | 13 | 14 |
| 1 | Livestock and livestock products | 0.00070 | 0.30678 |
| 2 | Other agricultural products | .00095 | .18453 |
| 3 | Forestry and fishery products | .00037 | .00679 |
| 4 | Agricultural, forestry, and fishery services | .00049 | .02815 |
| 5+6 | Metallic ores mining | .00485 | .00190 |
| 7 | Coal mining | .00355 | .00412 |
| 8 | Crude petroleum and natural gas | .00913 | .01686 |
| 9+10 | Nonmetallic minerals mining | .00121 | .00319 |
| 11+12 | Construction | .01631 | .01970 |
| 13 | Ordnance and accessories | .95565 | .00012 |
| 14 | Food and kindred products | .00243 | 1.23314 |
| 15 | Tobacco products | (*) | .00001 |
| 16 | Broad and narrow fabrics, yarn and thread mills | .00187 | .00227 |
| 17 | Miscellaneous textile goods and floor coverings | .00081 | .00121 |
| 18 | Apparel | .00072 | .00027 |
| 19 | Miscellaneous fabricated textile products | .00052 | .00079 |
| 20+21 | Lumber and wood products | .00460 | .00770 |
| 22+23 | Furniture and fixtures | .00068 | .00025 |
| 24 | Paper and allied products, except containers | .00779 | .03828 |
| 25 | Paperboard containers and boxes | .00284 | .02686 |
| 26A | Newspapers and periodicals | .01016 | .01371 |
| 26B | Other printing and publishing | .01067 | .02234 |
| 27A | Industrial and other chemicals | .01558 | .02347 |
| 27B | Agricultural fertilizers and chemicals | .00077 | .01697 |
| 28 | Plastics and synthetic materials | .00653 | .00899 |
| 29A | Drugs | .00017 | .00570 |
| 29B | Cleaning and toilet preparations | .00071 | .00275 |
| 30 | Paints and allied products | .00110 | .00134 |
| 31 | Petroleum refining and related products | .00810 | .01650 |
| 32 | Rubber and miscellaneous plastics products | .01663 | .02800 |
| 33+34 | Footwear, leather, and leather products | .00034 | .00035 |
| 35 | Glass and glass products | .00124 | .01661 |
| 36 | Stone and clay products | .00507 | .00314 |
| 37 | Primary iron and steel manufacturing | .03740 | .01462 |
| 38 | Primary nonferrous metals manufacturing | .04713 | .01655 |
| 39 | Metal containers | .00037 | .03428 |
| 40 | Heating, plumbing, and fabricated structural metal products | .00297 | .00186 |
| 41 | Screw machine products and stampings | .01136 | .00420 |
| 42 | Other fabricated metal products | .02272 | .00822 |
| 43 | Engines and turbines | .00236 | .00061 |
| 44+45 | Farm, construction, and mining machinery | .00122 | .00301 |
| 46 | Materials handling machinery and equipment | .00028 | .00027 |
| 47 | Metalworking machinery and equipment | .00656 | .00173 |
| 48 | Special industry machinery and equipment | .00068 | .00106 |
| 49 | General industrial machinery and equipment | .01130 | .00252 |

TABLE 4.15

| Tobacco products | Broad and narrow fabrics, yarn and thread mills | Miscellaneous textile goods and floor coverings | Apparel | Miscellaneous fabricated textile products | Lumber and wood products | Furniture and fixtures | Paper and allied products, except containers | Paperboard containers and boxes |
|---|---|---|---|---|---|---|---|---|
| 15 | 16 | 17 | 18 | 19 | 20+21 | 22+23 | 24 | 25 |
| 0.00301 | 0.01691 | 0.00587 | 0.00610 | 0.00713 | 0.00778 | 0.00261 | 0.00323 | 0.00200 |
| .08093 | .12372 | .03586 | .03547 | .04204 | .03045 | .00924 | .00646 | .00391 |
| .00040 | .00052 | .00070 | .00488 | .00082 | .09070 | .00843 | .00804 | .00381 |
| .00618 | .01010 | .00334 | .00383 | .00381 | .01673 | .00244 | .00227 | .00132 |
| .00038 | .00163 | .00238 | .00084 | .00116 | .00180 | .00456 | .00174 | .00180 |
| .00220 | .00868 | .00839 | .00436 | .00528 | .00486 | .00543 | .01327 | .00838 |
| .00666 | .02587 | .03180 | .01402 | .01734 | .02202 | .01498 | .02937 | .02649 |
| .00127 | .00451 | .00535 | .00190 | .00266 | .00300 | .00238 | .00755 | .00457 |
| .00777 | .02231 | .02038 | .01367 | .01666 | .02244 | .02163 | .02329 | .02072 |
| .00005 | .00010 | .00010 | .00009 | .00009 | .00048 | .00017 | .00013 | .00011 |
| .00203 | .00700 | .00803 | .00571 | .00743 | .00963 | .00560 | .01054 | .00700 |
| 1.16027 | .00001 | .00001 | .00001 | .00001 | - (*) | .00001 | .00031 | .00014 |
| .00116 | 1.34605 | .34258 | .35518 | .45718 | .00364 | .05777 | .01847 | .01004 |
| .00047 | .02614 | .97981 | .01007 | .07868 | .00427 | .02801 | .00716 | .00358 |
| .00020 | .00654 | .00277 | 1.22395 | .02405 | .00056 | .00101 | .00053 | .00040 |
| .00022 | .00319 | .00424 | .02801 | .90323 | .00123 | .00260 | .00060 | .00045 |
| .00479 | .00500 | .00680 | .00366 | .00944 | 1.39463 | .13191 | .10202 | .04803 |
| .00008 | .00021 | .00029 | .00017 | .00206 | .00423 | .98653 | .00060 | .00038 |
| .03971 | .01677 | .04054 | .01515 | .02511 | .01436 | .02154 | 1.17906 | .53800 |
| .04354 | .00990 | .01337 | .00848 | .01756 | .00887 | .01939 | .02051 | 1.00311 |
| .01266 | .00647 | .01154 | .00937 | .00823 | .00927 | .01072 | .00747 | .00544 |
| .02854 | .00895 | .01617 | .01113 | .01297 | .01183 | .01292 | .01997 | .01429 |
| .01074 | .12741 | .19206 | .05419 | .07834 | .03426 | .03635 | .08300 | .07250 |
| .00605 | .01270 | .00880 | .00439 | .00558 | .00847 | .00232 | .00638 | .00395 |
| .00403 | .17252 | .28064 | .07537 | .10007 | .01438 | .02826 | .04079 | .04225 |
| .00014 | .00091 | .00125 | .00053 | .00056 | .00031 | .00027 | .00068 | .00051 |
| .00152 | .00731 | .01133 | .00830 | .00591 | .00149 | .00173 | .00802 | .00462 |
| .00050 | .00215 | .00325 | .00112 | .00152 | .00858 | .01199 | .00220 | .00421 |
| .00737 | .02388 | .02737 | .01369 | .01579 | .02437 | .01470 | .02684 | .02757 |
| .00768 | .02372 | .03619 | .01787 | .04310 | .02204 | .04364 | .04423 | .03319 |
| .00007 | .00036 | .00035 | .00882 | .01624 | .00039 | .00368 | .00018 | .00027 |
| .00047 | .00897 | .00377 | .00286 | .00418 | .00563 | .00592 | .00178 | .00270 |
| .00112 | .00321 | .00535 | .00179 | .00262 | .01321 | .00876 | .00659 | .00424 |
| .00361 | .00533 | .00711 | .00358 | .00475 | .02095 | .06510 | .00819 | .01088 |
| .00226 | .00368 | .00482 | .00295 | .00375 | .01092 | .02830 | .00663 | .00923 |
| .00046 | .00140 | .00186 | .00085 | .00110 | .00139 | .00124 | .00126 | .00284 |
| .00076 | .00185 | .00204 | .00116 | .00158 | .01291 | .00503 | .00281 | .00217 |
| .00084 | .00200 | .00184 | .00133 | .00169 | .02177 | .01472 | .00338 | .00247 |
| .00972 | .00430 | .00632 | .00302 | .00445 | .03307 | .04498 | .01610 | .01626 |
| .00021 | .00062 | .00063 | .00037 | .00043 | .00091 | .00064 | .00068 | .00059 |
| .00093 | .00191 | .00122 | .00079 | .00096 | .00137 | .00108 | .00125 | .00091 |
| .00015 | .00072 | .00044 | .00033 | .00040 | .00108 | .00092 | .00041 | .00032 |
| .00079 | .00164 | .00190 | .00100 | .00144 | .00483 | .00464 | .00198 | .00217 |
| .00063 | .00514 | .01591 | .00462 | .00427 | .00226 | .00197 | .00640 | .00694 |
| .00092 | .00186 | .00251 | .00124 | .00147 | .00439 | .00483 | .00290 | .00233 |

**TABLE 4.15** Continued

| | | | |
|---|---|---|---|
| 50 | Miscellaneous machinery, except electrical | .00785 | .00221 |
| 51 | Computer and office equipment | .00334 | .00065 |
| 52 | Service industry machinery | .00058 | .00056 |
| 53 | Electrical industrial equipment and apparatus | .00432 | .00141 |
| 54 | Household appliances | .00017 | .00016 |
| 55 | Electric lighting and wiring equipment | .00132 | .00102 |
| 56 | Audio, video, and communication equipment | .02627 | .00037 |
| 57 | Electronic components and accessories | .04268 | .00189 |
| 58 | Miscellaneous electrical machinery and supplies | .00259 | .00206 |
| 59A | Motor vehicles (passenger cars and trucks) | .00173 | .00023 |
| 59B | Truck and bus bodies, trailers, and motor vehicles parts | .01504 | .00282 |
| 60 | Aircraft and parts | .13445 | .00047 |
| 61 | Other transportation equipment | .00093 | .00062 |
| 62 | Scientific and controlling instruments | .05339 | .00097 |
| 63 | Ophthalmic and photographic equipment | .00149 | .00112 |
| 64 | Miscellaneous manufacturing | .00149 | .00168 |
| 65A | Railroads and related services; passenger ground transportation | .00346 | .01385 |
| 65B | Motor freight transportation and warehousing | .01432 | .04443 |
| 65C | Water transportation | .00083 | .00358 |
| 65D | Air transportation | .01004 | .00671 |
| 65E | Pipelines, freight forwarders, and related services | .00208 | .00389 |
| 66 | Communications, except radio and TV | .01146 | .01104 |
| 67 | Radio and TV broadcasting | .00830 | .01104 |
| 68A | Electric services (utilities) | .01674 | .02199 |
| 68B | Gas production and distribution (utilities) | .00910 | .01321 |
| 68C | Water and sanitary services | .00093 | .00187 |
| 69A | Wholesale trade | .04535 | .11075 |
| 69B | Retail trade | .00184 | .00406 |
| 70A | Finance | .01122 | .01988 |
| 70B | Insurance | .00471 | .01421 |
| 71A | Owner-occupied dwellings | ............... | ............... |
| 71B | Real estate and royalties | .02051 | .04701 |
| 72A | Hotels and lodging places | .00385 | .00337 |
| 72B | Personal and repair services (except auto) | .00190 | .00294 |
| 73A | Computer and data processing services | .00449 | .00502 |
| 73B | Legal, engineering, accounting, and related services | .01883 | .01445 |
| 73C | Other business and professional services, except medical | .02941 | .03244 |
| 73D | Advertising | .00470 | .00625 |
| 74 | Eating and drinking places | .00634 | .00637 |
| 75 | Automotive repair and services | .00638 | .01134 |
| 76 | Amusements | .00387 | .00551 |
| 77A | Health services | .00002 | .00268 |
| 77B | Educational and social services, and membership organizations | .00182 | .00187 |
| 78 | Federal Government enterprises | .00324 | .00465 |
| 79 | State and local government enterprises | .00539 | .01032 |
| 82 | General government industry | ............... | ............... |
| 83 | Rest-of-the-world adjustment to final uses | ............... | ............... |
| 84 | Household industry | ............... | ............... |
| 85 | Inventory valuation adjustment | ............... | ............... |
| | **Total industry output multiplier** | **1.78562** | **2.58463** |

* Less than .000005.

## TABLE 4.15

| | | | | | | | | |
|---|---|---|---|---|---|---|---|---|
| .00143 | .00279 | .00301 | .00179 | .00251 | .00603 | .00389 | .00344 | .00345 |
| .00037 | .00077 | .00084 | .00056 | .00068 | .00072 | .00086 | .00086 | .00076 |
| .00025 | .00057 | .00058 | .00041 | .00064 | .00136 | .00211 | .00074 | .00063 |
| .00054 | .00153 | .00228 | .00102 | .00120 | .00297 | .00399 | .00191 | .00184 |
| .00006 | .00027 | .00021 | .00017 | .00029 | .00298 | .00082 | .00036 | .00023 |
| .00042 | .00098 | .00093 | .00065 | .00087 | .00334 | .00157 | .00141 | .00111 |
| .00020 | .00036 | .00038 | .00058 | .00053 | .00047 | .00082 | .00038 | .00037 |
| .00110 | .00210 | .00238 | .00181 | .00209 | .00211 | .00262 | .00204 | .00183 |
| .00074 | .00139 | .00105 | .00074 | .00087 | .00131 | .00099 | .00092 | .00086 |
| .00010 | .00015 | .00014 | .00012 | .00013 | .00074 | .00037 | .00026 | .00023 |
| .00153 | .00229 | .00193 | .00178 | .00230 | .00687 | .00344 | .00366 | .00338 |
| .00026 | .00050 | .00053 | .00042 | .00064 | .00063 | .00071 | .00067 | .00058 |
| .00014 | .00037 | .00045 | .00035 | .00075 | .00260 | .00090 | .00071 | .00071 |
| .00061 | .00125 | .00405 | .00339 | .00170 | .00144 | .00309 | .00329 | .00206 |
| .00080 | .00136 | .00174 | .00103 | .00129 | .00127 | .00149 | .00270 | .00178 |
| .00120 | .00156 | .00193 | .00819 | .00960 | .00276 | .00473 | .00173 | .00175 |
| .00287 | .00969 | .01190 | .00475 | .00676 | .01599 | .01018 | .01703 | .02018 |
| .01284 | .02704 | .03736 | .01790 | .02485 | .03401 | .02653 | .04110 | .04893 |
| .00062 | .00222 | .00427 | .00124 | .00193 | .00373 | .00199 | .00320 | .00276 |
| .00364 | .00544 | .00550 | .00571 | .00552 | .00586 | .00666 | .00931 | .00792 |
| ·.00136 | .00336 | .00418 | .00222 | .00272 | .00388 | .00288 | .00464 | .00483 |
| .00551 | .00965 | .01069 | .00836 | .00960 | .01013 | .01065 | .00954 | .01114 |
| .01002 | .00516 | .00939 | .00761 | .00657 | .00743 | .00867 | .00596 | .00427 |
| .00804 | .04671 | .03840 | .02482 | .02721 | .02627 | .02290 | .04081 | .03243 |
| .00407 | .02066 | .02766 | .01090 | .01465 | .01466 | .01241 | .02674 | .01967 |
| .00072 | .00277 | .00244 | .00133 | .00255 | .00204 | .00170 | .00700 | .00391 |
| .03913 | .09993 | .08389 | .08018 | .10227 | .10121 | .09419 | .08348 | .08815 |
| .00187 | .00341 | .00323 | .00247 | .00252 | .00468 | .00329 | .00460 | .00465 |
| .01250 | .01620 | .01678 | .01549 | .01943 | .02153 | .02293 | .01606 | .01390 |
| .00682 | .01060 | .00896 | .00697 | .00811 | .01176 | .00735 | .01018 | .00941 |
| ·············· | ·············· | ·············· | ·············· | ·············· | ·············· | ·············· | ·············· | ·············· |
| .01896 | .03187 | .02786 | .02458 | .02669 | .02493 | .02482 | .02211 | .02349 |
| .00153 | .00240 | .00260 | .00208 | .00303 | .00277 | .00279 | .00432 | .00300 |
| .00130 | .00897 | .00723 | .00620 | .00807 | .00261 | .00299 | .00375 | .00299 |
| .00288 | .00586 | .00605 | .00404 | .00495 | .00538 | .00591 | .00723 | .00653 |
| .00773 | .01972 | .02367 | .01444 | .01642 | .01842 | .02397 | .01542 | .01507 |
| .01740 | .03946 | .03556 | .02843 | .03416 | .03300 | .04465 | .02647 | .02826 |
| .00568 | .00292 | .00532 | .00431 | .00372 | .00421 | .00491 | .00337 | .00242 |
| .00324 | .00734 | .00766 | .00798 | .00817 | .00886 | .00949 | .00653 | .00721 |
| .00864 | .01167 | .01037 | .01217 | .00873 | .01918 | .01347 | .01873 | .01966 |
| .00448 | .00283 | .00444 | .00366 | .00335 | .00397 | .00463 | .00299 | .00231 |
| .00003 | .00017 | .00007 | .00007 | .00008 | .00008 | .00005 | .00004 | .00003 |
| .00133 | .00246 | .00238 | .00211 | .00706 | .00511 | .00366 | .00265 | .00200 |
| .00428 | .00628 | .00649 | .00649 | .00656 | .00531 | .00574 | .00556 | .00513 |
| .00366 | .01375 | .01281 | .00744 | .00980 | .01033 | .00860 | .02120 | .01505 |
| ·············· | ·············· | ·············· | ·············· | ·············· | ·············· | ·············· | ·············· | ·············· |
| ·············· | ·············· | ·············· | ·············· | ·············· | ·············· | ·············· | ·············· | ·············· |
| ·············· | ·············· | ·············· | ·············· | ·············· | ·············· | ·············· | ·············· | ·············· |
| 1.66430 | 2.50089 | 2.59678 | 2.30313 | 2.34455 | 2.35666 | 2.10133 | 2.18030 | 2.38394 |

**TABLE 4.15** Continued

| News-papers and periodicals | Other printing and publishing | Industrial and other chemicals | Agricultural fertilizers and chemicals | Plastics and synthetic materials | Drugs | Cleaning and toilet preparations | Paints and allied products | Petroleum refining and related products | Rubber and miscellaneous plastics products |
|---|---|---|---|---|---|---|---|---|---|
| 26A | 26B | 27A | 27B | 28 | 29A | 29B | 30 | 31 | 32 |
| 0.00127 | 0.00154 | 0.00286 | 0.00356 | 0.00261 | 0.00526 | 0.00686 | 0.00541 | 0.00085 | 0.00171 |
| .00173 | .00246 | .00667 | .00392 | .00979 | .00376 | .00533 | .00587 | .00107 | .00498 |
| .00127 | .00168 | .00124 | .00060 | .00077 | .00047 | .00067 | .00162 | .00040 | .00080 |
| .00075 | .00086 | .00125 | .00106 | .00134 | .00087 | .00109 | .00121 | .00114 | .00093 |
| .00060 | .00120 | .01063 | .00308 | .00511 | .00095 | .00242 | .00859 | .00087 | .00259 |
| .00300 | .00450 | .01070 | .00775 | .01129 | .00261 | .00379 | .00551 | .00404 | .00688 |
| .01093 | .01465 | .10073 | .10210 | .05287 | .01114 | .02574 | .03857 | .63987 | .02766 |
| .00162 | .00254 | .02785 | .08327 | .01060 | .00170 | .00434 | .00981 | .00599 | .00497 |
| .01325 | .01436 | .02496 | .02748 | .02245 | .01150 | .01342 | .01825 | .03730 | .01937 |
| .00017 | .00012 | .00012 | .00013 | .00010 | .00010 | .00009 | .00010 | .00008 | .00011 |
| .00447 | .00546 | .01676 | .01495 | .00909 | .01148 | .02633 | .02231 | .00293 | .00587 |
| .00004 | .00006 | .00004 | .00001 | .00002 | .00001 | .00001 | .00001 | (*) | .00001 |
| .00324 | .00655 | .00232 | .00173 | .05064 | .00141 | .00393 | .00699 | .00090 | .02805 |
| .00135 | .00260 | .00099 | .00086 | .00419 | .00066 | .00197 | .00090 | .00070 | .01376 |
| .00025 | .00036 | .00021 | .00029 | .00040 | .00017 | .00026 | .00020 | .00021 | .00077 |
| .00029 | .00045 | .00051 | .00040 | .00040 | .00019 | .00040 | .00026 | .00038 | .00084 |
| .01539 | .02036 | .00606 | .00557 | .00580 | .00333 | .00578 | .00383 | .00485 | .01004 |
| .00014 | .00022 | .00016 | .00018 | .00017 | .00014 | .00022 | .00014 | .00016 | .00111 |
| .16380 | .22155 | .02510 | .02286 | .03088 | .01869 | .03977 | .01454 | .00661 | .04396 |
| .00488 | .01052 | .00888 | .00921 | .01142 | .01213 | .03927 | .00509 | .00328 | .02004 |
| .97933 | .03990 | .00787 | .00975 | .00818 | .00591 | .01184 | .00590 | .00566 | .00693 |
| .15861 | 1.06242 | .01110 | .02633 | .01094 | .01367 | .02340 | .01095 | .00783 | .01060 |
| .02540 | .05052 | 1.06763 | .22956 | .46762 | .05175 | .14593 | .28191 | .03495 | .15451 |
| .00145 | .00222 | .02402 | 1.08740 | .01663 | .00888 | .00621 | .00724 | .00107 | .00503 |
| .00791 | .01601 | .06151 | .03033 | .90090 | .01022 | .03422 | .12428 | .00425 | .18759 |
| .00022 | .00041 | .00533 | .00711 | .00261 | 1.06797 | .02619 | .00150 | .00027 | .00115 |
| .00180 | .00297 | .02220 | .03851 | .01362 | .01111 | .98262 | .00998 | .00437 | .00554 |
| .00072 | .00142 | .00723 | .00220 | .00775 | .00074 | .00424 | .99498 | .00094 | .00337 |
| .01273 | .01569 | .10473 | .03866 | .04454 | .00907 | .02759 | .03944 | 1.04317 | .02437 |
| .01176 | .02981 | .02451 | .02275 | .05567 | .02816 | .06112 | .01691 | .00908 | 1.01398 |
| .00020 | .00052 | .00014 | .00013 | .00015 | .00012 | .00015 | .00009 | .00013 | .00032 |
| .00074 | .00105 | .00292 | .00451 | .00254 | .00835 | .00919 | .00214 | .00295 | .00722 |
| .00187 | .00252 | .00540 | .01019 | .00402 | .00199 | .00436 | .01513 | .00669 | .00741 |
| .00320 | .00472 | .01215 | .01042 | .00788 | .00492 | .01233 | .01956 | .00805 | .01506 |
| .00311 | .00740 | .01007 | .00751 | .00599 | .00416 | .01004 | .02222 | .00388 | .00994 |
| .00050 | .00075 | .00734 | .00709 | .00365 | .00512 | .01515 | .04760 | .00183 | .00209 |
| .00116 | .00147 | .00249 | .00281 | .00220 | .00136 | .00191 | .00168 | .00275 | .00523 |
| .00129 | .00177 | .00192 | .00261 | .00177 | .00311 | .00750 | .00222 | .00115 | .00827 |
| .00417 | .00749 | .01237 | .00929 | .00786 | .00464 | .01609 | .00759 | .00813 | .01767 |
| .00031 | .00038 | .00087 | .00112 | .00070 | .00025 | .00039 | .00055 | .00070 | .00074 |
| .00051 | .00056 | .00198 | .00277 | .00136 | .00049 | .00071 | .00104 | .00242 | .00104 |
| .00020 | .00022 | .00052 | .00120 | .00035 | .00018 | .00023 | .00031 | .00028 | .00040 |

TABLE 4.15

| Footwear, leather, and leather products | Glass and glass products | Stone and clay products | Primary iron and steel manufacturing | Primary non-ferrous metals manufacturing | Metal containers | Heating, plumbing, and fabricated structural metal products | Screw machine products and stampings | Industry number |
|---|---|---|---|---|---|---|---|---|
| 33+34 | 35 | 36 | 37 | 38 | 39 | 40 | 41 | |
| 0.05173 | 0.00104 | 0.00117 | 0.00086 | 0.00098 | 0.00101 | 0.00097 | 0.00087 | 1 |
| .03314 | .00199 | .00229 | .00131 | .00190 | .00161 | .00142 | .00136 | 2 |
| .00154 | .00208 | .00070 | .00068 | .00077 | .00070 | .00082 | .00055 | 3 |
| .00521 | .00088 | .00072 | .00073 | .00072 | .00070 | .00069 | .00061 | 4 |
| .00131 | .00155 | .00266 | .04278 | .10727 | .03788 | .01765 | .01487 | 5+6 |
| .00378 | .00620 | .01770 | .03668 | .01184 | .01349 | .01056 | .01242 | 7 |
| .01543 | .03566 | .02925 | .03423 | .03243 | .02545 | .01846 | .01853 | 8 |
| .00280 | .01691 | .09889 | .00777 | .00327 | .00340 | .00318 | .00315 | 9+10 |
| .01573 | .02298 | .02364 | .04090 | .02551 | .02510 | .02939 | .02844 | 11+12 |
| .00009 | .00013 | .00149 | .00052 | .00041 | .00028 | .00057 | .00044 | 13 |
| .16857 | .00375 | .00412 | .00309 | .00343 | .00390 | .00340 | .00303 | 14 |
| .00001 | .00001 | .00001 | (*) | (*) | .00001 | (*) | (*) | 15 |
| .05880 | .00182 | .00533 | .00120 | .00379 | .00209 | .00168 | .00293 | 16 |
| .02859 | .00076 | .00067 | .00060 | .00083 | .00064 | .00073 | .00086 | 17 |
| .00845 | .00042 | .00044 | .00043 | .00036 | .00038 | .00031 | .00043 | 18 |
| .00224 | .00030 | .00033 | .00030 | .00030 | .00035 | .00054 | .00350 | 19 |
| .00922 | .03259 | .00958 | .00900 | .01038 | .00871 | .01277 | .00710 | 20+21 |
| .00066 | .00599 | .00021 | .00093 | .00024 | .00087 | .00210 | .00072 | 22+23 |
| .02099 | .03500 | .02556 | .01060 | .01033 | .01955 | .01183 | .01136 | 24 |
| .01603 | .04933 | .00584 | .00393 | .00580 | .00996 | .00895 | .00729 | 25 |
| .01028 | .01070 | .00972 | .01410 | .00778 | .01389 | .00780 | .00863 | 26A |
| .01636 | .01276 | .01096 | .01463 | .01007 | .04073 | .00986 | .01079 | 26B |
| .06914 | .08169 | .05000 | .04187 | .04416 | .03664 | .02488 | .02733 | 27A |
| .00483 | .00231 | .00158 | .00131 | .00147 | .00126 | .00095 | .00095 | 27B |
| .03936 | .01105 | .01049 | .00575 | .02524 | .01373 | .00911 | .00803 | 28 |
| .00129 | .00046 | .00038 | .00026 | .00027 | .00025 | .00019 | .00020 | 29A |
| .00781 | .00220 | .00421 | .00136 | .00145 | .00186 | .00133 | .00145 | 29B |
| .00126 | .00277 | .00305 | .00159 | .00194 | .01805 | .00698 | .00378 | 30 |
| .01509 | .02014 | .02845 | .02211 | .02756 | .02008 | .01541 | .01400 | 31 |
| .04904 | .02170 | .01199 | .01054 | .02696 | .01615 | .02159 | .01550 | 32 |
| 1.18812 | .00026 | .00013 | .00016 | .00013 | .00014 | .00014 | .00019 | 33+34 |
| .00337 | 1.05830 | .00392 | .00105 | .00234 | .00131 | .01027 | .00299 | 35 |
| .00335 | .02999 | 1.10951 | .02382 | .01121 | .01066 | .01243 | .01168 | 36 |
| .00734 | .00811 | .01689 | 1.16568 | .04265 | .27469 | .24607 | .29902 | 37 |
| .00609 | .01126 | .00789 | .05416 | 1.41533 | .38202 | .12326 | .06612 | 38 |
| .00512 | .00100 | .00080 | .00070 | .00075 | 1.02369 | .00096 | .00109 | 39 |
| .00148 | .00332 | .00369 | .00525 | .00395 | .00444 | .98889 | .00924 | 40 |
| .00357 | .00587 | .00454 | .00782 | .00620 | .00828 | .03429 | .98177 | 41 |
| .01137 | .00584 | .01059 | .01970 | .02014 | .02433 | .04568 | .03114 | 42 |
| .00038 | .00064 | .00126 | .00196 | .00262 | .00134 | .00124 | .00167 | 43 |
| .00096 | .00111 | .00367 | .00514 | .00328 | .00232 | .00338 | .00437 | 44+45 |
| .00026 | .00040 | .00141 | .00110 | .00118 | .00073 | .00151 | .00066 | 46 |

**TABLE 4.15**   Continued

| | | | | | | | | | |
|---|---|---|---|---|---|---|---|---|---|
| .00090 | .00122 | .00239 | .00153 | .00191 | .00081 | .00160 | .00151 | .00135 | .00399 |
| .00327 | .00595 | .00546 | .00169 | .00317 | .00067 | .00157 | .00180 | .00041 | .00456 |
| .00113 | .00139 | .00284 | .00445 | .00371 | .00134 | .00320 | .00235 | .00258 | .00320 |
| .00145 | .00199 | .00259 | .00238 | .00296 | .00117 | .00236 | .00176 | .00158 | .00538 |
| .00187 | .00166 | .00096 | .00099 | .00088 | .00113 | .00060 | .00070 | .00066 | .00095 |
| .00043 | .00049 | .00082 | .00072 | .00068 | .00036 | .00089 | .00050 | .00063 | .00144 |
| .00108 | .00137 | .00256 | .00279 | .00211 | .00085 | .00129 | .00167 | .00230 | .00224 |
| .00013 | .00014 | .00020 | .00017 | .00017 | .00010 | .00013 | .00014 | .00017 | .00026 |
| .00071 | .00084 | .00107 | .00114 | .00114 | .00059 | .00077 | .00307 | .00122 | .00313 |
| .00045 | .00041 | .00038 | .00046 | .00038 | .00033 | .00032 | .00033 | .00038 | .00045 |
| .00327 | .00274 | .00249 | .00244 | .00215 | .00231 | .00190 | .00200 | .00147 | .00373 |
| .00166 | .00107 | .00088 | .00109 | .00138 | .00066 | .00075 | .00078 | .00073 | .00125 |
| .00015 | .00018 | .00014 | .00022 | .00014 | .00012 | .00020 | .00013 | .00012 | .00024 |
| .00220 | .00274 | .00193 | .00276 | .00201 | .00138 | .00172 | .00218 | .00185 | .00344 |
| .00142 | .00062 | .00075 | .00056 | .00063 | .00035 | .00048 | .00042 | .00037 | .00072 |
| .00032 | .00037 | .00056 | .00090 | .00062 | .00020 | .00033 | .00061 | .00034 | .00051 |
| .00207 | .00324 | .00207 | .00164 | .00151 | .00791 | .00385 | .00109 | .00073 | .00311 |
| .00379 | .00844 | .00434 | .00176 | .00223 | .00210 | .00157 | .00655 | .00088 | .00242 |
| .00189 | .00473 | .00202 | .00165 | .00157 | .00108 | .00382 | .00133 | .00101 | .00278 |
| .00719 | .00992 | .01590 | .02899 | .01882 | .00349 | .00840 | .01968 | .00409 | .01394 |
| .01806 | .02668 | .03340 | .09331 | .03399 | .01100 | .02625 | .04373 | .01243 | .04682 |
| .00098 | .00147 | .00419 | .00639 | .00532 | .00103 | .00257 | .00379 | .00887 | .00341 |
| .02585 | .00927 | .00685 | .00686 | .00645 | .00511 | .00634 | .00493 | .00484 | .00583 |
| .00392 | .00312 | .00803 | .00765 | .00547 | .00162 | .00345 | .00509 | .04489 | .00438 |
| .01188 | .00994 | .01002 | .01379 | .01066 | .01024 | .00944 | .01063 | .00927 | .01102 |
| .00825 | .00619 | .00631 | .00739 | .00659 | .00460 | .00945 | .00467 | .00454 | .00548 |
| .01319 | .01973 | .04307 | .04188 | .04232 | .01376 | .01773 | .02419 | .02571 | .03485 |
| .00640 | .00988 | .04796 | .06053 | .04234 | .01081 | .01617 | .02195 | .02087 | .02206 |
| .00138 | .00203 | .00256 | .00489 | .00277 | .00091 | .00148 | .00142 | .00149 | .00211 |
| .04217 | .06566 | .06867 | .09530 | .08044 | .05894 | .07782 | .06629 | .06543 | .08549 |
| .00273 | .00341 | .00330 | .00520 | .00326 | .00181 | .00230 | .00288 | .00275 | .00354 |
| .01694 | .01573 | .01635 | .03390 | .01516 | .01342 | .01194 | .01123 | .02444 | .01565 |
| .00664 | .00773 | .00800 | .00997 | .00839 | .00424 | .00620 | .00558 | .01069 | .00812 |
| ............ | ............ | ............ | ............ | ............ | ............ | ............ | ............ | ............ | ............ |
| .03454 | .02810 | .03799 | .04042 | .02933 | .02079 | .02201 | .02470 | .13417 | .02568 |
| .00512 | .00622 | .00296 | .00362 | .00335 | .00270 | .00270 | .00256 | .00269 | .00256 |
| .00303 | .00245 | .00213 | .00253 | .00396 | .00268 | .00220 | .00171 | .00189 | .00356 |
| .01937 | .00885 | .00663 | .00831 | .00685 | .00622 | .00436 | .00436 | .00597 | .00806 |
| .02960 | .01721 | .03495 | .02565 | .04461 | .05836 | .01815 | .02140 | .01927 | .02406 |
| .06300 | .03176 | .03586 | .05324 | .03595 | .05086 | .03302 | .02568 | .02748 | .03013 |
| .00467 | .00351 | .00357 | .00418 | .00373 | .00261 | .00535 | .00264 | .00257 | .00310 |
| .00908 | .01028 | .00695 | .00746 | .00675 | .00538 | .00603 | .00567 | .00577 | .00757 |
| .01420 | .01788 | .00895 | .01344 | .01033 | .00860 | .00846 | .00784 | .00781 | .01202 |
| .00396 | .00323 | .00313 | .00367 | .00327 | .00243 | .00437 | .00238 | .00235 | .00288 |
| .00005 | .00003 | .00004 | .00006 | .00004 | .00007 | .00008 | .00006 | .00002 | .00003 |
| .00353 | .00293 | .00204 | .00284 | .00300 | .00646 | .00277 | .00181 | .00252 | .00511 |
| .01810 | .01122 | .00581 | .00623 | .00560 | .00363 | .00463 | .00405 | .00433 | .00532 |
| .00668 | .00869 | .01498 | .02010 | .01511 | .00528 | .00762 | .00953 | .00996 | .01154 |
| ............ | ............ | ............ | ............ | ............ | ............ | ............ | ............ | ............ | ............ |
| **1.85432** | **1.94490** | **2.11738** | **2.47840** | **2.28005** | **1.66921** | **1.93174** | **2.13181** | **2.34805** | **2.13098** |

TABLE 4.15

| | | | | | | | | |
|---|---|---|---|---|---|---|---|---|
| .00166 | .00652 | .00286 | .00909 | .01375 | .00767 | .01390 | .03390 | 47 |
| .00160 | .00179 | .00081 | .00103 | .00094 | .00092 | .00183 | .00103 | 48 |
| .00134 | .00201 | .00435 | .01859 | .01794 | .00968 | .01419 | .00976 | 49 |
| .00263 | .00479 | .00300 | .00703 | .00603 | .00568 | .00794 | .01674 | 50 |
| .00062 | .00070 | .00087 | .00136 | .00093 | .00094 | .00089 | .00114 | 51 |
| .00048 | .00059 | .00069 | .00110 | .00129 | .00109 | .00575 | .00308 | 52 |
| .00107 | .00379 | .00300 | .01061 | .01139 | .00607 | .00776 | .00734 | 53 |
| .00025 | .00060 | .00015 | .00023 | .00032 | .00077 | .00090 | .00136 | 54 |
| .00081 | .00289 | .00197 | .00398 | .00198 | .00188 | .00215 | .00294 | 55 |
| .00034 | .00047 | .00072 | .00051 | .00087 | .00054 | .00066 | .00070 | 56 |
| .00200 | .00209 | .00221 | .00278 | .00515 | .00299 | .00292 | .00348 | 57 |
| .00085 | .00078 | .00090 | .00127 | .00774 | .00270 | .00155 | .00200 | 58 |
| .00015 | .00020 | .00023 | .00027 | .00029 | .00025 | .00102 | .01467 | 59A |
| .00172 | .00230 | .00291 | .00305 | .00465 | .00330 | .00478 | .01116 | 59B |
| .00050 | .00057 | .00067 | .00070 | .00069 | .00077 | .00219 | .00126 | 60 |
| .00033 | .00054 | .00065 | .00096 | .00063 | .00056 | .00171 | .00058 | 61 |
| .00131 | .00453 | .00149 | .00142 | .00158 | .00126 | .00381 | .00175 | 62 |
| .00125 | .00131 | .00124 | .00128 | .00120 | .00148 | .00123 | .00125 | 63 |
| .01466 | .00340 | .00271 | .00193 | .00152 | .00189 | .00215 | .00211 | 64 |
| .00777 | .01653 | .01982 | .02416 | .01378 | .01432 | .01106 | .01164 | 65A |
| .02951 | .02925 | .08528 | .03011 | .06070 | .04634 | .03497 | .03084 | 65B |
| .00190 | .00206 | .00552 | .00742 | .00310 | .00318 | .00260 | .00288 | 65C |
| .00716 | .00797 | .00523 | .00586 | .00731 | .00992 | .00731 | .00543 | 65D |
| .00301 | .00373 | .00603 | .00419 | .00515 | .00448 | .00342 | .00304 | 65E |
| .01076 | .01618 | .01604 | .01063 | .01139 | .01074 | .01107 | .01110 | 66 |
| .00832 | .00869 | .00769 | .01159 | .00625 | .01088 | .00625 | .00697 | 67 |
| .02049 | .03751 | .03795 | .05551 | .07190 | .04538 | .02963 | .03240 | 68A |
| .01160 | .05122 | .03095 | .04595 | .03403 | .02891 | .01991 | .02195 | 68B |
| .00144 | .00171 | .00223 | .00406 | .00174 | .00255 | .00164 | .00237 | 68C |
| .08812 | .06976 | .05476 | .09984 | .11455 | .13234 | .10220 | .09536 | 69A |
| .00252 | .00319 | .00506 | .00397 | .00480 | .00409 | .00402 | .00322 | 69B |
| .01821 | .01457 | .01925 | .01656 | .02021 | .01814 | .01634 | .01848 | 70A |
| .00719 | .00653 | .00800 | .00832 | .00959 | .00906 | .00769 | .00751 | 70B |
| ............ | ............ | ............ | ............ | ............ | ............ | ............ | ............ | 71A |
| .02625 | .02401 | .02478 | .02380 | .02643 | .02646 | .02376 | .02230 | 71B |
| .00780 | .00280 | .00294 | .00252 | .00338 | .00368 | .00682 | .00283 | 72A |
| .00380 | .00233 | .00304 | .00312 | .00413 | .00082 | .00260 | .00329 | 72B |
| .00497 | .00577 | .00778 | .00733 | .00668 | .00658 | .00544 | .00850 | 73A |
| .01540 | .01386 | .01483 | .01624 | .01704 | .01560 | .01592 | .03489 | 73B |
| .03383 | .02459 | .02877 | .04483 | .03645 | .03618 | .03425 | .03382 | 73C |
| .00471 | .00492 | .00436 | .00656 | .00354 | .00616 | .00354 | .00395 | 73D |
| .00826 | .00647 | .00706 | .00678 | .00742 | .00755 | .00814 | .00710 | 74 |
| .00879 | .01271 | .01400 | .01105 | .01907 | .01533 | .01444 | .01163 | 75 |
| .00400 | .00406 | .00364 | .00530 | .00322 | .00512 | .00325 | .00346 | 76 |
| .00047 | .00002 | .00003 | .00003 | .00003 | .00003 | .00003 | .00003 | 77A |
| .00186 | .00195 | .00243 | .00202 | .00276 | .00277 | .00245 | .00464 | 77B |
| .00777 | .00541 | .00558 | .00704 | .00777 | .00608 | .00524 | .00568 | 78 |
| .00840 | .01284 | .01374 | .01921 | .01657 | .01429 | .00980 | .01122 | 79 |
| ............ | ............ | ............ | ............ | ............ | ............ | ............ | ............ | 82 |
| ............ | ............ | ............ | ............ | ............ | ............ | ............ | ............ | 83 |
| ............ | ............ | ............ | ............ | ............ | ............ | ............ | ............ | 84 |
| ............ | ............ | ............ | ............ | ............ | ............ | ............ | ............ | 85 |
| ............ | ............ | ............ | ............ | ............ | ............ | ............ | ............ | |
| 2.27791 | 1.94175 | 1.99392 | 2.14775 | 2.47511 | 2.59246 | 2.16328 | 2.14177 | |

TABLE 4.15 Continued

| Industry number | Each entry represents the output required, directly and indirectly, of the industry named at the beginning of the row for each dollar of delivery to final demand of the commodity named at the head of the column | Other fabricated metal products | Engines and turbines |
|---|---|---|---|
| | Commodity number | 42 | 43 |
| 1 | Livestock and livestock products | 0.00095 | 0.00073 |
| 2 | Other agricultural products | .00148 | .00102 |
| 3 | Forestry and fishery products | .00082 | .00039 |
| 4 | Agricultural, forestry, and fishery services | .00066 | .00052 |
| 5+6 | Metallic ores mining | .01158 | .01356 |
| 7 | Coal mining | .00915 | .00906 |
| 8 | Crude petroleum and natural gas | .01840 | .01452 |
| 9+10 | Nonmetallic minerals mining | .00309 | .00239 |
| 11+12 | Construction | .02743 | .02236 |
| 13 | Ordnance and accessories | .00038 | .00099 |
| 14 | Food and kindred products | .00341 | .00255 |
| 15 | Tobacco products | .00001 | (*) |
| 16 | Broad and narrow fabrics, yarn and thread mills | .00237 | .00154 |
| 17 | Miscellaneous textile goods and floor coverings | .00122 | .00076 |
| 18 | Apparel | .00054 | .00044 |
| 19 | Miscellaneous fabricated textile products | .00087 | .00037 |
| 20+21 | Lumber and wood products | .01141 | .00492 |
| 22+23 | Furniture and fixtures | .00189 | .00036 |
| 24 | Paper and allied products, except containers | .01809 | .00970 |
| 25 | Paperboard containers and boxes | .01031 | .00511 |
| 26A | Newspapers and periodicals | .01021 | .00834 |
| 26B | Other printing and publishing | .01368 | .00988 |
| 27A | Industrial and other chemicals | .03829 | .01651 |
| 27B | Agricultural fertilizers and chemicals | .00125 | .00066 |
| 28 | Plastics and synthetic materials | .01391 | .00729 |
| 29A | Drugs | .00028 | .00013 |
| 29B | Cleaning and toilet preparations | .00161 | .00075 |
| 30 | Paints and allied products | .00844 | .00171 |
| 31 | Petroleum refining and related products | .01544 | .01162 |
| 32 | Rubber and miscellaneous plastics products | .03069 | .02116 |
| 33+34 | Footwear, leather, and leather products | .00017 | .00012 |
| 35 | Glass and glass products | .00384 | .00115 |
| 36 | Stone and clay products | .01084 | .01105 |
| 37 | Primary iron and steel manufacturing | .20138 | .19121 |
| 38 | Primary nonferrous metals manufacturing | .08136 | .09507 |
| 39 | Metal containers | .00118 | .00049 |
| 40 | Heating, plumbing, and fabricated structural metal products | .01273 | .01729 |
| 41 | Screw machine products and stampings | .02299 | .02357 |
| 42 | Other fabricated metal products | .94446 | .02405 |
| 43 | Engines and turbines | .00168 | .98256 |
| 44+45 | Farm, construction, and mining machinery | .00330 | .01765 |
| 46 | Materials handling machinery and equipment | .00095 | .00061 |
| 47 | Metalworking machinery and equipment | .01424 | .01139 |
| 48 | Special industry machinery and equipment | .00282 | .00112 |
| 49 | General industrial machinery and equipment | .01007 | .02950 |

TABLE 4.15

| Farm, construction, and mining machinery | Materials handling machinery and equipment | Metalworking machinery and equipment | Special industry machinery and equipment | General industrial machinery and equipment | Miscellaneous machinery, except electrical | Computer and office equipment | Service industry machinery | Electrical industrial equipmen and apparatus |
|---|---|---|---|---|---|---|---|---|
| 44+45 | 46 | 47 | 48 | 49 | 50 | 51 | 52 | 53 |
| 0.00090 | 0.00096 | 0.00089 | 0.00091 | 0.00086 | 0.00081 | 0.00092 | 0.00091 | 0.00094 |
| .00125 | .00121 | .00115 | .00116 | .00126 | .00099 | .00107 | .00135 | .00127 |
| .00058 | .00057 | .00046 | .00053 | .00047 | .00037 | .00036 | .00086 | .00055 |
| .00062 | .00059 | .00053 | .00055 | .00054 | .00051 | .00057 | .00061 | .00063 |
| .00863 | .00802 | .00727 | .00722 | .00883 | .00904 | .00369 | .01040 | .00946 |
| .00834 | .00663 | .00651 | .00556 | .00694 | .00592 | .00294 | .00601 | .00601 |
| .01480 | .01352 | .01328 | .01279 | .01325 | .01088 | .00907 | .01327 | .01650 |
| .00220 | .00198 | .00276 | .00208 | .00209 | .00208 | .00102 | .00300 | .00247 |
| .02348 | .02461 | .01835 | .01998 | .01980 | .02089 | .02003 | .01973 | .02129 |
| .00069 | .00065 | .00050 | .00068 | .00102 | .00063 | .00025 | .00058 | .00033 |
| .00319 | .00340 | .00321 | .00335 | .00295 | .00287 | .00330 | .00327 | .00332 |
| (*) | (*) | (*) | (*) | (*) | (*) | (*) | (*) | .00001 |
| .00187 | .00159 | .00180 | .00158 | .00352 | .00141 | .00178 | .00225 | .00220 |
| .00104 | .00100 | .00063 | .00089 | .00627 | .00135 | .00078 | .00104 | .00097 |
| .00042 | .00031 | .00034 | .00028 | .00040 | .00026 | .00030 | .00034 | .00035 |
| .00038 | .00054 | .00103 | .00027 | .00061 | .00024 | .00034 | .00039 | .00038 |
| .00757 | .00736 | .00569 | .00675 | .00595 | .00437 | .00401 | .01187 | .00680 |
| .00053 | .00196 | .00034 | .00041 | .00026 | .00035 | .00055 | .00102 | .00056 |
| .01146 | .01119 | .01021 | .01093 | .01152 | .00888 | .01183 | .01368 | .01920 |
| .00608 | .00438 | .00667 | .00534 | .00675 | .00532 | .00538 | .01009 | .00898 |
| .01218 | .01176 | .00897 | .00966 | .00855 | .00861 | .00499 | .01000 | .01441 |
| .01317 | .01704 | .01035 | .01127 | .01073 | .00994 | .00968 | .01134 | .01503 |
| .01841 | .01612 | .02237 | .02679 | .01621 | .01282 | .01422 | .02283 | .01987 |
| .00080 | .00069 | .00080 | .00088 | .00069 | .00059 | .00068 | .00090 | .00085 |
| .00945 | .00727 | .00871 | .00801 | .00991 | .00664 | .01054 | .01419 | .01291 |
| .00015 | .00014 | .00016 | .00018 | .00017 | .00011 | .00013 | .00030 | .00016 |
| .00084 | .00077 | .00094 | .00092 | .00083 | .00071 | .00073 | .00141 | .00095 |
| .00404 | .00256 | .00217 | .00125 | .00125 | .00114 | .00128 | .00440 | .00394 |
| .01217 | .01225 | .01242 | .01196 | .01143 | .00979 | .00992 | .01202 | .01711 |
| .03691 | .02401 | .03009 | .02584 | .02335 | .01218 | .03788 | .03232 | .03158 |
| .00024 | .00013 | .00018 | .00015 | .00011 | .00011 | .00015 | .00013 | .00013 |
| .00138 | .00133 | .00120 | .00426 | .00106 | .00113 | .00213 | .00250 | .00167 |
| .00932 | .00841 | .01950 | .00865 | .01059 | .01267 | .00345 | .01081 | .01516 |
| .17180 | .13906 | .11609 | .09713 | .13313 | .10012 | .02214 | .09684 | .07837 |
| .03790 | .04444 | .05456 | .05137 | .05790 | .07589 | .03830 | .09476 | .08963 |
| .00066 | .00058 | .00057 | .00055 | .00047 | .00040 | .00051 | .00072 | .00066 |
| .04474 | .03635 | .01877 | .02037 | .01465 | .01631 | .00630 | .01606 | .00784 |
| .02113 | .02460 | .04151 | .01273 | .01464 | .01495 | .00826 | .02753 | .02109 |
| .02787 | .04569 | .02143 | .02203 | .02412 | .02632 | .01565 | .02712 | .01980 |
| .04830 | .01118 | .00408 | .00550 | .00666 | .00538 | .00087 | .00333 | .01628 |
| 1.00820 | .02172 | .00639 | .00329 | .00632 | .00636 | .00061 | .00304 | .00209 |
| .00285 | .97644 | .00131 | .00454 | .00305 | .00149 | .00032 | .00118 | .00133 |
| .01381 | .01311 | .93896 | .02298 | .01599 | .02400 | .00281 | .00991 | .00889 |
| .00237 | .01040 | .00608 | .95573 | .00954 | .00337 | .00105 | .00629 | .00094 |
| .03920 | .04481 | .01278 | .04057 | 1.00047 | .02446 | .00256 | .02943 | .01079 |

**TABLE 4.15** Continued

| | | | |
|---|---|---|---|
| 50 | Miscellaneous machinery, except electrical | .01672 | .03899 |
| 51 | Computer and office equipment | .00159 | .00147 |
| 52 | Service industry machinery | .00212 | .00108 |
| 53 | Electrical industrial equipment and apparatus | .00939 | .04025 |
| 54 | Household appliances | .00054 | .00022 |
| 55 | Electric lighting and wiring equipment | .00409 | .00229 |
| 56 | Audio, video, and communication equipment | .00078 | .00081 |
| 57 | Electronic components and accessories | .00562 | .00552 |
| 58 | Miscellaneous electrical machinery and supplies | .00185 | .01244 |
| 59A | Motor vehicles (passenger cars and trucks) | .00047 | .00154 |
| 59B | Truck and bus bodies, trailers, and motor vehicles parts | .01170 | .05590 |
| 60 | Aircraft and parts | .00346 | .04306 |
| 61 | Other transportation equipment | .00063 | .00616 |
| 62 | Scientific and controlling instruments | .00539 | .00482 |
| 63 | Ophthalmic and photographic equipment | .00130 | .00114 |
| 64 | Miscellaneous manufacturing | .00405 | .00138 |
| 65A | Railroads and related services; passenger ground transportation | .00904 | .00721 |
| 65B | Motor freight transportation and warehousing | .02846 | .02374 |
| 65C | Water transportation | .00233 | .00209 |
| 65D | Air transportation | .00755 | .00735 |
| 65E | Pipelines, freight forwarders, and related services | .00310 | .00262 |
| 66 | Communications, except radio and TV | .01195 | .01116 |
| 67 | Radio and TV broadcasting | .00830 | .00676 |
| 68A | Electric services (utilities) | .03046 | .02756 |
| 68B | Gas production and distribution (utilities) | .01926 | .01646 |
| 68C | Water and sanitary services | .00155 | .00149 |
| 69A | Wholesale trade | .08911 | .09529 |
| 69B | Retail trade | .00342 | .00277 |
| 70A | Finance | .01786 | .01544 |
| 70B | Insurance | .00711 | .00674 |
| 71A | Owner-occupied dwellings | ............ | ............ |
| 71B | Real estate and royalties | .02140 | .01854 |
| 72A | Hotels and lodging places | .00279 | .00286 |
| 72B | Personal and repair services (except auto) | .00297 | .00350 |
| 73A | Computer and data processing services | .00712 | .00847 |
| 73B | Legal, engineering, accounting, and related services | .01644 | .01410 |
| 73C | Other business and professional services, except medical | .03379 | .03028 |
| 73D | Advertising | .00470 | .00383 |
| 74 | Eating and drinking places | .00756 | .00650 |
| 75 | Automotive repair and services | .01301 | .01157 |
| 76 | Amusements | .00396 | .00345 |
| 77A | Health services | .00003 | .00002 |
| 77B | Educational and social services, and membership organizations | .00245 | .00193 |
| 78 | Federal Government enterprises | .00527 | .00485 |
| 79 | State and local government enterprises | .00968 | .00862 |
| 82 | General government industry | ............ | ............ |
| 83 | Rest-of-the-world adjustment to final uses | ............ | ............ |
| 84 | Household industry | ............ | ............ |
| 85 | Inventory valuation adjustment | ............ | ............ |
| | **Total industry output multiplier** | **2.01018** | **2.13840** |

* Less than .000005.

TABLE 4.15

| | | | | | | | | |
|---|---|---|---|---|---|---|---|---|
| .03186 | .03139 | .03542 | .03083 | .02351 | .98897 | .00331 | .01594 | .01518 |
| .00116 | .00129 | .00165 | .00477 | .00329 | .00285 | 1.18959 | .00142 | .00458 |
| .00141 | .00346 | .00097 | .00308 | .00726 | .00193 | .00071 | 1.02208 | .00115 |
| .01143 | .04477 | .02867 | .06296 | .04828 | .00911 | .03207 | .07022 | .97115 |
| .00027 | .00036 | .00022 | .00108 | .00098 | .00033 | .00015 | .01111 | .00039 |
| .00194 | .00245 | .00220 | .00245 | .00183 | .00576 | .00650 | .00842 | .01236 |
| .00067 | .00157 | .00053 | .00076 | .00063 | .00067 | .02350 | .00095 | .00358 |
| .00333 | .00514 | .00425 | .01381 | .00656 | .00582 | .14252 | .00723 | .03982 |
| .00610 | .00761 | .00182 | .00217 | .00221 | .00267 | .00525 | .00185 | .00543 |
| .00151 | .00080 | .00172 | .00029 | .00038 | .00039 | .00022 | .00053 | .00045 |
| .01194 | .01250 | .00443 | .00335 | .01027 | .01592 | .00215 | .00606 | .00414 |
| .00306 | .00887 | .00193 | .00583 | .00198 | .00453 | .00182 | .00163 | .00782 |
| .00400 | .00109 | .00061 | .00078 | .00057 | .00107 | .00039 | .00111 | .00612 |
| .00177 | .00579 | .00185 | .00621 | .00581 | .00280 | .00939 | .01930 | .00756 |
| .00129 | .00133 | .00121 | .00380 | .00149 | .00124 | .00819 | .00133 | .00143 |
| .00201 | .00207 | .00179 | .00177 | .00171 | .00146 | .00153 | .00321 | .00243 |
| .00749 | .00638 | .00617 | .00587 | .00615 | .00547 | .00344 | .00664 | .00740 |
| .02679 | .02227 | .01968 | .01721 | .01834 | .01665 | .01153 | .02234 | .02100 |
| .00233 | .00174 | .00159 | .00142 | .00171 | .00141 | .00090 | .00177 | .00155 |
| .00921 | .00782 | .00771 | .01003 | .00956 | .00686 | .02078 | .01047 | .01416 |
| .00297 | .00258 | .00245 | .00253 | .00253 | .00207 | .00306 | .00282 | .00331 |
| .01161 | .01290 | .01102 | .01274 | .01209 | .01059 | .01231 | .01193 | .01438 |
| .00996 | .00947 | .00726 | .00784 | .00691 | .00696 | .00385 | .00813 | .01181 |
| .02470 | .02152 | .02358 | .02126 | .02504 | .02321 | .01633 | .02320 | .02523 |
| .01628 | .01333 | .01236 | .01173 | .01392 | .01082 | .00639 | .01290 | .01331 |
| .00147 | .00151 | .00110 | .00116 | .00118 | .00087 | .00094 | .00156 | .00119 |
| .10658 | .10372 | .06677 | .08968 | .08448 | .05775 | .11785 | .11005 | .09476 |
| .00298 | .00295 | .00296 | .00257 | .00258 | .00254 | .00246 | .00256 | .00285 |
| .01526 | .01599 | .01474 | .01749 | .01523 | .01818 | .01971 | .01400 | .02318 |
| .00743 | .00699 | .00670 | .00628 | .00630 | .00559 | .00594 | .00619 | .00670 |
| .01946 | .02148 | .02208 | .02263 | .01997 | .02293 | .02619 | .01948 | .02208 |
| .00337 | .00338 | .00369 | .00415 | .00384 | .00284 | .00982 | .00423 | .00784 |
| .00330 | .00321 | .00270 | .00270 | .00303 | .00253 | .00239 | .00259 | .00312 |
| .00807 | .00673 | .00612 | .00587 | .00695 | .00730 | .00567 | .00570 | .00688 |
| .01775 | .01755 | .01613 | .01692 | .01584 | .01669 | .01902 | .01607 | .01658 |
| .03259 | .03348 | .02897 | .03069 | .03045 | .04339 | .03035 | .02901 | .03345 |
| .00564 | .00536 | .00411 | .00444 | .00391 | .00394 | .00218 | .00461 | .00669 |
| .00789 | .00837 | .00805 | .00800 | .00767 | .00800 | .00927 | .00757 | .00823 |
| .00970 | .00988 | .01442 | .01080 | .01050 | .01169 | .01151 | .00941 | .01218 |
| .00471 | .00471 | .00347 | .00378 | .00360 | .00339 | .00238 | .00395 | .00544 |
| .00002 | .00003 | .00002 | .00002 | .00002 | .00003 | .00002 | .00002 | .00003 |
| .00189 | .00191 | .00223 | .00183 | .00219 | .00235 | .00221 | .00250 | .00211 |
| .00547 | .00533 | .00435 | .00515 | .00500 | .00426 | .00355 | .00402 | .00507 |
| .00817 | .00735 | .00719 | .00679 | .00747 | .00651 | .00565 | .00781 | .00790 |
| 2.08876 | 2.04968 | 1.81882 | 1.90336 | 1.91831 | 1.80296 | 2.04630 | 2.06395 | 1.95570 |

TABLE 4.15   Continued

| Household appliances | Electric lighting and wiring equipment | Audio, video, and communication equipment | Electronic components and accessories | Miscellaneous electrical machinery and supplies | Motor vehicles (passenger cars and trucks) | Truck and bus bodies, trailers, and motor vehicles parts | Aircraft and parts | Other transportation equipment |
|---|---|---|---|---|---|---|---|---|
| 54 | 55 | 56 | 57 | 58 | 59A | 59B | 60 | 61 |
| 0.00110 | 0.00100 | 0.00082 | 0.00089 | 0.00120 | 0.00132 | 0.00095 | 0.00080 | 0.00101 |
| .00208 | .00155 | .00118 | .00125 | .00162 | .00313 | .00161 | .00124 | .00233 |
| .00111 | .00064 | .00048 | .00039 | .00048 | .00071 | .00072 | .00035 | .00229 |
| .00072 | .00063 | .00054 | .00057 | .00067 | .00082 | .00067 | .00057 | .00100 |
| .00764 | .00837 | .00358 | .00631 | .01313 | .00652 | .01059 | .00757 | .00669 |
| .00688 | .00552 | .00311 | .00433 | .00478 | .00628 | .00717 | .00399 | .00541 |
| .01628 | .01486 | .00903 | .01178 | .01525 | .01760 | .01647 | .01041 | .01256 |
| .00279 | .00241 | .00119 | .00182 | .00205 | .00257 | .00251 | .00134 | .00218 |
| .01941 | .01885 | .01519 | .02156 | .01890 | .02169 | .02253 | .01786 | .03339 |
| .00055 | .00017 | .00387 | .00042 | .00380 | .00063 | .00057 | .05833 | .00049 |
| .00373 | .00353 | .00285 | .00323 | .00439 | .00407 | .00324 | .00268 | .00293 |
| .00001 | .00001 | (*) | (*) | .00001 | .00001 | (*) | (*) | (*) |
| .00697 | .00378 | .00300 | .00262 | .00317 | .01869 | .00428 | .00476 | .00838 |
| .00154 | .00092 | .00129 | .00111 | .00169 | .00705 | .00296 | .00229 | .00683 |
| .00044 | .00033 | .00057 | .00054 | .00047 | .00114 | .00042 | .00041 | .00060 |
| .00045 | .00042 | .00031 | .00033 | .00040 | .02470 | .00233 | .00252 | .00789 |
| .01531 | .00850 | .00623 | .00447 | .00536 | .00897 | .00960 | .00420 | .03423 |
| .00544 | .00126 | .01095 | .00102 | .00050 | .01301 | .00116 | .00079 | .00348 |
| .02503 | .01745 | .01451 | .01047 | .03643 | .01687 | .01304 | .00872 | .00943 |
| .02481 | .01836 | .00701 | .00545 | .01096 | .00766 | .00614 | .00247 | .00472 |
| .01163 | .00926 | .00980 | .00656 | .01298 | .01418 | .01354 | .01221 | .00733 |
| .01336 | .01113 | .01320 | .00871 | .01475 | .01563 | .01419 | .01286 | .00900 |
| .04000 | .03387 | .02227 | .03627 | .04416 | .03769 | .02685 | .01288 | .02307 |
| .00151 | .00127 | .00086 | .00118 | .00145 | .00144 | .00103 | .00067 | .00108 |
| .03613 | .03000 | .01557 | .01791 | .02076 | .02667 | .01888 | .00835 | .01766 |
| .00034 | .00023 | .00018 | .00026 | .00029 | .00031 | .00020 | .00024 | .00018 |
| .00160 | .00125 | .00121 | .00121 | .00148 | .00166 | .00125 | .00075 | .00100 |
| .00878 | .00299 | .00164 | .00121 | .00142 | .01490 | .00542 | .00280 | .00811 |
| .01365 | .01318 | .00876 | .01108 | .01519 | .01661 | .01489 | .01038 | .01222 |
| .05663 | .03535 | .05320 | .06151 | .05799 | .08865 | .05177 | .01919 | .02532 |
| .00019 | .00013 | .00021 | .00014 | .00015 | .00069 | .00019 | .00016 | .00028 |
| .01301 | .04013 | .00383 | .01006 | .00299 | .01296 | .00372 | .00148 | .01256 |
| .01062 | .00694 | .00354 | .00538 | .00560 | .00969 | .01199 | .00691 | .00888 |
| .11100 | .07517 | .01906 | .02272 | .03398 | .07921 | .11438 | .04112 | .08486 |
| .04891 | .07869 | .03872 | .07125 | .09667 | .04592 | .08723 | .08223 | .04911 |
| .00109 | .00072 | .00053 | .00059 | .00074 | .00143 | .00080 | .00046 | .00084 |
| .00721 | .00285 | .00486 | .00671 | .00790 | .01039 | .02283 | .00553 | .04164 |
| .02489 | .03407 | .01251 | .01253 | .01482 | .08957 | .04321 | .01707 | .01688 |
| .04277 | .02525 | .02189 | .04020 | .03012 | .03001 | .03187 | .01969 | .02988 |
| .00168 | .00102 | .00050 | .00059 | .00213 | .01974 | .00634 | .00796 | .04733 |
| .00356 | .00319 | .00060 | .00079 | .00109 | .00203 | .00256 | .00097 | .01594 |
| .00043 | .00034 | .00031 | .00031 | .00038 | .00067 | .00089 | .00029 | .00091 |

## TABLE 4.15

| Scientific and controlling instruments | Ophthalmic and photographic equipment | Miscellaneous manufacturing | Railroads and related services; passenger ground transportation | Motor freight transportation and warehousing | Water transportation | Air transportation | Pipelines, freight forwarders, and related services | Communications, except radio and TV | Industry number |
|---|---|---|---|---|---|---|---|---|---|
| 62 | 63 | 64 | 65A | 65B | 65C | 65D | 65E | 66 | |
| 0.00120 | 0.00102 | 0.00257 | 0.00099 | 0.00135 | 0.00204 | 0.00400 | 0.00207 | 0.00068 | 1 |
| .00189 | .00155 | .00523 | .00128 | .00991 | .00221 | .00304 | .00198 | .00081 | 2 |
| .00054 | .00080 | .00245 | .00083 | .00026 | .00051 | .00056 | .00057 | .00043 | 3 |
| .00062 | .00064 | .00130 | .00120 | .00091 | .00102 | .00107 | .00097 | .00075 | 4 |
| .00445 | .00223 | .00770 | .00131 | .00044 | .00218 | .00071 | .00045 | .00065 | 5+6 |
| .00363 | .00437 | .00443 | .00291 | .00224 | .00532 | .00214 | .00564 | .00120 | 7 |
| .00957 | .01211 | .01450 | .05101 | .04263 | .02912 | .07829 | .01459 | .00439 | 8 |
| .00135 | .00207 | .00283 | .00252 | .00091 | .00113 | .00125 | .00078 | .00105 | 9+10 |
| .01543 | .01475 | .01818 | .12308 | .01601 | .02263 | .02503 | .03320 | .05944 | 11+12 |
| .01572 | .00027 | .00021 | .00014 | .00008 | .00025 | .00270 | .00028 | .00014 | 13 |
| .00405 | .00368 | .00810 | .00323 | .00343 | .00710 | .01525 | .00750 | .00224 | 14 |
| .00001 | .00002 | .00009 | (*) | (*) | (*) | (*) | (*) | (*) | 15 |
| .00843 | .00346 | .02138 | .00126 | .00097 | .00527 | .00117 | .00306 | .00085 | 16 |
| .00614 | .00211 | .00342 | .00092 | .00048 | .00381 | .00046 | .00078 | .00041 | 17 |
| .00079 | .00027 | .00207 | .00060 | .00029 | .00564 | .00046 | .00033 | .00084 | 18 |
| .00037 | .00026 | .00541 | .00052 | .00047 | .00391 | .00075 | .00488 | .00018 | 19 |
| .00687 | .00952 | .03515 | .01153 | .00259 | .00414 | .00358 | .00558 | .00559 | 20+21 |
| .00165 | .00032 | .00151 | .00045 | .00010 | .00022 | .00014 | .00015 | .00038 | 22+23 |
| .01819 | .08910 | .03626 | .00657 | .00597 | .00878 | .00884 | .01022 | .00596 | 24 |
| .00714 | .01203 | .01806 | .00209 | .00205 | .00278 | .00195 | .00222 | .00134 | 25 |
| .01104 | .01091 | .01929 | .00347 | .00399 | .00776 | .01125 | .01137 | .00478 | 26A |
| .01285 | .01407 | .02596 | .00992 | .00826 | .01365 | .01460 | .01839 | .01130 | 26B |
| .02127 | .04590 | .04126 | .01072 | .00756 | .01042 | .00839 | .00549 | .00553 | 27A |
| .00091 | .00161 | .00203 | .00142 | .00123 | .00181 | .00071 | .00056 | .00040 | 27B |
| .01608 | .01754 | .03465 | .00401 | .00371 | .00534 | .00230 | .00265 | .00254 | 28 |
| .00390 | .00033 | .00037 | .00011 | .00009 | .00016 | .00016 | .00010 | .00049 | 29A |
| .00160 | .00171 | .00285 | .00104 | .00070 | .00100 | .00102 | .00052 | .00039 | 29B |
| .00150 | .00102 | .00549 | .00176 | .00087 | .00279 | .00066 | .00060 | .00159 | 30 |
| .00932 | .01198 | .01472 | .07886 | .06679 | .04271 | .12443 | .01224 | .00520 | 31 |
| .02977 | .03376 | .04259 | .01204 | .01532 | .01066 | .00578 | .00690 | .00802 | 32 |
| .00030 | .00012 | .00444 | .00015 | .00019 | .00028 | .00024 | .00046 | .00015 | 33+34 |
| .00469 | .00703 | .00251 | .00228 | .00103 | .00116 | .00105 | .00103 | .00083 | 35 |
| .00511 | .00677 | .00895 | .00871 | .00181 | .00331 | .00267 | .00254 | .00394 | 36 |
| .03442 | .01122 | .03011 | .01539 | .00455 | .01593 | .00527 | .00434 | .00569 | 37 |
| .03845 | .01946 | .08488 | .00934 | .00297 | .02122 | .00618 | .00336 | .00542 | 38 |
| .00076 | .00064 | .00102 | .00048 | .00037 | .00065 | .00080 | .00037 | .00024 | 39 |
| .00845 | .00209 | .00243 | .00803 | .00133 | .00315 | .00180 | .00213 | .00348 | 40 |
| .01583 | .00636 | .00733 | .00312 | .00229 | .00292 | .00229 | .00132 | .00383 | 41 |
| .02415 | .01812 | .01569 | .01160 | .00626 | .02576 | .00424 | .00391 | .00452 | 42 |
| .00071 | .00048 | .00070 | .00386 | .00088 | .00893 | .00073 | .00155 | .00207 | 43 |
| .00073 | .00061 | .00092 | .00126 | .00057 | .00148 | .00071 | .00062 | .00041 | 44+45 |
| .00028 | .00021 | .00035 | .00049 | .00017 | .00038 | .00018 | .00019 | .00022 | 46 |

**TABLE 4.15** Continued

| | | | | | | | | |
|---|---|---|---|---|---|---|---|---|
| .01335 | .00627 | .00331 | .00427 | .00658 | .01477 | .00911 | .01838 | .00624 |
| .00170 | .00080 | .00136 | .00402 | .00135 | .00147 | .00124 | .00104 | .00104 |
| .01188 | .00384 | .00303 | .00329 | .00604 | .01239 | .02691 | .00555 | .02641 |
| .00559 | .00559 | .00362 | .00480 | .01472 | .02527 | .04696 | .01413 | .01038 |
| .00183 | .00127 | .02236 | .03822 | .01678 | .00214 | .00148 | .00330 | .00100 |
| .02726 | .00291 | .00071 | .00067 | .00130 | .02638 | .01582 | .00057 | .00560 |
| .04179 | .03430 | .00979 | .00948 | .01595 | .00868 | .00794 | .00586 | .02134 |
| .95536 | .00082 | .00047 | .00019 | .00064 | .00057 | .00034 | .00016 | .00632 |
| .01399 | .96772 | .00954 | .00398 | .01303 | .00707 | .00610 | .00129 | .00566 |
| .01915 | .00360 | .97001 | .01204 | .00858 | .01090 | .00152 | .01694 | .00159 |
| .02145 | .01882 | .22148 | 1.04277 | .07712 | .01721 | .01097 | .02998 | .00477 |
| .00205 | .01206 | .01268 | .01141 | .92407 | .03498 | .02713 | .00275 | .00799 |
| .00042 | .00062 | .00028 | .00026 | .00045 | 1.00532 | .05817 | .00035 | .02677 |
| .00302 | .00557 | .00704 | .00195 | .02247 | .33205 | 1.02206 | .00288 | .02920 |
| .00134 | .00130 | .00254 | .00183 | .00671 | .00265 | .00137 | 1.16681 | .00712 |
| .00061 | .00066 | .00025 | .00029 | .00040 | .00113 | .00158 | .00062 | 1.00504 |
| .02120 | .00637 | .03490 | .01432 | .02866 | .01215 | .00291 | .03840 | .00574 |
| .00137 | .00161 | .00208 | .00130 | .00795 | .00176 | .00140 | .00131 | .00107 |
| .01025 | .00472 | .00201 | .00142 | .00208 | .00279 | .00209 | .00166 | .00352 |
| .00831 | .00704 | .00380 | .00482 | .00642 | .01098 | .00873 | .00393 | .00712 |
| .02770 | .02271 | .01404 | .01647 | .02277 | .04120 | .03301 | .01377 | .02360 |
| .00185 | .00144 | .00090 | .00117 | .00155 | .00207 | .00206 | .00107 | .00169 |
| .01032 | .01281 | .01071 | .01074 | .01617 | .01219 | .00841 | .02106 | .00720 |
| .00317 | .00313 | .00219 | .00248 | .00350 | .00412 | .00330 | .00317 | .00258 |
| .01122 | .01204 | .01196 | .01130 | .01306 | .01330 | .01426 | .01320 | .01200 |
| .00947 | .00747 | .00792 | .00523 | .01053 | .01160 | .01111 | .01001 | .00592 |
| .02431 | .02388 | .01697 | .02580 | .02503 | .02369 | .02697 | .01947 | .02083 |
| .01678 | .01445 | .00763 | .01020 | .01218 | .01572 | .01599 | .00892 | .01091 |
| .00155 | .00127 | .00080 | .00090 | .00157 | .00156 | .00137 | .00088 | .00130 |
| .11509 | .09897 | .08231 | .07558 | .10065 | .14639 | .10303 | .04962 | .09799 |
| .00265 | .00270 | .00190 | .00225 | .00312 | .00482 | .00471 | .00220 | .00352 |
| .01942 | .02101 | .01534 | .02058 | .02384 | .01926 | .01562 | .02410 | .01284 |
| .00667 | .00635 | .00565 | .00578 | .00760 | .01018 | .00853 | .00584 | .00629 |
| ............ | ............ | ............ | ............ | ............ | ............ | ............ | ............ | ............ |
| .02020 | .02177 | .02310 | .02258 | .02500 | .02298 | .02129 | .02123 | .03002 |
| .00464 | .00714 | .00347 | .00542 | .00860 | .00379 | .00302 | .01270 | .00279 |
| .00365 | .00297 | .00601 | .00245 | .00317 | .00401 | .00378 | .00244 | .00252 |
| .00636 | .00646 | .00683 | .00638 | .00726 | .00696 | .00643 | .00628 | .00488 |
| .01528 | .01753 | .01647 | .01713 | .02119 | .01820 | .01587 | .01757 | .01507 |
| .03251 | .03107 | .03199 | .03103 | .03628 | .03617 | .03400 | .03482 | .02960 |
| .00537 | .00423 | .00448 | .00296 | .00596 | .00657 | .00629 | .00567 | .00335 |
| .00770 | .00850 | .00705 | .00758 | .01077 | .00889 | .00783 | .00648 | .00640 |
| .00974 | .01158 | .00764 | .00786 | .01529 | .02449 | .04210 | .00730 | .01269 |
| .00452 | .00373 | .00398 | .00326 | .00630 | .00578 | .00523 | .00487 | .00360 |
| .00003 | .00003 | .00002 | .00002 | .00003 | .00003 | .00003 | .00002 | .00002 |
| .00348 | .00207 | .00320 | .00415 | .00273 | .00478 | .00322 | .00326 | .00192 |
| .00707 | .00486 | .00503 | .00430 | .00511 | .00631 | .00677 | .00511 | .00410 |
| .00862 | .00780 | .00541 | .00705 | .00853 | .00929 | .00913 | .00607 | .00722 |
| ............ | ............ | ............ | ............ | ............ | ............ | ............ | ............ | ............ |
| ............ | ............ | ............ | ............ | ............ | ............ | ............ | ............ | ............ |
| ............ | ............ | ............ | ............ | ............ | ............ | ............ | ............ | ............ |
| 2.13254 | 1.95934 | 1.93327 | 1.86771 | 2.05131 | 2.67846 | 2.24813 | 2.01852 | 2.07465 |

## TABLE 4.15

| | | | | | | | | | |
|---|---|---|---|---|---|---|---|---|---|
| .00476 | .00271 | .00461 | .00295 | .00076 | .00952 | .00161 | .00066 | .00065 | 47 |
| .00128 | .00134 | .00128 | .00043 | .00027 | .00083 | .00033 | .00036 | .00030 | 48 |
| .00467 | .00242 | .00366 | .00891 | .00194 | .02898 | .00155 | .00606 | .00311 | 49 |
| .00488 | .00365 | .00590 | .00557 | .00176 | .00977 | .00215 | .00175 | .00090 | 50 |
| .01515 | .00455 | .00177 | .00125 | .00091 | .00221 | .00169 | .00256 | .00269 | 51 |
| .00086 | .00046 | .00172 | .00221 | .00090 | .00088 | .00053 | .00064 | .00086 | 52 |
| .01891 | .00532 | .00417 | .01623 | .00195 | .00421 | .00137 | .00244 | .00327 | 53 |
| .00029 | .00020 | .00140 | .00053 | .00011 | .00087 | .00019 | .00017 | .00024 | 54 |
| .00410 | .00436 | .00189 | .00325 | .00140 | .00268 | .00092 | .00106 | .00185 | 55 |
| .01681 | .00145 | .00133 | .00077 | .00060 | .00065 | .00144 | .00106 | .01706 | 56 |
| .10356 | .08661 | .01666 | .00299 | .00190 | .00417 | .00506 | .00351 | .01931 | 57 |
| .00796 | .00291 | .00203 | .00290 | .00164 | .00289 | .00129 | .00264 | .00241 | 58 |
| .00035 | .00018 | .00053 | .00111 | .00071 | .00059 | .00013 | .00016 | .00014 | 59A |
| .00404 | .00183 | .00291 | .00992 | .00863 | .00298 | .00164 | .00243 | .00157 | 59B |
| .01208 | .00097 | .00109 | .00078 | .00042 | .00119 | .05232 | .00139 | .00042 | 60 |
| .00036 | .00028 | .00105 | .02116 | .00031 | .01724 | .00031 | .00065 | .00027 | 61 |
| .95889 | .01389 | .00212 | .00114 | .00049 | .00253 | .00258 | .00080 | .00147 | 62 |
| .00421 | .99892 | .00214 | .00100 | .00110 | .00167 | .00124 | .00167 | .00115 | 63 |
| .00304 | .00245 | 1.01047 | .00134 | .00107 | .00369 | .00196 | .00228 | .00152 | 64 |
| .00433 | .00561 | .00740 | .94368 | .00616 | .00406 | .00317 | .00561 | .00189 | 65A |
| .01426 | .01718 | .03234 | .01246 | 1.17942 | .01354 | .00965 | .01091 | .00684 | 65B |
| .00105 | .00175 | .00203 | .00244 | .00200 | 1.10822 | .00383 | .00195 | .00047 | 65C |
| .00835 | .00940 | .00683 | .00586 | .00602 | .00966 | 1.02442 | .02402 | .00401 | 65D |
| .00204 | .00248 | .00304 | .02851 | .04943 | .07693 | .09481 | 1.04047 | .00099 | 65E |
| .01255 | .00890 | .01030 | .00787 | .02423 | .01704 | .02466 | .03097 | 1.26341 | 66 |
| .00897 | .00883 | .01558 | .00249 | .00291 | .00603 | .00901 | .00890 | .00363 | 67 |
| .01808 | .01611 | .02191 | .01722 | .01529 | .03282 | .01400 | .04128 | .00706 | 68A |
| .00817 | .00943 | .01121 | .00706 | .00446 | .00638 | .00611 | .00636 | .00263 | 68B |
| .00106 | .00161 | .00226 | .00091 | .00069 | .00248 | .00090 | .00273 | .00374 | 68C |
| .06574 | .06309 | .08627 | .04993 | .03319 | .04916 | .05119 | .01963 | .01928 | 69A |
| .00224 | .00224 | .00336 | .01075 | .03558 | .00239 | .00303 | .00352 | .00337 | 69B |
| .01521 | .02116 | .01963 | .02015 | .01783 | .08554 | .04898 | .05804 | .02338 | 70A |
| .00605 | .00723 | .00773 | .01230 | .01933 | .00846 | .01413 | .03345 | .00372 | 70B |
| ......... | ......... | ......... | ......... | ......... | ......... | ......... | ......... | ......... | 71A |
| .02301 | .01872 | .02512 | .02954 | .04570 | .06254 | .04639 | .05607 | .03171 | 71B |
| .00319 | .00403 | .00307 | .00350 | .00327 | .00322 | .00311 | .00565 | .00186 | 72A |
| .00265 | .00201 | .00310 | .00231 | .00276 | .00287 | .00734 | .00357 | .00411 | 72B |
| .00541 | .00542 | .00538 | .01294 | .00821 | .02364 | .01743 | .02623 | .01672 | 73A |
| .01929 | .01564 | .02229 | .01914 | .01459 | .01673 | .01768 | .02855 | .01439 | 73B |
| .03148 | .03264 | .03626 | .03345 | .02812 | .09391 | .03601 | .03231 | .02007 | 73C |
| .00508 | .00500 | .00882 | .00141 | .00165 | .00341 | .00510 | .00504 | .00205 | 73D |
| .00763 | .00702 | .01194 | .00828 | .01002 | .00934 | .03621 | .02311 | .00632 | 74 |
| .01144 | .00980 | .01532 | .01953 | .04979 | .00881 | .00999 | .01453 | .00744 | 75 |
| .00456 | .00486 | .00733 | .00159 | .00184 | .00313 | .00509 | .00471 | .00850 | 76 |
| .00003 | .00003 | .00004 | .00003 | .00003 | .00007 | .00005 | .00003 | .00002 | 77A |
| .00338 | .00719 | .00581 | .00434 | .00181 | .00504 | .00212 | .00584 | .00269 | 77B |
| .00505 | .00403 | .00671 | .00441 | .00460 | .00660 | .00432 | .00790 | .00403 | 78 |
| .00600 | .00664 | .00876 | .11140 | .00670 | .01193 | .02335 | .01214 | .00756 | 79 |
| ......... | ......... | ......... | ......... | ......... | ......... | ......... | ......... | ......... | 82 |
| ......... | ......... | ......... | ......... | ......... | ......... | ......... | ......... | ......... | 83 |
| ......... | ......... | ......... | ......... | ......... | ......... | ......... | ......... | ......... | 84 |
| ......... | ......... | ......... | ......... | ......... | ......... | ......... | ......... | ......... | 85 |
| 1.81260 | 1.80504 | 1.99597 | 1.85843 | 1.82450 | 2.06097 | 1.94720 | 1.72170 | 1.68937 | |

**TABLE 4.15**   Continued

| Industry number | Each entry represents the output required, directly and indirectly, of the industry named at the beginning of the row for each dollar of delivery to final demand of the commodity named at the head of the column | Radio and TV broad- casting | Electric services (utilities) |
|---|---|---|---|
| | Commodity number | 67 | 68A |
| 1 | Livestock and livestock products | 0.00344 | 0.00050 |
| 2 | Other agricultural products | .00331 | .00075 |
| 3 | Forestry and fishery products | .00050 | .00051 |
| 4 | Agricultural, forestry, and fishery services | .00408 | .00089 |
| 5+6 | Metallic ores mining | .00054 | .00076 |
| 7 | Coal mining | .00160 | .11151 |
| 8 | Crude petroleum and natural gas | .00593 | .05256 |
| 9+10 | Nonmetallic minerals mining | .00076 | .00176 |
| 11+12 | Construction | .02807 | .07791 |
| 13 | Ordnance and accessories | .00014 | .00009 |
| 14 | Food and kindred products | .00963 | .00153 |
| 15 | Tobacco products | (*) | (*) |
| 16 | Broad and narrow fabrics, yarn and thread mills | .00178 | .00071 |
| 17 | Miscellaneous textile goods and floor coverings | .00047 | .00051 |
| 18 | Apparel | .00115 | .00023 |
| 19 | Miscellaneous fabricated textile products | .00072 | .00018 |
| 20+21 | Lumber and wood products | .00489 | .00719 |
| 22+23 | Furniture and fixtures | .00021 | .00023 |
| 24 | Paper and allied products, except containers | .00871 | .00373 |
| 25 | Paperboard containers and boxes | .00141 | .00116 |
| 26A | Newspapers and periodicals | .00793 | .00205 |
| 26B | Other printing and publishing | .01334 | .00476 |
| 27A | Industrial and other chemicals | .00736 | .00885 |
| 27B | Agricultural fertilizers and chemicals | .00122 | .00050 |
| 28 | Plastics and synthetic materials | .00275 | .00213 |
| 29A | Drugs | .00014 | .00010 |
| 29B | Cleaning and toilet preparations | .00082 | .00073 |
| 30 | Paints and allied products | .00054 | .00094 |
| 31 | Petroleum refining and related products | .00645 | .05098 |
| 32 | Rubber and miscellaneous plastics products | .00786 | .00577 |
| 33+34 | Footwear, leather, and leather products | .00070 | .00009 |
| 35 | Glass and glass products | .00099 | .00085 |
| 36 | Stone and clay products | .00223 | .00570 |
| 37 | Primary iron and steel manufacturing | .00338 | .00766 |
| 38 | Primary nonferrous metals manufacturing | .00472 | .00561 |
| 39 | Metal containers | .00046 | .00038 |
| 40 | Heating, plumbing, and fabricated structural metal products | .00187 | .00493 |
| 41 | Screw machine products and stampings | .00144 | .00260 |
| 42 | Other fabricated metal products | .00343 | .00397 |
| 43 | Engines and turbines | .00028 | .00501 |
| 44+45 | Farm, construction, and mining machinery | .00047 | .00539 |
| 46 | Materials handling machinery and equipment | .00018 | .00076 |
| 47 | Metalworking machinery and equipment | .00071 | .00133 |

TABLE 4.15

| Gas production and distribution (utilities) | Water and sanitary services | Wholesale trade | Retail trade | Finance | Insurance | Owner-occupied dwellings | Real estate and royalties | Hotels and lodging places |
|---|---|---|---|---|---|---|---|---|
| 68B | 68C | 69A | 69B | 70A | 70B | 71A | 71B | 72A |
| 0.00047 | 0.00142 | 0.00159 | 0.00218 | 0.00160 | 0.00231 | 0.00107 | 0.00166 | 0.00234 |
| .00062 | .00238 | .00150 | .00178 | .00158 | .00188 | .00172 | .00208 | .00331 |
| .00031 | .00174 | .00065 | .00044 | .00036 | .00039 | .00031 | .00054 | .00059 |
| .00103 | .00326 | .00101 | .00109 | .00084 | .00082 | .00819 | .00610 | .00611 |
| .00049 | .00184 | .00037 | .00031 | .00032 | .00023 | .00023 | .00038 | .00059 |
| .00225 | .00788 | .00178 | .00340 | .00158 | .00077 | .00031 | .00093 | .00508 |
| .49087 | .04801 | .01097 | .00896 | .00651 | .00381 | .00180 | .00474 | .01944 |
| .00115 | .00475 | .00058 | .00065 | .00054 | .00039 | .00093 | .00125 | .00199 |
| .04552 | .20297 | .01537 | .02564 | .01856 | .01378 | .05323 | .07124 | .05314 |
| .00006 | .00038 | .00016 | .00007 | .00014 | .00008 | .00003 | .00007 | .00016 |
| .00139 | .00408 | .00564 | .00805 | .00574 | .00858 | .00086 | .00395 | .00606 |
| (*) | (*) | .00001 | .00001 | (*) | (*) | ............. | (*) | .00001 |
| .00037 | .00178 | .00110 | .00078 | .00098 | .00061 | .00024 | .00057 | .00993 |
| .00026 | .00127 | .00051 | .00038 | .00036 | .00024 | .00026 | .00041 | .00201 |
| .00015 | .00075 | .00041 | .00024 | .00019 | .00015 | .00004 | .00014 | .00353 |
| .00014 | .00049 | .00061 | .00028 | .00074 | .00035 | .00009 | .00022 | .01210 |
| .00409 | .02486 | .00753 | .00400 | .00300 | .00216 | .00436 | .00646 | .00697 |
| .00014 | .00067 | .00011 | .00011 | .00010 | .00007 | .00014 | .00021 | .00032 |
| .00315 | .01094 | .01761 | .01697 | .01377 | .00933 | .00142 | .00655 | .01841 |
| .00067 | .00262 | .01129 | .00232 | .00122 | .00097 | .00039 | .00088 | .00303 |
| .00323 | .00464 | .00881 | .01730 | .00861 | .00798 | .00091 | .00690 | .00911 |
| .00469 | .00928 | .01877 | .01620 | .02702 | .02148 | .00191 | .01092 | .01950 |
| .00904 | .04333 | .00535 | .00454 | .00468 | .00332 | .00181 | .00410 | .01327 |
| .00061 | .00331 | .00061 | .00055 | .00056 | .00049 | .00228 | .00163 | .00701 |
| .00137 | .00744 | .00258 | .00196 | .00200 | .00137 | .00070 | .00161 | .00802 |
| .00007 | .00035 | .00010 | .00009 | .00011 | .00009 | .00004 | .00007 | .00036 |
| .00052 | .00174 | .00076 | .00059 | .00075 | .00045 | .00020 | .00053 | .00717 |
| .00063 | .00263 | .00058 | .00053 | .00045 | .00033 | .00052 | .00079 | .00098 |
| .01241 | .03761 | .01248 | .01038 | .00847 | .00481 | .00218 | .00544 | .01484 |
| .00283 | .01920 | .00695 | .00485 | .00573 | .00372 | .00174 | .00463 | .02567 |
| .00007 | .00025 | .00032 | .00021 | .00024 | .00028 | .00003 | .00012 | .00081 |
| .00050 | .00296 | .00116 | .00065 | .00085 | .00062 | .00029 | .00064 | .01354 |
| .00484 | .01546 | .00170 | .00195 | .00153 | .00110 | .00321 | .00456 | .00506 |
| .00599 | .01710 | .00332 | .00290 | .00243 | .00175 | .00252 | .00374 | .00500 |
| .00267 | .01175 | .00279 | .00229 | .00250 | .00173 | .00172 | .00278 | .00404 |
| .00019 | .00071 | .00192 | .00038 | .00034 | .00035 | .00010 | .00025 | .00061 |
| .00290 | .01195 | .00122 | .00157 | .00123 | .00089 | .00314 | .00399 | .00319 |
| .00085 | .00398 | .00142 | .00122 | .00105 | .00086 | .00044 | .00087 | .00183 |
| .00401 | .01818 | .00310 | .00354 | .00221 | .00155 | .00131 | .00240 | .00435 |
| .00043 | .00100 | .00029 | .00031 | .00029 | .00019 | .00011 | .00017 | .00049 |
| .00196 | .00378 | .00051 | .00044 | .00043 | .00025 | .00085 | .00051 | .00088 |
| .00018 | .00072 | .00077 | .00015 | .00017 | .00010 | .00016 | .00024 | .00032 |
| .00085 | .00177 | .00076 | .00049 | .00052 | .00032 | .00020 | .00040 | .00091 |

**TABLE 4.15**   Continued

| | | | |
|---|---|---|---|
| 48 | Special industry machinery and equipment | .00040 | .00023 |
| 49 | General industrial machinery and equipment | .00080 | .00384 |
| 50 | Miscellaneous machinery, except electrical | .00097 | .00193 |
| 51 | Computer and office equipment | .00322 | .00097 |
| 52 | Service industry machinery | .00084 | .00108 |
| 53 | Electrical industrial equipment and apparatus | .00184 | .00349 |
| 54 | Household appliances | .00029 | .00027 |
| 55 | Electric lighting and wiring equipment | .00128 | .00307 |
| 56 | Audio, video, and communication equipment | .00272 | .00048 |
| 57 | Electronic components and accessories | .03198 | .00182 |
| 58 | Miscellaneous electrical machinery and supplies | .00267 | .00214 |
| 59A | Motor vehicles (passenger cars and trucks) | .00011 | .00015 |
| 59B | Truck and bus bodies, trailers, and motor vehicles parts | .00137 | .00188 |
| 60 | Aircraft and parts | .00073 | .00055 |
| 61 | Other transportation equipment | .00081 | .00091 |
| 62 | Scientific and controlling instruments | .00184 | .00127 |
| 63 | Ophthalmic and photographic equipment | .00861 | .00071 |
| 64 | Miscellaneous manufacturing | .00267 | .00084 |
| 65A | Railroads and related services; passenger ground transportation | .00326 | .02952 |
| 65B | Motor freight transportation and warehousing | .00882 | .01013 |
| 65C | Water transportation | .00073 | .00477 |
| 65D | Air transportation | .01197 | .00458 |
| 65E | Pipelines, freight forwarders, and related services | .00193 | .00472 |
| 66 | Communications, except radio and TV | .01838 | .00582 |
| 67 | Radio and TV broadcasting | 1.02233 | .00152 |
| 68A | Electric services (utilities) | .01040 | .85344 |
| 68B | Gas production and distribution (utilities) | .00427 | .04944 |
| 68C | Water and sanitary services | .00108 | .00108 |
| 69A | Wholesale trade | .02157 | .03274 |
| 69B | Retail trade | .00244 | .00441 |
| 70A | Finance | .02505 | .02059 |
| 70B | Insurance | .00535 | .01196 |
| 71A | Owner-occupied dwellings | ............... | ............... |
| 71B | Real estate and royalties | .09383 | .02502 |
| 72A | Hotels and lodging places | .00452 | .00163 |
| 72B | Personal and repair services (except auto) | .01569 | .00265 |
| 73A | Computer and data processing services | .01199 | .01047 |
| 73B | Legal, engineering, accounting, and related services | .03583 | .01549 |
| 73C | Other business and professional services, except medical | .05481 | .01665 |
| 73D | Advertising | .00343 | .00086 |
| 74 | Eating and drinking places | .01487 | .00376 |
| 75 | Automotive repair and services | .00851 | .00843 |
| 76 | Amusements | .40593 | .00096 |
| 77A | Health services | .00006 | .00001 |
| 77B | Educational and social services, and membership organizations | .00798 | .00333 |
| 78 | Federal Government enterprises | .00623 | .05042 |
| 79 | State and local government enterprises | .00928 | .11696 |
| 82 | General government industry | ............... | ............... |
| 83 | Rest-of-the-world adjustment to final uses | ............... | ............... |
| 84 | Household industry | ............... | ............... |
| 85 | Inventory valuation adjustment | ............... | ............... |
| | **Total Industry output multiplier** | **2.02027** | **1.70058** |

* Less than .000005.

## TABLE 4.15

| | | | | | | | | |
|---|---|---|---|---|---|---|---|---|
| .00017 | .00064 | .00042 | .00033 | .00039 | .00026 | .00008 | .00022 | .00060 |
| .00169 | .00331 | .00084 | .00072 | .00104 | .00058 | .00040 | .00076 | .00129 |
| .00131 | .00769 | .00144 | .00096 | .00077 | .00055 | .00024 | .00068 | .00135 |
| .00155 | .00159 | .00092 | .00111 | .00461 | .00163 | .00022 | .00057 | .00142 |
| .00066 | .00330 | .00080 | .00077 | .00047 | .00034 | .00066 | .00097 | .00102 |
| .00171 | .00310 | .00080 | .00075 | .00106 | .00061 | .00049 | .00085 | .00144 |
| .00022 | .00122 | .00015 | .00019 | .00013 | .00010 | .00016 | .00042 | .00056 |
| .00125 | .00502 | .00080 | .00091 | .00091 | .00061 | .00094 | .00155 | .00435 |
| .00034 | .00140 | .00057 | .00058 | .00096 | .00082 | .00025 | .00053 | .00086 |
| .00131 | .00447 | .00250 | .00226 | .00605 | .00273 | .00050 | .00148 | .00372 |
| .00203 | .00346 | .00119 | .00102 | .00259 | .00200 | .00029 | .00091 | .00117 |
| .00008 | .00174 | .00021 | .00017 | .00014 | .00011 | .00003 | .00008 | .00020 |
| .00117 | .02960 | .00332 | .00266 | .00202 | .00170 | .00038 | .00110 | .00292 |
| .00029 | .00075 | .00077 | .00030 | .00077 | .00043 | .00008 | .00022 | .00041 |
| .00020 | .00078 | .00022 | .00021 | .00052 | .00026 | .00006 | .00018 | .00051 |
| .00069 | .01505 | .00062 | .00052 | .00074 | .00051 | .00024 | .00046 | .00171 |
| .00061 | .00163 | .00144 | .00128 | .00555 | .00316 | .00030 | .00095 | .00198 |
| .00071 | .00224 | .00206 | .00262 | .00284 | .00227 | .00032 | .00130 | .00468 |
| .00193 | .00718 | .00288 | .00368 | .00417 | .00391 | .00076 | .00250 | .00593 |
| .00530 | .12971 | .00751 | .00784 | .03166 | .01005 | .00271 | .00663 | .01506 |
| .00160 | .00320 | .00077 | .00062 | .00059 | .00036 | .00018 | .00041 | .00112 |
| .00408 | .00676 | .01349 | .00473 | .01348 | .00733 | .00108 | .00333 | .00576 |
| .00627 | .00762 | .00321 | .00148 | .00304 | .00201 | .00038 | .00127 | .00338 |
| .00501 | .01705 | .02202 | .01900 | .03491 | .03148 | .00249 | .00966 | .02612 |
| .00256 | .00348 | .00681 | .01422 | .00553 | .00598 | .00070 | .00554 | .00649 |
| .01501 | .05175 | .01111 | .02400 | .01049 | .00473 | .00138 | .00523 | .03412 |
| 1.10553 | .05609 | .00759 | .00595 | .00295 | .00189 | .00082 | .00302 | .02273 |
| .00076 | .39558 | .00070 | .00104 | .00090 | .00135 | .00013 | .00039 | .00427 |
| .01682 | .05234 | 1.03595 | .01516 | .01798 | .01195 | .00633 | .01200 | .03396 |
| .00282 | .01499 | .00395 | .99910 | .00306 | .00195 | .00318 | .00414 | .00503 |
| .01686 | .02938 | .02398 | .01804 | 1.20851 | .11255 | .01213 | .02814 | .06660 |
| .00592 | .06532 | .00516 | .00590 | .02292 | 1.40442 | .04374 | .02713 | .00776 |
| ............. | ............. | ............. | ............. | ............. | ............. | 1.00000 | | ............. |
| .09933 | .03702 | .03845 | .07446 | .05781 | .05680 | .04472 | 1.07371 | .05920 |
| .00165 | .00496 | .00868 | .00343 | .00560 | .01044 | .00099 | .00638 | 1.00308 |
| .00147 | .00705 | .00574 | .00620 | .00320 | .00291 | .00047 | .00303 | .01883 |
| .00696 | .01268 | .00619 | .01007 | .06143 | .01841 | .00198 | .00416 | .00946 |
| .01558 | .05401 | .01990 | .04402 | .04063 | .03368 | .01158 | .01669 | .03054 |
| .01462 | .03961 | .06608 | .03744 | .07363 | .03603 | .01068 | .03354 | .08971 |
| .00145 | .00197 | .00386 | .00805 | .00313 | .00339 | .00039 | .00314 | .00367 |
| .00361 | .00925 | .01722 | .01625 | .01815 | .02742 | .00197 | .01183 | .00624 |
| .00774 | .01595 | .02602 | .02068 | .01274 | .01083 | .00212 | .00745 | .02008 |
| .00130 | .00214 | .00629 | .00678 | .00346 | .00371 | .00053 | .00288 | .00376 |
| .00001 | .00003 | .00005 | .00004 | .00005 | .00004 | .00002 | .00003 | .00007 |
| .00124 | .00285 | .00286 | .00241 | .00491 | .00188 | .00034 | .00140 | .00527 |
| .00446 | .00845 | .00506 | .01043 | .03286 | .01730 | .00128 | .00577 | .01050 |
| .05390 | .54715 | .00546 | .01046 | .00551 | .00483 | .00152 | .02372 | .01915 |
| ............. | ............. | ............. | ............. | ............. | ............. | ............. | ............. | ............. |
| ............. | ............. | ............. | ............. | ............. | ............. | ............. | ............. | ............. |
| ............. | ............. | ............. | ............. | ............. | ............. | ............. | ............. | ............. |
| 2.03370 | 2.21174 | 1.53111 | 1.53953 | 1.84520 | 1.94722 | 1.26218 | 1.48199 | 1.85045 |

TABLE 4.15 Continued

| Personal and repair services (exc. auto) | Computer and data processing services | Legal, engineering, accounting, and related services | Other business and professional services, except medical | Advertising | Eating and drinking places | Automotive repair and services | Amusements | Health services |
|---|---|---|---|---|---|---|---|---|
| 72B | 73A | 73B | 73C | 73D | 74 | 75 | 76 | 77A |
| 0.00213 | 0.00118 | 0.00117 | 0.00186 | 0.00194 | 0.07242 | 0.00127 | 0.00737 | 0.00419 |
| .00268 | .00118 | .00108 | .00157 | .00224 | .05198 | .00140 | .00733 | .00355 |
| .00054 | .00037 | .00029 | .00038 | .00105 | .00816 | .00042 | .00085 | .00047 |
| .00120 | .00069 | .00087 | .00095 | .00166 | .00875 | .00089 | .01027 | .00154 |
| .00101 | .00076 | .00035 | .00072 | .00084 | .00068 | .00224 | .00052 | .00065 |
| .00364 | .00169 | .00116 | .00176 | .00272 | .00390 | .00329 | .00303 | .00221 |
| .01400 | .00677 | .00662 | .00901 | .00973 | .00940 | .02749 | .00888 | .01051 |
| .00145 | .00063 | .00048 | .00088 | .00144 | .00132 | .00129 | .00102 | .00134 |
| .02522 | .01379 | .01318 | .01680 | .01843 | .02227 | .02198 | .03171 | .01946 |
| .00014 | .00016 | .00022 | .00112 | .00014 | .00011 | .00019 | .00014 | .00040 |
| .00702 | .00424 | .00407 | .00618 | .00625 | .27883 | .00447 | .01811 | .01500 |
| .00001 | .00001 | (*) | (*) | .00003 | .00001 | (*) | (*) | (*) |
| .01086 | .00135 | .00088 | .00118 | .00295 | .00133 | .00196 | .00403 | .00290 |
| .00179 | .00061 | .00033 | .00050 | .00109 | .00079 | .00085 | .00083 | .00096 |
| .00690 | .00019 | .00027 | .00067 | .00052 | .00020 | .00189 | .00298 | .00161 |
| .00551 | .00021 | .00040 | .00031 | .00054 | .00043 | .00060 | .00169 | .00233 |
| .00562 | .00360 | .00276 | .00331 | .01237 | .00489 | .00439 | .00891 | .00351 |
| .00029 | .00017 | .00011 | .00015 | .00017 | .00015 | .00034 | .00031 | .00017 |
| .01831 | .02112 | .01226 | .01427 | .10644 | .02042 | .01023 | .01366 | .01339 |
| .00375 | .00225 | .00139 | .00198 | .00397 | .01148 | .00316 | .00190 | .00281 |
| .01142 | .00697 | .00409 | .00680 | .31207 | .01156 | .00875 | .01605 | .00417 |
| .02442 | .04562 | .01835 | .02305 | .24215 | .01539 | .01119 | .02253 | .01433 |
| .01782 | .00920 | .00487 | .01149 | .02341 | .01004 | .01308 | .00986 | .03008 |
| .00104 | .00059 | .00069 | .00225 | .00141 | .00500 | .00073 | .00261 | .00149 |
| .00979 | .00611 | .00233 | .00428 | .00746 | .00470 | .00677 | .00400 | .00792 |
| .00054 | .00013 | .00012 | .00050 | .00021 | .00137 | .00014 | .00021 | .02120 |
| .01464 | .00081 | .00110 | .00419 | .00150 | .00161 | .00113 | .00112 | .00377 |
| .00100 | .00053 | .00050 | .00127 | .00106 | .00066 | .00734 | .00073 | .00067 |
| .01302 | .00812 | .00841 | .01049 | .01091 | .00927 | .03702 | .00843 | .00991 |
| .02688 | .02526 | .00741 | .01455 | .01596 | .01613 | .02020 | .01157 | .02914 |
| .01301 | .00020 | .00042 | .00023 | .00033 | .00056 | .00034 | .00125 | .00021 |
| .00152 | .00106 | .00115 | .00192 | .00096 | .00584 | .00835 | .00106 | .00366 |
| .00679 | .00153 | .00127 | .00211 | .00216 | .00233 | .00409 | .00252 | .00284 |
| .00701 | .00419 | .00280 | .00602 | .00407 | .00533 | .02897 | .00390 | .00381 |
| .00832 | .00675 | .00272 | .00543 | .00626 | .00551 | .01588 | .00391 | .00324 |
| .00072 | .00034 | .00035 | .00156 | .00082 | .00798 | .00080 | .00077 | .00089 |
| .00203 | .00148 | .00101 | .00166 | .00145 | .00156 | .00400 | .00206 | .00150 |
| .00335 | .00191 | .00138 | .00258 | .00154 | .00176 | .03101 | .00182 | .00225 |
| .00777 | .00459 | .00306 | .00456 | .00428 | .00360 | .04107 | .00361 | .00351 |
| .00042 | .00033 | .00029 | .00106 | .00033 | .00040 | .00280 | .00044 | .00026 |
| .00064 | .00037 | .00053 | .00348 | .00049 | .00114 | .00103 | .00076 | .00045 |
| .00023 | .00015 | .00021 | .00141 | .00019 | .00021 | .00031 | .00024 | .00016 |

TABLE 4.15

| Educational and social services, and membership organizations | Federal Government enterprises | State and local government enterprises | Non-comparable imports | Scrap, used and second-hand goods | General government industry | Rest-of-the-world adjustment to final uses | Household industry | Inventory valuation adjustment | Industry number |
|---|---|---|---|---|---|---|---|---|---|
| 77B | 78 | 79 | 80 | 81 | 82 | 83 | 84 | 85 | |
| 0.00852 | 0.00122 | 0.00135 | | | | | | | 1 |
| .00684 | .00154 | .00236 | | | | | | | 2 |
| .00114 | .00023 | .00177 | | | | | | | 3 |
| .00521 | .00045 | .00467 | | | | | | | 4 |
| .00070 | .00030 | .00173 | | | | | | | 5+6 |
| .00305 | .00151 | .01148 | | | | | | | 7 |
| .01550 | .00940 | .04562 | | | | | | | 8 |
| .00166 | .00036 | .00628 | | | | | | | 9+10 |
| .06450 | .01055 | .31153 | | | | | | | 11+12 |
| .00020 | .00012 | .00018 | | | | | | | 13 |
| .02909 | .00436 | .00321 | | | | | | | 14 |
| .00001 | (*) | (*) | | | | | | | 15 |
| .00287 | .00211 | .00157 | | | | | | | 16 |
| .00093 | .00052 | .00133 | | | | | | | 17 |
| .00223 | .00020 | .00101 | | | | | | | 18 |
| .00085 | .00223 | .00044 | | | | | | | 19 |
| .01059 | .00159 | .02567 | | | | | | | 20+21 |
| .00027 | .00007 | .00084 | | | | | | | 22+23 |
| .03479 | .00514 | .00821 | | | | | | | 24 |
| .00378 | .00122 | .00212 | | | | | | | 25 |
| .01607 | .00178 | .00393 | | | | | | | 26A |
| .07941 | .01234 | .00801 | | | | | | | 26B |
| .01263 | .00321 | .03530 | | | | | | | 27A |
| .00226 | .00033 | .00445 | | | | | | | 27B |
| .00530 | .00161 | .00538 | | | | | | | 28 |
| .00118 | .00009 | .00026 | | | | | | | 29A |
| .00228 | .00114 | .00166 | | | | | | | 29B |
| .00130 | .00038 | .00324 | | | | | | | 30 |
| .01662 | .01280 | .03038 | | | | | | | 31 |
| .01490 | .00380 | .01048 | | | | | | | 32 |
| .00050 | .00058 | .00018 | | | | | | | 33+34 |
| .00312 | .00056 | .00204 | | | | | | | 35 |
| .00462 | .00099 | .02239 | | | | | | | 36 |
| .00601 | .00300 | .01640 | | | | | | | 37 |
| .00508 | .00231 | .01128 | | | | | | | 38 |
| .00114 | .00024 | .00065 | | | | | | | 39 |
| .00380 | .00092 | .01688 | | | | | | | 40 |
| .00279 | .00194 | .00289 | | | | | | | 41 |
| .00578 | .00227 | .00929 | | | | | | | 42 |
| .00037 | .00031 | .00111 | | | | | | | 43 |
| .00072 | .00027 | .00617 | | | | | | | 44+45 |
| .00030 | .00020 | .00100 | | | | | | | 46 |

# Appendix D

## List of Tables

**LIST OF TABLES**

## Appendices

# Appendix E

## List of Figures

### LIST OF FIGURES

# Appendix F

## Quick Reference Guide

**TABLE F.1  Quick Guide for Locating Financial and Investment Data[a]**

Sources (column key):

Standard & Poor's[b]
1. Analyst Handbook — REF MG 4519 .$772
2. Bond Guide — MG 4905 .$435
3. The Outlook — REF MG 4905 .$434
4. Stock Guide — MG 4915 .$67
5. Stock Reports — REF MG 4905 .$444
6. Statistical Service — REF MG 4921 .$72
7. Corporate Records — REF MG 4501 .$7663
8. Credit Week — MG 4501 .$7662
9. Industry Surveys — MC 106.6 .$74
10. Poor's Register — REF MG 4057 .A4

Other[c]
11. D&B Million $ Directory — REF MC 102 .D8
12. Predicasts F&S Index (US) — INDEX MG 491 .F8 2v.
13. Predicasts Forecasts — INDEX MC 101 .P7 Binder
14. Q File — REF MG 4028 .52 C7
15. Valueline — MG 4501 .v26
16. Cirr — INDEX MG 4001 .C53
17. U.S. Industrial Outlook — REF MC 106.5 .A17

Moody's[d]
18. Manual Industrial, OTC, S&F — REF MG 4961 .M7
19. Bond Survey — REF MG 4905 .M785
20. Handbook of Common Stocks — MG 4905 .M815
21. Bond Record — REF MG 4905 .M78

For company information:

| Data item | 1 | 2 | 3 | 4 | 5 | 6 | 7 | 8 | 9 | 10 | 11 | 12 | 13 | 14 | 15 | 16 | 17 | 18 | 19 | 20 | 21 |
|---|---|---|---|---|---|---|---|---|---|---|---|---|---|---|---|---|---|---|---|---|---|
| Headquarters address, phone | | | | | ● | | ● | | | ● | ● | | | ● | | | | ● | | ● | |
| Subsidiaries or divisions | | | | | ● | | ● | | | ● | ● | | | ● | | | | ● | | | |
| Number of employees | | | | | ● | | ● | | | ● | ● | | | ● | | | | ● | | | |
| Principal products, services or business | | | ● | ● | ● | | ● | | ● | ● | ● | ● | ● | ● | ● | | ● | ● | | ● | |
| Balance sheet data | ● | | | | ● | | ● | | | | | | | ● | | | | ● | | ● | |
| Income statement | ● | | | | ● | | ● | | | | | | | ● | | | | ● | | | |
| SIC classifications | | | | | | | ● | | | ● | ● | | | | | | | | | | |
| Officers and directors | | | | | ● | | ● | | | ● | ● | | | ● | | | | ● | | | |
| Corporation profile | | | | | ● | | ● | | | | | | | ● | | | | ● | | | |
| Comparative company financial ratios/data | ● | | | | ● | | ● | | | | | | | ● | | | | | | | |
| Corporate stock banking, prices | | | ● | ● | ● | ● | ● | ● | ● | | | | | ● | | | | ● | | ● | |
| Corporate bond ratings, prices | | ● | | | ● | ● | ● | ● | | | | | | ● | | | | ● | ● | | ● |
| News developments | | | ● | | | | | | ● | | | ● | | ● | ● | ● | | ● | | ● | |
| Transfer agents | | | | | | | ● | | | | | | | | | | | ● | | | |

Bond descriptions
Capitalization data
For Industry Information
News and trends
Industry company analysis
Marketing statistics/share of market
Per-share data by industry
Market action charts
Production indices
Balance sheet per-share-data
For Investment Information
Beta
Dividends amounts, dates
Earnings data
Stock action chart
Ticker symbol
Stock prices, ranges
S&P stock ranking
Price/earnings ratios
Institutional holdings
Stock trading volume
Stock advice/analysis
Mutual funds performance
Bond calls, redemptions, notices
Fixed-income interest payments
S&P or Moodys' bond ratings
Bond listing/analysis
Bond prices, ranges
New bond registrations

TABLE F.1 Continued

| For investment information (cont.) | Standard & Poor's[b] | | | | | | | | | | Other[c] | | | | | | | Moody's[d] | | | |
|---|---|---|---|---|---|---|---|---|---|---|---|---|---|---|---|---|---|---|---|---|---|
| | Analyst Handbook REF MG 4519 .$772 | Bond Guide MG 4905 .$435 | The Outlook REF MG 4905 .$434 | Stock Guide MG 4915 .$67 | Stock Reports REF MG 4905 .$444 | Statistical Service REF MG 4921 .$72 | Corporate Records REF MG 4501 .$7663 | Credit Week MG 4501 .$7662 | Industry Surveys MC 106.6 .$74 | Poor's Register REF MG 4057 .A4 | D&B Million $ Directory REF MC 102 .D8 | Predicasts F&S Index (US) INDEX MG 491 .F8 2v. | Predicasts Forecasts INDEX MC 101 .P7 Binder | Q File REF MG 4028 .52 C7 | Valueline MG 4501 .v26 | Cirr INDEX MG 4001 .C53 | U.S. Industial Outlook REF MC 106.5 .A17 | Manual Industrial, OTC, S&F REF MG 4961 .M7 | Bond Survey REF MG 4905 .M785 | Handbook of Common Stocks MG 4905 .M815 | Bond Record REF MG 4905 .M78 |
| New stock offerings | | | ■ | | | | | | | | | | | | | | | | | | |
| Bond indices | | ■ | ■ | | | | | | | | | | | | | | | | ■ | | └ |
| Dividend reinvestment data | | | | ■ | | ■ | | | | | | | | | | | | | | | |
| Tax status of dividends | | | | | ■ | | | | | | | | | | | | | | | | |
| Options data | | | ■ | ■,a | | | | | | | | | | | | | | | | | |
| S&P stock price indices | ■ | | ■ | | | ■ | ■ | ■ | ■ | | | | | | | | | | ■ | | └ |
| Commercial paper ratings | | | | | | | | | | | | | | | | | | | | | |

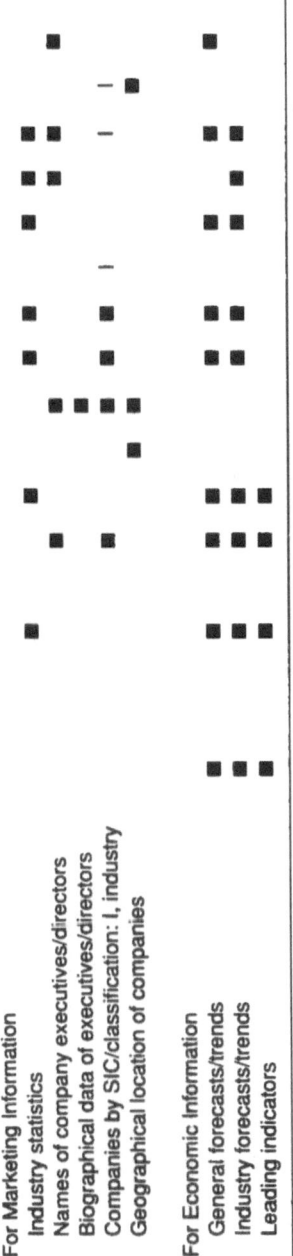

For Marketing Information
Industry statistics
Names of company executives/directors
Biographical data of executives/directors
Companies by SIC/classification: I, industry
Geographical location of companies

For Economic Information
General forecasts/trends
Industry forecasts/trends
Leading indicators

[a]This reference guide was prepared by the Reference Library staff, Mullins Library, University of Arkansas, Fayetteville, AR.
[b]Standard & Poor's Industrial Surveys, Standard and Poor's Corporation, Division of McGraw-Hill Companies, New York.
[c]Miscellaneous sources, such as Predicasts, Dun and Bradstreet, Value Line.
[d]Moody's Industrial Manual, Moody's Investors Service, Inc., New York.

# Index